# CARBON NANOTUBES - SYNTHESIS, CHARACTERIZATION, APPLICATIONS

Edited by **Siva Yellampalli**

**Carbon Nanotubes - Synthesis, Characterization, Applications**
Edited by Siva Yellampalli

**Published by InTech**

Janeza Trdine 9, 51000 Rijeka, Croatia



**Publishing Process Manager** Viktorija Zgela

**Technical Editor** Teodora Smiljanic

**Cover Designe**r InTech Design Team

Additional hard copies can be obtained from orders@intechopen.com

Carbon Nanotubes - Synthesis, Characterization, Applications, Edited by Siva Yellampalli
p. cm.
ISBN 978-953-307-497-9

For EU product safety concerns:
IN TECH d.o.o., Prolaz Marije Krucifikse Kozulić 3, 51000 Rijeka, Croatia,
info@intechopen.com or visit our website at intechopen.com.

# Meet the editor

Dr. Siva Yellampalli obtained his MS & Ph.D from Louisiana State University. He is currently with VTU Extension Centre, UTL Technologies Ltd. He worked on a broad range of research topics including Very Large Scale Integration (VLSI), mixed signal circuits/systems development, micro-electromechanical systems (MEMS), and integrated carbon nanotube based sensors. He has published a book, multiple journal and IEEE conference papers in these areas of research. In addition he has given many professional presentations including invited talks. He has been a consultant to a variety of industries and acts as a reviewer for technical journals and book publishers. He is a life member of Indian Society for Technical Education (ISTE), senior member, International Association of Computer Science and Information Technology (IACSIT) and member of IEEE, VLSI Society of India (VSI) and National Society of Collegiate Scholars (NSCS).

# Contents

# Preface

Carbon nanotubes are one of the most intriguing new materials with extraordinary properties being discovered in the last decade. Carbon nanotubes are typically considered as molecular-scale tubes of graphitic carbon. Depending on numbers of carbon layers, they are categorized as single-walled and multi-walled nanotubes. The unique structure provides nanotubes with extraordinary mechanical and electrical properties. The outstanding properties that these materials possess have opened new interesting research areas in nanoscience and nanotechnology.

Although nanotubes are very promising in a wide variety of fields, application of individual nanotubes for large scale production has been limited. The main roadblocks, which hinder its use, are limited understanding of its synthesis and electrical properties which lead to difficulty in structure control, existence of impurities, and poor processability.

This book focuses on understanding the synthesis mechanism behind the development of carbon nanotubes along with studying their properties and applications. The book has been divided into three parts, the first section focuses on the various synthesis and processing techniques, second part focuses on characterization of CNTs and the third part gives various applications of CNTs.

**Part 1. Synthesis and Processing**

*Chapter 1. Processing Carbon Nanotubes*
This chapter starts with an overview of the characterization techniques commonly used to follow surface and structural modification of CNTs upon chemical treatments; the respective sensitivity and the limits of each technique are also briefly discussed. The second part is dedicated to the description of the main kinds of CNT samples (obtained from different synthesis methods) and the question of their purification is in particular considered. In the following section, focus is on the treatments commonly used to induce the dispersion of the CNTs in a surrounding medium (solvent, monomer or polymer) and the methods leading to modify the CNT reactivity. The last part reports on the elaboration and the characterization of CNT-based composites taking into account their particular multi-scale character.

*Chapter 2. Initial Growth Process of Carbon Nanotubes in Surface Decomposition of SiC*
Understanding the growth mechanism is essential to control structural parameters of CNTs, which are necessary for realization of CNT devices. This chapter studies the formation process of carbon nanocaps, which are precursors of CNTs, determining the structures of CNTs.

*Chapter 3. Synthesis of Carbon Nanotubes by Microwave Irradiation*
The results of the synthesis of various carbon nanostructures under different preparation conditions using a domestic microwave oven as energy source have been presented. The chapter also presents a detailed description of the synthesis results in dependance of nanostructures on how the starting material is prepared and the time of exposure to microwaves.

*Chapter 4. Effects of Gravity and Magnetic Field On Production of Single Walled Carbon Nanotubes by Arc Discharge Method*
The production of SWNTs by the arc discharge method is investigated. As the reaction is strongly affected by gravity (heat convection) (Mieno, 2004) and the applied magnetic field (Lorentz force), the effects of gravity, heat convection and magnetic field on the production of SWNTs were experimentally studied.

*Chapter 5. Selective Growth of Carbon Nanotubes and Their Applications to Transparent Conductive Plastic Sheets and Optical Filters*
This chapter starts with explanation of radiation-heated CVD (RHCVD), which enables us to selectively synthesize SWCNT, DWCNT and TWCNT films on substrate followed by the synthesis of vertically aligned SWCNT, DWCNT and TWCNT films on substrates and the essence of synthesizing long CNTs on substrates. The optical performance of CNT forests is also explained in detail.

*Chapter 6. Direct Growth of Carbon Nanotubes on Metal Supports by Chemical Vapor Deposition*
Current developments of direct growth technique and its applications are summarized.

*Chapter 7. Flame Synthesis of Carbon Nanotubes*
This chapter summarizes the mechanisms of carbon nanotubes formation in flames followed by the summary of the types of flame configurations and chemistry that have been used in the synthesis process.

*Chapter 8. Carbon Nanotubes Synthesis and Growth Mechanism*
The material aspects of CNT synthesis by CVD is reviewed and the CNT growth mechanism in light of the latest progresses in the field has been discussed in this chapter.

*Chapter 9. Dielectrophoretic Deposition and Alignment of Carbon Nanotubes*
The selective deposition of CNTs - including both SWNTs and MWNTs - with dielectrophoresis to obtain aligned nanotubes in the forms of thin films, small bundles, and individual nanotubes has been examined by changing parameters in the dielectrophoresis process and the concentration of the CNT sample solutions.

*Chapter 10. Aligned Growth of Single-Walled and Double-Walled Carbon Nanotube Films by Control of Catalyst Preparation*
Growth of aligned CNT films using microwave plasma-enhanced CVD is described, with a particular emphasis on the pretreatment of substrate with catalytic metal nanoparticles. By preparing the Co nanoparticles in a controlled manner, aligned SWNT and DWNT films were fabricated on the Si substrate. Furthermore, area-selective growth of vertical CNTs to form organized SWNT and DWNT microstructures was demonstrated.

**Part 2. Characterization and Properties**

*Chapter 11. Study of Carbon Nanotube Based on Higher Order Cauchy-Born Rule*
A nanoscale continuum theory is established based on the higher order Cauchy-Born rule to study mechanical properties of carbon nanotubes. The theory bridges the microscopic and macroscopic length scale by incorporating the second-order deformation gradient into the kinematic description.

*Chapter 12. In-Situ Structural Characterization of SWCNTs in Dispersion*
A multi-scale characterization approach for a better understanding of the in-situ SWCNT structures in the dispersion, experimental techniques and methods such as viscosity and rheological measurements, scattering based techniques, sedimentation methods, spectroscopic techniques are reviewed in this chapter. For each of the methods, the underlying physical principles and their applications for the in-situ structural characterization of SWCNT dispersions were also discussed.

*Chapter 13. Microwave Absorption Characteristics of Carbon Nanotubes*
Microwave absorption characteristics of carbon nanotubes (CNTs) and aligned carbon nanotubes (ACNTs) were investigated in the frequency range of 2–18 GHz. Carbon nanotube (CNTs) and aligned carbon nanotubes were prepared by chemical vapor deposition (CVD). CNTs doppped by rare earth were also investigated for their microwave absorption properties in this chapter.

*Chapter 14. Structural Instability of Carbon Nanotube*
The present study employs MD simulations based on the Tersoff many-body potential function to perform a systematic and comprehensive investigation into the buckling behaviors of single-walled CNTs under uniaxial compressive displacement loading. In addition the applicability of the continuum buckling theory, which has been well developed for thin tubes, on predicting the buckling behavior of the CNT is also examined.

*Chapter 15. Molecular Dynamics Simulation Study on the Mechanical Properties and Fracture Behavior of Single-Wall Carbon Nanotubes*
The power of molecular dynamics simulation technique is exploited in investigating the mechanical characteristics of various types of defect-free as well as defective tubes with a varying number of Stone-Wales defects with different combinations. The effect

of interlayer interaction between different single-wall tubes in a bundle and the role of potential functions in the mechanical behavior of different carbon nanotubes are also presented here.

*Chapter 16. Microscopic Structure and Dynamics of Molecular Liquids and Electrolyte Solutions Confined by Carbon Nanotubes: Molecular Dynamics Simulation*
The influence of spatial confinements caused by internal space of single walled carbon nanotubes (SWCNTs) and multi walled carbon nanotubes (MWCNTs) on a microscopic structure and particle dynamics of the confined non-aqueous molecular liquids acetonitrile, methanol, dimethyl sulphoxide (AN, MeOH, DMSO) and infinitively diluted solutions of $Li^+$ in MeOH and solutions of $Et_4NBF_4$ of finite concentrations in AN are investigated by conducting molecular dynamics (MD) simulations on them.

*Chapter 17. Comparison of NQR of $O_2$, $N_2$ and CO on Surface of Single-Walled Carbon Nanotubes and Chemisorption of Oxygen-Doped on the Surface of Single-Walled Carbon Nanotubes: A DFT and NMR Computational Study*
The adsorption behavior of selected nitrogen, oxygen, and CO molecules on the surface of the single-walled carbon nanotubes (SWCNTs) was studied by the density functional theory (DFT) (B3LYP/6-311G*) using the Gaussian98 software. The nuclear quadrupole resonance (NQR) of the armchair (4, 4) SWCNTs with the optimal diameter of 5.6 Å and the length of 9.8 Å was also observed.

**Part 3. Applications**

*Chapter 18. Smart Materials and Structures Based on Carbon Nanotube Composites*
This section starts with a summary of the properties of CNTs and methods to control them. Then it proceeds to the description of the smart multifunctional applications of carbon nanotube composites.

*Chapter 19. Nonlinear Optical Properties of Graphene and Carbon Nanotube Composites*
In this chapter the excellent chemical activity of graphene and nanotubes which provides a broad platform for various functional counterparts, forming multi-component, multi-functional hybrid composites with wider spatial and temporal responses for OL has been studied.

*Chapter 20. Design and Demonstration of Carbon Nanotubes(CNTs)-Based Field Emission Device*
A large area, full colored field emission display (FED) based on CNT emitter is presented here along with a computer simulation of the field emission properties of CNT based on F-N theory, the issue of electron transmission efficiency and the design and experimental demonstration of a FED prototype.

*Chapter 21. Reinforced Thermoplastic Natural Rubber (TPNR) Composites with Different Types of Carbon Nanotubes (MWNTS)*
Thermoplastic natural rubber TPNR reinforced with two types of multi-walled carbon nanotubes (MWCNTs) nanocomposites were prepared by the melt blending

method which will disperse MWCNTs homogeneously in the TPNR matrix in an attempt to increase the mechanical properties of these nanocomposites. The effect of MWCNTs on the mechanical properties of TPNR nanocomposites is reported in this chapter.

*Chapter 22. Carbon Nanotubes and Semiconducting Polymer Nanocomposites*
This chapter presents a summary of the preparative methods, characterization data, and applications of conducting polymer/carbon nanotube composites. The electrical, thermal, mechanical and electrochemical properties of the composites are intermediate between pure polymer and CNT but vary depending on the method of preparation, type, purity, content of CNTs, the dispersion of CNTs in polymer matrix and the nature of the interaction between two components.

*Chapter 23. Carbon Nanotube Based Thin Flims: Synthesis and Properties*
A fundamental description of carbon nanotube (CNT)-based thin films, mainly concentrating on their synthesis and properties is presented. In the first part of this chapter, two synthesis methods, self assembling and electropolymerization are described, followed by another film developed consisting only of fullerene and CNTs, which can be called pure-carbon thin film. The chapter also discusses several important properties resulting from the exceptional structures of CNT films.

**Acknowledgments**

I would like to thank the authors of the chapters in this book for their excellent contributions and for the efforts placed in the publication of their work. I wish to thank them all and also my wife, Suma, for her patience and understanding.

**Dr. Siva Yellampalli**

*VTU Extension Centre*
*UTL Technologies Ltd*
*Banglaore, Karnataka*
*India*

# Part 1

## Synthesis & Processing

1

# Processing Carbon Nanotubes

Brigitte Vigolo and Claire Hérold
*Institut Jean Lamour, CNRS - Nancy Université – UPV MetzDépartement CP2SBP 70239, 54506 Vandoeuvre-lès-Nancy, France*

## 1. Introduction

Due to their combined superior chemical and physical properties, carbon nanotubes (CNTs) are recognized to have a huge potential in many fields of applications (Ajayan, 1999; Rao et al., 2001; Dai, 2002; Van Noorden, 2011). These molecular-scale tubes of graphitic carbon are one of the stiffest and strongest fibers known. Besides, they have remarkable electronic, optical, thermal and chemical properties. For these reasons their interest in both academic and industrial areas is unique. Nevertheless, the as-produced material is extremely difficult to process. Development of CNT-based devices or composites of interest for new applications has been consequently hindered. CNTs are hydrophobic and incompatible with a majority of solvents, including monomers and polymers; they indeed have a high tendency to agglomerate. Moreover, CNTs and especially single-walled carbon nanotubes (SWNTs) are assembled in bundles of generally several tens of tubes. Development of efficient processes and chemical treatments that are able to control the quality of the CNT samples and to induce both their dispersion and partial or complete debundling remains highly challenging.

CNTs can be produced using different methods that basically consist in heating carbon-containing solid or gas. On the contrary to the preparation of multi-walled carbon nanotubes (MWNTs), SWNT growing requires a metal catalyst. The characteristics of the samples depend on the control and the choice of the experimental parameters used for the synthesis. A better understanding of the growth mechanisms has permitted the development of mass production processes (Grobert, 2007). Nevertheless, their uniformity (length, diameter, chirality), the quality of their walls (number of defects) and also their purity are still partially controlled. The quality of the samples has to be improved in order to benefit of the exceptional properties of CNTs in new materials. Depending on the type and the synthesis method, the CNTs can differently behave through the applied chemical treatments. Whatever the synthesis method, CNT samples persistently contain several kinds of heterogeneities: (i) carbonaceous species like fullerenes, amorphous carbon, graphitic and carbon particles, …; (ii) impurities such as residual metallic catalyst often protected by more or less graphitized carbon shells or polyhedra; (iii) defects at the CNT surface or oxygenated grafted functions, (iv) dispersion in diameter, chirality and morphology (aspect ratio) and (v) aggregation into bundles. These heterogeneities represent a major obstacle for both the establishment of universal behaviors and the development of efficient processing methods.

Nano-scaled particles exhibit an enormous surface area being of several orders of magnitude larger than that of conventional fibers. This surface area can potentially act as a powerful interface but it is also responsible for the high tendency of CNTs to form agglomerates. It appears that the commonly used procedures for manufacturing composites with conventional fibers or other-carbon-form do not show the hoped results. Indeed, CNT samples particularly and unusually behave. Hence original chemical treatments and processes have to be proposed and optimized (Kuzmany et al., 2004; Tasis at al., 2006; Karousis et al., 2010); and efforts have to be made to disperse the CNTs prior to their incorporation within the chosen surrounding medium to obtain the desired device or material.

In this paper, we first give an overview of the characterization techniques commonly used to follow surface and structural modification of CNTs upon chemical treatments; the respective sensitivity and the limits of each technique are also briefly discussed. The second part is dedicated to the description of the main kinds of CNT samples (obtained from different synthesis methods) and the question of their purification is in particular considered. In the following section, after giving the parameters that are relevant regarding chemical treatments to process CNTs, we will focus on the treatments commonly used to induce the dispersion of the CNTs in a surrounding medium (solvent, monomer or polymer) and the methods leading to modify the CNT reactivity. The last part reports on the elaboration and the characterization of CNT-based composites taking into account their particular multi-scale character.

## 2. Characterization techniques

Characterization of CNT samples is a difficult task. Their inherent heterogeneity (discussed in the following part) is one of the main reasons for that statement. It is complex to obtain an unambiguous knowledge of their behavior based on the recorded data from one characterization technique. The usually and unavoidable employed approach is the use of several complementarily techniques. The most common analytical techniques used to characterize chemically modified CNTs are transmission electron microscopy (TEM), scanning electron microscopy (SEM), thermogravimetry analysis (TGA), Raman spectroscopy and XPS (X-ray photoelectron spectroscopy). Added to them, several other techniques can be used to specifically determine the nature of the attached chemical groups or their localization on CNT samples; for that purpose, TGA-MS (mass spectrometry) coupling technique, EXAFS (extended X-ray absorption fine structure) or adsorption volumetry are of interest.

The electron microscopy techniques (TEM and SEM) allow a qualitative and local examination of the morphology and the composition of the samples; by increasing the number of the observed zones, they can be reliable in the determination of the behavior of the CNT samples (Monthioux et al., 2001). The accuracy of electron microscopy being in constant progress, it permits to go further in the structural details of the analyzed species, including CNTs. EDS or EDX (energy-dispersive X-ray spectroscopy) for the elementary analysis at a specific location of the sample during observations can be used to determine the metal content after a purification process or the presence of one specific element belonging to the grafted functions. The main advantage of these techniques is that they allocate to separately analyze the behavior of the CNTs, the carbonaceous or catalytic impurities.

The TGA examines the weight lost of the CNT samples as a function of the temperature (usually from room temperature to 800°C-1000°C). In oxidative conditions (air or oxygen), the recorded weight loss corresponds to the combustion of the carbonaceous species of the samples. This oxidation treatment gives rise to their successive combustion as a function of their respective stability (Landi et al., 2005). At the end of the gasification process, only the oxidized catalysts remain. First, this technique is used to quantify the metal content in CNT samples after a purification process. In the example reported on Figure 1a (from Landi et al., 2005), the metallic oxides content is reduced from 9.6 wt% to 7.1 wt% after refluxing in a $NHO_3$/HCl acidic solution for 14h. The raw sample is named 'SWNT-HO' and the sample obtained after reflux is named 'SWNT-Reflux'. Second, the situation is a little bit complicated regarding the analysis of the nature and/or the content of the different carbonaceous species in the samples. Even if it is obvious that well-graphitized species will be more stable than amorphous carbon, the removal temperature range of CNTs and more- or less-ordered carbon species can be rather large and difficult to identify. However, TGA can be useful to evidence modification of CNT structure after treatments. For example, an increase of the CNT quality or concentration through a purification process or an introduction of defects through a chemical functionalization gives rise to modifications for both the recorded weight losses and the related temperature domains. It is commonly reported that CNTs decompose at higher temperature than amorphous carbon species. As illustration, Figure 1b (from Landi et al., 2005) shows the first derivative weight-loss curves for the samples previously mentioned. Raw SWNT-HO sample shows two major contributions of the weight loss with a prominent peak maximum at 427°C and a minor shoulder at 560°C, attributed to gasification of amorphous carbon and SWNTs, respectively. For 'SWNT-Reflux' sample, the two contributions are shifted to higher temperatures, namely, peak maxima at 503 and 598 °C, respectively.

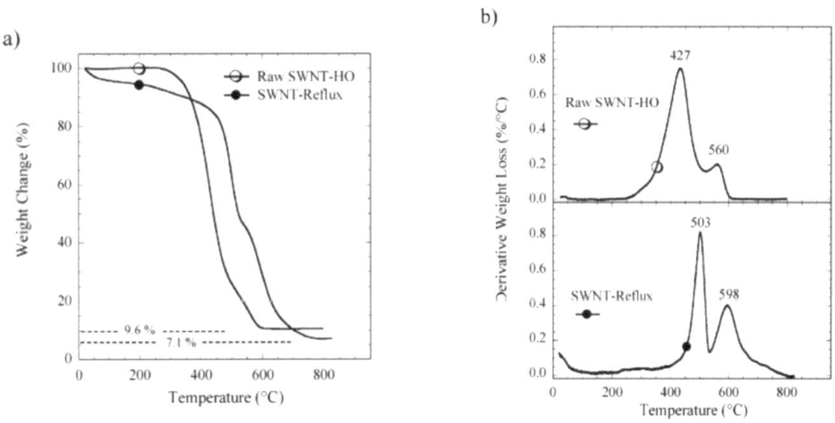


Fig. 1. TGA data for a raw SWNT (SWNT-HO) and SWNT-HO that has been treated using an acidic solution (SWNT-Reflux). SWNT-Reflux is obtained after refluxing raw SWNT-HO sample in a $NHO_3$/HCl acidic solution for 14h. The TGA was ramped at 10°C/min under air at a gas flow rate of 60 sccm. a) TGA data for raw SWNT-HO (O) and SWNT-Reflux (●) samples. b) TGA data from the first derivative analysis of the weight loss (%/°C) for raw SWNT-HO (O) and SWNT-Reflux (b) samples. The peak maxima for the prominent thermal decomposition features are labeled for clarity. From (Landi et al. 2005). Copyright 2005 by the American Chemical Society.

As it is reported in the work of Landi and coworkers (Landi et al., 2005), the ability to assign temperature regions of combustion for SWNTs and carbon impurities would be of great

importance for the optimization of purification processes. Nevertheless, a selective decomposition of carbon impurities if it exists remains an open question.
Under inert gas, TGA-MS coupling technique consists in analyzing the detachment of functions that have been initially grafted to the CNT surface (Chattopadhyay et al., 2005). The nature of the bonds created between the introduced functions and the CNTs can be identity from the release temperature domain (Lejosne et al., 2011). A molecule simply physisorbed or π-stacked will be detected at lower temperature than a group which is covalently linked to the sample surface. As a complementary analysis, MS investigation allows having a feedback of the nature of the functions that were effectively attached at the sample surface.
Raman spectroscopy is a widely used technique for the characterization of CNT samples (Burghard, 2005; Graupner, 2007). It is a powerful technique because the signal of CNTs is enhanced compared to that of the carbon impurities. Several features are modified upon chemical treatments and Raman spectroscopy allows probing the quality of CNT structure, the possible selectivity of reaction with respect to the electronic properties (metallic or semiconducting) (Dyke et al., 2005) or an induced electron transfer. A typical Raman spectrum of SWNTs shows three characteristic bands: the radial breathing mode RBM (100–400 $cm^{-1}$), the D mode (≈1350 $cm^{-1}$) and the tangential (C=C vibrations) stretching G mode (1500–1600 $cm^{-1}$).
At low frequency, RBM corresponds to the radial deformation of the carbon-carbon bonds. For SWNTs in bundle, the SWNT diameter can be calculated as follows (Jorio et al., 2003):

$$\omega_{RBM} = A/d_t + B \tag{1}$$

where the $A$ and $B$ parameters are determined experimentally. For typical SWNT bundles in the diameter range $dt$ = 1.5±0.2 nm, $A$ = 234 $cm^{-1}$ nm and $B$ = 10 $cm^{-1}$ has been found for SWNT bundles.
The intensity of the D-band is known to be related to introduction of defects in the CNT structure. The increase of the ratio of the intensity of the D band over the intensity of the G band, $I_D/I_G$, is commonly used to prove the covalent nature of the functionalization of CNTs since the attachment of the grafted groups leads to the breaking of C=C bonds (Dillon et al., 2005). The area of the D band is also reported to be sensitive to a presence of deposit of carbon layers on CNT surface. The removal of such deposit by oxidation for example possibly leads to a decrease of $I_D/I_G$ ratios (Osswald et al., 2005). Heating functionalized CNTs under vacuum or inert gas leads to the removal of functional groups and restores the initial low-defected structure. The obtained CNT samples after annealing show thus a reduced $I_D/I_G$. Figure 2 (from (Dyke & Tour, 2003)) shows three Raman spectra of raw SWNTs (A), covalently functionalized SWNTs (B) and of functionalized SWNTs after heating at 750°C under argon (C). $I_D/I_G$ increases after the functionalization process because of the attachment of functional groups on the SWNT sidewalls and it decreases after heating due to the detachment of the groups and the recovering of SWNT structure.
The G band corresponds to the tangential mode of vibration of the C=C bonds in CNTs (Jorio et al., 2002). The G band is mainly composed of two or three identifiable components even if it can be usually more complex to be fitted. A simple analysis can be carried out considering the two most intense peaks that basically originate from the symmetry breaking of the tangential vibration when the graphene sheet is rolled to make a cylindrically shaped tube. They are labeled $G^+$ for atomic displacements along the tube axis, and $G^-$ for modes with atomic displacement along the circumferential direction. The difference between

semiconducting and metallic SWNTs is evident in the lineshape of the $G^-$ feature, which is broadened for metallic SWNTs in comparison with the Lorentzian lineshape for semiconducting tubes. G band can be sensitive to chemical treatments. First, for defective SWNTs after functionalization, an additional contribution, referred as the G* band, at high wavenumber is added to the conventional G band (Cataldo, 2000; Vigolo et al., 2009a). Second, reaction of SWNTs with electron-donors or -acceptors induces a shift of the G band. The reaction between alkali metals (donors) and CNTs is accompanied by an electronic transfer giving rise to an electron enrichment of the CNT structure. The G band of the obtained reduced CNTs can be then down-shifted of several tens of $cm^{-1}$ (Sauvajol et al, 2003).

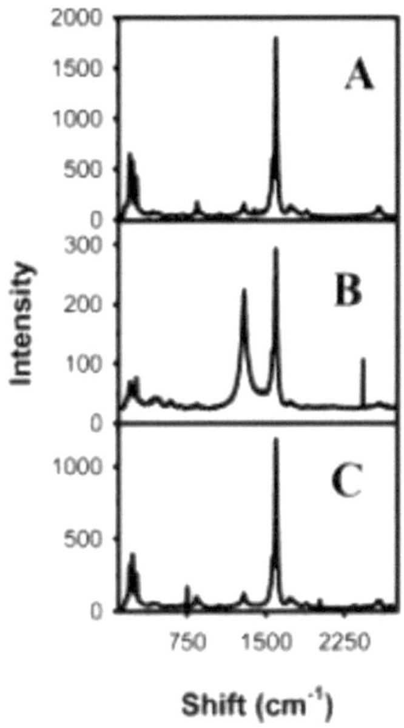


Fig. 2. Raman spectra (780.6 nm excitation) of (A) raw SWNTs, (B) functionalized SWNTs, and (C) functionalized SWNTs after TGA (10 °C/min to 750°C) in argon. From (Dyke & Tour, 2003). Copyright 2003 by the American Chemical Society.

Infrared (IR) spectroscopy is recognized to be useful for the study of SWNT sidewall chemistry. As a complementary technique of Raman spectroscopy, IR spectroscopy can be used to identify the functional groups and especially oxygenated functional groups added to the tube walls thanks to their characteristic vibrational modes (U.J. Kim et al., 2005a). The range expected for C-O stretching modes in ether, ester, alcohol, or phenol functions is around 1100 $cm^{-1}$ and in the 1700 $cm^{-1}$ domain, the bands can be assigned to carbonyl (C=O) stretching in ketone, aldehyde, or carboxylic acid groups. For aldehyde groups, the C-H stretching and bending vibration generally appears in the 2700-2900 $cm^{-1}$ range. The O-H stretching modes are found in the 3100-3600 $cm^{-1}$ range.

X-ray (XRD) and neutron diffraction techniques may help to ascertain the quality and crystalline nature of the treated CNT samples as opposed to amorphous carbon material. XRD, in particular, allows analyzing the state of oxidation of the catalyst through a purification process, for example (Vigolo et al., 2010a).

XPS (X-ray Photoelectron Spectroscopy) technique is widely used to determine atomic compositions and to qualitatively analyze different elements on the CNT surface especially upon oxidative treatments; it is also sensitive to the presence of structural defects on the nanotube surface. The C1s peak of non-treated CNTs shows a peak at 284.1 eV. The peak at 285.5 eV is attributed to defects on the nanotube structure (Datsyuk et al., 2008). Detection of oxygen-containing functional groups is evidenced by several contributions at 286.7, 288.3 and 290 eV. Finally, the π–π* transition loss peak is generally detected at 291.5 eV. The relative augmentation of the contributions related to the presence of oxygen with respect to that of pure carbon evidences the formation of C=O or C-O covalent bonds at the CNT surface. Figure 3 (from Liu et al., 2007a) shows that the C–O component after functionalization (SHR15) is increased compared to other components from the decomposition of C1s peak in agreement with the attached functional groups.

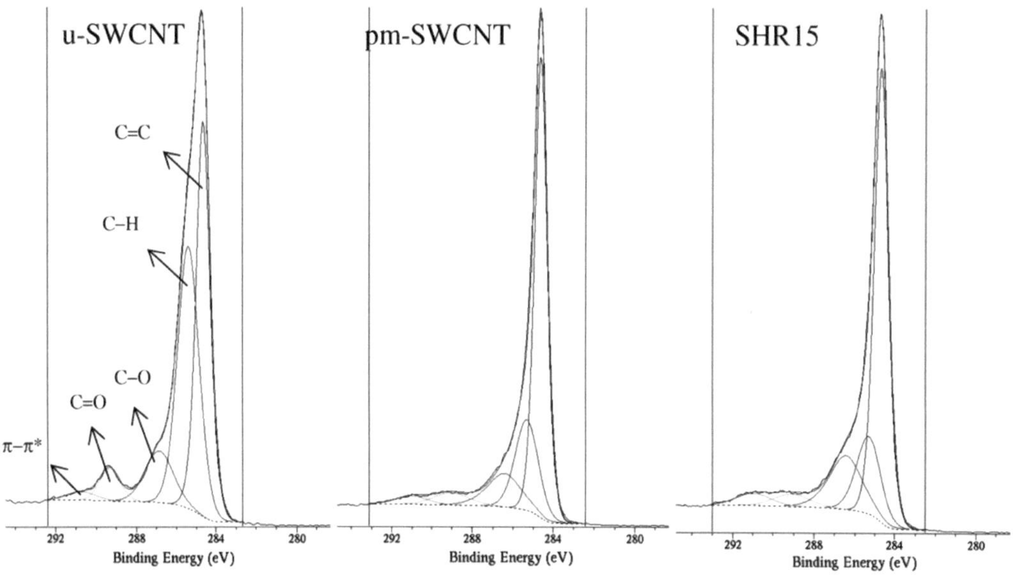


Fig. 3. Decomposition of C1s curves of as-produced SWNTs (u-SWCNT), product of control experiment (pm-SWCNT) and SWNTs that have been functionalized by methoxyphenyl (-$PhOCH_3$) functional groups through using a microwave-assisted reaction for 15 min designated as SHR15. From (Liu et al., 2007a) Copyright 2007 by Elsevier.

Near edge X-ray absorption fine structure (NEXAFS) can be used at the oxygen or carbon K edge to probe the bonding modification of the oxygen or carbon atoms in the sample upon a chemical treatment (Banerjee et al., 2004; Wang et al., 2010). CNT structure and chemical composition of oxygenated functional groups grafted to the CNT surface can be then analyzed using NEXAFS spectroscopy. In the case of C K-edge spectra, the CNTs are characterized by a sharp C–C σ* transition at 285.4 eV, three σ* transitions from 289.9–298 eV, and broad (σ+σ) transitions from 301–309 eV. The region between 287 and 290 eV mainly corresponds to the C–O bonding, with 287.6 eV assigned to C=O π* transition and 288.2 eV assigned to C–O σ* transition. The O K-edge spectra are usually also analyzed as complementary data.

As it can be commonly done for porous systems, the determination of adsorption-/desorption-curves with nitrogen at 77 K can be used to characterize the specific surface area of the CNTs according to the well-known Brunauer, Emmett, and Teller (BET-method). Since a chemical treatment can induce modifications of CNT surface, such analysis can provide a valuable feedback regarding the occurrence of the chemical treatments (C.M. Yang et al., 2002).

By using gas such as xenon or krypton, the adsorption isotherms exhibit specific profile with several steps and plateaus (Arab et al., 2007; Goudon & Lasjaunias, 2008). The steps are especially related to the existence of different adsorption sites related to CNT bundle morphology; each step being positioned at a characteristic pressure different from the characteristic pressure of the carbonaceous impurities of the samples. The isotherm of a CNT sample can be considered as a signature of the surface of the CNTs themselves. It is hence possible to selectively prove the occurrence of the chemical treatment on the CNT surface (Vigolo et al., 2009b).

## 3. Description of CNT samples

### 3.1 Synthesis methods and related characteristics of the samples

The CNT synthesis methods have been recently highly improved leading to the development of mass production processes (Sadeghian et al., 2009; Lehman et al., 2011). Although it is easier to produce significant quantities of MWNTs than SWNTs, their structure is less well understood than that of SWNTs because of their greater complexity and variety. Multitudes of exotic shapes and arrangements have also been observed under different processing conditions. The variety of forms may be interesting but also has a negative side because they diverge from the ideal $sp^2$ cylindrical structure and the CNT properties are consequently diminished.

CNTs can be synthesized using both high-temperature and low-temperature processes. CVD (Chemical Vapor Deposition) process is classified in the low-temperature methods. It is based on the reaction between a carbon containing flowing gas molecules and catalyst particles often above 1000°C. Arc discharge and laser ablation methods are based on sublimation of a graphite target which occurs at relatively high temperature (1000-3000°C).

In the case of CVD methods, the growth process involves heating a catalyst material to sufficient temperatures (550-1200°C) in a tubular furnace and flowing a hydrocarbon gas through the reactor for a period of time under vacuum. The growth mechanism is based on the dissociation of carbon containing molecules which is catalyzed by a transition metal (typically Ni, Fe or Co) (Gavillet et al., 2002; Deck & Vecchio, 2005; Esconjauregui et al., 2009). After dissolution in the metal particle, a precipitation phenomenon leads to the formation of tubular carbon solids in $sp^2$ structure. Materials grown over the catalyst, MWNTs or SWNTs, can be obtained as non-ordered soot, densely aligned bundles or as individual array deposited on the substrate (Maruyama et al., 2002; Bronikowski 2006; Singh et al. 2003; Vigolo et al., 2008). These advanced processes allow having a certain control of the diameter and the orientation of the CNTs (Yamada et al., 2006; Willems et al., 2000; N.S. Kim et al., 2002; Y. Chen et al., 2005).

For the synthesis of SWNTs in mass quantities, particular conditions are required in the HiPco process (Nikolaev et al., 1999). It allows producing high-quality and narrow-diameter SWNTs. The metal catalyst is formed *in situ* when $Fe(CO)_5$ or $Ni(CO)_4$ is injected into the reactor (900 to 1100°C) along with a stream of carbon monoxide (CO) gas and at a pressure of 30 to 50 atm. The reaction to make SWNTs is the 'disproportionation' of CO by nanometer-size metal catalyst particles.

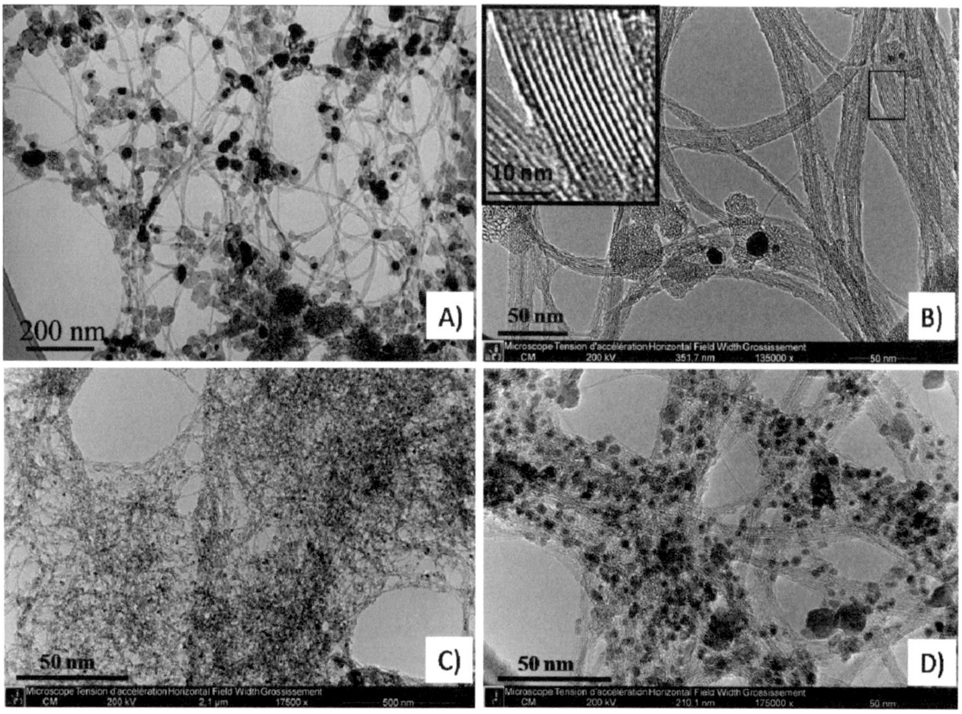


Fig. 4. TEM images at different magnifications showing arc-discharge as-produced SWNTs (A and B) (prepared in a home-made reactor) and SWNT synthesized by the HiPco process (C and D) (purchased from NanoIntergris Inc).

In principle, arc discharge and laser ablation are similar methods, both use a metal-impregnated or pure graphite target or electrode. The selection of which kind of CNT to be produced depends on the purity of graphite and the presence of catalyst. SWNTs could only be formed by adding metal catalysts (Fe, Ni, Y, Co) to graphite. MWNTs and also fullerenes can be synthesized when pure graphite is used instead. In a typical arc discharge synthesis, a low-voltage (~12 to 25 V) and high-current (50 to 120 A) power supply is used (Journet et al., 1997; Shi et al., 1999). An arc is produced across a 1-mm gap between two graphite electrodes of 5 to 20 mm in diameter. An inert gas such as He or Ar is used for the reaction, at a pressure of 100 to 1000 torr. The diameter distribution of SWNTs made by this method is roughly between 1.3 and 1.5 nm.

The characteristics of the CNTs (quality, defect amount, diameter distribution) and the nature of the impurities mainly depend on the synthesis method (Figure 4). Mass produced-SWNTs prepared by CVD (including HiPco) are usually more-defected and have a broader diameter distribution leading to less-ordered bundles than those obtained by high temperature synthesis processes (Figure 5 from U.J. Kim et al., 2005b). The carbonaceous nonnanotube impurities in the samples obtained by these latter are usually in larger concentration and show a larger variety of species compared to those included in the samples obtained by CVD. Both high temperature and CVD as-produced-CNT and especially SWNT samples contain metal residue coming from catalyst required for their growth. These catalysts are often protected by carbon shells and more or less graphitized carbon particles; and they are difficult to efficiently remove. This is especially the case for arc-discharge SWNT samples.

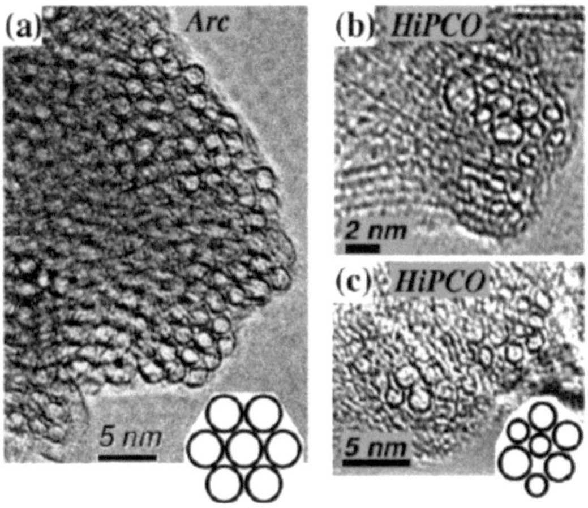


Fig. 5. TEM images of as-grown SWNT bundle cross-section produced by an arc-discharge method (a) and (b, c) SWNT bundle from HiPco materials. Insets are schematic representations of the spatial arrangement of SWNTs in the bundle for each material. From (U.J. Kim et al., 2005b) Copyright 2005 by the American Chemical Society.

### 3.2 Purification and purity

Even if the growing mechanisms are now better understood, it remains difficult to achieve both high-purity and high-yield CNT samples from the production methods. The as-produced soot persistently contains a part of nonnanotube species: (i) carbonaceous impurities showing large range of size and cristallinity from completely amorphous carbon to more or less ordered or well-graphitized particles and (ii) particles of residual metal catalyst required for the synthesis of SWNT-type. Numerous treatments and procedures for mass purification have been proposed, they are based on physical processes and/or chemical treatments (Hou et al., 2008; Cho et al., 2009). The physical methods are based on the difference in size, density, aspect ratio, magnetic properties between the impurities and the CNTs. They generally involve several steps of centrifugation (A. Yu et al., 2006) or filtration which prevent CNTs from severe damage. Most of the chemical procedures involve dry or wet oxidation process and/or an acid treatment. Such treatment is difficult to render selective towards the impurities because CNT are as well sensitive to the used oxidative process (Landi et al., 2005; Sen et al., 2003; Smith et al., 2003; Martinez et al., 2003;Vigolo et al., 2010a). Rigorous optimization of the experimental parameters has to be performed and adapted to the sample source. The carbonaceous impurities showing a large range of structural organization consequently lead to a large scale of stability. Moreover, inherent heterogeneities can be responsible for non controlled behaviors. The final quality of the CNTs (concentration and sidewall-defect amount) can be high but the yield consequent to the attack and the consumption of the CNTs is often disappointing. Subsequent high temperature annealing of the samples is required to restore the cristallinity of the CNTs and to remove the functions that have been grafted at their surface by the previous chemical treatments. Figure 6 (from Martinez et al., 2003) shows TEM images of SWNT samples at a raw state (a) and after each chemical treatment of the followed purification procedure. After nitric acid refluxing (b), SWNTs appear damaged due to the introduction of defects in their structure and a possible intercalation of $HNO_3$ molecules within the bundles. Oxidation treatments being also aggressive to the SWNTs, bundles (c) appear as well attacked. Annealing under Ar atmosphere at 950°C for 10 h (d) is able to remove the defects and the sidewall functions introduced through previous treatments and to restore the SWNT structure.

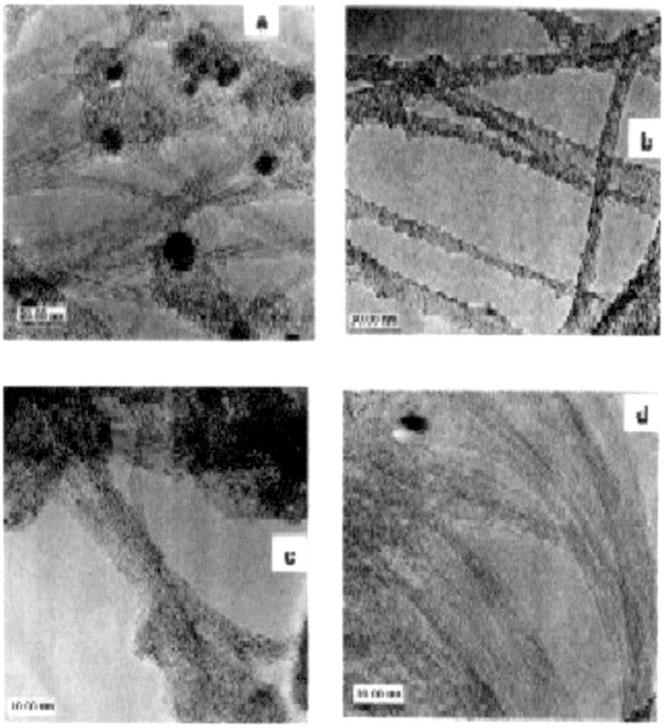


Fig. 6. TEM images of raw SWNT material (a), nitric acid-treated sample (b), air-oxidized sample (c), and annealed sample (d). From (Martinez et al. 2003) Copyright 2003 by Elsevier.

A long-standing issue involving the complete elimination of the metal catalysts from SWNT soot remains to be addressed. It is recognized that these metallic impurities (especially Ni, Fe, Co) can affect both magnetic and electric properties of the CNT samples; as well as defects that can be introduced in the CNT structure during the chemical treatments (Ellis & Ingham, 2006; Kolodiazhnyi & Pumera, 2008). Alternative nonconventional methods that combine selective elimination of catalytic impurities and weak sample-consumption have been developed. The sample is simply heated up under halogen gas (usually chlorine) combined or not- with an oxidation treatment (Zimmerman et al., 2000; Vigolo et al., 2010b). The efficiency of such one-pot process is due to the favored formation of metallic chlorides that are able to induce a mechanical stress on the protecting carbon shells leading to their fracture. In the meantime, the formed metallic chlorides being highly volatile at the used conditions, they are spontaneously eliminated from the sample by sublimation; they simply deposit out of sample at a colder location.

The assessment of CNT purity is really challenging (Arepalli et al., 2004); it does not exist a dedicated characterization technique allowing determining selectively the concentration of CNTs. The currently used techniques for the characterization of the treated samples are

MET, TGA and Raman spectroscopy. TEM allows a qualitative description of the SWNT concentration and the degree of their wall damaging. TGA carried out under oxidative atmosphere is supposed to lead to an assignment and a quantification of the present carbonaceous species from the observed temperatures of removal. Nevertheless, it is difficult to discriminate from the temperature of elimination of CNTs to that of other carbonaceous species. Raman spectroscopy can help in characterizing the damaging of the tubes upon the used treatment.

## 4. Modification of CNT surface properties

CNTs are often entangled according to the production process and they have high tendency to rapidly re-aggregate if no special surface agent or treatment is used to maintain them in a dispersed state. Several means can be used to modify the CNT surface properties, indispensable step for their characterization, their manipulation, their processing or their incorporation in materials (Figure 7 from Hirsh, 2002). It can be achieved by the use of surfactants (Vigolo et al., 2000; Dror et al., 2005), polyelectrolytes (Grunlan et al., 2006), biological molecules (Qiao & Ke, 2006)... that are able to reduce the interfacial energy between the CNT sidewalls and a solvent (Niyogi et al., 2002; Hirsh, 2002; Karousis et al.,

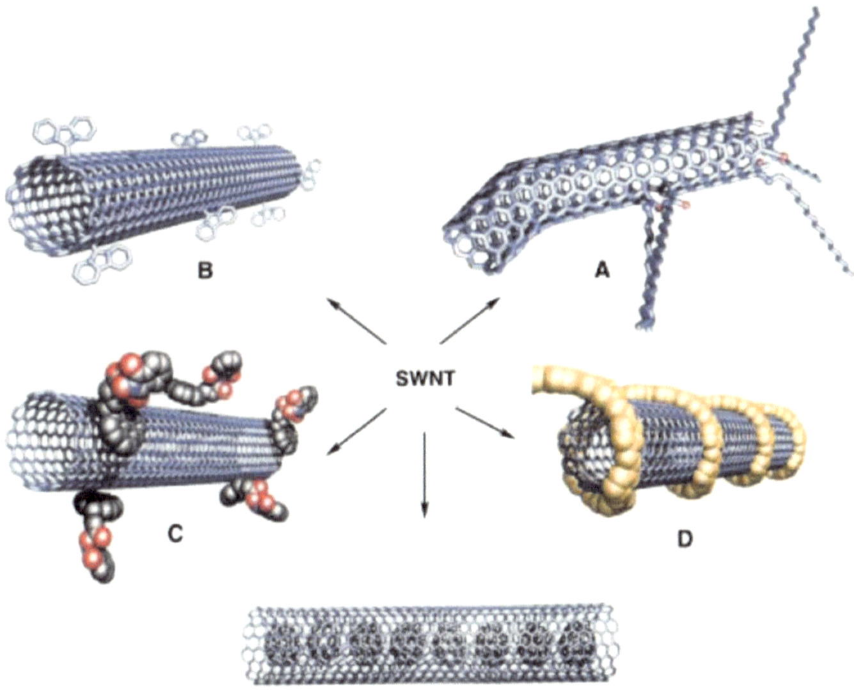


Fig. 7. Possible functionalization approaches for SWNTs. Functionalization possibilities for SWNTs: A) defect-group functionalization, B) covalent sidewall functionalization, C) noncovalent functionalization with surfactants, D) noncovalent functionalization with polymers, and E) filling tube cavity of SWNT, for example, $C_{60}$. From (Hirsch, 2002) Copyright 2002 by John Wiley and Sons.

2010). These physically adsorbed coatings are indeed able to counter-balance the van der Waals attractive forces between the CNT bundles and can also lead to the debundling and CNT individualization (Grossiord et al., 2005). Two others approaches are usually proposed to increase the affinity of the CNTs towards a surrounding media. Covalent functionalization which consists in the attachment of a chemical group, generally having a hydrophilic character, is recognized to be an efficient way to obtain well-quality CNT dispersions. An alternative soft-chemistry route which is based on an electron transfer between an alkali metal and the CNTs allows obtaining high-stable CNT dispersion. This process avoids any introduction of defects in CNT walls since only their electronic structure is modified. These two last mentioned processes are described in more details in the following sections.

### 4.1 Covalent functionalization

The use of surfactant molecules or polymers which are physically adsorbed onto the CNT surface has the advantage of not altering the CNT surface and they can facilitate their manipulation. However, for manufacturing CNT-based composites, their removal during the process is difficult and their presence in the final materials can be responsible for diminishing the composite properties. Indeed, since the remaining molecules are situated at the CNT surface, they can drastically reduce the interaction between the CNTs and the polymer matrix. The chemical functionalization which consists in covalently grafting functional groups on the CNT surface is commonly used to induce both the dispersion of CNTs in solutions of monomers or polymers and good CNT-polymer interaction.

Covalent functionalization can be realized by either modification of surface-bound carboxylic groups situated on the CNTs or direct addition of reagents to the CNT sidewalls by radical attack for example (Sun et al., 2002; Dyke & Tour 2004b). In the first category, CNTs are simply submitted to an oxidation process using $HNO_3$ for example (H. Yu et al., 2008). Oxygen-containing groups including carboxylic acid functions are either directly formed on intrinsic defects or are added at the CNT surface (Zhang et al., 2003). The treated CNTs can be easily dispersed in many solvents (Rosca et al., 2005; Tchoul et al., 2007) and the attached acid functions can be used as sites to attach a variety of functional groups (Niyogi et al., 2002; Wepasnick et al., 2011). In the second category, the functional groups are directly added on the CNT sidewalls without using a preceding acid attack. The developed procedures are often based on the generation of radicals that open the C=C bonds of the CNT structure. In that case, the pre-existing defects are not the favored sites but the addition mechanism rather involves an introduction of additional defects. (Dyke & Tour, 2004a; Liang et al., 2004; Mickelson et al., 1998; Ying et al., 2003). Various functional groups such as alkyl, aryl or fluorine can be covalently attached at the CNT sidewalls. Dispersability of CNTs in various solvents and in polymers can be successfully increased by using a functional group having a good affinity towards the surrounding medium. Changes in the affinity of the functionalized SWNTs towards the solvent can be merely evidenced by dispersion tests. Figure 8 shows photographs of dispersions (in DMF) of SWNT samples that have been submitted to a functionalization process in three steps. Functionalized samples are well dispersed after steps 1 and 3 (dark solution); on the contrary, the dispersion quality of SWNT-PhOH is much reduced after step 2 of the chemical process.

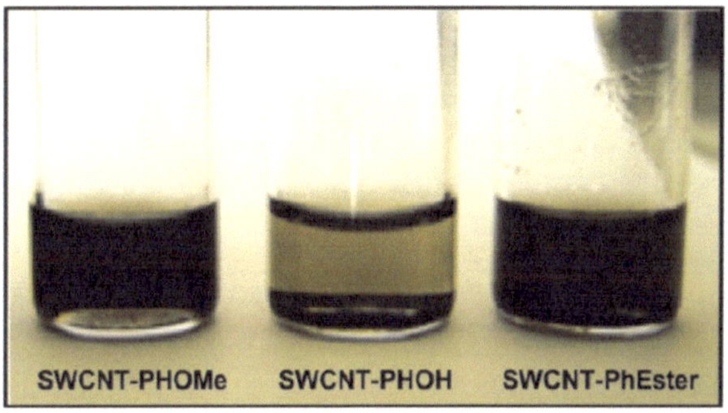


Fig. 8. Photographs of dispersions of functionalized SWNTs in DMF after each step of a chemical procedure. The photographs have been taken two weeks after the dispersion preparation. From (B. Vigolo et al., 2009b) Copyright 2009 by Elsevier.

| methodology | addend | characterization techniques | degree of functionalization | highest solubility |
|---|---|---|---|---|
| diazonium[23,27,29,31] | aryl | UV/vis/NIR, Raman, TGA, XPS, ATR-IR, AFM, TEM | 1 addend in every 10 carbons in SDS/water and 1 addend in every 25 carbons in organic solvent or neat | 0.8 mg/mL in DMF |
| diazonium in oleum[30] | aryl | UV/vis/NIR, Raman, TGA, XPS, ATR-IR, AFM, TEM | 1 addend in every 20 carbons | 0.25 mg/mL in $H_2O$ |
| fluorination[34] | fluorine | ATR-IR, AFM, STM, VTP-EIMS, UV/vis/NIR, Raman, TGA | 1 addend in every 2 carbons | 1 mg/mL in 2-propanol |
| radical chemistry[37] | alkyl | Raman, ATR–IR, UV/vis/NIR, TGA | 1 addend in every 6 carbons | not given |
| dissolving metal reduction (Billups reaction)[38] | alkyl, aryl | TEM, AFM, Raman, ATR-IR, TGA-MS | 1 addend in every 17 carbons | not given |
| azomethine ylides (Prato reaction)[39] | pyrrolidine | UV/vis/NIR, Raman, TEM | 1 addend in every 100 carbons | 50 mg/mL in $CHCl_3$ |
| nitrene[40] | aziridene | XPS, AFM, TEM, NMR, UV/vis/NIR, Raman | 1 addend in every 50 carbons | 1.2 mg/mL in DMSO |
| Bingel reaction[41] | cyclopropane | AFM, $^{19}F$ NMR, XPS | 1 addend in every 50 carbons | not given |
| dichlorocarbene[42] | cyclopropane | mid-IR, Raman, EDS, UV/vis/NIR, AFM, thermal | 1 addend in every 25 carbons | not given |

Table 1. Common SWNT sidewall functionalization methodologies. From (Dyke & Tour 2004b) Copyright 2004 by the American Chemical Society.

Numerous chemical routes have been developed; they are able to attach various functional groups at the CNT sidewalls. They are based on the use of a highly reactive intermediate which is required to attack the carbon nanotubes. As example, table 1 gives the mainly used methods. The aim here is not to enter into details for each method of functionalization but focus the discussion on the related functionalization levels: pertinent parameter for CNT-based materials. Based on its high reactivity with graphite, fluorination was chosen for initial studies (Mickelson et al. 1998). In that case, the very high functionalization level can be found since one C-F function is present every 2 carbon atoms on the CNT. The second methodology involves the well-known substitution by benzenediazonium salts; it leads obtaining CNT being less functionalized with 1 function every 10-20 carbons (Dyke & Tour, 2004a). Arylation and alkylation of CNTs often used as preliminary step for numerous functionalization procedures, can be also obtained by radical reactions for which the degree of functionalization depends on the used process for the generation of the radicals as we will see (Ying et al. 2003; Liu et al., 2007a). SWNTs can be as well modified using cyclization reactions using reactive carbene and nitrene reagents to attack the SWNT walls (Holzinger et al., 2003). Cyclopropanation of SWNTs under Bingel reaction conditions has also been reported (K.S. Coleman et al., 2003). In the case of the functionalization process developed by Billups and coworkers, the reaction leads to ultrahighly lithiated SWNTs (1 lithium atom per 2.2 carbon atoms) that can be further treated with numerous electrophiles including alkyl halides, aryl halides, and even vinyl monomer. Interestingly, functionalized SWNTs are obtained in an individualized state (Liang et al., 2004).

As we have just seen, depending on the chemical mechanism and procedure, the obtained levels of functionalization can be relatively elevated. Integration of covalently functionalized CNTs in polymer matrix could induce good stress–strain transfer between nanotubes and polymer guarantying interesting mechanical properties in composite materials (c.f. section 5). Nevertheless, the breaking of CNT conjugated $\pi$ system may have negative impact on properties (conductivity, in particular) of the obtained composites (Garg & Sinnott, 1998; Byrne & Gun'ko, 2010; Bose et al., 2010). This is the reason why, for composite processing, the functionalization levels have to be controlled and maintained relatively low in order to avoid a strong alteration of the CNT structure. However, grafting degrees are not easy to master and they mainly depend on the involved mechanism of reaction and the means used to facilitate the reaction (Syrgiannis et al., 2010). Chemical reactions assisted by micro-wave are recognized to lead to higher functionalization degree than those obtained by thermally-assisted reactions(Liu et al., 2007b). The chemical procedure we have developed is based on the direct attack of the $sp^2$ carbon on the CNT surface. It advantageously allows having a certain control of the yield of functionalization without the introduction of a large number of defects (Liu et al., 2006; Vigolo et al., 2009b). The obtained low yield of functionalization is efficient enough to modify the surface properties of the CNTs but preserve their structural integrity (Dossot et al., 2007). The other main difficulty regarding the integration of functionalized-CNTs in polymer matrix concerns the homogeneity of the functionalization degree on the CNT surface over the several milligrams of the used CNT sample for composite elaboration (Vigolo et al., 2009c). Because of the high tendency of CNTs to form aggregates, accessibility of reactants to CNT surface has to be improved by using pre-dispersion process (usually done by ultrasounds). Depending on the used solvent which is conducted by the functionalization treatment itself, the CNTs are often poorly dispersed.

### 4.2 CNT reduction: dispersion and debundling

Development of soft chemistry processes such as intercalation reactions is highly challenging for both dispersion and debundling of the SWNT bundles. Indeed, CNTs have demonstrated an amphoteric character since they can be doped or intercalated either by electron-donors or -acceptors (Duclaux, 2002). These reactions are accompanied by an electronic transfer that has been evidenced by means of several techniques such as transport measurements (Grigorian 'et al., 1998; Fischer, 2002) or various spectroscopies and especially Raman spectroscopy (Bendiab et al., 2001; G. Chen et al., 2005). This electronic transfer could as well play a major role in the dispersion process of SWNT bundles. To our knowledge, only donor-type reactions with SWNTs have been successfully used for this purpose. Donor-type reactions are carried out with the strongest reducing metals: the alkali metals. Three routes are possible. The chemical reduction can be thermally assisted (i) in vapor phase or conducted (ii) in liquid phase at room temperature; or, (iii) based on an electrochemically process. The electrochemical intercalation that was mainly studied with lithium is known to induce damages of the SWNTs by progressive solvent co-intercalation.

In-situ X-ray diffraction evidences an irreversible loss of the "triangular" lattice of SWNTs (Fischer, 2002). Routes (i) and (ii) do not show any alteration of SWNT structure. Light alkali metals, lithium and sodium were firstly chemically intercalated into SWNTs using the liquid phase method in THF solutions with several radical anions (naphtalene, benzophenone, anthraquinone, benzoquinone) at ambient temperature (Petit et al., 1999). Heavy alkali metals such as potassium could also be successfully intercalated by means of the same method. In that case, within the obtained compounds, intercalated alkali cations are surrounded by solvent molecules (THF, for example). Figure 9 gives a scheme of two ternary intercalation compounds and reveals that the intertube distance depends on the size of the alkali metal.

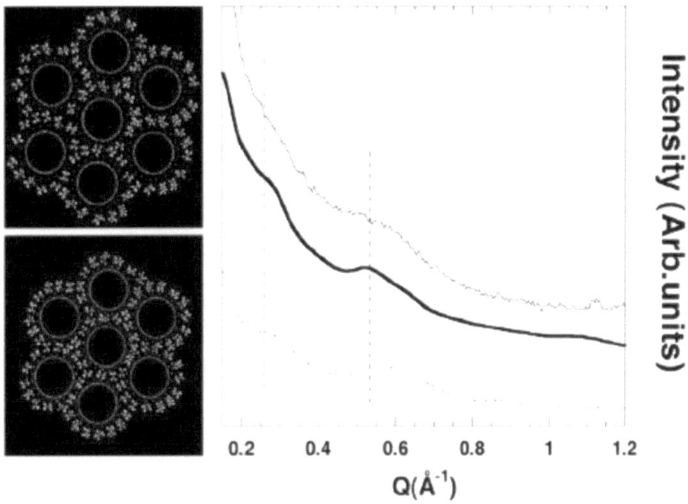


Fig. 9. Left part: Structure of the $KC_{5.88}$-THF (top) and the $LiC_{5.88}$-THF (bottom) compound after energy minimization. Right part: Experimental Neutron Diffraction patterns of the $KC_{5.88}$-THF sample (upper line) and the $LiC_{5.88}$-THF (lower dotted line, shown for comparison). The bold line is a calculated Neutron Diffraction pattern of the $KC_{5.88}$ sample involving the structural parameters described in the text. From (Cambedouzou et al., 2005) Copyright 2005 by the American Physical Society

In the case of intercalation in liquid phase, the electronic transfer was estimated from optical absorption response of the formed radical ion. Heavy alkali metals could be intercalated into SWNTs using vapor phase method (Duclaux, 2002; Duclaux et al., 2003; Vigolo et al., 2009c). The reaction temperature depends on the used alkali metal, more precisely on its vapor pressure. Contrary to the previous method, the vapor-phase process allows preparing binary compounds. The structure of the intercalated materials strongly depends on the quality and the structural parameters of the SWNT sample. For heavy alkali metals, an expansion of the 2D lattice is often observed (Duclaux, 2002). This expansion is reported to be due to the occupation of the interstitial channels. However, the precise location of the alkali metal atoms and the type of sites preferentially occupied are difficult to determine. Moreover, from different observations and measurements, it is possible to claim the absence of intercalation stages in SWNT system contrary to that could be observed in graphite intercalation compounds. The level of chemical intercalation is then difficult to control. The excess of reagent with respect to carbon materials during the chemical reactions leads generally to "saturated" materials. However, the intercalation process itself remains poorly understood because it is not a straightforward process. Indeed, it involves some modifications of both the SWNT electronic structure upon a reduction reaction and the structural parameters subsequent to the intercalation of guest species within the host structure. Two main forces oppose each other throughout the chemical process: favorable forces resulting from the electron transfer and non favorable mechanical forces acting against an increase of the 2D lattice. As for graphite intercalation compounds, the obtained structure mainly results from the balance between these two opposite forces.

Fig. 10. AFM height image of electric arc SWNTs deposited on mica from solution after debundling process of ternary intercalation compounds prepared by a liquid medium intercalation method. Height measurements show a height of about 1 nm on all nanotubes measured consistent with the presence of isolated SWNTs in the solution. From (Pénicaud et al., 2005) Copyright 2005 by the American Chemical Society.

Beyond the precise understanding of the intercalation process itself, the preparation of SWNT-intercalation compounds is of great interest since the obtained reduced SWNTs are able to be dispersed and partially or completely debundled. Indeed, these intercalation compounds can be assimilated to salts capable to dissolve in aprotic polar solvents such as NMP (1-méthyl-2pyrrolidone), DMF (dimethyl formamide) or DMSO (dimethyl sulfoxide) (Pénicaud et al., 2005; Vigolo, et al. 2009c). Such dissolution-like mechanism is spontaneous avoiding the need of sonication or other high energy mechanical mix methods which can damage CNTs. The obtained dispersions are highly stable as long as they are kept in inert atmosphere. Polarity is a key parameter for minimizing the mixing energy of particles in a solvent. The electron transfer between carbon atoms on the CNT surface and the intercalated metal, increasing the surface reactivity, is very helpful for the dispersion of CNTs in polar solvents. Regarding the question of the preferentially occupied sites, the dispersion stability indicates that alkali metal atoms certainly occupy the external sites and the groove sites of the bundles. Moreover, locally the presence of alkali metal atoms in the interstitial sites (between two tubes within a bundle) will favor the debundling process since the increase of

the 2D lattice has already taken place. The dispersion process is then expected to lead to the debundling of the nanotubes between which alkali metals were inserted. The degree of debundling will then depend on the level of filling of interstitial channels. Indeed, if the interstitial channels are not occupied at all, preserved bundles should be observed after dispersion. On the contrary, in the case of a complete occupation of the interstitial sites, complete debundling of SWNTs leading to their individualization is observed. This is the case of SWNTs reduced by alkali metals in liquid phase (Pénicaud et al., 2005). It is then possible to prepare solution containing individualized SWNTs (Figure 10).
In the case of binary compounds prepared by vapor phase, only partial debundling is observed, explained by a non compete filling of the interstitial sites especially favored at the ending and the external part of the bundles. This partial debundling leads to hyperbranched structures of CNTs (Figure 11).

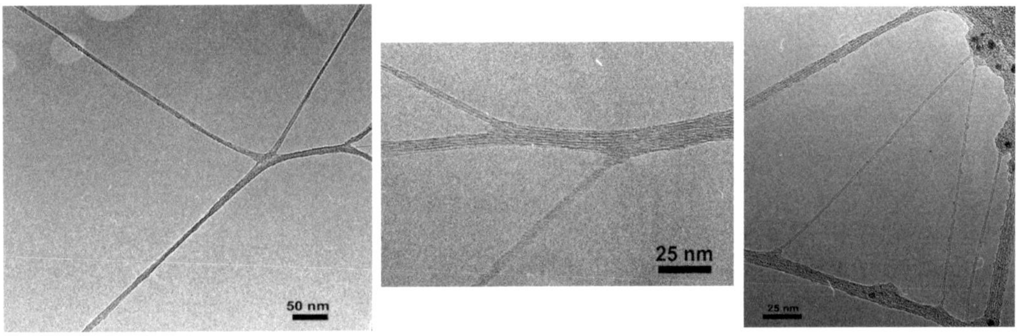


Fig. 11. TEM images of SWNT bundles after the debundling process of binary alkali metal-SWNT compounds prepared through a vapor-phase intercalation process.

This method involving a CNT reduction reaction is of great interest since it allows the formation of individual CNTs or hyperbranched structures of CNTs reducing that way the size of the usually obtained bundles and increasing the developed interfacial surface. Obtaining transparent and conducting composites or antistatic coatings are some of the most challenging applications of CNTs. High-quality dispersions of debundled non-defected SWNTs can be of great interest for manufacturing such new CNT based composites.

## 5. CNT-based composites: multi-scale materials

CNTs are recognized to be the ideal filler to obtain superior composites (J.N. Coleman et al., 2006). Their high aspect ratio combined to their lightness is one of the fundamental requirements for reinforcement of polymer matrices. The challenge consists in successfully transfer both the remarkable mechanical and conductive intrinsic properties of the CNTs at the macroscopic scale of the composite materials. Dispersion is known to be certainly the most fundamental issue. Efficient load transfer cannot be achieved if CNT aggregates remain within the polymer matrix; CNTs have to be randomly dispersed in bundled or isolated state. Their alignment is also reported to enhance the mechanical properties along the CNT axis (Xie et al., 2005). High interfacial strength between CNT surface and the polymer matrix is also an indispensable requirement to induce efficient stress transfer within the composites (Cadek et al., 2004; Gorga et al., 2006).

Various procedures have been developed for the preparation of CNT-based composites (Fiedler et al., 2006; Ma et al., 2010; Spitalsky et al. 2010). Functionalization is widely used to achieve both the dispersion of CNTs and a good wetting between CNTs and polymer chains. Indeed, the procedures that consist in simply mixing non-treated CNTs in the polymer system (either solubilized in a solvent or used in a melting state) present the advantage to be compatible with large-scale production but re-agglomeration phenomenon is still difficult to avoid. Two main approaches for the incorporation of functionalized CNTs in polymer matrices have been developed: *grafting to* and *grafting from*. Their principle is here briefly recalled. The "grafting to" method consists in the attachment of already preformed end-functionalized polymer molecules to functional groups on the CNT surface (Z. Yang et al., 2005). The main limitation of this technique is that the initial binding sterically prevents diffusion of additional macromolecules to the surface. The "grafting from" method consists in first grafting monomer-like functional groups on the CNT surface; the polymerization reaction being further initiated in the monomer solution, the formed polymer chains are directly bound to the CNTs. This method requires a strict control of the amount of the initiator but allows efficient polymer-CNT binding (Jia et al., 1999). Such *in situ* polymerization method in the presence of the functionalized CNTs has been successfully developed (Velasco-Santos et al., 2003; Putz et al., 2004; Vigolo et al. 2009a). The main advantage of this method is the possibility of creating covalent bonds between a modified CNT surface and the polymer matrix.

| Nanotube/Polymer Composite | Nanotube Functionality | Preparation Techniques | $Y_{Poly}$ [GPa] | $Y_{Max}$ [GPa] | NT content [wt%] | $dY/dV_f$ [GPa] | Reference and Year |
|---|---|---|---|---|---|---|---|
| CVD-MWNT/methyl and ethyl methacrylate P(MMA-co-EMA) | Octadecylamine | Solution casting | 1.64 | 2.62 | 10 | 15 | [131] 2004 |
| SWNT/epoxy | Large organic groups | Solution casting and curing | 2.02 | 3.4 | 4 | 95 | [77] 2004 |
| SWNT/polyamide 6 (PA6) | PA6 | Solution casting and melt spinning | 0.44 | 1.2 | 1.5 | 120 | [132] 2005 |
| SWNT/PVA | Hydroxyl | Solution casting | 2.4 | 4.3 | 0.8 | 305 | [133] 2005 |
| CVD-MWNT/PVA | Ferritin protein | Solution casting | 3.4 | 7.2 | 1.5 | 380 | [134] 2005 |
| MWNT/PC | Epoxide | Solution casting and injection molding | 2.0 | 3.8 | 5 | 95.5 | [135] 2006 |
| MWNT/PMMA | PMMA | Melt mixing and extrusion | 2.7 | 2.9 | 3 | 18.15 | [136] 2006 |
| CVD-MWNT/PVC | CPP | Solution casting | 0.56 | 0.9 | 1 | 115 | [129] 2006 |
| CVD-MWNT/PS | CPP | Solution casting | 1.48 | 2.63 | 1 | 304 | [129] 2006 |
| SWNT/copolymers of styrene and vinyl phenol (PSVPh) | Carboxylic acid | Solution casting | 1.5 | 2.1 | 5 | 24.25 | [44] 2006 |
| CVD-MWNT/PU | Carboxylic acid | Solution casting | 0.05 | 0.42 | 20 | 4.5 | [46] 2006 |
| SWNT/biosteel (synthetic spider silk) | Octadecylamine | Solution casting | 1.6 | 1.9 | 0.125 | 381 | [122] 2007 |
| SWNT/PAMAM | Epoxy | Solution casting and curing | 2.76 | 3.49 | 1 | 153 | [56] 2008 |
| CVD-MWNT/Kevlar | PVC | Solution casting | 1.5 | 2.5 | 2 | 300 | [64] 2008 |
| CVD-MWNT/epoxy | TETA | Cast molding | 1.56 | 2.4 | 0.6 | 355 | [126] 2008 |
| CVD-MWNT/PS | Butyl | Solution casting | 1.29 | 1.63 | 0.25 | 433 | [130] 2008 |

Table 2. Mechanical properties of polymer composites containing functionalized CNTs. From (Byrne & Gun'ko, 2010) Copyright 2010 by John Wiley and Sons.

Both conductivity and mechanical properties (see table 2,) of composites containing functionalized CNTs cover quite a large domain (Byrne & Gun'ko, 2010). Mechanical performance of the obtained materials showing the highest Young's modulus of 7.2 GPa (Table 2) is to a certain extent disappointing compared to Young's modulus of the CNTs being more than 100 times higher. It is however essential to remind, that the measured properties depend on the CNT type, the treatment used to modify CNT reactivity and the used preparation technique. It is indeed difficult to directly compare the values of Young's modulus to estimate the effectiveness of a given CNT-polymer system. Moreover, the rate of increase of the Young's modulus ($dY/dV_f$) as a function of the concentration of CNTs within the matrix is reported to be a most pertinent parameter for a quantitative evaluation of the induced reinforcement (Cadek et al., 2004). The interfacial area in CNT-based-materials plays a fundamental role in the reinforcement mechanism and $dY/dV_f$ indeed quantifies the efficiency of the interfacial area to reinforce the obtained materials (table 2).

Micromechanical mechanisms for the reinforcement phenomenon in CNT-based composites are not easy to ascertain due to the multi-scale character of the materials. The interfacial stress transfer could be analyzed by means of fiber pullout model. If adhesion between CNT surface and polymer matrix is efficient, crack propagation tends to be inhibited by a bridging phenomenon increasing the toughness of the material. Reinforcement is thus provided by the positioning of elongated nanoparticles perpendicular to the cracks reinforcing the brittle polymer zone by CNT bridges. Resulting CNT pullout could be useful to characterize mechanical reinforcement in CNT-polymer composites (Hwang et al., 2004; Cooper et al., 2002).

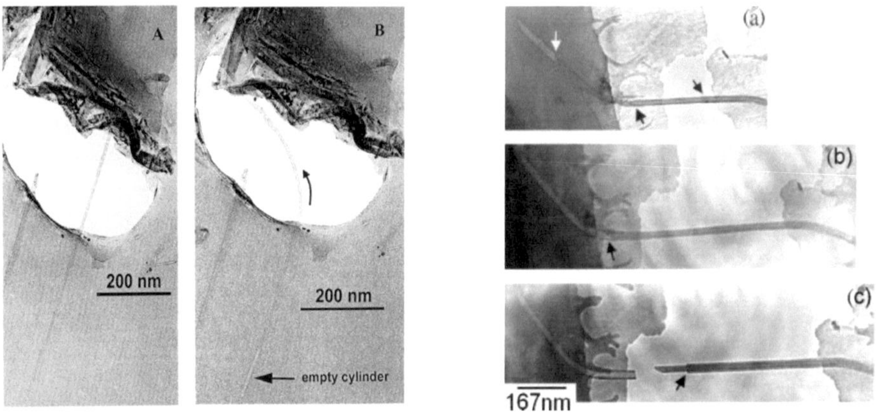


Fig. 12. TEM images of a MWNT crossing a hole in CNT polymer composites: the fiber pullout model.On the left side (from Cooper et al., 2002) (a) TEM image of a nanotube bridging a matrix hole. The bridging nanotube in this image has a diameter of 8.2 nm. (b) TEM image of same specimen following partial pull-out by means of a scanning probe microscope (SPM) tip. The direction of the tip movement was from right to left across the hole; The larger arrow shows the direction of the tip movement; the small arrow indicates the empty cylindrical hole left behind after partial pullout. Copyright 2002 by the American Institute of Physics. On the right side (from Hwang et al., 2004), TEM images of MWNT bridges in PMMA (poly(methyl methacrylate) matrix taken at different times a) t=0, b) t=4, and t=10 min. Copyright 2004 by John Wiley and Sons

Added to the development of tailored preparation procedures, studying CNT-based materials requires tools, techniques and methods that are able to relate the behavior at the molecular scale to the composite properties at a macroscopic scale (Wagner & Vaia, 2004). Beyond the experimental parameters or the nature of the CNT-polymer system, the mechanisms that take place at the CNT-polymer interface have to be better understood; they are indeed recognized to be the key for the CNT-based composite processing.

## 5. Conclusion

To summarize, significant progress has been achieved in the area of CNT processing. A range of new chemical treatments that are tailored in agreement with the desired application of the CNTs have demonstrated interesting results. Among them, covalently functionalized CNTs have been shown to improve both dispersion and polymer-CNT interaction. Avoiding damaging of the CNT structure upon chemical functionalization or purification procedure is still challenging. Further, the exact influence of the pre-treatments on the chemical and physical properties of the CNTs is difficult to assess. Regarding CNT-based composites, the prevailing problems of dispersion and stress transfer are not completely being overcome. However progress in this area has to be continued for the development of selective and innovative chemical treatments that will hopefully help for manufacturing of CNT-based devices and materials.

## 6. Acknowledgment

We would like to thank the Region Lorraine for financial support.

## 7. References

Ajayan, P.M. (1999). Nanotubes from Carbon. *Chemical Reviews*, Vol. 99, pp. 1787-1799

Arab, M.; Picaud, F.; Ramseyer, C.; Babaa, M. R.; Valsaque, F.; McRae, E. (2007). Characterization of single wall carbon nanotubes by means of rare gas adsorption. *Journal Chemical Physics*, Vol. 126(5), 054709/1- 054709/10

Arepalli, S.; Nikolaev, P.; Gorelik, O.; Hadjiev, V.G.; Holmes, W.; Files, B.; Yowell, L. (2004) Protocol for the characterization of single-wall carbon nanotube material quality. *Carbon*, Vol. 42, pp. 1783–1791

Banerjee, S.; Hemraj-Benny, T.; Balasubramanian, M., Fischer, D.A.; Misewich, J.A.; Wong, S.S. (2004). Surface Chemistry and Structure of Purified, Ozonized, Multiwalled Carbon Nanotubes Probed by NEXAFS and Vibrational Spectroscopies. *ChemPhysChem*, Vol. 5, pp.1416-1422

Bendiab, N.; Spina, L.; Zahab, A.; Poncharal, P.; Marlière, C.; Bantignies, J.L.; Anglaret, E.; Sauvajol, J.L. (2001). Combined *in situ* conductivity and Raman studies of rubidium doping of single-wall carbon nanotubes. *Physical Review B*, Vol. 63, pp. 153407

Bose, S.; Khare, R.A.; Moldenaers. P. (2010) Assessing the strengths and weaknesses of various types of pre-treatments of carbon nanotubes on the properties of polymer/carbon nanotubes composites: A critical review. *Polymer*, Vol. 51, pp. 975-993

Bronikowski, M.J. (2006). CVD growth of carbon nanotube bundle arrays. *Carbon*, Vol. 44, pp. 2822-2832

Burghard, M. (2005). Electronic and vibrational properties of chemically modified single-wall carbon nanotubes. *Surface Science Reports*, Vol. 58, pp. 1-109

Byrne, M.T. & Gun'ko Y.K. (2010). Recent Advances in Research on Carbon Nanotube–Polymer Composites. *Advanced Matererials*, Vol. 22, pp. 1672-1688

Cadek, M.; Coleman, J. N.; Ryan, K. P.; Nicolosi, V.; Bister, Fonseca, G. A.; Nagy, J. B.; Szostak, K.; Béguin, F.; Blau W.J. (2004). Reinforcement of Polymers with Carbon Nanotubes: The Role of Nanotube Surface Area. *Nanoletters*, Vol. 4, pp. 353-356

Cambedouzou, J.; Rols, S.; Bendiab, N.; Almairac, R.; Sauvajol, J.L.; Petit, P.; Mathis C.; Mirebeau I.; Johnson, M. (2005). Tunable intertube spacing in single-walled carbon nanotube bundles. *Physical Review B*, Vol. 72, 041404(R)

Cataldo, F. (2000). A Raman study on radiation-damaged graphite by g-rays. *Carbon*, Vol. 38, pp. 623-641

Chattopadhyay, J.; Sadana A.K.; Liang F.; Beach J.M.; Xiao Y.; Hauge R. H.; and Billups W.E. (2005). Carbon Nanotube Salts. Arylation of Single-Wall Carbon Nanotubes. *Organic Letters*, Vol. 7, pp. 4067-4069

Chen, G.; Furtado, C.A.; Kim, U.J.; Eklund, P.C. (2005). Alkali-metal-doping dynamics and anomalous lattice contraction of individual debundled carbon nanotubes. *Physical Review B*, Vol. 72, 155406

Chen, Y.; Yu, J. (2005). Growth direction control of aligned carbon nanotubes. Carbon, Vol. 43, pp. 3183-3194

Cho, H.G.; Kim, S.W.; Lim, H.J.; Yun C.H.; Lee, H.S.; Park C.R. (2009). A simple and highly effective process for the purification of single-walled carbon nanotubes synthesized with arc-discharge. *Carbon*, Vol. 47, pp. 3544-3549.

Coleman, K.S.; Bailey, S.R.; Fogden, S.; Green M.L.H. (2003). Functionalization of Single-Walled Carbon Nanotubes via the Bingel Reaction. *Journal of American Chemical Society*, Vol. 125, pp. 8722-8723

Coleman, J.N.; Khan, U.; Blau, W.J.; Gun'ko Y.K.. (2006). Small but strong: A review of the mechanical properties of carbon nanotube–polymer composites. *Carbon*, Vol. 44, pp. 1624-1652.

Cooper, C.A.; Cohen, S.R.; Barber, A.H.; Wagnera, H.D. (2002). Detachment of nanotubes from a polymer matrix. *Applied Physics Letters*, Vol. 81, pp. 3873-3875

Dai, H. (2002). Carbon Nanotubes: Synthesis, Integration, and Properties. *Acc. Chem. Res.* Vol. 35, pp. 1035-1044

Datsyuk, V.; Kalyva, M.; Papagelis, K.; Parthenios, J.; Tasis, D.; Siokou, A.; Kallitsis I.; Galiotis, C. (2008). Chemical oxidation of multiwalled carbon nanotubes. *Carbon*, Vol. 46, pp. 833-840

Deck, C.P. & Vecchio, K. (2005). Growth mechanism of vapor phase CVD-grown multi-walled carbon nanotubes. Carbon, Vol. 43, pp. 2608-2617

Dillon, A.C.; Parilla, P.A.; Alleman, J.L.; Gennett, T. Jones, K.M.; Heben, M.J. (2005). Systematic inclusion of defects in pure carbon single-wall nanotubes and their effect on the Raman D-band. *Chemical Physics Letters*, Vol 401, pp. 522–528

Dossot, M.; Gardien, F.; Mamane, V.; Fort, Y.; Liu, J.; Vigolo, B.; Humbert, B.; McRae. E. (2007). Optical Parameter to Reveal the Interplay between Covalent Functionalization and the State of Aggregation of Single-Walled Carbon Nanotubes. *Journal of Physical Chemistry C*, Vol. 111, pp.12199-12206

Dror, Y.; Pyckhout-Hintzen, W.; Cohen. Y. (2005). Conformation of Polymers Dispersing Single-Walled Carbon Nanotubes in Water: A Small-Angle Neutron Scattering Study. *Macromolecules*, Vol. 38, pp. 7828-7836

Duclaux, L. (2002). R eview of the doping of carbon nanotubes (multiwalled and single-walled). *Carbon*, Vol. 40, pp. 1751-1764

Duclaux, L.; Salvetat, J.P. ; Lauginie, P.; Cacciaguera, T.; Faugère, A.M.; Goze-Bac, C.; Bernier P. (2003). Synthesis and characterization of SWNT-heavy alkali metal intercalation compounds, effect of host SWNTs materials. *Journal of Physics and Chemistry of Solids*, Vol. 64, pp. 571–581

Dyke, C.A.; Stewart, M.P.; Tour J.M. (2005). Separation of Single-Walled Carbon Nanotubes on Silica Gel. Materials Morphology and Raman Excitation Wavelength Affect Data Interpretation. *Journal of American Chemical Society*, Vol. 127, pp. 4497-4509

Dyke C.A. & Tour, J.M. (2003). Solvent-Free Functionalization of Carbon Nanotubes. *Journal of American Chemical Society*, Vol. 125, pp. 1156-1157

Dyke, C.A. & Tour, J.M. (2004a). Overcoming the Insolubility of Carbon Nanotubes Through High Degrees of Sidewall Functionalization. *Chemistry - A European Journal*, Vol. 10, pp. 812-817

Dyke C.A. & Tour, J.M. (2004b). Covalent Functionalization of Single-Walled Carbon Nanotubes for Materials Applications. *The Journal of Physical Chemistry A*, Vol. 108, pp. 11151-11159

Ellis, A.V. & Ingham B. (2006). Magnetic properties of multiwalled carbon nanotubes as a function of acid treatment. *Journal of Magnetism and Magnetic Materials*, Vol. 302, pp. 378–381

Esconjauregui, S.; Whelan, C.M.; Maex K. (2009). The reasons why metals catalyze the nucleation and growth of carbon nanotubes and other carbon nanomorphologies. *Carbon*, Vol. 47, pp. 659-669

Fiedler B.; Gojny, F.H.; Wichmann, M.H.G.; Nolte, M.C.M.; Schulte, K. (2006). Fundamental aspects of nano-reinforced composites. *Composites Science and Technology*, Vol. 66, pp. 3115-3125

Fischer J.E. (2002). Chemical Doping of Single-Wall Carbon Nanotubes. *Accounts of Chemical Research*, Vol. 35, pp. 1079-1086

Garg, A. & Sinnott, S.B. (1998). Effect of chemical functionalization on the mechanical properties of carbon nanotubes. *Chemical Physics Letters*, Vol. 295, pp. 273–278

Gavillet, J. ; Loiseau, A.; Ducastelle, F.; Thair, S.; Bernier, P.; Stéphan, O.; Thibault, J.; Charlier, J.C. (2002) Microscopic mechanisms for the catalyst assisted growth of single-wall carbon nanotubes. *Carbon*, Vol.40, pp. 1649-1663

Gorga, R.E.; Lau, K.K.S.; Gleason, K.K.; Cohen R.E. (2006). The Importance of Interfacial Design at the Carbon Nanotube/Polymer Composite Interface. *Journal of Applied Polymer Science*, Vol. 102, pp. 1413–1418

Goudon V. & Lasjaunias J.C. (2008). Characterization of two single-wall carbon nanotubes samples by Ar and Kr adsorption isotherms. *Adsorption*, Vol. 14, pp. 1-9

Graupner R. (2007). Raman spectroscopy of covalently functionalized single-wall carbon nanotubes. *Journal of Raman Spectroscopy*, Vol. 38, pp. 673-683

Grigorian, L.; Sumanasekera, G.U.; Loper, A.L.; Fang, S.; Allen, J.L.; Eklund, P.C. (1998) Transport properties of alkali-metal-doped single-wall carbon nanotubes. *Physical Review B*, Vol. 58, R4195

Grobert, N. (2007). Carbon nanotubes –becoming clean. *MaterialsToday*, Vol.10, pp. 28-35

Grossiord, N.; Regev, O.; Loos, J.; Meuldijk, J.; Koning, C.E. (2005). Time-Dependent Study of the Exfoliation Process of Carbon Nanotubes in Aqueous Dispersions by Using UV-Visible Spectroscopy. *Analytical Chemistry*, Vol. 77, pp. 5135-5139

Grunlan, J.C.; Liu, L.; Kim, Y.S. (2006). Tunable Single-Walled Carbon Nanotube Microstructure in the Liquid and Solid States Using Poly(acrylic acid). *Nano Letters*, Vol. 6, pp. 911-915

Hirsch A. (2002). Functionalization of Single-Walled Carbon Nanotubes. *Angew. Chem. Int. Ed.* Vol. 41, pp. 1853-1859

Holzinger, M.; Vostrowsky, O.; Hirsch, A.; Hennrich, F.; Kappes, M.; Weiss, R.; Jellen, F. (2001). Sidewall Functionalization of Carbon Nanotubes. *Angew. Chem. Int. Ed.*, Vol. 40, pp. 4002-4005

Holzinger, M.; Abraham, J.; Whelan, P.; Graupner, R.; Ley, L.; Hennrich, F.; Kappes, M.; Hirsc, A. (2003). Functionalization of Single-Walled Carbon Nanotubes with (R-)Oxycarbonyl Nitrenes. *Journal of American Chemical Society*, Vol. 125, pp. 8566-8580

Hou, P.X.; Liu, C.; Cheng, H.M. (2008). Purification of carbon nanotubes. *Carbon*, Vol. 46, pp. 2003-2025

Hwang G.L.; Shieh, Y.T.; Hwang, K.C. (2004). Efficient load transfer to polymer-grafted multiwalled nanotubes in polymer composites. *Advanced Functional Materials*, Vol. 14, pp. 487-491

Jia, Z.; Wang, Z.; Xu, C.; Liang, J.; Wei, B.; Wu, D.; Zhu, S. (1999). Study on poly(methyl methacrylate):carbon nanotube composites. *Materials Science and Engineering*, A271, pp. 395-400

Jorio, A.; Souza Filho, A.G.; Dresselhaus, G.; Dresselhaus, M.S.; Swan, A. K.; Ünlü, M.S.; Goldberg, B.B.; Pimenta, M.A.; Hafner, J.H.; Lieber, C.M.; Saito R. (2002). *G*-band resonant Raman study of 62 isolated single-wall carbon nanotubes. *Physical Review B*, Vol. 65, 155412

Jorio, A.; Pimenta, M.A.; Souza Filho, A.G.; Saito, R.; Dresselhaus, G.; Dresselhaus, M.S. (2003). Characterizing carbon nanotube samples with resonance Raman scattering. *New Journal of Physics*, Vol. 5, 139.1–139.17

Journet, C.; Maser, W.K.; Bernier, P.; Loiseau, A.; Lamyde la Chapelle, M.; Lefrant, S.; Deniard, P.; Leek, R.; Fischer J.E. (1997). Large-scale production of single-walled carbon nanotubes by the electric-arc technique. *Nature*, Vol. 388, pp. 756-758

Karousis, N.; Tagmatarchis, N.; Tasis, D. (2010). Current Progress on the Chemical Modification of Carbon Nanotubes. *Chemical Reviews*, Vol. 110, pp. 5366-5397

Kim, U.J.; Furtado, C.A.; Liu, X.; Chen, G.; Eklund, P.C. (2005a). Raman and IR Spectroscopy of Chemically Processed Single-Walled Carbon Nanotubes. *Journal of American Chemical Society*, Vol. 127, pp. 15437-15445

Kim, U.J.; Gutiérrez, H. R.; Kim, J.P.; Eklund, P.C. (2005b). Effect of the Tube Diameter Distribution on the High-Temperature Structural Modification of Bundled Single-Walled Carbon Nanotubes. *Journal of Physical Chemistry B, Vol. 109, pp.* 23358-23365

Kim, N.S.; Lee, Y.T.; Park, J.; Ryu H.; Lee, H.J.; Choi S.Y.; Choo, J. (2002). Dependence of the Vertically Aligned Growth of Carbon Nanotubes on the Catalysts. *Journal Physical Chemistry B*, Vol. 106, pp. 9286-9290

Kolodiazhnyi T. & Pumera, M. (2008). Towards an Ultrasensitive Method for the Determination of Metal Impurities in Carbon Nanotubes. *Small*, Vol. 4, pp. 1476 1484

Kuzmany, H.; Kukovecz, A.; Simona, F.; Holzweber, M.; Kramberger, Ch.; Pichler, T. (2004). Functionalization of carbon nanotubes. *Synthetic Metals*, Vol. 141, pp. 113-122

Landi, B.J.; Cress, C.D.; Evans, C.M.; Raffaelle R.P. (2005). Thermal Oxidation Profiling of Single-Walled Carbon Nanotubes. *Chemistry of Materials*, Vol. 17, pp. 6819-6834

Lehman, J.H.; Terrones, M.; Mansfield E.; Hurst, K.E.; Meunier, V. (2011). Evaluating the characteristics of multiwall carbon nanotubes. *Carbon*, Vol. 49, pp. 2581-2602

Lejosne, J. ; Mercier, G. ; Mamane, V. ; Fort, Y. ; Marêché, J.F.; McRae, E. ; Valsaque, F. ; Vigolo. B. (2011). Low degree of functionalization of Single-Walled Carbon Nanotubes probed by highly sensitive characterization Techniques. *Carbon*, Vol. 49, pp. 3010-3018

Liang, F.; Sadana, A.K.; Peera, A.; Chattopadhyay, J.; Gu, Z.; Hauge, R.H.; Billups. W. E. (2004). A Convenient Route to Functionalized Carbon Nanotubes. *NanoLetters*, Vol. 4, pp. 1257-1260

Liu, J.; Rodriguez i Zubiri, M.; Dossot, M.; Vigolo, B.; Hauge, R.H.; Fort, Y.; Ehrhardt, J.J.; McRae, E. (2006). Sidewall functionalization of single-wall carbon nanotubes (SWNTs) through aryl free radical addition. *Chemical Physics Letters*, Vol. 430, pp. 93-96

Liu, J.; Rodriguez i Zubiri, M.; Vigolo, B.; Dossot, M.; Fort, Y.; Ehrhardt, J.J.; McRae. E. (2007a). Efficient microwave-assisted radical functionalization of single-wall carbon nanotubes. Carbon, Vol. 45, pp. 885-891

Liu, J.; Rodriguez i Zubiri, M.; Vigolo, B.; Dossot, M.; Humbert, B.; Fort, Y.; McRae, E. (2007b). Microwave-Assisted Functionalization of Single-Wall Carbon Nanotubes Through Diazonium. *Journal of Nanoscience and Nanotechnology*, Vol. 7, pp. 3519-3523

Ma, P.C.; Siddiqui, N.A.; Marom, G.; Kim J.K. (2010). Dispersion and functionalization of carbon nanotubes for polymer-based nanocomposites: A review. *Composites: Part A*, Vol. 41, pp. 1345-1367

Martinez, M.T.; Calleja, M.A.; Benito, A.M.; Cochet, M.; Seeger, T.; Anson, A.; Schreiber, J.; Gordon, C.; Marhic, C.; Chauvet, O.; Fierro, J.L.G.; Maser, W.K. (2003). Sensitivity of single wall carbon nanotubes to oxidative processing: structural modification, intercalation and functionalization. *Carbon*, Vol. 41, pp. 2247-2256

Maruyama, S.; Kojima, R.; Miyauchi, Y.; Chiashi, S.; Kohno, M. (2002). Low-temperature synthesis of high-purity single-walled carbon nanotubes from alcohol. *Chemical Physics Letters*, Vol. 360, pp. 229-234

Monthioux, M.; Smith, B.W., Burteaux, B.; Claye, A.; Fischer, J.E.; Luzzi, D.E. (2001). Sensitivity of single-wall carbon nanotubes to chemical processing: an electron microscopy investigation. *Carbon*, Vol.39, pp. 1251-1272

Mickelson, E.T.; Huffman, C.B.; Rinzler, A.G.; Smalley, R.E.; Hauge, R.H.; Margrave, J.L. (1998). Fluorination of single-wall carbon nanotubes. Chemical Physics Letters, Vol. 296, pp. 188-194

Nikolaev, P.; Bronikowski, M.J.; Kelley Bradley, R.; Rohmund, F.; Colbert, D.T.; Smith, K.A.; Smalley, R.E. (1999). Gas-phase catalytic growth of single-walled carbon nanotubes from carbon monoxide. *Chemical Physics Letters*, Vol. 313, pp. 91-97

Niyogi, S.; Hamon, M.A.; Hu, H.; Zhao, B.; Bhowmik, P.; Sen, R.; Itkis, M.E.; Haddon. R.C. (2002). Chemistry of Single-Walled Carbon Nanotubes. *Accounts of Chemical Research*, Vol. 35, pp. 1105-1113

Osswald, S.; Flahaut, E.; Ye, H.; Gogotsi, Y. (2005). Elimination of D-band in Raman spectra of double-wall carbon nanotubes by oxidation. *Chemical Physics Letters*, vol. 402, pp. 422-427

Pénicaud, A. ; Poulin, P.; Derré, A.; Anglaret,E.; Petit, P. (2005). Spontaneous Dissolution of a Single-Wall Carbon Nanotube Salt. *Journal of American Chemical Society*, Vol. 127, pp. 8-9

Petit, P.; Mathis, C.; Journet, C.; Bernier, P. (1999). Tuning and monitoring the electronic structure of carbon nanotubes. *Chemical Physics Letters*, Vol. 305, pp. 370–374

Putz, K.W.; Mitchell, C.A.; Krishnamoorti, R.; Green, P.F. (2004). Elastic Modulus of Single-Walled Carbon Nanotube/Poly(methyl methacrylate) Nanocomposites. *Journal of Polymer Science: Part B: Polymer Physics*, Vol. 42, pp. 2286–2293

Qiao, R. & Ke, P.C. (2006). Lipid-Carbon Nanotube Self-Assembly in Aqueous Solution. *Journal of American Chemical Society*, Vol. 128, pp. 13656-13657

Rao, C. N. R.; Satishkumar, B.C.; Govindaraj, A.; Nath, M. (2001). Nanotubes. *ChemPhysChem*, Vol. 2, pp. 78-105

Rosca, I.D.; Watari, F.; Uo, M.; Akasaka, T. (2005). Oxidation of multiwalled carbon nanotubes by nitric acid. *Carbon*, Vol. 43, pp. 3124–3131

Sadeghian, Z. (2009). Large-scale production of multi-walled carbon nanotubes by low-cost spray pyrolysis of hexane. *New Carbon Materials*, Vol. 24, pp. 33-38

Sauvajol, J.L.; Bendiab, N.; Anglaret, E. ; Petit, P. (2003). Phonons in alkali-doped single-wall carbon nanotube bundles. *Comptes Rendus Physique*, Vol. 4, pp. 1035-1045

Sen, R.; Rickard, S.M.; Itkis, M.E.; Haddon, R.C. (2003). Controlled Purification of Single Walled Carbon Nanotube Films by Use of Selective Oxidation and Near-IR Spectroscopy. *Chemistry of Materials*, Vol. 15, pp. 4273-4279

Shi, Z.; Lian, Y.; Zhou, X.; Gu, Z. ; Zhang, Y.; Iijim S.; Zhou, L.; Yue, K.T.; Zhang, S. (1999). Mass-production of single-wall carbon nanotubes by arc discharge method. *Carbon*, Vol. 37, pp. 1449-1453

Singh, C.; Shaffer, M.S.P.; Windle, A.H. (2003). Production of controlled architectures of aligned carbon nanotubes by an injection chemical vapour deposition method. *Carbon*, Vol. 41, pp. 359-368

Smith, Jr.M.R.; Hedges, S.W.; LaCount, R.; Kern, D.; Shah, N.; Huffman, G.P.; Bockrath, B. (2003). Selective oxidation of single-walled carbon nanotubes using carbon dioxide. *Carbon*, Vol. 41, pp. 1221-1230

Spitalsky, Z.; Tasis, D.; Papagelis, K.; Galiotis, C. (2010). Carbon nanotube–polymer composites: Chemistry, processing, mechanical and electrical properties. *Progress in Polymer Science*, Vol. 35, pp. 357-401

Sun, Y.P.; Fu, K.; Lin, Y.; Huang, W. (2002). Functionalized Carbon Nanotubes: Properties and Applications. *Accounts of Chemical Research*, Vol. 35, pp.1096-1104

Syrgiannis, Z.; Gebhardt, B.; Dotzer, C.; Hauke, F.; Graupner, R.; Hirsch, A. (2010). Reductive Retrofunctionalization of Single-Walled Carbon Nanotubes. *Angew. Chem. Int. Ed.*, Vol. 49, pp. 3322 -3325

Tasis, D.; Tagmatarchis, N.; Bianco, A.; Prato, M. (2006). Chemistry of Carbon Nanotubes. Chemical Reviews, Vol. 106, pp. 1105-1136

Tchoul, M.N.; Ford, W.T.; Lolli, G.; Resasco, D.E.; Arepalli, S. (2007). Effect of Mild Nitric Acid Oxidation on Dispersability, Size, and Structure of Single-Walled Carbon Nanotubes. *Chemistry of Materials*, Vol. 19, pp. 5765-5772

Van Noorden, R. (2011). The Trials of new carbon. Nature, Vol. 469, pp. 14-16

Velasco-Santos, C.; Martínez-Hernández, A.L.; Fisher, F.T.; Ruoff, R.; Castaño, V.M. (2003). Improvement of Thermal and Mechanical Properties of Carbon Nanotube Composites through Chemical Functionalization. *Chemistry of Materials*, Vol. 15, pp. 4470-4475

Vigolo, B.; Coulon, C.; Pénicaud, A.; Bernier, P.; Journet, C.; Sauder, C.; Pailler R.; Poulin, P. (2000). Macroscopic Fibers and Ribbons of Oriented Carbon Nanotubes. *Science*, Vol. 289, pp. 1331-1334

Vigolo, B.; Cojocaru, C.S.; Faerber, J.; Arabski, J.; Gangloff, L.; Legagneux, P.; Lezec, H.; Le Normand, F. (2008). Localized CVD growth of oriented and individual carbon nanotubes from nanoscaled dots prepared by lithographic sequences. *Nanotechnology*, vol. 19, 135601

Vigolo, B.; Vincent, B.; Eschbach, J.; Bourson, P.; Marêché, J.F.; McRae, E.; Müller, A.; Soldatov, A.; Hiver, J.M.; Dahoun, A.; Rouxel, D. (2009a). Multiscale Characterization of Single-Walled Carbon Nanotube/Polymer Composites by Coupling Raman and Brillouin Spectroscopy. *Journal of Physical Chemistry C*, Vol. 113, pp. 7624-7628

Vigolo, B.; Mamane, V.; Valsaque, F.; Le, T.N.H.; Thabit, J.; Ghanbaja, J.; Aranda, L.; Fort, Y.; McRae, E. (2009b). Evidence of sidewall covalent functionalization of single-walled carbon nanotubes and its advantages for composite processing. *Carbon*, vol. 47, pp. 411-419

Vigolo, B.; Hérold, C.; Marêché, J.F.; Bourson, P.; Margueron, S.; Ghanbaja, J.; McRae E. (2009c). Direct Revealing of the Occupation Sites of Heavy Alkali Metal Atoms in Single-Walled Carbon Nanotube Intercalation Compounds. *Journal of Physical Chemistry C*, Vol. 113, pp. 7624–7628

Vigolo, B.; Hérold, C. ; Marêché, J.F. ; Ghanbaja, J. ; Gulas, M.; Le Normand, F. ; Almairac, R. ; Alvarez, L. Bantignies, J.L. (2010a). A comprehensive scenario for commonly used purification procedures of arc-discharge as-produced single-walled carbon nanotubes. *Carbon*, Vol. 48, pp. 949-963

Vigolo, B.; Mercier, G; Hérold, C.; Marêché, J.F.; Cahen, S. (2010b). Purification Method of Carbon Nanotubes. *French Patent Application* CNRS-UHP-INPL n° FR 1058904

Wagner, H.D. & Vaia, R.A. (2004). Nanocomposites: issues at the interface. *MaterialsToday*, 38-42

Wang, X.; Li, N.; Webb, J.A.; Pfefferle, L.D.; Haller, G.L. (2010). Effect of surface oxygen containing groups on the catalytic activity of multi-walled carbon nanotube supported Pt catalyst. *Applied Catalysis B: Environmental*, Vol. 101, pp. 21-30

Wepasnick, K.A.; Smith, B.A.; Schrote, K.E.; Wilson, H.K.; Diegelmann, S.R.; Fairbrother, D.H. (2010). Surface and structural characterization of multi-walled carbon nanotubes following different oxidative treatments. *Carbon*, Vol. 49, pp. 24-36

Willems, I.; Kónya, Z.; Colomer, J.F.; Van Tendeloo, G.; Nagaraju, N.; Fonseca, A.; Nagy, J.B. (2000). Control of the outer diameter of thin carbon nanotubes synthesized by catalytic decomposition of hydrocarbons. *Chemical Physics Letters*, Vol. 317, pp. 71-76

Xie, X.L.; Mai, Y.W.; Zhou, X.P. (2005). Dispersion and alignment of carbon nanotubes in polymer matrix: A review. *Materials Science and Engineering*, R 49, pp. 89–112

Yamada, T.; Namai, T.; Hata, K.; Futaba, D.N.; Mizuno, K.; Fan, J.; Yudasaka, M.; Yumura, M.; Iijima, S. (2006). Size-selective growth of double-walled carbon nanotube forests from engineered iron catalysts. *Nature Nanotechnology*, Vol. 1, pp. 131-136

Yang, Z.; Pu, H.; Yin, J. (2005). Covalent functionalization of multiwalled carbon nanotubes by polyvinylimidazole. *Materials Letters*, Vol. 59, pp. 2838-2841

Yang, C.M.; Kaneko, K.; Yudasaka, M.; Iijima, S. (2002). Effect of Purification on Pore Structure of HiPco Single-Walled Carbon Nanotube Aggregates. NanoLetters, Vol. 2, pp. 385-388

Ying, Y.; Saini, R.K.; Liang, F.; Sadana, A.K.; Billups, W.E. (2003). Functionalization of Carbon Nanotubes by Free Radicals. *Organic Letters*, Vol. 5, pp. 1471-1473

Yu, A.; Bekyarova, E.; Itkis, M.E.; Fakhrutdinov, D.; Webster, R.; Haddon, R.C. (2006). Application of Centrifugation to the Large-Scale Purification of Electric Arc-Produced Single-Walled Carbon Nanotubes. *Journal of American Chemical Society*, Vol. 128, pp. 9902-9908

Yu, H.; Jin, Y.; Peng, F.; Wang, H.; Yang, J. (2008). Kinetically Controlled Side-Wall Functionalization of Carbon Nanotubes by Nitric Acid Oxidation. *Journal of Physical Chemistry C*, Vol. 112, pp. 6758-6763

Zhang, J.; Zou, H.; Qing, Q.; Yang, Y.; Li, Q.; Liu, Z.; Guo, X.; Du, Z. (2003). Effect of Chemical Oxidation on the Structure of Single-Walled Carbon Nanotubes. *Journal of Physical Chemistry* B, Vol. 107, pp. 3712-3718

Zimmerman, J.L.; Kelley Bradley, R.; Huffman, C.B.; Hauge, R.H.; Margrave, J.L. (2000). Gas-Phase Purification of Single-Wall Carbon Nanotubes. Chemistry of Materials, Vol. 12, pp. 1361-1366

2

# Initial Growth Process of Carbon Nanotubes in Surface Decomposition of SiC

Takahiro Maruyama and Shigeya Naritsuka
*Department of Materials Science and Engineering, Meijo University*
*Japan*

## 1. Introduction

### 1.1 Basic property of carbon nanotube

Carbon nanotubes (CNTs) are cylinders of graphene which is a single planar sheet of $sp^2$-bonded carbon atoms. From their unique structures, they exhibit several novel properties, such as high strength, high performance in electrical and thermal conductors, etc. Therefore, since the discovery by Iijima in 1991 (Iijima, 1991), CNTs have been one of the hottest topics in both physics and material science. Mainly, their structures are specified with four parameters; the number of walls, diameter, length and chirality. Dependent of the number of walls, CNTs are categorized as single-walled nantoubes (SWNTs) and multi-walled nanotubes (MWNTs) and SWNT structure can be specified uniquely with a chiral vector $\boldsymbol{C}_h = n\boldsymbol{a}_1 + m\boldsymbol{a}_2$ ($n$, $m$ are integers, $0 \leq |m| \leq n$) defined relative to the two-dimensional graphene sheet (Saito et al., 1998). Figure 1 shows the graphene honeycomb lattice. The unit cell is spanned by the two vectors $\boldsymbol{a}_1$ and $\boldsymbol{a}_2$ and contains two carbon atoms, where the basis vectors of length $a_0 = 0.246$ nm form an angle of 60°. The chiral vector $\boldsymbol{C}_h$ can be expressed by these two vectors $\boldsymbol{a}_1$ and $\boldsymbol{a}_2$, and it corresponds to a section of the SWNT perpendicular to the nanotube axis. In Fig. 1, the chiral vector $\boldsymbol{C}_h = 6\boldsymbol{a}_1 + 3\boldsymbol{a}_2$ of an (6, 3) tube is shown. The translation vector, $\boldsymbol{T}$, which is parallel to the nanotube axis and is normal to the chiral vector $\boldsymbol{C}_h$ in the unrolled honeycomb lattice is also shown and it can be expressed in terms of the basis vectors $\boldsymbol{a}_1$ and $\boldsymbol{a}_2$ as, $\boldsymbol{T} = 4\boldsymbol{a}_1 - 5\boldsymbol{a}_2$. The electronic structure of a SWNT is determined by these two indices $(n, m)$ and the SWNTs could be semiconductors and metals, depending on them. For example, metallic CNTs show the relation, $n - m = 3q$, with $q$ being an integer. The diameter of a SWNT, $d$, is also specified with these chiral indices as,

$$d = \left(\frac{a_0}{\pi}\right)\sqrt{\left(n^2 + nm + m^2\right)} \tag{1}$$

The band gap of semiconducting SWNTs is inversely proportional to the diameter. Therefore, it is important to control and obtain SWNTs with unique chirality at will, for realization of CNT devices.

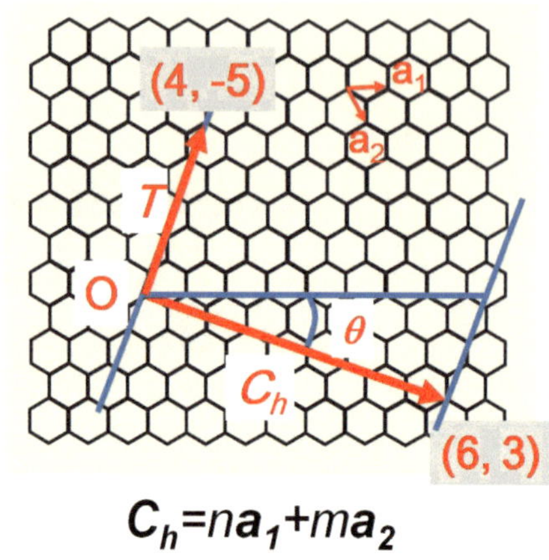


Fig. 1. Graphene honeycomb lattice with the lattice vectors $a_1$ and $a_2$. The chiral vector $C_h = 6a_1 + 3a_2$ of the *(6, 3)* tube and the translation vector $T = 4a_1 - 5a_2$ are shown.

### 1.2 Carbon nanotube growth by surface decomposition of SiC

There have been three main methods to produce CNTs; arc discharge, laser ablation and chemical vapour deposition (CVD). Arc discharge and laser ablation utilize the evaporation of a graphite target to create gas-phase carbon fragments which recombine to form CNTs. Although both produce CNTs with high crystallinity due to the high growth temperature, it is difficult to control the structure of CNTs. Recently, CVD has been widely used for CNT growth because it is suitable for organizing CNTs over a large surface. CVD utilize metal catalyst particles in the gas phase or on surfaces, decomposing a carbon containing feedstock gas, such as acetylene or ethanol. These procedures can be applied to control the structure of CNTs through the size and shape of catalyst. So far, the diameters of CNTs could be controlled to some extent through the control of catalyst particle size (Jeong et al., 2007). However, there still remains some distribution in the CNT diameter, and also the control of chirality has never been achieved.

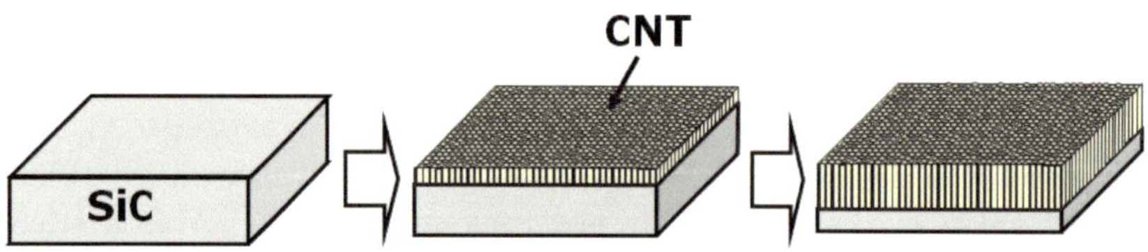


Fig. 2. Schematic illustration of CNT growth by surface decomposition of SiC.

As distinct from these conventional growth techniques, surface decomposition of SiC has been reported as a novel CNT growth technique during recent years. It was first discovered by M. Kusunoki et al. in 1997, who found that CNTs grew on SiC(-1-1-1) surface after heating a 3C-SiC single crystal at 1700°C using YAG laser during a transmission electron microscope (TEM) observation (Kusunoki et al., 1997). They observed that CNTs were mostly oriented along the [111] direction. Later, they reported that CNTs could be easily produced on 6H-SiC(000-1) surface only after heating in a vacuum electric furnace (Kusunoki et al., 2000). The schematic picture of the growth process of CNTs by surface decomposition of SiC is shown in Fig. 2. As heating temperature rises, Si atoms are desorbed and CNTs grew into the SiC substrate the axes of which are kept perpendicular to the substrate surface. This growth method has an exclusive characteristics because it needs no catalyst, and self-aligned CNTs were grown with desorption of Si atoms.

Also, they pointed out that CNT growth was strongly dependent on the polar faces of SiC. Figures 3(a) and (b) shows schematic cross-sectional and top view of 6H-SiC, respectively, and SiC (000-1) and (0001) faces are shown at the top layer and bottom in Fig. 3(a). On the (000-1) face (carbon–face), there are dangling bonds from carbon atoms which are perpendicular to the substrate surface, while dangling bonds from Si atoms are seen on the (0001) face (Si-face). Kusunoki et al. pointed out that CNT growth was observed only on carbon-faces of SiC single crystals, such as (000-1) face of 6H-SiC and (-1-1-1) face of 3C-SiC, while graphite sheets were formed on the surface after heating Si-faces (Kusunoki et al., 2000).

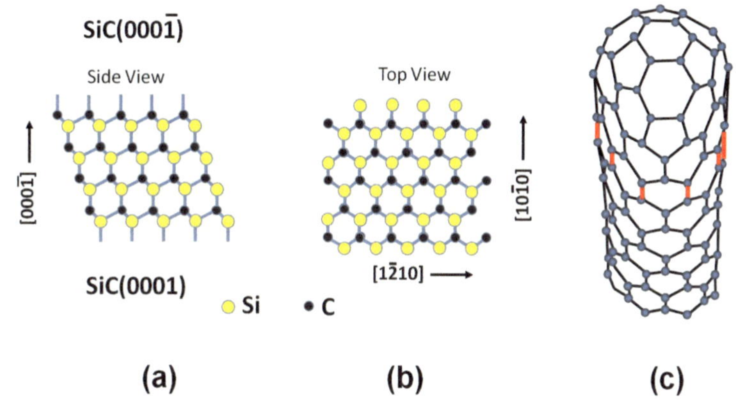


Fig. 3. Schematic representations of (a) cross-sectional view and (b) top view of 6H-SiC, and (c) zigzag-type of a SWNT. The top side and bottom side of (a) correspond to (000-1) and (0001) faces, respectively.

So far, through high-resolution TEM (HRTEM) images, they revealed several characteristics of CNTs grown by this method: (a) most CNTs are two to five layered 2-5 nm in diameter, (b) The CNTs are grown perpendicular to the SiC surface, (c) The density of CNTs is above $10^{12}$ cm$^{-2}$, which is higher than those grown by other methods, (d) The CNTs are atomically bonded with the SiC crystal at the interface and (e) most of the CNTs have unique chirality, that is, zigzag-type (Fig. 3(c)) (Kusunoki et al., 2002). Although there are still several problems to be resolved for device fabrication, such as purity of the CNT film and homogeneity in the number of walls and diameter, CNTs grown by surface decomposition of SiC have several advantages to realize CNT devices. Among these useful properties, the unique chirality of grown CNTs is the most important and mysterious, because the growth of CNTs with unique chirality has never been attained by other growth techniques. Therefore, surface decomposition method might play an important role in fabrication of CNT devices. Also, elucidation of the mechanism of selectively chirality alignment will be invaluable and useful to control and align chirality of CNTs grown with other methods.

### 1.3 Carbon nanocap

The formation process in the initial stage of CNT growth by surface decomposition of SiC has been reported by several groups. TEM studies by Kusunoki's group showed that arced sheets of carbon with an outer diameter of 5 nm and height of 1-2 nm appeared on the SiC(000-1) face after heating at 1250°C for 30 min (Kusunoki et al., 2000). After heating at 1300°C, two-, to three-layered carbon nanocaps 3-5 nm in diameter and 3-5 nm in height were generated in a dense formation on the surface. After heating at higher temperature, CNTs grew below these carbon nanocaps with desorption of Si atoms indicating that these carbon nanocaps correspond to the top of CNTs. This means that the structure of CNTs is closely related to the carbon nanocaps, and elucidation of the formation mechanism of the carbon nanocaps is a key point to clarify unique chirality of CNTs grown by this method.

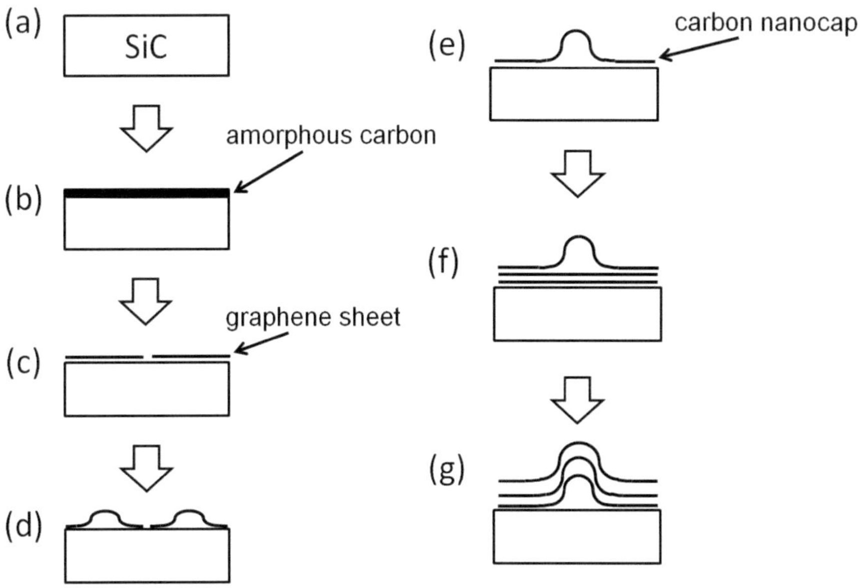


Fig. 4. Schematic illustration of the initial growth process of carbon nanocaps and CNTs proposed by Watanabe's group.

Except for pioneering and essential investigation by Kusunoki's group, few studies have been reported about the formation mechanism of carbon nanocaps. Watanabe et al. carried out TEM and scanning tunnelling microscope (STM) observation for CNT growth on 3C-SiC(-1-1-1) and proposed a model for carbon nanocap formation. The schematic figure of their model is shown in Fig. 4 (Watanabe et al., 2001). In their model, at first, amorphous carbon was formed on the SiC surface with the evaporation of Si atoms. Then, they were crystallized and graphene flakes were formed and aligned parallel to the SiC surface with the temperature rise. Finally, lifting of part of the graphene, carbon nanocaps were formed. This model is significant as first attempt to explain this unique phenomenon. However, intuitively, their model seems to have some discrepancy against the observed experimental results; in their model, the diameter of a CNT would be enlarged, however, TEM observation showed that the diameter was maintained during the growth. Also, it seems to be difficult to explain the reason why enormously high-density CNTs are grown. To clarify the real growth process, analysis at the atomic level is essential, including the alignment and direction of C-C bonding during the crystallization

In this study, we investigated the details of formation process of carbon nanocaps using various experimental techniques; STM, X-ray photoemission (XPS) and near edge X-ray absorption fine structure (NEXAFS) spectroscopies. Based on these experimental results, we attempted to propose a formation mechanism of carbon nanocaps.

## 2. Formation mechanism of carbon nanocap

### 2.1 Formation process of carbon nanocap

To clarify the details of formation process of carbon nanocaps, it is essential to clarify the surface structure and composition. However, few studies have been reported with combing XPS measurement and STM observation for formation process of carbon nanocaps. Here, we carried out both XPS measurements and STM observations for carbon nanocaps formed on 6H-SiC(000-1) (Maruyama et al., 2006, 2007). Figure 5(a) and (b) show XPS spectra of 6H-SiC(000-1) surface before and after heating at 1100°C in UHV, respectively, where the spectrum of Fig. 5(a) is that from the sample after etching with 10% HF for 15 min. To grow carbon nanocaps, the temperature was slowly increased, keeping the heating rate below 2°C/min in the temperature region above 400°C. Both spectra were measured at room temperature with a laboratory photon source ($Mg\ K_{\alpha}$ : 1253.6 eV) at the take-off angle of 20° from the surface plane, to enhance the surface sensitivity. As expected, no elementary carbon was observed before heating. It should be noted that some oxides such as $Si_xC_yO_z$ are main components on the surface and the SiC component is weak. Taking into account that the escape depth of photoelectron was about only one nanometer in the measurement condition, most of SiC(000-1) surface was covered with these oxides even after HF etching. After heating at 1100°C, a component of elementary carbon derived from $sp^2$ bonding became dominant, corresponding to both desorption of Si. The component from the SiC substrate is still observed, which suggests that the carbon layer accumulated on the SiC surface was very thin and about a few nanometers. Relative intensity of each component in XPS spectra was estimated as the sum should be one, and its temperature dependence is shown in Fig. 5(c). After heating above 1000°C, the component of elementary carbon ($sp^2$) appeared and it gradually increased, as the heating temperature rose. On the other hand, those related from SiC and $Si_xC_yO_z$ gradually decreased with the temperature, although a small amount of oxides still remained at 1200°C. The increase of the component of carbon $sp^2$ bonding correspond to the progress in the desorption of Si atoms and the accumulation of carbon atoms on the SiC surface, which leads to carbon nanocap formation.

Surface structure change with the heating temperature is shown in Fig. 6. These STM images were observed at room temperature after heating at each temperature in UHV. After heating at 800°C, step-like structures appeared on some portion of the surface (a). The area of each terrace was small and several tens of nanometers. After heating at 1100°C, densely packed grains, around 1-2 nm in size, were observed over the entire surface (b). These nanoparticles formed clusters at 1150°C, keeping the sizes of individual particles unchanged (c). The cluster size ranged from around 3 to 5 nm. Taking into account the result of XPS measurements, the nanoparticles were elemental carbon which were accumulated after desorption of Si. After heating at 1200°C, the SiC surface was covered by a large number of grains ranging in size from around 3 to 5 nm, as shown in STM images of Fig. 6(d). The blowup in Fig. 6(d) shows that their surfaces were covered with cobweb-like patterns, indicating that the carbon nanocaps were formed at this temperature.

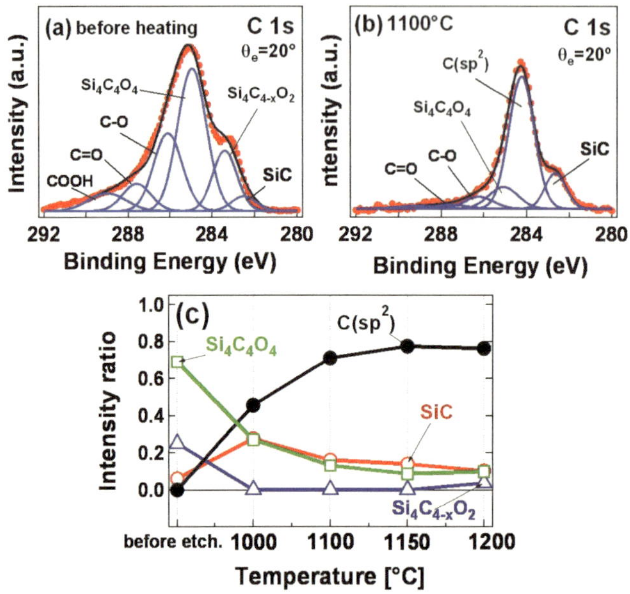


Fig. 5. XPS spectra of SiC(000-1) surface of (a) before heating (after HF etching), and (b) after heating at 1100°C in UHV. Both spectra were measured at the take-off angle of 20°from the surface plane. (c) Temperature dependence of relative intensities of each component in XPS spectra.

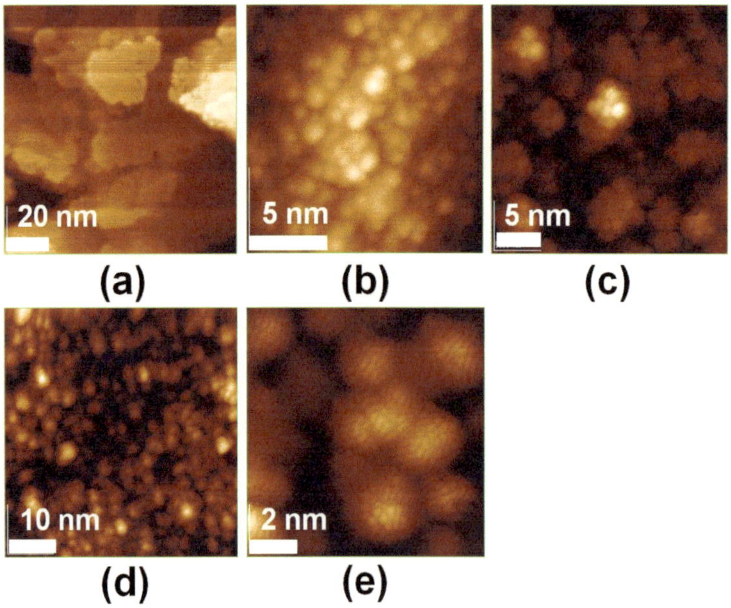


Fig. 6. STM images of 6H-SiC(000-1) substrate obtained in subsequent stages of the annealing process under UHV: (a) after annealing at 800°C, (b) 1100°C, (c) 1150°C, and (d) and (e) 1200°C. (e) is a blowup of (d).

Figure 7 shows another STM images for carbon nanocaps formed on SiC(000-1) which were annealed at 1250ºC in a vacuum electric furnace (Bang et al., 2006). Carbon nanocaps formed on the entire surface were clearly shown in (a). Figure 7(b) shows an atomic-resolution STM images of the blowup of a cap structure produced on a 6H-SiC(000-1) substrate. They showed convex structures, 3-5 nm in diameter and 1-2 nm in height, exhibiting cobweb-like patterns, as shown in Fig. 6(e).

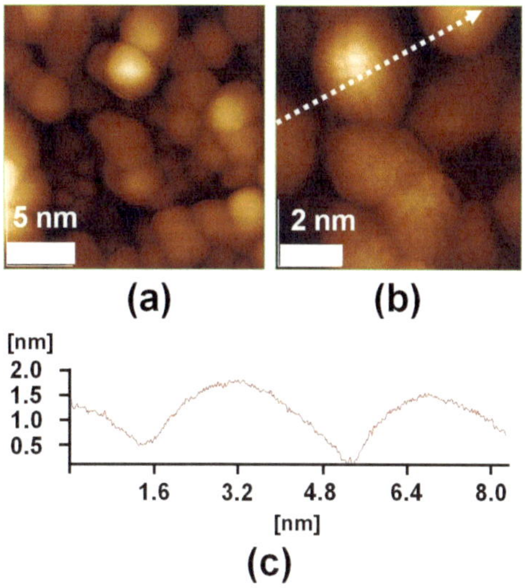


Fig. 7. (a) and (b) STM image of the cap structures produced on 6H-SiC(000-1) substrates after heating at 1250°C in a vacuum electric furnace. (c) Height profile of cap structures in the direction marked by an arrow in (b).

Figure 8(a) shows one portion of the surface of a carbon nanocap formed on 6H-SiC(000-1) after annealing in a vacuum electric furnace and (b) and (c) show magnified feature of STM images of (a). Figure (b) and (c) are the same except for the markers shown by open circles and a square. (Bang et al., 2006). Figure 8(b) clearly reveals the cobweb-like pattern, which corresponds to carbon networks, on the surface of carbon nanocaps. In Fig. 8(c), one large circle, one large square, and nine small circles can be seen. There are six small open circles around the right large open circle, which confirms that this pattern is a carbon network composed of hexagons. In contrast, the left large square is surrounded by only five small circles, which indicates that the portion of network represented by the left large open square is a pentagon on the nanocap surface. It is well known that a convex carbon network contains several pentagons, as observed on the surface of fullerenes. Given the convex shape of the nanocap, it is reasonable to observe a pentagon on the surface. These STM observations support the assumption that the carbon nanocap had been already crystallized when it was formed on the SiC surface.

Crystallization during the formation of carbon nanocaps was directly confirmed by *in situ* NEXAFS measurements. The left part of Fig. 9 shows carbon *K* edge NEXAFS spectra of SiC(000-1) surfaces at 1190 and 1250°C obtained using a synchrotron light source. All spectra were measured in Auger electron yield detection mode, with the sample maintained at the heating temperature to carry out *in situ* measurements. The arrangement of the sample and the synchrotron radiation source is shown schematically in the right part of Fig. 9, where the angle between the incident X-rays and the electron energy analyzer was fixed at 60°, and the incident angle, $\theta$, is defined as the angle between the light source direction and the sample surface. In Fig. 9, NEXAFS spectra obtained with different $\theta$ values are shown for each temperature and are normalized to the highest peaks. The light from the synchrotron source was linearly polarized, so that when $\theta$ = 90°, the electric field vector, $E$, is parallel to the sample surface, whereas when $\theta$ = 30°, $E$ contains components both parallel and perpendicular to the surface.

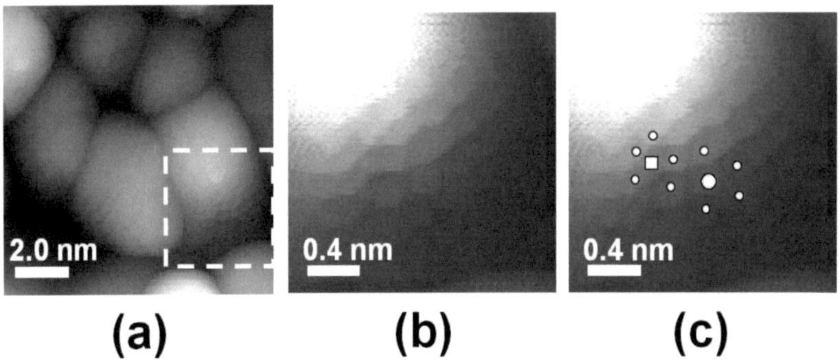


Fig. 8. (a) STM image of another region of the sample shown in Fig. 7. (b) and (c) Magnified feature of STM images of the cap structures in (a). (b) and (c) are exactly the same image, except for the open circle and square markers in (c). The large circle on the right is surrounded by six small circles, while the large square on the left is surrounded by five small circles.

As seen in Fig. 9, for both incident angles, and irrespective of the temperature, sharp C-C π* resonance peaks were observed at approximately 285 eV in all spectra. In addition, at 1250°C, broad σ* resonances were observed in the region from 290 to 315 eV, accompanied by a sharp σ* bound exciton at ~291.5 eV ((a) and (b)). These two peaks are characteristic of carbon materials composed of well-ordered graphene sheets. Therefore, the spectra in Figs. 9(a) and (b) are expected to be associated with carbon nanocaps. At 1190°C, slight differences were observed between the spectra for the two incident angles. At $\theta$=30°, the spectrum was quite similar to that at 1250°C. On the other hand, at $\theta$=90°, the relative intensity of the σ* to the π* resonance increased, and the σ* bound exciton peak became blurred. It has been reported that for amorphous carbon, the NEXAFS spectrum exhibits a structureless σ* resonance and the relative intensity of the σ* to the π* resonance is higher than that for graphite (Comeli et al., 1988). Therefore, the spectra at 1190°C were considered to be derived from a mixture of ordered graphene layers and amorphous carbon. Taking the STM results into account, the spectra at 1190°C are associated with crystallization during the production of carbon nanocaps. It is well known that the π* resonance peak is strongly enhanced when the electric field vector of the incident X-rays is parallel to the π* orbitals (Banerjee et al., 2005). Therefore, the increased relative intensity of the π* resonance at $\theta$=30° indicates that the majority of π* orbitals were nearly perpendicular to the surface, that is, most C-C bonds were directed nearly parallel the surface at the beginning of carbon nanocap formation. On the other hand, the carbon nanocaps are dome-like and they consist of graphene sheets with almost the same amount parallel and perpendicular to the SiC surface (Kusunoki et al., 2000, Bang et al., 2006). That is why there was no remarkable difference in the spectra at 1250°C for the two incident angles.

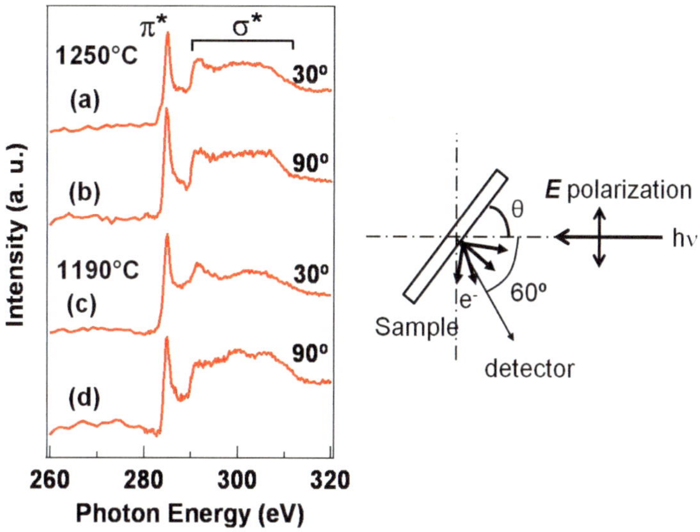


Fig. 9. Left part: NEXAFS spectra of 6H-SiC(000-1) during heating at 1250°C (a) and (b), and 1190°C(c) and (d). The spectra of (a) and (c) were measured at an incident angle of light source was 30°, while those of (b) and (d) were at 90°. Right part: Scheme of registration of angular dependence of NEXAFS, where $E$ and $\theta$ is electric field vector and incidence angle of X-ray beam. The incidence angle, $\theta$, is measured as an angle between the direction of X-ray and the sample surface and the angle between the incident X-ray and the electron energy analyzer was fixed at 60°.

### 2.2 Mechanism of chirality alignment to zigzag-type

One of the most significant and important characteristics of CNT films grown by surface decomposition of SiC is their unique chirality. As stated above, Kusunoki et al. pointed out that the CNTs mostly exhibit zigzag-type chirality (Kusunoki et al., 2002). Taking into account that the carbon nanocaps were the tips of CNTs and that CNTs grow from below the carbon nanocaps, this suggests that they have chemical bonds perpendicular to the SiC(000-1) surfaces at its periphery, connected with zigzag-type CNTs (Fig. 3(c)). On the other hand, considering that aligned CNTs were apt to form on the carbon-face of SiC, Irle et al. pointed out that dangling bonds of carbon-face play a key role to form cap structures, based on quantum chemical molecular dynamics simulations (Irle et al., 2006).

Our XPS and STM results show that amorphous carbon nanoparticles are accumulated on SiC surface before the carbon nanocap formation (Maruyama et al., 2006). These nanoparticles assemble and coalesce as the temperature rises and finally are crystallized to carbon nanocaps. From the NEXAFS results, majority of C-C bonds are directed toward perpendicular to the surface at the beginning of crystallization. Therefore, utilizing dangling bonds on SiC surface, crystallization occurred, which serves to form carbon nanocaps. Considering that the final structure is a dome-like, the bond alignment perpendicular to the surface should occur circumferentially.

On the basis of the various experimental results, we propose a model for carbon nanocap formation by surface decomposition of SiC. Schematic images are shown in Fig. 10. (1) At first, carbon nanoparticles are accumulated as Si atoms are desorbed from SiC surface. (~1100°C) (2) Upon heating, carbon particles start migration on the SiC surface as a result of the thermal energy, forming clusters of several tens of carbon nanoparticles to reduce the surface energy further (~1150°C) (Fig. 10(b)). Then, the redistribution of carbon atoms begins within each nanoparticle cluster, and the nanoparticles begin to coalesce (Fig. 10(c)). At around 1200°C, coalescence proceeds within the clusters, and crystallization occurs, forming CNT nanocaps. In the crystallization, the size of clusters is presumably almost the same as that of the formed nanocaps. Thus, we believe that the size of clusters determines the nanocap size. (5) Above 1200°C, CNT growth into SiC proceeds by the sublimation of Si atoms from the SiC(000-1) surface. The NEXAFS spectra which suggest that the C-C bonds are directed nearly parallel to the surface at the beginning of crystallization, indicates that the direction of C-C bonds were changed during the crystallization. (Some difference might be seen in the measurement temperature between STM and NEXAFS because of the detection limitation of a pyrometer.) Taking into account the final structure of carbon nanocaps, we speculate that crystallization occurs at the periphery of amorphous carbon dome, and formation of C-C bonding utilize the dangling bonds from carbon atoms on the SiC(000-1) surface (Fig. 10). From the standpoint of crystal growth, this means that nucleation starts to occur at the edge in the carbon nanocaps.

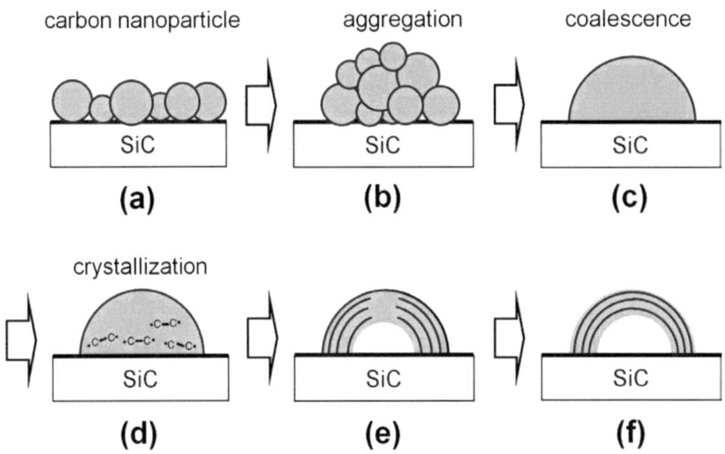


Fig. 10. Schematic model of the carbon nanocap formation process during surface decomposition of SiC. (a) formation of carbon nanoparticles by surface decomposition of SiC, (b) aggregation of carbon nanoparticles, (c) coalescence, (d) start of crystallization utilizing the dangling bonds, (e) progress of crystallization, and (f) complete formation of carbon nanocap. (The portion consisting of elementary carbon is shown as grey area.)

Recently, graphene growth from SiC(000-1) carbon-face by surface decomposition has been reported and a lot of researchers have been focusing on it because of its superiority in electronic property (Jernigan et al., 2009, Tedesco et al., 2009). From the point of view of growth technique, the method of graphene growth on SiC(000-1) is remarkably similar to that of CNT growth by surface decomposition, but one significant difference is the heating rate. In the growth of graphene, in general, the temperature of SiC is quickly increased to the growth temperature, typically, about several minutes to ~1400°C (Gamara et al., 2008). On the other hand, in the CNT growth by surface decomposition, the heating rate is typically 1-15°C/min (Kusunoki et al. 1999, 2002) and much slower than that in the graphene growth. This low heating rate might be necessary to enhance the migration of carbon atoms, assembling and leading to formation of carbon nanocaps.

## 3. Effect of ambient oxygen gas

In early studies, Kusunoki et al. proposed a chemical reaction process for CNT growth by surface decomposition of SiC as follows (Kusunoki et al. 1999):

$$SiC(s)+1/2O_2(g) \rightarrow C(s) + SiO(g) \quad (2)$$

In their CNT growth, heating of SiC was carried out using a vacuum electric furnace whose pressure was about $10^{-2}$ Pa. In this circumstance, residual gases, such as $O_2$, $H_2O$, CO etc., should exist in the furnace during the growth. They considered that these residual gases reacted with Si atoms on SiC surface, leading to the desorption.

Fig. 11. shows C 1s XPS spectra for 6H-SiC at 1270°C under various oxygen pressures. The main peak at around 284.3 eV and the shoulder peak at around 282.7 eV are derived from carbon $sp^2$ and SiC components, respectively, and these spectra are normalized with the main peaks. It is clearly shown that the ratio of the peak intensity of the SiC component to that of the carbon $sp^2$ component decreased as the oxygen partial pressure increased, confirming that oxygen enhances the decomposition of the SiC surface and promoting the carbon nanocap formation. Thus, surface reaction represented with eq. (2) seems to be reasonable. In the case of CNT growth in a vacuum electric furnace, the main reaction may be replaced by the reactions as follows, since residual gases are mainly composed of CO and $H_2O$.

$$SiC(s)+CO(g) \rightarrow 2C(s) + SiO(g) \quad (3)$$

or

$$SiC(s)+H_2O(g) \rightarrow C(s)+ SiO(g)+ H_2(g) \quad (4)$$

As for the reaction of oxygen with SiC surface, thermal oxidation of SiC surface has been an interesting theme from a fundamental surface science perspective, since it is chemically much more complex than that of Si. So far, three distinct regions have been observed in the phase diagram describing the SiC+$O_2$ interaction; "passive oxidation", "active oxidation" and "surface segregation of carbon" (Song & Smith, 2002). In the passive oxidation, which occurs under high oxygen pressure and low temperature, $SiO_2$ layer grows on the surface, and, in the active oxidation, which occurs under low oxygen pressure and high temperature, SiC surface is etched by the reaction with oxygen. In addition, under the lower oxygen pressure and the higher temperature, surface segregation of carbon has been observed. The CNT growth by surface decomposition of SiC should occur under the third mode where segregation of carbon atoms proceeds on the surface. However, it has not been well investigated between the reaction with oxygen gas and the CNT growth in surface decomposition of SiC. To clarify these, it is necessary to minutely control oxygen pressure and investigate the relation among the pressure, temperature and surface decomposition of SiC during the CNT growth.

Fig. 12 shows a phase diagram for the interaction of $O_2$ with the SiC surface, where formation of carbon nanocaps and $SiO_2$ are separately shown (Maruyama et al., 2007). All data were obtained by heating the 6H-SiC(000-1) in a vacuum electronic furnace and in a UHV chamber where the oxygen gas was introduced by controlling the pressure, and the heating rates were kept below 1°C/min which were low enough to form carbon nanocaps. In the figure, the surface covered with carbon nanocaps, elemental carbon (not formed nanocaps), $SiO_2$, and tiny amount of $SiO_2$ (weak $SiO_2$ peak was observed in XPS spectra) are shown as open circles, open squares, solid squares and solid triangles, respectively. In addition, the boundary between passive oxidation (P. O.) and active oxidation (A. O.) and that between active oxidation (A. O.) and "surface segregation of carbon (S. C) " which were obtained by extrapolating the experimental results for thermal oxidation of 6H-SiC(000-1) reported by Song and Smith (Song & Smith 2002) to wider ranges in both oxygen pressure and temperature. In their data, the critical oxygen pressure at which the active-transition occurred was estimated to be,

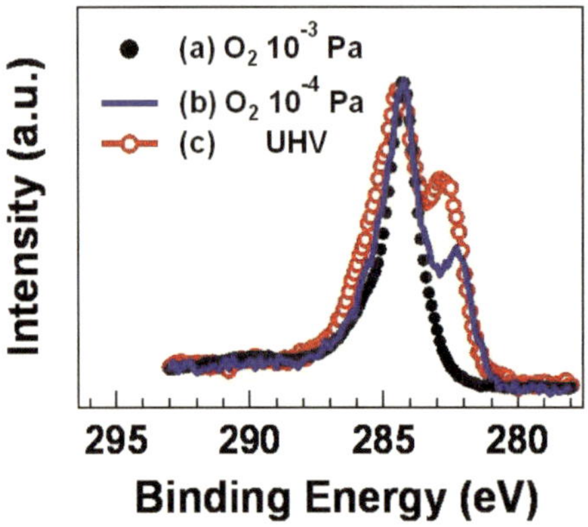


Fig. 11. C 1s XPS spectra for 6H-SiC(000-1) after heating at 1270°C for 30 min under UHV and various oxygen partial pressures ($10^{-4}$ and $10^{-3}$ Pa).

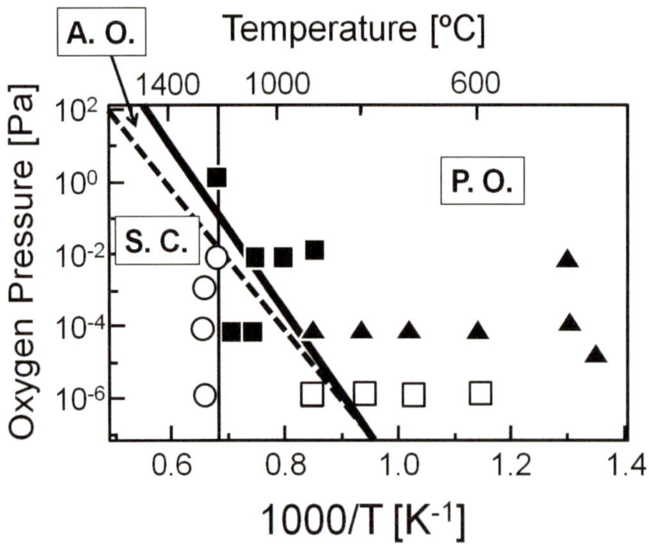


Fig. 12. Phase diagram for the interaction of $O_2$ with the 6H-SiC(000-1) face. Our experimental results are indicated: the open circles, open squares, solid squares and solid triangles correspond to the surfaces covered with carbon nanocaps, elemental carbon (not formed nanocaps), $SiO_2$, and tiny amount of $SiO_2$. The active-passive oxidation (A. O. –P. O.) and active oxidation (A. O.) - "surface segregation of carbon" (S. C.) boundaries reported by Ref. (Song and Smith, 2002) are also shown as thick solid line and dashed line, respectively.

$$P_c(O_2, T) = P_0 \exp(-\Delta E/k_B T) \qquad (5)$$

where $P_0=2.7x10^{15}$ *Pa* and $\Delta E = 4.8$ *eV*. It should be noted that carbon nanocaps were formed in the "surface segregation of carbon" region, and formation of carbon nanocap was observed above 1200°C, irrespective of the ambient pressure. This indicates that the heating above 1200°C is necessary to coalesce and crystallize amorphous carbon layer to form nanocaps.

Previous studies reported that graphitization of amorphous carbon occurs just above 1000°C (Botti, 2005; Chadderton & Chen, 1999). In the case of CNT growth by surface decomposition of SiC, both accumulation of sufficient carbon atoms and carbon migration are important to form carbon nanocap. This might lead to necessity of the higher temperature. On the other hand, $SiO_2$ layers were formed on the SiC surface in the passive oxidation region, except for the UHV region, where slight elementary carbon was seen because of insufficient oxygen gas. Although, in our experimental result, data were not obtained in the just active oxidation region, shown between the dashed line and solid line, it was clearly shown that carbon nanocaps were formed above 1200°C in the "surface segregation of carbon" region.

## 4. Effect of ambient hydrogen gas

Effect of hydrogen gas on carbon nanocap and CNT formation is another interesting theme, since reduction of residual oxides on SiC surfaces has been reported after annealing SiC in ambient hydrogen gases (Seyller, 2004) and this may affect the formation of carbon nanocaps. Figure 13 shows survey XPS spectra for SiC(000-1) surfaces after annealing at 1250°C for 30 min in a hydrogen atmosphere at various pressures (Ueda et al., 2010). The O 1s peak was not observed for the samples annealed at $10^{-3}$ and $10^{-2}$ Pa, whereas the O 1s peak was clearly observed for that annealed at $10^{-4}$ Pa. Figures 14(a) and (b) shows Si 2p and C 1s spectra for 6H-SiC(000-1) after annealing at 1250°C in a hydrogen atmosphere of $10^{-2}$ Pa.

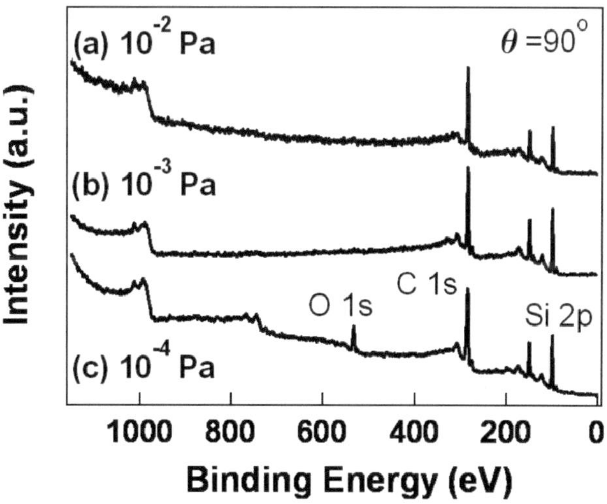


Fig. 13. XPS survey spectra for 6H-SiC(000-1) substrates after annealing at 1250°C in a hydrogen atmosphere of: (a) $10^{-2}$, (b) $10^{-3}$ and (c) $10^{-4}$ Pa.

For comparison, Si 2p spectrum of the sample after annealing at 1250°C in a UHV is also shown in (c). In the Si 2p spectrum for the sample after annealing in a hydrogen atmosphere, only SiC component was observed and the component of $Si_xC_yO_z$ related oxides were negligible, while there still remain some oxides after annealing in a UHV. These results confirm removal of residual $Si_xC_yO_z$ oxides by annealing a hydrogen atmosphere of sufficient high pressure, as reported by Seyller. Figure 14(b) shows a C 1s XPS spectrum of the same sample. Except for a peak derived from adsorbed CO molecules, the spectrum consists of only a SiC-component at 282.7 eV and carbon $sp^2$ hybridization bonding at 284.3 eV, and no $Si_xC_yO_z$ related oxides components were observed, consistent with the Si 2p spectrum in (a).

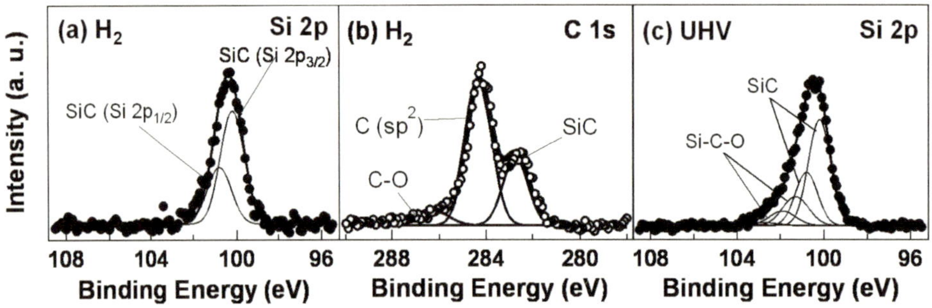


Fig. 14. (a) Si 2p and (b) C 1s XPS spectra for 6H-SiC(000-1) substrates after annealing at 1250°C in a hydrogen atmosphere of $10^{-2}$ Pa. (c) Si 2p XPS spectrum for 6H-SiC(000-1) substrates after annealing in a UHV. All spectra were deconvoluted into a set of Gaussian peaks after implementing background subtraction that accounted for previous XPS spectra of SiC (Hornetz et al., 1995).

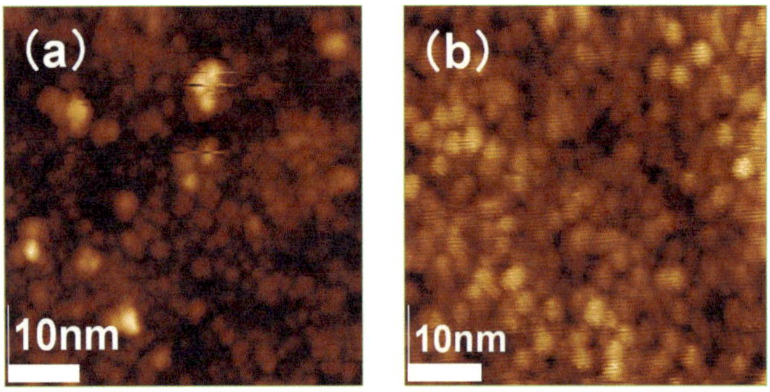


Fig. 15. STM images of 6H-SiC(000-1) for 30 min after annealing at 1200°C in (a) a UHV and (b) a hydrogen atmosphere of $10^{-2}$ Pa.

STM images of SiC(000-1) substrates after annealing at 1200°C for 30 min in the UHV and a hydrogen atmosphere of $10^{-2}$ Pa are shown in Fig. 15 (Ueda et al., 2010). The images reveal that convexities corresponding to carbon nanocaps were presented over the entire surface, independent of the annealing conditions. On the sample surface annealed in a UHV, the resulting carbon nanocaps were mainly in the diameter range of 1-3 nm, but nanocaps of more than 4 nm were observed. There were also some areas where no carbon nanocaps were formed. On the other hand, carbon nanocaps whose diameters were mainly distributed between 3 and 4 nm were formed on the SiC surface annealed in a hydrogen atmosphere, as shown in Fig. 15(b). The change in the diameter distribution is summarized in Fig. 16, which shows histograms of carbon nanocap diameter after annealing in a hydrogen atmosphere of $10^{-2}$ Pa. They indicates that the distribution of carbon nanocap diameter after annealing in the hydrogen atmosphere was narrower than that found in the UHV, demonstrating that the homogeneity of carbon nanocaps can be improved by reducing the native oxide from the SiC(000-1) surface.

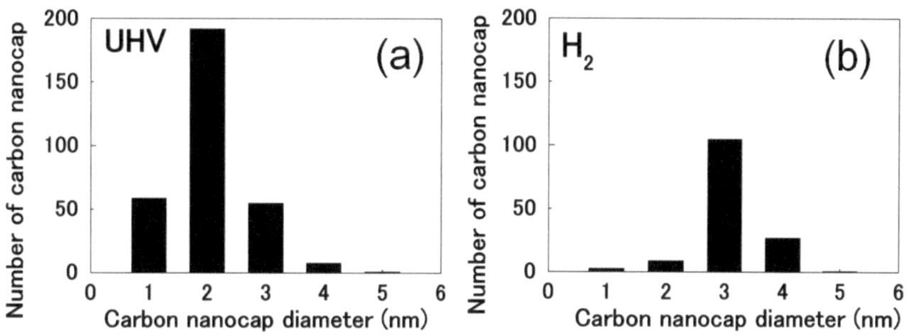


Fig. 16. Histograms, (a) and (b), of carbon nanocap diameter correspond to the samples shown in Fig. 15(a) and (b), respectively.

XPS result suggests that native $Si_xC_yO_z$ oxide on 6H-SiC still remain even after carbon nanocap formation without pre-annealing in $H_2$ ambient (Fig. 5(b)). Therefore, carbon nanocap formation should be disturbed among the areas covered with the oxides, because those areas should be suppressed sublimation of Si atoms. Such a localized suppression will induce an inhomogeneous distribution of an amorphous carbon layer, which, in turn, will have a deleterious effect on the size uniformity of the resulting carbon nanocaps. On the other hand, if there is no native oxide on the surface, simultaneous desorption of Si atoms would occur and homogeneous distribution of amorphous carbon layer would be formed. This process would produce carbon nanocaps with a uniform size.

## 5. Conclusion

CNT growth by surface decomposition of SiC is not only an interesting method as peculiar crystal growth but also has various characteristics useful for device applications. Development of this growth technique should directly lead to realization of CNT devices in future, replaced by conventional semiconductor devices. There still remain several problems to be solved for fabrication of CNT devices. To understand its growth mechanism is essential to control structural parameters of CNTs, which is absolutely necessary for realization of CNT devices. To clarify the formation process of carbon nanocaps, which are precursors of CNTs, determining the structures of CNTs, is the first step to reach this objective.

## 6. Acknowledgment

The authors are grateful to Prof. Amemiya for his help during NEXAFS measurements and also thank Dr. Ito, Mr. Kawamura, Ms. Fujita and Mr. Tanioku for assisting with our study on STM observation. A part of this work was supported by the Japan Society for the Promotion of Science (JSPS), a Grant-in-Aid for Scientific Research (C) 17560015 and 21510119, and by the Ministry of Education, Culture, Sports, Science and Technology (MEXT) through the 21st century COE program "Nano Factory". We are grateful to Professor T. Yokoyama and Dr. T. Nakagawa of the Institute for Molecular Science (IMS), Okazaki, for providing the XPS facility through the Nanotechnology Support Project in Central Japan, financially supported by the Nanotechnology Network. The NEXAFS measurements were performed under the approval of Photon Factory Program Advisory Committee (Proposal No. 2009G530). We would also like to to thank the Academic Research Institute of Meijo University for their financial support.

## 7. References

Banerjee, S., Hemrj-Benny, T., Sambasivan, S., Fisher, D. A., Misewich, J. A. & Wong, S. S. (2005). Near-Edge X-ray Absorption Fine Structure Investigations of Order in Carbon Nanotube-Based Systems, *The Journal of Physical Chemistry B*, Vol. 109, No. 17, (October 2004), pp. 8489-8495, ISSN 1520-6106.

Bang, H., Ito, Y., Kawamura, Y., Hosoda, E., Yoshida, C., Maruyama, T., Naritsuka, S. & Kusunoki, M. (2006). Observation of Nanosized cap Structurese on 6H-SiC(000-1) Substrates by Ultrahigh-Vacuum Scanning Tunneling Microscoy, *Japanese Journal of Applied Physics*, Vol.45, No.1A, (January 2006), pp.372-374, ISSN 0021-4922.

Botti, S. Ciardi, R., Fabbri, F., Larciprete, R., Goldoni, A., Gregoratti, L., Kaulich, B. & Kiskinova, M. (2005). Electron microscopy and photoelectron spectromicroscopy study of catalyst-free transformation of carobn nanoparticles into nanotubes, *Journal of Applied Physics*, Vo.98, No.8, (October 2005), pp.084307-1-084307-6, ISSN 0021-8979.

Camara, N., Rius, G., Huntzinger, J. –R., Tiberj, A., Magaud, L., Mestres, N., Godignon, P. & Camassel, J. (2008). Early stage formation of graphene on the C face of 6H-SiC, *Applied Physics Letters*, Vol.93, No.26, (December 2008), pp.263102-1-263102-3, ISSN 0003-6951.

Chadderton, L. T. & Chen, Y. (1999). Nanotube growth by surface diffusion, *Physcis Letters A*, Vol.263, No.4-6, (December 1999), pp.401-405, ISSN 0375-9601.

Comelli, G., Stohr, J., Robinson, C. J. & Jark, W. (1988). Structural studies of argon-sputtered amorphous carbon films by means of extended x-ray-absorption fine structure, *Physical Review B*, Vol. 38, No. 11, (October 1988), pp. 7511-7519, ISSN 1098-0121.

Hornetz, B., Michel, H.-J. & Halbritter, J. (1995). Oxidation and 6H-SiC-$SiO_2$ interfaces, *J. Vac. Sci. Technol. A* Vol. 13, No. 3, (June 1995), pp. 767-771, ISSN 0734-2101.

Iijima, S. (1991). Helical microtubules of graphitic carbon, *Nature*, Vol.354, No.6348, (November 1991) pp.56-58, ISSN 0028-0836.

Irle, S., Wang, Z., Zheng, G. & Morokuma, K. (2006). Theory and experiment agree : Single-walled carbon nanotube caps grow catalyst-free with chirality preference on a SiC surface, *The Journal of Chemical Physics*, Vol.125, No.4 (July 2006), pp.044702-1-044702-4, ISSN 0021-9606.

Jeong, G. H., Suzuki, S., Kobayashi, Y., Yamazaki, A., Yoshimura H. & Homma, Y. (2007). Size control of catalytic annoparticles by thermal treatmnet and its application to diameter control of single-walled carbon nanotube, *Applied Physcs Letters*, Vol.90, No.4, (Janualy 2007), pp.043108-1-043108-3, ISSN 0003-6951.

Jernigan, G. G., VanMil, B. L., Tedesco, J. L., Tischler, J. G., Glaser, E. R., Davidson, III, A., Cambell, P. M., & Gaskill, D. K. (2007). Comparison of Epitaxial Graphene on Si-face and C-face 4H SiC Formed by Ultrahigh Vacuum and RF Furnace Production, *Nano Letters*, Vol.9, No.7, (September 2009), pp.2605-2609, ISSN 1530-6984.

Kusunoki, M., Rokkaku, M. & Suzuki, T. (1997). Epitaxial carbon nanotube film self-organizaed by sublimation decomposition of silicon carbide, *Applied Physics Letters*, Vol.71, No.18, (September 1997), pp.2620-2622, ISSN 0003-6951.

Kusunoki, M., Suzuki, T., Kaneko, K. & Ito, M. (1999). Formation of self-aligned carbon nanotubes films by surface decomposition of silicon carbide, *Philosophical Magazine Letters*, Vo.79, No.4, (April 1999), pp.153-161, ISSN 0950-0839.

Kusunoki, M., Suzuki, T., Hirayama, T. & Shibata, N. (2000). A formation mechanism of carbon nanotube films on SiC(0001), *Applied Physics Letters*, Vol.77, No.4, (July 2000), pp.531-533, ISSN 0003-6951.

Kusunoki, M., Suzuki, C. Honjo, T., Hirayama, T. & Shibata, N. (2002). Selective synthesis of zigzag-type aligned carbon nanotubes on SiC(000-1) wafers, *Chemical Physics Letters*, Vol.366, No.5-6, (December 2002), pp.458-462, ISSN 0009-2614.

Maruyama, T., Bang, H., Kawamura, Y., Fujita, N., Tanioku, K., Shiraiwa, T., Hozumi, Y., Naritsuka, S. & Kusunoki, M. (2006). Scanning-tunneling-microscopy of the formation of catobn nanocaps on SiC(000-1), *Chemical Physics Letters*, Vol.423, No.4-6, (June 2006), pp.317-320, ISSN 0009-2614.

Maruyama, T., Bang, H., Fujita, N., Kawamura, Y., Naritsuka, S. & Kusunoki, M. (2007). STM and XPS studies of early stages of carbon nanotube growth by surface decomposition of 6H-SiC(000-1) under various oxygen pressures, *Diamond & Related Materials*, Vol.16, No.4-7, (January 2007), pp.1078-1081, ISSN 0925-9635.

Saito, R., Dresselhaus, G. & Dresselhaus, M. S. (1998). Structure of a Single-Wall Carbon Nanotube, In : *Physical Properties of Carbon Nanotubes*, pp. 35-58, Imperial College Press, ISBN 1-86094-223-7, London.

Seyller, T. (2004). Passivation of hexagonal SiC surfaces by hydrogen termination, *Journal of Physics : Condensed Matter*, Vo.16, No. 17 (May 2004), pp.S1755-S1782, ISBN 0953-8984.

Song, Y. & Smith, F. W. (2002). Phase diagram for the interaction of oxygen with SiC, *Applied Physics Letters*, Vol.81, No.16, (October 2002) pp.3061-3063, ISSN 0003-6951.

Tedesco, J. L., VanMil, B. L., Myers-Ward, R. L., McCrate, J. M., Kitt, S. A., Campbell, P. M., Jernigan, G. G., Culbertson, J. C., Eddy, Jr., C. R. & Gaskill, D. K. (2009). Hall effect mobility of epitaxial graphene grown on silicon carbide, *Applied Physics Letters*, Vol.95, No.12, (September 2009), pp.122102-1-122102-3, ISSN 0003-6951.

Ueda, K., Iijima, Y., Maruyama, T. & Naritsuka, S. (2010). Effect of Annealing in Hydrogen Atomosphere on Carbon Nanocap Formation in Surface Decomposition of 6H-

SiC(000-1), *Journal of Nanoscience and Nanotechnology*, Vol.10, No.6, (June 2010), pp.4054-4059, ISSN 1550-7033.

Watanabe, H., Hisada, Y., Mukainakano, S., & Tanaka, N. (2001). *In situ* observation of the initial growth process of carbon nanotubes by time-resolved high resolution transmission electron microscopy, *Journal of Microscopy*, Vol.203, No.1, (July 2001) pp.40-46, ISSN 1365-2818.

# 3

# Synthesis of Carbon Nanostructures by Microwave Irradiation

J. Vivas-Castro[1], G. Rueda-Morales[1], G. Ortega-Cervantez[1], J. Ortiz-López[1], L. Moreno-Ruiz[2] and M. Ortega-Avilés[2]
[1]*Escuela Superior de Física y Matemáticas, Instituto Politécnico Nacional,*
[2]*Centro de Nanociencias y Micro-Nanotecnologías, I.P.N.,*
*Mexico*

## 1. Introduction

Carbon nanotubes (CNT) have been synthesized with various techniques of which the most common ones are laser ablation, electric arc discharge, and chemical vapor deposition. These methods produce CNTs with different characteristics, sometimes involving complex experimental setups that add to their cost of production. It is of current general interest the development of new techniques for the efficient and selective synthesis of CNTs and other carbon nanostructures at the cheapest possible cost. One such possibility is the use of microwave radiation, which over the past few years has played an important role as a thermal tool in organic synthesis due to considerable advantages over conventional methods (Lidström, et al., 2001). The use of microwave radiation in the synthesis and functionalization of carbon nanotubes or other nanostructures is advantageous because it provides a fast and uniform heating rate that can be selectively directed towards a targeted area. The first report of the production of carbon nanostructures with microwaves was made by Ikeda et al (Ikeda et al., 1995), who synthesized fullerenes from microwave-induced naphthalene-nitrogen plasma at atmospheric pressure inside a cylindrical coaxial cavity. O. Kharissova has reported the synthesis of vertically aligned carbon nanotubes using a domestic microwave oven (Kharissova, 2004).

Graphite is a good microwave radiation absorber. It has been used in military applications as radar-absorbing material and in anti-electromagnetic interference coatings for civil purposes. Milled flake graphite and carbon nanotubes have microwave absorption maxima in the 10-15 GHz frequency range (Fan et al., 2009). Microwave radiation can heat or cause arcing in many objects and powdered samples can absorb such radiation and be heated efficiently. Short-time direct exposure to microwave irradiation has been used to produce exfoliated graphite as well as to reduce graphite oxide (Zhu et al., 2010).

In graphite powder, absorbed microwave radiation is converted into heat via dielectric loss and conductive loss mechanisms. Graphite powder is oxidized by long exposure to ambient air and may become partly electrically insulating. Microwaves are absorbed with energy dissipation through the coupling of the radiation electric field with local electric dipoles associated with structural defects in graphite powder particles such as particle edges, dangling bonds, C-O bonds, impurities and others. The electric field of microwaves also drives electric currents with efficient generation of heat due to the highly diffusive transport of $\pi$ electrons within small defective graphite particles loosely interconnected between each other.

In this work, we present results on the synthesis of various carbon nanostructures under different preparation conditions using a domestic microwave oven as energy source. Starting material is a mixture of graphite powder with iron acetate which is then subjected to microwave irradiation under different conditions. Heating of mixed powders under microwave irradiation decomposes iron acetate leaving small Fe particles that act as catalysts for the synthesis of carbon nanostructures. Various nanostructures are obtained depending on the way the starting material is prepared and exposed to microwaves as well as on the time of exposure. We make a detailed description of the synthesis results in dependence of those parameters. With this technique, we have obtained blocks of dense arrays of aligned multiwall carbon nanotubes, disordered nanotubes intercalated in between graphite planes, long iron-filled multiwall carbon nanotubes, and other peculiar formations. In addition, we propose mechanisms to explain how the synthesis of these nanostructures takes place.

## 2. Experimental methods and materials

All the material obtained in this work was produced with the aid of a conventional domestic microwave oven operating at 2.45 GHz (12 cm wavelength) with 1000 W power. Starting material was prepared by mixing 30 wt% of iron (II) acetate (ferrous acetate, $Fe(CO_2CH_3)_2$) from Sigma-Aldrich (99.999% purity) with 70 wt% graphite powders (99.99% purity, particle size less than 75 μm). The powders were then mixed and placed inside quartz ampoules (8 mm inner diameter, 10 cm long) which were then sealed under vacuum ($10^{-4}$ torr).

Temperature of microwave-irradiated ampoules rapidly rises and stabilizes within an elapsed time (few seconds) that depends on sample mass. The final temperature correlates linearly with total mass. It has been reported that 0.25 g of graphite powder irradiated inside a 700 W microwave oven reaches about 1000 °C in 12 seconds (Curling et al., 2009). Microwave irradiation decomposes iron acetate into small Fe particles and at the same time produces thermal swelling of graphite particles inside the ampoules. Various molecular species resulting from thermal decomposition of the iron acetate molecule participate in the chemical reduction of the oxide layer that forms on the surface of graphite and metallic particles and make them catalytically active. Once formed, small Fe particles will also absorb microwave radiation (Liu et al., 2006) and participate in further heating the powder mixture. At initial stages of irradiation, metallic particles are small and eventually coalesce into larger ones. At the same time, they will combine with graphite to form iron carbide from which various nanostructures are generated. Evaporation of the catalytic metal will occur in the case of direct exposure to microwaves because inside the ampoules the mixed powders reach high enough temperatures. Evaporated iron will ultimately condense on the ampoule walls and act as catalyst for the growth of dense arrays of aligned nanotubes.

Quite different products may be obtained by variations in the preparation conditions. We have applied the following: (a) quartz ampoules containing the mixed powders may be sealed under ambient air or under vacuum; (b) the ampoules can be partially submerged in water inside the oven to avoid excessive heating and exposure to microwaves, and (c) direct exposure of the ampoules to microwaves during short periods.

The resulting material was analyzed with scanning electron microscopy (SEM) using a FEI-Sirion instrument operated at 5 kV with secondary electrons. Scanning-transmission electron microscopy (STEM) was performed with a FEI Quanta Dual Beam instrument. Transmission electron microscopy (TEM) was also employed for the analysis using a JEOL 200 C (200 kV) instrument as well as a FEI-Titan 80-300 with spherical aberration correction (Cs TEM). For determination of the crystalline structure of metal particles inside MWNTs, a double-tilt TEM sample holder was used. For examination with Raman scattering spectroscopy we used two instruments: a Perkin-Elmer Raman Station 400F and a Horiba Jobin Yvon LabRam HR800. For SEM and Raman studies, samples were analyzed 'as grown' with no special preparation. For TEM analysis, suspensions were prepared either in dichloroethane or in deionized water with 0.5 wt% SDBS (dodecylbenzenesulfonic acid) surfactant. To optimize dispersion of sample particles, the suspensions were sonicated in an ultrasonic bath or using a 750 watt ultrasonic tip (Sonics) and later were centrifuged at 16,000 rpm. Samples for STEM and TEM analysis were taken from the supernatant material and deposited by dripping in 300 mesh holey carbon TEM copper grids.

## 3. Description of synthesized material

### 3.1 Irradiation of quartz ampoules partially submerged in water

To retard sample reactions, we avoided direct microwave irradiation of the mixed powders by partially submerging the prepared ampoules in water. In this way, most of microwave radiation is absorbed by water and only a fraction of the radiation acts on the sample. The consequence of this is that the material temperature rises more slowly and reactions in the synthesis process will take longer to occur. We estimate that in our case, with our 1000 W oven, the material reaches 1000 °C after 30 minutes. We prepared four samples (all of them evacuated ampoules) with 30, 60, 90 and 120 minutes irradiation time at full power with the turning plate of the oven in operation. Material produced under direct irradiation is distinctly different and will be described separately.

Carbon nanotubes (CNT) grow disorderly on the surface of graphite particles after 30 min irradiation, as can be seen in the particle indicated with the lower arrow in Fig. 1(a). Fig. 1(b) is a magnified view of the same particle where CNTs are clearly seen having curled geometry and a wide distribution of diameters. Statistical analysis in the zone of the sample corresponding to Fig 1(b) reveals that 3% of the tubes have diameters in the 4-10 nm range, 40% in the 10-20 nm range, 25% in the 20-30 nm range and the rest are wider than 30 nm. At this stage, some graphite particles have not completely reacted with iron particles and remain with flat surfaces, as shown with the upper arrow in Fig. 1(a).

In samples with 60 min exposure (Fig. 2), CNTs are more abundant and other types of structures appear. This is the case of Fig. 2(a) where a disordered arrangement of 'worm-like' strips about 0.5 μm wide is observed. Fig. 2(b) shows a magnified view of the zone marked with the rectangle in Fig. 2(a). The worm-like features seem to be formed by small-area graphite layers that have slipped and displaced with respect to each other. Analysis of this sample with transmission electron microscopy reveals the presence of graphite nano particles, graphene layers, encapsulated iron particles inside multiwall carbon nanotubes (MWNT) in different stages of growth, iron-filled graphitic onions and peapod-like structures. Two examples of these structures are shown in Fig. 2(c) and (d). In Fig. 2(c) the arrow marks a series of free graphene layers as a result of exfoliated graphite particles, and Fig. 2(d) marks with an arrow a peapod-like structure inside a double-walled nanotube.

For the sample with 90 min exposure in Fig. 3, curled and disordered MWNTs dominate the type of structures observed as seen in Fig. 3(a). TEM image in Fig. 3(b) demonstrates that these nanotubes are indeed (defective) MWNTs, with irregular number of wavy walls separated 0.34 nm.

(a) (b)

Fig. 1. SEM images of material obtained in an evacuated ampoule submerged in water after 30 min microwave irradiation: a) view at low magnification; b) view at higher magnification of the surface of particle indicated by lower arrow in (a).

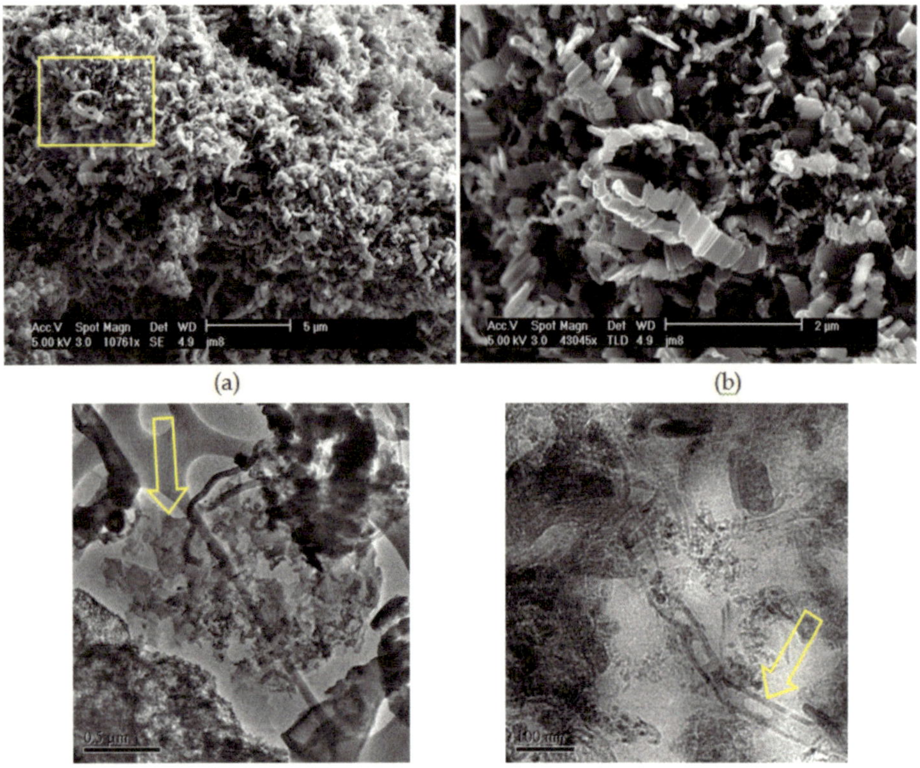


(a) (b)

(c) (d)

Fig. 2. SEM images of material obtained in an evacuated ampoule submerged in water after 60 min microwave irradiation: a) view at low magnification; b) higher magnification of the rectangular zone marked in (a); c) TEM image of graphene layers marked with an arrow; d) TEM image of a peapod-like structure marked with an arrow.

After 120 min irradiation, the majority of the original graphite powder has reacted with iron catalytic particles and has been converted into MWNT. Large (more than 100 μm) and thin (about 2 μm) blocks of disordered MWNTs can be collected on the ampoule walls as seen in Fig. 4(a). The distribution of MWNT diameters falls in the 10-30 nm range. The magnified image of Fig. 4(b) shows a cross sectional view of a fractured block revealing that the blocks are highly porous and have the appearance of metal sponge.
Raman spectroscopy was performed for all samples prepared in evacuated ampoules submerged in water. Their spectra do not display great differences with respect to the irradiation time. Fig. 4(c) shows the Raman response (785 nm excitation, $E_{laser}$=1.58 eV) of the sample irradiated 120 min which illustrates the typical spectrum observed in other samples. The signal is dominated by a large and wide D-band (1100-1400 $cm^{-1}$) with a shoulder around 1176 $cm^{-1}$ and a smaller G-band (1500-1700 $cm^{-1}$). The spectrum has features typical of nano graphite (Pimenta et al., 2007), but in view of our SEM and TEM observations, the signal of a large amount of disordered and defective MWNTs should be also present. A lorentzian line shape analysis reveals some possible component bands as shown in Fig. 4(c). The measured G band has two components, the first one at 1578 $cm^{-1}$, associated with tangential modes in graphite or MWNT and the second one at 1602 $cm^{-1}$, to the D′ defect induced band in $sp^2$ carbons (graphite and MWNTs). The D band can be decomposed in two main components at 1281 and 1310 $cm^{-1}$ as is usual in disordered graphite and MWNTs (Dresselhaus et al., 2005). We identify the shoulder at 1176 $cm^{-1}$ as the manifestation of a weak dispersive mode (sometimes called T-mode) observed in disordered MWNTs, graphite, and other non-planar $sp^2$ carbons in the 1084-1100 $cm^{-1}$ range (Kawashima, 1995; Li, 1997; Tan, 2004). As seen in our SEM and TEM images, all those types of carbon structures may coexist in our highly inhomogeneous samples.

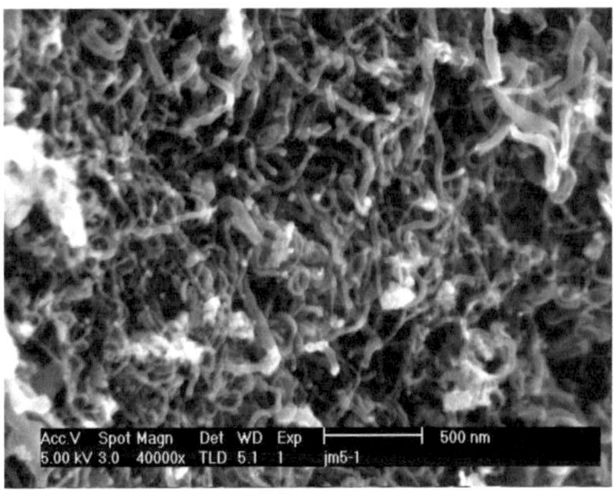


(a)

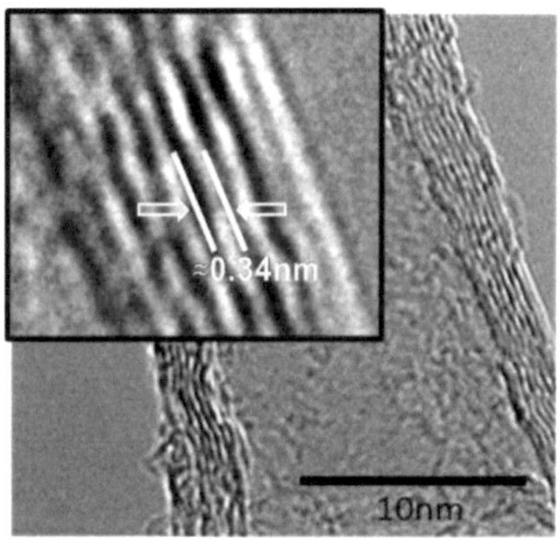


(b)

Fig. 3. Material obtained in an evacuated ampoule submerged in water after 90 min microwave irradiation: a) SEM image of a group of disordered nanotubes; b) TEM image showing that the observed tubes are multiwall; the insert shows a magnified view indicating a 0.34 nm interwall separation.

In samples obtained from ampoules submerged in water, small iron particles are attached to the tips of the observed MWNTs, as shown in the TEM image of Fig. 5. For samples prepared under direct irradiation, higher temperatures and stronger temperature gradients induce complete filling of MWNTs interior with metallic iron as described in next section. In

Fig. 5(a) we observe an elongated iron particle (16 by 28 nm in size) attached to the end of a MWNT and covered by several graphitic layers. Few new layers are seen forming inside the tube as the MWNT grows behind the tip. For the same sample, Fig. 5(b) shows of a group of MWNTs of various diameters, some of them as wide as 50 nm and others as thin as 10 nm, the latter ones with an iron particle at their tip.

### 3.2 Direct irradiation of quartz ampoules

With direct irradiation, temperature of the mixed powders rises rapidly in few seconds. Reactions are very fast and may become so violent that can cause explosion of the ampoules. Care must be taken to avoid accidents for this reason. To perform these experiments we previously determined the sites inside the microwave oven with maximum radiation intensity. The ampoules (all of them evacuated) were placed horizontally, so that their bottom (filled with the powder mixture) coincided with one of these sites.

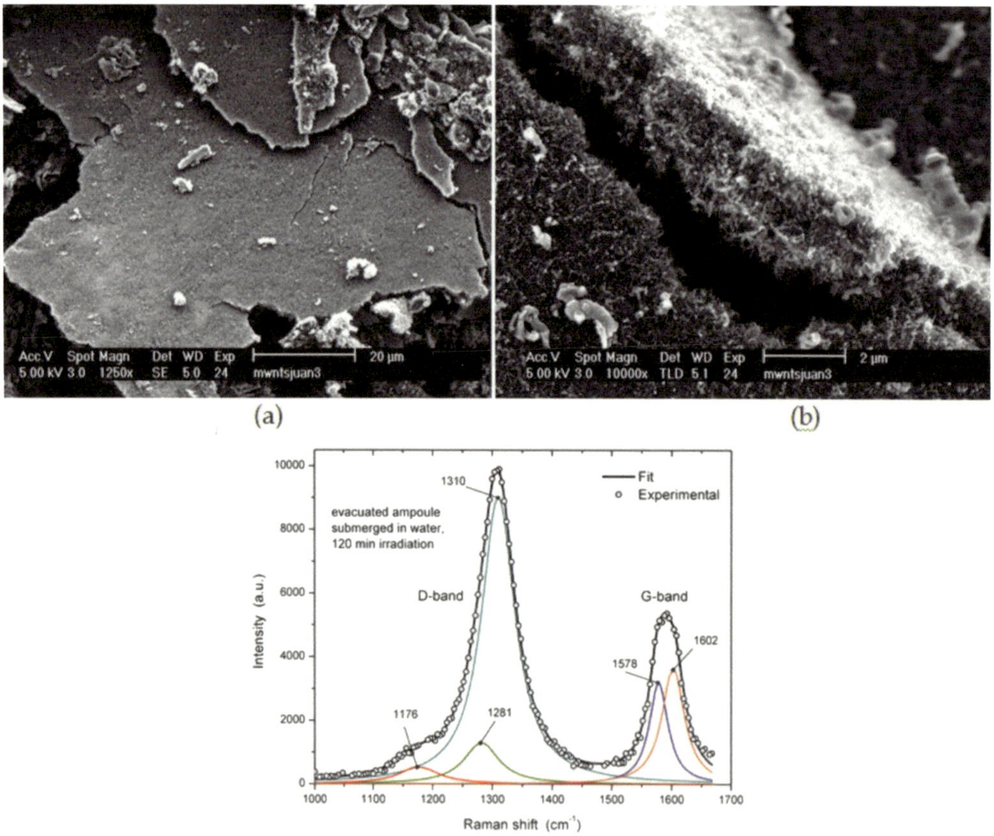


(c)

Fig. 4. SEM images of material obtained in an evacuated ampoule submerged in water after 120 min microwave irradiation: a) large blocks of disordered nanotubes collected from the ampoule walls; b) cross sectional view of a fractured block showing that it is entirely composed of disordered MWNTs; (c) Typical Raman spectrum at 785 nm excitation of material obtained after 120 min microwave exposure.

Fig. 6(a) illustrates SEM images of material collected from the ampoule inner walls after 10 min exposure to microwaves. The material is formed by blocks 7 to 10 μm thick of aligned MWNTs also known as CNT 'forests' or 'carpets'. Fig. 6(b), a magnified view of the zone marked with the rectangle in Fig. 6(a), indicates a diameter distribution within the 40-60 nm range. These CNT forests grow from iron particles produced by thermal decomposition of the acetate that accumulate on the inner walls of the ampoules. In Fig. 6(c) we present a Raman spectrum of the same sample. Defect D and tangential G bands are the strongest features of the measured spectrum. The D-band is almost as large as the G-band indicative of a sample with rather defective MWNTs and graphitic particles. At low energy there is a weak feature around 319 $cm^{-1}$ whose nature is unclear but could be attributed either to iron oxide, iron carbide or nano Fe crystals which become Raman active at small size. Another weak and narrow feature can be found at about 1450 $cm^{-1}$, in between D and G bands, which is identified as an LO or defective mode (Dresselhaus, 2005; Gupta et al., 2006). Other weak features in the Raman spectra of Fig. 6(c) include two small bands, one around 1750 $cm^{-1}$ (M-band) and the other around 1950 $cm^{-1}$ (iTOLA-band). The M-band consists of two components, 1745 and 1788 $cm^{-1}$, and it has been assigned to an overtone of the infrared-active 'out-of-plane' mode (oTO) at 864 $cm^{-1}$ in $sp^2$ carbon materials (Brar et al., 2002). The iTOLA-band at 1950 $cm^{-1}$ is identified as a combination of the in-plane transverse optic (iTO) and longitudinal acoustic (LA) modes by the same authors. Other three notorious features are identified as second harmonics and combination modes: (i) at 2427 $cm^{-1}$, the G*-band which is combination of defective D and T modes; (ii) at 2672 $cm^{-1}$, the G'-band which is overtone of the D defective mode band, and (iii) at 2927 $cm^{-1}$, the superposition of the

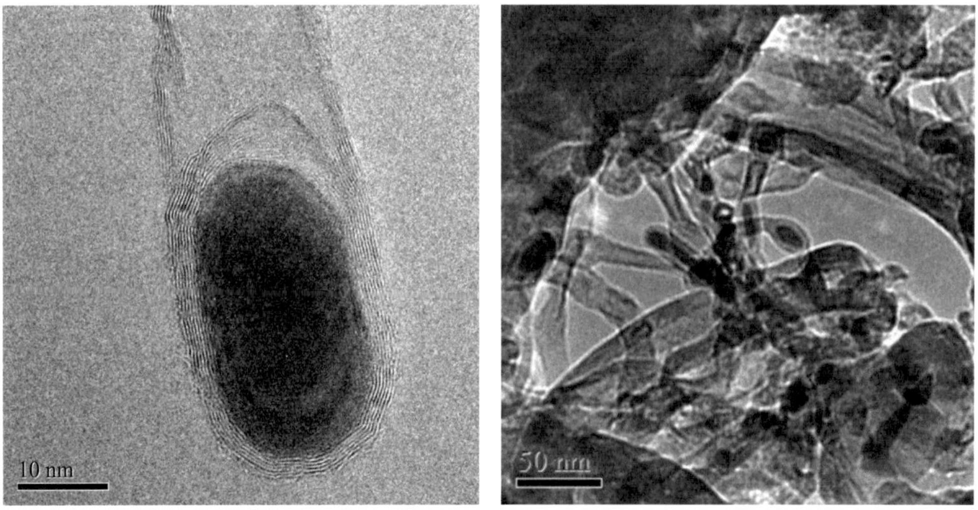


(a) (b)

Fig. 5. TEM images of material obtained in an evacuated ampoule submerged in water after 120 min microwave irradiation a) image of an iron particle at the tip of a MWNT; b) image of a group of MWNTs of various diameters, the narrow ones with an iron particle at their tip.

combined G+D modes and the overtone of the LO mode (Dresselhaus et al., 2005; Kawashima, 1995; Shimada et al., 2005; Sveningsson et al., 2001; Tan, 2004). We note that the 2427 $cm^{-1}$ (G*) band has also been assigned to other combination modes (Maultzsch et al.,

2004) and is usually a weak feature in most $sp^2$ carbons (Tan, 2004; Yoon, 2009) but in our sample it is rather intense.

Another type of structure that can be obtained is iron-filled MWNTs as shown in the STEM image of Fig. 7(a). The MWNT shown is of about 60 nm external diameter, while the iron filling (in dark trace) is about 2 µm long and 15 nm diameter in its thinnest part. This result was obtained with iron acetate catalyst and direct intermittent exposure to microwaves for an effective time of 15 min. In Fig. 7(b) we present powder X-ray diffraction of the sample corresponding to the image in Fig. 7(a). The identified reflections show the presence of graphite particles (and MWNTs), α-Fe (cubic bcc) and γ-Fe (cubic fcc) crystalline iron as well as iron carbide $Fe_3C$ (cementite), in agreement with nanostructured carbon material synthesized with other techniques (Chen et al., 2004).

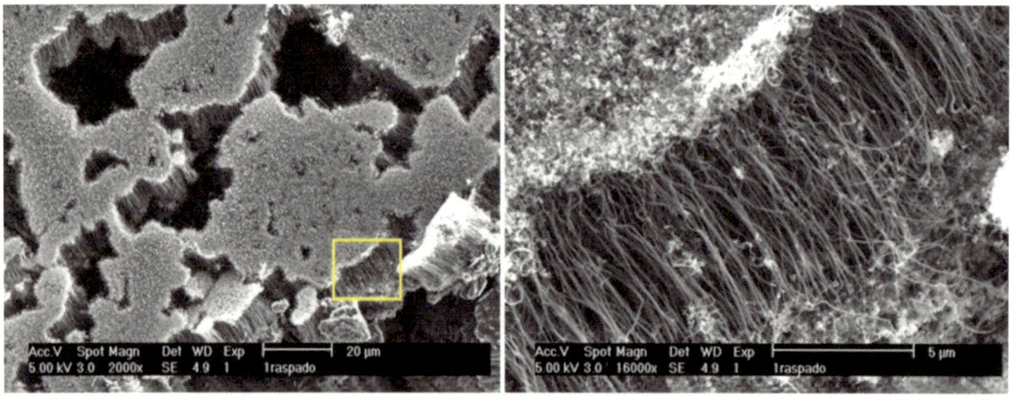


(a) (b)

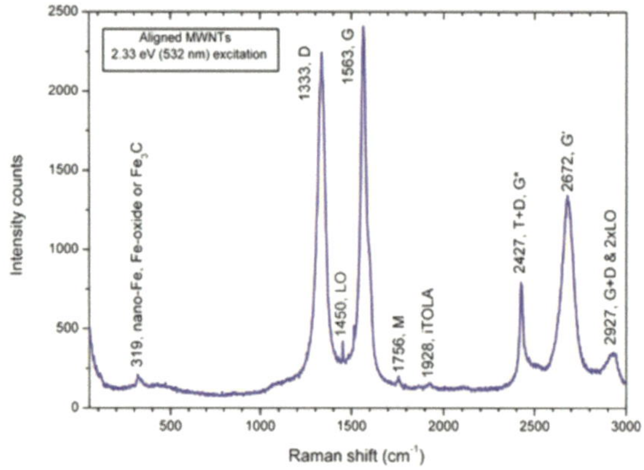


(c)

Fig. 6. Material obtained in an evacuated ampoule directly exposed to microwaves for 10 min: a) SEM image at low magnification of parallel arrays of aligned MWNTs; b) higher magnification of the rectangular zone marked in (a); c) Raman spectrum at 532 nm excitation of the same sample.

When the graphite/iron acetate powder mixture is finely ground and thoroughly mixed with mortar and pestle, the time needed of microwave exposure for the synthesis of nanostructures can be reduced substantially. Seven minutes of irradiation suffice for the appearance of MWNTs and almost complete reaction of the powders. TEM analysis of a sample prepared in this way shows abundant iron filled MWNTs as well as iron particles attached to MWNT tips like the one shown in Fig. 8(a). We chose that particular MWNT for structural identification of the metal particle at the tip using a double-tilt TEM sample holder. The metal particle is relatively large (70 nm wide, 150 nm long) and, as we found from its diffraction pattern, it consisted of several smaller ones of different composition and structure. After few minutes of exposure to the electron beam, the large particle split into smaller ones allowing determination of their individual diffraction patterns from spots close to thin edges. Fig. 8(b) shows the situation after splitting of the original particle and tilting of the sample holder to optimize crystalline orientation respect to the electron beam. Various spots around particle edges were analyzed in detail. The specific spot marked with the square in Fig. 8(b) is shown magnified in Fig. 8(c), in which crystalline planes of the metal particle are clearly resolved. The area marked with the square in this image is further magnified and shown in the inset at the upper left corner in Fig. 8(c). Below this magnified view, its Fourier transform is also shown as inset in Fig. 8(c). The measured interplanar distance is of 2.12 Å in the high resolution image of the inset, and the Fourier transform identifies the crystalline structure as bcc oriented in [110] direction. From this we determine that the particle corresponds to an α-Fe nanoparticle. From analysis of other spots we also find γ-Fe nanoparticles (fcc structure) oriented in [111], as well as $Fe_3C$ nanoparticles (orthorhombic structure). These results are in agreement with chemical analysis by energy

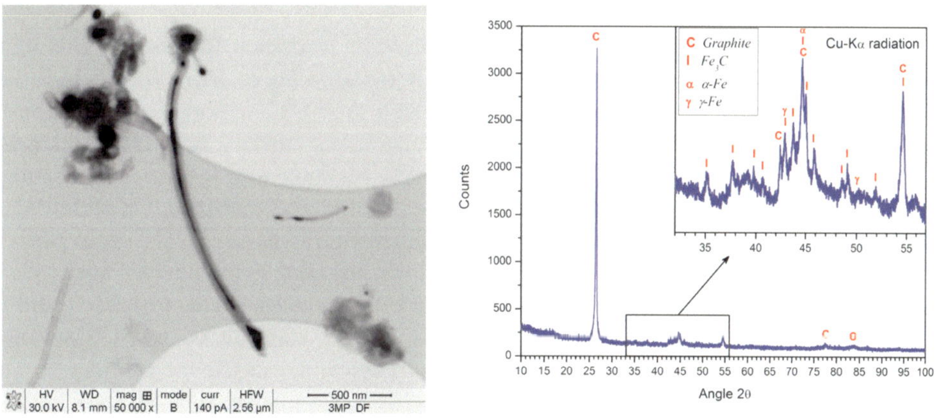


(a) (b)

Fig. 7. a) STEM image of iron-filled MWNTs from samples obtained by direct microwave irradiation; b) X-ray diffraction pattern of sample corresponding to the image in a) in which graphite C, α-Fe, γ-Fe and $Fe_3C$ phases are identified.

dispersive spectroscopy (EDS) which always showed presence of Fe and C and with the X-ray diffraction pattern presented in Fig. 7(b).

## 4. Mechanisms of nanostructure formation

Both methods of irradiation described in this work result in somewhat different synthesized material. Ampoules submerged in water are dynamically exposed to microwaves with temperature gradients constantly changing because the ampoules themselves are set in random rocking motion due to the surrounding boiling water. The resulting material in this case are curled MWNTs for long exposure times. In the case of direct irradiation, exposure to microwaves and temperature gradients are static, resulting in the growth of well oriented and aligned MWNTs arrays. In spite of these contrasts, it is possible to elaborate models of the mechanisms involved in the formation of the observed nanostructures.

For initial stages, when temperature is not too high, microwave absorption by graphite particles thermally expands its layers, facilitating the intercalation of foreign species between them (red particles), whilst exfoliation of outermost layers is also taking place. This is described schematically in Fig. 9(a). At the same time, ferrous acetate in our mixed powders is pyrolyzed above 200 °C [1]. Similar as reported for nickel acetate (Afsal et al., 1991), pyrolysis of ferrous acetate proceeds according to the following reactions:

$$Fe(CO_2CH_3)_2 \rightarrow FeO + CH_3COCH_3 + CO_2 \quad (1)$$

$$2\,Fe(CO_2CH_3)_2 \rightarrow 2\,FeO + 2\,CH_3COCH_3 + 2\,CO + O_2 \quad (2)$$

In both reactions, thermal decomposition yields FeO (wustite), and volatile and gaseous components such as acetone $CH_3COCH_3$, carbon monoxide, carbon dioxide and oxygen. Acetone itself can reduce FeO into metallic Fe which will eventually combine with graphite to form iron carbide $Fe_3C$ (cementite) from which carbon nanotubes and other nanostructures grow. A possible route for this to happen is the following:

$$4\,FeO + CH_3COCH_3 \rightarrow Fe + Fe_3C + 2\,CO + 3\,H_2O \quad (3)$$

Graphite particles in our powder mixture will provide feedstock for the growth of the observed nanostructures either from reaction with metallic Fe or from the carbide $Fe_3C$ obtained from the above reaction. At temperatures above 590 °C, acetone from reactions (1) and (2) may decompose into methane $CH_4$, ethylene $C_2H_2$, carbon monoxide and ketene $CH_2CO$, as reported as early as 1929 (Rice et al., 1929). Methane and ketene may also act as reduction agents of FeO. In relation to pyrolysis of acetone and methane, we should draw attention to the early work of E. L. Evans and collaborators (Evans et al., 1973) who reported the growth of filamentary carbonaceous deposits in the presence of iron, nickel and stainless steel surfaces. In the case of acetone pyrolysis on iron surfaces, these authors found the formation of lamellar $\gamma$-$Fe_2O_3$ at temperatures as low as 400 °C. This last observation indicates that pyrolysis and redox reactions of all chemical species involved may be quite more complex than those implied by the simple reactions (1) - (3).

Metallic particles will interact differently with neighboring graphite grains depending on their relative position respect to the graphite planes (graphene layers). It is expected that catalytic reactivity will be facilitated when the metallic particle sits close to the edges of graphite planes because it will readily interact with carbon end atoms. Evidence of this is seen in the SEM image of Fig. 9(b) where a graphite grain has reacted with iron particles mostly along its flanks (right arrow), while the uppermost layers look unaffected (left arrow). When this occurs, carbon and metallic iron will combine into metallic carbide and a carbon nanostructure will grow from the carbide particle as it becomes saturated with carbon atoms. This growth mechanism for carbon nanotubes is well known in the literature (Saito, 1995). If the metallic particles are small enough (less than 0.7 nm), they will react only

[1]Sigma-Aldrich, material safety data sheet of iron (II) acetate

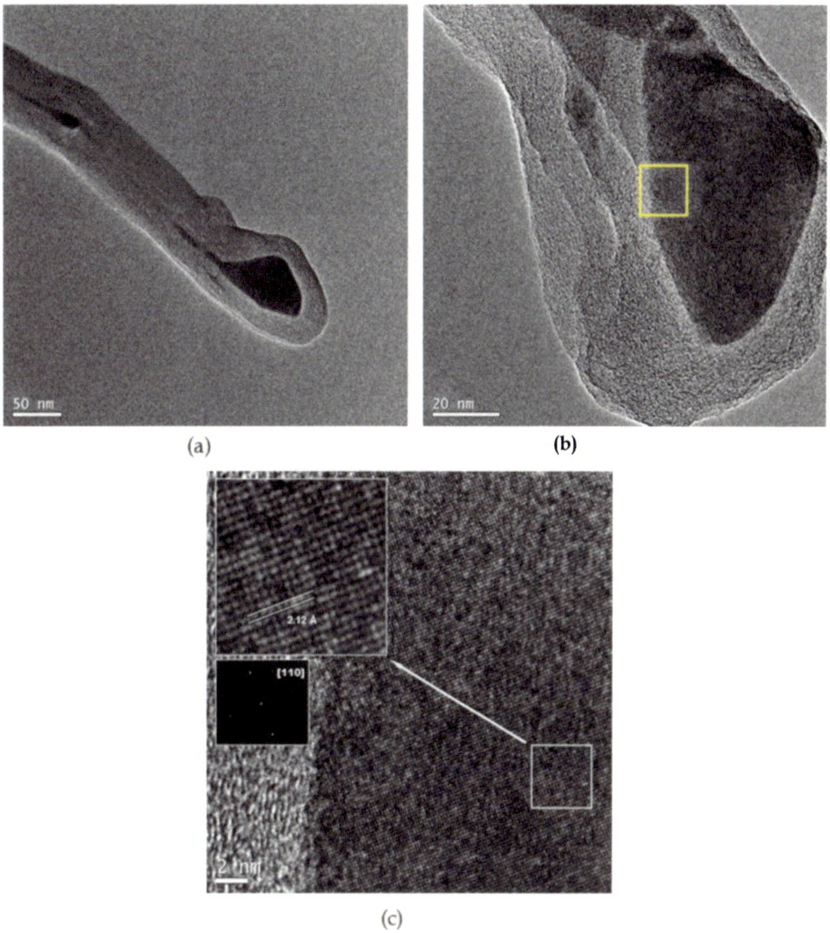


Fig. 8. TEM analysis of metal particle at the tip of a MWNT: a) image of the analyzed particle at low magnification; b) image at higher magnification after splitting of the original particle and tilting of the sample holder; c) higher magnification of the spot marked with the square in (b); inset at the upper left corner is an amplified view of the square marked at the lower right hand side, with its Fourier transform shown underneath.

with a single graphene layer of the graphite particle and may result in an assortment of carbon nanotubes intercalated in between planes of the graphite particle structure. This is schematically illustrated in the drawing of Fig. 9(c) and supported by the SEM image of Fig. 9(d) where a series of disordered nanotubes are seen protruding from amid graphite planes which are normal to the plane of the image and run across the vertically direction. These effects occur at short times of microwave exposure when metallic particles are small. For longer times, metallic particles coalesce into larger ones (red particles in upper drawing of Fig. 9(e), carbon nanostructures emanating from these particles are not drawn) and can interact with more than one graphene layer, leaving unaffected zones of the graphite grains in the form exfoliated graphitic structures, consisting of small-area layers with notorious slippage between each other as indicated in the lower drawing in Fig. 9(e). The SEM image of Fig. 8(f) shows this type of 'worm-like' strips which were already described in Fig. 2(b).

## 5. Conclusions

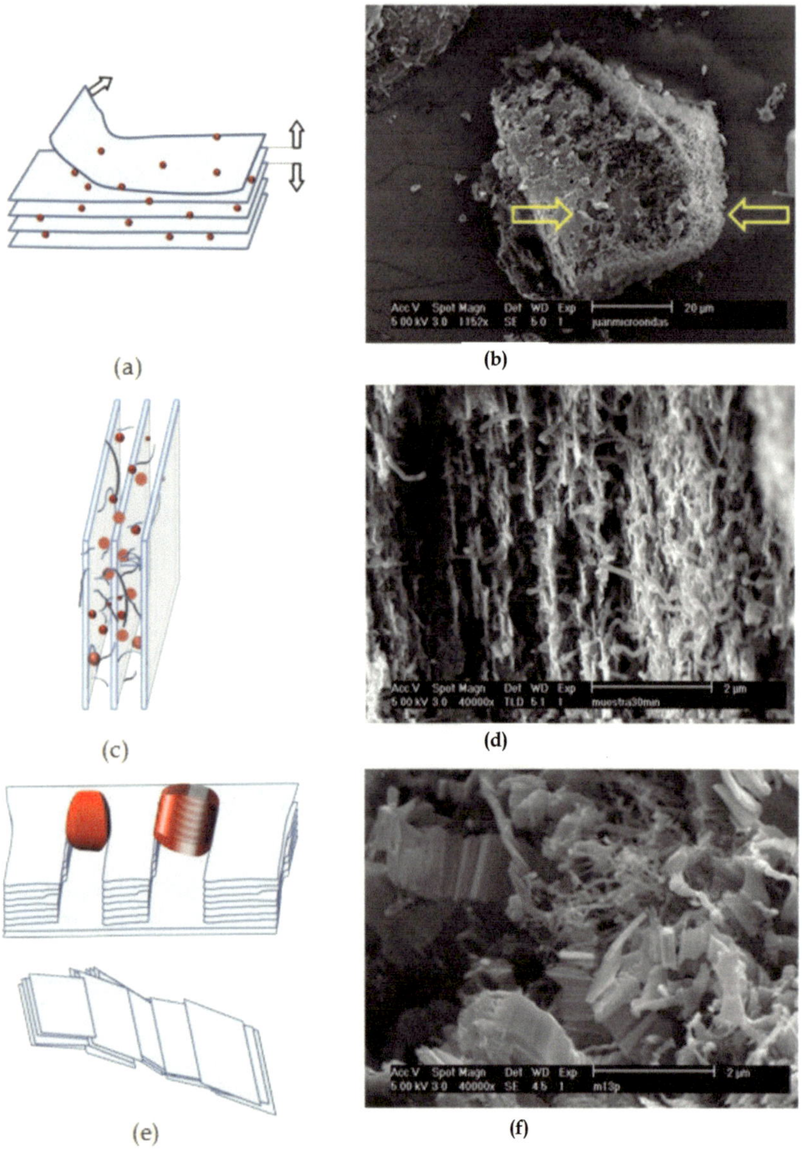


Fig. 9. a) Schematic drawing of expansion, intercalation and exfoliation of graphite particles as consequence of microwave irradiation; b) SEM image showing preferential reaction with iron catalytic particles along the flanks of a graphite particle (right arrow), whilst uppermost layers are almost unaffected (left arrow); c) reaction of small iron particles with individual graphene layers; d) SEM image supporting illustration in (c); e) upper drawing: reaction of large iron particles (red color) with several graphene layers at a time; lower drawing: leftovers of later mechanism leaving 'worm-like' strips of small-area graphene layers with notorious slippage between each other; f) SEM image of 'worm-like' strips illustrated in (e).

The method of synthesis of carbon nanotubes and other carbon nanostructures by microwave irradiation is simple and low-cost. With this technique and under various preparation conditions we obtained nanostructured material from a graphite/iron acetate powder mixture using a commercial microwave oven as energy source. Microwave absorption by the powder mixture results in pyrolysis of iron acetate. Decomposition of the acetate provides metallic iron nanoparticles that act as catalysts in the synthesis of nanotubes and other carbon nanostructures. Different type of nanostructured carbon can be obtained by variation of preparation conditions. Direct irradiation of (vacuum sealed) quartz ampoules, and attenuated irradiation by partial submerging the ampoules in water provide examples of the variety of nanostructures that can be obtained at different stages of microwave exposure. With information collected from these two cases we provide a general explanation of the observed phenomena with a model based on the reaction and exfoliation of graphite powder grains by metallic iron particles. This reaction occurs preferentially on graphene layer edges where carbon end atoms have non saturated dangling bonds. Depending on the size of the iron particle (which depends on effective irradiation time), the interaction will be established with single graphene layers when the particle is small, or with several layers when the particle is large. These conditions will result in the formation of different carbon nanostructures as irradiation time evolves.

## 6. Acknowledgments

We are thankful to Ing. Daniel Ramirez Gonzalez of IPICYT-Mexico, for TEM analysis of some samples. GRM, GOC and JOL acknowledge support from COFAA-IPN scholarships. GRM, GOC and MOA acknowledge support from EDI-IPN scholarships. JOL acknowledges support from EDD-IPN scholarship. This work was supported by CONACYT-Mexico grant No. 57262 and by SIP-IPN through GRM, GOC and JOL research projects.

## 7. References

Afzal, M.; Butt, P.K.; Ahrnad, H. (1991). Kinetics of thermal decomposition of metal acetates. *Journal of Thermal Analysis*, Vol. 37, No.5, pp.1015-1023, ISSN 1388-6150.

Brar, V.W.; Samsonidze, Ge.G.; Dresselhaus, M.S.; Dresselhaus, G.; Saito, R.; Swan, A.K.; Ünlü, M.S.; Goldberg, B.B.; Souza Filho, A.G.; Jorio, A. (2002). Second-order harmonic and combination modes in graphite, single-wall carbon nanotube bundles, and isolated single-wall carbon nanotubes. *Physical Review B*, Vol.66, No.15, pp. 155418-155427, ISSN 1098-0121.

Chen, Y.; Conway, M.J.; Fitz Gerald, J.D.; Williamsand, J.M. & Chadderton, L.T. (2004). The nucleation and growth of carbon nanotubes in a mechano-thermal process. *Carbon*, Vol.42, No.8-9, pp.1543-1548, ISSN 0008-6223.

Curling, M.; Collins, A.; Dima, G.; Proud, W. (2009). Progress towards microwave ignition of explosives. *Shock Compression of Condensed Matter-2009*, AIP Conference Proceedings 1195, pp. 486-489, ISBN 978-0-7354-0732-9.

Dresselhaus, M.S.; Dresselhaus, G.; Saito, R. & Jorio, A. (2005). Raman spectroscopy of carbon nanotubes. *Physics Reports*, Vol.409, No. 2, pp. 47-99, ISSN 0370-1573.

Evans, E.L.; Thomas J.L.; Thrower, P.A.; Walker, P.L. (1973). Growth of filamentary carbon on metallic surfaces during the pyrolysis of methane and acetone, *Carbon*, Vol.11, No.5, pp. 441-445, ISSN 0008-6223.

Fan, Y.; Yang, H.; Li, M.; Zou, G. (2009). Evaluation of the microwave absorption property of flake graphite. *Materials Chemistry and Physics,* Vol.115, No.2-3, pp.696-698, ISSN 0254-0584.

Gupta, A.; Chen, G.; Joshi, P.; Tadigadapa, S.; Eklund, P.C. (2006). Raman Scattering from High-Frequency Phonons in Supported n-Graphene Layer Films. A. *Nano Letters,* Vol.6, No.12, pp. 2667-2673, ISSN 1530-6984.

Ikeda, T.; Kamo, T. & Danno, M. (1995). New synthesis method of fullerenes using microwave-induced naphthalene-nitrogen plasma at atmospheric pressure, *Applied Physics Letters,* Vol.67, No.7, pp. 900-902, ISSN 0003-6951.

Kawashima, Y.; Katagiri, G. (1995). Fundamentals, overtones, and combinations in the Raman spectrum of graphite. *Physical Review B,* Vol.52, No.14, pp. 10053-10059, ISSN 1098-0121.

Kharissova, O. (2004). Vertically aligned carbon nanotubes fabricated by microwaves, *Reviews on Advanced Materials Science,* Vol.7, No.1, pp. 50-54, ISSN 1605-8127.

Li, W.; Zhang, H., Wang C.; Zhang, Y.; Xu, L.; Zhu, K.; Xie, S. (1997). Raman characterization of aligned carbon nanotubes produced by thermal decomposition of hydrocarbon vapor. *Applied Physics Letters,* Vol.70, No.20, pp. 2684-2686, ISSN 0003-6951.

Lidström, P.; Tierney, J.; Whatey, B. & Westman J. (2001). Microwave assisted organic synthesis–a review. *Tetrahedron,* Vol.57, pp. 9225-9283, ISSN 0040-4020.

Liu, J.R.; Itoh, M.; Machida, K. (2006) Magnetic and electromagnetic wave absorption properties of α-Fe/Z-type Ba-ferrite nanocomposites. *Applied Physics Letters* Vol.88, No.6, pp. 062503-1-3, ISSN 0003-6951.

Maultzsch, J.; Reich, S.; Thomsen, C.; Requardt, H.; Ordejon, P. (2004). Phonon dispersion in graphite. *Physical Review Letters,* Vol. 92, No. 7, pp. 075501-075504 , ISSN 0031-9007.

Pimenta, M.A.; Dresselhaus, G.; Dresselhaus, M.S.; Cançado, L.G.; Jorio, A.; Saito. R. (2007) Studying disorder in graphite-based systems by Raman spectroscopy. *Physical Chemistry Chemical Physics,* Vol.9, No.11, pp. 1276–1291, ISSN 1463-9076.

Rice, F.O.; Vollrath, R. E. (1929). The thermal decomposition of acetone in the gaseous state. *Proceedings of the National Academy of Sciences,* Vol.15, No.9, pp. 702-705, ISSN 0027-8424.

Saito, Y. (1995). Nanoparticles and filled nanocapsules. *Carbon,* Vol.33, No.7, pp. 979-988, ISSN 0008-6223.

Shimada, T.; Sugai, T.; Fantini, C.; Souza, M.; Cançado, L.G.; Jorio, A.; Pimenta, M.A.; Shinohara, R.; Saito, Y.; Grüneis, A.; Dresselhaus, G.; Dresselhaus, M.S.; Ohno, Y.; Mizutani, T. & Shinohara, H. (2005). Origin of the 2450 $cm^{-1}$ Raman bands in HOPG, single-wall and double-wall carbon nanotubes. *Carbon,* Vol.43, No.5, pp. 1049-1054, ISSN 0008-6223.

Sveningsson, M.; Morjan, R.E.; Nerushev, O.A.; Sato Y.; Bäckström, J.; Campbell, E.E.B.; Rohmund, F. (2001). Raman spectroscopy and field-emission properties of CVD-grown carbon-nanotube films. *Applied Physics A,* Vol.73, No.4, pp. 409–418, ISSN 0947-8396.

Tan, P.H.; Dimovski, S.; Gogotsi, Y. Raman scattering of non-planar graphite: arched edges, polyhedral crystals, whiskers and cones (2004). *Philosophical Transactions of the Royal Society A,* Vol.362, No. 1824, pp. 2289–2310, ISSN 1471-2962.

Yoon, D.; Moon, H.; Cheong, H.; Sik Choi, J.; Ae Choi, J.; Ho Park, B. (2009). Variations in the Raman Spectrum as a Function of the Number of Graphene Layers, *Journal of the Korean Physical Society,* Vol.55, No.3, pp. 1299-1303, ISSN 0374-4884.

Zhu, Y.; Murali, S.; Stoller, M.D.; Velamakanni, A.; Piner, R.D.; Ruoff, R.S. (2010). Microwave assisted exfoliation and reduction of graphite oxide for ultracapacitors, *Carbon,* Vol.48, No.7, pp. 2118-2122, ISSN 0008-6223

# 4

# Effects of Gravity and Magnetic Field on Production of Single-Walled Carbon Nanotubes by Arc-Discharge Method

Tetsu Mieno and GuoDong Tan
*Department of Physics, Shizuoka University*
*Japan*

## 1. Introduction

Since the breakthrough enabling the mass production of single-walled carbon nanotubes (SWNTs) (Iijima 1993), many researchers in institutes and companies around the world have been developing efficient production methods for SWNTs (Harris, 1999). Various applications of this new and stable carbon nanomaterial with unique properties have been proposed (Jorio et al., 2008). However, insufficient control of the production of SWNTs is a major problem in developing applications of SWNTs. The mass production of long high-quality, defect-free SWNTs such as those with a length of 10 cm and a diameter of 1 nm is a major research target. Quality control in the production of SWNTs in terms of their diameter, chirality and defect density is also an important research target. Through the development of SWNTs, it is hoped that they can be used in strong and lightweight carbon wires and lightweight but strong composite bodies for many types of vehicles. Therefore, a basic study on the production process of SWNTs is very important for establishing new methods of producing high-quality SWNTs. In this study, the production of SWNTs by the arc discharge method is investigated. This is one of most popular methods of producing SWNTs, and it is essential to carry out the reaction in a hot helium gas atmosphere. As the reaction is strongly affected by gravity (heat convection) (Mieno, 2004) and the applied magnetic field (Lorentz force), the effects of gravity, heat convection and magnetic field on the production of SWNTs were experimentally studied. The process was examined under zero gravity, normal gravity and high gravity. As there are large differences among them, the authors discusses experimental results in comparison with reaction models and fluid simulation results. An investigation of the effect of applied magnetic field on inducing the *JxB* force in the arc plasma is also reported, and the effect of such a field on the production of SWNTs is clarified.
The effect of zero gravity was first examined using a vertical swing tower (VST), that repeatedly produces 1.1 s of zero gravity. Then, a series of parabolic-flight experiments were carried out with the support of Japan Space Forum, in which 10-20 periods of 20 s of zero gravity were obtained per flight. The results were compared with those of the laboratory experiment. A higher-gravity experiment was carried out using a rotating acceleration generator, which produces gravity of 1-3 $g_0$ ($g_0$: normal gravitational acceleration). The produced SWNTs were measured by TEM, Raman spectrometry and other methods.

## 2. Theoretical model of production process under selected reaction conditions

Figure 1 shows a reaction model of SWNTs fabricated by the arc-discharge method. Sublimated carbon molecules and metal atoms diffuse in the region of hot He gas, where the metal atoms fuse to form nanosize catalytic particles, and carbon atoms diffuse into the catalyst particles or diffuse on them. During the cooling process of the hot gas, SWNTs grow on the catalyst particles, which is a self-organizing process, and no special mould is used. The reaction is strongly affected by the gas temperature, the diffusion speed, the cooling rate of the particles, the carbon density and the catalyst particle density.

When spherical carbon clusters of the same diameter and mass are produced in a globe, which is located at the centre of spherical coordinates and has radius of $r_0$, and the clusters diffuse continuously and isotropically with diffusion coefficient $D_C$ without any external forces or chemical reactions, the density profile of the carbon clusters $n_C(r)$ and radial cluster flux $J_C$ $(r)$ can be calculated by Fick's diffusion laws (Cussler, 1984; Mieno & Takeguchi, 2006). Under the steady flow condition,

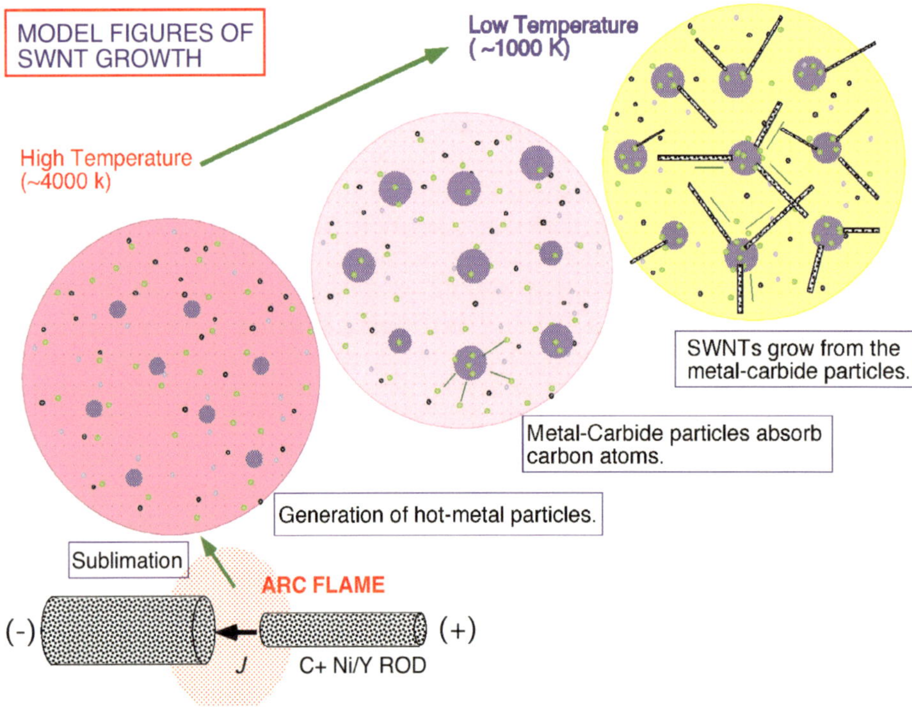


Fig. 1. Model of production process of SWNTs in He gas phase fabricated by the arc discharge method, in which metal catalyst particles play an important role.

$$n_C(r) = \frac{r_1 r_0}{r_1 - r_0}\left(n_0 - n_1\right)\frac{1}{r} + \frac{r_1 n_1 - r_0 n_0}{r_1 - r_0} \tag{1}$$

$$J_C(r) = D_C \frac{r_1 r_0}{r_1 - r_0}\left(n_0 - n_1\right)\frac{1}{r^2} \tag{2}$$

where $r_1 > r > r_0$. $n_0$ and $n_1$ are the cluster densities at $r= 0$ and $r= r_1$, respectively. When the carbon clusters diffuse in a noble gas, according to classical diffusion theory (Bird et al., 1960; Mieno, 2006) their diffusion coefficient $D_{CN}$ is written as

$$D_{CN} = \frac{4}{3}\frac{\sqrt{2}}{\pi^{3/2}}\left(\frac{1}{m_C}+\frac{1}{m_N}\right)\frac{(\kappa_B T)^{3/2}}{p(d_C + d_N)^2} \tag{3}$$

where $m_C$ and $m_N$ are the cluster mass and noble atom mass, respectively. $l_B$, $T$, $p$, $d_C$ and $d_N$ are Boltzmann's constant, the gas temperature, the gas pressure, and the diameters of the cluster and noble-gas atom, respectively. From these equations, the ideal radial diffusion velocities of the clusters can be calculated. Figure 2 shows the thermal diffusion velocities of carbon clusters $C_2$, $C_{60}$, $C_{1000}$ and $C_{10000}$ as a function of He gas pressure, where the clusters are assumed to be spherical, the He temperature is $T= 5000$ K and $r= 3.0$ cm from the centre (the centre of the arc flame). These boundary conditions are based on the experimental cylindrical vessel (102 mm in diameter and 184 mm high).

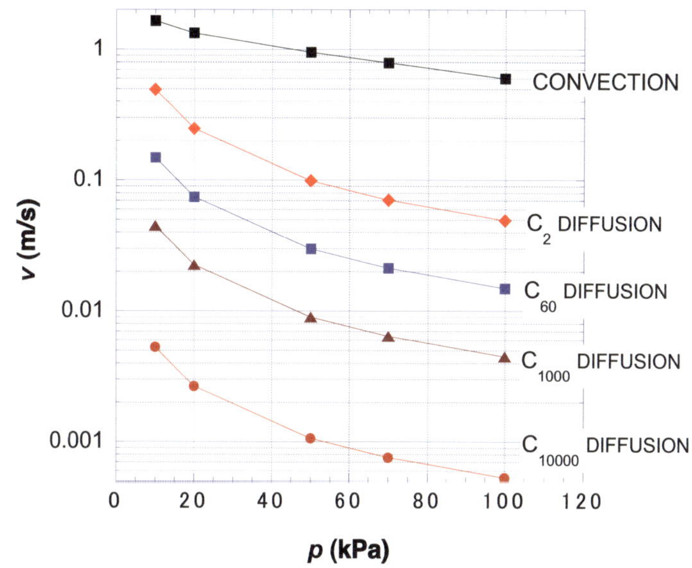


Fig. 2. He pressure dependence of thermal diffusion velocities of carbon clusters and helium convection velocity. Calculation at $T= 5000$ K.

Under the normal gravity, the gravitational force causes strong heat convection, in which all the carbon clusters flow upward with the He gas under a collisional pressure condition. This convection velocity can be calculated using a set of fluid equations. Here, a simulation by the simplified marker and cell (SMAC) method (Payret & Taylor, 1983) (Mizuho Information & Research Institution Inc., Fuji-RIC/Alfa-Flow) was used. The heat convection velocity versus He gas pressure under normal gravity is shown in figure 2. The heat convection velocity is much larger than the diffusion velocity of the clusters. Therefore, it can be expected that under zero gravity, the diffusion velocity of the clusters dramatically decreases. From equation 3, it is expected that larger and heavier clusters diffuse more slowly in He gas. In the case of SWNTs, the structure is chainlike, which decreases their mobility in the gas. From the measurement of the mobilities of fullerenes, chainlike carbon clusters and ring-type carbon clusters, it was shown that the diffusion velocity of long-chain-type clusters is about half that of spherical clusters with the same mass (von Helden et al., 1993). Therefore, SWNTs are estimated to have half the diffusion velocity of spherical carbon clusters with the same mass. In the case of arc discharge, carbon atoms diffuse from the arc-plasma region. Therefore, this large suppression of the diffusion speed under zero-gravity should realize high-temperature long-time hot reaction for synthesizing SWNTs.

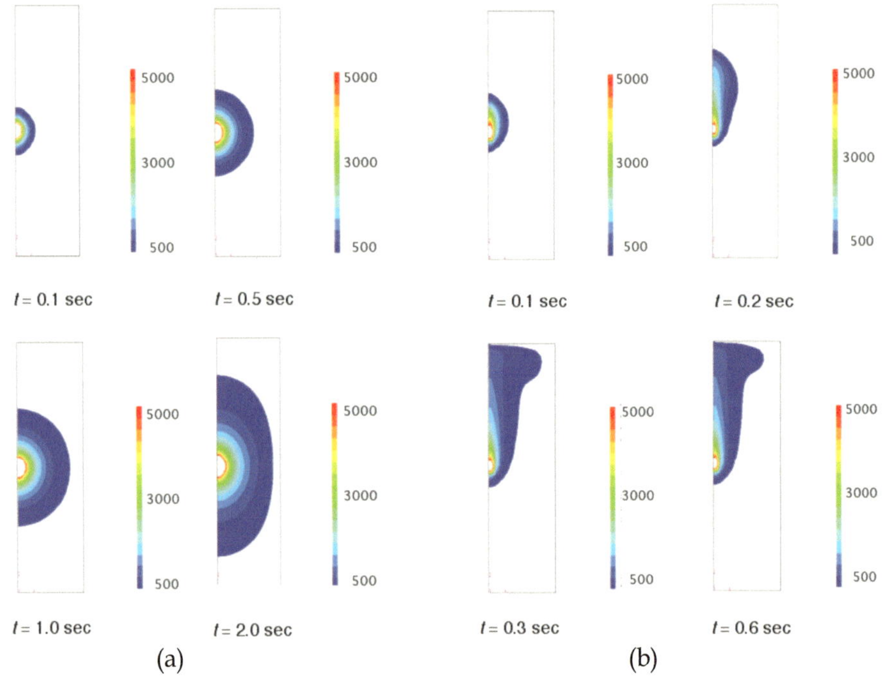


Fig. 3. Simulation results of temperature contours ($T$= 500-5000 K). (a) zero gravity, (b) normal gravity. $p$(He)= 40 kPa, and the arc starts at $t$= 0.

If the heat convection is suppressed, the gas atoms and clusters isotropically diffuse, resulting in the high-temperature gas region around the arc plasma becoming spherical and the volume of this region becoming much larger. Using the SMAC simulation method, the time evolution of the He gas temperature contours around the plasma was calculated and is shown in figures 3(a) and 3(b) (Mieno, 2006). These figures show the profiles of He gas temperature contours under (a) zero gravity and (b) normal gravity, and the right half of the reaction chamber is visualized. Here, $p$(He)= 40 kPa, the input electric power is $P_{IN}$= 1.0 kW and $T$= 500-5000 K. Under zero gravity, the temperature contours isotropically and monotonically expand until the contour arrives at the side wall ($t$~ 2 s). It takes more than 1 min for the temperature distribution to reach a steady-state condition. In contrast, under normal gravity, the contours expand only in the upward direction and attain a steady-state condition within 0.6 s. Thereafter, the temperature contours exhibit almost no change with time, which means that the produced thermal energy is continuously transported to the upper wall under this condition.

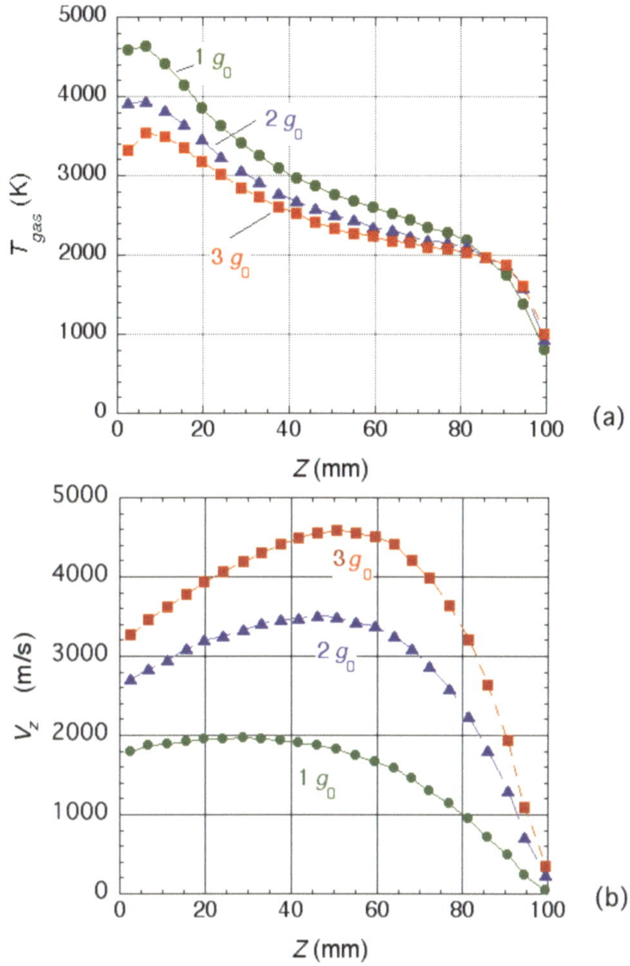


Fig. 4. (a) Vertical ($z$) direction distributions of gas temperatures for gravitational strengths of 1.0 $g_0$, 2.0 $g_0$ and 3.0 $g_0$, where the centre of the arc flame is set to $z$= 0 cm. $r$= 1.1 cm, $p$(He)= 50 kPa and $P_{IN}$= 1.0 kW. (b) vertical direction distributions of gas velocity $V_z$ for the three gravitational strengths.

When the gravity is increased from 1.0 $g_0$ to 3.0 $g_0$, the temperature contours and gas velocities exhibit distinct properties. Using the SMAC simulation method, the temperature contours were calculated (Tan & Mieno, 2010b). The vertical distributions of the gas temperature along the $z$ direction at $r$= 1.1 cm for three gravitational strengths are graphed as shown in figure 4(a), where $p$(He) = 50 kPa and $P_{IN}$=1.0 kW. The corresponding vertical distributions of the gas velocity $V_z$ were calculated and are shown in figure 4(b). The gas temperature above the plasma decreases with increasing gravity. Whereas $V_z$ above the plasma increases with increasing gravity. From these calculations, the vertical distribution of the gas cooling rate can be evaluated. The results for the three gravity conditions corresponding to figures 4(a) and (b) are shown in figure 5. At around $z$= 20 mm, the highest cooling rate is observed for these gravities. Increasing gravity (3 $g_0$) increases this cooling rate at $z$= 20 - 80 mm. These simulation results are easily explained by the simple fluid model.

When a steady-state magnetic field is applied to the arc plasma, electrons and ions are accelerated by the Lorentz force ($F$= $J$x$B$) as shown in figure 6. At first, electrons are accelerated in the $J$x$B$ direction, and then ions are accelerated in the same direction by an ambipolar electric field. By the Lorentz force, the plasma and neutral particles including many carbon atoms and metallic atoms are jetted out in the $J$x$B$ direction, and sublimated carbon atoms are efficiently transformed from the electrode region, by which the redeposition of carbons in the gas phase to the cathode is supressed (Aoyama & Mieno, 1999), and thus an efficient reaction to produce SWNTs is expected.

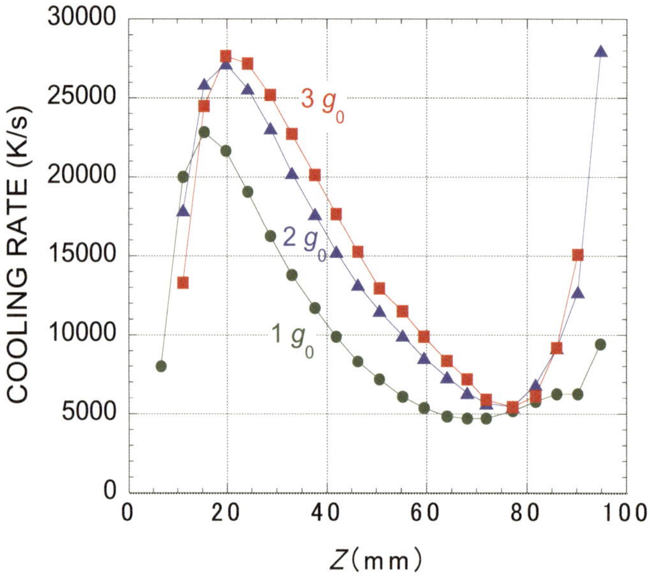


Fig. 5. Vertical distributions of calculated cooling rates of the gas for the three gravitational strengths. $p$(He)= 50 kPa and $P_{IN}$= 1.0 kW.

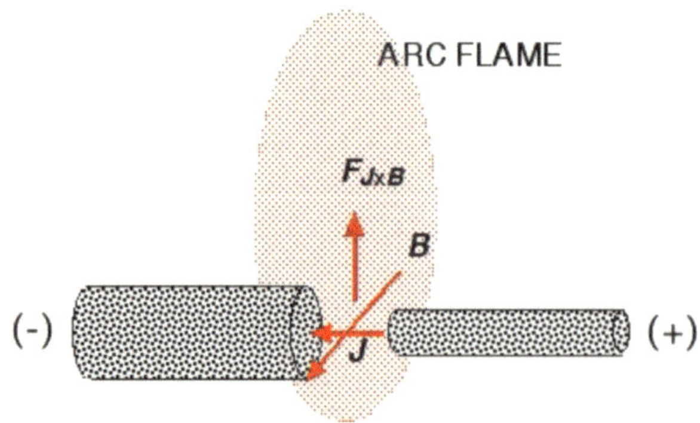


Fig. 6. Model figure of the $J$x$B$ arc discharge.

## 3. Production of SWNTs by arc-discharge method

### 3.1 Gravity-free production by means of parabolic flight

At first, we constructed a 12-m-high vertical swing tower (VST) in the university campus supported by the Institute of Space & Astronautical Science, Japan (ISAS/JAXA). A photograph and the schematic of the VST are shown in figure 7. Suspended by a thin stainless wire and a 2-m-long rubber rope, an arc reactor was swung by force from an air cylinder, and a constant-amplitude swing of the reactor was realized (Mieno, 2004). The amplitude is about 4 m peak to peak, the period was 2.3 s and the gravity-free time was 1.1 s. Synchronous with this swing, the arc-discharge current was pulse-time-modulated. After about 30 min of swinging, 15 min of integrated gravity-free production was realized. By this method, SWNTs were produced (Kanai et al., 2001). The result shows that amount of produced carbon soot increased by about 13.5 times that produced under normal gravity. However, 1.1 s of gravity-free time was not sufficient to form a large and spherical gas region for synthesizing SWNTs.

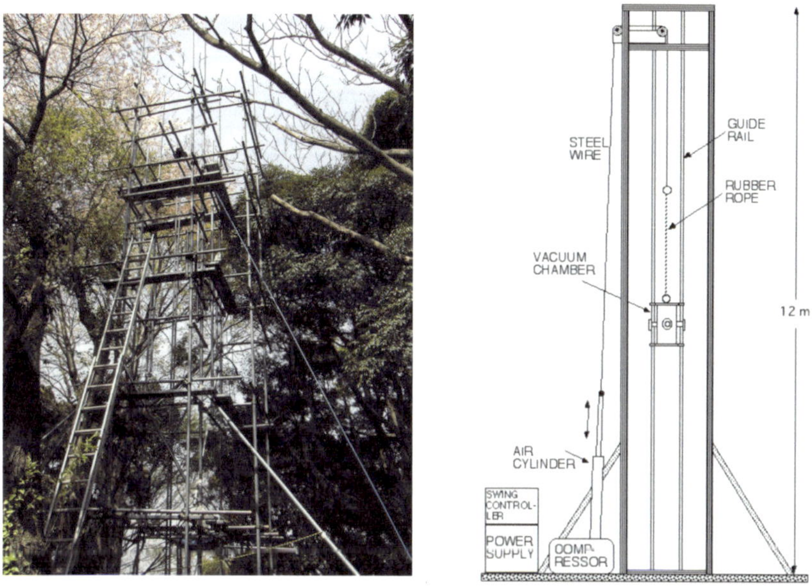


Fig. 7. Photograph and schematic of 12-m-high vertical swing tower (VST).

Then, I had the chance to perform an experiment using a specially prepared jet plane in Japan, in which a repetitive series of experiments under 20 s of microgravity condition was conducted using parabolic flights. I tried to examine the microgravity effects in the production of SWNTs by this method (Mieno, 2006, Mieno & Takeguchi, 2006). We developed special equipment for this purpose (which includes a small arc reactor, a DC power supply, a pumping system, a gas-feeding system, diagnostic systems such as video cameras, thermocouples, and a Mie scattering unit.) as shown in figure 8. The equipment was installed in the jet plane (Grumman G-II), which was operated to make repeated parabolic flights (Mieno, 2003). In one flight, 10-20 parabolic flights were carried out, by which 200-400 s of the integrated gravity-free production of SWNTs were realized. The reactor is a cylindrical metal chamber 16 cm in diameter and 20 cm high, in which a carbon rod anode 6.0 mm in diameter including Ni/Y metal particles and a carbon rod cathode 10.0 mm in diameter were set.

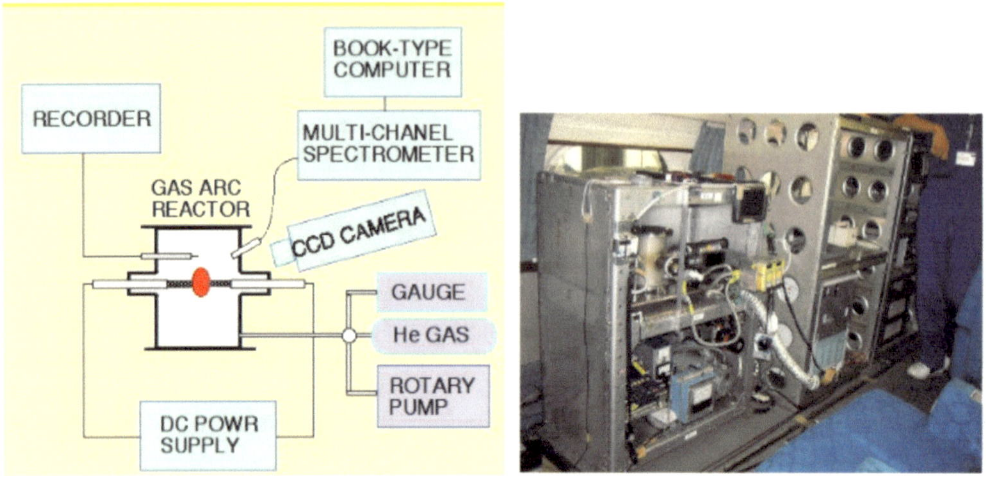


Fig. 8. Schematic and photograph of the reactor installed in the jet plane.

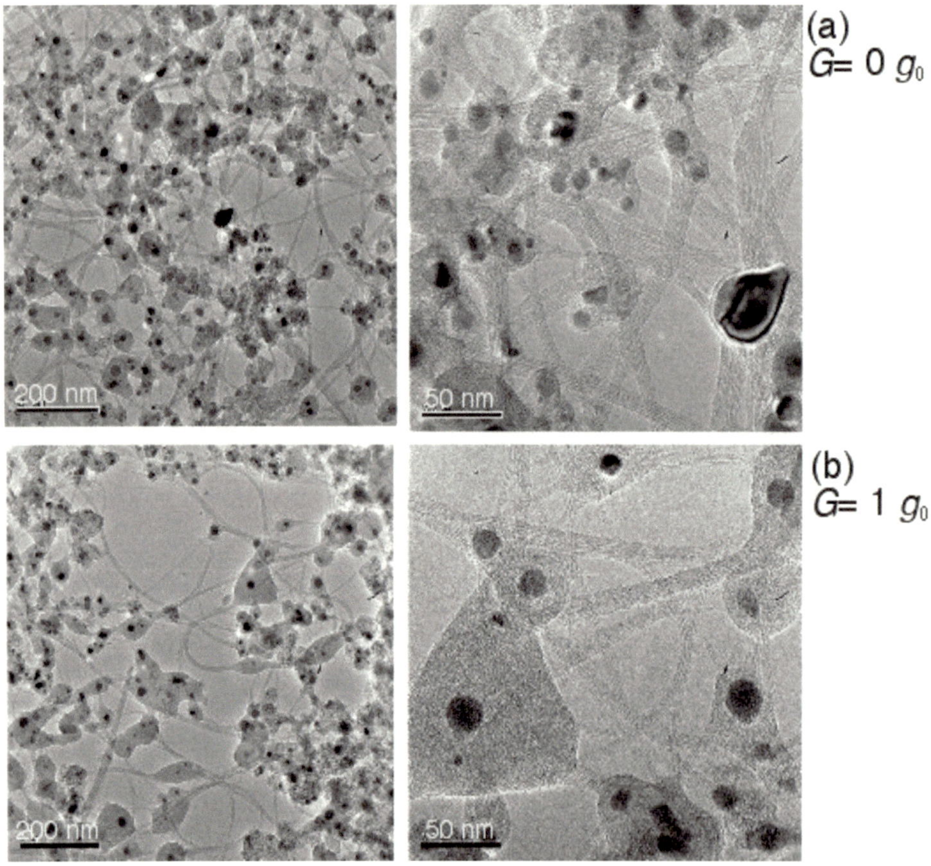


Fig. 9. Typical TEM images of SWNTs produced under two gravity conditions. $p$(He)= 65 kPa, $I_d$~ 50 A.

After evacuating the reactor with a rotary pump, He gas was introduced in the reactor and the reactor was closed. A DC power supply (Daihen Co., VRTP-200) was used to continue DC arc discharge under a constant current.

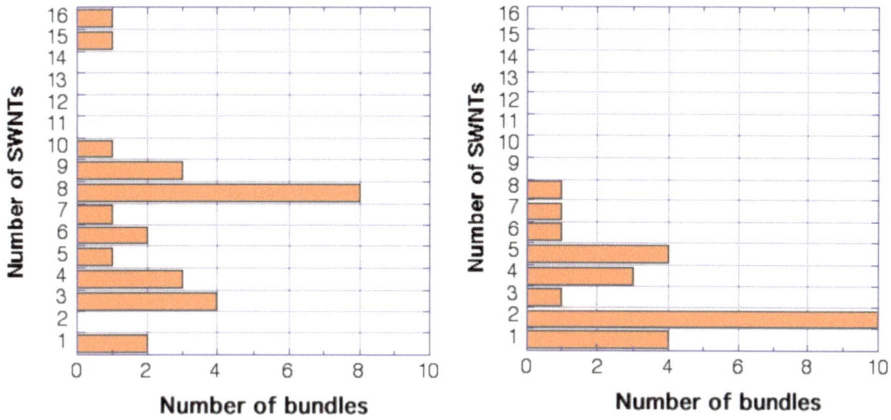


(a) $G$= 0$g_0$ (b) $G$=1$g_0$

Fig. 10. Number of observable SWNTs in a bundle counted from figure 9. Shaded SWNTs in each bundle are not counted.

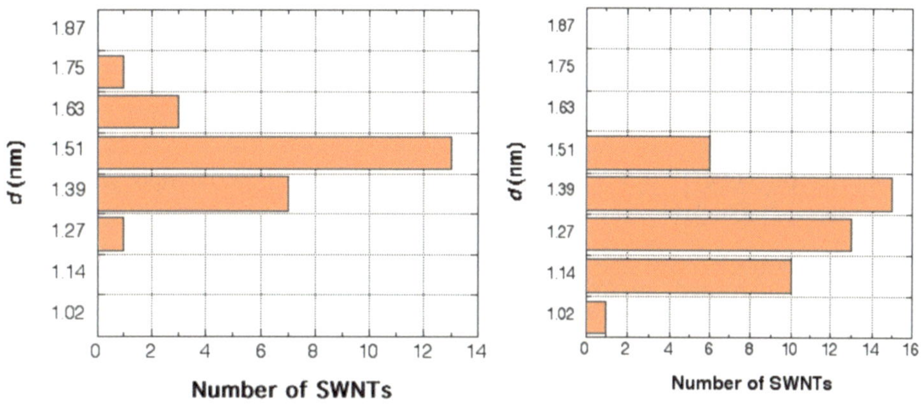


(a) $G$= 0$g_0$ (b) $G$=1$g_0$

Fig. 11. Diameter distributions of SWNTs in figure 9.

To clarify the gravity effect, the production was carried out on the ground under the same discharge condition of zero gravity. After the production, the produced carbon soot was carefully collected and analysed. The TEM images of the produced SWNTs are shown in Fig. 9, where $p$(He)= 65 kPa and discharge current $I_d$= 50 A. Under zero gravity, thicker bundles of SWNTs with higher content were produced than those under normal gravity. The diameter is about 1.5 nm. From the TEM images, the number of observable SWNTs in a bundle was measured, and its histograms are shown in figure 10. From this figure it was confirmed that thicker bundles of SWNTs were produced under zero gravity. Also the diameter distribution of SWNTs in figure 9 was measured, and its histograms are shown in figure 11. Under zero gravity, thicker SWNTs (average diameter of 1.5 nm) were produced. For the gravity-free experiment, the main target is production of much longer SWNTs. Although the SWNTs shown in figure 9(a) appear to be more than 100 nm, longer than those produced under the normal gravity, it is difficult to confirm the length of each nanotube. The length of most of the SWNTs did not exceed 1 mm. It is conjectured that 20 s of gravity-free time is not enough to produce very long SWNTs. Further analysis and a search for better methods are now under way.

### 3.2 High-gravity production using rotating acceleration generator

To clarify the gravity effects, SWNTs were produced under high gravity. In this case, stronger heat convection is expected and the reactor time decreases. For this purpose, the rotating acceleration generator of JAXA, Japan was used.

The photograph of the generator is shown in figure 12. The generator has a 6.5-m-long arm rotating at a constant speed to reach an acceleration of 0 – 490 $m/s^2$. (Tan & Mieno, 2010a) On one end of the arm, the experimental set up was installed, and by remote control with a motor drive and video cameras, arc discharge under a continuous high gravity was carried out. Under 2-$g_0$ and 3-$g_0$, stronger heat convections were clearly observed by the passive Mie scattering method, in which direct light from the arc plasma was cut by a metal plate, and scattered lights from the clusters are recorded. (Mieno, 2006b) Figure 13(a) shows a Mie scattering image under 3 $g_0$ (side view), in which the white clouds are carbon clusters flowing from the plasma region forming one large swirl, where $p$(He)= 50 kPa, $I_d$= 40 A and $t$= 4.0 min. Figure 13(b) shows corresponding velocity distribution calculated by the SMAC method. A large swirl (cluster flow) from the inner side to the outer side is clearly obtained. After the experiment, the produced soot was carefully collected and analysed by Raman spectrometry and TEM. Figure 14 shows the produced soot yield versus applied gravity G, where $p$= 50 kPa and $I_d$= 50 A. During the discharge, the discharge voltage was stable and should little change with gravity. However, the soot yield gradually increased with gravity.

Fig. 12. Photograph of the rotation acceleration generator of JAXA.

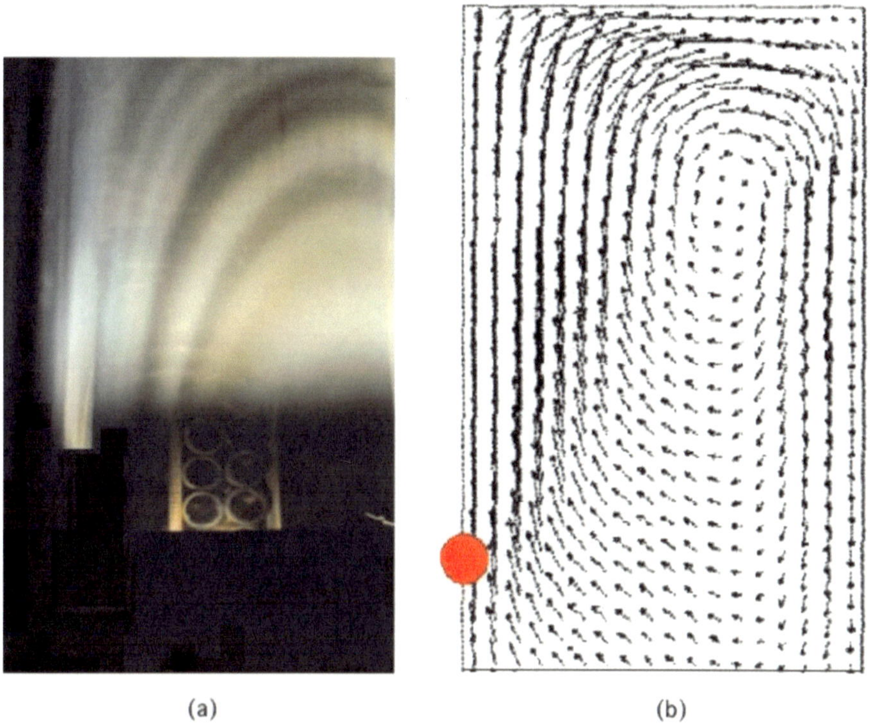


Fig. 13. (a) Mie scattering image of the carbon-cluster flow from the plasma region under 3 $g_0$ (side view). $p$(He)= 50kPa, $I_d$= 40 A, $t$= 4.0 min. (b) Corresponding calculation of the gas velocity by the SMAC method.

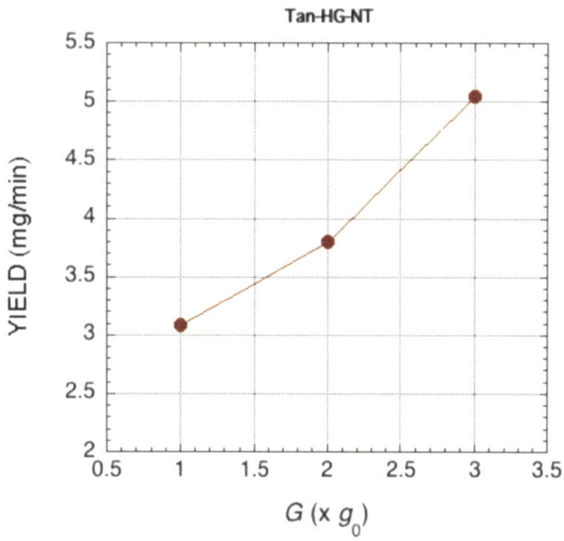


Fig. 14. Yield of carbon soot versus applied gravity $G$. $p$(He)= 50 kPa and $I_d$= 50 A.

To determine the production site of SWNTs, a cylindrical carbon collector array (each cylinder is 0.8 cm in diameter and 2.0 cm long) was installed 1.0 cm from the arc centre (Tan & Mieno, 2010a). The soot is collected at $z$= 0, 10, 20, 30, 40, 50, 60, 70 and 80 mm. The Raman spectra of the collected soot at 5 vertical positions under 1 $g_0$ are shown in figure 15,

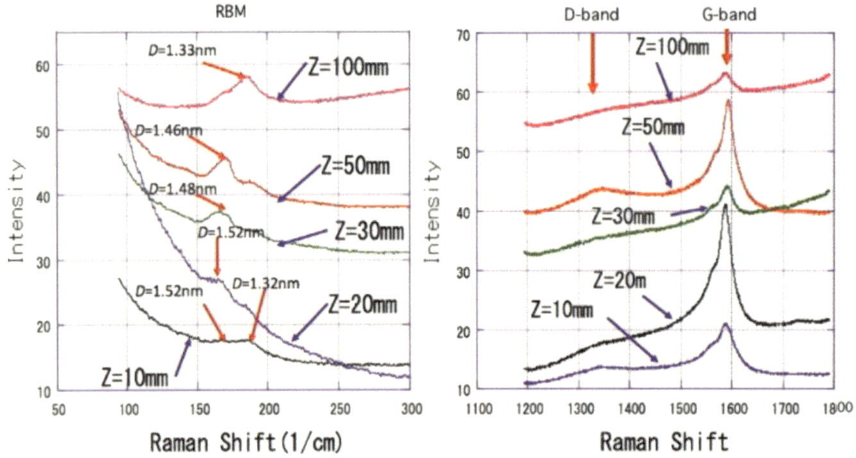


Fig. 15. Raman spectra of the soot produced under 1.0 $g_0$. The RBM, G-band and D band of soot collected at five positions are shown.

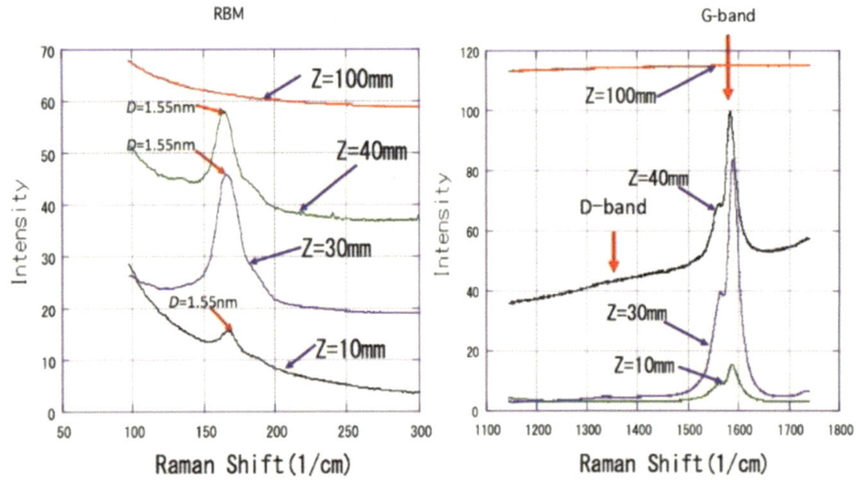


Fig. 16. Raman spectra of the soot produced under 3.0 $g_0$. The RBM, G-band and D band of soot collected at several positions are shown.

where the radial breathing modes (RBM) of SWNTs are shown on the left side. The G-band and D-band of carbon are shown on the right side. Vertical direction variations of these signals can be confirmed. In the same way, the Raman spectra of the carbon soot under 3 $g_0$ are shown in figure 16. Under this condition, the RBM and G-band are clearly obtained only at $z$= 10, 30 and 40 mm. At higher positions, no clear peaks from SWNTs were obtained. As expected from the theory, the higher the gravity, the higher convection speed, which reduces the reaction time for producing SWNTs in a hot gas atmosphere. This model supports the results shown by the Raman spectra in figures 15 and 16.

From the Raman spectra, the vertical distributions of G/D ratio for the soot produced under the three gravitational conditions were evaluated and are shown in figure 17 (Tan & Mieno, 2010a). This G/D ratio indicates the relative content of SWNTs in the produced soot, because the disordered carbon impurity emits the D-band signal. We can find that the G/D ratios under 3 $g_0$ at $z$= 10, 20 and 30 mm are much higher than those under 1 $g_0$, in spite of the stronger heat convection. This phenomenon can be explained from the model calculation. From the Mie scattering image and calculation in figure 13, it can be conjectured that the large swirl observed in the chamber flows down the carbon particles from the top wall region to the hot gas region, which makes the repetitive heating of carbon particles crucial for producing good SWNTs.

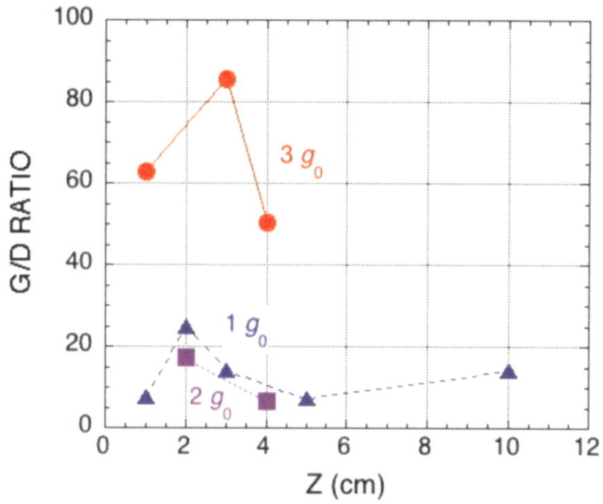


Fig. 17. Vertical distributions of G/D ratio for the soot produced under the three gravitational conditions.

### 3.3 Production under steady-state magnetic field

After considering the control of gas flow around the arc plasma, it was found that a steady-state magnetic field applies a *JxB* force to the plasma and changes the flow of sublimated carbon molecules (Aoyama & Mieno, 1999). The plasma flow affected by the magnetic field can be observed from a viewing port of the arc reactor, as shown in figure 18 (Matsumoto & Mieno, 2003). A strong upper flow of the plasma along the *JxB* direction was clearly observed (figure 18 (b)). To confirm this effect, an arc discharge experiment with a magnetic field was carried out (Mieno et al., 2006). Depending on the field, the production rate of the soot including SWNTs increased, as shown in figure 19, where $p$(He)= 60 kPa and $I_d$= 50 A. By the Raman spectrometry and TEM, the characteristics of produced SWNTs are determined (Mieno et al., 2006). The Raman spectra of the soot including SWNTs under the two magnetic fields are shown in figure 20, where $p$(He)= 60 kPa and $I_d$= 50 A. From the RBMs in the left graphs, the diameter of SWNTs, $d$, can be estimated and is written on the top of the figure. By applying the magnetic field, the diameter distribution is narrower and the main diameter is 1.4 nm. In the right graphs, the peaks of the G-band and D-band of carbon are shown. A good G/D ratio was obtained and the D-band signal was very small with application of a magnetic field.

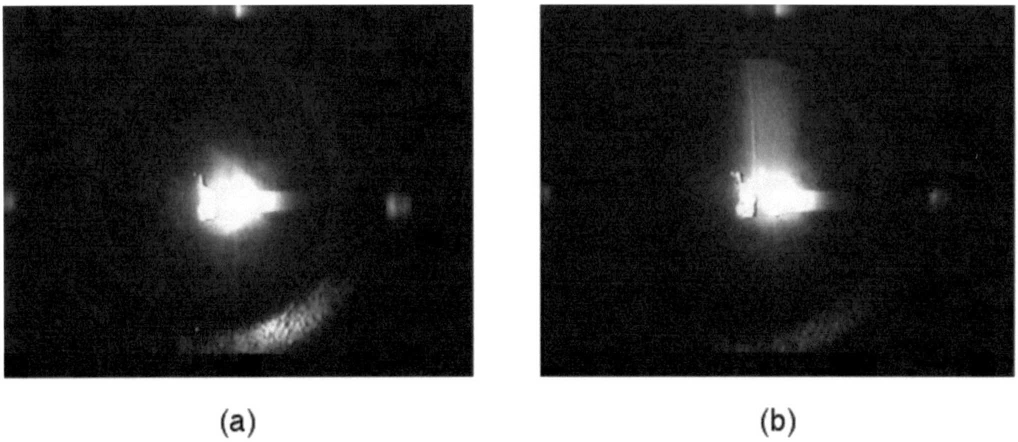


Fig. 18. Typical images (side views) of the arc flame. (a) No magnetic field and (b) $J$x$B$ arc discharge.

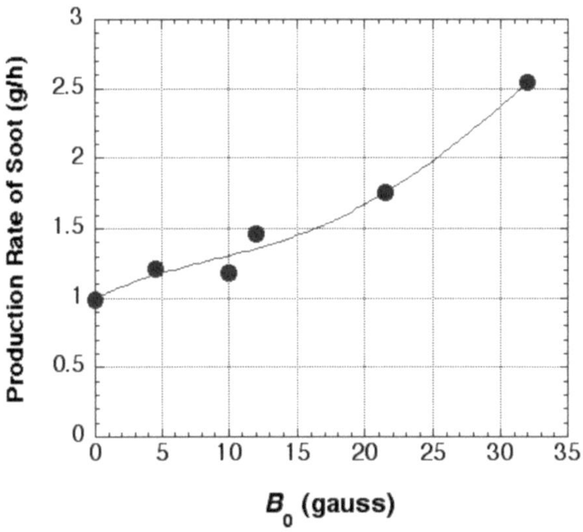


Fig. 19. Production rate of soot icluding SWNTs vs applied magnetic field.

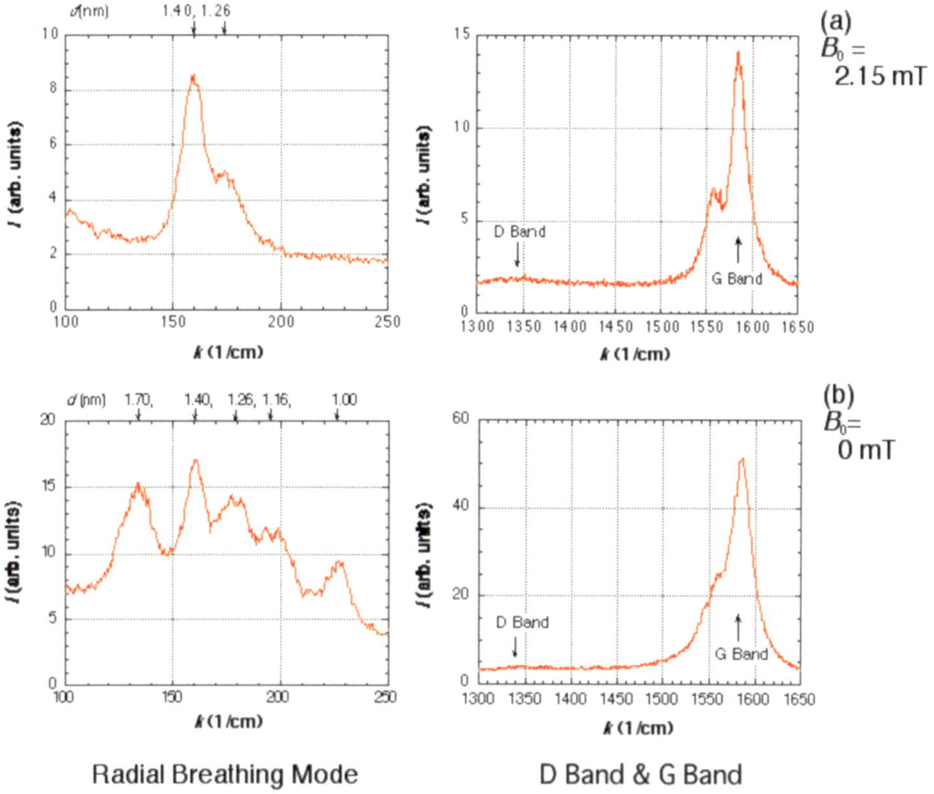


Fig. 20. Raman spectra of the soot including SWNTs under the two magnetic fields. $p$(He)= 60 kPa and $I_d$= 50 A. From the RBMs in the left graphs, the diameter of SWNTs, $d$, can be estimated and is written on the top of the figure. In the right graphs, the peaks of the G-band and D-band of carbon are shown.

## 3. Conclusion

Control of the reaction conditions in the hot gas phase is important for the production of high-quality SWNTs, because the analysis and control of the production process of SWNTs have not yet been investigated. Now, to improve the production of longer and high-quality SWNTs, a basic study of the production process is necessary. Here, the effects of zero-gravity, high-gravity and the *J*x*B* arc jet condition for the production of SWNTs were investigated. Under the zero-gravity and the *J*x*B* arc jet condition, considerably excellent production was obtained. However, the mass production of long SWNTs (more than 1 cm) could not be achieved. Therefore, further experimental and simulation studies are necessary. This challenge would lead to breakthroughs in the worldwide use of SWNTs.

## 4. Acknowledgments

I would like to thank Mr. N. Matsumoto for his collaboration as a graduate student. This study was partly supported by Japan Space Forum (JSF), ISAS/JAXA and The Ministry of Education, Culture, Sports, Science &Technology (MEXT), Japan.

## 5. References


Aoyama, A. & Mieno, T. (1999) Effect of Gravity and Magnetic Field in Production of $C_{60}$ by a DC Arc Discharge, *Japanese Journal of Applied Physics*, Vol. 38, No. 3A, pp. L267-L269

Bird, R. B. et al. (1960). Transport Phenomena, Wiley, New York, p. 337

Cussler, E. L. (1984). Diffusion: Mass Transfer in Fluid Systems, Cambridge University Press, New York, p. 19

Harris, P. J. F. (1999). Carbon Nanotubes and Related Structures, Cambridge University Press, ISBN 0-521-00533-7, Cambridge

Iijima, S. (1993). Single-Shell Carbon Nanotubes of 1-nm Diameter, *Nature*, Vol. 363, pp. 603-605

Jorio, A. et al. (eds.) (2008). Carbon Nanotubes, Springer-Verlag, ISBN 978-3-540-72864-1, Berlin

Kanai, M. et al. (2001). High-yield synthesis of single-walled carbon nanotubes by gravity-free arc discharge, *Applied Physics Letters*, Vol. 79, No. 18, pp. 2967-2969

Matsumoto, N. & Mieno, T. (2003). Characteristics of heat flux of *JxB* gas-arc discharge for the production of fullerenes, *VACUUM*, Vol. 69, pp. 557-562

Mieno, T. (2003). Gas Temperature Evolution of the Gravity-Free Gas Arc Dicharge under a Parabolic Flight of Jet Plane, *Japanese Journal of Applied Physics*, Vol. 42, No. 8A, pp. L960-L963

Mieno, T. (2004). Characteristics of the gravity-free gas-arc discharge and its application to fullerene production, *Plasma Physics and Controlled Fusion*, Vol. 46, pp. 211-219

Mieno, T. & Takeguchi, M. (2006). Thermal motion of carbon clusters and production of carbon nanotubes by gravity-free arc discharge, *Journal of Applied Physics*, Vol. 99, pp. 104301-1-5

Mieno, T. (2006). Diffusion and Reaction of Carbon Clusters in Gas Phase for Production of Carbon Nanotubes, *New Diamond and Frontier Carbon Technology*, Vol. 16, No. 3, pp. 139-150

Mieno, T.; Matsumoto N. & Takeguchi, M. (2006). Efficient Production of Single-Walled Carbon Nanotubes by *JxB* Gas-Arc Method. *Japanese Journal of Applied Physics*, Vol. 43, No. 12A, pp. L1527-1529

Payret, R. & Taylor, T. D. (1983). Computational Methods for Fluid Flow, Springer, New York, p. 160

Tan, G. & Mieno, T. (2010a). Synthesis of single-walled carbon nanotubes by arc-vaporisation under high gravity condition, *Thin Solid Films*, Vol. 518, pp. 3541-3545

Tan, G. & Mieno, T. (2010b). Experimental and Numerical Studies of Heat Convection in the Synthesis of Single-Walled Carbon Nanotubes by Arc Vaporisation, *Japanese Journal of Applied Physics*, Vol. 49, pp. 045102-1-6

Von Helden, G. et al. (1993). Carbon cluster cations with up to 84 atoms: structures, formation mechanism, and reactivity, *Journal of Physical Chemistry*, Vol. 97, pp. 8182-8192

# 5

# Selective Growth of Carbon Nanotubes, and Their Applications to Transparent Conductive Plastic Sheets and Optical Filters

Yusuke Taki, Makiko Kikuchi, Kiyoaki Shinohara,
Yosuke Inokuchi and Youhei Takahashi
*Smart Materials Research Section, Materials & Advanced Research Laboratory, Research & Development Headquarters, Core Technology Center, Nikon Corporation*
*Japan*

## 1. Introduction

Carbon nanotubes (CNTs) have incomparable physical properties and their applications in various fields have been examined. In particular, CNTs composed of a few cylindrical walls are useful for optoelectronic applications. The electronic properties of CNTs, however, significantly change depending on their chirality and the number of graphene walls. Therefore, first of all, the selective growth of graphene walls is required, and ultimately, chiral selection technology should be established. In single-walled CNTs (SWCNTs) and double-walled CNTs (DWCNTs), both semiconducting and metallic characteristics exist according to the chirality of the CNTs. On the other hand, it was predicted that all triple-walled CNTs (TWCNTs) have semimetal characteristics. Therefore, TWCNTs may be used in electronic applications without the need for chiral selection. Moreover, because TWCNTs is the finest multi-walled CNTs (MWCNTs), it is academically interesting to clarify the formation mechanism and various properties of TWCNTs.

Nowadays, as-grown SWCNT films are well synthesized on substrates by several types of chemical vapor deposition (CVD) (Hata et al., 2004; Murakami et al., 2004; Zhong et al., 2005). In addition, a synthetic process of producing high-purity SWCNT powder was established at the end of the last century (Nikolaev et al., 1999). As-grown DWCNT films on substrates have also been reported (Hiramatsu et al., 2005; Yamada at al., 2006). From the viewpoint of obtaining DWCNT powder, the CVD of DWCNT powder with supporting material was reported (Muramatsu et al., 2005). The combustion removal of SWCNTs from a mixture of SWCNTs and DWCNTs was also reported (Ramesh et al., 2006). The combustion removal method required a post-treatment to obtain high-yield DWCNT powder. On the other hand, the synthesis of TWCNTs has not reported at all, regardless of the form, for example, films and powder. For the purpose of obtaining CNT films with high DWCNT and TWCNT contents, post-treatments after CNT synthesis are not suitable, because CNTs have almost same chemical properties not related to the number of graphene walls. Post-treatments are only effective for removing amorphous carbon and metallic catalysts from CNTs. Therefore, it is necessary to develop an as-grown synthetic process for DWCNT and TWCNT films.

It is necessary to prevent the aggregation of catalyst particles on a substrate and to control the catalyst diameter distribution on the substrate for the entire period from catalyst deposition to the beginning of CNT growth, because catalyst particles are considered to function as growth nuclei for CNTs. It is necessary for DWCNT and TWCNT syntheses to enlarge the average diameters of catalyst particles more than SWCNT synthesis and to maintain narrow catalyst diameter distributions. In the techniques reported thus far, when catalyst particles are deposited on substrates, catalyst diameter distributions are narrow. While temperature rises slowly up to the CNT growth temperature, the catalyst diameter distributions broaden because of aggregation. As a result, CNTs that consist of various numbers of graphene walls are synthesized involuntarily.

In Section 2.1, it is explained about radiation-heated CVD (RHCVD), which enables us to selectively synthesize SWCNT, DWCNT, and TWCNT films on substrates (Taki et al., 2008a). The number of graphene walls of CNTs has a close relationship to catalyst diameter. It is very important to maintain narrow catalyst diameter distributions for a selective growth of graphene walls. RHCVD is characterized by its use of IR radiation, which enables the entire reactor to be heated rapidly. Therefore, RHCVD has the capability of maintaining narrow catalyst diameter distributions for the entire period from catalyst deposition to the beginning of CNT growth. The principle of RHCVD is explained as follows. First, catalyst particles whose diameters are several nanometers are deposited onto a substrate. Second, an entire reactor in hydrogen atmosphere is heated using IR radiation with an originally developed heating procedure. Catalyst particles should be reduced at a relatively low temperature to prevent their aggregation. Third, after reduction, the reactor is rapidly heated up to a prescribed temperature. During this period, it is considered that alloy particles are created within a narrow diameter distribution. Fourth, hydrocarbon gas is introduced into the reactor. The gas adheres to the catalyst particles, hydrogen is dissociated, and CNTs grow from the catalyst particles. Most of CNTs grown are composed of the same number of graphene walls because the catalyst diameter distribution is controlled within a narrow range.

In addition, in Section 2.2, it is mentioned about the synthesis of vertically aligned SWCNT, DWCNT, and TWCNT films on substrates by a combination of RHCVD and long-throw sputtering as a catalyst particle deposition process (Taki et al., 2008b). For obtaining vertically aligned SWCNT, DWCNT, and TWCNT films, it is necessary to deposit catalyst particles on substrates uniformly with a high population density, prior to RHCVD. When the population density of catalysts on substrates is very high, particles aggregate and the selective growth of graphene walls becomes impossible. In contrast, when the population density is very low, SWCNTs, DWCNTs, and TWCNTs do not grow vertically. In order to solve this dilemma, it is very effective to first deposit non-catalyst particles on a substrate, and secondarily deposit catalyst particles onto non-catalyst particles on the substrates. The diameters of non-catalyst particles must be slightly larger than those of catalyst particles. Even though the population density of catalyst particles on the substrate is quite high, catalyst particles may not aggregate each other during pre-heating process. Because each catalyst particle is located on each non-catalyst particle, a catalyst particle is not able to aggregate with a nearest catalyst particle on another non-catalyst particle.

Section 2.3 explains the essence of synthesizing long CNTs on substrates. It is necessary to delay the inactivation of metallic particles during a CNT growth period. Using Al particles as a non-catalyst, and lowering growth temperature are effective for delaying the inactivation of metallic catalysts.

Transparent conductive plastic sheets containing CNTs have also been studied by some groups. The plastic sheets reported so far have isotropic conductivity (Dan et al., 2009; Hecht et al., 2009; Wu et al., 2004). In contrast, anisotropic conductivity has been achieved in this study. In Section 3, it is explained how to fabricate an anisotropically coductive sheet and comparative study of electrical features using several kinds of CNTs. Anisotropically conductive and transparent plastic sheet is expected for both optical and electronic new applications.

In Section 4, the excellent optical performance of CNT forests is explained in detail. In the field of laser optics, stray light is significant matter to be solved. For example, the reflection on an inner wall of a laser device box induces stray light problem and causes an optical sensor in the box not to work well. Therefore, extremely low-reflection optical elements with a wide incident angle range are necessary for solving stray light. CNT forests, which grow vertically on substrates with a high population density, are effective for such optical applications. By using CNT forests, extremely low-reflection absorbance and ND filter were fabricated and characterized for high power ArF excimer laser equipment.

## 2. CNT growth

### 2.1 Selective growth of SWCNTs, DWCNTs, and TWCNTs through precise control of catalyst diameter by radiation-heated chemical vapor deposition

Dip solutions were prepared by dissolving both $Co(CH_3COO)_2 \cdot 4H_2O$ and $(Mo(CH_3COO)_2)_2$ in ethanol. The Co and Mo concentrations in the dip solutions were accurately adjusted to 0.01, 0.02, 0.03, 0.04, 0.05, 0.06, 0.07, 0.08, 0.09, and 0.10 wt%. The Co / Mo mole ratio was maintained at 1/1 for each dip solution. Quartz glass substrates were cleaned ultrasonically, and then hydrophilically modified by UV irradiation in ambient atmosphere. Hydrophilically modified substrates were dipped in the Co / Mo solutions and pulled up at a regular speed. Catalyst diameters on the substrates were controlled by changing the concentrations of the dip solutions. Several substrates which were dip-coated with each concentration were simultaneously placed on a quartz glass holder in a RHCVD reactor. The holder temperature was controlled with a thermocouple and a proportional-integral-derivative controller. The RHCVD reactor was heated up and kept at an intermediate temperature in air. The dip-coated films on the substrates were oxidized and volatile gases were exhausted from the films. After this period, the reactor was evacuated and filled with hydrogen gas at the same temperature. The dip-coated films were reduced and they became metallic particles. The intermediate temperature was as low as the temperature at which catalyst particles scarcely aggregate, and as high as the temperature at which oxidation and reduction reactions progress sufficiently. After that, the reactor was rapidly heated up to 800°C in hydrogen atmosphere. As soon as the temperature reached 800°C, ethanol vapor diluted with argon gas was introduced into the reactor. CNTs grew from the metallic catalyst particles on the substrates. During a CNT growth period of 1 h, the substrate temperature and total pressure were kept at 800°C and 1.7 kPa, respectively.

The characteristics of catalyst particles and the morphology of as-grown CNT films on substrates were observed by scanning electron microscopy (SEM). Two hundred CNTs scratched from each film were observed by transmission electron microscopy (TEM) in order to evaluate the number of graphene walls and the diameter distributions of as-grown CNTs. The chemical bonding states of as-grown CNT films were investigated using Raman spectroscopy.

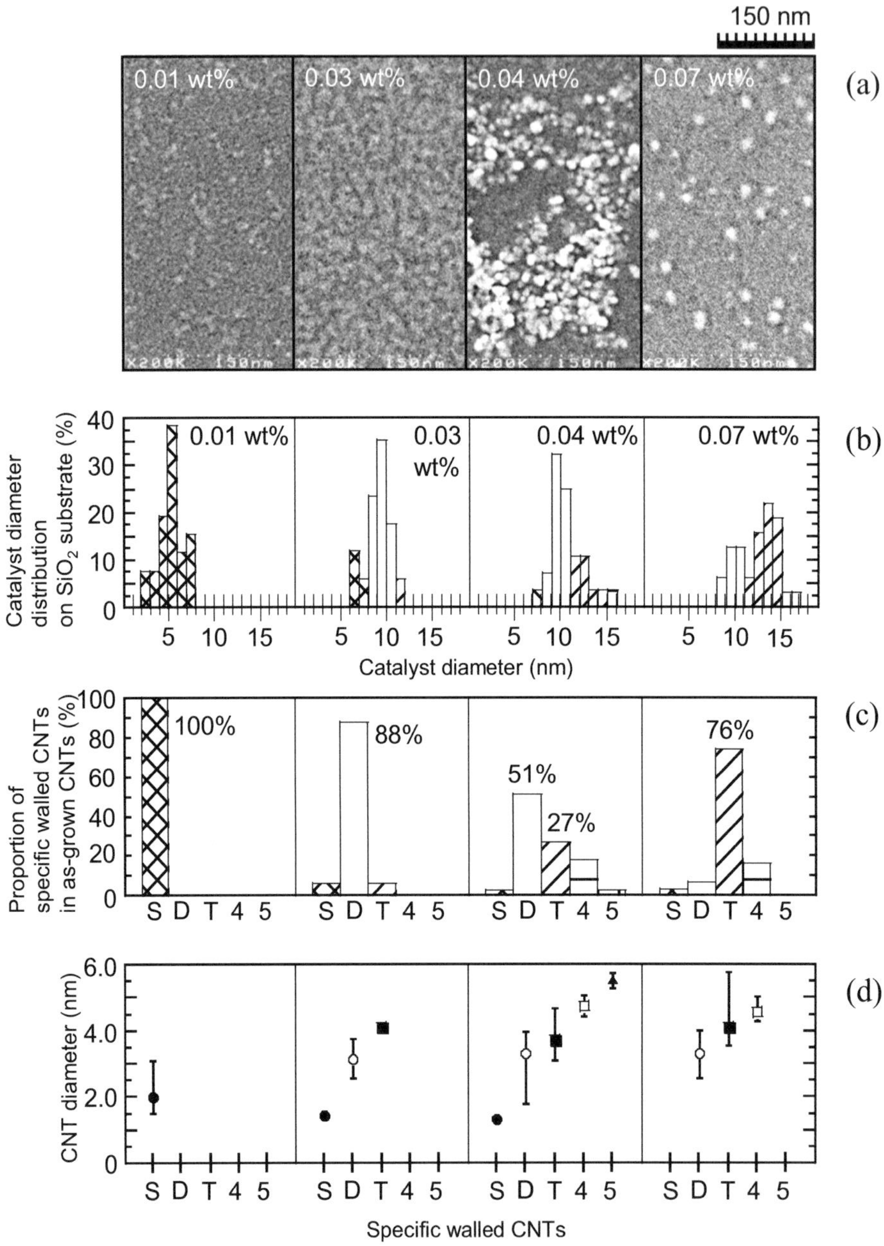


Fig. 1. Relationship between catalyst conditions on the eve of CNT growth and CNTs synthesized by RHCVD. (a) top-view SEM images of catalyst particles on quartz glass substrates dip-coated with 0.01, 0.03, 0.04, and 0.07 wt% solutions, (b) histograms of catalyst diameter distributions on the substrates, (c) histograms of specific walled CNTs, and (d) outer diameters of the specific walled CNTs.

It is important to understand the catalyst diameter distributions on substrates at the beginning of CNT growth. The CNT growth procedure explained above was executed until the beginning of CNT growth. As soon as the substrate temperature reached 800°C, substrates were rapidly cooled down to room temperature without introducing ethanol vapor. The substrates obtained thus have similar appearances to the catalyst particles on the eve of CNT growth. Figure 1(a) shows top-view SEM images of the substrates on the eve of CNT growth, which were dip-coated with 0.01, 0.03, 0.04, and 0.07 wt% solutions and then reduced. The white regions are catalyst particles. Catalyst diameters were measured and are summarized in Fig. 1(b) in the histogram. Each histogram shows the sharp distribution within several nm. When a catalyst concentration in a dip solution is raised, catalyst diameters on a substrate increase. Therefore, catalyst diameters are able to be controlled by adjusting a concentration of catalyst in a dip solution.

Figure 2(a) shows an SEM image of the as-grown CNT film on the substrate that was dip-coated with a 0.01 wt% solution. The entire area of the substrate is covered with nonaligned CNTs. The CNTs gathered from the substrate shown in Fig. 2(a) were observed by TEM. The proportion of SWCNTs in the as-grown CNTs was 100%. Figure 3(a) shows TEM images of the as-grown SWCNTs. Figure 4(a) shows the Raman spectrum of the as-grown film shown in Fig.2(a). The graphitic (G) band shape has a typical feature of SWCNTs, and the radial breathing mode (RBM) is clearly observed. Moreover, the disorder (D) band that arises from the disorder in the carbon arrangement is detected. The intensity ratio of the G band to the D band is 11 to 1.

Figure 2(b) shows an SEM image of the as-grown CNT film on the substrate that was dip-coated with a 0.03 wt% solution. The entire area of the substrate is covered with nonaligned CNTs. The CNTs gathered from the substrate shown in Fig. 2(b) were observed by TEM. The proportion of DWCNTs in the as-grown CNTs was 88%, and the remainder was composed of SWCNTs and TWCNTs. Figure 3(b) shows TEM images of the as-grown DWCNTs. Figure 4(b) shows the Raman spectrum of the as-grown film shown in Fig.2(b). The D band is strengthened compared with that of the SWCNT film shown in Fig. 4(a). The detected RBM is attributed to the inner walls of the DWCNTs or SWCNTs.

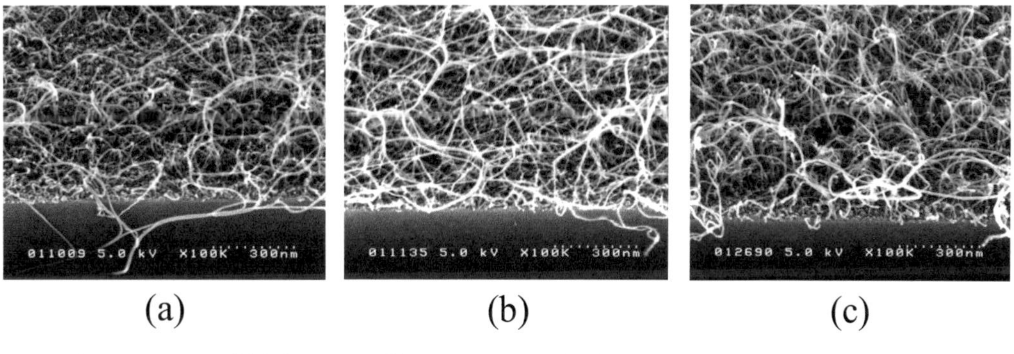


Fig. 2. SEM images of (a) SWCNT, (b) DWCNT, and (c) TWCNT films synthesized by RHCVD.

Figure 2(c) shows an SEM image of the as-grown CNT film on the substrate that was dip-coated with a 0.07 wt% solution. The entire area of the substrate is covered with nonaligned CNTs. The CNTs gathered from the substrate shown in Fig. 2(c) were observed by TEM. The proportion of TWCNTs in the as-grown CNTs was 76%, and the remainder was composed of DWCNTs and 4WCNTs. Figure 3(c) shows TEM images of the as-grown TWCNTs. Figure

4(c) shows the Raman spectrum of the as-grown film shown in Fig.2(c). The RBM is not detected. This result confirms that SWCNTs and DWCNTs with diameters smaller than 2.5nm are not included in this film. In addition, the G band becomes broader and the relative strength of the D band is higher. It is suggested that the amount of defects in CNTs increases as the number of graphene walls increases.

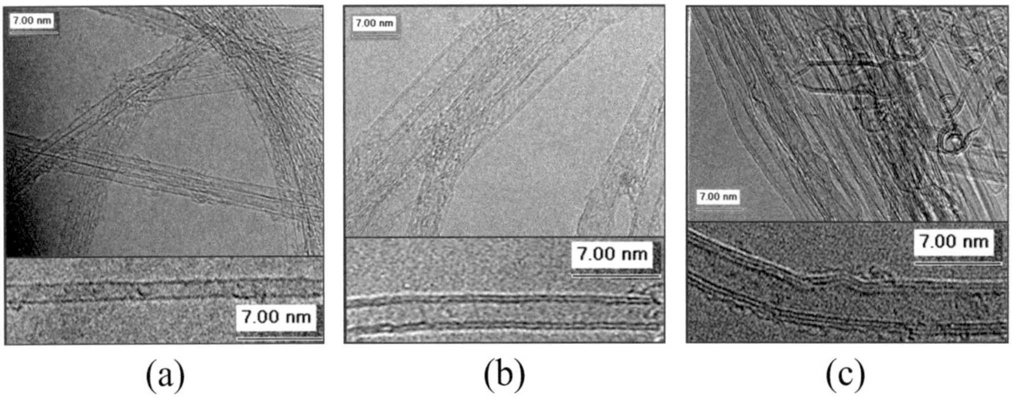


Fig. 3. TEM images of (a) SWCNTs, (b) DWCNTs, and (c) TWCNTs synthesized by RHCVD.

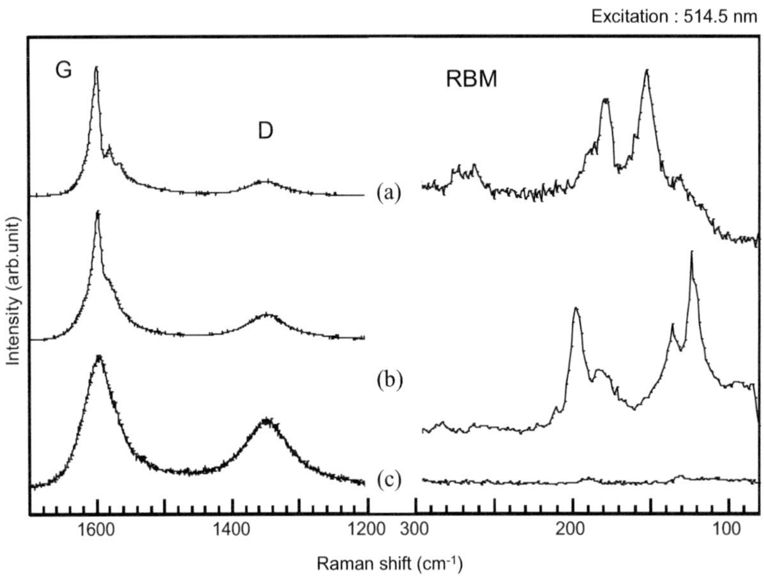


Fig. 4. Raman spectra of (a) SWCNT, (b) DWCNT, and (c) TWCNT films synthesized by RHCVD.

The discussion will now return to Fig. 1. CNTs were grown by RHCVD on each substrate with the catalyst diameter distribution shown in Fig. 1(b). The number of graphene walls was evaluated by TEM. The results are summarized in Fig. 1(c). Dominant CNTs are changed from SWCNTs and DWCNTs to TWCNTs as catalyst diameters increase. By comparing Figs. 1(b) and 1(c) carefully, it seems that there are effective catalyst diameter ranges suitable for SW, DW, and TWCNT growth. SWCNTs grow when catalyst diameters are in the range of 2-8 nm. DWCNTs grow from catalysts with diameters of 8-11nm, and TWCNTs grow from catalysts with diameters of 11-15 nm. Figure 1(d) shows the outer diameters of CNTs measured by TEM. The number of graphene walls shows the tendency to increase as the outer diameters of CNTs increase. In addition, the outer diameters of CNTs are invariably smaller than the catalyst diameters.

A comparative study between RHCVD and conventional CVD (Murakami et al., 2004) was executed. The difference between the two processes is the heating rate used before the beginning of CNT growth. A dip solution with a concentration of 0.06wt% was prepared, and two quartz glass substrates were continuously dip-coated with the solution. Therefore, in the period before heating, the catalyst conditions of both substrates must be equivalent to each other. One substrate was heated using the conventional CVD procedure and the other substrate was heated using the RHCVD procedure. As soon as the temperature reached 800°C, both substrates were rapidly cooled down to room temperature without introducing ethanol vapor. Figure 5(a) shows histograms of the catalyst diameter distributions on both substrates measured by SEM. It is obvious that the catalyst diameter distribution obtained with the RHCVD procedure is comparatively narrow and has a sharp peak at 11-12 nm, while the distribution obtained with the conventional CVD procedure is broad and does not have a distinct peak.

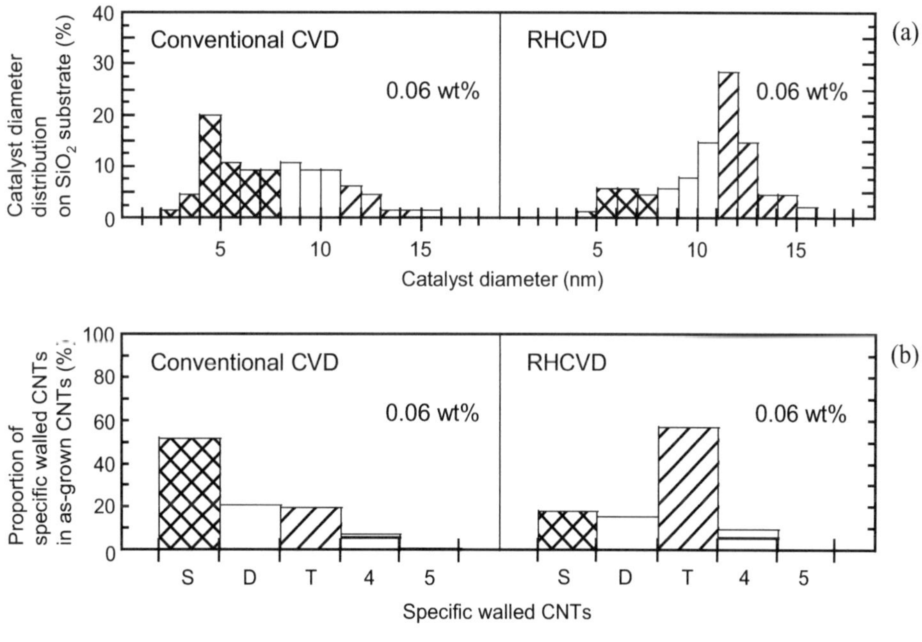


Fig. 5. Comparative study between RHCVD and conventional CVD. (a) histograms of the catalyst diameter distributions on quartz glass substrates dip-coated with a 0.06 wt% solution and heated up to 800°C, and (b) histograms of specific walled CNTs.

CNT growth was performed using substrates equal to those shown in Fig. 5(a). Figure 5(b) shows the results. In the case of conventional CVD, there are more SWCNTs than any other

CNTs. In the case of RHCVD, contrastingly, TWCNTs grow preferentially. From this comparative experiment, it has been confirmed that RHCVD is an effective process to selectively synthesize CNTs that consist of a specific number of graphene walls.

## 2.2 Selective growth of vertically aligned SWCNTs, DWCNTs, and TWCNTs by RHCVD

Quartz glass substrates were cleaned ultrasonically and then hydrophilically modified by UV irradiation in ambient atmosphere. A multicathode RF long-throw sputtering system was used for catalyst particle deposition. $Al_2O_3$ particles as a non-catalyst were first deposited on the substrates with slightly larger diameter prior to metallic catalyst deposition. Secondarily, both Co and Fe particles were deposited on the substrates as catalysts for SWCNT synthesis. Both Fe and Mo particles were deposited as catalysts for DWCNT and TWCNT syntheses. For a comparative study, MWCNT films were also synthesized with the catalysts of Fe particles. Catalyst diameters were adjusted by changing sputter time. The conversion film thicknesses of catalyst particles were calculated from sputter time. In this study, conversion film thicknesses were used as substitutes for catalyst diameters. After sputter deposition, the substrates for SW, DW, TW, and MWCNT syntheses were placed on a quartz glass holder in an RHCVD reactor simultaneously. The holder temperature was controlled with a thermocouple and a proportional-integral-derivative controller.

The RHCVD reactor was heated up and kept at an intermediate temperature in hydrogen atmosphere. Sputtered particles were reduced and they became metallic particles. The intermediate temperature is as low as the temperature at which catalyst particles scarcely aggregate and as high as the temperature at which the reduction reaction progresses sufficiently. Subsequently, the reactor was rapidly heated up to 800 °C in hydrogen atmosphere. As soon as the temperature reached 800 °C, ethanol vapor diluted with argon gas was introduced into the reactor. CNTs grew from the metallic catalyst particles on the substrates. During a CNT synthesis period of 1 h, the substrate temperature and total pressure were kept at 800 °C and 1.7 kPa, respectively.

The morphologies and lengths of the as-grown CNT films on substrates were observed by SEM. One hundred CNTs included in each film were observed by TEM in order to evaluate the number of graphene walls and the diameter distributions of the as-grown CNTs. In addition, the chemical bonding states of the as-grown SWCNTs were investigated by Raman spectroscopy. The photoluminescence (PL) properties of SWCNTs were also measured. PL gives the chirality distributions of semiconducting SWCNTs included in total SWCNTs.

SWCNT films are synthesized on the substrates on which both Co and Fe particles are deposited. When the total conversion film thickness of both particles is less than 0.40 nm and the Fe / Co thickness ratio ranges from 3 to 1/3, SWCNTs are preferentially synthesized. When coming off from this condition, MWCNTs are dominantly grown. The diameters of SWCNTs are proportional to the total film thickness of catalysts. Figure 6(a) shows an SEM image of the vertically aligned CNT film grown on a Co 0.15 nm / Fe 0.15 nm / $SiO_2$ glass substrate. In this case, non-catalyst particles are not used. Figure 7(a) shows the TEM images of the CNTs extracted from the CNT film shown in Fig. 6(a). The CNTs are composed of SWCNTs (87 %), DWCNTs (10 %), and TWCNTs (3 %). The diameters of the SWCNTs are distributed from 1.2 to 4.0 nm, and the average diameter is 2.8 nm. Most SWCNTs form bundles, whereas some SWCNTs with relatively large diameters are naturally isolated.

Figure 8 shows the Raman spectra of the CNT film shown in Fig. 6(a). The radial breathing mode (RBM) appears clearly in Fig. 8(a). In Fig.8(b), G band shape has a typical feature of SWCNTs and D band appears with a weak intensity. The intensity ratio of the G band to the D band is 16 to 1. The CNT film shown in Fig. 6(a) was removed from the substrate and then was ultrasonically dispersed in an aqueous solution of sodium dodecylbenzenesulfonate. After centrifugation, the upper solution was used for PL measurement. Figure 9 shows the PL properties of the SWCNTs synthesized by RHCVD and purified HiPco as a reference. These PL properties indicate the chirality distributions of small-diameter semiconducting SWCNTs included in both specimens. Although both have some common features, the most intense peaks are (7,5) and (8,3) in HiPco, while the most intense peak is (8,6) in the SWCNTs synthesized by RHCVD. From Raman and PL characterizations, it is clear that the as-grown SWCNTs synthesized by RHCVD have similar features to general SWCNTs synthesized by conventional CVD processes.

DWCNT films are synthesized on the substrates on which both Fe and Mo particles are deposited. When the total conversion film thickness of both particles ranges from 0.30 to 0.50 nm and the Fe / Mo thickness ratio ranges from 4 to 1, DWCNTs are preferentially synthesized. Figure 6(b) shows an SEM image of the vertically aligned CNT film grown on an Fe 0.30 nm / Mo 0.10 nm / $Al_2O_3$ 5 nm / $SiO_2$ glass substrate. Figure 7(b) shows TEM images of the CNTs extracted from the CNT film. The CNTs shown in Fig. 7(b) are composed of DWCNTs (83 %) and TWCNTs (17 %). The outer diameters of the DWCNTs are distributed from 2.9 to 5.0 nm, and the average diameter is 4.1 nm.

TWCNT films are also synthesized on the substrates on which both Fe and Mo particles are deposited. When the total conversion film thickness of both particles ranges from 0.30 to 0.80 nm and the Fe / Mo thickness ratio ranges from 6 to 3, TWCNTs are preferentially synthesized. Figure 6(c) shows an SEM image of the vertically aligned CNT film grown on an Fe 0.25 nm / Mo 0.05 nm / $Al_2O_3$ 5 nm / $SiO_2$ glass substrate. Figure 7(c) shows TEM images of the CNTs extracted from the CNT film. The CNTs shown in Fig. 7(c) are composed of DWCNTs (26 %), TWCNTs (62 %), and 4WCNTs (12 %). The outer diameters of the TWCNTs are distributed from 3.8 to 4.8 nm, and the average diameter is 4.4 nm.

MWCNT films are synthesized on the substrates on which only Fe particles are deposited. The diameters of MWCNTs are proportional to the conversion film thickness of Fe. Figure 6(d) shows an SEM image of the vertically aligned MWCNT film grown on an Fe 0.50 nm / $Al_2O_3$ 5 nm / $SiO_2$ glass substrate. TEM observation revealed that the outer diameters of the MWCNTs are distributed from 5.3 to 9.5 nm, and the average diameter is 7.1 nm.

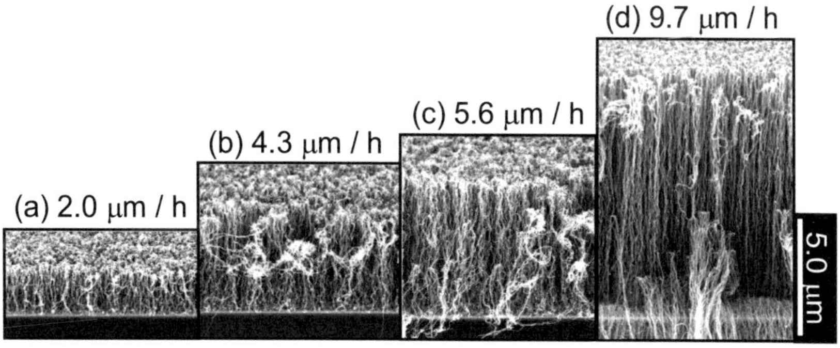


Fig. 6. SEM images of vertically aligned CNT films synthesized by RHCVD under optimal catalyst conditions for (a) SWCNTs, (b) DWCNTs, (c) TWCNTs, and (d) MWCNTs.

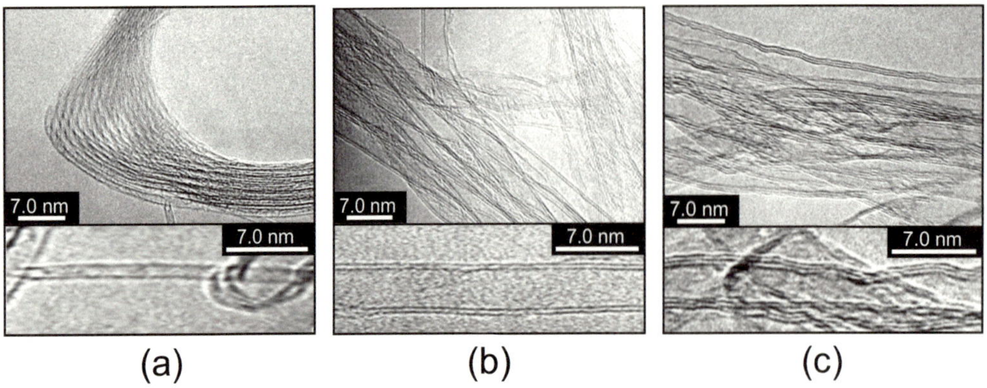


Fig. 7. TEM images of (a) SWCNTs, (b) DWCNTs, and (c) TWCNTs extracted from the CNT films shown in Figs. 6(a) - 6(c), respectively.

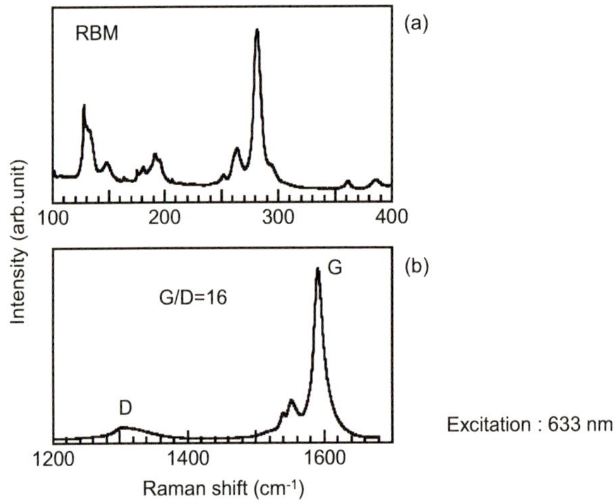


Fig. 8. Raman spectra of (a) RBM and (b) G-D regions of the CNT film shown in Fig. 6(a).

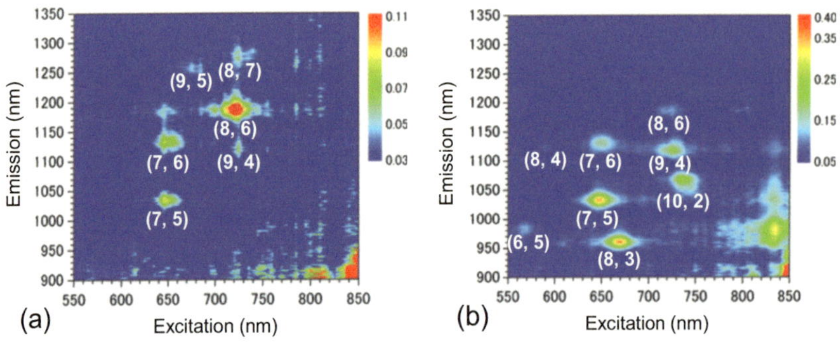


Fig. 9. PL properties of dispersed specimens of (a) the CNT film shown in Fig. 6(a) and (b) purified HiPco as a reference.

In this study, there are three important points concerning CNT growth, namely, the selection of graphene walls, vertical alignment, and lifetime of catalysts. These are sequentially discussed.

Figure 10 shows the proportions of specific walled CNTs / as-grown CNTs and the average outer diameters of the as-grown CNT films. The number of graphene walls of CNTs increases as the outer diameters of CNTs increase. Table 1 shows the total conversion film thicknesses and film thickness ratios of the main catalysts over the subcatalysts, which are used in order to synthesize the CNT films shown in Figs. 6(a)-6(d). Fe and Co are the main catalysts. Mo is considered to behave as the subcatalyst, which becomes an alloy with the main catalysts during heating, preventing the aggregation between the main catalyst particles. In the case of our long-throw sputtering, SWCNT films are not grown when both Fe/Mo and Co/Mo catalysts are used. From Fig. 10 and Table 1, it is clarified that the thickness ratio is important rather than the total thickness in achieving the selective growth of graphene walls. By alloying main catalyst particles with high melting-point metal particles, main catalyst particles are kept small and uniform in size.

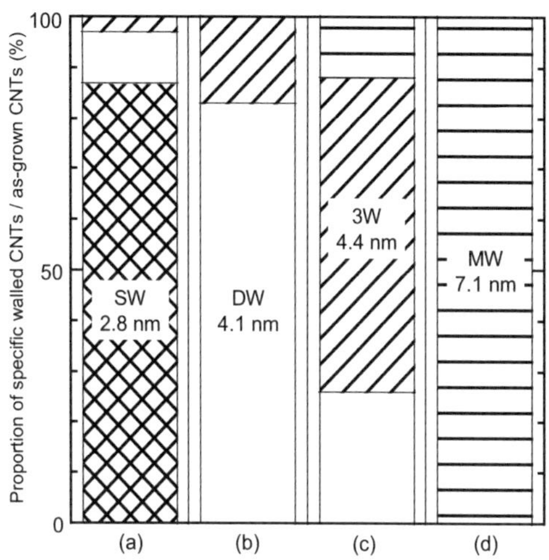


Fig. 10. Proportions of specific walled CNTs / as-grown CNTs and average outer diameters of the CNT films shown in Figs. 6(a) - 6(d). This result was obtained by TEM evaluation.

It is estimated from SEM observations that the population density of CNTs grown on a quartz glass substrate ranges from $1.1\times10^{9}$ to $7.1\times10^{10}$ bundles/cm$^2$. The population density of CNTs on a substrate depends on the population density of non-catalyst particles adjusted by sputtering. When alloyed metallic particles are individually placed on each different non-catalyst particles and the population density of non-catalyst particles on a substrate is high enough, the alloyed metallic particles never aggregate and finally CNTs are able to vertically grow with the specific number of graphene walls. This phenomenon is similar to that observed in the case of a jam-packed train. In contrast, when the population density of alloyed catalyst particles without non-catalyst particles on a substrate is very low, CNTs do not grow vertically like the case of Section 2.1.

In Figs. 6(a)-6(d), as the outer diameters and the number of graphene walls of CNTs increase, CNTs grow longer. It is inferred that this result is due to the difference in lifetime between metallic catalysts. In short, the metallic catalysts become inactivated by intermediates decomposed from ethyl alcohol during a CNT growth period of 1 h. It needs longer time to make larger metallic particles inactivated completely. It is considered that the larger the metallic catalyst diameter, the longer the catalyst lifetime.

| | (a) | (b) | (c) | (d) |
|---|---|---|---|---|
| Selective growth | SWCNTs | DWCNTs | TWCNTs | MWCNTs |
| Total conversion film thickness | 0.3 | 0.4 | 0.3 | 0.5 |
| Film thickness ratio | Co/Fe=1 | Fe/Mo=3 | Fe/Mo=5 | Fe only |

Table 1. Total conversion film thicknesses and film thickness ratios of main catalysts over subcatalysts, which are used in order to synthesize the CNT films shown in Figs. 6(a) - 6(d).

### 2.3 Vertical growth of long CNTs by RHCVD

In order to make CNTs grow longer, it is necessary to delay the inactivation of metallic particles during a CNT growth period. Changing a non-catalyst from $Al_2O_3$ to Al particles, and lowering a growth temperature from 800 °C to 650 °C are effective for delaying the inactivation of metallic catalysts. In Section 2.3, the vertically long TWCNT growth is explained with detailed data.

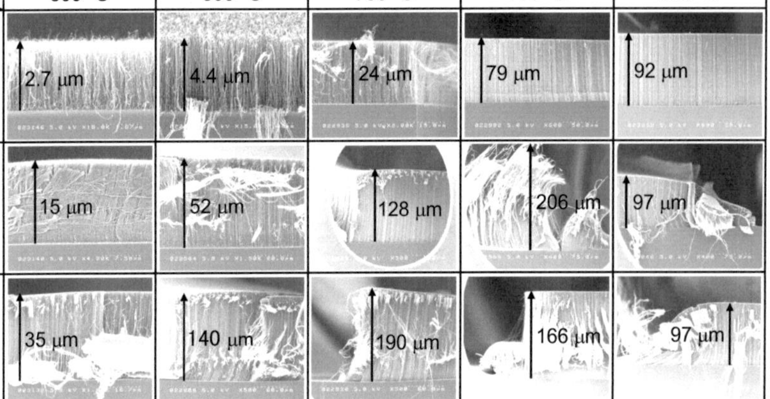

| | 800 °C | 800 °C | 750 °C | 700 °C | 650 °C |
|---|---|---|---|---|---|
| Center | 2.7 µm | 4.4 µm | 24 µm | 79 µm | 92 µm |
| Side | 15 µm | 52 µm | 128 µm | 206 µm | 97 µm |
| Corner | 35 µm | 140 µm | 190 µm | 166 µm | 97 µm |
| Co | 1 nm | 1 nm | 1 nm | 1 nm | 1 nm |
| Fe | 1 nm | 1 nm | 1 nm | 1 nm | 1 nm |
| Non-catalyst | $Al_2O_3$ 5 nm | Al 5 nm | Al 5 nm | Al 5 nm | Al 5 nm |

Fig. 11. SEM images of vertically aligned TWCNT films which locate at centers, sides, and corners of substrates, synthesized at various growth temperatures and catalyst conditions.

Figure 11 shows SEM images of CNTs grown from the alloy particles of Fe and Co on Al or $Al_2O_3$ particles on quartz glass substrates. The growth time was 30 min. CNTs which grow at a center, a side, and a corner of each substate with a size of 20 mm$^2$ were observed by SEM. These CNTs were also confirmed to consist of mainly TWCNTs by TEM observation. In Fig.11, when comparing two kinds of CNTs synthesized at 800 °C, it is found that using Al particles instead of $Al_2O_3$ can develop CNTs longer. Al is considered to react with intermediates easier than both Fe and Co. Al particles play a role of a getter catching intermediates. In contrast, Because $Al_2O_3$ itself is inert against intermediates, the lifetimes of Fe and Co become shorter and consequently CNTs grow shorter. Next, when comparing between CNTs synthesized at 800, 750, 700, and 650 °C, CNTs grow much longer at a lower temperature. This result attributes to delaying chemical reaction between metallic particles and intermediates at a lower temperature. Moreover, the difference of CNT length between a center, a side, and a corner in each substrate becomes smaller when lowering a growth temperature. This tendency is considered to the relatively larger distribution of Al particles in size since Al is low-melting point material. Lowering a growth temperature is also effective for synthesizing CNTs with uniform length in a wide area. Vertically long growth of SWCNTs and DWCNTs have been similarly accomplished though this chapter explained only TWCNTs.

## 3. Fabrication of transparent and anisotropically conductive plastic sheets by using CNTs

### 3.1 Behavior of CNTs in an alternating electric field

An anisotropically conductive CNTs sheet is fabricated by using dielectrophoresis phenomena. In general, when particles are placed in an alternating electric field, polarization is created inside the particles. As a result, two kinds of forces generated inside particles depending on an alternating frequency. Under a relatively low frequency, particles move to the region of higher electric field. This is called positive dielectrophoresis. In contrast, when non-conductive particles are under a relatively high frequency, non-conductive particles move to the region of lower electric field. This is called negative dielectrophoresis. Positive dielectrophoresis phenomenon is utilized to fabricate anisotropically conductive CNT sheets. At first, the distribution of electric field intensity was calculated by finite element method. Figure 12 shows the cross-sectional distribution of electric field intensity between both electrodes. The distance between both electrodes is 100 μm and the medium is pure water. The electric field intensity becomes much stronger near both electrodes, especially, at the edges of electrodes, the electric field intensity is maximum. On the other hand, in the intermediate region between both electrodes, the electric field intensity is minimum. Therefore, when particles are influenced by positive dielectrophoresis, the particles move quickly toward both electrodes.

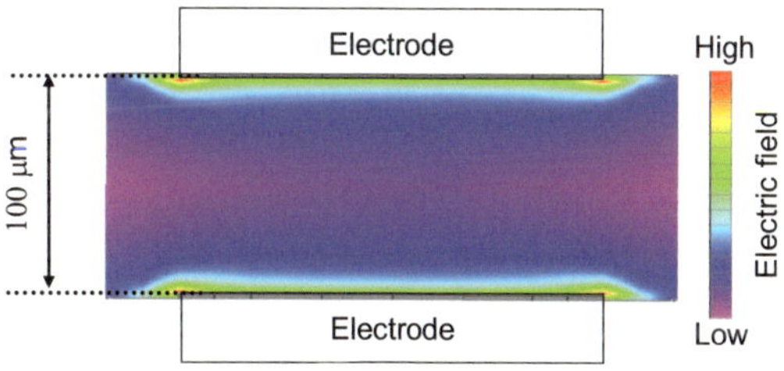


Fig. 12. Cross-sectional distribution of electric field intensity between both electrodes.

In order to understand the behavior of CNTs in dielectrophoresis, a basic experiment was performed. The circuit as illustrated in Fig 12 was prepared. ITO films were deposited on a pair of glass substrates as both electrodes, because particles could be observed to move during dielectrophoresis with a microscope. SWCNTs were ultrasonically treated in a 5 mol/L-nitric acid solution for cutting the lengths of CNTs. Treated CNTs were dispersed in pure water with a concentration of 1 mg/mL, and the dispersion liquid was droped on the circuit. When an AC voltage of 4 Vp-p was applied with a frequency of 100 kHz, SWCNTs were observed to move quickly to both surfaces of ITO electrodes. Figure 13 shows an SEM image of an area near one electrode in a few seconds just after starting to apply an AC voltage. In Fig 13, first reached SWCNTs are observed to align perpendicularly to the surface of the electrode. Secondarily reached SWCNTs connect to the edges of first reached SWCNTs, and Tertiarily reached SWCNTs connect to the edges of secondarily reached SWCNTs. Because CNTs have such a fiber-like shape, polarization is created between both edges and center of CNTs. As a result, CNTs always move and align as shown Fig 13. After applying an AC voltage between both electrodes, CNTs finally connect between both electrodes and an electrical circuit is closed.

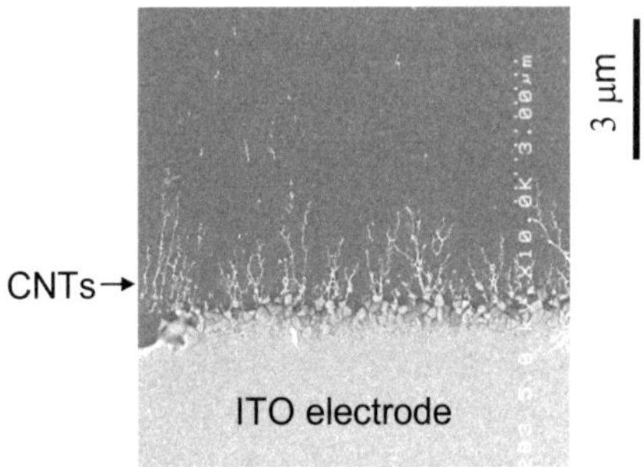


Fig. 13. SEM image of CNTs aligned perpendicularly at an ITO electrode in the early stage of positive dielectrophoresis.

### 3.2 Transparent and anisotropically conductive plastic sheets containing CNTs

By utilizing this fundamental and interesting behavior of CNTs in dielectrophoresis, an anisotropically conductive CNT sheet can be fabricated. CNTs are treated in a 5 mol/L-nitric acid solution for cutting the lengths of CNTs, linsed with pure water and ethanol, and then dispersed in a polymer medium by ultrasonic homogenizer for 20 min. Low viscosity acrylic acid resin R-604 is used as a polymer medium. The viscosity is 500 cP and the contraciton percentage is approximately 7 %. The molecular structure of R-604 is illustrated in Fig 14. A pair of flat glass substrates are prepared, and Au films with a thickness of 50 nm are deposited on one side of each substrate as electrodes. A pair of Au electrodes are faced and fixed with a space of 100 μm. The space is easily adjusted with spacers. The polymer medium in which CNTs are dispersed is introduced into the space between the electrodes. When applying an alternating voltage of 20 Vp-p with a frequency of 1 kHz to 100 kHz, CNTs move toward the electrodes, align perpendicularly to the surfaces of electrodes, finally connect between both electrodes in the polymer medium. Positive dielectrophoresis takes place same as the case of water medium. During positive dielectrophoresis, the curcuit current is continuously monitored. When the current rapidly increases, the circuit is irradiated with UV light while applying the voltage. After polymer becomes solid, irradiating with UV light and applying the voltage are finished. Molded plastic sheet is easily removed from a pair of substrates, because Au electrode films also play a role of mold lubricant. If necessary, after demolding, both sides of the plastic sheet are coated with transparent material such as ITO, ZnO and so on. This consecutive procedure is shown in Fig 15.

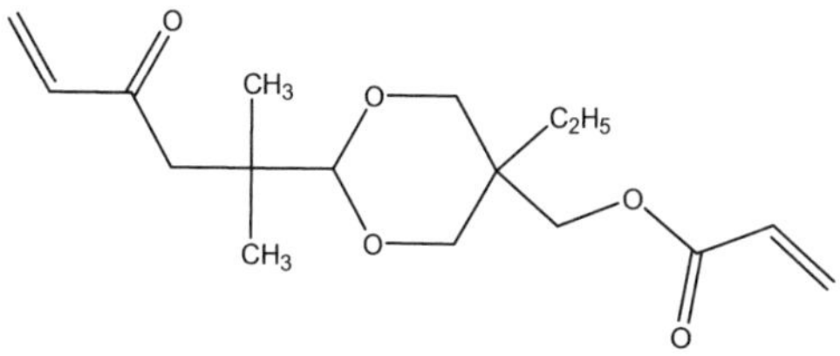

Fig. 14. Molecular structure of acrylic acid resin R-604.

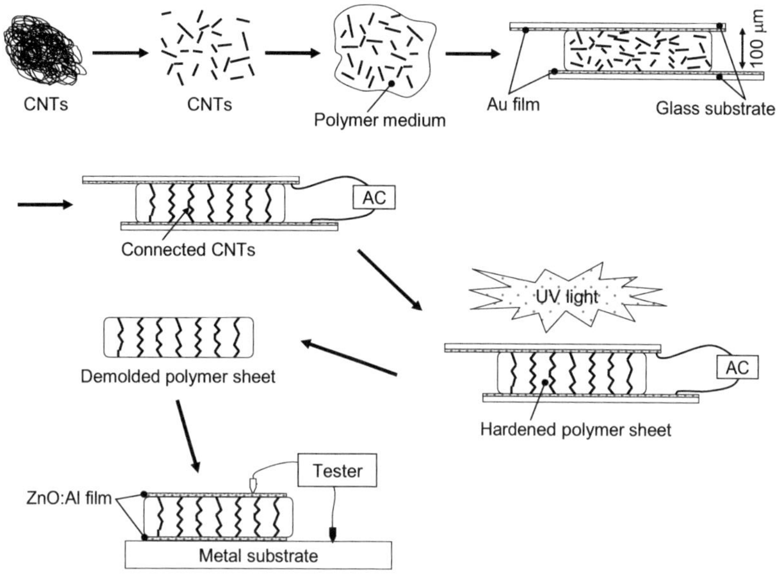


Fig. 15. Procedure for fabricating a transparent and anisotropically conductive plastic sheet containing CNTs.

Figure 16 shows typical appearances of a plastic sheet conatining CNTs and an original plastic sheet without CNTs. Because CNTs are aligned perpendicularly to a surface of a plastic sheet, a CNT-containing plastic sheet is completely transparent in visible light, as shown in Fig 16(a).

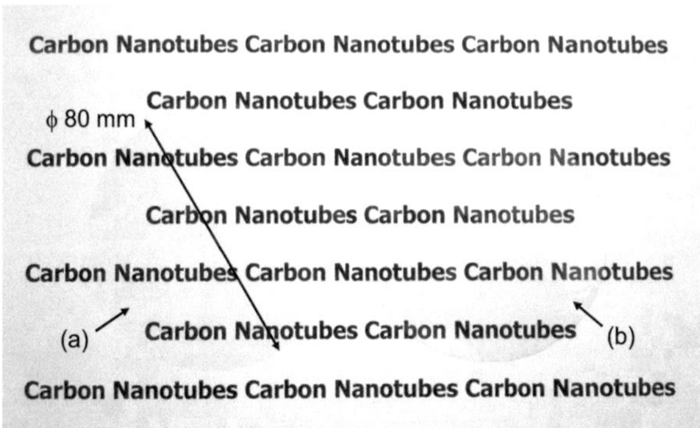


Fig. 16. Appearances of (a) an transparent and anisotropically conductive plastic sheet conatining CNTs, and (b) an original plastic sheet without CNTs.

For comparative study, SWCNTs, MWCNTs, and TWCNTs were used in order to fabricate conductive plastic sheets. Table 2 shows providers, the number of graphene walls, approximate average diameters and lengths of CNTs. SWCNTs were cut by nitric acid treatment. MWCNTs were purified by alkaline and then cut by nitric acid treatment. On the other hand, as mentioned in Section 2.3, a vertically long TWCNT film with 90 μm long was synthesized on a quartz glass substrate, and then was scratched from the glass substrate. The TWCNTs were used without nitric acid treatment. By using these CNT sources at conditions of 1 kHz and 20Vp-p, conductive plastic sheets were fabricated. The results are summarized in Table 3. Electrical resistivity was measured two times, before demolded, and after demolded and coated with ZnO:Al film. All four plastic sheets have conductivity in perpendicular direction, in contrast, high resistivity in in-plane direction. This indicates that CNTs are completely oriented and connected perpendicularly in molded plastic sheets. The lengths of TWCNTs are mcuh longer than the other CNTs. Therefore, the number of connecting points between each CNT in a sheet is considered less than the other CNTs. This is the main reason why the sheet containing TWCNTs have much lower resistivity than the other sheets containing SWCNTs and MWCNTs. Figure 17 shows transmittance and reflectance spectra of anisotropically conductive plastic sheets containing three kinds of CNTs with different concentrations, and an insulating plastic sheet without CNTs as a reference. These anisotropically conductive plastic sheets give high enough transparency through the visible region.

| CNTs | The number of graphene wall | CNT diameter (nm) | CNT length (μm) |
|---|---|---|---|
| SWCNTs (CNI, HiPco) | 1 | about 1 | 0.2 After nitric acid treatment |
| MWCNTs (Meijo Nano Carbon) | about 20 | about 20 | 0.2 After nitric acid treatment |
| TWCNTs (Nikon) | Mainly 3 | 3 ~ 6 | 90 As-grown |

Table 2. The number of graphene walls, diameters, and lengths of SWCNTs, TWCNTs, and MWCNTs used for fabricating transparent and anisotropically conductive plastic sheets.

| CNTs | CNT content (wt%) | Sheet thickness (μm) | Sheet transmittance @ 500 nm (%) | Resistivity (Ω) | | |
|---|---|---|---|---|---|---|
| | | | | Before demolding | | After demolding |
| | | | | perpendicular | parallel | perpendicular |
| SWCNTs (CNI, HiPco) | 0.01 | ~100 | 90.3 | 920 | >$10^{14}$ | 867 |
| MWCNTs (Meijo Nano Carbon) | 0.01 | ~100 | 91.0 | 1,500 | >$10^{14}$ | 900,000 |
| TWCNTs (Nikon) | 0.01 | ~100 | 81.9 | 58 | >$10^{14}$ | 58 |
| TWCNTs (Nikon) | 0.005 | ~100 | 86.9 | 104 | >$10^{14}$ | 111 |

Table 3. Transmittances and electrical resistivities of transparent and anisotropically conductive plastic sheets containing SWCNTs, TWCNTs, and MWCNTs.

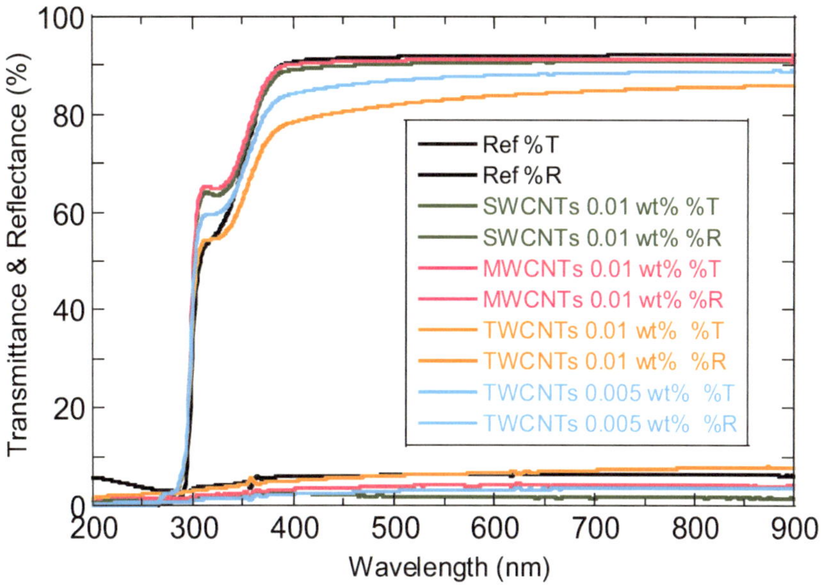


Fig. 17. Transmittance and reflectance spectra of transparent and anisotropically conductive plastic sheets containing SWCNTs, TWCNTs, and MWCNTs, and an insulating plastic sheet without CNTs.

## 4. Optical filters by using CNT forests

A vertically aligned CNT film, in other words, a CNT forest was synthesized on a quartz glass substrate by RHCVD as stated in Section 2.3. Figure 18 shows an appearance and a cross-sectional SEM image of a typical CNT forest. In Fig.18, the CNT forest is observed to be black enough without reflection, and identified to grow vertically on a substrate with a high population densitiy. The absolute transmittance and reflectance of the CNT forest are T=0.0002 % (optical density, OD = 5.7) and R=0.0018 % (at an incident angle of 5 deg.) at 193.4 nm of ArF laser wavelength, measured with Cary 5 (Varian) and U-4000 (Shimadzu) respectively. The reflectance of the CNT forest is less than 0.002 % at an incident angle of 5 deg. in the wide wavelength range from ultraviolet to infrared. In addition, the reflectance of the CNT forest was also measured under wide incident angles of 10 to 80 deg. at 193.4 nm with VUV-VASE (J.A.Woollam), as shown in Fig 19. The reflectance of the CNT forest is less than 0.01 % up to incident angles of 75 deg., and less than 0.35 % up to an incident angles of 80 deg. Extremely low-reflection feature is achieved at such a wide incident angle range. This CNT forest performs as an anti-reflective black filter in a wide incident angle range and wide wavelength range. The ArF laser marathon test of the CNT forest shown in Fig.18 was performed. The CNT forest was irradiated with ArF laser up to doses of 7.5 * 10E+8 mJ/cm$^2$ under nitrogen gas purge environment. Figures 20 & 21 show the OD value and reflectance as a function of the dose of ArF laser. The OD value merely decreases from 5.7 to 5.5 during the marathon test. This degradation can be small enough and neglibile. The reflectance is maintained less than 0.002 % through the test. This results have proven practical stability of the CNT forest as an anti-reflective black filter for ArF laser.

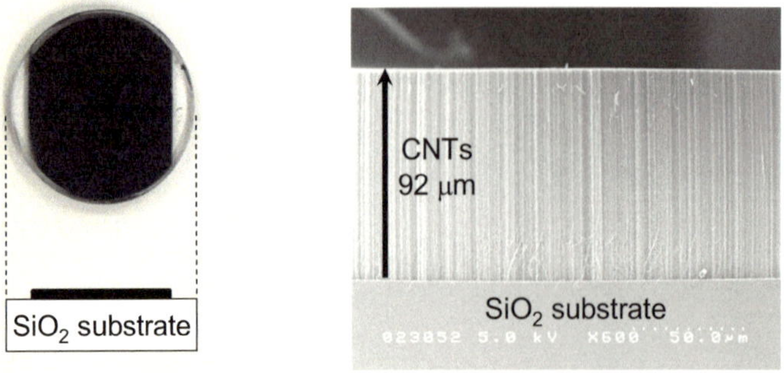


(a) (b)

Fig. 18. (a) an appearance and (b) a cross-sectional SEM image of a typical CNT forest.

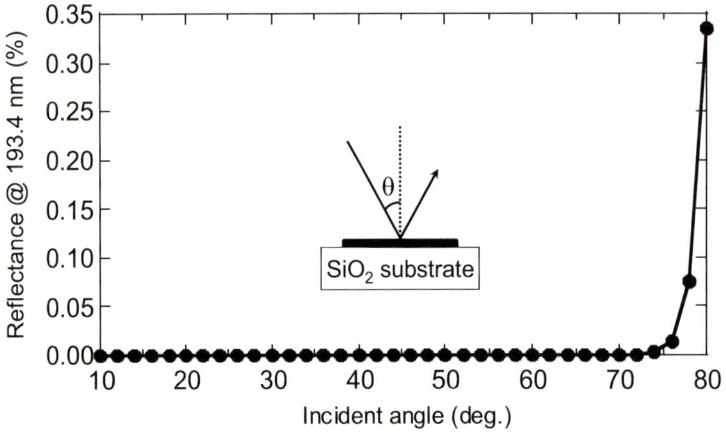


Fig. 19. Incident angle dependence of the reflectance at 193.4 nm of a CNT forest.

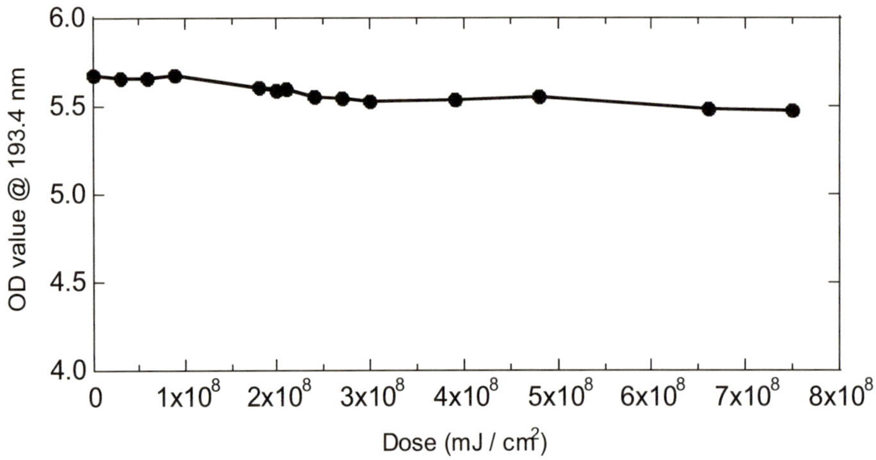


Fig. 20. OD value changes of a CNT forest duirng long-time ArF laser irradiation.

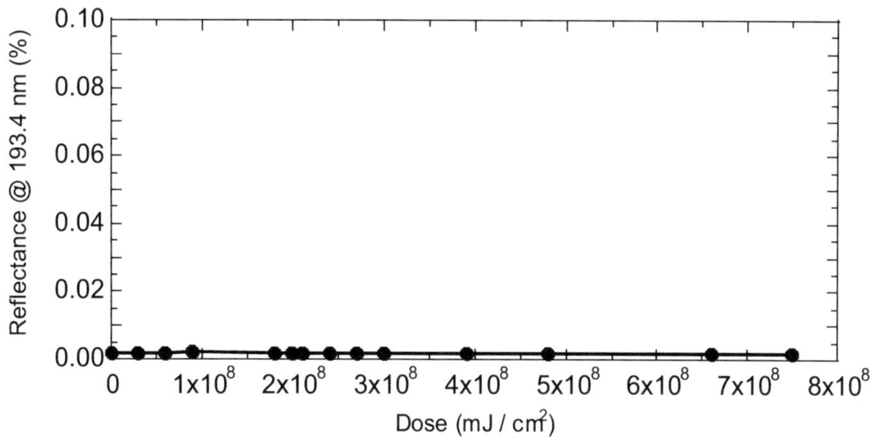


Fig. 21. Reflectance changes of a CNT forest duirng long-time ArF laser irradiation.

Detailed investigation regarding to the thickness of CNT forests revealed that CNT forests more than 10 μm long on quartz glass substrates perform as anti-reflective black filters. On the other hand, CNT forests of a few microns long are useful for ND filters. By changing catalyst diameters, a CNT growth time, and a CNT growth temperature, the OD value is adjusted sequentially. Figure 22 shows top-view appearances, OD values, and reflectances at 193.4 nm of CNT forests deposited by different growth conditions. In addition, the easier way to adjust an OD vaule of a ND filter is shown in Fig 23. Stacking of a metallic Cr film and a CNT forest of a few microns long performs as an anti-reflective ND filter. First, a Cr film is deposited on a quartz glass substrate, secondarily, a CNT forest is grown on the Cr film. The optical density of the ND filter is adjusted by the thickness of the Cr film. The CNT forest plays a role of anti-reflection. A CNT forest is capable of a unique optical filter.

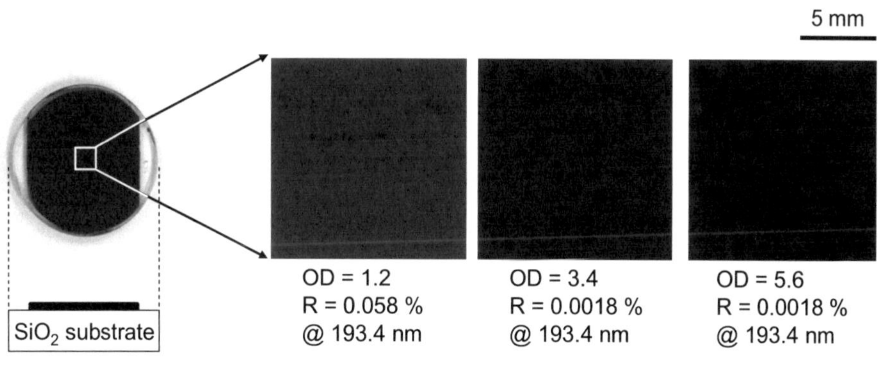


Fig. 22. Colors, OD values and reflectances at 193.4 nm of CNT forests with various lengths.

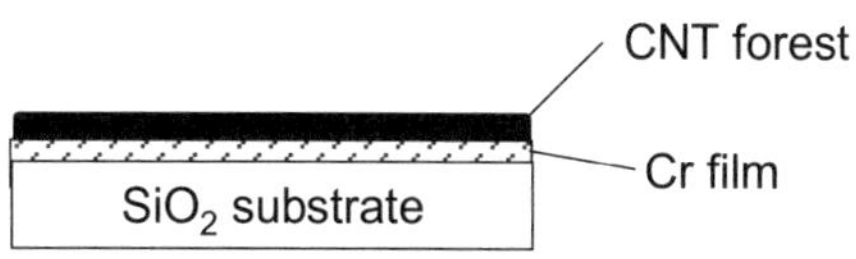


Fig. 23. Illustration of stacking of a Cr film and a CNT forest for an anti-reflective ND filter.

## 5. Conclusions

RHCVD is a newly developed process that enables the maintenance of narrow catalyst diameter distributions until CNTs start growing, and the syntheses of SWCNT (SWCNT / as-grown CNT ratio of 100%), DWCNT (DWCNT / as-grown CNT ratio of 88%), and TWCNT (TWCNT / as-grown CNT ratio of 76%) films by changing catalyst diameters. It is clarified that catalyst diameters have a close relationship to CNT diameters and the number of graphene walls of CNTs.

The syntheses of vertically aligned SWCNT, DWCNT, and TWCNT films have been achieved by a combination of RHCVD and the particle-arranged substrates where alloyed catalyst particles are deposited on non-catalyst particles. The diameters of non-catalyst particles must be slightly larger than those of catalyst particles. Because each catalyst particle is located on each non-catalyst particle, a catalyst particle is not able to aggregate with a nearest catalyst particle on another non-catalyst particle during pre-heating process. Therefore, CNTs are able to grow vertically on a substrate with a high population density.

Metallic catalysts become inactivated by intermediates decomposed from ethyl alcohol as a carbon source during a CNT growth period. In order to make CNTs grow longer, it is necessary to delay the inactivation of metallic particles during a CNT growth period. By using Al particles as a non-catalyst, and lowering a growth temperature up to 650 $^{o}$C, vertically long SWCNT, DWCNTs, and TWCNT growth has been accomplished. Al particles play a role of a getter catching intermediates. Lowering a growth temperature delays chemical reaction between metallic particles and intermediates.

The new method to fabricate a transparent and conductive plastic sheet with CNTs has been developed. CNTs are dispersed in a polymer medium. While casting a plastic sheet with the polymer medium, positive dielectrophoresis of CNTs is executed in the polymer medium. CNTs align perpendicularly to both surfaces of the sheet, and finally connect between both surfaces of the sheet. By irradiating with UV light to solidify the sheet, an anisotropically conductive CNT sheet is fabricated. A remarkable feature of this transparent sheet is its conductivity in perpendicular direction and high resistivity in-plane direction.

By using a CNT forest, a unique optical filter has been developed. A CNT forest more than 10 μm long on a quartz glass substrate performs as an anti-reflective black filter. The reflectance of a CNT black filter is nearly zero in the wide wavelength range from ultraviolet to infrared and the wide incident angles from 0 deg to 75 deg. A CNT black filter has been proven to be capable of ArF laser applications. On the other hand, a CNT forest of a few microns long performs as an anti-reflective ND filter by stacking of a metallic Cr film and a CNT forest.

## 6. References

Dan B., Irvin G.C., & Pasquali M. (2009). Continuous and Scalable Fabrication of Transparent Conducting Carbon Nanotube Films. *ACS Nano*, Vol.3, No.4, (April 2009), pp. 835–843

Hata K., Futaba D. N., Mizuno K., Namai T., Yumura M., & Iijima S. (2004). Water-Assisted Highly Efficient Synthesis of Impurity-Free Single-Walled Carbon Nanotubes. *Science*, Vol. 306, No. 5700, (November 2004), pp. 1362-1364

Hecht D.S., Thomas D., Hu L., Ladous C., Lam T., Park Y., Irvin G., & Drzaic P. (2009). Carbon-nanotube film on plastic as transparent electrode for resistive touch screens. *Journal of the Society for Information Display,* Vol. 17, No. 11, (November 2009), pp. 941-946, ISSN 10710922

Hiramatsu M., Nagao H., Taniguchi M., Amano H., Ando Y., & Hori M. (2005). High-Rate Growth of Films of Dense, Aligned Double-Walled Carbon Nanotubes Using Microwave Plasma-Enhanced Chemical Vapor Deposition. *Japanese Journal Applied Physics,* Vol. 44, No. 22, (May 2005), pp. L 693 - L 695

Murakami Y., Chiashi S., Miyauchi Y., Hu M., Ogura M., Okubo T., & Maruyama S. (2004). Growth of Vertically Aligned Single-Walled Carbon Nanotube Films on Quartz Substrates and Their Optical Anisotropy. *Chemical Physics Letters,* Vol. 385, No. 3-4 (February 2004), pp. 298-303

Muramatsu H., Hayashi T., Kim Y. A., Endo M., Terrones M., & Dresselhaus M. S. (2005). Growth of Double-Walled Carbon Nanotubes Using a Conditioning Catalyst. *Journal of Nanoscience and Nanotechnology,* Vol. 5, No. 3, (March 2005), pp. 404-408, ISSN 15334880

Nikolaev P., Bronikowski M. J., Bradley R. K., Rohmund F., Colbert D. T., Smith K. A., & Smalley R. E. (1999). Gas-Phase Catalytic Growth of Single-Walled Carbon Nanotubes from Carbon Monoxide. *Chemical Physics Letters,* Vol. 313, No. 1-2, (November 1999), pp. 91-97

Ramesh P., Okazaki T., Sugai T., Kimura J., Kishi N., Sato K., Ozeki Y., & Shinohara H. (2006). Purification and Characterization of Double-Wall Carbon Nanotubes Synthesized by Catalytic Chemical Vapor Deposition on Mesoporous Silica. *Chemical Physics Letters,* Vol. 418, No. 4-6, (February 2006), pp. 408-412

Taki Y., Shinohara K., Kikuchi M., & Tanaka A. (2008). Selective Growth of Single-, Double-, and Triple-Walled Carbon Nanotubes through Precise Control of Catalyst Diameter by Radiation-Heated Chemical Vapor Deposition. *Japanese Journal Applied Physics,* Vol. 47, No. 1, (January 2008), pp. 725 - 729

Taki Y., Kikuchi M., Shinohara K., & Tanaka A. (2008). Selective Growth of Vertically Aligned Single-, Double-, and Triple-Walled Carbon Nanotubes by Radiation-Heated Chemical Vapor Deposition. *Japanese Journal Applied Physics,* Vol. 47, No. 1, (January 2008), pp. 721 - 724

Wu Z., Chen Z., Du X., Logan J.M., Sippel J., Nikolou M., Kamaras K., Reynolds J.R., Tanner D.B., Hebard A.F., Rinzler A.G. (2004). Transparent, conductive carbon nanotube films. *Science,* Vol. 305, No. 5688, (August 2004), pp. 1273-1276, ISSN 00368075

Yamada T., Namai T., Hata K., Futaba D. N., Mizuno K., Fan J., Yudasaka M., Yumura M., & Iijima S. (2006). Size-Selective Growth of Double-Walled Carbon Nanotube Forests from Engineered Iron Catalysts. *Nature Nanotechnology,* Vol. 1, (November 2006), pp. 131-136

Zhong G., Iwasaki T., Honda K., Furukawa Y., Ohdomari I., & Kawarada H. (2005). Low Temperature Synthesis of Extremely Dense and Vertically Aligned Single-Walled Carbon Nanotubes. *Japanese Journal Applied Physics,* Vol. 44, No. 4A, (April 2005), pp. 1558 - 1561

6

# Direct Growth of Carbon Nanotubes on Metal Supports by Chemical Vapor Deposition

Naiqin Zhao[1] and Jianli Kang[2]
[1]*School of Materials Science and Engineering, Tianjin University, China*
*Tianjin Key Laboratory of Composite and Functional Materials, Tianjin,*
[2]*School of Mechanical Engineering, Tianjin Polytechnic University,*
*WPI Advanced Institute for Materials Research, Tohoku University,*
[1]*China*
[2]*Japan*

## 1. Introduction

Since their discovery (Iijima, 1991), carbon nanotubes (CNTs) have been attracted much attention due to their excellent mechanical, physical and chemical properties. In the past thirty years, thousands of articles have been published to discuss their growth, properties, and applications. For the CNT growth, there are three main methods: arc discharge, laser ablation and chemical vapour deposition (CVD). Compared with the first two methods, CVD was regarded as the most promising one for industrial application with low cost, although it usually introduces more defects in CNTs during the growth process. Furthermore, CVD is the preferred choice to grow patterned CNTs on substrates for nanoelectronic applications. Typically, nanotubes are grown by CVD on metal catalyst particles or islands that are deposited on top of semiconducting or insulating materials, such as alumina, silicon and silicon oxide. These non-conducting substrates assist in the formation of small islands or nanoparticles of metal catalyst on their surfaces, which is necessary for the CNT growth. However, for many applications, such as displays, cell electrodes, gigascale interconnects, high electrical and thermal conductivities are required, usually needing a conductive substrate to connect with CNTs. Furthermore, minimization of the contact resistance between metal substrate and CNTs is also a major challenge in nano-electronics. Apart from geometrical factors, contact resistance depends mostly on alignment of Fermi energy levels of CNT and substrate. Since multi-walled CNTs are predominantly metallic (Single-walled CNTs can be adjusted to be metallic by controlling their structure), a metallic substrate to connect with CNTs is expected to have least contact resistance. In order to assemble the CNTs on metal substrate with fine patterned architectures, the obvious and effective way would be to grow CNTs directly on metal substrates. Herein, the current development of direct growth technique and its applications were summarized.

## 2. Direct growth of CNTs on buck metal substrates

With respect to isolating ceramic materials, an important drawback of metal substrate is that the elevated temperature involved in CNT growth by CVD may activate the diffusion of the catalyst material into the metallic substrate, which inhibits its activity. To overcome this problem, one possible strategy is to employ pure catalytic metal or alloys containing at least one of these well known catalyst materials as substrates. However, if the substrate selection is constrained to specific metals not including the known catalyst materials, other solutions must be explored. An alternative effective choice is to deposit a thin barrier layer that restrains the diffusion reaction between catalyst and substrate, while minimizing the effect of this layer on the property of the device.

### 2.1 Pure catalytic metal or alloys containing at least one of these well known catalyst materials as substrates

It is accepted that the most common and effective catalysts are Fe, Co, Ni and their alloys. In order to deposit CNTs on these metal foils, nucleation site formation and density on the foil surface are crucial. Du et. al investigated pure Ni as substrate to grow CNTs without any pretreatment (Du & Pan, 2005). It was found that nickel nanoparticles as nucleation sites for CNT growth come from the fragmentation of nickel surface. However, in most reports, the grown CNTs on pure metal substrates were spaghetti-like, showing low density and weak control of the nanotube growth knetics, even though modifying the surface properties by wet-chemical etching treatments (Mata, et al. 2010), polishing with different roughness (Reddy, et al. 2006). Compared with single catalytic metal, using alloy foils containing at least one of these well known catalyst materials, such as stainless steel (Masarapu & Wei, 2007; Baddour, et al. 2008), mumetal (Benito & Lefferts, 2010) and inconel (Talapatra, et al. 2006), as substrates have generated a higher density of CNTs, which may be related to other experiments that an alloy or intermetallic compound containing catalytic and non-catalytic elements assists the formation of a high density of nucleation sites on silicon substrates (Cui, et al. 2003; Zhang, et al. 2006). As it is known that, the role of catalyst and its interaction with the substrate is a key issue of CNT growth. The chemical composition and thermal stability of the substrate at the growth temperature should aquire important consideration for CNT growth. The chemical nature of the substrate under growth conditions should be such that it neither allows the catalyst to diffuse into its bulk, nor spreads out and forms a continuous film. For some Ni-, Co-, Fe- based superalloys, in situ formation of a conductive passivation oxide layer on the surface of these superalloys played an important role in stabilizing nanoclusters of Fe catalyst under typical growth conditions (Bult, et al. 2009 & Pal, et al. 2010). On the other hand, introducing an insulating thin buffer layer (such as $Al_2O_3$) between the extra catalyst and metallic substrate, where the buffer layer was too thin to form an insulating barrier on the metallic plate, was suggested to obtain relative higher density and quality of CNT forest (Gao, et al. 2008 & Martinez-Latorre, et al. 2009 & Lepro, et al. 2010). Moreover, single- and double-walled carbon nanotube forests could be synthesized successfully on various alloys spaning many standard metals, such as Inconel 601, YEF 426, NiCr, YEF 50, and SUS 310S, covering a wide range of Ni-Fe-Cr compositions, as shown in Fig. 1 (Hiraoka, 2006). Apart from the chemical compostion of substrate, pretreatment of the alloy substrates, especially for that no any other catalyst introduced, significantly affected the efficiency of CNT growth on the metallic substrates. It was reported that surface roughness variation of the alloy by polishing might change the surface chemistry, resulting in inhomogeneous growth of CNTs (Oye, et al. 2010).

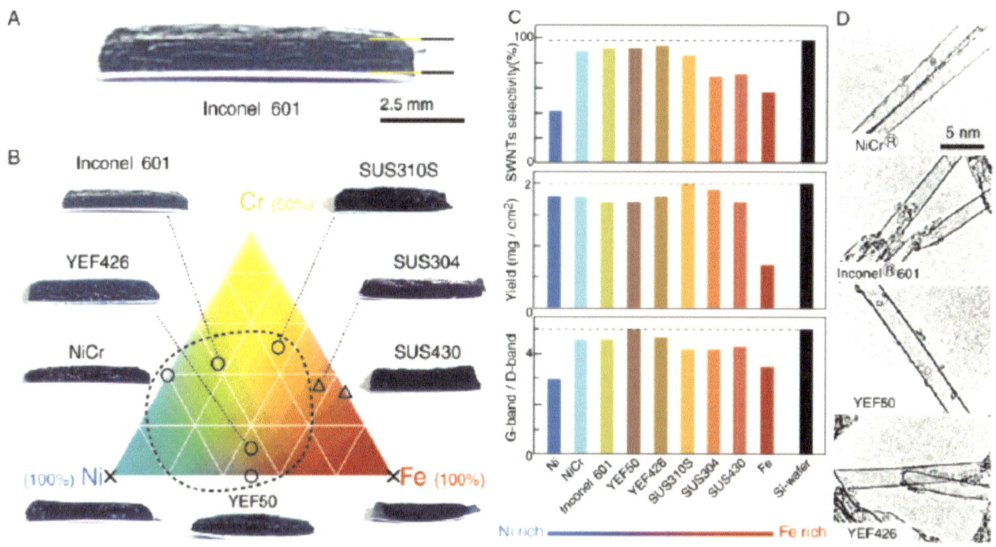


Fig. 1. Direct growth of CNT forests on Ni-Cr-Fe alloys with water-assisted CVD. (A) A photograph of a SWNT forest synthesized on Inconel 601. (B) Correlation between components of Ni-Cr-Fe alloys and photographs of CNT forests grown on them. Symbols indicate the level of SWNT selectivity (O high selectivity; ≥ medium selectivity; · low selectivity). (C) A family of histograms showing selectivity of SWNTs, yield per cm$^2$, and G-band/D-band ratio of Raman spectra. (D) TEM images of SWNTs synthesized on Ni-based alloys. (Hiraoka, 2006)

## 2.2 Other metals as substrates

In current electrical and electronic industries, Cu is the most popular raw material due to their high electric and thermal conductivity with relatively low cost. Thus, it is very interesting to directly grow CNTs on Cu substrate. It is known that Cu can act as effective catalyst for CNT growth under certain conditions (Gan, et al. 2001 & Qin, et al. 2004). However, the grown carbon nanostructures catalyzed by Cu are spaghetti-like and the catalytic activity of copper is accepted to be much lower than Fe, Co and Ni (Zhang, 2010). Therefore, it seems necessary to introduce other more active catalyst to grow dense aligned CNTs on copper substrate. Directly deposition of catalyst on Cu substrate is an obvious choice. However, as discussed above, Cu is not an efficient catalyst for CNT growth and easy to form a solid solution with the common catalysts (especially for Ni) at high temperature. The diffusion between catalyst and copper substrate may induce the loss of catalyst. Thus, the catalyst layer should be thick enough to guarantee enough unaffected catalyst residues to catalyze CNT growth (Singh, 2002 & Yin, 2008), meanwhile, some authors reported that thicker catalyst layers resulted in larger diameter of grown CNTs (Atthipalli, 2011). Therefore, adding a diffusion barrier layer between catalyst and copper substrate is a promising and preferable choice for high-quality aligned CNT growth, although this may increase the contact resistance between CNTs and Cu substrate in some content. A thin layer of alumina has been used as diffusion barrier in most literatures. Lin et al. investigated the effect of the deposition method and thickness of alumina layer on the quality of CNT growth systematically, infering that a well-controlled conformal alumina support layer played an important role for the aligned CNT growth with high quality and reproducibility (Lin, et al., 2010). Other ceramic materials, such as TiN, $In_2O_3$:Sn (ITO), and $SiN_x$, were also examined as effective barrier, inferring that the extent of CNT coverage depended on the barrier layer quality (Garcı´a-Ce´spedes, et al., 2009). Apart from ceramic materials, some metals, such as Ti (Sung, et al., 2008), Cr, Au, Ta (Nessim, et al., 2009 & 2010), Pa (Nessim, et al., 2010), Al (Burt, et al., 2009 & Kavian, et al., 2011), and so on (Wang, et al., 2003 & Kim & Gangloff, 2009), were also used as efficient barrier layers or substrates directly. Although CNT growth on these substrates or underlayers was affected by many factors, understanding the coupled influences of barrier layer thickness, grain size, and catalyst dewetting is an important step in development of a more complete understanding of CNT growth mechanisms.

### 2.3 Applications of CNTs directly grown on bulk metal substrate

Due to a high aspect ratio geometry, small tip radius, good chemical stability, low coefficient of thermal expansion, high thermal and electrical conductivity and mechanical strength, CNTs have attracted great interest as thermal management materials for microelectronic packaging, electrode materials for use in energy storage cells and supercapacitors, field emission (FE) electron sources for use in flat panel displays, e-beam lithography equipment, x-ray sources, and vacuum microwave amplifiers, and so on. For these applications, as discussed above, high electrical and thermal conductivities are required, which usually needs a conductive substrate to connect with CNTs. Compared with the external connections, the biggest benefit for direct growth technique of CNTs on metal substrate is that can reduce the contact resistance and form an ohmic connection between CNTs and metal substrate, which is crucial for electronic devices. Thus, it is expected to have a significant advance toward the goal of replacing external metal conections with CNTs, such as CMOS-compatible processes, by direct growth technique (Nessim, et al., 2009 & Kim, et al., 2010). Experimental results indicated electrical contact through the CNT carpet to the metallic substrate (Cu) by direct growth with an approximate resistance of 35 kΩ for multiwall CNT carpets taller than two micrometers (Nessim, 2009). Talapatra et al. further certified that the average total contact resistance of aligned CNTs directly grown on inconel by a vapour-phase catalyst delivery CVD method, measured over many samples (typical pattern size = 70μm×70μm, height=50μm), was even as low as about 500Ω. The power density for the double-layer capacitors fabricated using the Inconel substrate with directly grown aligned nanotubes as electrodes was about 7 kW $kg^{-1}$ even at a high scan rate of 1000 mV $s^{-1}$ (Talapatra, et al., 2006). Due to the good bonding and excellent conductivity by interface-controlled growth of CNTs directly on Cu current collectors, Li ion battery using the novel directly grown binder-free CNTs structure as anode showed very high specific capacity, almost three times as that of graphite, excellent rate capability even at a charging/discharging rate of 3 C, and no capacity degradation up to 50cycles (Lahiri, ea al., 2010). The CNT-based field emitters, prepared on metallic substrates by direct thermal chemical vapor deposition, exhibited a very low turn-on field, high emission current, long time stability and good resistance to degradation in high-field, long-time exposure (Mauger, et al., 2004 & Sung, et al., 2008 & Mahanandia, et al., 2009 & Yi & Yang, 2010 & Li, et al., 2010 & Lahiri, et al, 2010). Furthermore, Zhang, et al. indicated that coiled carbon nanosturctures with more defects by direct growth exhibited better field emission properties than straight CNTs (Zhang, et al., 2010). Several studies have revealed that CNTs have unusually high thermal conductivity in their axial direction (Berber, et al., 2000 & Kim, et al., 2010). Thus, one promising application of CNTs in microelectronics is to use vertically aligned CNT arrays as novel thermal interface materials (TIMs). Direct synthesis of vertically aligned CNT TIM layer on the backside of a silican chip is not compatible with current electronic packaging systems. Instead vertically aligned CNT synthesis on metallic substrate, especially for copper, is preferred. The TIM, made up of a thin copper foil covered with Cr-Au-MWCNT on both sides, showed a thermal resistance as low as 12 $mm^2$ K/W (Wang, et al., 2007). Recently, Lin, et al. developed a novel assembling process of incorporating carbon nanotubes as TIM for heat dissipation by synthesizing vertically aligned CNTs on a copper substrate and chemically bonding the carbon nanotubes to a silicon surface. Experimental results indicated that such an interface modification improved the effective thermal diffusivity of the carbon nanotube-mediated thermal interface by an order of magnitude and conductivity by almost two orders of magnitude (Lin, et al., 2010). Due to the retained mechanical flexibility of the metallic substrate (such as stainless steel) after CNT forest synthesis and its renewability and low cost, direct growth of CNTs on metallic substrate becomes a promising approach for industrial production of CNTs, CNT yarns and sheets (Baddour, et al., 2010 & Lepro´, et al., 2010 & Iijima, 2011).

## 3. Direct growth of CNTs on metal powders

Since their discovery, carbon nanotubes have been regarded as an ideal reinforcement for composites to overcome the performance limits of conventional materials, due to their excellent mechamical and physical properties. Many research efforts have dealt with CNT/polymer composites, which exhibit a tremendous strengthening effect for the composites (García, et al., 2007). However, the results in metals are not as encouraging as those in polymer. This is mainly due to the difficulties in achieving homogeneous dispersion of CNTs in metal matrix and good interfacial bonding between CNTs and metal matrix. So far, nearly all the efforts were made through powder metallurgy route, which involved

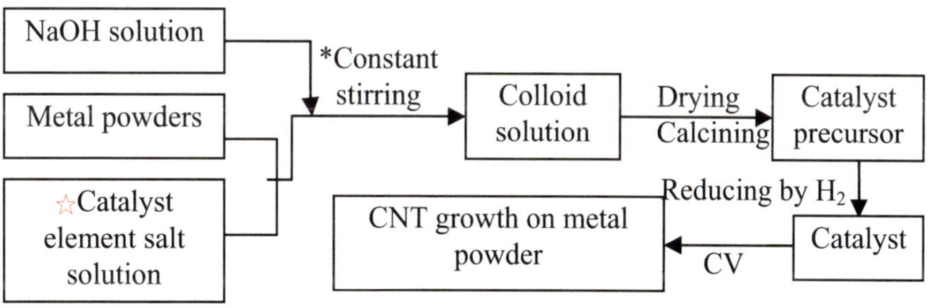


☆For Ni catalyst, the catalyst element salt is $Ni(NO_3)_2 \cdot 6H_2O$; For Ni/Y catalyst, the catalyst element salt is a mixture of $Ni(NO_3)_2 \cdot 6H_2O$ and $Y(NO_3)_3 \cdot 6H_2O$

*For Mg supported catalyst, metal powders mixed with NaOH solution firstly. The salt solution was added to the mixed solution dropwise under constant stirring

Fig. 2. Schematic diagram of the CNT growth directly on metal powders mixing of CNTs with metal powders by ball milling or chemical process (Cha, et al., 2005 & Kim, et al., 2006). In the above methods, CNTs was inclined to be damaged and lose their initial properties during the mechanical milling or their functionalization process. Alternatively direct synthesis of CNTs in metal matrix composites may overcome the problems associated with conventional methods and achieve homogeneous distribution of CNTs and good interfacial bonding between CNTs and metal matrix. In this section, we will summarize our recent progress in this field. Several metal powders, such as Al, Cu, Mg, Ag, have been proved as effective support for CNT growth. The catalyst dispersed on these metal powders was prepared by deposition-precipitation. The experimental procedure was described in Fig. 2. CNT growth was performed by a simple CVD process.

### 3.1 Aluminum powder

As mentioned above, aluminum can be used as effective underlayer or substrate for catalyst deposition, probably due to its easy formation of a passivation oxide layer on the surface of Al (Liu, et al., 2008). In our study, Ni nanoparticles were homogeneously deposited on Al particle surface by a direct deposition-precipitation method. Fig. 3 displays transmission electron microscopy (TEM) images of representative Ni-Al catalyst powders. Almost all Al powders are evenly decorated by several Ni nanoparticles. It is found that catalytic particles of different sizes can be formed by varying the reduction temperature, which thus controls the diameter of the growing tubes. When reducing the catalyst precursor at 400 °C for 2 h, the diameter of the Ni particles ranged from 5 to 20 nm. It can be clearly seen from Figure 3b that Ni nanoparticles with even diameters are homogeneously dispersed on the surface of the Al powders.

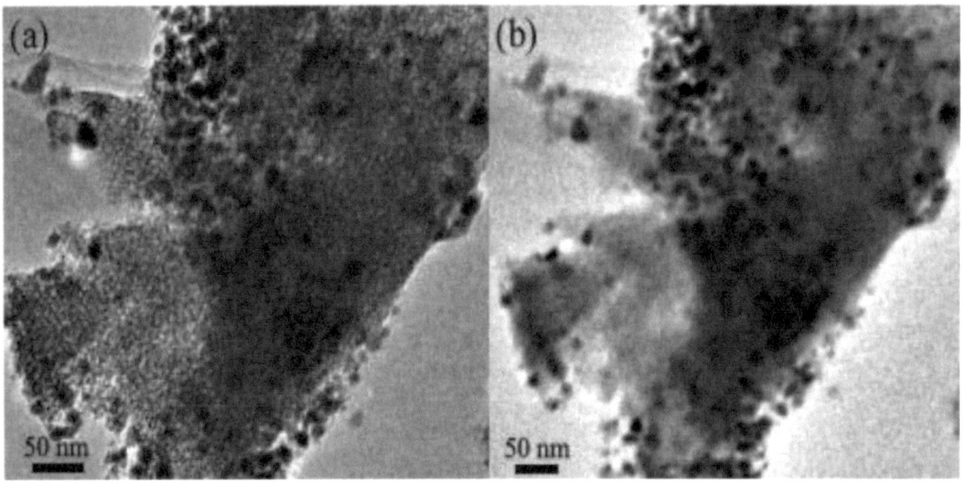


Fig. 3. Microstructures of the Ni-Al catalyst powders, in which the Ni nanoparticles with a narrow diameter distribution are homogeneously dispersed on the surface of the Al powder. a) TEM image of a Ni-Al catalyst powder, obtained by reducing the catalyst precursor at 400℃ for 2 h (scale bar: 50 nm). b) TEM image of a Ni-Al catalyst powder, showing that the gray Al powder is evenly decorated by several black Ni nanoparticles (scale bar: 50 nm).

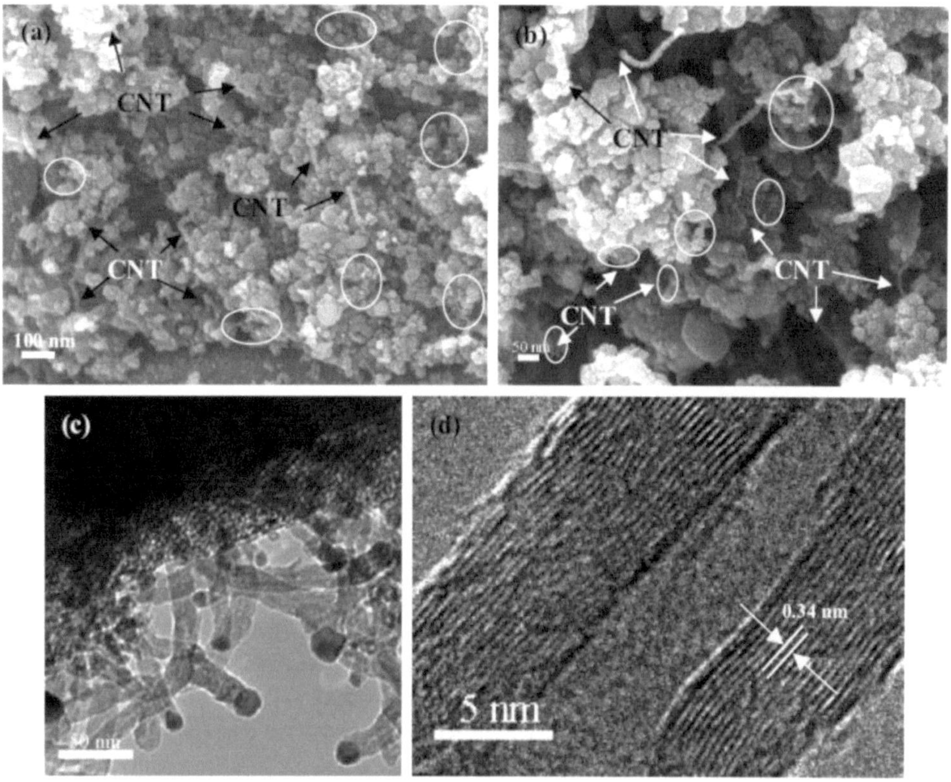


Fig. 4. a) Scanning electron microscopy (SEM) image of a CNT(Ni)-Al composite powder (scale bar: 100 nm). b) Enlarged SEM image of several typical CNTs (indicated by arrows) dispersed in the Al powders, the locations indicated with ellipses show the network structures between CNTs and Al powders (scale bar: 50 nm). c) Low-magnification HRTEM image of a CNT-Al composite powder (scale bar: 50 nm). d) High-resolution TEM (HRTEM) image of a typical CNT, showing a well-graphitized multiwalled nanotube (scale bar: 5 nm).

Fig. 4a shows a low-magnification scanning electron microscopy (SEM) image of homogeneously dispersed CNTs within the Al powders. The surfaces of the walls of the as-grown CNTs are clean, and their diameters range from 5 to 25 nm. In addition, metallic nanoparticles are visible at the tips of the CNTs. The most important feature of our process is that the CNTs are synthesized into the Al powders in situ. The morphology of the CNT(Ni)–Al powders shows an ideal composite microstructure, displaying spherical morphologies with CNTs homogeneously dispersed into the powders. It is also found that the density and length of the CNTs, or the CNT content of the composite powders, can be tuned by adjusting the experimental parameters, such as the growth time and the Ni content in the Ni–Al catalyst. With a shorter time and lower Ni content, shorter and sparser CNTs can be obtained in the Al powders. Figure 4b is an enlarged SEM image of several typical CNTs (indicated by arrows) dispersed in the Al powders. It can be distinctly observed that the CNTs are not agglomerated at all. Moreover, we have noticed that some locations, marked with ellipses in Figure 4b, form network structures between the CNTs and the Al powders.

The microstructure details of the CNT(Ni)–Al composite powders were investigated with high-resolution TEM (HRTEM). The low-magnification HRTEM image in Figure 4c further verifies that the tips of the CNTs encapsulate metallic particles. An HRTEM image of a typical CNT, as seen in Figure 4d, demonstrates that the CNTs are well-graphitized multiwalled nanotubes. The graphitic sheets of the CNTs (see Fig. 4d) are apparent, and the interlayer spacing between the sheets is 0.34 nm, consistent with the ideal graphitic interlayer space (0.34 nm). To determine the length of the as-grown CNTs, dilute nitric acid was used to remove the Al from the composite powders. TEM analysis indicates that the CNTs are easily entangled after removing Al as a result of strong van der Waals forces between them, and their lengths range from 1500 to 2500 nm.
The CNT(Ni)-Al bulk composite was prepared by pressing and sintering the CNT(Ni) composite powders directly. The microstructure of the composite was investigated by TEM. In the low magnification TEM image of Figure 5a, we see only some black Ni nanoparticles (indicated by arrows) distributed evenly within the Al grains. When investigating the area around the black Ni nanoparticles in detail by high-magnification TEM, it is found that the CNTs with encapsulated Ni nanoparticles are dispersed very homogeneously within the Al matrix (as seen in the inset of Fig. 5a). In particular, the TEM image (Fig. 5a, inset) shows that the interfaces of the CNTs and Al bond well, and that no reactant is formed by the present technique. Furthermore, the Al grains reinforced by CNTs present a very low location density. Figure 5b is an SEM image of the composite fracture surface, which shows CNTs with an obviously tubular structure (Fig. 5b, inset). Moreover, it can be observed that the CNTs are dispersed very well into the Al matrix, and some CNTs are pulled-out or broken (marked with triangles), which indicates that the load transfer from the matrix to the nanotubes is sufficient to fracture the nanotubes.

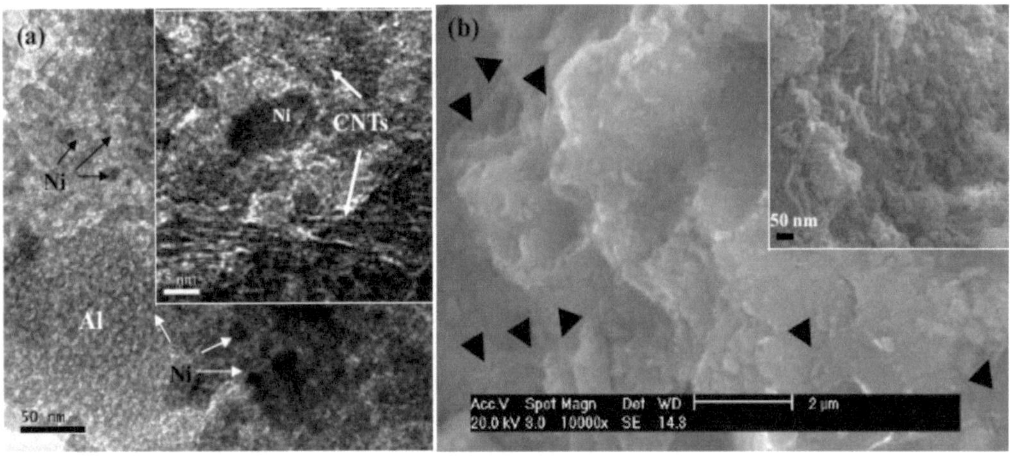


Fig. 5. a) TEM image of a CNT(Ni)-Al matrix bulk composite, showing the homogeneously dispersed CNT(Ni) within the Al matrix (scale bar: 50 nm, scale bar of inset: 5 nm). b) SEM image of the fractured surface of a CNTs(Ni)-Al matrix bulk composite, showing the existence of pulled-out and broken CNTs (scale bar of inset: 50 nm).

Table 1 presents a comparative study of hardness and tensile strength values of the composite and two other typical materials that were fabricated by the same procedure (i.e., pressing, sintering, and re-pressing). The density of the CNT(Ni)–Al bulk composite (c) was measured by the Archimedes principle as 2.50 g $cm^{-3}$, which is about 96% of the theoretical density (2.6 g $cm^{-3}$) of Al with a 5 wt% CNT reinforcement, and about 93% of the matrix density. The hardness (0.65 GPa) and tensile strength (398 MPa) of the CNT(Ni)/Al composites (c) are 4.3 and 2.8 times that of the pure Al matrix (Table 1). To further verify the strengthening effect of the in situ synthesized CNTs for the bulk composites, the same composite was also prepared by a traditional method, which involved the preparation of CNT(5 wt %)–Ni (1 wt %)–Al composite powders by ball-milling the CNTs and the Ni–Al powders, pressing, and sintering of the composite powders. According to Table 1, the hardness and tensile strength of the in situ synthesized CNT(Ni)–Al composites (c) are 2.0 and 1.8 times, respectively, that of the composites with a similar composition (b). This remarkable strengthening is caused by the dispersion strengthening of the homogeneously dispersed CNTs and Ni nanoparticles. Besides, the retention of the perfect structure gives the CNTs their extreme hardness (62–150 GPa), whereas molecular-level homogenous mixing between CNTs and Al powders brings about strong interfacial strength between CNTs and Al powders. Both factors contribute to the tremendous enhancement of the overall hardness and strength of the composites. The strong interfacial strength between CNTs and Al powders that resulted from very good homogeneous dispersion of CNTs in Al powders is especially important to improve the composite performance, because it can cause high load translation during tensile processes (as suggested by pulling out and broken of CNTs in Fig. 5b) and thus raise the fracture energy and the tensile strength of the composites. As for the CNT–Al matrix composite produced by a traditional method, the interfacial strength between CNTs and Al powders can not be expected to be high because of the mere mixing of the CNTs and the Al matrix. Moreover, the high-temperature generated by high-energy ball milling and plasma spraying damages the perfect structure of CNTs. As a result, the reinforcement effect of the CNTs on the composites is not very outstanding (Kuzumaki, et al., 1998 & Laha, et al., 2004 & George, et al., 2005). Compared to reinforcements such as $Al_2O_3$ (Huang, et al., 2003 & Kang, et al., 2004), SiC (Moreno, et al., 2006), $TiB_2$ (Huang, et al., 2005), aluminum borate whiskers (Zhu & Lizuka, 2003), TiN (Shyua, et al., 2002), and others used for Al matrix composites (Zambona, et al., 2004 & Tang, et al., 2003), the strengthening effect of the CNT reinforcement is the strongest ever reported.

| Material | Theoretical density [g $cm^{-3}$] | Measured density [g $cm^{-3}$] | Hardness [GPa] | Tensile strength at RT [MPa] |
|---|---|---|---|---|
| Pure Al | 2.7 | 2.69 | 0.15 | 140 |
| CNT(5 wt %)-Ni (1wt %)-Al matrix composite [a] | 2.6 | 2.48 | 0.32 | 213 |
| CNT(5 wt %)-Ni (1 wt %)-Al matrix composite [b] | 2.6 | 2.50 | 0.65 | 398 |

[a] Obtained by ball-milling CNTs and Ni–Al powders. [b] Obtained by in situ CVD synthesis.

Table 1. Comparative study of the CNTs(Ni)-Al matrix bulk composite and two typical materials with respect to density, hardness and tensile strength

### 3.2 Copper powder

From the binary phase diagram, it is known that Cu is inclined to form solid solution with Ni, Co and Fe. Experimental results showed that there were almost no carbon deposits synthesized by CVD using Ni catalyst nanoparticles deposited on Cu particle surface (prepared by deposition-precipitation) at any conditions. XRD analysis indicated that Ni and Cu had alloyed completely by diffusion at even lower than 500℃. It is also impossible to deposit a diffusion barrier layer on Cu particle surface, like that on buck Cu substrate. Thus, the only alternative way is to prepare a new kind of stable and effective catalyst, which is not affected by the underlayer materials. Herein, a novel Y stabilized Ni catalyst was successfully prepared by deposition-precipitation. XRD examination of the Ni-Y catalyst reduced at 500℃ indicated that there were no reactions between Ni and Cu. However, we did not found any phase peaks that possibly contained Y element (Fig. 6a). Fig. 6b shows the TEM images of the Ni-Y catalyst with a weight ratio of 4:1 (Ni:Y) reduced at 500℃. It can be seen that the crystal lattice structure of the catalyst nanoparticle is very complex. The basic structure of the nanoparticle is Ni crystal lattice, while, some very small crystals with larger interplanar spacing (about 0.3nm, which is similar to that of the (111) plane of $Y_2O_3$) seems to insert in the basic crystal lattice, as indicated by arrows in Fig. 6b. Thus, the whole particle demonstrates a mixture of Ni and $Y_2O_3$ (very small $Y_2O_3$ crystals embedded in Ni, which may be responsible to that why the XRD cannot detect any phases containing Y). The more Y content in the composite catalyst, the more stable the catalyst is. However, Y ($Y_2O_3$) itself does not have any catalytic activity for CNT growth. Too Y loading will degrade the general activity of the novel composite catalyst.

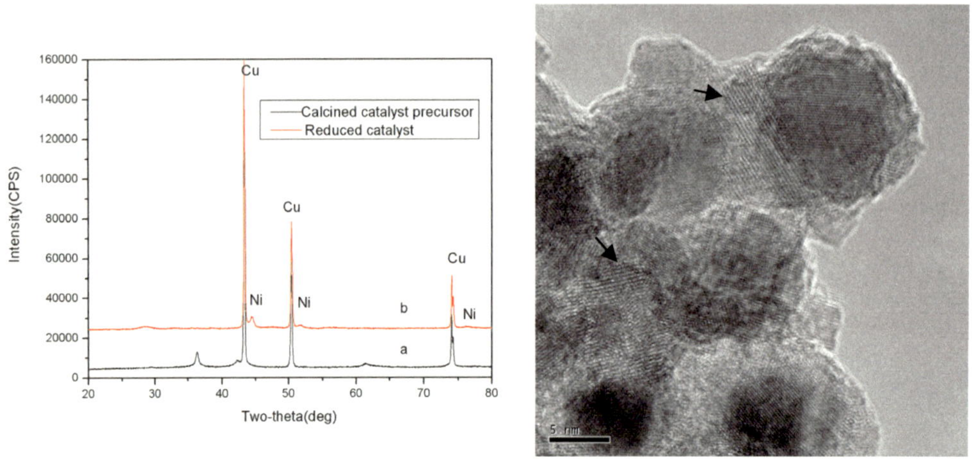


Fig. 6. (a) XRD analysis of the Ni/Y catalyst (Ni:Y =4:1) precursor calcined at 300℃ and 400℃ for 2h respectively under $N_2$ atmosphere and catalyst reduced at 500℃ for 2h under $H_2$ atmosphere (b) TEM image of the catalyst after reduction

In order to prepare Cu composite with homogeneous CNT dispersion, the first step is to obtain homogeneously dispersed active catalyst on the Cu particle surface (Fig. 7a) and then grow CNTs with controllable content and quality by CVD (Fig. 7.b). The resulting powder was termed as in situ CNT(Ni/Y)-Cu composite powder. Since some ceramic materials are suggested to be suitable as catalyst support, the in situ CNT synthesis has been widely used in the fabrication of ceramic matrix composite, which develops homogeneous CNT dispersion in the matrix (Peigney, et al., 2002). However, most of the CNTs are located on the surface of ceramic powders, which inhibits the diffusion of matrix materials across or along the powder surfaces; thus, sintering cannot easily proceed without damaging the CNTs or removing them from the powder surface. Even if sintering is successful, CNTs are mostly located at grain boundaries of the matrix and are insignificant in improvement of material performance. Furthermore, the CNTs should be short enough to avoid entanglement (Balani, et al., 2008). In order to overcome these shortcomings, we introduce the third step effectively to implant CNTs into the Cu powders. The in situ synthesized CNT(Ni/Y)-Cu composite powders and copper ions from the copper salt ($Cu(NO_3)_2 \cdot 2.5H_2O$) were added into a minimal amount of ethanol and the solution was heated under constant magnetic stirring until the ethanol was vaporized completely (Fig. 7c). Because the direct grown CNTs are attached on the Cu particles by metallurgical bond, CNTs can flow with the Cu particles without separation during the solution mixing by magnetic stirring, which prevents the agglomeration of CNTs. Thus, this unique process combining in situ synthesis of CNTs and a solution mixing can easily obtain a high dispersion of CNTs in metal matrix and a clean interfacial bond between the CNTs and matrix materials. The quantity of copper ions added is equal to that of the in situ CNT(Ni/Y)-Cu composite powders. The powders generated in the third step are generally a mixture, where the CNT(Ni/Y)-Cu composite powders were encapsulated by the basic Cu salts from decomposition of $Cu(NO_3)_2 \cdot 2.5H_2O$ (Fig. 7d). The fourth is the calcination and reduction process to obtain chemically stable crystalline powders (Figure 1e). During this process, the powders become CNT(Ni/Y)-Cu-CuO($Cu_2O$) by heating at 300 °C in air and are then reduced to CNT(Ni/Y)-Cu composite powders under a hydrogen atmosphere. Finally, the composite powers are hot pressed at 500 °C for 30min in a vacuum of $10^{-6}$ torr with an applied pressure of 50MPa.

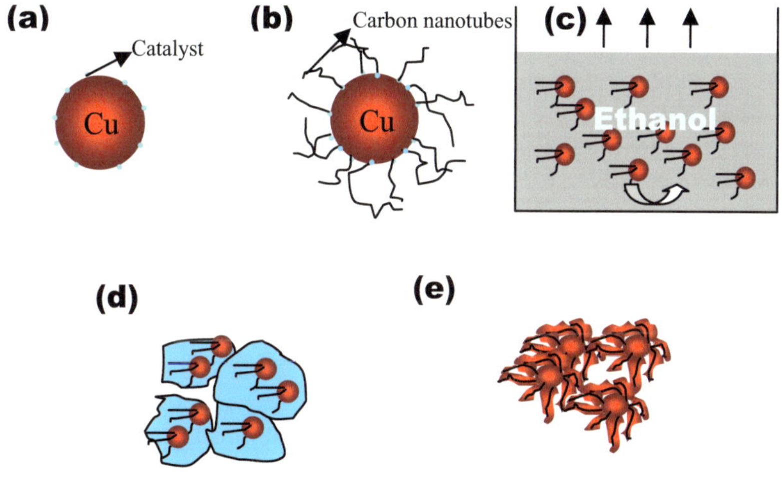


Fig. 7. Schematic illustration of the fabrication of CNT(Ni/Y)-Cu composite powders: a) formation of active catalyst nanoparticles scattered homogeneously on the surface of Cu powder by deposition-precipitation followed by calcination and reduction processes, b) in situ synthesis of CNTs in the matrix by CVD, c) mixing a Cu salt and the in situ CNT(Ni/Y)-Cu composite powders in ethanol, d) vaporization of the solvent by heating under constant magnetic stirring, e) calcination and reduction to obtain CNT-implanted Cu matrix composite powders

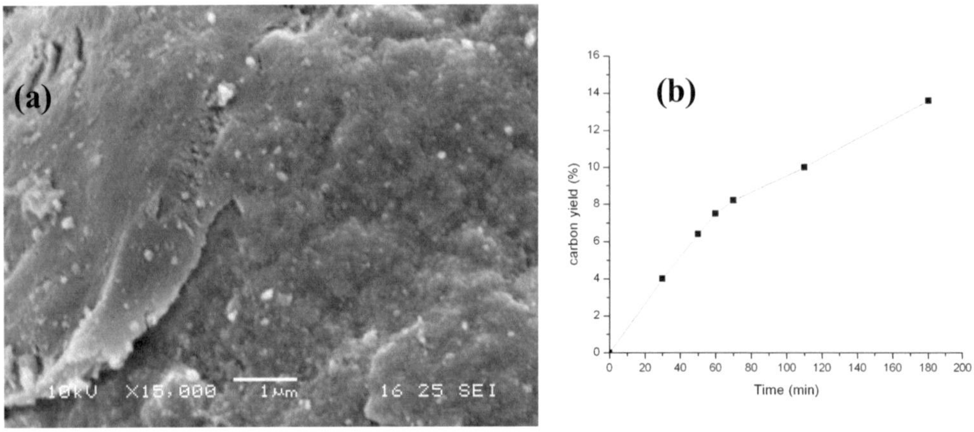


Fig. 8. (a) SEM image of the Ni/Y-Cu catalyst, showing a homogeneous dispersion of Ni/Y catalyst on the surface of Cu powder; (b) variation of carbon yield with growth time by CVD at 500 °C using 2%Ni1%Y-Cu catalyst

SEM images show that almost all of the Cu powders are evenly decorated by Ni/Y composite nanoparticles after the first processing step. Figure 8a displays a high-magnification SEM image of representative Ni/Y catalyst supported on Cu powders (containing 2wt.%Ni and 1wt.%Y). It can be seen that the catalyst (Ni/Y) nanoparticles with a diameter of about 20-40nm have been homogeneously dispersed on the surface of the Cu powders. The density and length of the CNTs, or the CNT fraction in the composite powders, can be tuned by adjusting the experimental parameters, such as growth time. Figure 8b is a typical growth time-carbon yield curve of the composite powders by CVD at 500°C using 2%Ni1%Y-Cu catalyst. With longer time, the slope of the time-carbon yield curve is almost stable at first and then decreases, which is similar with previous reports (Venegoni, et al., 2002).

Figure 9a shows a low-magnification SEM image of the in-situ CNT(Ni/Y)-Cu composite powders obtained by CVD. Almost all Cu particles are decorated by CNTs, displaying spherical morphologies with CNTs highly dispersed into the powders. The most important feature of our process is that the CNTs are synthesized into the Cu powders in situ, resulting in a relatively firm bonding between CNTs and Cu powders. From Figure 9b, we can observe the direct growth of CNFs on the surface of Cu powder (indicated by white arrows). The diameter of CNFs ranges from 20-50nm. A high-resolution transmission electron microscopy (HRTEM) image of a typical CNT, as inserted in Figure 9b, demonstrates the well-graphitized herringbone structure. The graphitic sheets of CNTs are apparent, and the interlayer space between the sheets is similar with the ideal graphitic interlayer spacing (0.34nm). It is also found that the surface of the CNTs is not smooth, displaying a zigzag structure.

During the third mixing process, the in-situ composite powder at micrometer level is easy to disperse highly in the solution under constant stirring, and the fluid force can not separate the CNTs and force them to move with the Cu powders. Furthermore, the Cu ions at the atomic level are mixed with CNTs in the solution and thus contact with CNTs very well. Figure 10 shows the microstructures of CNT(Ni/Y)-CuO and CNT(Ni/Y)-Cu composite powders. It can be seen that CNTs are located within the powders rather than on the powder surface. The morpholgies of the CNT(Ni/Y)-CuO and CNT(Ni/Y)-Cu composite powders show an ideal composite microstructure, which displays a network with CNTs implanted in the Cu powders. It is also interesting that Cu is directly synthesized at the surface of neat CNTs, producing CNT-Cu chains instead of agglomerating together to form large particles (see Figure 10b), thus, realizing excellent wetting of the tubes by Cu, separation of the tubes, and excellent dispersion in the matrix.

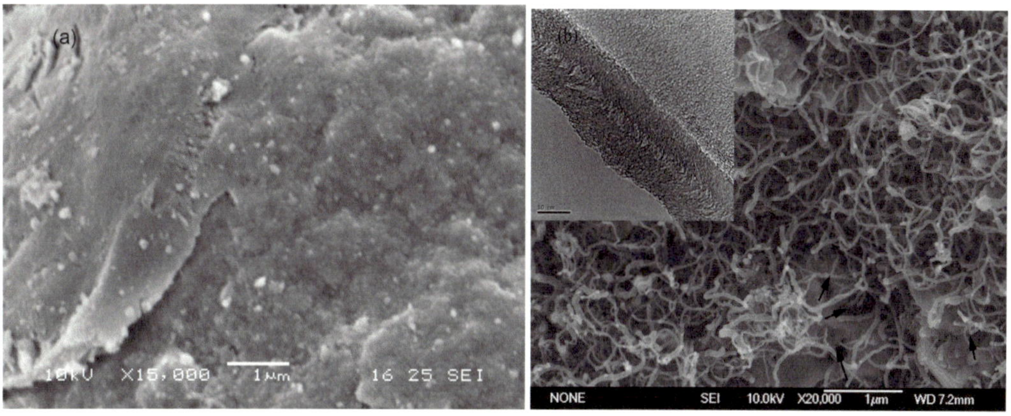


Fig. 9. (a) SEM image of the in-situ CNT(Ni/Y)-Cu composite powders; (b) High-magnification SEM image of a representative composite powder, the inserted is a HRTEM image of a typical CNT, showing a multi-walled nanotube with herringbone structure

Fig. 10. Microstructures of CNT(Ni/Y)-CuO (a) and CNT(Ni/Y)-Cu (b) composite powders, where the CNTs are homogeneously covered with copper

The final composite powders were consolidated into a bulk CNT(Ni/Y)-Cu composite by a vacuum hot pressing process. Fig. 11a shows a low-magnification SEM micrograph of the CNT(Ni/Y)-Cu composite surface after chemical etching. There are many dark gray particles highly dispersed into the light gray area. High-magnification SEM images indicated that there are many twin striations in the dark gray particles, which is the typical character of crystalline copper (Fig. 11b). These dark gray particles are the original copper particles in the in situ CNT(Ni/Y)-Cu powders, which have been encapsulated completely and not oxidized during the calcination in air. XRD analysis of the calcined powders also indicates the existence of Cu. The inserted picture in Fig. 11b shows a typical boundary between the dark gray particles and the gray matrix area. It can be seen that the CNTs are still connected with the Cu particles, inferring that the CNTs are not separated from the in situ CNT(Ni/Y)-Cu powders during the mechanical stirring or the bonding strength between the CNTs and Cu support is strong enough to endure the flowing force. The flowing trace of CNTs with the solution can be observed (indicated by white arrow and ellipse). The gray area shows a highly homogeneous distribution of CNTs within the Cu matrix (Fig. 11c). Particularly, the CNTs form a 3-dimensional (3-D) network within the Cu grains instead of locating at the grain boundary (Fig. 11c insert).

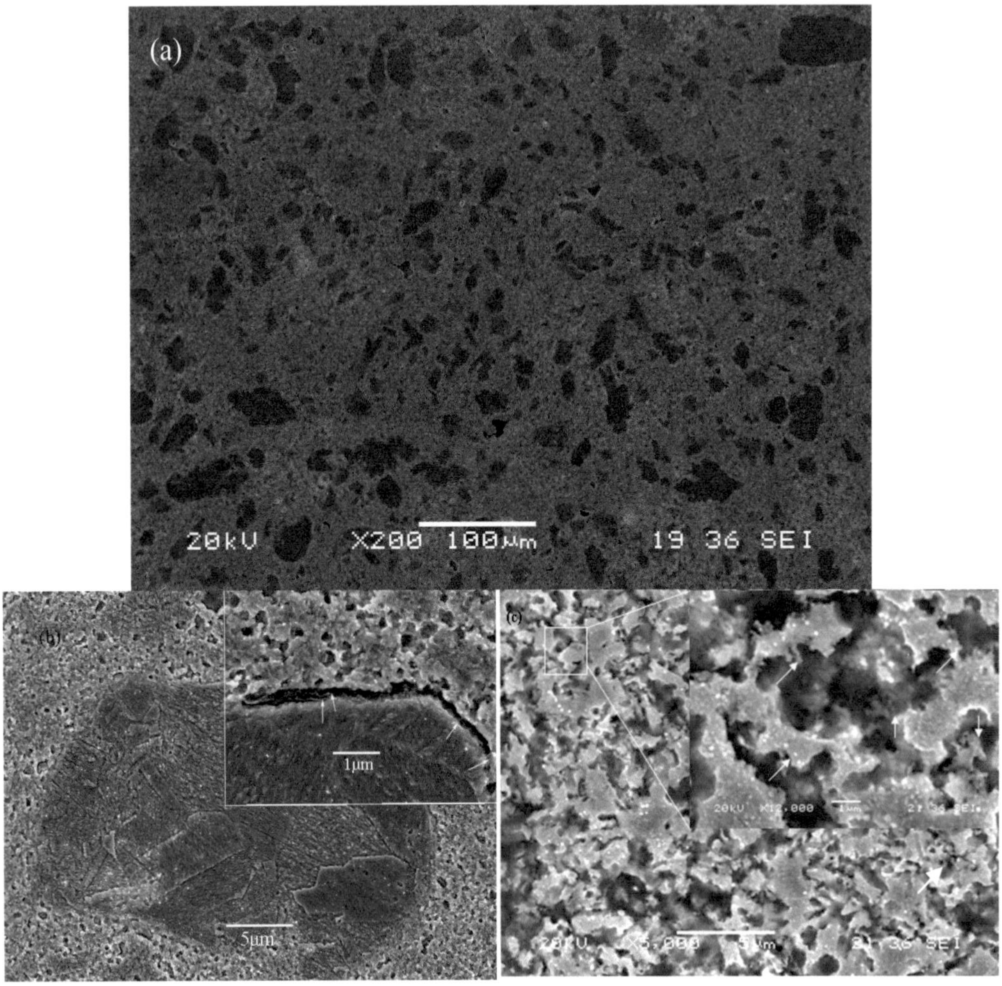


Fig. 11. SEM images of the CNT(Ni/Y)-Cu composite surface after chemical etching, showing homogeneous 3-D distribution of CNTs in the composite.

The mechanical properties of the CNT(Ni/Y)-Cu composite were characterized using compressive tests. As shown in Fig. 12a, the compressive yield strengths of CNT(Ni/Y)-Cu composite were much higher than that of the Cu matrix, which was fabricated by the same hot pressing process without adding CNT(Ni/Y). A 3.4wt.% CNT-reinforced Cu composite showed a yield strength of 581MPa, which is more than 3.6 times higher than that of Cu.

Even with as much as 5.7wt.% CNTs, the yield strength was still 448MPa, which is 2.8 times higher than that of Cu. The reduction of the properties of the composites with high fraction of CNTs may be due to the local agglomeration of CNTs and low relative density. However, compared with that obtained by traditional methods, the yield strength of the composites with 5.7%CNTs was still higher, meaning that the agglomeration of CNTs is not serious. By reducing the powder size of Cu used for catalyst and post treatment of the composites, such as hot extrusion, it has high potential to improve the properties of the composites further. Such research is in the processing. The linear coefficient of thermal expansion (CTE, 30-200 $^{o}$C) of the composite with 5.7wt.% CNTs reduced to $10.1*10^{-6}/ {}^{o}C$, about 57.7% the CTE measured on the Cu. As shown in Table 2, the reinforcement efficiency for reduction of CTE of composites, i.e., the effect of a given volume percentage of reinforcement on the matrix, is more than two times that of W, Mo, SiC and diamond (Table 2). This indicates that the composites with high CNT content have potential as advanced heat sink materials, which requires low CTE, high thermal conductivity and machinability.

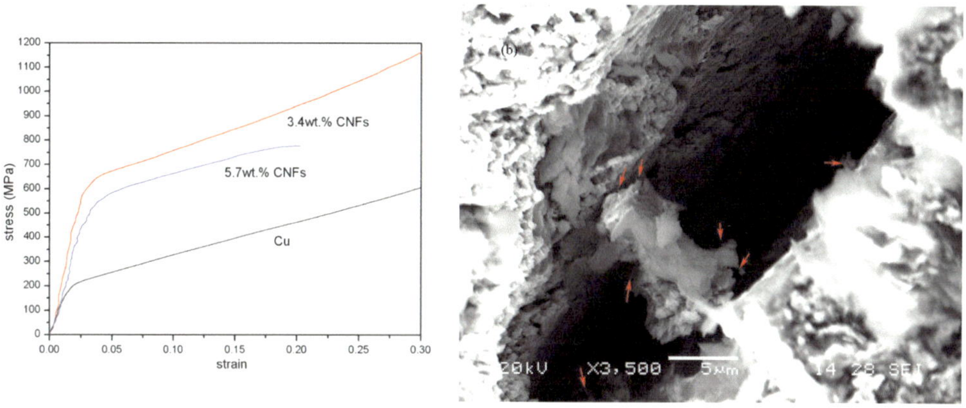


Fig. 12. a) stress-strain curves of CNF(Ni/Y)-Cu composites obtained by compressive testing, and b) SEM image of the crack structure of the composite under compressive testing.

| matrix | reinforcement | Vol.% | CTE σ ($10^{-6}$/K) | Efficiency of the reinforcement R[a] |
|---|---|---|---|---|
| Cu | $SiC_p$ (Mo) | 40 | 11.5 | 0.81 |
| Cu | Diamond (0.8Cr) | 42 | 11 | 0.84 |
| Cu | AlN | 40 | 9.6 | 1.09 |
| Al | $SiC_p$ | 50 | 10.8 | 1.06 |
| Cu | Mo | 60-85 | 6.27-9 | <0.8 |
| Cu | W | 80-90 | 5.6-9.1 | <0.8 |
| Cu | CNFs | 23[b] | 10.1 | 1.76 |

[a] The efficiency of the reinforcement $R=(\sigma_c-\sigma_m)/V_f\sigma_m$.

[b] The density of CNFs is estimated as $1.8g/cm^{-3}$.

Table 2. The efficiencies of several reinforcement materials for reduction of CTE of composites (Schubert, et al., 2008 & Wu, et al., 2006 & Geffroy, et al., 2007 & Lee, et al., 2007)

Such excellent strengthening by CNT reinforcement was due to the high dispersion of CNTs and the high load-transfer efficiency of CNTs in the metal matrix. The fracture surface of the CNT(Ni/Y)-Cu composite under bending test indicates that almost all dimples, inferring ductile fracture mechanism of the composite, are evenly decorated by CNT tips, further confirming the high dispersion of CNTs. Moreover, all the CNTs that pull out are very short, meaning that some CNTs may be broken during the fracture and thus indicating that the load transfer from the matrix to the nanotube is sufficient to break the CNTs. Fig. 12b shows the fracture surface after a compression test. It can be observed that there are composite particles instead of single CNTs that pull out, also inferring the strong interfacial bonding strength between CNTs and Cu. bridging CNTs between the pull-out particles and bulk composite can be observed (indicated by circle in Fig. 12b). As is known, there are three main load transfer mechanisms that control the full operation of the stress transfer, including micro-mechanical interlock, chemical bonding (interaction) and a weak van der Waals attractive force. Because Cu cannot react with untreated CNTs, chemical bonding between CNTs and Cu in our experiments is impossible. The weak van der Waals attractive force cannot transfer high loads. Thus, the possible main mechanism is micro-mechanical interlock. As observed by SEM, the nanotubes have physically contacted well with the Cu from the copper salt by mixing in a solution. Furthermore, the herringbone structure and uneven wall of CNTs also increase the micro-mechanical interlock strength.

### 3.3 Other metal powders

Apart from Al and Copper, other popular metal powders, such as Ag and Mg, were also examined as catalyst support for CNT growth. For the Ni/Ag catalyst, the CVD processes were performed at temperatures among 500-700℃. The weight of the catalyst before and after CVD was increased only a very little amount (<1wt.%). TEM and SEM analysis showed that there was almost no carbon deposit observed in the product obtained at 500℃, while, only a few amorphous carbon or graphite fragments was detected in the product obtained at 700℃ and the diameter of most catalyst particles became micro-size level (Fig. 13), meaning that the activity and selectivity of the catalyst was impaired seriously even there was only a little diffusion of Ag into Ni, apart from the aggregation of the catalyst.

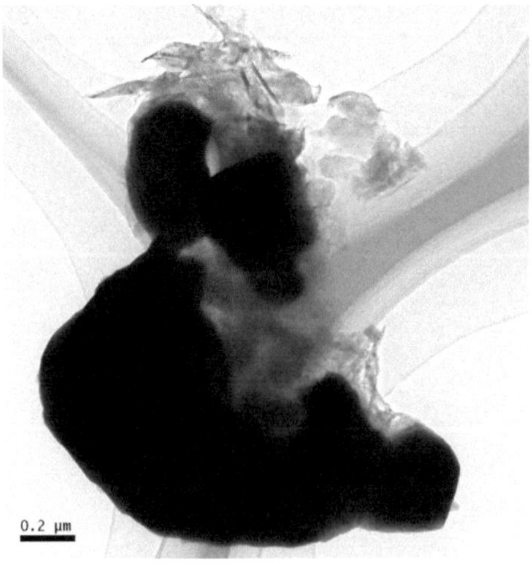


Fig. 13. TEM image of the products obtained by CVD using Ni/Ag catalyst at 700℃

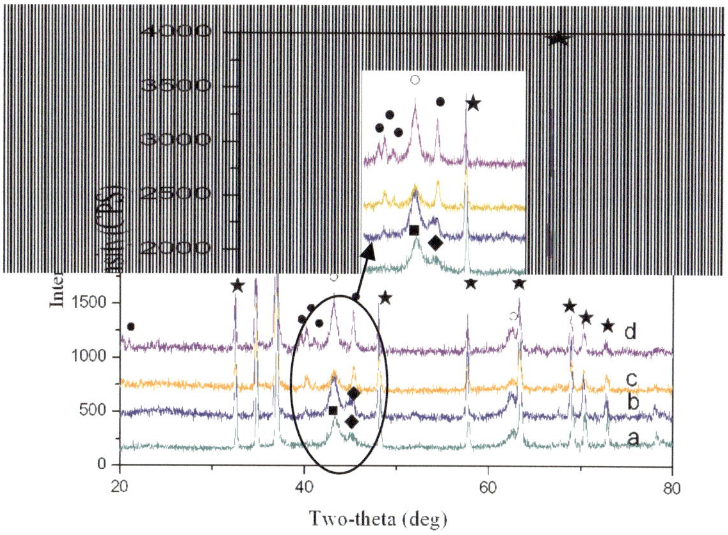


Fig. 14. XRD analysis of the 10%Ni/Mg catalyst precursor after calcination (a) and reaction products by CVD at 400℃(b); 450℃ (c) and 500℃ (d)

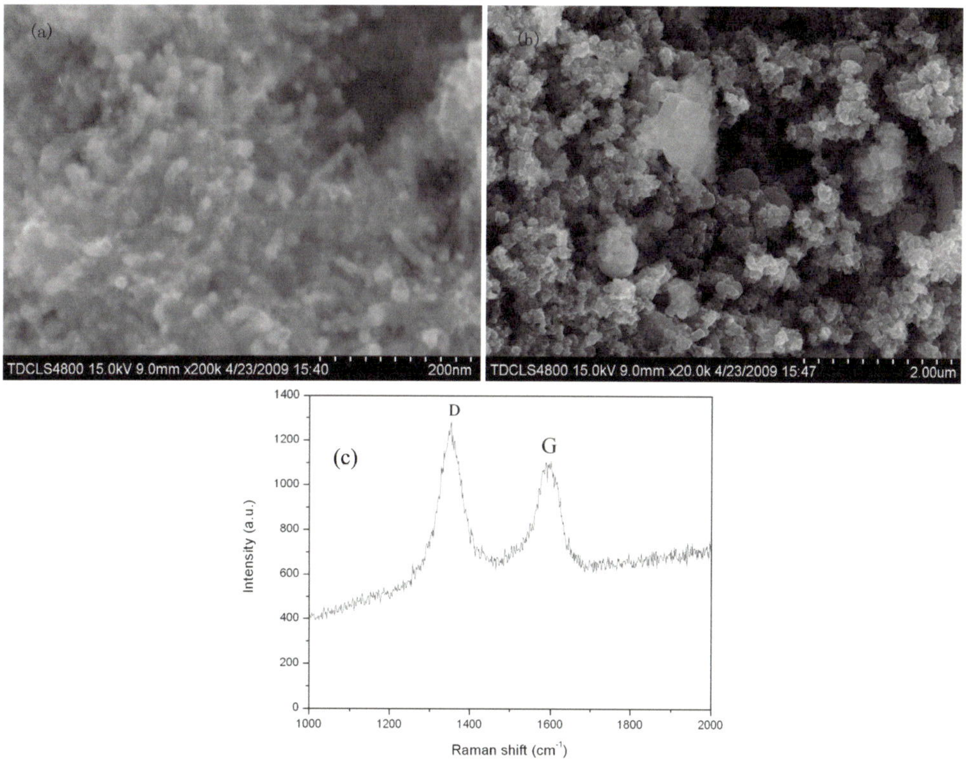


Fig. 15. SEM images and Raman spectra of the products obtained by CVD using Ni/Mg catalyst at 400℃ (a, c); 450℃ (b)

Fig. 14 shows the XRD analysis of the products obtained by CVD using Ni/Mg at various reaction temperatures. Both Ni and MgO peaks are detected in the product obtained at 400℃. However, MgO can not resist the reactions between Ni and Mg at higher temperature, inferring that MgO is not compact and some Ni particles may be attached directly on the Mg surface. From Fig. 14b, it is observed that $Mg_2Ni$ begins to be formed when the temperature increases to 450℃. SEM image and Raman spectrum show that short and thin CNTs are obtained at 400℃ (Fig. 15a and c). When the reaction temperature is above 400℃, there are almost no carbon deposits formed as shown in Fig. 15b, indicating that $Mg_2Ni$ can not catalyze the decomposition of $CH_4$.

Based on the above analysis, it seems that the chemical reaction or diffusion between catalyst and substrate impairs the activity of the catalyst seriously. When the catalyst reacts with substrate, the catalyst will poison. The novel Ni/Y catalyst was also applied in Ag an Mg powders. Experimental results indicated that CNTs grown on Ag particles using Ni/Y catalyst was similar with that on Cu particles. Ni/Y catalyst supported on Mg, which presents a higher stability and catalytic property than Ni catalyst, was also successful to catalyze CNT growth directly on Mg particles. Those results infer that different metal supports have relative little effect on the activity of the catalyst if the reaction between catalyst and support is critically controlled, meaning that all metal powders are possible to be used as catalyst carriers for CNT growth by CVD. The works using these in situ CNT-metal composite powders to fabricate buck composites are in progress.

## 4. Conclusion

In summary, carbon nanotube integration with metal substrate by direct growth technique has been explored as a new approach for electronic device design and CNT-based composite fabrications. Many teams demonstrated solutions to the key challenge of growing high-quality CNTs directly on various metal substrates. The performance of electronic devices by direct growth of CNTs, such as field emitters, supercapacitors, lithium ion batteries and ITMs, has been validated to be more excellent than that by external connection of CNTs. Direct growth of CNTs in metal powders can greatly improve the dispersion of CNTs in matrix and interfacial properties between CNTs and metal matrix, thus promoting the performance of the CNTs/metal composites. The next milestone will be the commercialization of this technology.

## 5. Acknowledgment

This work was sponsored by National Basic Research Program of China (2010CB934700), National Natural science Foundation of China (No.51071107 and No. 51001080) and the open project of The State Key Laboratory of Metal Matrix Composites (No. mmc-kf09-03).

## 6. References

Atthipalli, G.; Epur, R.; Kumta, R.; Allen, B.; Tang, Y.; Star, A.; Gray, J.; (2011), The effect of temperature on growth of carbon nanotubes on copper foil using a nickel thin film as catalyst, *Thin Solid Films*, in Press (doi:10.1016/j.tsf.2011.02.046)

Baddour, C.; Fadlallah, F.; Nasuhoglu, D.; Mitra, R.; Vandsburger, L.; Meunier, J. (2008), A simple thermal CVD method for carbon nanotube synthesis on stainless steel 304 without the addition of an external catalyst, *Carbon*, 47, 313-347

Baddour, C.; Upham, D. & Meumier, J. (2010). Direct and pepetitive growth cycles of caron nanotubes on stainless steel particles by chemical vapor deposition in a fluidized bed, *Carbon*, 48, 2652-2656.

Benito, S.; Lefferts L. (2010), The production of a homogeneous and well-attached layer of carbon nanofibers on metal foils, *Carbon*, 48, 2862-2872

Berber, S.; Kwon, Y. & Tománek, D. (2000). Unusually high thermal conductivity of carbon nanotubes, *Physics Review Letters*, 84, 4613-4616.

Bult, J.; Sawyer, W.; Ajayan P.; Schadler L. (2009) Passivation oxide controlled selective carbon nanotube growth on metal substrates, *Nanotechnology*, 20, 085302(7pp)

Burt, D.; Whyte, W.; Weaver, J.; Glidle, A.; Edgeworth, J.; Macpherson, J.; Dobson, P.; (2009), Effects of metal underlayer grain size on carbon nanotube growth, *Journal of Physics and chemistry: C*, 2009, 113, 15133-151399

Cha, S.; Kim, T.; Arshad, S.; Mo, C. & Hong, S. (2005), Extraordinary strengthening effect of carbon nanotubes in metal matrix nanocomposites processed by molecular-level mixing, *Advanced Materials*, 17, 1377-1381.

Cui, H.; Eres, G.; Howe, J.; Puretkzy, A.; Varela, M.; Geohegan, D.; Lowndes, D. (2003), Growth behavior of carbon nanotubes on multilayered metal catalyst film in chemical vapor deposition, *Chemical Physics Letter*, 374(3–4), 222–228.

Du, C.; Pan, N. (2005) CVD growth of carbon nanotubes directly on nickel substrate, *Materials Letters*, 59, 1678-1682

Gan, B.; Ahn, J.; Zhang, Q.; Rusli; Yoon, S.F.; Yu, J. (2001) Y-junction carbon nanotubes grown by in situ evaporated copper catalyst, *Chem. Phys. Lett.* 333, 23–28

Gao, L. ; Peng, A. ; Wang, Z. ; Zhang, Hao. ; Shi, Z. ; Gu, Z. ; Cao, G. ; Ding, B. (2008), Growth of aligned carbon nanotube arrays on metallic substrate and its application to supercapacitors, *Solid Communications*, 146, 380-383

García, E.; Hart, A.; Wardle, B. & Slocum, A. (2007). Fabrication and nanocompression testing of aligned carbon-nanotube-polymer nanocomposites, *Advanced Materials*, 19, 2151-2156.

García-Ce´spedes, J.; Thomasson, S.; Teo, K.; Kinloch, I.; Milne, W.; Pascual, E.; Bertran, E.; (2009), Efficient diffusion barrier layers for the catalytic growth of carbon nanotubes on copper substrates, *Carbon*, 47, 613-621.

Geffroy, P.; Chartier, T. & Silvain, J. (2007). Preparation by tape casting and hot pressing of copper carbon composites films, *Journal of the European Ceramic Society*, 27, 291-299.

George, R.; Kashyap, K.; Rahul, R. & Yamdagni, S. (2005). Strengthening in carbon nanotube/aluminium (CNT/Al) composites, *Scripta Materialia*, 53, 1159-1163.

Hiraoka, T. ; Yamada, T. ; Hata, K. ; Futaba, D. ; Kurachi, H. ; Uemura, S. ; Yumara M. ; Iijima, S. (2006), Synthesis of single- and double-walled carbon nanotube forests on conducting metal foils, *Journal of American Chemical Society*, 128, 13338-13339

Huang, M. ; Li, X. ; Yi, H. ; Ma, N. & Wang, H. (2005). Effect of in situ $TiB_2$ particle reinforcement on the creep resistance of hypoeutectic Al-12Si alloy, *Journal of Alloy and Compounds*, 389, 275-280

Huang, Z. ; Yang, B. ; Cui, H. & Zhang, J. (2003). Study on the fabrication of Al matrix composites strengthened by combined in-situ alumina particle and in-situ alloying elements, *Materials Science and Engineering A*, 351, 15-22

Iijima, S. (1991), Helical microtubules of graphitic carbon, *Nature*, 354 (6348), 56-58.

Iijima, S. (2011). Science and industrial applications of nano-carbon materials such as fullerence, carbon nanotubes and graphenes, *The 2011 WPI-AIMR Annual Workshop*, Sendai, Japan.

Kang, Y. & Chan, S. (2004). Tensile properties of nanometrix Al2O3 particulate-reinforced aluminum matrix composites, *Materials Chemistry and Physics*, 85, 438-443.

Kavian, R.; Vicenzo, A.; Bestetti, M.; (2011), Growth of carbon nanotubes on aluminium foil for supercapacitors electrodes, Journal of Materials Science, 46, 1487-1493.

Kim, B.; Chung H.; Chu, K.; Yoon, H.; Lee, C.; Kim, W.; (2010), Synthesis of vertically-aligned carbon nanotubes on stainless steel by water-assisted chemical vapor deposition and characterization of their electrochemical properties, *Synthetic Metals*, 160, 584-587.

Kim, K.; Cha, S.; Hong, S. & Hong, S. (2006). Microstructures and tensile behavior of carbon nanotube reinforced Cu matrix nanocomposites, *Materials Science and Engineering A*, 430, 27-33.

Kim, P.; Shi, L.; Majumdar, A. & McEuen, P. (2001). Thermal transport measurements of Individual multiwalled nanotubes, *Physics Review Letters*, 87, 215502 (4pp).

Kim, S.; Gangloff, L.; (2009), Growth of carbon nanotubes (CNTs) on metallic underlayers by diffusion plasma-enhanced chemical vapour deposition (DPECVD), *Physica E*, 41, 1763-1766

Kuzumaki, T.; Miyazawa, K.; Ichinose, H. & Ito, K. (1998). Processing of carbon nanotube reinforced aluminum composite, *Journal of materials Research*, 13, 2445-2449.

Laha, T.; Agarwal, A.; Mckechnie, T. & Seal, S. (2004). Synthesis and characterization of plasma spray formed carbon nanotube reinforced aluminum composite, *Materials Science and Engineering A*, 381, 249-258.

Lahiri, I.; Oh, S.; Hwang, J.; Cho, S.; Sun, Y.; Banerjee, R.; Choi, W.; (2010), High capacity and excellent stability of lithium ion battery anode using interface-controlled binder-free multiwall carbon nanotubes grown on copper, *ACS Nano*, 4, 3440-3446.

Lahiri, I.; Seelaboyina, R.; Hwang, J.; Banerjee, R.; Choi, W.; (2001), Enhanced field emission from multi-walled carbon nanotubes grown on pure copper substrate, *Carbon*, 48, 1531-1538

Lee, K.; Oh, D.; Choi, W.; Weissgärber, T. & Kieback, B. (2007), Thermomechanical properties of AIN-Cu composite materials prepared by solid state processing, *Journal of Alloy and Compounds*, 434-435, 375-377.

Lepro´ X. ; Lima, M. ; Baughman, R. (2010), Spinnable carbon nanotube forests grown on thin, flexible metallic substrates, *Carbon*, 48, 3621-3627

Li, C. ; Zhang, Y. ; Mann, M. ; Hasko, D. ; Lei, W. ; Wang, B. ; Chu, D. ; Pribat, D. ; Amaratunga, G. ; Milne, W. ; (2010), High emission current density, vertically aligned carbon nanotube mesh, field emitter array, *Applied Physics Letters*, 97, 113107 (3pp)

Li, X. ; Zhang, X. ; Ci L. ; Shah R. ; Wolfe, C. ; Kar, S. ; Talapatra, S. ; Ajayan, P. ; (2008) Air-assisted growth of ultra-long carbon nanotube bundles, *Nanotechnology*, 19, 455609(7pp)

Lin, W. ; Zhang, R.; Moon, K. ; Wong, C. ; (2010), Synthesis of high-quality vertically aligned carbon nanotubes on buck copper substrate for thermal management, *IEEE Transactions on Advanced Packaging*, 33, 370-376.

Lin, W. ; Zhang, R.; Moon, K. & Wong, C. (2010). Molecular Phonon couplers at carbon nanotube/substrate interface to enhance interfacial thermal transport, *Carbon*, 48, 107-113.

Liu, H.; Zhang, Y.; Arato, D.; Li, R.; Merel, P. & Sun, X. (2008) Aligned multi-walled carbon nanotubes on different substrates by floating catalyst chemical vapor deposition: Critical effects of buffer layer, *Surface Coating Technology*, 202, 4114-4120

Mahanandia, P. ; Arya, V. ; Bhotla, P. ; Subramanyam, S. ; Schneider, J. ; Nanda, K. ; (2009), Excellent field emission from semialigned carbon nanofibers grown on cylindrical copper surface, *Applied Physics Letters*, 95, 083108 (3pp).

Martinez-latorre, L.; Ruiz-Cebollada, P.; Monzon, A.; Garcia-Bordeje, E. (2009), Preparation of stainless steel microreactors coated with carbon nanofiber layer: impact of hydrocarbon and temperature, *Catalysis Today*, 1475, 587-593

Masarapu C. ; Wei B. (2007) Direct growth of aligned multiwalled carbon nanotubes on treated stainless steel substrates, *Langmuir*, 23, 9046-9049

Mata, D.; Ferro, M.; Fernandes, A.J.S. ; Amaral, M. ;Oliveira, F.J. ; Costa, P. ; Silva, R. (2010) Wet-etched Ni foil as active catalysts towards carbon nanofiber growth, *Carbon*, 48, 1939-2854

Mauger, M. ; Binh, V. ; Levesque, A. ; Guillot, D. ; (2004), Freestanding vertically aligned arrays of individual carbon nanotubes on metallic substrates for field emission cathodes, *Applied Physics Letters*, 85, 385-387.

Moreno, M. & Oliver, C. (2006). Compression creep of PM aluminum matrix composites reinforced with SiC short fibers, *Materials Science and Engineering A*, 418, 172-181.

Nessim, G. ; Seita, M. ; O'Brien, K. ; Hart, A . ; Bonaparte, R. ; Mitchell, R. ; Thompson, C. ; (2009), Low temperature synthesis of vertically aligned carbon nanotubes with electrical contact to metallic substrates enabled by thermal decomposition of the carbon feedstock, *Nano Letter*, 9, 3398-3405.

Nessim, G. ; Seita, M. ; O'Brien, K. ; Speakman, S. ; (2010), Dual formation of carpets of large carbon nanofibers and thin crystalline carbon nanotubes from the same catalyst-underlayer system, *Carbon*, 48, 4519-4526.

Nessim, G. ; Acquaviva, D. ; Seita, M. ; O'Brien, K. ; Thompson, C. ; (2010), The critical role of the underlayer material and thickness in growing vertically aligned carbon nanotubes and nanofibers on metallic substrates by chemical vapor deposition, *Advanced Functional materials*, 20, 1306-1312.

Oye, M. ; Yim, S. ; Fu, A. ; Schwanfelder, K. ; Meyyapan, M. ; Nguyen, C. (2010), Surface smoothness effect for the direct growth of carbon nanotubes on bulk FeCrAl metal substrates, *Journal of Nanoscience and Nanotechnology*, 10, 4082-4088

Pal, S. ; Kar, S. ; Lastella S. ; Kumar, A. ; Vajtai, R. ; Talapatra, S. ; Borca-Tasciuc, T. ; Ajayan, P. ; (2010), Importance of $Cr_2O_3$ layer for growth of carbon nanotubes on superalloys, *Carbon*, 48, 844-853

Peigney, A. ; Flahaut, E. ; Laurent, C. ; Chastel, F. & Rousset, A. (2002). Aligned carbon nanotubes in ceramic-matrix nanocomposites prepared by high-temperature extrusion, *Chemical Physics Letter*, 352, 20-25

Qin, Y.; Zhang, Q.; Cui, Z.L.; (2004) Effect of synthesis method of nanocopper catalysts on the morphologies of carbon nanofibers prepared by catalytic decomposition of acetylene, *J. Catal.* 223, 389–394.

Reddy, N. ; Meunier J. ; Coulombe, S. (2006) Growth of carbon nanotubes directly on a nickel surface by thermal CVD, *Matterials Letter*, 60, 3761-3765

Schubert, T. ; Trindade, B. ; Weißgärber, T. & Kieback, B. (2008). Interfacial design of Cu-based composites prepared by powder metallurgy for heat sink applications, *Materials Science and Engineering A*, 475, 39-44.

Shyu, R. ; Weng, F. & Ho, C. (2002). In situ reacted titanium nitride-reinforced aluminum alloy composite, *Journal of Materials Processing Technology*, 122, 301-304.

Singh, M. ; Singh, P. Titus, E. ; Misra, D. ; LeNormand, F. ; (2002), High density of multiwalled carbon nanotubes observed on nicke electroplated copper substrates by microwave plasma chemical vapor deposition, *Chemical Physics Letters*, 354, 331-336.

Sung, W. ; Kim, W. ; Lee, H. ; Kim, Y. ; (2008), Field emission characteristics of carbon nanofibers grown on copper micro-tips at low temperature, *Vacuum*, 82, 551-555.

Talapatra, S. ; Kar, S. ; Pal, S. ; Vajtai, R. ; Ci, L. ; Victor, P. ; Shaijumon, M. ; Kaur, S. ; Nalamasu, O. ; Ajayan. (2006), Direct growth of aligned carbon nanotubes on bulk metals, Nature *Nanotechnology*, 1, 112-116

Tang, F. ; Anderson, I. & Biner, S. (2003). Microstructures and mechanical properties of pure Al matrix composites reinforced by Al-Cu-Fe alloy particles, *Materials Science and Engineering A*, 363, 20-29.

Venegoni, D. ; Serp, P. ; Feurer, R. ; Kihn, Y. ; Vahlas, C. & Kalck, P. (2002). Parametric study for growth of carbon nanotubes by catalystic chemical vapor deposition in a fluidized bed reactor, *Carbon*, 40, 1799-1807.

Wang, H. ; Feng, J. ; Hu, X. & Ng, K. (2007). Synthesis of aligned carbon nanotubes on double-sided metallic substrate by chemical vapor deposition, *Journal of Physics and Chemistry : C*, 111, 12617-12624.

Wang, B. ; Liu, X. ; Liu, H. ; Wu, D. ; Wang, H. ; Jiang, J. ; Wang, X. ; Hu, P. ; Liu, Y. ; Zhu, D. ; (2003), Cotrollable preparation of patterns of aligned carbon nanotubes on metals and metal-coated silicon substrates, *Journal of Materials Chemistry*, 13, 1124-1126.

Wu, H. ; Wang, Z. ; Zheng, Q. & Zhou, J. (2006) The review and prospect of copper-matrix composites for electronic packaging, *China Molybdenum Industry*, 03, 30-32

Yi, W. & Yang, Q. (2010), CVD growth and field electron emission of aligned carbon nanotubes on oxidized inconel plates without addition of catalyst, *Diamond & Related Materials*, 19, 870-874.

Yin, X. ; Wang, Q. ; Lou, C. ; Zhang, X. & Lei, W. (2008). Growth of multi-walled CNTs emitters on an oxygen-free copper substrate by chemical-vapor deposition, *Applied surface Science*, 254, 6633-6636

Zambon, A. ; Badan, B. & Maddalena, A. (2004). Production, microstructural and mechanical characterizaition of spray formed Al-6 wt.% Fe alloy and pre-mixed Al-6 wt.% Fe/$SiC_p$ composite, *Materials Science and Engineering A*, 375-377, 645-650.

Zhang, L., Tan, Y. & Resasco, D. (2006). Controlling the growth of vertically oriented single-walled carbon nanotubes by varying the density of Co–Mo catalyst particles, *Chemical Physics Letter*, 422(1–3), 198–203.

Zhang, Z.; He, P.; Sun, Z.; Feng, T.; Chen, Y.; Li, H. & Tay, B. (2010) Growth and field emission property of coiled carbon nanostructure using copper as catalyst, *Applied Surface Science*, 256, 4417-4422

Zhu, S. & Lizuka, T. (2003). Fabrication and mechanical behaviou of Al matrix composites reinforced with porous ceramic of in situ grown whisker framework, Materials Science and Engineering A, 354, 306-314.

7

# Flame Synthesis of Carbon Nanotubes

Jay P. Gore and Anup Sane
*Purdue University*
*USA*

## 1. Introduction

Carbon is one of the most abundant elements found in nature. It is also one of the elements that form the fundamental building blocks of life on earth. Carbon, in solid, liquid and gaseous forms, is a major contributor in many human activities. Solid forms of carbon have a diverse phase diagram ranging from very soft graphitic to very hard diamond structures. These forms are produced through various processes ranging from coking to crystalline condensation from gaseous precursors. These diverse processes and the variables that define them lead to many different forms of solid carbon including coke, carbon blacks, graphitized carbon, pyrocarbon, glossy carbon, active carbon, diamond, fullerenes, carbon fibers and the newest identified form- carbon nanotubes. Carbon nanotubes (CNTs) are one of the most elegant arrangements of solid carbon that are elongated structures of C60 molecules (also known as fullerenes (Kroto, 1987; Kroto et al., 1985)). CNTs were first observed by Baker et al. in the 1970s (Baker et al., 1972; , 1973; Oberlin et al., 1976). However, these findings did not spark as much interest in the scientific community as that sparked by the re-discovery of CNTs by Iijima(Iijima, 1991). For the last two decades carbon nanotubes and their remarkable properties have been extensively studied and different synthesis methods have been developed for their production.

Carbon nanotubes can be classified by the number of concentric walls. A single wall carbon nanotube (SWNT) can be visualized as a flat graphene sheet of fullerene (C60) molecules rolled up to make a seamless cylinder. C60 molecules are tiny spherical structures (diameter of ~ 7 Å) of 60 carbon atoms connected together forming 20 hexagons and 12 pentagons. Two halves of the fullerene molecule can be visualized to form the end caps of the nanotube. SWNTs exhibit extraordinary physical properties including very high thermal (1750-5800 W/m-k) and electrical conductivity (resistivity equals ~ $10^{-6}$ Ω-m) in the axial direction (Hone et al., 1999; Thess et al., 1996) and equally remarkable structural properties such as high Young's modulus (1054 GPa) (Yu et al., 2000). A multi wall carbon nanotube (MWNT) can be visualized as a structure with concentric cylinders of increasing diameters with correspondingly larger hemispherical end caps terminating them at each end. MWNTs are more commonly found in products of synthesis than the SWNTs. SWNTs are produced by carefully controlling the condensation process. Diameters of SWNTs are, typically a few nanometers, and those of MWNTs are generally of tens of nanometers. Another classification of CNTs is based on the orientation of the hexagonal (six-member carbon ring) in a honeycomb lattice with respect to the axis of the nanotube. Three possible structures include armchair, zigzag and chiral, as shown in figure 1. Armchair and zigzag structures are achiral, that is, their mirror image is identical to the original object. On the other hand, chiral structures exhibit spiral symmetry, whose mirror images cannot be superimposed on the original one.

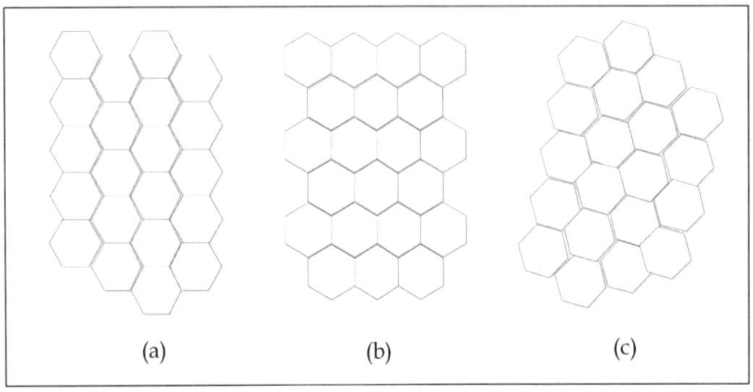


Fig. 1. Various configurations of carbon nanotubes (a) Armchair, ( b) Zigzag and (c) Chiral

Catalytic synthesis of carbon nanotubes is typically a multiple length-scale multi-step process in which carbon is deposited in solid form. This conversion occurs at the nano scale via endothermic reactions occurring at surface of catalyst nano-particle. Therefore, a synthesis reactor generally consists of three essential components: (i) source of gaseous carbon (ii) source of heat and (iii) catalyst particles that provide the reaction sites. A mixture of carbon containing gaseous precursor species at appropriate temperature is maintained throughout the reactor. The size of the reactor defines the largest scale. The catalyst is added either in a form of nano-scale depositions on a substrate or freely floating nano-particle aerosol embedded in the bulk phase. In the past, carbon nanotubes have been produced using methods such as plasma arc discharge, pulsed laser vaporization (PLV), chemical vapor deposition (CVD), Plasma Enhanced (PECVD) and hydrocarbon flame synthesis. Schematic representation of these synthesis processes is shown in figure 2.

Iijima (Iijima, 1991) used the arc discharge method for production of CNTs. The process involved condensation of carbon atoms generated from evaporation of a solid carbon source. In this method, high electric current (~50 - 120 A) is passed through graphite electrodes placed at a distance of approximately 1 mm in the synthesis chamber that causes material from the cathode to sublimate and the nanotubes to form on the anode. The arc discharge process is difficult to control because of the very high temperature (~3200 K) in the electrode gap. The method is also cost and energy intensive and unwanted byproducts such as polyhedron graphite particles contaminate the relatively low yield of CNTs.

In the pulsed laser vaporization or laser ablation method, a high energy laser is directed to ablate a carbon target that contains some nickel and cobalt in a tube furnace, at the temperature of ~ 1400 K. A flow of inert gas is passed through the chamber to carry the CNTs downstream, to a collector surface. Single walled carbon nanotubes, mostly in the form of ropes, at a 1- 10g scale have been formed by this method. The CNTs formed by the laser ablation method are of a higher quality than those produced by the arc discharge method. However, the production rate is low, and the pulsed laser vaporization or laser ablation method is both capital and energy intensive.

The chemical deposition method (CVD) is an alternative method in which CNTs are grown using catalysis. This method involves decomposition of a hydrocarbon gas over a transition metal catalyst and initiation of CNT synthesis by some of the resulting carbon atoms. CVD growth mechanism generally involves the dissociation of hydrocarbon molecules and saturation of carbon atoms in the catalyst metal nano-particles. The precipitation of carbon from the saturated metal particles leads to the formation of carbon nanotubes. Use of catalysis reduces the need for high temperatures. Hydrogen from the decomposition process, and supplemented by that carried with the bulk phase, contributes to the activation and reactivation of the catalytic surface. The CVD method has a better CNT yield and is potentially scalable to commercial manufacturing.

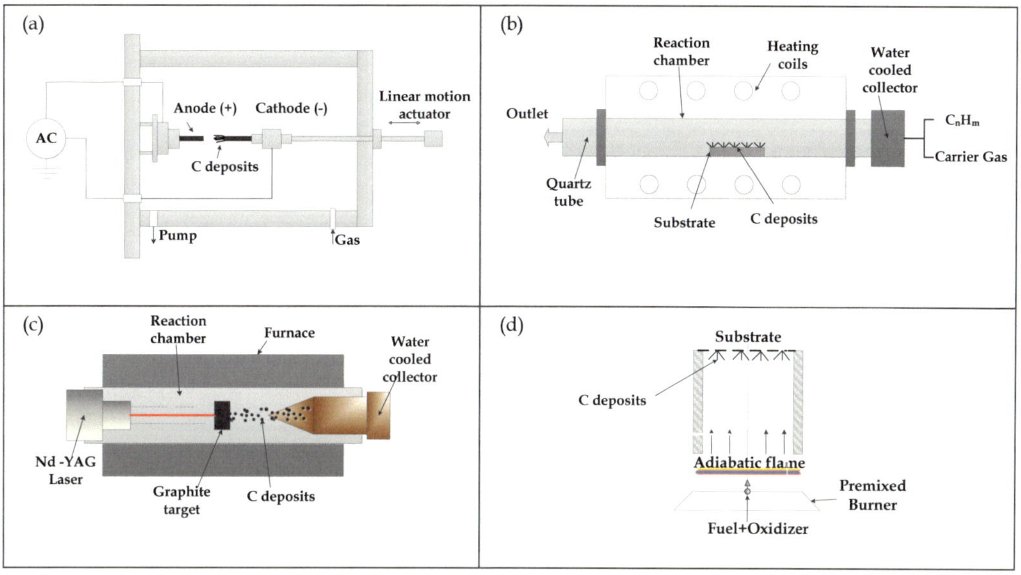


Fig. 2. Schematic representation of methods used for carbon nanotube synthesis (a) Arc discharge (b) Chemical vapor deposition (c) Laser ablation (d) hydrocarbon flames

Hydrocarbon flames provide a unique combination of the chemical and catalytic factors that are conducive to initiation and growth of carbon nanotubes. Gases ($CO$, $CH_4$, $C_2H_2$, $C_2H_4$ and $C_2H_6$) present in the post flame environment form a diverse source of gaseous carbon. The chemical energy released in the form of heat in the flame supports the endothermic carbon deposition reactions. Catalysts in appropriate form (substrate or aerosol) provide the reaction sites for deposition of solid carbon. Growth mechanisms similar to those observed in the CVD process govern the growth of nanotubes in flames. The geometry and characteristics of the catalysts play an important role in the structural properties of the carbon nanotubes. Flames are scalable and are commercially used for the production of solid carbon forms such carbon black and printing ink. Appropriately tailored flame conditions may provide an ideal environment for growth of CNTs on a large commercial scale.

The rest of this chapter summarizes the mechanisms of carbon nanotubes formation in flames followed by a summary of the types of flame configurations and chemistry that have been used in the synthesis process. The process parameters such as equivalence ratio, temperature, pressure are discussed next. Computational models for exploring the parameter space for optimization of the synthesis process are discussed next and the chapter ends with a few comments about future trends.

## 2. Mechanism of carbon nanotube formation in a catalytic synthesis process

The inception and growth mechanisms of CNTs have been studied extensively but a consensus on a single mechanism has not emerged. In fact, more than one mechanism may be involved in the inception and growth of CNTs depending on the specifics of the gaseous precursors, catalysts and operating parameters. One of the most popular descriptions involves the carbon dissolution-diffusion-precipitation mechanism proposed by (Baker et al., 1972). Catalytic nano-particles from transitional metal/metal alloys (e.g. Fe, Ni, and Co) are assumed to be spherical or pear-shaped and are either floating or supported on a substrate. The catalytic decomposition of the carbon precursor molecules (e.g. CO, $CH_4$, $C_2H_2$, $C_2H_4$ and $C_2H_6$) is conjectured to occur on one half of the nano-particle surfaces (the lower curvature side for the pear shaped particles). The released carbon atoms diffuse into the catalyst nano-particles along the concentration gradients until carbon super-saturation at the particle temperature occurs (Moisala et al., 2003). Post super saturation of the catalyst particle, carbon atoms precipitate in solid carbon form on the opposite half of the catalyst particle around and below the bisecting diameter. This description is similar to the Vapor-Liquid-Solid (VLS) process suggested by (Tibbetts, 1984). As per this process, the solid carbon fibers grow from a super-saturated molten liquid catalyst droplet. Decomposition of the gas phase hydrocarbon molecules provides the carbon necessary for saturation of the molten catalyst. The possibility of gas phase and surface decomposition of the hydrocarbon molecules exists. Melting of metal catalyst particles at the normal synthesis temperature (900 - 1200 K) is plausible only as a result of non-equilibrium processes within the thin surface layer of the particle. Reduction in size of the particles also leads to increased carbon solubility within available process time.

The specific physical form (e.g. MWNT, SWNT, amorphous carbon and particle-encapsulated graphite cell) of the precipitated solid carbon depends on several factors; including catalyst particle size and precipitation rate (Moisala et al., 2003). When the precipitation rate is in equilibrium with or less than the carbon diffusion rate, graphitic layers are formed surrounding the catalytic nano-particles resulting in the thermodynamically most stable carbon forms. When the precipitation rate is larger than the carbon diffusion rate can CNT growth occur. Generally only catalytic nano-particles that are sufficiently small (<20nm) are active for CNT nucleation and growth, with the tube diameter corresponding to that of the catalytic nano-particle. For particles smaller than 20 nm, solid carbon atoms do not precipitate from the apex of the hemisphere but from a circular ring close to one of the diameters of the spherical particle. This accounts for the hollow core characteristic of CNTs with diameter approximately corresponding to that of the catalyst particles. For supported catalysts, formation occurs either by "extrusion" (also known as base growth as shown in Figure 3 (a)), in which CNTs grow upward from the nano-particles that remain attached to the substrate, or by lifting of the catalyst nano-particles by the growing CNT (tip growth as shown in Figure 3 (b)).

As summarized in Figure 3 (c) below, catalyst particle diameter plays an important role in defining the synthesized carbon nano-structure. Particles of the order of 1 nm diameter predominantly form SWNTs (Rao et al., 2001). MWNTs are formed with catalyst particle diameters in the range of 10~50 nm with the number of layers increasing with the particle diameter. Particles with diameters larger than 50 nm are covered with amorphous graphitic sheets often given another visually descriptive name "nano-onion."

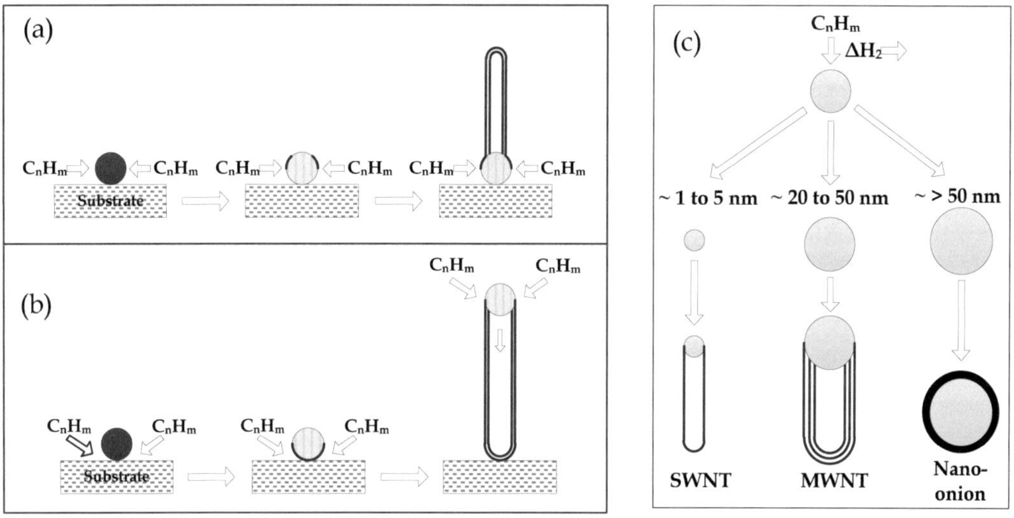


Fig. 3. Mechanisms for carbon nanotube growth (a) Base growth (b) tip growth (c) Structural dependence on catalyst particle size

Dai et al. provided a visually descriptive "yarmulke (Yiddish for skull cap) mechanism," name to the observed growth of MWNTs (Dai et al., 1996). A key characteristic of this mechanism involves nucleation on the catalyst material (e.g. Fe, Ni, and Co) surface by decomposition of gaseous hydrocarbon molecules followed by diffusion of the hydrogen away from the surface. Once the nano-particle is supersaturated with the carbon atoms, the linking of the carbon atoms together in the form of a hexagonal sheet that conforms to the curvature of the particle is energetically favored (Tibbetts, 1984). For experiments in which the nano-particles are supported on a substrate the carbon "yarmulke" grows approximately half the particle diameter towards the substrate and then starts growing longitudinally with the particle remaining at the root. Newly arriving carbon atoms are integrated into the network and the tube grows longer. For experiments involving suspended catalytic nano-particles, the strain of the carbon sheet curving around the nano-particle is higher and conjectured to result in smaller diameter tube formation. A second yarmulke can form underneath the first with approximate spacing between the two equal to the interlayer spacing of graphite crystals (~0.34 nm). As additional yarmulkes grow, one beneath the other, older yarmulkes lift up to form the MWNT. The open ends of these structures remain chemisorbed to the catalytic particle. As the strain resulting from increasing curvature exceeds a certain value, nucleation of new inner walls ceases defining the diameter of the innermost tube. The pre-nucleation step involving saturation of the catalyst particle by dissolved carbon defines the maximum diameter of the catalyst nano-particles for CNT growth.

## 3. Flame synthesis of carbon nanotubes

### 3.1 Experimental investigation

The first experimental observation and conjecture of the formation of filamental carbon in flames (Singer & Grumer, 1959) came long before CNTs were discovered. Formation of CNTs by plasma arc discharge method was first reported in 1991 (Iijima, 1991). In the same year, formation of elongated carbonecous structures on the surface of a probe inserted in methane air diffusion flames (Saito et al., 1991) was reported and presence of C60 and C70 fullerens was detected in premixed flames (Howard et al., 1991). A year later, co-existence of MWNTs with soot like structures was detected in premixed flames (Howard et al., 1992). Prior to these reports, the highly ordered carbon cylindrical structures had been produced only by very energetic processes such plasma and laser vaporization but not in flames. These discoveries prompted scientist to investigate the potential of various types of hydrocarbon flames for synthesis of CNTs.

Flame is a unique synthesis medium that provides both energy and chemical species for the synthesis of CNTs and other carbon nano-forms. Schematics of various flame synthesis processes are shown in figure 4.

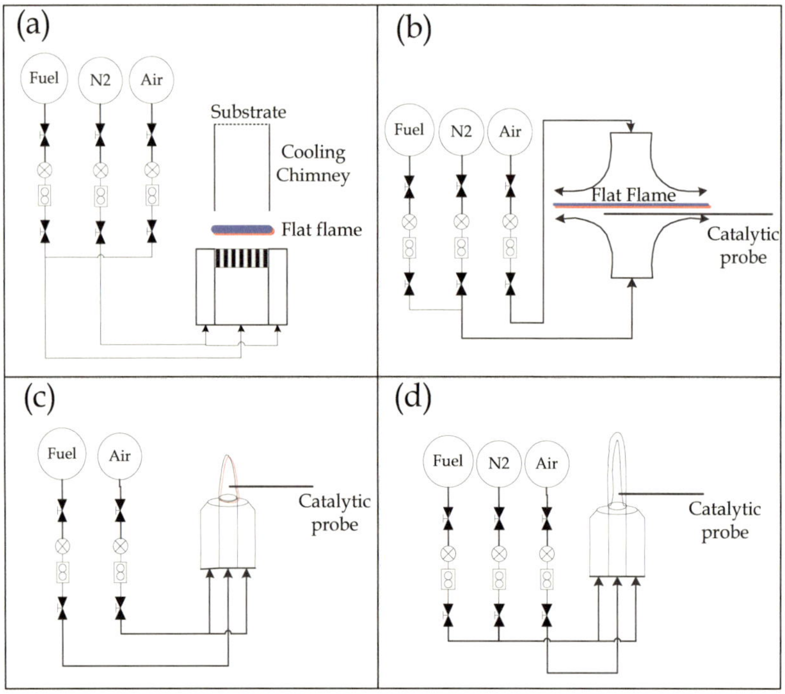


Fig. 4. Schematic for flame synthesis of carbon nanotubes (a) premixed flame (Gopinath & Gore, 2007; Height et al., 2004; Vander Wal et al., 2002) (b) counter-flow diffusion flame (Li et al., 2007; Merchan-Merchan et al., 2003; Saveliev, 2003; Xu et al., 2007)(c) co-flow diffusion flame (Unrau & Axelbaum, 2010; Vander Wal, 2000; Xu, 2007; Yuan et al., 2001; Yuan et al., 2001)(d) inverse diffusion flame(Lee et al., 2004; Unrau et al., 2007; Xu et al., 2006)

In a flame, fuel (generally hydrocarbons such as methane ($CH_4$), ethylene ($C_2H_4$), and acetylene ($C_2H_2$) etc.) reacts with oxidizer ($O_2$ from air) to produce gaseous mixture that includes carbon dioxide ($CO_2$), water vapor ($H_2O$), carbon monoxide (CO), hydrogen ($H_2$), saturated and unsaturated hydrocarbons($C_2H_2$, $C_2H_4$, $C_2H_6$ etc.) and radicals. Hydrocarbons and carbon monoxide constitute the gaseous precursor mixture that is the source of solid carbon deposited on catalyst particles to form carbon nanostructures. Metal catalysts, inserted in the flame either in the form of a substrate coating or as aerosol particles, provide the necessary reaction sites for deposition of solid carbon. Nanotubes are believed to form on the catalyst particle via the same carbon dissolution-diffusion-precipitation mechanism discussed in section 2 above. The structure of the formed carbon nanotube (MWNTs and/or SWNT) depends on the catalyst particle size and carbon deposition rate. Post flame gas phase chemistry, temperature at the surface of the catalyst particle and the structure and type of catalyst particle are the key controlling parameters for growth of nanotubes in the flame synthesis process.

In comparison with the other processes, flame synthesis is an auto-thermal process that is capable of providing temperature optimal for achieving desired synthesis conditions. Flame medium is characterized by a complex homogeneous gas phase kinetics that involves reactions between the fuel and the oxidizer with formation of water, carbon dioxide, and partial oxidation products and fuel pyrolysis with formation of secondary hydrocarbon species such as single and multi-chained hydrocarbons, polycyclic aromatic hydrocarbons, and soot precursors. In comparison with CVD, a flame medium is rich in intermediate radicals that are formed in high concentrations during intense homogeneous gas phase reactions. This homogeneous gas phase kinetics is closely coupled with the heterogeneous kinetics of gas-surface interactions leading to the formation of nano-structured solid carbon.

Post flame gas phase chemistry and temperature are a complex function of fluid dynamics, mass transfer and heat transfer phenomena at play inside a flame. Fuel and oxidizer primarily impact the gas phase composition and the maximum temperature in the flame. Hydrocarbon fuels such as $CH_4$, $C_2H_4$ and $C_2H_6$ when used with oxygen/air as the oxidizer result in unique product gas phase compositions and maximum temperatures (~ 2000 K). The degree of mixing between fuel and the oxidizer that is identified by the parameter of equivalence ratio ($\phi$), generally determines the extent to which the chemical oxidation reactions are complete to stable product gases such as $CO_2$ and $H_2O$. Intermediate product gases such as CO, $C_2H_4$, and $C_2H_2$ and in general $C_nH_m$ provide compositions supportive of solid carbon formation. At steady state equivalence ratio is a spatially varying quantity within the diffusion flame structure. The zone of maximum temperature within the flame is also known as flame front is formed at the location where the local $\phi$ is equal to 1.

Configuration of the flame plays an important role in establishing the fluid dynamics, the mass and energy transfer and the chemistry in the flames. Flames are classified mainly as premixed, non-premixed (diffusion), and partially premixed. Diffusion flames are further characterized by the orientation of the reactant nozzles into co-flow diffusion, inverse diffusion, and counter flow diffusion. All of these flame configurations have been used for carbon nanotube growth. A premixed flame is defined as a flame where the oxidizer and fuel are completely mixed before burning (e.g. Bunsen flame). A co-flow jet burner establishes a diffusion flame with the fuel issued from an inner tube and the oxidizer is injected from an outer tube. When the fuel and the oxidizer are inverted in a co-flow jet burner, an inverse diffusion flame is formed. A counter-flow flame is established from two converging nozzles arranged in an opposed flow configuration with a fixed distance, where oxidizer issued from one nozzle impinges onto the fuel flow issued from the other. Independent of the flame type used, it must provide a source of carbon to form the graphite layers, utilize the catalytic metal nano-particles to form the solid graphitic layers from gas-phase carbon containing molecules, and provide a heat source for forming and activating the catalytic nano-particles. The flow structure of the flame can be laminar or turbulent

based on the Reynolds number. However, only laminar flames have been used for synthesis of carbon nanotubes due to the uniform structure. In the further discussion only laminar flames have been addressed unless explicitly otherwise stated.
Use of inert diluents also affects the flame chemistry and temperature. Nitrogen ($N_2$) and Argon (Ar) have been used as the diluents in many carbon nanotube flame synthesis experiments. The ambient conditions of pressure and temperature also impact the flame and hence the synthesis conditions for carbon nanotubes.
Metal catalysts in the form of both substrate and aerosol have been used for growing carbon nanotubes. Typical catalysts include transition metals such as Iron (Fe), Nickel (Ni), and Cobalt (Co). Alloys of transition with other metals like chromium (Cr), copper (Cu) and zinc (Zn) have been used. In the substrate method, a substrate coated with a catalyst layer is positioned at the appropriate location inside the flame. Catalyst nano-particles are formed on the substrate as a result of flame-substrate interactions. These particles further act as the nucleation site for nanotube growth. Stationary substrates have been typically used for synthesis of MWNTs because of the larger size of catalyst nano-particle (~20 nm). Catalyst can be injected inside a flame in the form of a vapor aerosol. Generally, nitrates of transition metals and metallocenes have been used in flame synthesis. Catalyst particles of the size of approximately ~5nm are formed due the condensation of the catalyst vapor, that are suitable for growth of SWNTs. Catalysts have been found to be very active towards a particular gas phase precursor of solid carbon, and hence structure and type of the catalyst play an important role in determining the growth rate as well as the structure of the CNTs.

### 3.2 Synthesis of MWNTs using flames

MWNTs and larger forms of nanotubes have been successfully synthesized using flames. Substrate type catalysts have been used in most of the experiments because of the inherently closer control over the catalyst formation processes. A variety of flame configurations mentioned above have been used with $CH_4$, $C_2H_4$, $C_2H_2$, $C_3H_8$ and alcohols as the fuels and air or $O_2$-$N_2$ and $O_2$-Ar mixtures as the oxidizer species.

#### 3.2.1 Premixed flame synthesis

Premixed flames offer distinct advantages for CNT synthesis when compared to non-premixed flames. As the mixing of fuel and air occurs before ignition, equivalence ratios can be easily controlled by varying the mass flow rate of the fuel and/or the oxidizer. Premixed burners with a flat radial profile and variation only in the axial direction (e.g. McKenna burner) have been used for nanotube synthesis. The flame temperature can be reduced to an appropriate value by the use of chimneys. Uniform gas flow composition can be obtained by appropriate arrangements in the burner.
The first evidence for filamentous carbon growth in flames was established using premixed flames (Singer & Grumer, 1959). In the last two decades premixed flat flames have been extensively studied for synthesis of carbon nanotubes (Diener et al., 2000; Gopinath & Gore, 2007; Grieco et al., 2000; Howard et al., 1992; Howard et al., 1991; Vander Wal et al., 2002, 2002). Formation of C60 and C70 fullerenes was first observed in premixed flames by Howard et al. In their studies, sooty discharge from premixed laminar flames of benzene, oxygen and argon at low pressures (1.60 to 13.35 kPa) were analyzed using electron impact mass spectroscopy. The results showed presence of C60 and C70 fullerenes that were confirmed by FTIR (Fourier Transform Infrared Spectroscopy). The yields of C60, and C70

and the C70/C60 ratio were found to depend on temperature, pressure, carbon/oxygen ratio, and residence time in the flame. The amount of fullerenes formed in the flame was very low (0.009% to 0.03 % of the soot mass) as compared to that formed in graphite vaporization (1% to 14%). Nonetheless, this finding motivated combustion scientists to pursue the synthesis of CNTs using flames as the precursors (C60 and C70 fullerenes) were found in flames.

Vander Waal et al. carried out a comprehensive study of the MWNT synthesis in premixed flames. Premixed flat flame McKenna burner was used with SS chimney for cooling. Methane ($CH_4$), Ethane ($C_2H_6$), Propane ($C_3H_8$), Ethylene ($C_2H_4$) and Acetylene ($C_2H_2$) were used as fuels with air as the oxidizer. At the top of the chimney, a circular molybdenum ring held the mesh (Stainless Steel) supported catalyst (cobalt) within the post-flame gases. The flame equivalence ratio was varied, adjusting the fuel flow rate to the burner while maintaining a constant air flow rate. The post-flame gas temperature was recorded to be (~1100 K). Meshes were retained in the flame gases for 12 minutes, measured from the time of insertion to extraction. Results of chemical equilibrium calculations were correlated with the experimental measurements to determine the optimal gas phase chemistry for the growth of CNTs. CO was identified as the main gas precursor. Both SEM and HRTEM imaging were used to correlate the nanotube morphology and internal structure to the reaction gas composition. The variations observed were understood in light of the gas composition and the interaction of the reactive components with both the deposited Co catalyst particles and supporting metal substrate. Coated and uncoated (with Co catalyst) meshes were subjected to post flame gases. The uncoated SS meshes resulted in a dense random CNT growth because of formation of catalysts through surface break up. However coated meshes showed a uniform and dense growth confirming the dominance of catalyzed CNT formation. With $C_2H_2$ flames high deposition was found even on uncoated meshes. Most significantly, catalyst particles are observed at many tips using the uncoated SS mesh, characteristic of surface breakup processes.

Gopinath and Gore further investigated the carbon containing gas phase species responsible for deposition of carbon during the synthesis of MWNTs. Premixed flames of ethylene and air were established in a flat flame McKenna burner. Due to the flat flame profile, radial gradients were ignored. This assumption significantly simplified the computational analysis of the flame and post flame chemistry. The yellowish high carbon region just above the flame was found to be best suited for the gas phase chemistry required to encourage CNT growth. In order to enable the required gas phase chemistry and temperature to exist around the catalyst substrate, a chimney was placed just above the flame. The chimney served two functions, (a) it prevented the optimal gas phase chemistry of the near flame zone from dissipating and (b) it provided the necessary wall losses to quench the post flame environment to attain the ideal temperature for rapid CNT growth. A cooling chimney was used to cool the flame products and obtain appropriate post flame temperatures for CNT synthesis. A $N_2$ co-flow stream was employed to stabilize the flat flame. The air flow rate was held constant at 11.5 lpm and the equivalence ratio ($\phi$) was varied by changing the fuel flow rate. A 2 nm thin layer of cobalt catalyst was deposited commercially, using a physical vapor deposition technique, on SS304 200 mesh standard TEM grids. The experimental arrangement is shown in figure 5 (a). The gas phase temperature and substrate temperature were measured using a thermocouple and found to be within 10 K of each other at 1100 K under steady state conditions confirming the observations by Vander Wal.

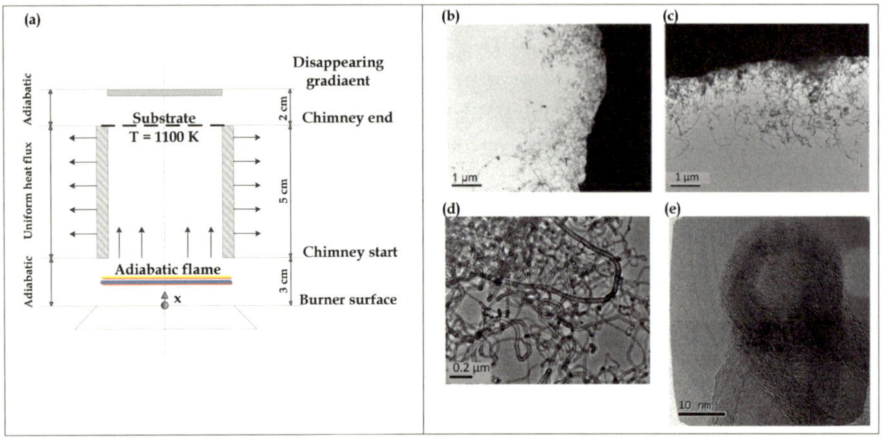


Fig. 5. Experimental synthesis of carbon nanotubes using premixed flames (a) Experimental arrangement, (b) & (c) TEM images of nanotube growth on cobalt catalyst at $\phi$ = 1.55 (d) HRTEM image of MWNTs at $\phi$ = 1.55 (e) HRTEM image of closed MWNT (Gopinath & Gore, 2007)

Figure 5 shows the TEM images of the carbon nanotubes synthesized during the experiment at the equivalence ratio of 1.55. Figure 5 (b) and (c) show the low magnification images of the CNTs. Optimum yield of nanotubes was produced at this particular value of equivalence ratio. As shown in figure 5 (d) and (e), the magnified images reveal the multi-walled structure of nanotubes. The well-graphitized structure is evident from this image. These observations were found to be consistent with the experiments performed by Vander Wal.

### 3.2.2 Diffusion flame synthesis

Formation of fibrous carbon in diffusion flames was first observed by Saito et al. (Saito et al., 1991; Saito et al., 1986) while conducting soot characterization studies on methane air diffusion flames. The growth was observed beyond a certain height above the burner with associated color change from brown to black. Later, Yuan et al. completed detailed characterization studies using methane- air (Yuan et al., 2001), ethylene-air and $N_2$ diluted ethylene-air flame synthesis (Yuan et al., 2001) of CNTs. The catalytic supports (Ni-Cr) used were in the form of wires and grids undergoing an oxidation process, and grids pre-loaded with Co nano-particles.

In the experiment done with methane by Yuan and co-workers, SEM images suggested that most nanotubes have a particle attached at the base near the substrate. Since the particles were not seeded in the flame and the images indicated their presence it is apparent that the particles are lifted from the surface. The base location of most of the particles supports the base growth model for the CNTs. Nevertheless, some particles were found at the tip of the nanotube indicating catalyst surface breakup. This phenomenon has been reported by Vander Waal et al. in their reports on premixed flame synthesis especially with an uncoated stainless steel mesh. Soot was formed when only Ni-Cr wire was held in the flame without any support mesh. This indicated that the stainless steel mesh may be essential for the formation of CNTs. Soot was found to grow over a broad range of conditions in the flame whereas CNTs grow in a narrower region in the presence of a catalyst. The optimum harvest conditions were observed within non-dimensional physical locations between h/H = 0.2 to 0.3 and r/R = 0.6 to 0.9. The temperature of nanotube formation was found to be around 1520 K. The CNTs were collected on Ni-Cr wire whereas brown deposits were formed on the stainless steel mesh later identified as iron oxide. CNT formation was observed even at low catalyst concentration indicating that soot formation and CNT growth may be competing phenomenon in the flame. Optimum region for CNT growth was found to be in the region of minimum oxygen concentration. However, the rate of catalyst particle formation was found to be low. Thus, it was suggested that oxygen might play an important role in the formation of metal catalysts. However, high concentrations of oxygen might lead to oxidation of the CNT precursors and incipient CNTs. The zone of temperature was also determined to be a critical parameter for the synthesis of CNTs.

With the ethylene flames Yuan and coworkers tried to study growth mechanism for the carbon nanotubes inside the flame. The effect of $N_2$ addition to the flame on the growth of nanotubes was assessed. Soot instead of CNT formation occurred when a bare stainless steel mesh was used as the substrate. However, when pre-oxidized substrate was used CNT growth was visible indicating the criticality of formation of metal oxides for CNT growth. When Ni-Cr wire similar to methane experiment was placed in the ethylene flame amorphous carbon growth was visible that signified the difference between formation mechanism of CNTs in methane and ethylene flames. The deposition rate for gray material was found to be more than 3 mg/min. With increase in sampling time the production rate of amorphous carbon increased leading to eventual solidification of the tube due to deposition of carbon on the nanotube walls. The increase in the thickness was attributed to deposition of pyrolytic carbon on to the carbon nanotube walls. Addition of $N_2$ to the flame resulted in decreased synthesis temperature (from 1820 K to 1517 K) and carbon gas concentration that led to fewer nanotubes. However, more uniform nanotubes resulted in the presence of $N_2$. Cobalt coated grids resulted in well aligned and uniform CNTs with diameters well correlated to the catalyst particle. Similar observations related to the optimum temperature for nanotube synthesis were made by Lee et al. (Lee et al., 2004) for an ethylene air inverse diffusion flame. It was observed that when the gas temperature was varied from 1400 to 900 K, well-aligned MWNTs with diameters ranging from 20 to 60 nm were formed on the probe's surface. Ni was used as catalyst in the form of $Ni(NO)_3$ particles pre-loaded on the substrate. Reduction up to 60% in melting temperature of the transition metal due to small particle size of the bulk value has been reported (Moisala et al., 2003; Petroski et al., 1998). This fact was used to explain the formation of active catalyst particles in the temperature well below the bulk melting temperature of Ni (~1726 K).

Xu et al. (Xu et al., 2006)examined the effect of different types of catalysts on growth of carbon nanotubes in a methane air inverse diffusion flame. They tried to correlate composition of the catalyst with the observed morphology of the carbon nanotube. A methane air inverse diffusion flame of total height of 15 mm was established. Temperatures and concentrations were determined, at various radial locations at particular heights from the burner, using spontaneous Raman spectroscopy. Optimum range for nanotube growth was found to be at a height of Z = 12 mm and radius r between 2 - 4 mm. Peak concentrations of CO and $H_2$ were found in the optimum synthesis range. Ideal temperature range for CNT growth was reported to be in between 1200 - 1400 K.

Counter flow diffusion flames are been increasingly used for synthesis of MWNTs due to their 1-D geometry and convenience in positioning the catalyst substrate in the flame (Hou et al., 2009; Li et al., 2007; Merchan-Merchan et al., 2003; Merchan-Merchan et al., 2004; Merchan-Merchan et al., 2002; Merchan-Merchan et al., 2009; Saveliev, 2003; Xu et al., 2007). Merchan-Merchan et al. (Merchan-Merchan et al., 2002) recorded the formation of CNTs in a methane oxygen counter diffusion flame without any catalysts. They employed an atmospheric, opposed flow burner with $N_2$ co-flow in which the oxidizer was enhanced to 50% oxygen or greater. High resolution SEM and TEM images revealed soot like structure with presence of carbon nano-particles and nanotubes however, no catalyst particles were found embedded in the soot like structure. The tube diameter and length were approximately 20 and 320 nm, respectively. The distribution of the sizes of nano-particles and nanotubes was found to be bimodal, indicating that both structures originated in similar sized solid carbon precursor seed. Presence of nano-particles and nanotubes inside soot like structure pointed towards a similar mechanism responsible for formation of all three structural forms. Currently, oxy-flames are being pursued for CNT synthesis (Hou et al., 2009; Merchan-Merchan et al., 2009) due to the high temperature and radical concentration obtained at the flame location.

### 3.3 Flame synthesis of SWNTs

Similar to the synthesis of MWNTs, a combustion system tailored with an ideal source of carbon, heat source, and appropriate catalytic material, can result in the production of single-walled carbon nanotubes. In the flame method, the catalytic precursors are generally introduced into the flame system in the gas-phase and nucleate and condense to solidify into spherical metallic nanoparticles. Flame parameters can be used to obtain an appropriate flame environment that would allow the formation of ideal sizes of catalytic particles for carbon nanotube inception and growth. The available literature on the flame synthesis of SWNTs is scarce, in contrast to flame synthesis of MWNTs, consisting of only a handful of experiments that have been conducted on the synthesis of SWNTs.

To some extent all products obtained in the SWNT synthesis experiment have common morphological trends; even though they are synthesized in flames formed using different burner configurations and conditions. These morphological trends include (Merchan-Merchan et al., 2010): (i) SWNTs always coexist with metallic and/or soot particles, (ii) particles often appear to be poisoned; even when ultra small catalytic particles, ideal for SWNT inception, can be achieved, they can be heavily encapsulated with amorphous carbon becoming inactive as catalysts for nanotubes, (iii) the presence of larger metallic particles with very short SWNTs.

Vander Wal studied the effect of catalysts in aerosol form on the growth of CNTs (Vander Wal, 2002). Primarily SWNTs were grown on aerosol catalyst particles using an acetylene air flame. Same flame configuration was used except the catalyst in form of Fe(III) nitrate ($Fe(NO)_3$) vapor dissolved in a solvent was introduced through a nebulizer. Absolute ethanol was found to be the optimum solvent for the catalyst. The experiment was directed towards identifying the correct precursor for the SWNT growth by introduction of pyrolysis gas mixtures ($CO/H_2/He$ and $C_2H_2/H_2/He$) and studying the effect of catalyst particle size on the growth of SWNT. Higher CO concentrations led to metal particles becoming encapsulated within amorphous carbon. There appears to be a minimum limit for presence of CO and $H_2$ and maximum limit for presence of $H_2O$ for the production of SWNT synthesis. Increase in catalyst vapor concentration led to increased particle size, making them ineffective for fullerenic growth. Therefore, a need for appropriate gas phase precursor and catalyst particle size was identified for SWNT growth. $C_2H_2$ was found responsible for poisoning of catalysts and presence of $H_2$ was deemed essential for etching of the catalyst particle.

Height et al. (Height et al., 2004) studied the transitional conditions between soot formation and CNT formation and the effect of operating conditions on structure of nanotubes. Optimum zone for equivalence ratio was identified that was required for formation of SWNTs. A premixed $C_2H_2/O_2$ flame with argon dilution of 15 molar percent, cold gas feed velocity of 30 cm/s, and burner pressure of 6.7 kPa formed the basis of the experiments. Iron penta-carbonyl ($Fe(CO)_5$) vapor was used as the catalyst. Carbon nanotubes were formed as the distance above the burner surface is increased. A nanotube formation window for equivalence ratio was anticipated with upper and lower limits determined by sooting and carbon availability factors. Flames with equivalence ratios between 1.4 and 2.0 were examined, with samples extracted at 70 mm HAB (approx. 53 ms). Multistep mechanism for nanotube formation in flames was recognized. Post flame gas chemistry and formation of appropriate size catalyst particles were identified as the most critical steps. An order of magnitude growth-rate for the nanotubes in this interval is between 10 and 100 μm per second. Optimal condition for SWNT growth is around $\phi$ of 1.6 and appropriate size of catalyst particles.

The growth mechanism for SWNTs has been found to be very similar to the mechanism for other forms of solid carbon like soot. It has been well known that the precursors for soot are Polyaromatic Hydrocarbons (PAH) that are formed through the breakdown of $C_2H_2$. However, presence of high concentration of $C_2H_2$ causes the catalyst particle to be coated with amorphous carbon inhibiting the growth of SWNT. An earlier abundance of carbon species might poison the particle and prevent CNT inception earlier on in the flame volume. Therefore, following occurrences can affect the formation of SWNTs in a flame (Diener et al., 2000): (i) soot formation begins at a time where the metal particles have not yet grown large enough to act as a SWNT catalyst; (ii) catalytic particles with suitable size are synthesized but the large amount of acetylenic species poison the catalytic particles preventing their activation and inception of SWNTs.

Even though all the above mentioned experiments were conducted with fixed flame parameters and single catalyst material, the synthesized forms of carbon nano-materials is found to change dramatically. It is observed that change of flame position induces variation in macro-morphology and in the microstructure of the formed carbon nano-materials. The modification of growth conditions is directly related to variation of the flame environment pertinent to the specific flame location. Temperature, radical and hydrocarbon concentrations are strong functions of axial position in the flame. Availability of specific hydrocarbons at given flame location alters the growth mechanism leading to the selective production of various nanoforms. Hence, there is a need for more fundamental study related to the establishment of optimum growth region and the associated structure of carbon nanotubes inside a flame environment.

## 4. Growth controlling parameters

As mentioned previously, gas phase composition, temperature and the catalyst are the three major factors that determine the optimum region for carbon nanotube growth inside a flame. Careful control of these variables can result in a high yield rate of pure carbon nanotubes when compared to other synthesis methods. In this section, effect of each of these variables on the carbon nanotube growth is outlined.

### 4.1 Gas phase composition inside a flame

Carbon nanotubes are formed when carbon in gaseous form is deposited in form of the structured solid on to a catalyst particle. The concentration of gaseous precursors and the resulting deposition rate play an important role in determining the structure of the nanotube that is dependent on the concentration of gaseous precursor. These gaseous precursors are formed through the complex phenomena that occur inside a flame.

#### 4.1.1 Fuel type and equivalence ratio

At steady state, the concentration of gaseous precursors in a flame is a function of the type of fuel, configuration of the flame and the local equivalence ratio ($\phi$). Fuel rich flames ($\phi > 1$) are utilized for nanotube growth. In a premixed flat flame the equivalence ratio is uniform. Therefore, premixed flames have been studied to understand the effect of equivalence ratio on nanotube growth (Gopinath & Gore, 2007; Height et al., 2004; Vander Wal, 2000; Vander Wal et al., 2002).

Vander Waal et al. characterized the equivalence ratio range that may be ideal for CNT growth for different fuels. They found that methane did not produce any significant CNT growth. Ethane produced various nanostructures for equivalence range of $1.52 < \phi < 1.9$. For ethylene the growth started at $\phi = 1.50$ and best results are obtained at equivalence ratio of 1.62. Acetylene provided high growth of CNTs at all equivalence ratios. However, the growth was non-uniform and MWNTs with large diameter were obtained.

Gopinath and Gore observed that CNTs with maximum yield and best morphology were produced with ethylene flame for the equivalence ratio range of $1.5 < \phi < 1.6$. For a richer equivalence ratio ($1.62 < \phi < 1.75$), the yield of CNTs fell substantially. For a leaner equivalence ratio ($1.47 < \phi < 1.49$), the yield of CNTs was less than the maximum yield range, even for longer residence times. For $\phi < 1.45$, no CNTs were observed to grow and for $\phi > 1.75$, non- CNT structures of low yield were found to be predominant.

In the experiments carried out by Height et al. with $C_2H_2/O_2/Ar$ flames for growth of SWNTs, they examined the flame for equivalence ratio range of $1.4 < \phi < 2.0$. Nanotubes were observed to form between $1.5 < \phi < 1.9$. For low $\phi$ ($1.4 < \phi < 1.5$) the condensed material (particles and nanotubes) in the flame was dominated by discrete particles. For $\phi$ of 1.9 and higher, soot-like structures were found to dominate with clustered networks of primary particles ranging in size from 5 to 20 nm.

From these observations it is seen that for ethylene and acetylene there is a general agreement on the range of equivalence ratio ($1.5 < \phi < 1.8$) optimal for CNT growth. However for other fuels like (methane) there is no consensus. This suggests that the growth of CNTs is a function of gas phase concentrations of carbon containing gases which is in turn a combined function of equivalence ratio and type of fuel used.

In case of diffusion flames the mixing and hence the equivalence ratio is determined by the mass transfer due to diffusion between fuel and oxidizer streams. Hence the equivalence ratio is a function of spatial location and is difficult to measure experimentally. Thus the CNT growth region cannot be directly related to the equivalence ratio. However, when similar conditions were used by Yuan et al. (Yuan et al., 2001; Yuan et al., 2001) for CNT growth with ethylene and methane, ethylene deposited amorphous carbon whereas methane deposited CNTs. This observation suggests that, also in case of diffusion flames CNT growth gas composition in the flame.

Vander Wal et al. (Vander Wal et al., 2000) also noted that dilution of fuel with an inert like $N_2$ or Ar might be critical to the nanotube synthesis. Absence of diluent resulted in soot formation and encapsulation of the catalyst nano-particle with amorphous carbon. Yuan et al. (Yuan et al., 2001)observed that addition of diluent ($N_2$) reduced the temperature in the synthesis region that further resulted in reduced but more uniform yield of CNTs. Addition of diluents leads to altered flame structure which in turn can affect the gas phase composition.

From the discussion above, it is clear that, there remains a need for characterizing different flames for similar gas phase composition that is favorable for CNT growth.

### 4.2.2 Gas phase precursors: CO and $C_2H_2$

In a flame environment, various carbon containing gas phase species are formed that are responsible for deposition of solid carbon. The two main contributors are hydrocarbons ($C_nH_m$) and carbon monoxide (CO).

Hydrocarbons decompose at high temperature to form solid carbon. With increase in number of carbon atoms ($CH_4$, $C_2H_6$, $C_3H_8$) these compounds become unstable. Methane is the most stable hydrocarbon that begins to decompose at 1200 K. Un-saturated hydrocarbons such as $C_2H_2$ and $C_2H_4$ are very susceptible to decomposition due to presence of disruptive $\pi$ bonds. $C_2H_2$ is found to be unstable even at room temperature. In fact, $C_2H_2$ is one of the main precursors for soot formation inside flame. It also contributes primarily to the formation of carbon nanotubes if sooting conditions are avoided. Very fast decomposition of acetylene is the main cause of catalyst deactivation due to encapsulation by amorphous carbon.

Carbon monoxide (CO) participates in deposition of solid carbon via the Boudard (CO disproportionation) reaction and the hydrogenation reaction that are shown in equation (1) and (2) respectively.

$$2CO_{(g)} \leftrightarrow C_{(s)} + CO_{2(g)} \qquad \Delta H = -171\, kJ/mol \tag{1}$$

$$CO + H_2 \leftrightarrow C_{(s)} + H_2O \quad \Delta H = -131\, kJ/mol \tag{2}$$

The decomposition rate of CO disproportionation reaction is found to be low when compared with the acetylene decomposition, making it the ideal precursor for SWNT formation. Based on thermodynamic equilibrium, temperature range of 800 - 1100 K has been found to be ideal for CO disproportionation at normal pressure (Moisala et al., 2003). However, this range may not be ideal for catalyst particle formation, carbon dissolution and carbon precipitation. CNT yield is found to increase with increase in the CO pressure.

$C_2H_2$ and CO exhibit preferential activity towards certain catalysts. Comparative studies between these two gaseous precursors (Vander Wal, 2002; Vander Wal & Hall, 2001) in pyrolysis flames indicated that CO reacts with Fe based catalyst through carbide formation whereas $C_2H_2$ is active towards Ni based catalysts. Particle size plays a critical role towards determining the catalytic activity. Fe nano-particles of all sizes are generally inactive (toward nanotube synthesis) within $C_2H_2$ mixtures.

To assess the effect of carbon precursor on nanotube growth, gas phase chemistry in a premixed flame was studied by Vander Wal et al. (Vander Wal et al., 2002). Concentrations of various gases in the post flame environment were determined using gas phase equilibrium calculations. Water gas shift reaction was assumed to be at equilibrium. Based on the experimental results and calculations a strong relation emerged between the optimum CNT synthesis conditions and the concentration of CO and $H_2$. An optimum window for nanotube synthesis based on CO and $H_2$ concentrations was deduced as shown in figure 6 (a). The concentration of $C_2$ species was found to be negligible in comparison to CO based on detailed chemistry calculations. Hence, CO was considered to be the main carbon source. However, the study lacked a comparison between the amount of solid carbon deposited and the gas phase carbon present in various species, making the above assumption speculative. The post flame temperatures were found to be constant irrespective of variation in the adiabatic flame temperature. With an identical $H_2$ concentration, there was a dramatic increase in the CNT yield with increases in CO concentration. At very high CO concentrations (in case of $C_2H_2$ flames) PAH and soot formation may result in the encapsulation of the catalyst material and reduction in CNT yields. At very high CO concentrations, surface carbon builds up to form an inactive layer on the catalyst surface (coking layer) without a carbon removal mechanism. Once formed, such a layer prevents further contact with carbon gas-phase species and thus stops the carbon atom supply.

Detailed chemistry calculations instead of equilibrium were performed by Gopinath and Gore (Gopinath & Gore, 2007) for a similar premixed flame arrangement. The hypothesis of water gas equilibrium at the CNT synthesis conditions was assessed. The effect of variation in equivalence ratio on the substrate temperature was found to be negligible confirming the observations by Vander Wal. The effect of changing equivalence ratios on CNT yield was interpreted, based on gas phase chemistry, using chemical kinetics computations. A one-dimensional premixed flame code with a post flame heat loss model, including detailed chemistry, was used to estimate the gas phase chemical compositions in the region of interest. The CNTs formed were in a very small amount even at the highest yield location. Hence it was difficult to quantitatively relate the change in the gas phase chemistry to CNT synthesis. Comparison of variation of concentration with equivalence ratio for different gas phase species was done to assess the effect of gas phase chemistry. Significant rise of up to 10 orders of magnitude was found to occur for $C_2$ hydrocarbons and up to 6 orders of magnitude of $CH_4$ relative to hydrogen mole fraction near the maximum yield equivalence ratio. On the other hand, the trends for CO and $H_2$ concentration showed monotonic variation with equivalence ratio. Rapid departures from the partial equilibrium of the water gas shift reaction and rapid changes in mole fraction ratios of $C_2$ unburned hydrocarbon to $H_2$ were observed in the range of equivalence ratios suitable for CNT growth. Based on this observation, it was argued that $C_2$ species might play an important role in carbon deposition as compared to CO. The slow kinetics of the CO disproportionation reaction at the experimental conditions was found to be in favor of the argument. Based on chemical kinetics calculations the optimum window was found to shift slightly to the lower side (as shown in Figure 6 (b) with maximum CO concentration ~10 %). Water gas shift reaction was found to diverge from the equilibrium at the CNT synthesis conditions. Hence the need of detailed chemistry calculations was established to correctly assess the effect of gas phase chemistry on the synthesis of CNTs in flames.

From the discussion it is evident that CO and $C_2H_2$ both contribute to the nanotube synthesis. However, further experimental and computational studies are required to examine the competing effects of $C_2H_2$ and CO on solid carbon formation in a flame.

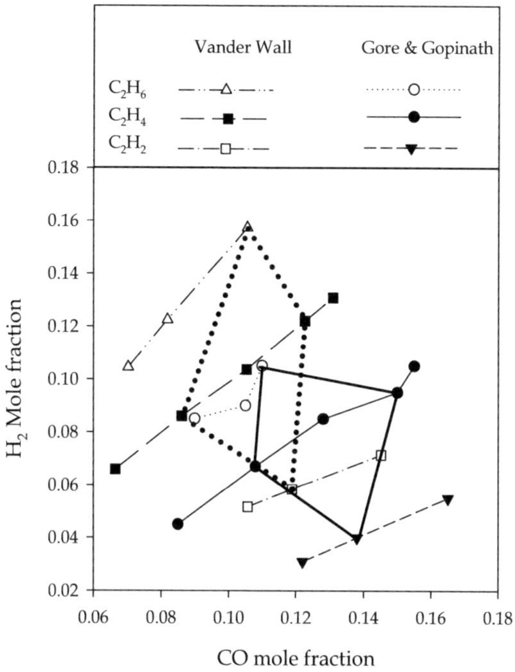


Fig. 6. Optimal CO and H2 conditions for CNT synthesis based on partial equilibrium calculations ........ (Vander Wal et al., 2002) and detailed chemistry calculations _______ (Gopinath & Gore, 2007)4.2.3 Effect of other gas phase constituents: $H_2$, $CO_2$ and $H_2O$

Other gas phase constituents of the flame chemistry also have significant effect on CNT growth. $H_2$ acts as the primary etching agent to remove surface-adsorbed carbon. Higher $H_2$ concentrations greatly facilitate the catalysis of uniform and graphitic CNTs. However, very high concentrations of $H_2$ could compete for adsorption sites on the catalyst surface, thereby slowing the rate of CO adsorption and subsequent carbon supply to the CNT. $H_2O$ and $CO_2$ participate indirectly in the CNT synthesis by altering the water-gas shift equilibrium. $CO_2$ is not known to contribute to the surface reactions nor can to interfere with adsorption of $H_2$ or CO. $H_2O$ interacts with the adsorbed carbon on the particle surface. Gasification of surface carbon by $H_2O$ at elevated temperatures can also hinder CNT growth. Thus CNT synthesis is hindered at lean conditions due to higher concentrations of $H_2O$, $CO_2$.

### 4.3 Temperature

Temperature is one of the important parameter that governs the growth of carbon nanotubes inside a flame. Evolution of temperature field and gas phase chemistry occurs simultaneously inside a flame, because of the coupled energy and mass transfer phenomena. The gas-solid interactions responsible for deposition of carbon are endothermic and thus are favored by high temperature (~ 1000 K). Flame environment provides an inherent source of heat and thus high temperature which makes it one of the ideal candidates for CNT growth. However, the temperature field inside the flame shows a large variation ranging from ~ 2000 K near the flame front to the cooler regions of ~ 800 K. Thus, appropriate regions of the flame need to be probed for the growth of carbon nanotubes. Li et al. (Li et al., 2009) tried to characterize the effect of temperature on the CNT growth in counter-flow flames. According to their study the ideal range for CNT growth was found to be 1000-1200 K which is also comparable to the temperature range of CVD synthesis method. At this location in the flame the concentration of various carbon precursors is also found to be high. Similar observations have been reported by Vander Wal et al. (Vander Wal et al., 2002) and Gopinath and Gore (Gopinath & Gore, 2007) in case of premixed flames. Xu et al. also reported similar temperature range around ~1200 K for synthesis of MWNTs in an inverse diffusion flame.

The gas phase composition and temperature range discussed above are typically measured at scales much larger than those corresponding to that of the C60 molecule inception, growth and organization. Therefore, the ranges defined above can be deceptively broad and yet support a narrower range at the micro scale. In situ measurements of carbon growth species such as $C_2H_2$ and abstraction and addition processes involving H and C atoms are needed to develop nano scale and micro scale models.

### 4.4 Catalyst

Mainly, transition metals such as Iron (Fe), Nickel (Ni), Cobalt (Co) have been employed for CNT synthesis in a flame. Alloys of these metals with metals like Copper (Cu) and Zinc (Zn) have also been used. Properties of these catalyst materials can be found in literature (Moisala et al., 2003). Fe has the highest melting point (~1800 K) amongst the transition metal catalysts and is found to be reactive towards CO as compared to $C_2H_2$ (Baker et al., 1972; Baker et al., 1973; Vander Wal, 2002, 2002). It is found to be active even at smaller particle diameters (~1 nm) and hence has been successfully used for synthesis of SWNTs. It supports formation of MWNTs at sufficiently high temperature and CO concentration (Xu et al., 2006). MWNT formation has been extensively reported on stainless steel (Baker et al., 1973; Soneda & Makino, 2000; Vander Wal et al., 2002). Nickel has the lowest melting point amongst the transition metals (~1725 K) and is active towards $C_2H_2$ as compared to CO. It has a diametrical size threshold of ~5 nm above which it becomes active towards solid carbon deposition. Ni has been used for growth of MWNTs in flames with other carbon forms. Nickel when used as a substrate inside a flame undergoes surface break-up similar to Fe. Usually, with nickel the catalyst particles are found on the tip of the nanotubes indicating tip growth through surface growth. The MWNTs produced by nickel are well aligned as compared to Fe, however when subjected to high concentration of $C_2H_2$ nickel particles are poisoned due amorphous carbon deposition.

In case of substrate catalyst, the required catalytic nano-particles responsible for CNT growth are formed inside the flame primarily through the probe surface breakup induced by surface carbonization. The surface breakup occurs due lattice stress mismatch which a result of oversaturation of metal surface with solid carbon. This formation mechanism of catalyst nano-particles generally creates a wide variety of sizes and geometries (Baird et al., 1974; Moisala et al., 2003; Soneda et al., 2002; Soneda & Makino, 2000) which are determined by various factors, such as temperature, chemical species, and carbon solubility of the metal. Two other possible mechanisms for the direct formation of nano-particles on a metal probe surface in flames are OH oxidation-hydrogen reduction and evaporation–condensation. The inherent presence of oxygen-bearing species (e.g. OH) near the flame front on the fuel side can lead to local oxidization of the metal probe. However, these two mechanisms contribute little to nano-particle formation due to limited oxygen content and lower temperature. Furthermore, carbon-bearing species are overwhelmingly dominant in the local flame structure such that the surface breakup mechanism dominates for nano-particle formation.

In case of aerosol form, vapor molecule containing catalyst particle (nitrates, carbonyls or metallocenes of transition metals) undergoes rapid decomposition inside the flame to form the atoms of transition metal catalyst. These atoms coagulate further downstream in the flame giving rise to metal catalyst particle distribution that form the active sites for nanotube growth. The particle sizes generated in this case are of the order of ~ 5nm that are suitable for SWNT growth. In fact, simultaneous growth of SWNT and the catalyst particle has been reported (Vander Wal, 2002). Ferrocene and nickelocene were used as the catalyst precursors for formation of catalyst particles. It was observed that it was difficult to form catalyst particle sizes above 5nm by the aerosol method due to large number of particles needed for coagulation (5 nm particle corresponds to ~$10^4$ atoms). It has been observed that Nickel becomes active towards nanotube synthesis for particle sizes above 5 nm. Thus it has been difficult to synthesize SWNTs using Ni.

## 5. Mathematical models for growth of CNTs in flames

Growth of carbon nanotubes in a flame environment is a multistep multi-scale phenomenon. Mathematical modeling of the various processes occurring during the synthesis is essential for gaining predictive capability and minimizing the number of experiments. As shown in figure 7 various processes occur either simultaneously or consecutively at different length scales. The two main spatial scales can be identified as (i) the bulk scale that includes the entire flame ~ 10 cm (as shown in figure 7 (a)) and (ii) the catalyst particle scale where the deposition of solid carbon occurs at small-scale ~ 100 nm (as shown in figure 7 (c)). The growth process can be broken down into following steps, (a) establishment of the flame that acts as the source for gaseous precursors and heat (b) simultaneous formation of catalyst particles and growth of carbon nanotube occurring at nano-scale. Each of the above mentioned process requires mathematical modeling along with a model to couple the models at various length scales.

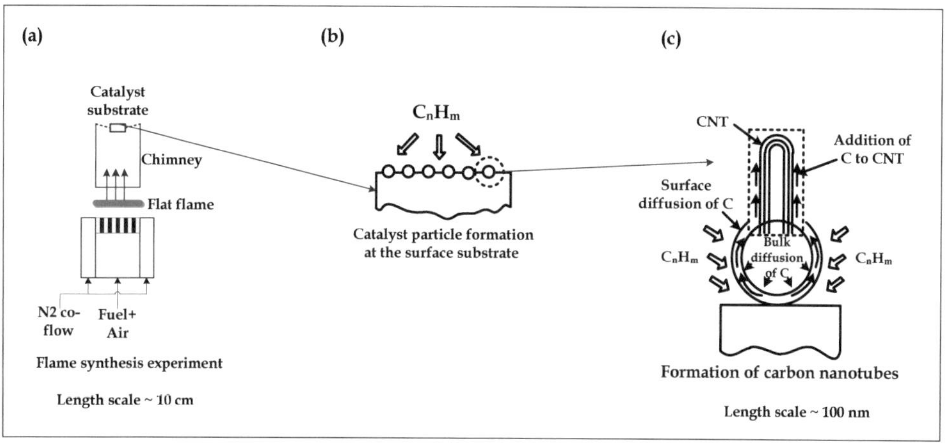


Fig. 7. Formation of carbon nanotubes in a flame environment

Accurate mathematical description of advection, diffusion and chemical processes is necessary in order to predict the gas phase composition and temperature in the flame environment. The fluid dynamics of the flame is captured through the solution of Navier-Stokes (NS) equations. Energy and species equation need to be solved simultaneously with the NS equations to completely capture the mass and energy transfer. Most of the reported experiments have used laminar flames for nanotube growth. Thus, a 2-D axisymmetric CFD calculation may be sufficient for capturing the flow in most of the cases. As described in section 3, in case of premixed flames, or counter flow diffusion flames, flat flame profiles have been employed in CNT synthesis. For these flames, even a 1-D formulation can provide acceptable solutions, where in which 1-D continuity and momentum equation are solved along with energy and species equations. The equation of state provides the required relation between density, pressure and temperature. CHEMKIN packages based on this philosophy (e.g., premix and oppdiff) and other similar ones have been extensively used to model premixed and counter diffusion flames. The chemistry is modeled via detailed chemical kinetics mechanisms. However, in some of the growth experiments these flame environments have been modified by use of cooling chimneys (Gopinath & Gore, 2007; Vander Wal et al., 2002) in case of premixed flames and by the insertion catalyst substrates in case of counter flow flames (Li et al., 2007; Merchan-Merchan et al., 2003; Xu et al., 2007). In these situations, 2-D simulation that accurately captures the development of boundary layer along the walls needs to be used for better prediction of quantities inside the flame and post flame environment. In case of co-flow and inverse flow diffusion flames 2-D computations coupled with mixture fraction formulation comprise a good solution strategy. As the flow is inherently 2-D in these flames, one dimensional solution does not fare well. Axisymmetric combustion codes like UNICORN have been successfully used to model axisymmetric diffusion flames (Katta et al., 2005; Unrau et al., 2010). Complete mathematical description of the flame environment provides the required information of the gas phase composition and temperature at the bulk reactor scale. This information is further used to calculate various parameters related to the growth of carbon nanotubes.

Growth of carbon nanotubes at the catalyst particle surface through the deposition of solid carbon constitutes the next length scale i.e. the particle scale. The literature related to the flame synthesis of CNTs/CNFs is overwhelmingly experimental and provides observations rather than fundamental explanations and models (Naha et al., 2007). However, there have been few good attempts of describing the growth process of carbon nanotubes inside a flame (Celnik et al., 2008; Naha & Puri, 2008; Naha et al., 2007; Unrau et al., 2010; Wen et al., 2008; Zhang & Smith, 2005).

Development of gas phase chemistry and formation of catalyst particles occur simultaneously in a flame environment as shown in figure 7 (b). Catalyst particle coagulation in the aerosol form needs to be mathematically described along with the gas phase chemistry (Celnik et al., 2008; Kuwana & Saito, 2005; Wen et al., 2008). Kuwana and Saito (Kuwana & Saito, 2005) have described nano-particle growth from ferrocene in which they provided a two-step catalytic reaction model for the formation of Fe nano-particles. Wen et al. (Wen et al., 2008; Wen et al., 2007) employed a sectional method developed on simultaneous particle and molecule modeling (SPAMM) approach developed by Pope and Howard (Pope & Howard, 1997). In this approach, catalyst particle formation is modeled as reactions, that can be incorporated into a gas-phase reaction mechanism and allow for the simultaneous modeling of the gas-phase chemistry and the nano-particle formation processes.

Growth of carbon nanotube at the catalyst particle can be visualized as a collection of following process, (a) diffusion of carbon precursor gas species and their conversion to solid carbon (b) diffusion of the solid carbon over the surface of the catalyst particle and through the bulk, (c) encapsulation of the catalyst particle by the solid carbon and (d) nucleation and growth of the carbon nanotube. A schematic of these processes occurring together is shown in figure 7 (c).Diffusion of gaseous carbon species to the catalyst particle followed by the deposition of carbon is usually assumed to occur at the leading face of the particle. In most of the studies, this combined process has been modeled in terms of impingement rate based on the kinetic theory (Naha & Puri, 2008; Naha et al., 2007; Wen et al., 2008). It can also be modeled in terms of reaction kinetics for the dissociation reaction of the individual gas species as mentioned in section 4.1. The deposited carbon is also desorbed, which can be described in an Arrhenius type equation. The solid carbon remaining at the leading face of the particle can now diffuse either along the surface of the catalyst nano-particle or through the particle. Surface diffusion leads to coating of the particle by the solid carbon. This process can be captured through a rate equation (Naha et al., 2007; Zhang & Smith, 2005). On the other hand the diffusion of solid carbon through the bulk of the particle establishes a concentration gradient across the particle. The concentration of solid carbon inside the catalyst particle can be obtained by solving the unsteady mass diffusion equation over the volume of the particle. At the trailing face of the catalyst particle, CNT nucleation can be considered to occur due to precipitation of the bulk diffused solid carbon. The nucleation and growth of the CNTs can be modeled based on a critical threshold cluster density number similar to the soot growth mechanism.

In conclusion, the mathematical description of the CNT growth process, involves large temperature and species variation occurring at the scale of flame environment, whereas the processes at the nano-scale are determined by the mass and energy transfer across the boundary layer. Thus, a careful multi-scale modeling approach is necessary to capture the essence of the nanotube growth.

## 6. Conclusions and future trends

From this review it is evident that flames have emerged as a viable method for bulk synthesis of CNTs and related nanostructures. The synthesis of CNTs and related nanostructures is affected by the flame environment's temperature and chemical species, geometry, burner configuration, and catalyst composition. Flames provide the chemical species and the thermal energy necessary for driving the synthesis process. Flame synthesis processes are scalable and large scale production of SWNTs has been reported (Richter et al., 2008). However, control over the complex processes occurring in a flame is deemed critical for generation of appropriate condition for CNT growth. A large variety of hydrocarbons have been used as fuels for flames. Thus identification of the optimum ingredients, fuel and oxidizer, for the development of industrial processes is essential. Measurements of carbon nanotube growth rates under well defined micro-scale gas phase environment are scarce. Development of boundary layer around the catalyst particle or substrate inside a flame environment needs to be carefully examined. More sophisticated models are required for capturing mass and energy transfer across the boundary layer. Multi-scale modeling approach is essential for these computations. Spectroscopic diagnostic techniques at the micro-scale need to be applied in the boundary layer region to obtain insitu measurement of species concentration and temperature near the catalyst surface. These measurements can provide the missing link of data between the large scale and the nano-scale processes occurring simultaneously during the flame synthesis of CNTs. It has been observed that careful control of the synthesis process is essential to preserve the properties of the CNTs and to avoid any impurities. However, flame environment is characterized by presence of large number of chemical species and varying temperature field. Thus control techniques are necessary for pure and uniform yield of CNTs in the flame environment. Electric biasing of the catalyst substrate has been found helpful in alignment of CNTs in flames. Better control over the flame temperature with use of chimneys has led to uniform growth of

CNTs. Addition of diluents to the flame has resulted in temperature reduction and more uniform growth of CNTs. Injection of pyrolysis gases in a non-hydrocarbon flame ($H_2$-$O_2$) flame has also been attempted for improved control.
In summary, future work in flame synthesis of CNTs is needed for the definition of a standard set of reactants based on nano-scale and micro-scale measurements and computations of the optimum growth environment. Macro-scale reactor geometries and bulk material composition that lead to desirable product quality and yield need to be defined. Finally, cost versus quality tradeoffs depending on the specific needs of an application will decide the method for synthesis of carbon nanotubes.

## 7. References


Baird, T., Fryer, J. R., & Grant, B. (1974). Carbon formation on iron and nickel foils by hydrocarbon pyrolysis - reactions at 700 C. *Carbon,* Vol. 12, No. 5, pp. 591-602, ISSN 0008-6223

Baker, R. T. K., Barber, M. A., Waite, R. J., Harris, P. S., & Feates, F. S. (1972). Nucleation and growth of carbon deposits from nickel catalyzed decomposition of acetylene. *Journal of Catalysis,* Vol. 26, No. 1, pp. 51-&, ISSN 0021-9517

Baker, R. T. K., Harris, P. S., Thomas, R. B., & Waite, R. J. (1973). Formation of filamentous carbon from iron, cobalt and chromium catalyzed decomposition of acetylene. *Journal of Catalysis,* Vol. 30, No. 1, pp. 86-95, ISSN 0021-9517

Celnik, M., West, R., Morgan, N., Kraft, M., Moisala, A., Wen, J., Green, W., & Richter, H. (2008). Modelling gas-phase synthesis of single-walled carbon nanotubes on iron catalyst particles. *Carbon,* Vol. 46, No. 3, pp. 422-433, ISSN 0008-6223

Dai, H. J., Rinzler, A. G., Nikolaev, P., Thess, A., Colbert, D. T., & Smalley, R. E. (1996). Single-wall nanotubes produced by metal-catalyzed disproportionation of carbon monoxide. *Chemical Physics Letters,* Vol. 260, No. 3-4, pp. 471-475, ISSN 0009-2614

Diener, M. D., Nichelson, N., & Alford, J. M. (2000). Synthesis of single-walled carbon nanotubes in flames. *Journal of Physical Chemistry B,* Vol. 104, No. 41, pp. 9615-9620, ISSN 1089-5647

Gopinath, P., & Gore, J. (2007). Chemical kinetic considerations for postflame synthesis of carbon nanotubes in premixed flames using a support catalyst. *Combustion and Flame,* Vol. 151, No. 3, pp. 542-550, ISSN 00102180

Grieco, W. J., Howard, J. B., Rainey, L. C., & Vander Sande, J. B. (2000). Fullerenic carbon in combustion-generated soot. *Carbon,* Vol. 38, No. 4, pp. 597-614, ISSN 0008-6223

Height, M., Howard, J. B., Tester, J. W., & Vander Sande, J. B. (2004). Flame synthesis of single-walled carbon nanotubes. *Carbon,* Vol. 42, No. 11, pp. 2295-2307, ISSN 00086223

Hone, J., Whitney, M., Piskoti, C., & Zettl, A. (1999). Thermal conductivity of single-walled carbon nanotubes. *Physical Review B,* Vol. 59, No. 4, pp. R2514, ISSN

Hou, S.-S., Chung, D.-H., & Lin, T.-H. (2009). High-yield synthesis of carbon nano-onions in counterflow diffusion flames. *Carbon,* Vol. 47, No. 4, pp. 938-947, ISSN 00086223

Howard, J. B., McKinnon, J. T., Johnson, M. E., Makarovsky, Y., & Lafleur, A. L. (1992). Production of C-60 and C-70 fullerenes in benzene oxygen flames. *Journal of Physical Chemistry,* Vol. 96, No. 16, pp. 6657-6662, ISSN 0022-3654

Howard, J. B., McKinnon, J. T., Makarovsky, Y., Lafleur, A. L., & Johnson, M. E. (1991). Fullerenes C60 and C70 in flames. *Nature,* Vol. 352, No. 6331, pp. 139-141, ISSN 0028-0836

Iijima, S. (1991). Helical microtubes of graphitic carbon. *Nature,* Vol. 354, No. 6348, pp. 56-58, ISSN 0028-0836

Katta, V. R., Blevins, L. G., & Roquemore, W. M. (2005). Dynamics of an inverse diffusion flame and its role in polycyclic-aromatic-hydrocarbon and soot formation. *Combustion and Flame,* Vol. 142, No. 1-2, pp. 33-51, ISSN 0010-2180

Kroto, H. W. (1987). The stability of the fullerenes C-24, C-28, C-32, C-36, C-50, C-60 and C-70. *Nature,* Vol. 329, No. 6139, pp. 529-531, ISSN 0028-0836

Kroto, H. W., Heath, J. R., Obrien, S. C., Curl, R. F., & Smalley, R. E. (1985). C-60 - Buckminsterfullerene. *Nature,* Vol. 318, No. 6042, pp. 162-163, ISSN 0028-0836

Kuwana, K., & Saito, K. (2005). Modeling CVD synthesis of carbon nanotubes: Nanoparticle formation from ferrocene. *Carbon,* Vol. 43, No. 10, pp. 2088-2095, ISSN 0008-6223

Lee, G., Jurng, J., & Hwang, J. (2004). Formation of Ni-catalyzed multiwalled carbon nanotubes and nanofibers on a substrate using an ethylene inverse diffusion flame. *Combustion and Flame,* Vol. 139, No. 1-2, pp. 167-175, ISSN 00102180

Li, T. X., Kuwana, K., Saito, K., Zhang, H., & Chen, Z. (2009). Temperature and carbon source effects on methane–air flame synthesis of CNTs. *Proceedings of the Combustion Institute,* Vol. 32, No. 2, pp. 1855-1861, ISSN 15407489

Li, T. X., Zhang, H. G., Wang, F. J., Chen, Z., & Saito, K. (2007). Synthesis of carbon nanotubes on Ni-alloy and Si-substrates using counterflow methane–air diffusion flames. *Proceedings of the Combustion Institute,* Vol. 31, No. 2, pp. 1849-1856, ISSN 15407489

Merchan-Merchan, W., Saveliev, A., & Kennedy, L. A. (2003). Carbon nanostructures in opposed-flow methane oxy-flames. *Combustion Science and Technology,* Vol. 175, No. pp. 2217-2236, ISSN

Merchan-Merchan, W., Saveliev, A., & Kennedy, L. A. (2004). High-rate flame synthesis of vertically aligned carbon nanotubes using electric field control. *Carbon,* Vol. 42, No. 3, pp. 599-608, ISSN 00086223

Merchan-Merchan, W., Saveliev, A., Kennedy, L. A., & Fridman, A. (2002). Formation of carbon nanotubes in counter-flow, oxy-methane diffusion flames without catalysts. *Chemical Physics Letters,* Vol. 354, No. 1-2, pp. 20-24, ISSN 0009-2614

Merchan-Merchan, W., Saveliev, A. V., Kennedy, L., & Jimenez, W. C. (2010). Combustion synthesis of carbon nanotubes and related nanostructures. *Progress in Energy and Combustion Science,* Vol. 36, No. 6, pp. 696-727, ISSN 03601285

Merchan-Merchan, W., Saveliev, A. V., & Nguyen, V. (2009). Opposed flow oxy-flame synthesis of carbon and oxide nanostructures on molybdenum probes. *Proceedings of the Combustion Institute,* Vol. 32, No. 2, pp. 1879-1886, ISSN 15407489

Moisala, A., Nasibulin, A. G., & Kauppinen, E. I. (2003). The role of metal nanoparticles in the catalytic production of single-walled carbon nanotubes - a review. *Journal of Physics-Condensed Matter,* Vol. 15, No. 42, pp. S3011-S3035, ISSN 0953-8984

Naha, S., & Puri, I. K. (2008). A model for catalytic growth of carbon nanotubes. *Journal of Physics D: Applied Physics,* Vol. 41, No. 6, pp. 065304, ISSN 0022-37271361-6463

Naha, S., Sen, S., De, A. K., & Puri, I. K. (2007). A detailed model for the flame synthesis of carbon nanotubes and nanofibers. *Proceedings of the Combustion Institute,* Vol. 31, No. 2, pp. 1821-1829, ISSN 15407489

Oberlin, A., Endo, M., & Koyama, T. (1976). Filamentous growth of carbon through benzene decomposition. *Journal of Crystal Growth,* Vol. 32, No. 3, pp. 335-349, ISSN 0022-0248

Petroski, J. M., Wang, Z. L., Green, T. C., & El-Sayed, M. A. (1998). Kinetically controlled growth and shape formation mechanism of platinum nanoparticles. *Journal of Physical Chemistry B,* Vol. 102, No. 18, pp. 3316-3320, ISSN 1089-5647

Pope, C. J., & Howard, J. B. (1997). Simultaneous particle and molecule modeling (SPAMM): An approach for combining sectional aerosol equations and elementary gas-phase reactions. *Aerosol Science and Technology,* Vol. 27, No. 1, pp. 73-94, ISSN 0278-6826

Rao, C. N. R., Satishkumar, B. C., Govindaraj, A., & Nath, M. (2001). Nanotubes. *Chemphyschem,* Vol. 2, No. 2, pp. 78-105, ISSN 1439-4235

Richter, H., Treska, M., Howard, J. B., Wen, J. Z., Thomasson, S. B., Reading, A. A., Jardim, P. M., & Vander Sande, J. B. (2008). Large Scale Combustion Synthesis of Single-Walled Carbon Nanotubes and Their Characterization. *Journal of Nanoscience and Nanotechnology,* Vol. 8, No. 11, pp. 6065-6074, ISSN 15334880

Saito, K., Gordon, A. S., Williams, F. A., & Stickle, W. F. (1991). A study of the early history of soot formation in various hydrocarbon diffusion flames. *Combustion Science and Technology,* Vol. 80, No. 1-3, pp. 103-119, ISSN 0010-2202

Saito, K., Williams, F. A., & Gordon, A. S. (1986). Strucutre of laminar co-flow methane air diffusion flames. *Journal of Heat Transfer-Transactions of the ASME,* Vol. 108, No. 3, pp. 640-648, ISSN 0022-1481

Saveliev, A. (2003). Metal catalyzed synthesis of carbon nanostructures in an opposed flow methane oxygen flame. *Combustion and Flame,* Vol. 135, No. 1-2, pp. 27-33, ISSN 00102180

Singer, J. M., & Grumer, J. 1959. Paper read at Seventh symposium (international) on combustion, at Pittsburgh.

Soneda, Y., Duclaux, L., & Beguin, F. (2002). Synthesis of high quality multi-walled carbon nanotubes from the decomposition of acetylene on iron-group metal catalysts supported on MgO. *Carbon,* Vol. 40, No. 6, pp. 965-969, ISSN 0008-6223

Soneda, Y., & Makino, M. (2000). Formation and texture of carbon nanofilaments by the catalytic decomposition of CO on stainless-steel plate. *Carbon,* Vol. 38, No. 3, pp. 478-480, ISSN 0008-6223

Thess, A., Roland, L., Nikolaev, P., Dai, H., Petit, P., Robert, J., Xu, C., Lee, Y. H., Seong Gon, K., Rinzler, A. G., Colbert, D. T., Scuseria, G. E., Tománek, D., Fischer, J. E., & Smalley, R. E. (1996). Crystalline Ropes of Metallic Carbon Nanotubes. *Science,* Vol. 273, No. 5274, pp. 483-487, ISSN 00368075

Tibbetts, G. G. (1984). Why are carbon filaments tubular? *Journal of Crystal Growth,* Vol. 66, No. 3, pp. 632-638, ISSN 0022-0248

Unrau, C. J., & Axelbaum, R. L. (2010). Gas-phase synthesis of single-walled carbon nanotubes on catalysts producing high yield. *Carbon,* Vol. 48, No. 5, pp. 1418-1424, ISSN 00086223

Unrau, C. J., Axelbaum, R. L., Biswas, P., & Fraundorf, P. (2007). Synthesis of single-walled carbon nanotubes in oxy-fuel inverse diffusion flames with online diagnostics. *Proceedings of the Combustion Institute,* Vol. 31, No. 2, pp. 1865-1872, ISSN 15407489

Unrau, C. J., Katta, V. R., & Axelbaum, R. L. (2010). Characterization of diffusion flames for synthesis of single-walled carbon nanotubes. *Combustion and Flame,* Vol. 157, No. 9, pp. 1643-1648, ISSN 00102180

Vander Wal, R. L. (2000). Flame synthesis of substrate-supported metal-catalyzed carbon nanotubes. *Chemical Physics Letters,* Vol. 324, No. 1-3, pp. 217-223, ISSN 0009-2614

Vander Wal, R. L. (2002). Fe-catalyzed single-walled carbon nanotube synthesis within a flame environment. *Combustion and Flame,* Vol. 130, No. 1-2, pp. 37-47, ISSN 0010-2180

Vander Wal, R. L. (2002). Flame synthesis of Ni-catalyzed nanofibers. *Carbon,* Vol. 40, No. 12, pp. 2101-2107, ISSN 0008-6223

Vander Wal, R. L., & Hall, L. J. (2001). Flame synthesis of Fe catalyzed single-walled carbon nanotubes and Ni catalyzed nanofibers: growth mechanisms and consequences. *Chemical Physics Letters,* Vol. 349, No. 3-4, pp. 178-184, ISSN 0009-2614

Vander Wal, R. L., Hall, L. J., & Berger, G. M. (2002). The chemistry of premixed flame synthesis of carbon nanotubes using supported catalysts. *Proceedings of the Combustion Institute,* Vol. 29, No. pp. 1079-1085, ISSN 0082-0784

Vander Wal, R. L., Hall, L. J., & Berger, G. M. (2002). Optimization of flame synthesis for carbon nanotubes using supported catalyst. *Journal of Physical Chemistry B,* Vol. 106, No. 51, pp. 13122-13132, ISSN 1520-6106

Vander Wal, R. L., Ticich, T. M., & Curtis, V. E. (2000). Flame synthesis of metal-catalyzed single-wall carbon nanotubes. *Journal of Physical Chemistry A,* Vol. 104, No. 31, pp. 7209-7217, ISSN 1089-5639

Wen, J. Z., Celnik, M., Richter, H., Treska, M., Vander Sande, J. B., & Kraft, M. (2008). Modelling study of single walled carbon nanotube formation in a premixed flame. *Journal of Materials Chemistry,* Vol. 18, No. 13, pp. 1582, ISSN 0959-9428 1364-5501

Wen, J. Z., Goldsmith, C. F., Ashcraft, R. W., & Green, W. H. (2007). Detailed kinetic mdeling of ion nanoparticle synthesis from the decomposition of $Fe(CO)_5$. *The Journal of Physical Chemistry C,* Vol. 111, No. 15, pp. 5677-5688, ISSN 1932-7447

Xu, F. 2007. Investigating flame-based synthesis of carbon nanotubes and metal-oxide nanowires. Ph.D.

Xu, F. S., Liu, X. F., & Tse, S. D. (2006). Synthesis of carbon nanotubes on metal alloy substrates with voltage bias in methane inverse diffusion flames. *Carbon,* Vol. 44, No. 3, pp. 570-577, ISSN 0008-6223

Xu, F. S., Zhao, H., & Tse, S. D. (2007). Carbon nanotube synthesis on catalytic metal alloys in methane/air counterflow diffusion flames. *Proceedings of the Combustion Institute,* Vol. 31, No. pp. 1839-1847, ISSN 1540-7489

Yu, M.-F., Lourie, O., Dyer, M. J., Moloni, K., Kelly, T. F., & Ruoff, R. S. (2000). Strength and Breaking Mechanism of Multiwalled Carbon Nanotubes under Tensile Load. *Science,* Vol. 287, No. 5453, pp. 637-640, ISSN 00368075

Yuan, L. M., Saito, K., Hu, W. C., & Chen, Z. (2001). Ethylene flame synthesis of well-aligned multi-walled carbon nanotubes. *Chemical Physics Letters,* Vol. 346, No. 1-2, pp. 23-28, ISSN 0009-2614

Yuan, L. M., Saito, K., Pan, C. X., Williams, F. A., & Gordon, A. S. (2001). Nanotubes from methane flames. *Chemical Physics Letters,* Vol. 340, No. 3-4, pp. 237-241, ISSN 0009-2614

Zhang, Y., & Smith, K. (2005). A kinetic model of $CH_4$ decomposition and filamentous carbon formation on supported Co catalysts. *Journal of Catalysis,* Vol. 231, No. 2, pp. 354-364, ISSN 00219517

# 8

# Carbon Nanotube Synthesis and Growth Mechanism

Mukul Kumar
*Department of Materials Science & Engineering*
*Meijo University, Nagoya 468-8502*
*Japan*

## 1. Introduction

A carbon nanotube (CNT) is a tubular structure made of carbon atoms, having diameter of nanometer order but length in micrometers. Right from its discovery, we have been listening exciting quotations about CNT, viz.

- "CNT is 100 times stronger than stainless steel and six times lighter..."
- "CNT is as hard as diamond and its thermal capacity is twice that of pure diamond..."
- "CNT's current-carrying capacity is 1000 times higher than that of copper..."
- "CNT is thermally stable up to 4000K..."
- "CNT can be metallic or semiconducting, depending on their diameter and chirality..."

However, it is important to note that all those superlative properties were predicted for an atomically-perfect *ideal* CNT which is far from the CNTs we are *practically* producing today. Despite a huge progress in CNT research over the years, we are still unable to produce CNTs of well-defined properties in large quantities by a cost-effective technique. The root of this problem is the lack of proper understanding of the CNT growth mechanism. There are several questions at the growth level awaiting concrete answer. Till date no CNT growth model could be robustly established. Hence this chapter is devoted to review the present state of CNT synthesis and growth mechanism.

There are three commonly-used methods of CNT synthesis. Arc-discharge method, in which the first CNT was discovered, employs evaporation of graphite electrodes in electric arcs that involve very high (~4000°C) temperatures (Iijima, 1991). Although arc-grown CNTs are well crystallized, they are highly impure; about 60–70% of the arc-grown product contains metal particles and amorphous carbon. Laser-vaporization technique employs evaporation of high-purity graphite target by high-power lasers in conjunction with high-temperature furnaces (Thess et al., 1996). Although laser-grown CNTs are of high purity, their production yield is very low (in milli gram order). Thus, it is obvious that these two methods score too low on account of efficient use of energy and resources. Chemical vapor deposition (CVD), incorporating catalyst-assisted thermal decomposition of hydrocarbons, is the most popular method of producing CNTs; and it is truly a low-cost and scalable technique for mass production of CNTs (Cassell et al., 1999). That is why CVD is the most popular method of producing CNTs nowadays. Here we will review the materials aspects of CNT synthesis by CVD and discuss the CNT growth mechanism in the light of latest progresses in the field.

## 2. CNT synthesis

Figure 1 shows a schematic diagram of the experimental set-up used for CNT growth by CVD method in its simplest form. The process involves passing a hydrocarbon vapor (typically 15–60 min) through a tubular reactor in which a catalyst material is present at sufficiently high temperature (600–1200°C) to decompose the hydrocarbon. CNTs grow on the catalyst in the reactor, which are collected upon cooling the system to room temperature. In the case of a liquid hydrocarbon (benzene, alcohol, etc.), the liquid is heated in a flask and an inert gas is purged through it, which in turn carries the hydrocarbon vapor into the reaction zone. If a solid hydrocarbon is to be used as the CNT precursor, it can be directly kept in the low-temperature zone of the reaction tube. Volatile materials (camphor, naphthalene, ferrocence etc.) directly turn from solid to vapor, and perform CVD while passing over the catalyst kept in the high-temperature zone. Like the CNT precursors, also the catalyst precursors in CVD may be used in any form: solid, liquid or gas, which may be suitably placed inside the reactor or fed from outside. Pyrolysis of the catalyst vapor at a suitable temperature liberates metal nanoparticles in-situ (such a process is known as floating catalyst method). Alternatively, catalyst-coated substrates can be placed in the hot zone of the furnace to catalyze the CNT growth.

CNT synthesis involves many parameters such as hydrocarbon, catalyst, temperature, pressure, gas-flow rate, deposition time, reactor geometry. However, to keep our discussion compact, here we will consider only the three key parameters: hydrocarbon, catalyst and catalyst support.

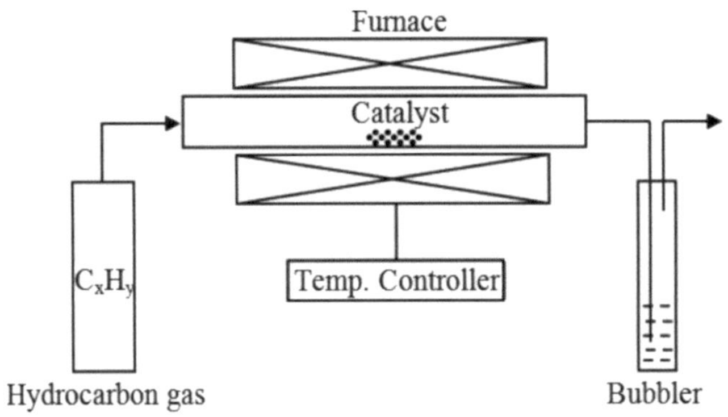


Fig. 1. Schematic diagram of a CVD setup in its simplest form.

### 2.1 CNT precursors

Most commonly used CNT precursors are methane, ethylene, acetylene, benzene, xylene and carbon monoxide. Among the early reports of CVD, MWCNTs were grown from the pyrolysis of benzene at 1100°C (Endo et al., 1991) and from acetylene at 700°C (Jose-Yacaman et al., 1993). In those cases, iron nanoparticles were used as the catalyst. Later, MWCNTs were also grown from many other precursors including cyclohexane (Li et al., 2007) and fullerene (Nerushev et al., 2003). On the other hand, SWCNTs were first produced from the disproportionation of carbon monoxide at 1200°C, in the presence of molybdenum nanoparticles (Dai et al., 1996). Later, SWCNTs were also produced from benzene, acetylene, ethylene, methane, cyclohexane, fullerene etc. by using various catalysts. In 2002, a low-temperature synthesis of high-purity SWCNTs was reported from alcohol CVD on Fe-Co-impregnated zeolite support (Maruyama et al., 2002); and since then, ethanol became the most popular CNT precursor in the CVD method worldwide. Special feature of ethanol is that ethanol-grown CNTs are almost free from amorphous carbon, owing to the etching effect of OH radical. Later, vertically-aligned SWCNTs were also grown on Mo-Co-coated quartz and silicon substrates (Murakami et al., 2004). Recently, it has been shown that intermittent supply of acetylene in ethanol CVD significantly assists ethanol in preserving the catalyst's activity and thus enhances the CNT growth rate (Xiang et al., 2009).

The molecular structure of the precursor has a detrimental effect on the morphology of the CNTs grown (K. Ghosh et al., 2009). Linear hydrocarbons such as methane, ethylene, acetylene, thermally decompose into atomic carbons or linear dimers/trimers of carbon, and generally produce straight and hollow CNTs. On the other hand, cyclic hydrocarbons such as benzene, xylene, cyclohexane, fullerene, produce relatively curved/hunched CNTs with the tube walls often bridged inside.

General experience is that low-temperature CVD (600–900°C) yields MWCNTs, whereas high-temperature (900–1200°C) reaction favors SWCNT growth. This indicates that SWCNTs have a higher energy of formation (presumably owing to small diameters; high curvature bears high strain energy). Perhaps that is why MWCNTs are easier to grow (than SWCNTs) from most of the hydrocarbons, while SWCNTs grow from selected hydrocarbons (viz. carbon monoxide, methane, etc. which have a reasonable stability in the temperature range of 900–1200°C). Commonly efficient precursors of MWCNTs (viz. acetylene, benzene, etc.) are unstable at higher temperature and lead to the deposition of large amounts of carbonaceous compounds other than the nanotubes.

In 2004, a highly-efficient synthesis of impurity-free SWCNTs was reported by water-assisted ethylene CVD on Si substrates (Hata et al., 2004). It was proposed that controlled supply of steam into the CVD reactor acted as a weak oxidizer and selectively removed amorphous carbon without damaging the growing CNTs. Balancing the relative levels of ethylene and water was crucial to maximize the catalyst's lifetime. Recently, however, it has been shown that a reactive etchant such as water or hydroxyl radical is not required at all in cold-wall CVD reactors if the hydrocarbon activity is low (Zhong et al., 2009). These studies emphatically prove that the carbon precursor plays a crucial role in CNT growth. Therefore, by proper selection of CNT precursor and its vapor pressure, both the catalyst's lifetime and the CNT-growth rate can be significantly increased; and consequently, both the yield and the quality of CNTs can be improved.

### 2.2 CNT catalysts

For synthesizing CNTs, typically, nanometer-size metal particles are required to enable hydrocarbon decomposition at a lower temperature than the spontaneous decomposition temperature of the hydrocarbon. Most commonly-used metals are Fe, Co, Ni, because of two main reasons: (i) high solubility of carbon in these metals at high temperatures; and (ii) high carbon diffusion rate in these metals. Besides that, high melting point and low equilibrium-vapor pressure of these metals offer a wide temperature window of CVD for a wide range of carbon precursors. Recent considerations are that Fe, Co, and Ni have stronger adhesion with the growing CNTs (than other transition metals do) and hence they are more efficient in forming high-curvature (low diameter) CNTs such as SWCNTs (Ding et al., 2008).

Solid organometallocenes (ferrocene, cobaltocene, nickelocene) are also widely used as a CNT catalyst, because they liberate metal nanoparticles in-situ which catalyze the hydrocarbon decomposition more efficiently. It is a general experience that the catalyst-particle size dictates the tube diameter. Hence, metal nanoparticles of controlled size, pre-synthesized by other reliable techniques, can be used to grow CNTs of controlled diameter. Thin films of catalyst coated on various substrates are also proven good in getting uniform CNT deposits (Fan et al., 1999). The key to get pure CNTs is achieving hydrocarbon decomposition on the catalyst surface alone and prohibiting the aerial pyrolysis. Recently, a high-yield CNT growth has been observed from acetylene decomposition on a stainless steel sheet at 730°C, without using any additional catalyst (Camilli et al., 2011). This study proves that the catalyst as a whole does not necessarily need to be a nanoparticle. Even a bulk metal can catalyze the CNT growth provided that the surface roughness is on nanometer scale. Moreover, alloys are proven to have a higher catalytic activity than pure metals. In 2008, *gigas growth* of CNT was reported from Fe-Co catalyst on zeolite support resulting in a weight gain of 1000% and volume gain of 10,000%, relative to the zeolite bed (Kumar et al., 2008). Hence, by combining different metals in different ratios and carefully controlling the catalyst calcination conditions, it is possible to evolve new crystallographic phases that could exhibit much higher catalytic activity toward CNT growth. Very recently, highly active crystallographic phases of Co-Mo and Ni-Mo have been achieved on MgO support, yielding ~3000 wt% CNT growth (Nunez et al., 2011).

Apart from the popular transition metals (Fe, Co, Ni), other metals of this group, such as Cu, Au, Ag, Pt, Pd were also found to catalyze CNT growth from various hydrocarbons (Moisala et al., 2003). On the role of CNT catalysts, it is worth mentioning that transition metals are proven to be efficient catalysts not only in CVD but also in arc-discharge and laser-vaporization methods. Therefore, it is likely that these apparently different methods might inherit a common growth mechanism of CNT, which is not yet clear. Hence this is an open field of research to correlate different CNT techniques in terms of the catalyst's role in entirely different temperature and pressure range.

### 2.3 CNT catalyst supports

The same catalyst works differently on different support materials. Commonly used substrates in CVD are quartz, silicon, silicon carbide, silica, alumina, alumino-silicate (zeolite), $CaCO_3$, magnesium oxide, etc. For an efficient CNT growth, the catalyst-substrate interaction should be investigated with utmost attention. Metal–substrate reaction (chemical bond formation) would cease the catalytic behavior of the metal. The substrate material, its surface morphology and textural properties greatly affect the yield and quality of the resulting CNTs. Zeolite substrates with catalyst in its nanopores have resulted significantly high yields of CNTs with a narrow diameter distribution (Kumar et al., 2005). Alumina materials are reportedly a better catalyst support than silica owing to stronger metal-support interaction in the former, which allows high metal dispersion and thus a high density of catalytic sites. Such interactions prevent metal species from aggregating and forming unwanted large clusters that lead to graphite particles or defective MWCNTs. Recent in-situ XPS analysis of CNT growth from different precursors on iron catalyst supported on alumina and silica substrates have confirmed these theoretical assumptions. Thin Alumina flakes (0.04–4 μm thick) loaded with iron nanoparticles have shown high yields of aligned CNTs of high aspect ratio (Mattevi et al., 2008). Latest considerations are that the oxide substrate, basically used as a physical support for the metal catalyst, might be playing some chemistry in the CNT growth (Noda et al., 2007). Accordingly, the chemical state and structure of the substrate are more important than that of the metal.

The crystallographic orientation of the exposed substrate surface governs the CNT growth direction. CNTs preferentially grow at 90° to Si (100) surface, but at 60° to Si (111) surface (Su et al., 2000). While α-plane sapphire leads to CNT growth normal to (001), no orientation is observed on c-plane or m-plane sapphire (Han et al., 2005). Similarly, single-crystal MgO (001) substrate shows preferential growth of CNTs along [110] direction (Maret et al., 2007). Very recently, an orthogonal CNT growth has been observed on micro-sized alumina particles (Fig. 2) by CVD of ferrocene-xylene mixture at 600°C (He et al., 2011). This observation proves that the grain boundary or crystalline steps of the support material play a crucial role in CNT growth and orientation.

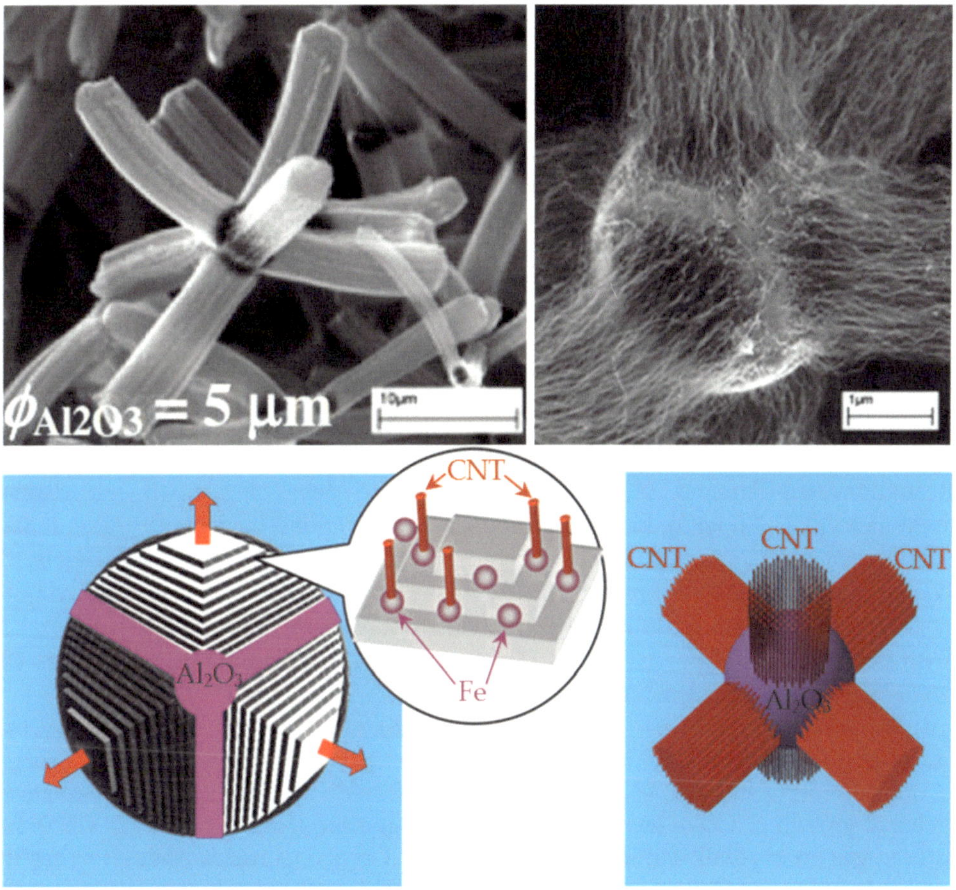


Fig. 2. Orthogonal growth of CNTs from an $Al_2O_3$ substrate. The crystalline steps of $Al_2O_3$ microparticle guide the direction of CNTs growing from the Fe nanoparticles lying on those steps. (He et al., 2011. Carbon 49, 2273-2286. © Elsevier Science).

### 2.4 New CNT catalysts

Recent developments in the nanomaterials synthesis and characterization have enabled many new catalysts for the CNT growth. Apart from popularly used transition metals (Fe, Co, Ni), a range of other metals (Cu, Pt, Pd, Mn, Mo, Cr, Sn, Au, Mg, Al) has also been successfully used for horizontally-aligned SWCNT growth on quartz substrates (Yuan et al.,

2008). Unlike transition metals, noble metals (Au, Ag, Pt, Pd etc.) have extremely low solubility for carbon, but they can dissolve carbon effectively for CNT growth when their particle size is very small (<5nm). Recently, a controlled growth of SWCNTs was reported on Au nanoparticles deposited on the atomic steps of Si (Takagi et al., 2008). It was found that the active catalyst is Au-Si alloy with about 80 at% Au.

Although copper is a transition metal, it showed insignificant catalytic effect on the CNT growth in the past (Vanderwal et al., 2001). In fact, it had been considered as an adverse contaminant. As a recent development, however, Cu has been found to catalyze the CNT growth efficiently. Methane and ethanol decomposition at 825–925°C on Cu nanoparticles supported on silicon wafers produced high densities of well-crystallined SWCNTs up to 1 cm in length (Zhou et al., 2006). The Cu nanoparticles were synthesized by the reduction of $CuCl_2$ in the presence of $Cu_2O$ nanoparticles produced by thermolysis of cupric formates in coordinating solvents. This implies that the novelty lies in the catalyst-preparation method.

Rhenium (Re) is a rare catalyst used for CNT growth. Diamagnetic SWCNTs and MWCNTs were reported from methane decomposition on Re catalyst (Ritschel et al., 2007). The current scenario is that, just as any carbon-containing material can yield CNT, any metal can catalyze the CNT growth, provided that the experimental conditions are properly optimized. Accordingly, there is a huge scope in exploring new metals as CNT catalyst.

### 2.5 Metal-free CNT growth

Recently, nanodiamond particles (5 nm) were shown to catalyze the CNT growth (Takagi et al., 2009). Ethanol suspension of nanodiamond particles was spread on graphite plates and dried in air at 600°C. This resulted in isolated diamond particles, monolayers of diamond, and multilayered-diamond stacks on the substrate, depending upon the diamond concentration (0.01–1.0 wt%). Ethanol CVD over these diamond-loaded substrates at 850°C produced isolated CNTs, layered CNTs and high-density CNT mats, respectively. The nanodiamond particles do not fuse even after high-temperature CVD process, implying that they remain in solid state during CVD. Nanodiamond is therefore said to act as a CNT growth seed. This result proves that CNT growth is possible without metal catalyst. Does nanodiamond act as a catalyst? If it does, how? These are open questions.

In many studies, oxygen was noticed to activate the CNT growth. Recent studies have revealed that many metals, which do not exhibit catalytic activity in pure-metal form, do well in oxide form (Rummeli et al., 2005). Does metal oxide act as a catalyst? Template-free directional growth of CNTs has been achieved on sapphire (Han et al., 2005). CNTs have also been grown on semiconductors such as Si and Ge nanoparticles (though C has little solubility in bulk Si or Ge), provided that the nanoparticles are heated in air just before CVD (Takagi et al., 2007). Similarly, CNT growth on SiC substrates takes place only when some oxygen is present in the chamber (Kusunoki et al., 2000). Porous $Al_2O_3$ has already been shown to facilitate CNT growth without any catalyst (Schneider et al., 2008). Catalyst-free CNT growth is also possible in oxy-fuel flames (Merchan-Merchan et al., 2002). And oxide, typically used as a catalyst support in CVD, is itself capable of forming graphene layers (Rummeli et al., 2007a). All these examples resoundingly indicate that oxygen plays a key role in CNT growth. The question is whether oxygen is a catalyst! HRTEM investigation of the CNTs grown from cyclohexane pyrolysis over iron nanoparticles supported on thin $Al_2O_3$ layers shows that CNT keeps on growing even when the metal is completely

encapsulated in the tube center (Rummeli et al., 2007b). The authors propose that the metal only helps to initiate the CNT precipitation at the nucleation stage. Once the CNT head is created, metal becomes non functional; subsequent carbon addition to the CNT base periphery is facilitated from the substrate's oxide layer. This concept is radically different from the existing concept that the metal must remain exposed (either on the CNT tip or base) to keep the growth on. Hence more careful in-situ observation and robust theoretical support are required to establish the oxide's direct role as a catalyst.
More recent development of the field is even more exciting: CNT growth is possible with no metal at all; the non-metallic substrate itself acts as the CNT catalyst. Liu et al. (2009) passed methane and hydrogen (1:1) over an $SiO_2$-sputtered Si wafer at 900°C for 20 min and got dense SWCNTs grown on it. In the same CVD condition, thermally-grown $SiO_2$ films did not result CNTs. The success lies in the in-situ transformation of the sputtered $SiO_2$ film (30 nm) into isolated Si particles (1.9 nm) which efficiently catalyzed methane decomposition due to small-size effect. Similar $SiO_2$ nanoparticle generation and subsequent CNT growth was reported by another group from ethanol decomposition on annealed $SiO_2$/Si substrates (Liu et al., 2010). On the other hand, Huang et al. (2009) simply scratched the existing $SiO_2$/Si wafers by a diamond blade and passed ethanol over it at 900°C for 10 min. Bunch of SWCNTs grew on the scratched portions. Random scratches on thin $SiO_2$ films protrude some nanoparticles mechanically. These developments raise many new questions and compel us to reconsider the existing CNT-growth models. $SiO_2$ has no carbon solubility; how does it assist hydrocarbon-to-CNT conversion? Does it act as a solid-state catalyst like nanodiamond? Or does it melt at 900°C as usual metal nanoparticles do? If it is in molten state, Si and O atoms might have some mobility thus creating a vacancy or dislocation which would attract hydrocarbon and cause dehydrogenation. If it is in solid state, it would be strained enough (high curvature at small particle size) and could possibly interact with hydrocarbon. It is also likely that it is in a quasi-liquid (or semi solid) state; slight distortion in its overall shape would develop some polarity on Si and O atoms, which could possibly facilitate dehydrogenation; and its fluctuating shape could act as a template for tubular graphite formation. These speculations evoke serious discussion. $SiO_2$ has a number of distinct crystalline forms. Si-O bond length and Si-O-Si bond angle vary significantly in different crystal forms (e.g., 154–171 pm, 140°–180°). Ab-initio calculations indicate that CNT-cap nucleation is influenced by solid-surface curvatures (Reich et al., 2006). More theoretical considerations and experimental verifications are sought for proper understanding of CNT growth on $SiO_2$ nanoparticles. It will take due time to come up with a convincing model; nevertheless, there is no doubt that metal-free CNT synthesis is a major breakthrough in CNT research, and it has opened a new avenue in nanotechnology.

### 2.6 New CNT precursors

Apart from the popular hydrocarbons mentioned in the section 2.1, CNTs have also been synthesized from many other organic compounds, especially from polymers. Carbonization (prolonged pyrolysis in vacuum to convert organic compounds into solid carbon) of polyacrylonitrile (Parthsarathy et al., 1995) and poly-furfuryl-alcohol (Kyotani et al., 1996) within nanoporous alumina templates resulted in thick CNTs. Reported in 1995-96, this was a multi-step tedious process requiring chemically-controlled monomer initiators to achieve polymerization. The field has matured enough and nowadays super-aligned highly-uniform CNTs can be produced from readily available polymers without taking pains for chemical initiators or catalysts. Recently, several polymer precursors, loaded on commercially-available alumina templates of well-defined pore size, were carbonized (400–600°C for 3h) to obtain MWCNTs of desired diameter (Han et al., 2009). N-doped MWCNTs obtained from carbonization of polypyrrole within alumina and zeolite membranes have shown better hydrogen-storage capacity than pristine MWCNTs obtained from polyphenyl acetylene in the same conditions (Sankaran et al., 2008). As for SWCNT, pyrolysis of tripropylamine within the nanochannels (0.73 nm) of aluminophosphate crystals (AFI) resulted in the narrowest nanotubes (0.4 nm) (Tang et al., 1998). Later, several carbon precursors were pyrolyzed within the AFI channels and tetrapropylammonium hydroxide was found to yield high densities of 4Å CNTs with better crystallinity (Zhai et al., 2006). It is suggested that the number of carbon atoms in the precursor molecule influences the SWCNT packing density in the template channels.

Among other organic compounds, amino-dichloro-s-triazine, pyrolyzed on cobalt-patterned silica substrates, resulted in highly pure CNTs (Terrones et al., 1997) Almost contemporary, organometallic compounds such as metallocene (ferrocene, cobaltocene, nickelocene) (Sen et al., 1997) and nickel phthalocyanine (Yudasaka et al., 1997) were used as the carbon-cum-catalyst precursor; however, as-grown CNTs were highly metal-encapsulated and the yield was very low. Later, pyrolysis of thiophene with metallocene led to the formation of Y-junction CNTs (Satishkumar et al., 2000). Recently, high-temperature pyrolysis (1300°C) of simple saccharides (from table sugar (sucrose) to lactose) resulted in straight as well as helical MWCNTs (Kucukayan et al., 2008).

In 2001, high yield of CNTs was obtained from camphor, a tree product (Kumar et al., 2001). Since then the authors remained involved with this environment-friendly source of CNTs and established the conditions for growing MWCNTs (Kumar et al., 2002; 2003a), SWCNTs (Kumar et al., 2003b) and vertically-aligned CNTs on quartz and silicon substrates (Kumar et al., 2003c; 2004) by using ferrocene catalyst. Later, using Fe-Co catalyst impregnated in zeolite support, mass production of CNTs was achieved by camphor CVD (Kumar et al., 2008). MWCNTs were grown at a temperature as low as 550°C, whereas SWCNTs could be grown at relatively high (900°C) temperature. Because of very low catalyst requirement with camphor, as-grown CNTs are least contaminated with metal, whereas oxygen atom present in camphor helps in oxidizing amorphous carbon in-situ (Kumar et al., 2007). These features of camphor stimulated more in-depth, basic and applied research worldwide.

Camphor-grown CNTs were used as the anode of secondary lithium battery (Sharon et al., 2002). Andrews et al. (2006) investigated the effect of camphor's molecular structure on the CNT growth and quality. Yamada et al. (2006) studied camphor CVD with different ways of catalyst feeding and addressed catalyst activation/deactivation process for the synthesis of highly-dense aligned CNT arrays. Parshotam (2008) studied the effect of carrier gases (nitrogen, argon, argon-hydrogen mixture) as well as catalyst-support materials ($SiO_2$, $Al_2O_3$ and MgO) on the quality of camphor-grown CNTs. Antunes et al. (2010) carried out thermal annealing and electrochemical purification of camphor-grown CNTs. Tang et al. (2010) synthesized tree-like multi-branched CNTs from camphor and reported the effects of temperature, argon flow rate and catalyst concentration on the structure of as-grown carbon nanotrees. Musso et al. (2007) got 2.3 mm thick CNT mats at a high deposition rate of 500 nm/sec. Later, the same group published fluid-dynamic analysis of the carrier-gas flow for camphor-CVD system (Musso et al., 2008) and hydrogen-storage analysis of camphor-grown CNTs (Bianco et al., 2010). Thus, camphor has emerged as a promising and the most-efficient CNT precursor amongst the new/unconventional ones. Moreover, it has opened up a new avenue of exploring other botanical products as a CNT precursor. Appreciable efforts have been made by Sharon and his coworkers who investigated the pyrolysis of a range of plant-based materials for this purpose (Sharon et al., 2006). Recently, high yields of aligned and non-aligned CNTs have also been reported from other plant-derived cheap raw materials such as turpentine (Afre et al., 2005) and eucalyptus oils (P. Ghosh et al., 2009). Most recently, Zhao et al. (2011) used sesame seeds as a CNT catalyst precursor. Sesame seed consists of uniform microcells containing an Fe-complex. During CVD at 800°C, this Fe-complex releases uniformly-distributed Fe nanoparticles, which efficiently catalyze the CNT growth.

Apart from the well-defined chemical reagents described above, CNTs have also been successfully and systematically synthesized from domestic fuels such as kerosene (Pradhan et al., 2002), liquefied petroleum gas (Qian et al., 2002) and coal gas (Qiu et al., 2006). More interestingly, there are scientific reports of CNT production from green grasses. Grass contains dense vascular bundles mainly composed of cellulose, hemicellulose and lignin. Rapid heat treatment of grass (600°C) in controlled-oxygen ambience dehydrates and carbonizes the vascular bundles into CNTs (Kang et al., 2005). Thus, now it is almost certain that any carbon-containing material may be a CNT precursor under suitable experimental conditions. The point is: can we reproduce the product quality and quantity from those materials of inconsistent composition? Certainly not. Depending upon the chemical composition of the raw material, one will have to change the experimental conditions every now and then; and the impurity elements of the raw material would greatly contaminate the resulting CNTs which would ultimately be of no practical importance. Hence, seemingly novel and interesting research of CNT production from abundant materials, such as waste plastics or domestic garbage, would be a too-long-term project. To meet the immediate need of mass production of CNTs, it is advisable to choose a raw material of consistent chemistry, which is abundant and regenerative too; so that it could lead to a reproducible as well as sustainable industrial technique.

## 3. CNT growth mechanism

CNT growth mechanism has been debatable right from its discovery. Based on the reaction conditions and post-deposition product analyses, several groups have proposed several possibilities which are often contradicting. Therefore, no single CNT growth mechanism is well established till date. Nevertheless, widely-accepted most-general mechanism can be outlined as follows. A hydrocarbon vapor when comes in contact with the "hot" metal nanoparticles, first decomposes into carbon and hydrogen species; hydrogen flies away and carbon gets dissolved into the metal. After reaching the carbon-solubility limit in the metal at that temperature, as-dissolved carbon precipitates out and crystallizes in the form of a cylindrical network having no dangling bonds and hence energetically stable. Hydrocarbon decomposition (being an exothermic process) releases some heat to the metal's exposed zone, while carbon crystallization (being an endothermic process) absorbs some heat from the metal's precipitation zone. This precise thermal gradient inside the metal particle keeps the process on.

Now there are two general cases. (Fig. 3a) When the catalyst-substrate interaction is weak (metal has an acute contact angle with the substrate), hydrocarbon decomposes on the top surface of the metal, carbon diffuses down through the metal, and CNT precipitates out across the metal bottom, pushing the whole metal particle off the substrate (Fig. 3a(i)). As long as the metal's top is open for fresh hydrocarbon decomposition, the concentration gradient exists in the metal allowing carbon diffusion, and CNT continues to grow longer and longer (Fig. 3a(ii)). Once the metal is fully covered with excess carbon, its catalytic activity ceases and the CNT growth is stopped (Fig. 3a(iii)). This is known as "tip-growth model".

In the other case, (Fig. 3b) when the catalyst-substrate interaction is strong (metal has an obtuse contact angle with the substrate), initial hydrocarbon decomposition and carbon diffusion take place similar to that in the tip-growth case, but the CNT precipitation fails to push the metal particle up; so the precipitation is compelled to emerge out from the metal's apex (farthest from the substrate, having minimum interaction with the substrate). At first, carbon crystallizes out as a hemispherical dome (the most favorable closed-carbon network on a spherical nanoparticle) which then extends up in the form of seamless graphitic cylinder. Subsequent hydrocarbon decomposition takes place on the lower peripheral surface of the metal, and as-dissolved carbon diffuses upward. Thus CNT grows up with the catalyst particle rooted on its base; hence, this is known as "base-growth model".

However, there are several points of discord in the above-mentioned *general* CNT growth mechanism. We are not sure that, during the CNT growth, whether the metal is in solid or liquid state, whether the carbon diffusion in metal is volume diffusion or surface diffusion, whether the actual catalyst for CNT growth is the pure metal or metal carbide, etc. Let us now review some important in-situ electron microscopic studies on these aspects.

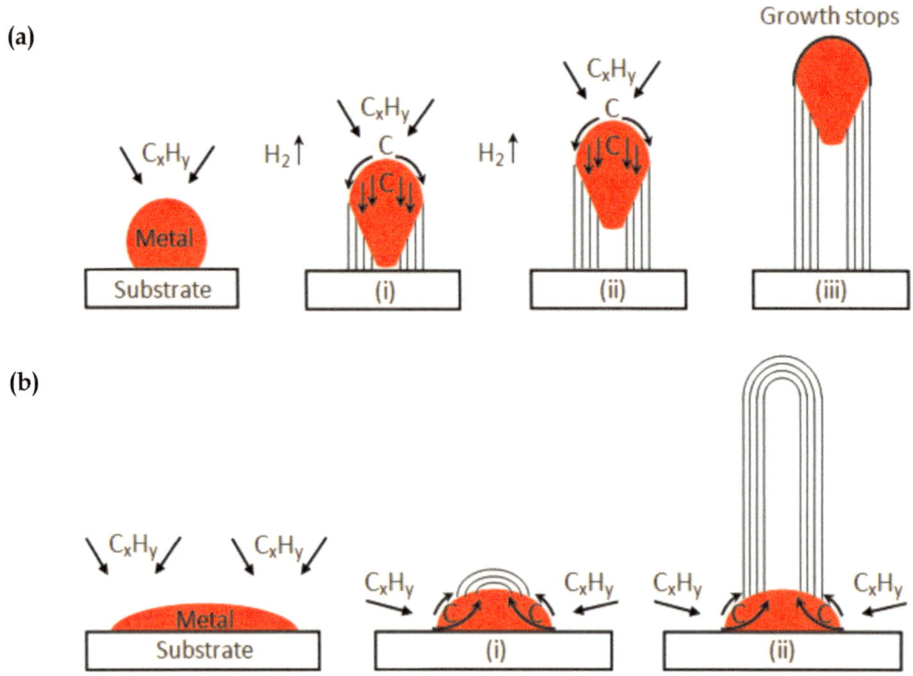


Fig. 3. Widely-accepted growth mechanisms for CNTs: (a) tip-growth model, (b) base-growth model.

### 3.1 Physical state of the catalyst

The first effort to observe the carbon filament growth process in-situ was made by Baker et al. (1972). By installing a gas-reaction cell in the TEM specimen chamber, they were able to perform carbon fiber growth in a temperature range of 600–1200°C at different gas pressures up to 225 torr (maximum), while the TEM column was maintained at sufficiently low temperature and pressure suitable for electron microscopy. For acetylene decomposition on nickel catalyst supported on silica and graphite supports at 600°C, they clearly observed that the metal particles changed it shape and moved up with a trail of carbon deposit (30–50 nm diameter). From the changing shape of the metal particle during fiber growth, they assumed that the catalyst was in liquid phase. The activation energy calculated for this growth was nearly same as the activation energy of carbon diffusion in liquid nickel; hence they suggested that carbon diffuses through the bulk metal and the fiber growth rate is diffusion-controlled. Similar tip-growth process was observed with Fe, Co and Cr catalysts (Baker et al., 1973). But in the case of acetylene decomposition on bimetallic (Pt-Fe) catalyst, the catalyst was observed to remain static on the substrate, while the carbon filament went on growing up. This led them to enunciate a base-growth model (Baker et al., 1975). It was explained that strong interaction between Pt-Fe and $SiO_2$ substrate kept the metal particle anchored to the substrate surface, and carbon precipitation occurred from the free upper face of the particle. Temperature and concentration gradients were thought to be the main driving forces for the continued growth dynamics. The filament growth was seen to be ceased when the particle was fully covered with the carbon cloud, but it could be re-activated by exposure to either hydrogen or oxygen at higher temperatures (Baker et al., 1972). Later, however, many scientists reported base-grown CNTs from Fe and Co catalysts on Si and $SiO_2$ substrates (Li et al., 1999; Bower et al., 2000). This indicates that the same set of hydrocarbon, catalyst and substrate may act differently in slightly different experimental conditions (temperature, pressure, etc.).

In 1984, Tibbetts explained why catalytically-grown carbon nanofibers were tubular. Because the surface free energy of the (002) basal plane of graphite is exceptionally low, the free energy required for a filament growth is minimum when graphite is in the form of a seamless cylinder circumfering the metal. And the inner core is hollow because inner cylindrical planes of small diameter would be highly strained, energetically unfavorable to form. He also explained the CNT growth mechanism with a vapor-liquid-solid (VLS) model, originally formulated for Si, Ge whiskers and many other crystals (Wagner et al., 1965). Although this model is convincing and acceptable to a great extent, it is often doubted how Fe, Co, Ni etc. (normal melting point ~1500°C) could be in liquid state within 600–900°C, the growth temperature of typical CNTs in CVD. Here it is important to note that the melting point of nanoparticles below 10 nm falls abruptly (Fig. 4). For instance, an 8-nm Fe and Au particle (or 4-nm Ni particle) can melt at about 800°C. Typical CNT growth temperature range is 700–900°C, implying that in some cases (>800°C) the catalyst metal may be in liquid state, while in some cases (<800°C) it may be in solid state. Also, in any experiment, all particles are not strictly of the same size. So, it is still hard to say on the metal's state authoritatively. However, recalling that hydrocarbon decomposition on metal surface is an exothermic reaction, it is likely that the extra heat generated during hydrocarbon decomposition helps metal liquefaction to some extent. Hence the opinion of active catalyst being in liquid phase wins, as reported by many scientists for SWCNT growth (Ding et al., 2004; Harutyunyan et al., 2005). But then, what about the case of MWCNTs which usually grow on bigger (>20 nm) metal particles? Bigger particles must be in solid phase; and in turn, MWCNT would involve a different growth mechanism than that of SWCNT!!

Another reasonable disagreement between the SWCNT and MWCNT growth is on the existence of temperature gradient inside the metal catalyst. Baker's explanation of temperature-gradient driven fiber growth might be applicable to MWCNTs which involve big catalyst particles. In the case of SWCNTs, however, it is very hard to imagine a significant temperature gradient within a particle of 1-2 nm. Hence SWCNT growth must be driven by the carbon concentration gradient during the process. A molecular dynamics simulation study suggests the possibility of SWCNT growth without any temperature gradient in the metal (Ding et al., 2006).

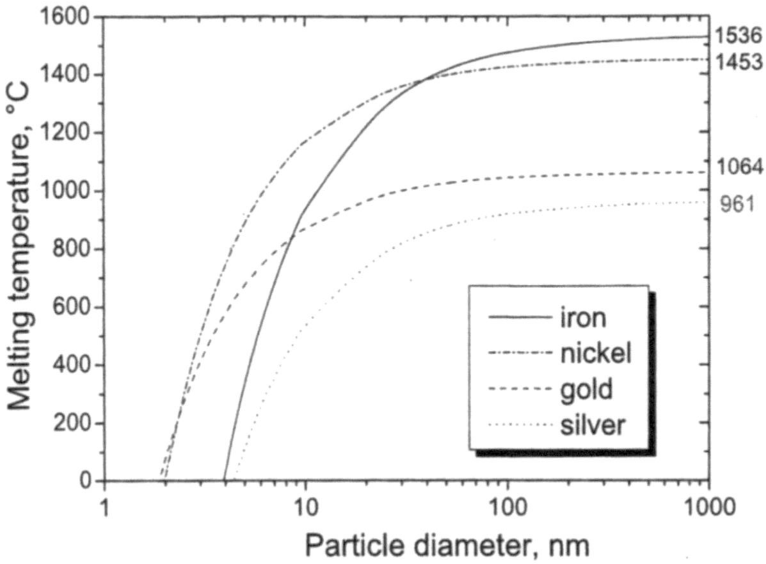


Fig. 4. Melting temperature of selected metals as a function of particle diameter. (Moisala et al., 2003. *J. Phys.: Condens. Mater.* 15, S3011-S3035. © IOP Publishing Ltd.)

### 3.2 Mode of carbon diffusion

Another highly-debatable question is whether the so-called diffusion of carbon species through metal particle is surface diffusion or bulk (volumetric) diffusion. Endo's group who extensively carried out benzene decomposition on iron catalyst at 1100°C (Oberlin et al., 1976), argued that hollow fiber could form only by surface diffusion on the metal particle, as earlier proposed by Baird et al. (1971). In 2004, Helveg et al. succeeded in observing MWCNT growth from methane decomposition at 500°C on Ni catalyst in a high-resolution TEM. They noted that, throughout the growth process, the nickel cluster remained crystalline with well-faceted shapes (Fig. 5). The graphite layers were found to grow as a consequence of dynamic interaction between carbon and nickel atoms. 'Surface atoms' of the nickel cluster moved up and down, in and out (continuously changing the metals' surface texture) as if they were knitting a graphene sheet out of the surrounding carbon atoms. The nanocluster shape was periodically changing its shape from spherical to cylindrical, aligning the graphene layers around them. The authors proposed that the mono-atomic steps on the cluster boundary played a key role in anchoring carbon atoms and knitting the graphene network. This observation reveals that the catalyst is in solid phase and the carbon diffusion is a surface diffusion around the catalyst.

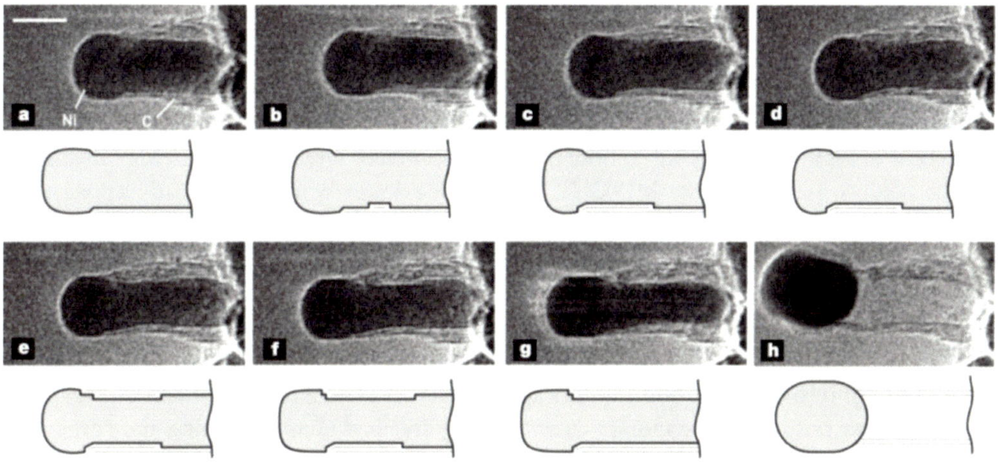


Fig. 5. In situ HRTEM image sequence of a growing carbon nanofiber. Images (a–h) illustrate one cycle in the elongation/contraction process. Drawings are included to guide the eye in locating the positions of mono-atomic Ni step-edges at the graphene–Ni interface. Scale bar = 5 nm. (Helveg et al., 2004. *Nature* 427, 426-429. © Nature Publishing Group)

Later, Raty et al. (2005) reported a molecular dynamics simulation study of the early stages of SWCNT growth on metal nanoparticles. They showed that carbon atoms diffuse only on the outer surface of the metal cluster. At first, a graphene cap is formed which floats over the metal, while the border atoms of the cap remain anchored to the metal. Subsequently, more C atoms join the border atoms pushing the cap up and thus constituting a cylindrical wall (Fig. 6).

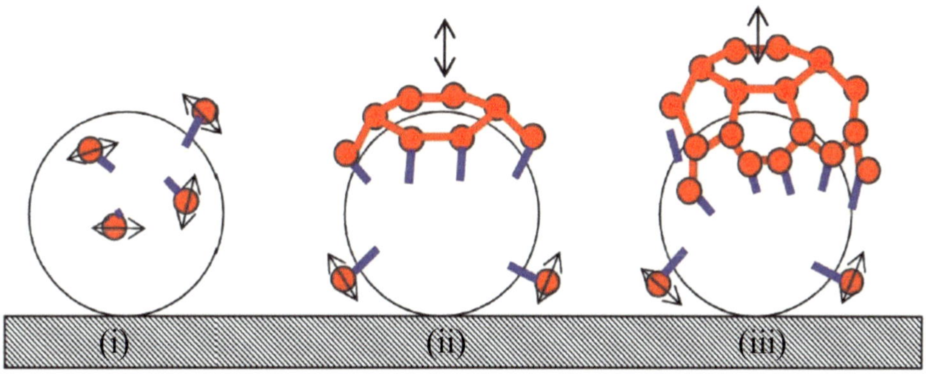


Fig. 6. Schematic representation of the basic steps of SWCNT growth on a Fe catalyst, as observed in *ab initio* simulations. (i) Diffusion of single C atoms (red spheres) on the surface of the catalyst. (ii) Formation of an $sp^2$ graphene sheet floating on the catalyst surface with edge atoms covalently bonded to the metal. (iii) Root incorporation of diffusing single C atoms. (Raty et al., 2005. *Phys. Rev. Lett.* 95, 096103. © American Physical Society)

In 2007, Hofmann et al. reported an atomic-scale environmental TEM observation of CNT growth from acetylene decomposition on Ni particles at 480°C. Figure 7 shows the growth stage at different times (as mentioned therein) captured from a continuous video recording.

Initially, the Ni cluster had a round shape which transformed into an elongated shape (perpendicular to the substrate) surrounded by a thin carbon layer. This elongation was in contact with the substrate up to 0.8s and suddenly (at 0.87s), the Ni cluster left the substrate contact and contracted upward taking a round shape and leaving behind a hollow carbon tube. This elongation and contraction of Ni re-occurred alternately, moving ahead and leaving behind a bamboo-like MWCNT grown. The inner walls of the tube appeared to emerge from step-like stages of Ni cluster (see HRTEM and corresponding model in Fig. 7), suggesting that carbon atoms also diffuse deep inside the Ni cluster and crystallize in the form of inner tube walls when the Ni cluster moves up (contracts back to round shape at the CNT tip). This was a MWCNT observation via tip growth model. In another experiment at 615°C, the authors observed SWCNT formation from a small Ni cluster via base growth model (Fig. 8). Initially, a carbon cap emerged with a diameter smaller than the Ni cluster. Then, the apex portion of the cluster assumed a cylindrical shape, pushing the carbon cap off the cluster and forming a SWCNT. Finally, the CNT network expanded upward by itself. These evidences also explain the general experience that small nanoparticles are crucial for SWCNT formation. Small metal clusters (1–2 nm) have sharp edges (atomic steps); hence they possess high catalytic activity and are capable to form high-strain SWCNTs. With the increasing cluster size, the sharpness of the atomic steps at the cluster boundary decreases and so does their catalytic activity. Therefore, bigger metal clusters (5–20 nm) form less-strained MWCNTs. Too big clusters (viz. 100 nm) acquire almost spherical boundary with no sharp steps; that is why they do not form CNTs at all.

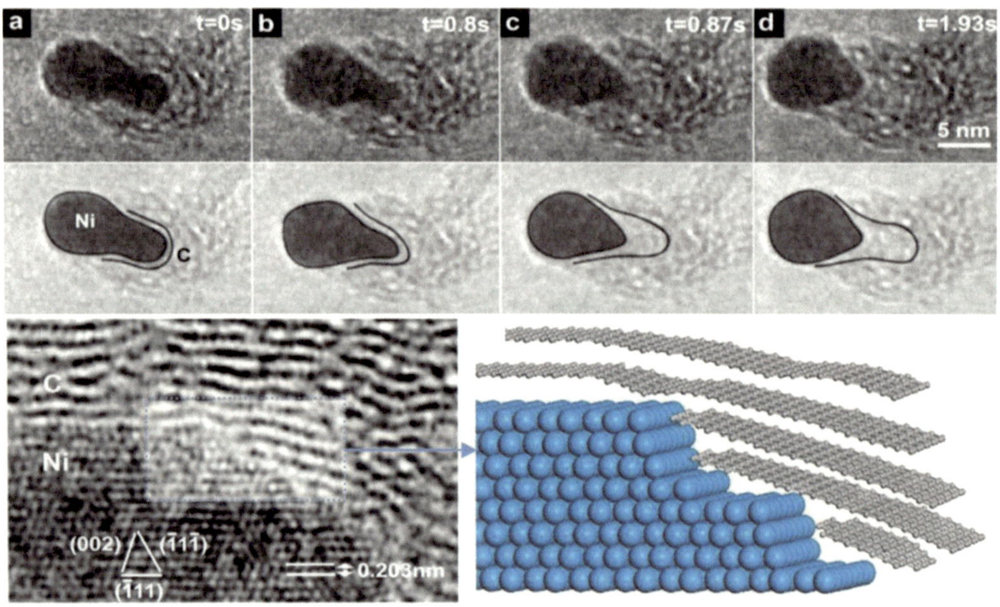


Fig. 7. In-situ TEM image sequence showing a MWCNT nucleation and growth, and corresponding growth scheme. (Hofmann et al., 2007. Nano Lett. 7, 602-608. © American Chemical Society)

Quite intriguingly, however, two months after Hofmann's (2007) report, Rodriguez-Manzo et al. (2007) reported an exciting observation of CNT formation in an HRTEM by simply holding a metal-encapsulated MWCNT at 600°C under electron beam (300kV) for 90 min. Carbon atoms from the side walls (the existing graphite layers around the encapsulated metal) got injected into the metal bulk and emerged in the form of new SW, DW and MWCNTs of smaller diameters coaxial to the original MWCNT (Fig. 9). Such a prolonged observation of the CNT-growth dynamics (atom-by-atom) clearly evidences that carbon diffuses through the metal bulk (volume diffusion). Nevertheless, we should note that this observation was an exclusive case of rearrangement of the carbon-iron ensemble inside a constrained nanoreactor (the original MWCNT) under high-energy electron-beam irradiation, a situation far away from usual CVD conditions. Hence such bulk diffusion cannot be conceptualized as a general CNT growth mechanism.

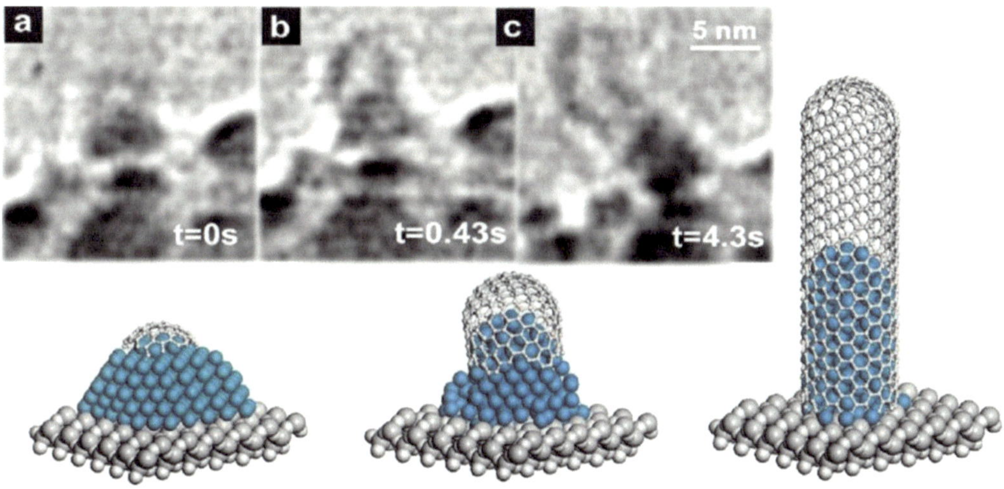


Fig. 8. In-situ TEM image sequence showing a SWCNT nucleation and growth, and corresponding growth scheme. (Hofmann et al., 2007. *Nano Lett.* 7, 602-608. © American Chemical Society)

In the context of the changing metal shape during CVD, it is pertinent to mention another aspect of the CNT growth. Many a time we encounter CNTs with their graphene layers inclined to the tube axis (herringbone or stacked-cup structure). It is puzzling to think how they form. Keeping in mind that graphite layers grow preferentially on selected crystal planes of metal, this can be understood as follows. The shape of the catalyst metal cluster acts as a template for the surrounding graphene layers. Nanoclusters (say, 10–20 nm Ø) under suitable thermodynamic conditions, tend to form an elongated cylindrical shape (viz., 3 nm Ø) so that CNTs grow with the graphene layers parallel to the tube axis. Under certain (different) thermodynamic conditions the metal clusters tend to become pear-shaped, giving birth to graphene layers parallel to their inclined facets. This usually happens with bigger clusters (say, 100 nm) or for alloy catalysts (Kim et al., 1992). When it comes to explain how such open-edged nanographenes are energetically stable, scientists suggest that those dangling bonds at the edges of the stacked graphite platelets are stabilized with the hydrogen atoms expelled from the hydrocarbon or from the $H_2$ supply (Nolan et al., 1998). Such a fiber is known as graphite whisker, and many people do not consider that as a CNT.

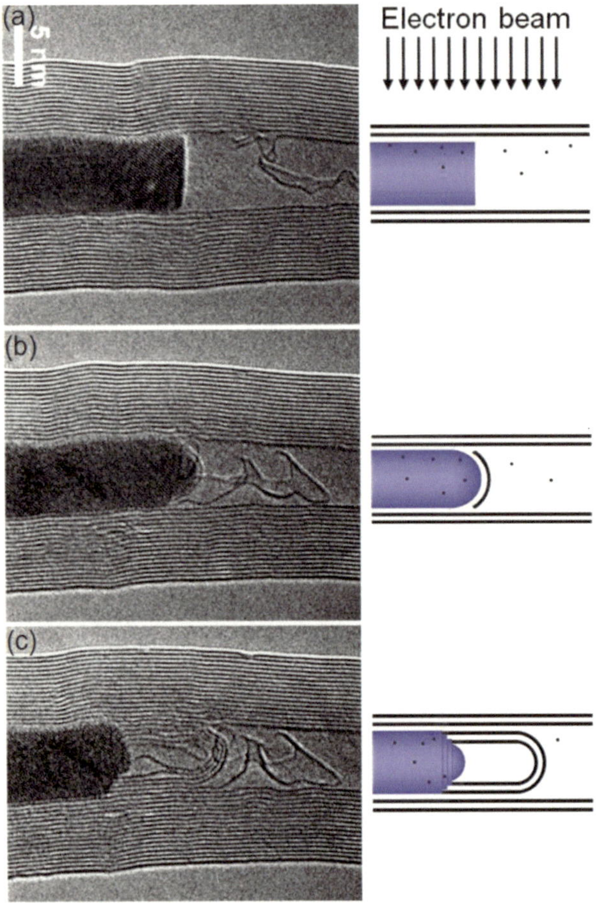


Fig. 9. In-situ observation of CNT growth under HRTEM. (a) Electron beam knocks some carbon atoms from the MWCNT side walls into the encapsulated metal cluster. (b) The metal cluster reshapes its flat cross section into a convex dome and a carbon cap appears over the dome. (c) At the base of the metal dome, atomic steps develop and new MWCNTs emerge coaxial to the original MWCNT. (Image courtesy: M. Terrones)

### 3.3 Chemical state of the catalyst

Another frequently debated point in the CNT growth mechanism is about the chemical state of the active catalyst. Most common concept is that the starting catalyst material (pre-deposited on substrates) is usually in oxide form. Even if we deposit fresh metal nanoparticles on a substrate, the nanoparticles get quickly oxidized when exposed to oxygen during the substrate transfer to the CVD reactor. During CVD, hydrogen gas is supplied to reduce the metal oxide into pure metal upon which hydrocarbon decomposition and subsequent diffusion leads to the CNT growth. Even when no hydrogen is supplied externally, the hydrogen atoms liberated from the hydrocarbon decomposition on the catalyst surface are likely to serve the same. However, there are many conflicting reports right from the early-stage CVD experiments. Baker and many others proposed that pure metal is the active catalyst (Baker et al., 1982; Yang et al., 1986), while Endo and many others detected the encapsulated particles (in the CNTs) to be iron carbide (Oberlin et al., 1976; Ducati et al., 2004).

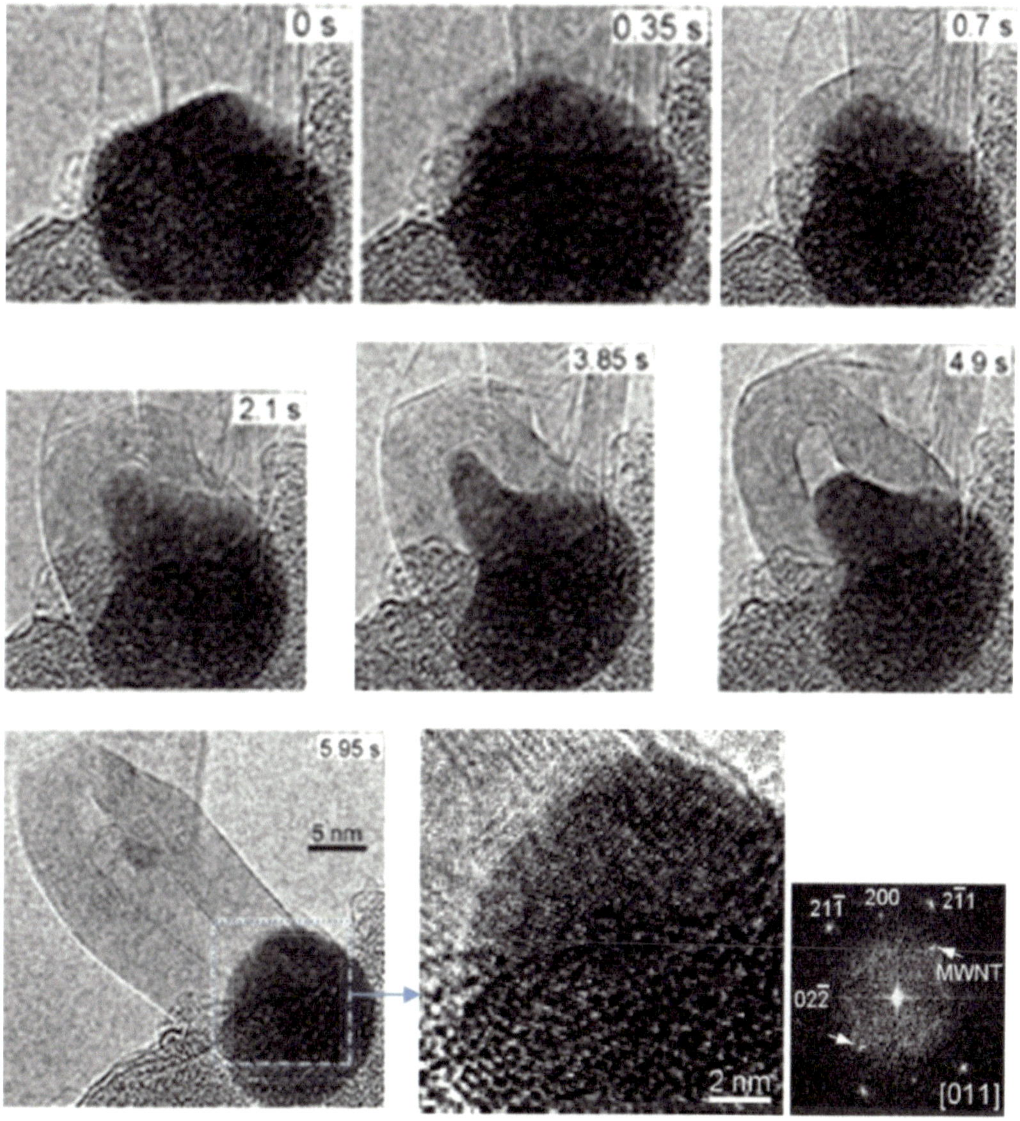


Fig. 10. In-situ TEM observation of MWCNT nucleation and growth as a consequence of acetylene decomposition on iron catalyst at 600°C. Initially, graphene layers appear around the metal cluster. Subsequently, the metal cluster assumes a conical shape, elongates upward (expelling a CNT), and finally, comes back to round shape. Fourier transform of the enlarged image of the metal cluster at 5.96s suggests it to be $Fe_3C$. (Yoshida et al., (2008). *Nano Lett.* 8, 2082-2086. © American Chemical Society)

In 2008, Yoshida et al. performed atomic-scale in-situ observation of acetylene decomposition on Fe catalyst at 600°C and $10^{-2}$ torr. Both SWCNT and MWCNT were clearly observed to be growing from metal particles rooted on the substrate (base-growth model). Electron diffraction analysis of the metal clusters in each frame was reported to match with that of iron carbide in cementite ($Fe_3C$) form (Fig. 10). Accordingly, the authors concluded that the active catalyst was in 'fluctuating solid state' of 'iron carbide'; the carbon diffusion was volumetric; and all layers of the MWCNTs grew up simultaneously, at the same growth rate.

However, Wirth et al. (2009), based on their in-situ electron microscopy and XPS analyses, emphatically advocate that the catalyst exists in pure metallic form: right from the CNT nucleation to the growth termination (Fig. 11). When the CNT growth ceases due to catalyst poisoning with excess carbon, that supersaturated metal-carbon assembly crystallizes in carbide form upon cooling. Confusion persists because lattice constants of pure metal and their carbide or oxide are very close. For instance, 2Å reflection is possible from fcc Ni(111) or $Ni_3C$(113) or $Ni_2O_3$(200). Moreover, for 'nano' particles, some distortion in the lattice constants is expected due to the small-size effect. The authors also suggest that the catalyst particle undergoes severe mechanical re-shaping during the tip growth of multi-wall nanotubes. This looks like the metal is in liquid state. However, this shape distortion occurs due to relative displacement of different atomic layers (in solid state) due to the large forces exerted by the surrounding CNT in the growth stage.

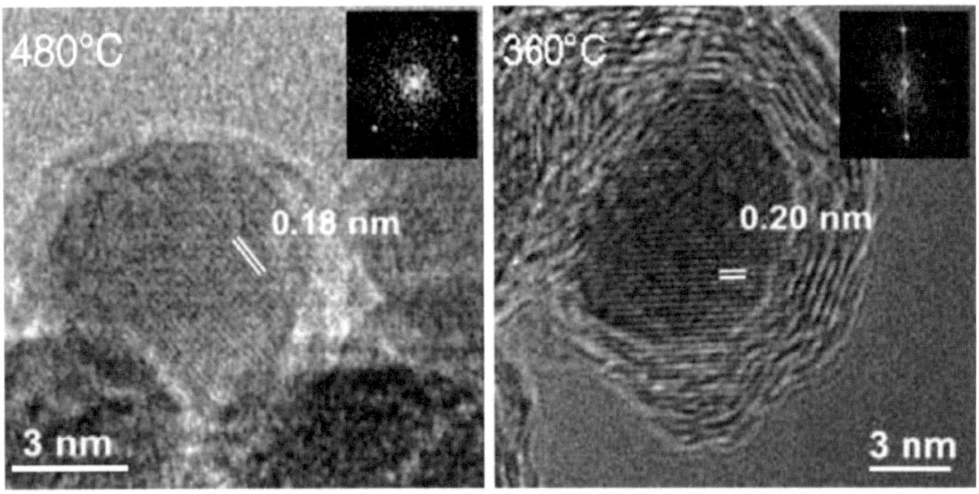


Fig. 11. In-situ TEM images of CNT growth from acetylene decomposition on Ni particles at 480 °C and 360°C. Clear observation of lattice planes suggests the metal to be in crystalline (solid) state. (Wirth et al., 2009. *Diamond & Related Mater.* 18, 940-945. © Elsevier Science)

Among the latest literature, Chen et al. (2011) performed in-situ resonance Raman spectroscopy and XRD photoelectron spectroscopy of ethanol CVD directly on $SiO_2$ nanoparticles without any metal. Their results suggest that, during the CVD, $SiO_2$ does not form any SiC phase and the SWCNT nucleation and growth occur by carbon diffusion on solid $SiO_2$ particles, following a vapor-solid-solid (VSS) model, instead of the conventional vapor-liquid-solid (VLS) model (Fig. 12).

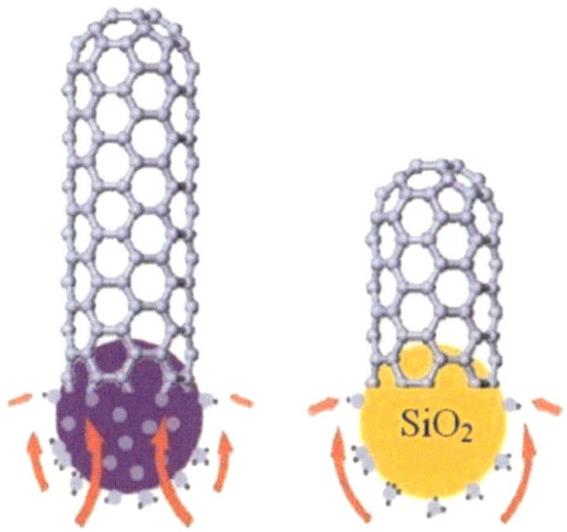


(a) VLS model (b) VSS model

Fig. 12. (a) VLS model for SWCNT growth from a metal nanoparticle: volume diffusion of C atoms (vapor) in metal (liquid) to crystallize a CNT (solid). (b) VSS model for the SWCNT growth from an $SiO_2$ nanoparticle: surface diffusion of carbon atoms (vapor) on $SiO_2$ (solid) resulting in a CNT (solid). (Chen et al., 2011. *Carbon* 49, 3316-3324. © Elsevier Science)

These diversities of observation unambiguously reflect that we have not matured enough to understand the world inside the nanotubes. Hence more research is required to establish a concrete growth mechanism. Then only we would be able to make CNTs of specific properties, and in turn, CNT would be able to fulfill the expectations and needs of the society.

## 4. Existing challenges and future directions

In the foregoing sections, we raised several questions on the roles of precursor, catalyst, catalyst support and growth mechanism, and indicated possible directions. In addition to those basic issues, other growth-related challenges are briefly outlined below.

1. Researchers have succeeded in minimizing the diameter distribution of SWCNTs up to some extent. However, synthesis of SWCNTs of a given diameter is yet to be achieved. It would be possible only when we have the catalyst particles all of exactly the same diameter (say, 0.5 nm).
2. Chirality control is even more challenging. Re-growth from ordered arrays of open-ended SWCNTs may help up to some extent. Alternatively, we have to develop proper separation methods that could first sort out CNTs according to metallic or semiconducting tubes and then select tubes of specific chirality.
3. In MWCNTs, control on the number of walls is another big challenge. Synthesis of thin MWCNTs (3–6 walls) is a better choice than thick MWCNTs.
4. Growth of isolated CNTs has not yet reached a mature stage. There are indications that even a single CNT does not possess the same diameter and chirality over the entire length. How can we solve it?
5. Researchers have succeeded in growing CNTs from almost all metals. However, we do not know how different metals affect the physical, chemical, electronic and optical and magnetic properties of as-grown CNTs. If we could discover any correlation between

the type of the metal used and the property of the CNT grown, we would be able to grow CNTs of selective properties.

6. What is/are the determining steps in CNT nucleation and growth? Having known these steps, the CNT growth rate could be increased for mass production.
7. The exact role of $H_2$, $O_2$ and $H_2O$ in CNT growth is yet to be clarified. Simultaneous presence of reducing as well as oxidizing agents in the reaction zone makes it ambiguous whether amorphous carbon is etched by atomic hydrogen, oxygen or water. Are they really essential?
8. Researchers have succeeded in bringing down the CNT growth temperature to ~400°C in low-pressure CVD. However, low-pressure CVD greatly reduces the growth rate and yield. Low-temperature CNT growth must be devised at atmospheric pressure for high yields of CNTs.
9. Many technological applications are looking for room-temperature CNT growth which is still a dream, so far as thermal CVD is concerned.
10. Mass-produced CNTs usually contain catalyst particles or support materials as impurity. Post-deposition purification greatly reduces the CNT quality and final output.
11. CVD-grown CNTs (especially low-temperature MWCNTs) have poor crystallinity. With a suitable combination of different catalysts, it should be possible to get better-crystallinity CNTs.
12. Recent metal-free oxygen-assisted CNT growth is a breakthrough. It must be scaled up to mass production of high-purity CNTs.
13. Carbon-metal phase diagram needs to be reconstructed, especially for 1–5 nm range, relevant to CNT growth.
14. All extraordinary properties of CNTs are predicted for atomically-perfect CNTs. To make those predictions true, it is of prime importance to develop new techniques to monitor and remove defects during the growth.
15. Lack of the quality control and assessment of the CNTs synthesized by different groups by different methods does not allow us to get the correct product details. Analytical sampling of CNTs obtained from different sources at an authorized standard laboratory would reveal exact merits and demerits of different techniques, which would in turn help us explore combinations of techniques toward high-yield, high-purity and low-cost mass production.

## 5. Conclusion

As we have seen in the foregoing sections, despite extensive progress over the years, there are many basic issues concerning the CNT growth mechanism which are still not clear. Contradictory observations of CNT growth under electron microscopy by different groups suggest that the mechanism is extremely sensitive to each parameter such as carbon precursor, metal catalyst, particle size, temperature, pressure. Even a minor change in any of these parameters leads the growth in critically different directions. Catalysis is the main stem of CVD-CNT technique; and it seems that we have not yet utilised the best of catalysis in this field. New nano-catalyst materials are needed to be developed and investigated in more detail. In principle, with the use of a suitable catalyst, the CVD temperature can be brought down to room temperature. By identifying the growth-limiting steps it should be possible to control the diameter and chirality of the resulting CNTs. To comply with the environmental concerns, renewable materials should be explored as CNT precursors. In view of the expected giant demand of CNTs in the near future, industrial production of CNTs should be carried out with far-sighted thoughts for long-term sustainability. Fossil-fuel based CNT-production technology would not be sustainable. The unanswered questions about growth mechanism and the existing problems concerning the growth control will keep the CNT researchers engaged for a long time. Thus, there is no doubt that CNT research will continue to remain a hot topic, a prospective research area.

## 6. References

Afre R.A., Soga T., Jimbo T., Kumar M., Ando Y. and Sharon M. (2005). *Chem. Phys. Lett.* 414, 6-10.

Andrews R.J., Smith C.F. and Alexander A.J. (2006). *Carbon* 44, 341-347.

Antunes E.F., Almeida E.C., Rosa C.B.F., Medeiros L.I., Pardini L.C., Massi M. and Corat E.J. (2010). *J. Nanosci. Nanotechnol.* 10, 1296-1303.

Baird T., Fryer J.R. and Grant B. (1971). *Nature* 233, 329-330.

Baker R.T.K., Barber M.A., Harris P.S., Feates F.S. and Waite R.J. (1972). *J. Catalysis* 26, 51-62.

Baker R.T.K., Harris P.S., Thomas R.B. and Waite R.J. (1973). *J. Catalysis* 30, 86-95.

Baker R.T.K. and Waite R. J. (1975). *J. Catalysis* 37, 101-105.

Baker R.T.K., Alonzo J.R., Dumesic J.A. and Yates D.J.C. (1982). *J. Catalysis* 77, 74-84.

Bianco S., Giorcelli M., Musso S., Castellino M., Agresti F., Khandelwal A., Russo S.L., Kumar M., Ando Y. and Tagliaferro A. (2010). *J. Nanosci. Nanotechnol.* 10, 3860-3866.

Bower C., Zhou O., Zhu W., Werder D.J. and Jin S. (2000). *Appl. Phys. Lett.* 77, 2767-2769.

Camilli L., Scarselli M., Gobbo S.D., Castrucci P., Nanni F., Gautron E., Lefrant S. and Crescenzi M.D. (2011). *Carbon* 49, 3307-3315.

Cassell A.M., Raymakers J.A., Kong J. and Dai H., (1999). *J. Phys. Chem. B* 103, 6484-6492.

Chen Y. and Zhang J. (2011). *Carbon* 49, 3316-3324.

Dai H., Rinzler A.G., Nikolaev P., Thess A., Colbert D.T. and Smalley R.E. (1996). *Chem. Phys. Lett.* 260, 471-475.

Ding F., Bolton K. and Rosen A. (2004). *J. Chem. Phys. B* 108, 17369-17377.

Ding F., Bolton K. and Rosen A. (2006). *Comput. Mater. Sci.* 35, 243-246.

Ding F., Larsson P., Larsson J.A., Ahuja R., Duan H., Rosen A. and Bolton K. (2008). *Nano Lett.* 8, 463-468.

Ducati C., Alexandrou I., Chhowalla M., Robertson J. and Amaratunga G.A.J. (2004). *J. Appl. Phys.* 95, 6387-6391.

Endo M., Fujiwara H. and Fukunaga E. (1991). 18th Meeting Japanese Carbon Society Saitama (Dec 1991) p34.

Fan S., Chapline M., Frankline N., Tombler T., Cassell A.M. and Dai H. (1999). *Science* 283, 512-514.

Ghosh K., Kumar M., Maruyama T. and Ando Y. (2009). *Carbon* 47, 1565-1575.

Ghosh P., Soga T., Tanemura M., Zamri M., Jimbo T., Katoh R. and Sumiyama K. (2009). *Appl. Phys. A* 94, 51-56.

Han C.Y., Xiao Z.L., Wang H.H., Lin X.M., Trasobares S. and Cook R.E. (2009). *J. Nanomaterials* 2009, 562376(11pages).

Han S., Liu X. and Zhou C. (2005). *J. Am. Chem. Soc.* 127, 5294-5295.

Harutyunyan A.R. and Tokune T. (2005). *Appl. Phys. Lett.* 87, 51919-15921.

Hata K., Futaba D.N., Mizuno K., Namai T., Yumura M. and Iijima S. (2004). *Science* 306, 1362-1364.
He D., Li H., Li W., Haghi-Ashtiani P., Lejay P. and Bai J. (2011). *Carbon* 49, 2273-2286.
Helveg S., Lopez-Cartes C., Sehested J., Hansen P.L., Clausen B.S., Rostrup-Nielsen J.R., Abild-Pedersen F. and Nørskov J.K. (2004). *Nature* 427, 426-429.
Hofmann S., Sharma R., Ducati C., Du G., Mattevi C., Cepek C., Cantoro M., Pisana S., Parvez A., Cervantes-Sodi F., Ferrari A.C., Dunin-Borkowski R., Lizzit S., Petaccia L., Goldoni A. and Robertson J. (2007) *Nano Lett.* 7, 602-608.
Huang S., Cai Q., Chen J., Qian Y. and Zhang L. (2009). *J. Am. Chem. Soc.* 131, 2094-2095.
Iijima S. (1991). *Nature* 354, 56-58.
Jose-Yacaman M., Miki-Yoshida M., Rendon L. and Santiesteban J. G., (1993). *Appl. Phys. Lett.* 62, 657-659.
Kang Z., Wang E., Mao B., Su Z., Chen L. and Xu L. (2005). *Nanotechnology* 16, 1192-1195.
Kim M.S., Rodriguez N.M. and Baker R.T.K. (1992) *J. Catalysis* 134, 253-268.
Kucukayan G., Kayacan S., Baykal B. and Bengu E. (2008). *Mater. Res. Soc. Symp. Proc. (Symposium-P)* Vol. 1081E, P05-14.
Kumar M., Zhao X. and Ando Y. (2001). *International Symposium on Nanocarbons, Nagano, Japan (November 14-16, 2001) Extended Abstract*: 244-245.
Kumar M., Zhao X., Ando Y., Iijima S., Sharon M. and Hirahara K. (2002). *Mol. Cryst. Liq. Cryst.* 387, 117-121.
Kumar M. and Ando Y. (2003a). *Diamond Related Mater.* 12, 998-1002.
Kumar M. and Ando Y. (2003b). *Diamond Related Mater.* 12, 1845-1850.
Kumar M. and Ando Y. (2003c). *Chem. Phys. Lett.* 374, 521-526.
Kumar M., Kakamu K. Okazaki T. and Ando Y. (2004). *Chem. Phys. Lett.* 385, 161-165.
Kumar M. and Ando Y. (2005). *Carbon* 43, 533-540.
Kumar M., Okazaki T. Hiramatsu M. and Ando Y. (2007). *Carbon* 45, 1899-1904.
Kumar M. and Ando Y., (2008). *Defence Science Journal* 58, 496-503.
Kusunoki M., Suzuki T., Hirayama T. and Shibata N. (2000). *Appl. Phys. Lett.* 77, 531-533.
Kyotani T., Tsai L.F. and Tomita A. (1996). *Chem. Mater.* 8, 2109-2113.
Li J., Papadopoulos C., Xu J.M. and Moskovits M. (1999). *Appl. Phys. Lett.* 75, 367-369.
Li N., Chen X., Stoica L., Xia W., Qian J., Abmann J., Schuhmann W. and Muhler M. (2007). *Adv. Mater.* 19, 2957-2960.
Liu B., Ren W., Gao L., Li S., Pei S., Liu C., Jiang C. and Cheng H.M. (2009). *J. Am. Chem. Soc.* 131, 2082-2083.
Liu H., Takagi D., Chiashi S. and Homma Y. (2010). *Carbon* 48, 114-122.
Maret M., Hostache K., Schouler M.C., Marcus B., Roussel-Dherbey F., Albrecht M. (2007). *Carbon* 45, 180–187.
Maruyama S., Kojima R., Miyauchi Y., Chiashi S. and Kohno M. (2002). *Chem. Phys. Lett.* 360, 229-234.
Mattevi C., Wirth C.T., Hofmann S., Blume R., Cantoro M., Ducati C., Cepek C., Knop-Gericke A., Milne S., Castellarin-Cudia C., Dolafi S., Goldoni A., Schloegl R. and Robertson J. (2008). *J. Phys. Chem. C* 112, 12207-12213.
Merchan-Merchan W., Savaliev A., Kennedy L.A. and Fridman A. (2002). *Chem. Phys. Lett.* 354, 20-24.
Moisala A., Nasibulin A.G. and Kauppinen E.I. (2003). *J. Phys. Condens. Mater.* 15, S3011-S3035.

Murakami Y., Chiashi S., Miyauchi Y., Hu M., Ogura M., Okubo T. and Maruyama S. (2004). *Chem. Phys. Lett.* 385, 298-303.
Musso S., Porro S., Giorcelli M., Chiodoni A., Ricciardi A.C. and Tagliaferro A. (2007). *Carbon* 45, 1133-1136.
Musso S., Porro S., Rovere M., Giorcelli M. and Tagliaferro A. (2008). *J. Cryst. Growth* 310, 477-483.
Nerushev O.A., Dittmar S., Morjan R.E., Rohmund F. and Campbell E.E.B. (2003). *J. Appl. Phys.* 93, 4185-4190.
Noda S., Hasegawa K., Sugime H., Kakehi K., Zhang Z., Maruyama S. and Yamaguchi Y. (2007). *Jpn. J. Appl. Phys.* 46, L399-L401.
Nolan P.E., Lynch D.C. and Cutler A.H. (1998). *J. Phys. Chem. B* 102, 4165-4175.
Nunez J.D., Maser W.K, Mayoral M.C., Andres J.M. and Benito A.M. (2011). *Carbon* 49, 2483-2491.
Oberlin A., Endo M. and Koyama T. (1976). *J. Cryst. Growth* 32, 335-349.
Parshotam H. (2008). *M.Tech. Thesis, Chem. Dept., University of Johannesburg.*
Parthasarathy R.V., Phani K.L.N. and Martin C.R. (1995). *Adv. Mater.* 7, 896-897.
Pradhan D. and Sharon M. (2002). *Mater. Sci. Engg. B* 96, 24-28.
Qian W., Yu H., Wei F., Zhang Q. and Wang Z. (2002). *Carbon* 40, 2968-2970.
Qiu J., Li Q., Wang Z., Sun Y. and Zhang H. (2006). *Carbon* 44, 2565-2568.
Raty J.Y., Gygi F. and Galli G. (2005). *Phys. Rev. Lett.* 95, 096103(4pages).
Reich S., Li L. and Robertson J. (2006). *Chem. Phys. Lett.* 421, 469-472.
Ritschel M., Leonhardt A., Elefant D., Oswald S. and Buchner B. (2007). *J. Phy. Chem. C* 111, 8414-8417.
Rodriguez-Manzo J.A., Terrones M., Terrones H., Kroto H.W., Sun L. and Banhart F. (2007). *Nature Nanotechnology* 2, 307-311.
Rummeli M.H., Borowiak-Palen E., Gemming T., Pichler T., Knupfer M., Kalbac M., Dunsch L., Jost O., Silva S.R.P., Pompe W. and Buchner B. (2005). *Nano Lett.* 5, 1209-1215.
Rummeli M.H., Kramberger C., Gruneis A., Ayala P., Gemming T., Buchner B. and Pichler T. (2007a). *Chem. Mater.* 19, 4105-4107.
Rummeli M.H., Schaffel F., Kramberger C., Gemming T., Bachmatiuk A., Kalenczuk R.J., Rellinghaus B., Buchner B. and Pichler T. (2007b). *J. Am. Chem. Soc.* 129, 15772-15773.
Sankaran M. and Viswanathan B. (2008). *Indian J. Chem.* 47A, 808-814.
Satishkumar B.C., Thomas P.J., Govindaraj A. and Rao C.N.R. (2000). *Appl. Phys. Lett.* 77, 2530-2532.
Schneider J.J., Maksimova N.I., Engstler J., Joshi R., Schierholz R. and Feile R. (2008). *Inorganica Chimica Acta* 361, 1770-1778.
Sen R., Govindaraj A. and Rao C.N.R. (1997). *Chem. Phys. Lett.* 267, 276-280.
Sharon M., Hsu W.K., Kroto H.W., Walton D.R.M., Kawahara A., Ishihara T. and Takita Y. (2002). *J. Power Sources* 104, 148-153.
Sharon M. and Sharon M. (2006). *Synthesis and Reactivity in Inorganic, Metal-Organic and Nano-Metal Chemistry* 36, 265-279.
Su M., Li Y., Maynor B., Buldum A., Lu J.P., Liu J. (2000). *J Phys Chem B* 104, 6505–6508.
Takagi D., Hibino H., Suzuki S., Kobayashi Y. and Homma Y. (2007). *Nano Lett.* 7, 2272-2275.
Takagi D., Kobayashi Y., Hibino H., Suzuki S. and Homma Y. (2008). *Nano Lett.* 8, 832-835.
Takagi D., Kobayashi Y. and Homma Y. (2009). *J. Am. Chem. Soc.* 131, 6922-6923.

Tang J., Jin G., Wang Y. and Guo X. (2010). *Carbon* 48, 1545-1551.
Tang Z.K., Sun H.D., Wang J., Chen J. and Li G. (1998). *Appl. Phys. Lett.* 73, 2287-2289.
Terrones M., Grobert N., Olivares J., Zhang J.P., Terrones H., Kordatos K., Hsu W.K., Hare J.P., Townsend P.D., Prassides K., Cheetham A.K., Kroto H.W. and Walton D.R.M. (1997). *Nature* 388, 52-55.
Thess A., Lee R., Nikolaev P., Dai H., Petit P., Robert J., Xu C. and Smalley R. (1996). *Science* 273, 483-487.
Tibbetts G.G. (1984). *J. Cryst. Growth* 66, 632-638.
Vanderwal R.L., Ticich T.M. and Curtis V.E. (2001). *Carbon* 39, 2277-2289.
Wagner R.S. and Ellis W.C. (1965). *Trans. Metallurg. Soc. AIME* 233, 1053-1064.
Wirth C.T., Hofmann S. and Robertson J. (2009). *Diamond Related Mater.* 18, 940-945.
Xiang R., Einarsson E., Okawa J., Miyauchi Y. and Maruyama S., (2009). *J. Phys. Chem. C*, 113, 7511-7515.
Yamada K., Abe K., Mikami M., Saito M. and Kuwano J. (2006). *Key Engg. Mater.* 320, 163-166.
Yang K.L. and Yang R.T. (1986). *Carbon* 24, 687-693.
Yoshida H., Takeda S., Uchiyama T., Kohno H. and Homma Y. (2008). *Nano Lett.* 8, 2082-2086.
Yuan D., Ding L., Chu H., Feng Y., McNicholas T.P. and Liu J. *Nano Lett.* (2008). 8, 2576-2579.
Yudasaka M., Kikuchi R., Ohki Y. and Yoshimura S. (1997). *Carbon* 35, 195-201.
Zhai J., Tang Z., Sheng P. and Hu X. (2006). *IEEE Conf. 2006: (ICONN-2006)* 155-158.
Zhao J., Guo X., Guo Q., Gu L., Guo Y. and Feng F. (2011). *Carbon* 49, 2155-2158.
Zhong G., Hofmann S., Yan F., Telg H., Warner J.H., Eder D., Thomsen C., Milne W.I. and Robertson J. (2009). *J. Phys. Chem. C* 113, 17321-17325.
Zhou W., Han Z., Wang J., Zhang Y., Jin Z., Sun X., Zhang Y., Yan C. and Li Y. (2006). *Nano Lett.* 6, 2987-2990.

# 9

# Dielectrophoretic Deposition and Alignment of Carbon Nanotubes

Wei Xue and Pengfei Li
*Washington State University, Vancouver, WA*
*U.S.A.*

## 1. Introduction

The carbon nanotube (CNT) is a unique form of carbon material. Since its discovery, CNT has been intensively studied due to its remarkable electrical, mechanical, thermal, and chemical properties (Iijima, 1991; Katz & Willner, 2004). Based on the structures and dimensions, CNTs can be divided into two groups: single-walled carbon nanotubes (SWNTs) and multi-walled carbon nanotubes (MWNTs). An SWNT is one tube of graphene capped at both ends and it consists of only surface carbon atoms; its diameter is in the range of 1-2 nm. An MWNT is composed of multiple coaxial tubes of SWNTs and its diameter is often in the range of 10-150 nm. CNTs have a high potential in a broad range of applications, especially in nanoelectronics and biomedical sensors. A wide variety of electronic devices based on individual CNTs or CNT thin films have been developed and used as sensors (Boul et al., 2009; Wang et al., 2009; Xue & Cui, 2008b), field-effect transistors (Xue & Cui, 2009; Xue et al., 2006), conductive interconnects (Robertson et al., 2008; Xue & Cui, 2008a), and energy storage systems (Hu et al., 2009; Kaempgen et al., 2009). A critical step to obtain these practical devices is to deposit well-organized and highly aligned CNTs in desired locations. Recently, researchers have developed a number of methods to align CNTs: using moving fluids to organize nanotubes (S. Li et al., 2007), introducing gas flows in reactors or channels (Liu et al., 2009), withdrawing microfluidic channels from solutions (Tsukruk et al., 2004), spin coating nanotube dispersions with controlled speeds (LeMieux et al., 2008; Roberts et al., 2009), and magnetic capturing of nanotubes (Shim et al., 2009). However, many of these techniques have limitations and restrictions because they require either intensive preparation processes or assisting materials with special properties. Therefore, their applications are relatively limited.

By comparison, dielectrophoresis, a simple but versatile method, has proven to be effective in aligning CNTs in small and large scales (Gultepe et al., 2008; Mureau et al., 2006). This method can be conducted at room temperature with low voltages. In addition, a number of parameters such as solution concentration, deposition time, alternating current (AC) amplitude, and frequency can be adjusted to optimize the quality of the aligned CNTs. More importantly, dielectrophoresis can be easily incorporated into device fabrication and has the potential to be used in wafer-level deposition for the mass production of CNT-based devices (Monica et al., 2008; Stokes & Khondaker, 2008; Xiao & Camino, 2009).

Recently, devices based on dielectrophoresis-aligned CNTs have been developed and used in various applications such as biocompatible substrates for cell growth (Yuen et al., 2008), bacteria capturing chips (Zhou et al., 2006), gas sensors (Lim et al., 2010), and memory devices (Di Bartolomeo et al., 2010). Numerical studies have also been performed to provide theoretical support of the process (Dimaki & Bøggild, 2004; Padmaraj et al., 2009). However, most of these research efforts are focused on the alignment of CNT thin films. Even though the controlled assembly of single CNT bundles has been studied by using various voltage magnitudes and types (Seo et al., 2005), a thorough investigation into the electrode geometry and the solution concentration is still necessary to achieve the deposition and alignment of individual CNTs and small nanotube bundles.

In this chapter, we examine the selective deposition of CNTs—including both SWNTs and MWNTs—with dielectrophoresis to obtain aligned nanotubes in the forms of thin films, small bundles, and individual nanotubes. These different results are obtained by changing the parameters in the dielectrophoresis process and the concentration of the CNT sample solutions. Pristine CNTs are treated with acids for surface functionalization and then diluted with deionized (DI) water to obtain different concentrations. The CNT thin films are deposited and aligned using a large-width electrode design; the alignment of nanotube bundles is achieved using a "teeth"-like electrode design; and the alignment of individual nanotubes is achieved using "teeth"-like electrodes with sharp tips. The "teeth"-like electrodes are used to generate a concentrated and highly directional electric field in between two opposite "teeth". The electric field induces electric dipoles in CNTs and applies dielectrophoretic forces on them. Consequently, the dielectrophoretic forces move and rotate the CNTs to follow the electric field lines and eventually they land on the substrates to cover the electrodes. The electrodes are fabricated with optical lithography and wet etching; expensive equipment such as electron-beam writer—commonly used in the fabrication of individual nanotube devices—is avoided. The dielectrophoresis experiments are conducted at room temperature. Scanning electron microscopy (SEM) inspection shows that the CNTs are aligned in desired locations. The deposited SWNTs show better alignment than the MWNTs. Electrical characterization of the SWNT devices demonstrates that they have linear current-voltage (*I-V*) curves and their resistance is dependent on the SWNT solution concentration. For the MWNT devices, however, the *I-V* curves are less linear. The material preparation, electrode design, fabrication process, dielectrophoresis process, quality of the aligned nanotubes and thin films, microscope observation, and electrical characterization of the thin-film devices are described and discussed in this chapter.

## 2. Materials and experiments

The CNTs—both SWNTs and MWNTs—used in our experiments are purchased from SES Research (Houston, TX). The pristine CNTs are in the form of powder; the specifications for the SWNTs are: outer diameter < 2 nm, length 5-15 μm, purity > 90%; the specifications for the MWNTs are: outer diameter 60-100 nm, length 5-15 μm, purity > 95%. Because dielectrophoresis is a solution-based method, it requires that the deposited materials used in the process are free to move and rotate in a medium. However, it is well known that the pristine CNTs suffer from poor solubility in most solvents. Therefore, to increase the solubility and processability of CNTs, defects such as terminal groups can be intentionally introduced to their sidewalls and open ends with various chemical oxidation methods (Banerjee et al., 2005).

In our study, the pristine CNTs are chemically functionalized with a mixture of sulfuric and nitric acids (3:1 $H_2SO_4$:$HNO_3$) at 110°C for 45 min. The mixture of acids cuts the CNTs into short tubes with openings at both ends, as shown in Fig. 1. The acid treatment introduces the covalent attachment of carboxylic (-COOH) groups on the surfaces and open ends of CNTs. As a result, the CNTs can be uniformly dispersed in DI water and remain stable for a long period of time (10-12 months). In addition, the nitric acid in the mixture can purify the CNTs by removing amorphous carbon, carbon particles, and other impurities.

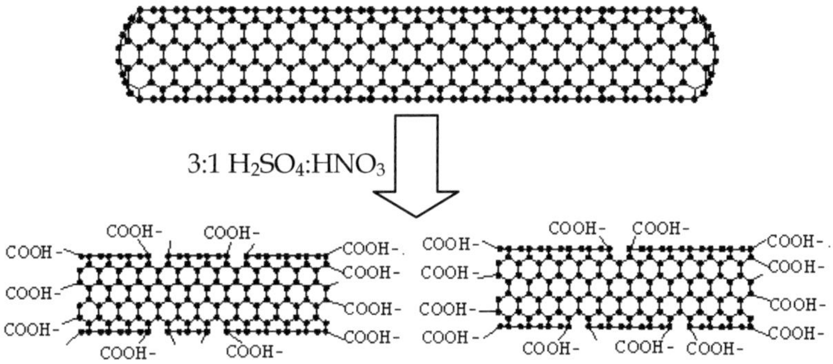


Fig. 1. Scheme of the chemical functionalization of a SWNT. The SWNT is cut into short tubes with carboxylic groups covalently attached to the sidewalls and open ends.

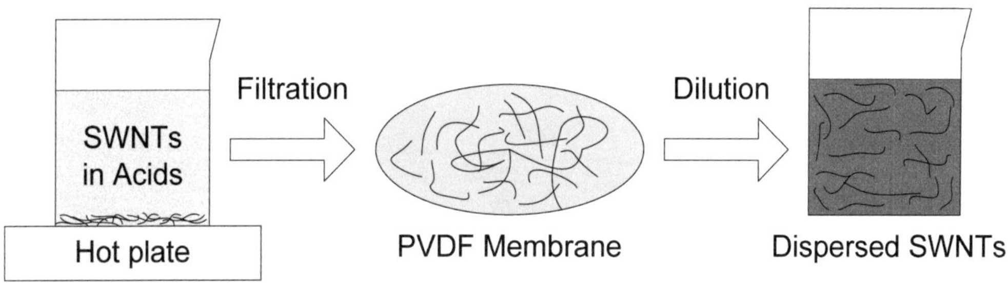


Fig. 2. Surface functionalization process of the CNTs with acid treatment, filtration, and dilution.

After the acid treatment, the functionalized CNTs are diluted with DI water and filtered with a polyvinylidene fluoride (PVDF) filtration membrane (with an average pore diameter of 0.22 μm) repeatedly for 5-6 times until the pH value of the dispersion reaches five. Next, the purified and functionalized CNTs are collected from the PVDF membrane and dispersed in DI water. The CNT dispersion is treated with an ultrasonic process for 30 min to obtain a more uniform solution. Last, the dispersed CNTs are diluted with DI water to obtain different concentrations: 0.2, 0.1, 0.05, 0.025, and 0.0125 mg/ml for SWNT solutions and 0.1, 0.05, 0.025, 0.0125, and 0.00625 mg/ml for MWNT solutions. These concentrations are selected because the samples can show clear differences after the nanotube deposition. Figure 2 shows the process flow to obtain the diluted CNT solutions.

To investigate the deposition and alignment of CNTs in various configurations, two types of electrodes are designed and fabricated (P. Li & Xue, 2010a). The first electrode design is used for the deposition of thin films. In this design, the electrodes have a large width of 400 μm and a gap of 5 μm. The schematic structure of the design and an SEM image of the fabricated electrodes are shown in Fig. 3. When an AC signal is applied on the device, an alternating electric field with parallel lines is generated in between the electrodes. The CNTs in the dispersion can be attracted to this area by the dielectrophoretic force. They are re-oriented to follow the electric field lines and evenly distributed across the width of the electrodes after deposition.

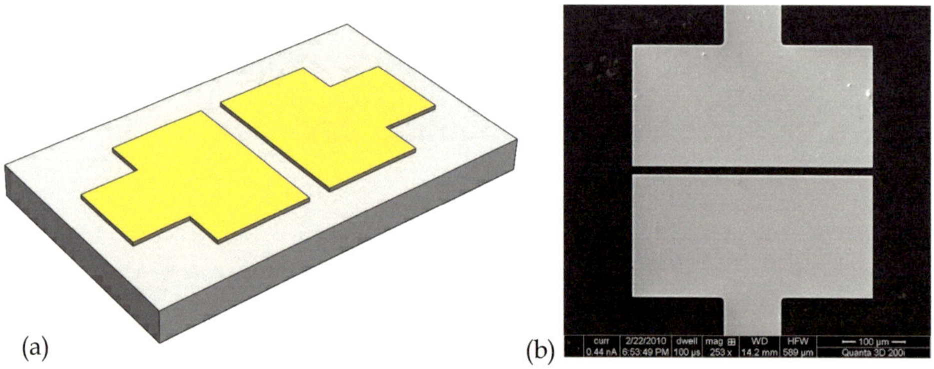


Fig. 3. (a) Schematic structure of the wide electrode design. The width and the gap of the electrodes are 400 and 5 μm, respectively. (b) An SEM image of the fabricated electrodes. Reprinted with permission from P. Li & Xue, 2010a. @ 2010 Springer.

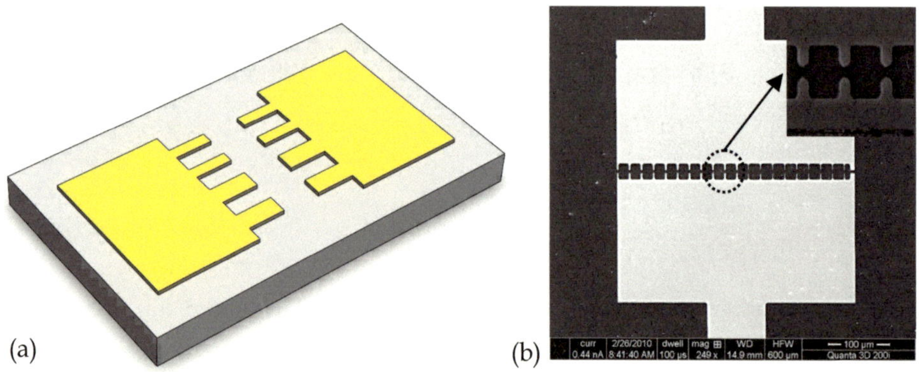


Fig. 4. (a) Schematic structure of the "teeth"-like electrode design. The electrodes have two different widths of 5 and 3 μm. The gap between the electrodes is 3 μm. (b) An SEM image of the fabricated electrodes. The enlarged area shows the electrode pairs with sharp tips. Reprinted with permission from P. Li & Xue, 2010a. @ 2010 Springer.

The second electrode design is used for the deposition and alignment of sparsely distributed nanotubes, nanotube bundles, and individual nanotubes. The electrodes are designed as "teeth"-like structures with a 3-μm gap in between, as shown in Fig. 4. The "teeth"-like electrode design enables a highly concentrated electric field in desired regions. In order to control the electric field strength and distribution profile, the electrodes are designed to have two different widths of 5 and 3 μm. The 5-μm-wide electrodes are used for the alignment of sparsely distributed nanotubes; the 3-μm-wide electrodes are used for the alignment of nanotube bundles and individual nanotubes.

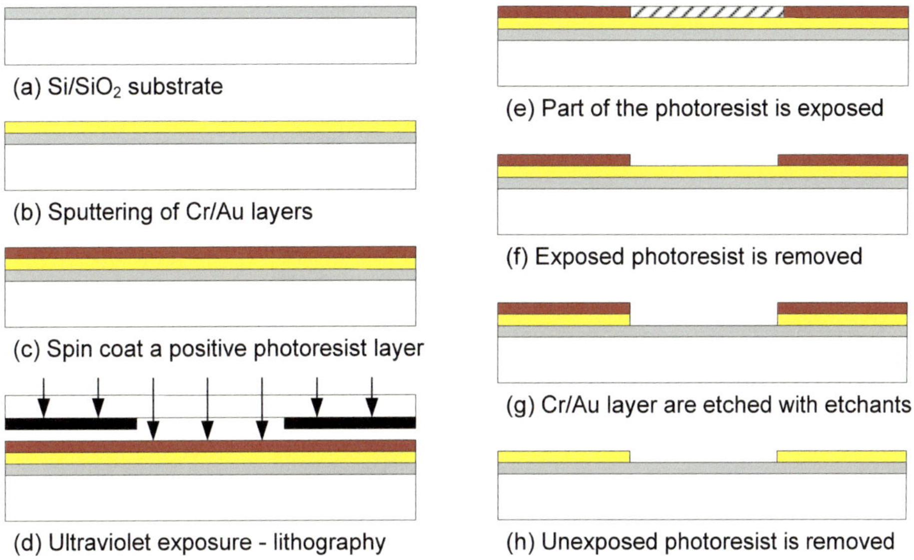


Fig. 5. Fabrication process of the electrodes using optical lithography and wet etching.

The electrodes are fabricated on 4-inch silicon wafers. The wafers are covered with a 200-nm-thick thermal-grown $SiO_2$ insulating layer. Metal layers of Cr (100 nm, adhesion material) and Au (200 nm, electrode material) are coated on the wafer surface with sputtering. One of the biggest advantages of dielectrophoresis in nanotube alignment is its high potential in wafer-level deposition, which is compatible with the parallel micro/nano-fabrication processes used in the semiconductor industry. Therefore, series (and often expensive) nanofabrication processes using equipment such as electron-beam lithography and scanning-probe lithography should be avoided. In our investigation, the entire fabrication process is compatible with the traditional microfabrication technology, as shown in Fig. 5. We use optical lithography with a hard contact aligner (OAI 200 Mask Aligner), a positive photoresist (Shipley S1813), and controlled wet etching to obtain electrodes, which can generate electric field with desired strength and distribution. The parameters for the etching steps are 120 sec for Cr etching using a standard chromium mask etchant and 20 sec for Au etching using a gold etchant (Type TFA from Transene Inc.). After the controlled etching steps, the actual widths of the electrodes are reduced to 2-3 (designed width: 5 μm) and 0.5-1 μm (designed width: 3 μm). Figure 6 shows an optical image of a fabricated device. The dimensions of the device are 75 mm × 75 mm, smaller than a dime. The center of the device contains the electrodes used for the dielectrophoresis process. The electrodes are elongated to the edge of the device as probing pads for the application of the AC signal.

## 3. Dielectrophoresis

When dielectric particles are exposed to a non-uniform electric field, charges including electrons (-) and protons (+) are moved away from their initial balanced positions and redistributed in these particles. The charge redistribution creates electric dipole moments which force these particles to rotate along the electric field lines. The induced effective dipole moment for an ellipsoidal particle can be expressed as (Peng et al., 2006):

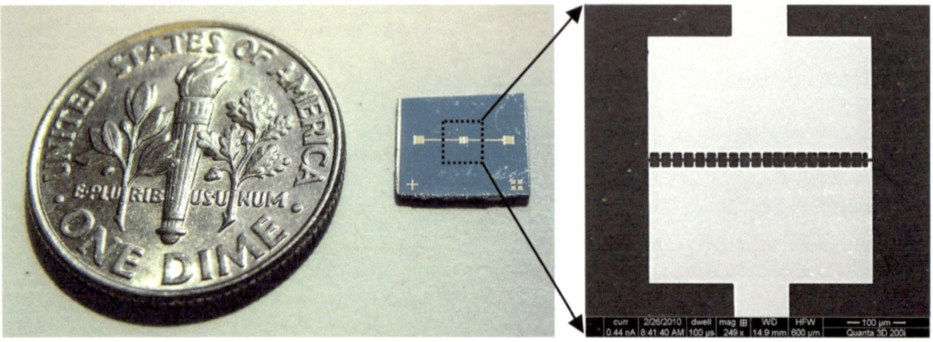

Fig. 6. Optical image of a fabricated device next to a dime.

$$p_{eff} = \frac{V(\varepsilon_p - \varepsilon_m)E}{[1 + (\varepsilon_p - \varepsilon_m) / \varepsilon_m]A_L} \quad (1)$$

where $V$ is the volume of the particle, $\varepsilon_p$ and $\varepsilon_m$ are the permittivities of the particle and the medium, $E$ is the external electric field strength, $A_L \approx 4r^2/l^2[\ln(l/r)-1]$ is the depolarization factor, and $r$, $l$ are the radius and length of the ellipsoidal particle, respectively.
Similarly, an external non-uniform electric field can induce dipole moments in rod-like objects including nanowires and carbon nanotubes. The dipole moment parallel to the tube axis is much stronger than that in the perpendicular direction. Therefore, the polarized nanotube, if free to move in a medium, is subject to a net force and can be aligned to follow the electric field direction, as shown in Fig. 7.
The net force exerted on the nanotube is expressed as (Dimaki & Bøggild, 2004; Raychaudhuri et al., 2009):

$$F = \frac{\pi r^2 l}{6} \varepsilon_m Re[f_{cm}] \nabla E_{rms}^2 \quad (2)$$

where the term $r^2l/6$ is a geometry factor that contains the volume information of the nanotube, $Re[f_{cm}]$ is the real number part of the Clausius-Mossotti factor $f_{cm}$, and $\nabla E_{rms}$ is the root mean square of the gradient of the external electric field. When the long axis of the nanotube is aligned along the electric field line, the Clausius-Mossotti factor $f_{cm}$ can be derived from (Ahmed et al., 2009; Dimaki & Bøggild, 2004):

$$f_{cm(long\ axis)} = \frac{\varepsilon^*_n - \varepsilon^*_m}{(\varepsilon^*_n - \varepsilon^*_m)A_L + \varepsilon^*_m} \quad (3)$$

where $A_L$ is the depolarization factor along the long axis, the subscripts $n$ and $m$ represent the nanotube and the medium, respectively, and $\varepsilon^*$ is the complex permittivity that contains the information of the physical permittivity $\varepsilon$, the conductivity $\sigma$, and the angular velocity of the external electric field $\omega$. The term $\varepsilon^*$ is defined as:

$$\varepsilon^* = \varepsilon - j\frac{\sigma}{\omega} \quad (4)$$

These equations indicate that the dielectrophoresis of CNTs is affected by many factors including the dimensions of the nanotubes, the properties of the medium, and the strength of the electric field. In our investigation, the following parameters are adjusted to control the deposition and alignment of the nanotubes: bias voltage, frequency, deposition time, width of the electrodes, and nanotube solution concentration.

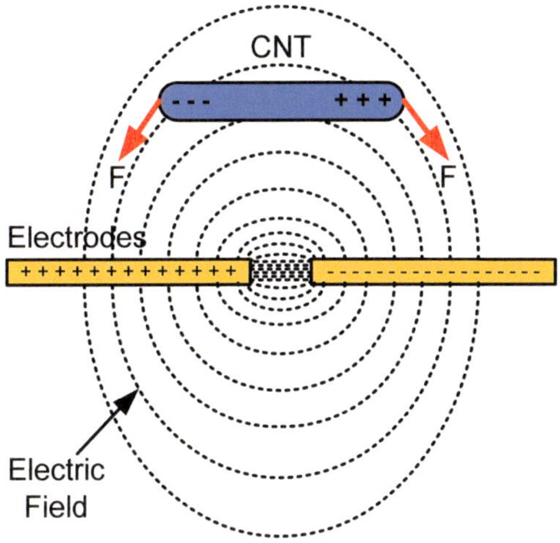


Fig. 7. Principle of dielectrophoresis deposition and alignment of a carbon nanotube.

Early research demonstrates that the polarization along the longitudinal direction is much higher than that along the transverse direction for metallic CNTs, but comparable for semiconducting CNTs (Padmaraj et al., 2009). This is because the metallic CNTs have a larger $Re[f_{cm}]$ and the dielectrophoretic force exerted on them is much stronger than that experienced by the semiconducting CNTs. Therefore, we expect that the metallic CNTs will dominate the movement of SWNT bundles in the dielectrophoresis process in our experiments.

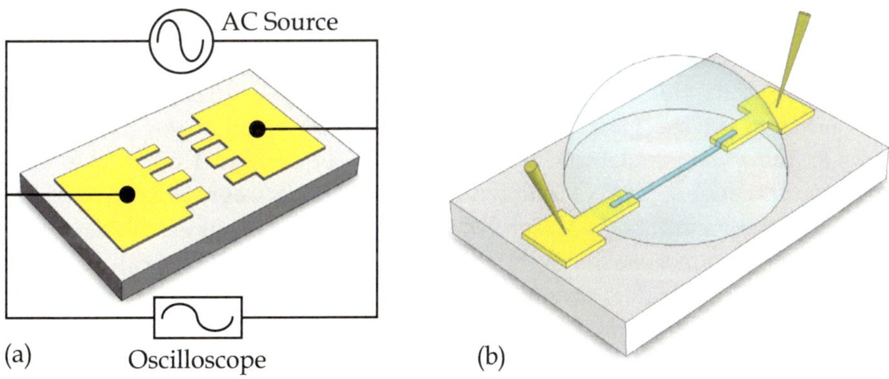


Fig. 8. (a) Experimental system for the dielectrophoresis of CNTs. (b) Schematic diagram of the CNT alignment. Reprinted with permission from P. Li & Xue, 2010a. @ 2010 Springer.

Figure 8a shows the configuration of the experimental system for the dielectrophoresis of the CNTs. A function generation (Agilent Technologies 81150A) is used as the AC signal source and it is connected to the electrodes through two metal probes. The potential drop across the electrode gap is monitored with an oscilloscope (Agilent Technologies MSO 7054A), which provides the voltage value in real time. After the instruments are set up, a droplet of the CNT solution is carefully placed in the area between the electrodes with a syringe. Next, the AC signal source is switched on. An electric field is generated in between the electrode "teeth". The electric field exerts dielectrophoretic forces on the CNTs and forces them to rotate along the field lines. The CNTs can be deposited on the substrate with this orientation, as shown in Fig. 8b. After 30 sec of dielectrophoresis, the AC signal is switched off and the CNT solution is removed with another syringe. The experiments are conducted at room temperature with an applied AC signal with a peak-to-peak voltage of $V_{pp}$ = 10 V and frequency of $f$ = 5 MHz.

## 4. Results and discussion

In the dielectrophoresis process, the dielectrophoretic forces drag the CNTs to the gap of the electrodes where the electric field has the highest magnitude. In reality, however, the dielectrophoresis process and the deposition results of the CNTs are affected by a number of factors including the dimensions of the nanotubes, the properties of the medium, and the strength of the electric field. In this research, the alignment results of the two forms of CNTs—SWNTs and MWNTs—are investigated and compared. In addition, the electrical properties of these resulting CNTs are studied.

### 4.1 SWNTs

As described earlier, the fabricated devices can be divided into three groups: 400-μm-wide electrodes, 5-μm-wide electrodes, and 3-μm-wide electrodes. For each group, the electrodes are exposed to the SWNT solutions with different concentrations, ranging from 0.2 to 0.00125 mg/ml. An SEM (FEI Quanta 3D 200i) is used to inspect the deposited SWNTs after the dielectrophoresis process. A semiconductor device analyzer (Agilent Technologies B1500A) is used to characterize the electrical properties of the aligned SWNTs.

#### 4.1.1 Deposition and alignment of the SWNTs

Figure 9 shows the SEM images of the deposited SWNTs on the substrates with the 400-μm-wide electrodes. This electrode design enables an evenly distributed electric field with parallel field lines in the gap. The SWNTs are stretched and aligned in between the electrodes to follow the parallel field lines. The density of the SWNT thin film is dependent on the concentration of the SWNT solution used in the dielectrophoresis process. A higher solution concentration induces a denser SWNT film. The alignment experiments are highly repeatable when used to fabricate SWNT thin films. However, it is difficult to obtain small nanotube bundles or individual nanotubes using this group of devices.

In order to explore the possibility of producing aligned nanotube bundles and individual nanotubes, narrower electrodes are used. Figure 10 shows the SEM images of the aligned SWNTs on the devices with the 5-μm-wide electrodes. Figure 10a demonstrates a dense film of SWNTs covering the electrode gap, where most SWNTs are aligned to follow the direction of the electric field lines. As the solution concentration decreases, fewer SWNTs are attracted by the dielectrophoretic force and deposited on the substrate. In addition, the SEM inspection shows that there are no SWNTs observed outside the electrode gap area.

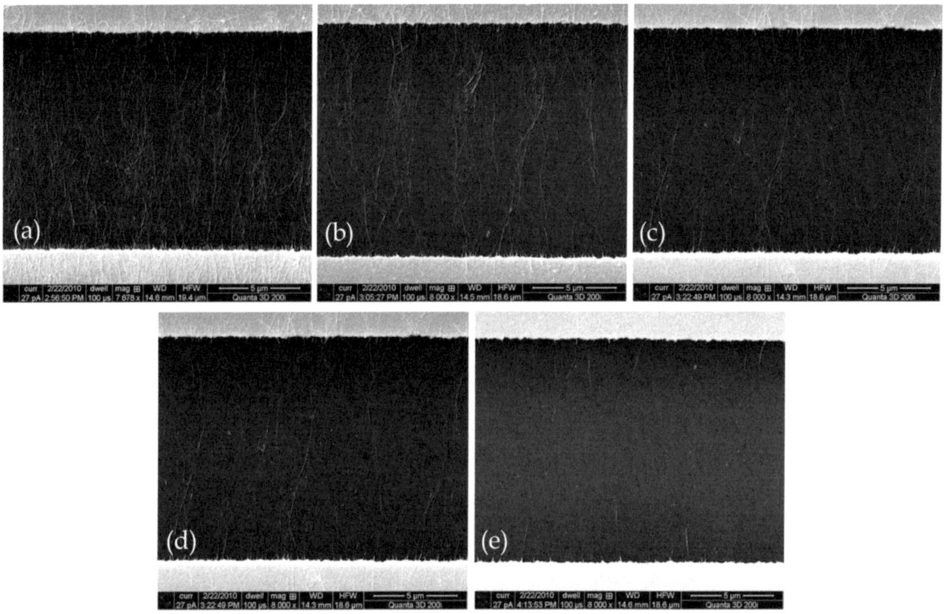


Fig. 9. SEM images of aligned SWNTs with the 400-μm-wide electrodes and solutions with different concentrations: (a) 0.2 mg/ml, (b) 0.1 mg/ml, (c) 0.05 mg/ml, (d) 0.025 mg/ml, and (e) 0.0125 mg/ml. Reprinted with permission from P. Li & Xue, 2010a. @ 2010 Springer.

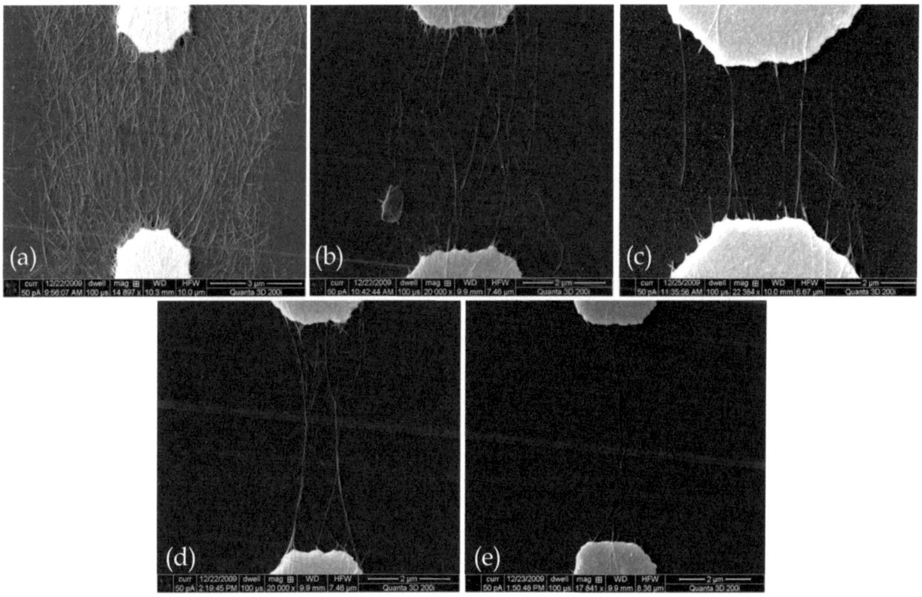


Fig. 10. SEM images of aligned SWNTs with the 5-μm-wide electrodes and solutions with different concentrations: (a) 0.2 mg/ml, (b) 0.1 mg/ml, (c) 0.05 mg/ml, (d) 0.025 mg/ml, and (e) 0.0125 mg/ml. Reprinted with permission from P. Li & Xue, 2010a. @ 2010 Springer.

This proves that the dielectrophoresis process is a selective deposition technique which only moves SWNTs to areas with strong electric fields. Figure 10b-d show sparsely distributed nanotube bundles and Fig. 10e shows a small bundle of SWNTs. Compared with the devices with wider electrodes (Fig. 9), the devices with 5-μm-wide electrodes can generate a more concentrated electric field in the gap and force the SWNTs to land on the substrate to cover a smaller area. The bundling phenomenon of the SWNTs in the solution is caused by their high aspect ratios and high flexibility. Based on Fig. 10b-e, as the solution concentration decreases, the bundles become thinner due to the lower availability of SWNTs in the solution.

For the narrowest electrodes with the width of 3 μm, the quality of the deposition and alignment of SWNTs is similar to that for the electrodes with 5-μm width, as shown in Fig. 11. However, there are three major differences comparing the results from the two designs. First, the amount of the aligned SWNTs on the 3-μm-wide electrodes is smaller. This is because the area with a strong and concentrated electric field, generated by the narrower electrodes, is smaller. Second, the bundles deposited on the 3-μm-wide electrodes are thinner and contain smaller numbers of nanotubes. Third, and most importantly, individual nanotubes can be observed in the dielectrophoresis experiments using low-concentration solutions and 3-μm-wide electrodes (Fig. 11e). In this case, the narrow electrode design ensures that the electric field in between the electrodes is extremely concentrated and highly directional. The low solution concentration ensures that in the adjacent area there is only one nanotube available in the solution to be attracted by the electric field. Consequently, this individual nanotube is deposited on the substrate and aligned in between the electrodes.

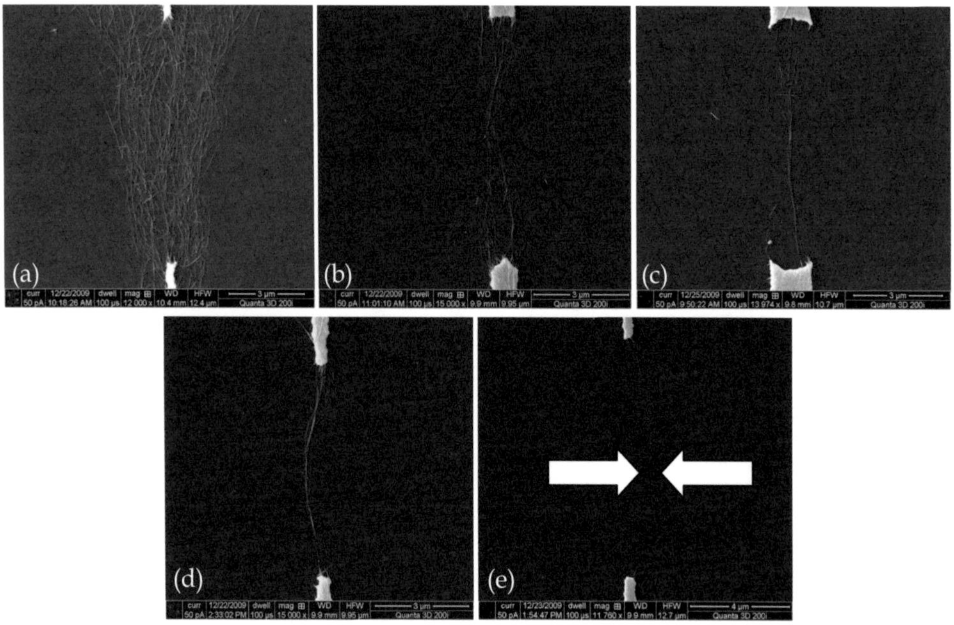


Fig. 11. SEM images of aligned SWNTs with the 3-μm-wide electrodes and solutions with different concentrations: (a) 0.2 mg/ml, (b) 0.1 mg/ml, (c) 0.05 mg/ml, (d) 0.025 mg/ml, and (e) 0.0125 mg/ml. Reprinted with permission from P. Li & Xue, 2010a. @ 2010 Springer.

### 4.1.2 Electrical characteristics of the aligned SWNTs

After the dielectrophoresis process, the gap of the electrodes are covered with the deposited and aligned SWNTs, which allows the transmission of electric charges across the electrodes. As a result, these devices can be measured as regular resistors with the electrodes as the testing pads and the aligned SWNTs as the conductive path (P. Li & Xue, 2010b). To gain a better understanding of the effects of the solution concentration on the deposition and alignment of SWNTs, the electrical properties of the devices after the dielectrophoresis process are characterized with the semiconductor device analyzer. Based on the design of the electrodes, the devices are divided into two groups: the first group contains devices with the wide electrodes (width: 400 μm) and the second group includes devices with the "teeth"-like electrodes (width: 5 μm and 3 μm). Each group consists of five different SWNT devices corresponding to the solutions used in the dielectrophoresis process.

The electrical properties of the wide-electrode devices are illustrated in Fig. 12a. The width and the gap of the electrodes are 400 and 5 μm, respectively. These devices are measured as regular resistors with the current-voltage plots as the output characteristics. In the measured range of -3 to 3 V, all plots are highly linear. The resistances of these devices are calculated accordingly and plotted in Fig. 12b. At a low SWNT solution concentration of 0.0125 mg/ml, which corresponds to the device shown in Fig. 9e, only a few SWNTs are captured. The resistance is measured as approximately 35.015 kΩ. As the concentration increases, the resistance quickly decreases to a lower value and remains relatively constant in a low-resistance region. The calculated resistances are 23.641 kΩ (SWNT solution: 0.025 mg/ml), 0.832 kΩ (SWNT solution: 0.05 mg/ml), 0.099 kΩ (SWNT solution: 0.1 mg/ml), and 0.035 kΩ (SWNT solution: 0.2 mg/ml).

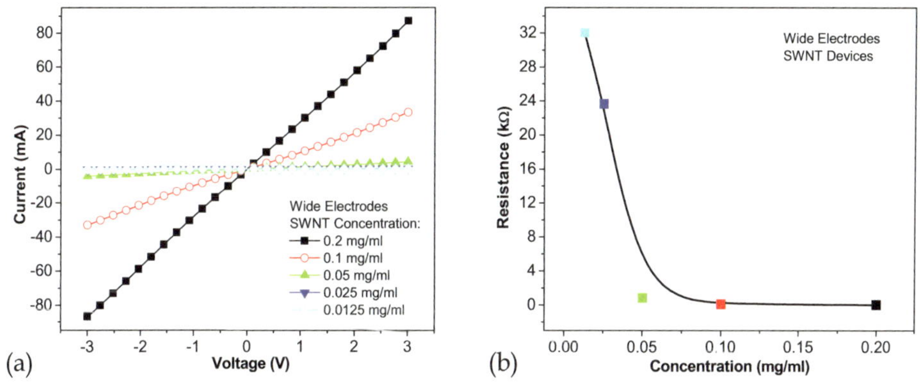


Fig. 12. (a) Current-voltage plots of five devices with the 400-μm-wide electrodes using SWNT solutions with different concentrations. (b) Resistance as a function of the concentration for the aligned SWNTs. The resistances are calculated from (a). Reprinted with permission from P. Li & Xue, 2010b. @ 2010 American Society of Mechanical Engineers.

The electrical characteristics of the devices with the "teeth"-like electrodes are obtained using the same method. The devices are based on the second electrode design which contains electrodes with variable widths (Fig. 4). Each device consists of eleven 5-μm-wide electrode pairs and nine 3-μm-wide electrode pairs. Figure 13a shows the *I-V* plots of five devices using SWNT solutions with five different concentrations. In the measured range of -3 and 3 V, all five *I-V* plots are highly linear. Figure 13b shows the calculated resistance as a

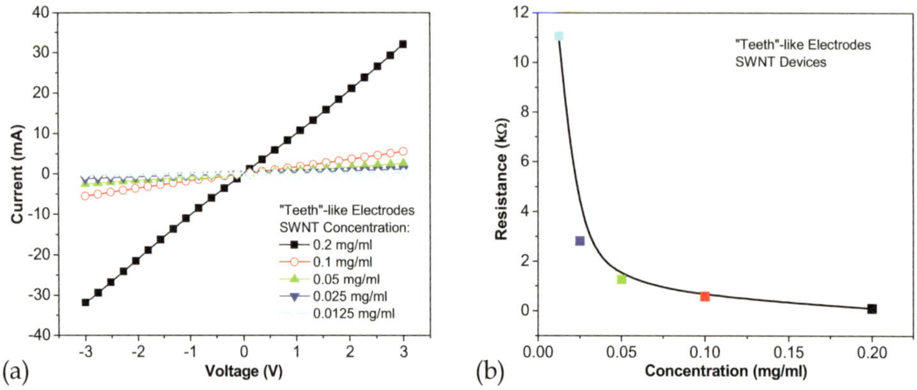


Fig. 13. (a) Current-voltage plots of five devices with the "teeth"-like electrodes using SWNT solutions with different concentrations. (b) Resistance as a function of the concentration for the aligned SWNTs. The resistances are calculated from (a). Reprinted with permission from P. Li & Xue, 2010a. @ 2010 Springer.

function of the solution concentration. The resistance of the device starts as 11.055 kΩ at the lowest concentration (SWNT solution: 0.0125 mg/ml) and quickly decreases to 2.806 kΩ (SWNT solution: 0.025 mg/ml). It continues to decrease to lower values of 1.256 kΩ (SWNT solution: 0.05 mg/ml), 0.571 kΩ (SWNT solution: 0.1 mg/ml), and 0.097 kΩ (SWNT solution: 0.2 mg/ml). Although this group of devices shows a similar resistance-concentration relationship to the first group of devices with wide electrodes, the resistance range is different. For example, the largest obtained resistances are 11.055 and 32.015 kΩ for the "teeth"-like and wide electrodes, respectively. This difference is caused by the different amount of aligned SWNTs on the substrates. The devices with wide electrodes are able to capture more SWNTs, providing more conductive paths for the charges to move across the electrodes. As a result, these devices show lower resistance.

The characterization of these devices verifies the results from the SEM inspection and proves that the higher-concentration solutions lead to denser films with more deposited SWNTs. The high linearity of the *I-V* plots suggests that the aligned SWNTs are mostly metallic nanotubes instead of semiconducting nanotubes. These experimental results fit well with the prediction described in Section 3: Dielectrophoresis.

### 4.2 MWNTs

The deposition, alignment, and electrical characteristics of the MWNTs are investigated using the same procedures as those for the SWNTs. The experiment and measurement results of the aligned MWNTs are described; the main differences between the deposited MWNTs and SWNTs are discussed.

#### 4.2.1 Deposition and alignment of the MWNTs

The alignment results of the MWNTs are similar to those for the SWNTs - the solutions with higher MWNT concentrations generate denser films, as shown in Fig. 14. However, most MWNTs are tangled together and are not fully stretched in between the electrodes. This can be explained by the special dimensions and structures of the MWNTs. First, the typical diameter of the MWNTs used in our experiments is 60-100 nm, which is much larger than

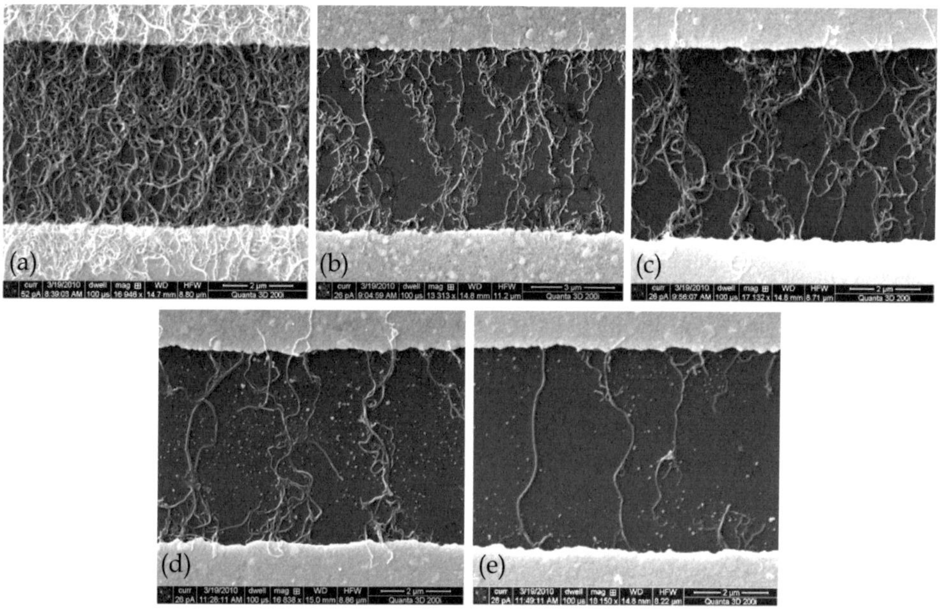


Fig. 14. SEM images of aligned MWNTs with the 400-μm-wide electrodes and solutions with different concentrations: (a) 0.1 mg/ml, (b) 0.05 mg/ml, (c) 0.025 mg/ml, (d) 0.0125 mg/ml, and (e) 0.00625 mg/ml.

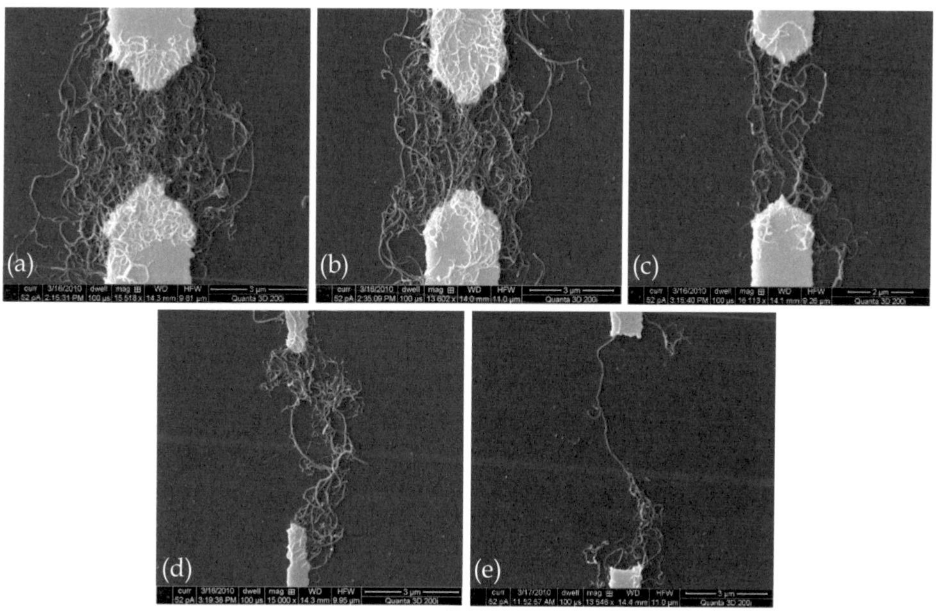


Fig. 15. SEM images of aligned MWNTs with the "teeth"-like electrodes and solutions with different concentrations: (a) 0.1 mg/ml, (b) 0.05 mg/ml, (c) 0.025 mg/ml, (d) 0.0125 mg/ml, and (e) 0.00625 mg/ml.

that of the SWNTs (2 nm). Second, the multi-walled structure of the MWNTs makes their mechanical properties different from the SWNTs – the MWNTs are heavier and stiffer. As a result, when the MWNTs are exposed to an external electric field, the generated dielectrophoretic forces are not strong enough to fully stretch the MWNTs. Furthermore, the thin films generated from the solutions with higher MWNT concentrations have a higher degree of randomness. As the concentration decreases to a lower level, the MWNTs are suspended in the solution more homogeneously and their interaction with each other becomes weaker. Therefore, the resulting films and bundles are more organized with better quality of alignment.

The "teeth"-like electrodes are also used for the dielectrophoretic assembly of the MWNTs, as shown in Fig. 15. Unlike the wide electrodes, the "teeth"-like electrodes only generate high-gradient electric field between two opposite "teeth". Consequently, the MWNTs are only deposited in these locations. Similar to the experimental results of the SWNT samples, the MWNTs only deposit in between the electrodes. In addition, we are able to achieve not only the deposition of narrow-width thin films, but also the assembly of a controlled amount of MWNTs or even individual MWNTs, as demonstrated in Fig. 15e.

### 4.2.2 Electrical characteristics of the aligned MWNTs

The electrical characterization of the MWNTs deposited on the electrodes is performed after the dielectrophoresis process. The devices are divided into two groups: the first group contains devices with the wide electrodes (width: 400 μm) and the second group includes devices with the "teeth"-like electrodes. Each group consists of five different MWNT devices corresponding to the solutions used in the dielectrophoresis process.

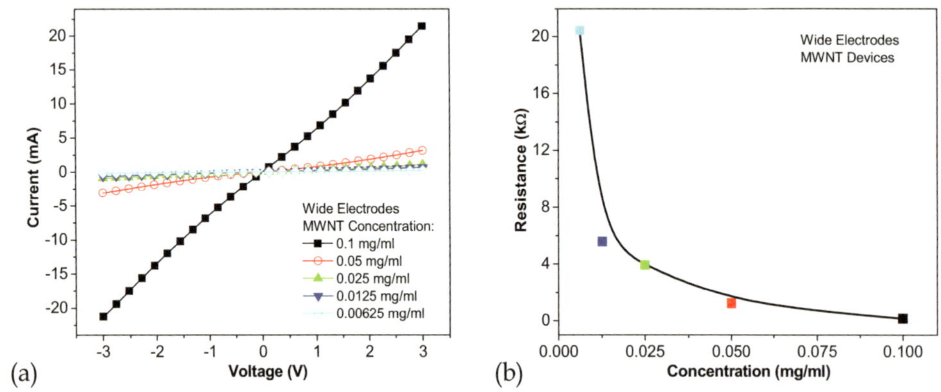


Fig. 16. (a) Current-voltage plots of five devices with the 400-μm-wide electrodes using MWNT solutions with different concentrations. (b) Resistance as a function of the concentration for the aligned MWNTs. The resistances are calculated from (a). Reprinted with permission from P. Li & Xue, 2010b. @ 2010 American Society of Mechanical Engineers.

Figure 16 shows the *I-V* plots and the resistance-concentration function of the five MWNT devices with the wide electrodes. The MWNT thin films behave in a similar way to the SWNT thin films. At a low MWNT concentration of 0.00625 mg/ml, the resistance is measured as 20.433 kΩ. As the solution concentration increases, the resistance decreases to

5.576 kΩ (MWNT solution: 0.0125 mg/ml), 3.926 kΩ (MWNT solution: 0.025 mg/ml), 1.220 kΩ (MWNT solution: 0.05 mg/ml), and 0.154 kΩ (MWNT solution: 0.1 mg/ml).

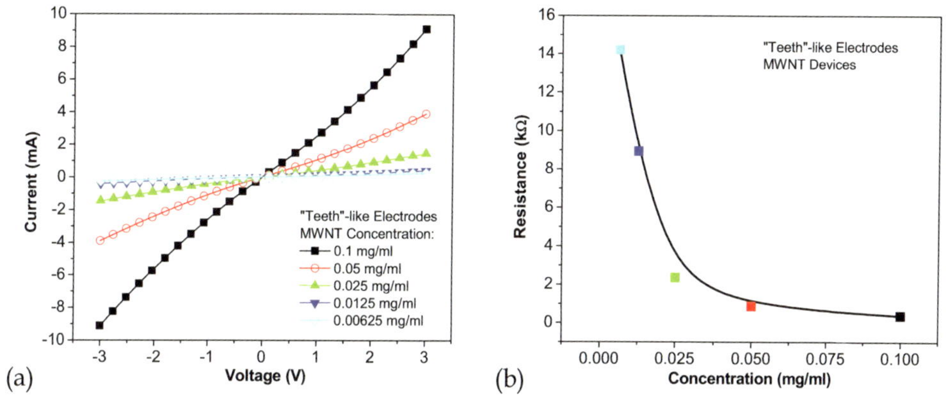


Fig. 17. (a) Current-voltage plots of five devices with the "teeth"-like electrodes using MWNT solutions with different concentrations. (b) Resistance as a function of the concentration for the aligned MWNTs. The resistances are calculated from (a).

The devices with the "teeth"-like electrodes demonstrate similar results, as shown in Fig. 17. The resistance of the devices decreases from 14.201 kΩ (MWNT solution: 0.00625 mg/ml) to 8.938 kΩ (MWNT solution: 0.0125 mg/ml), 2.356 kΩ (MWNT solution: 0.025 mg/ml), 0.890 kΩ (MWNT solution: 0.05 mg/ml), and 0.374 kΩ (MWNT solution: 0.1 mg/ml).

However, the *I-V* plots of the MWNT devices are not as linear as those of the SWNT devices. The plots show slight curvature in the measurement range of -3 to 3 V. One possible reason for the nonlinearity is that the MWNTs tend to tangle with each other. Therefore, many semiconducting MWNTs are mixed in the bundles and deposited on the substrates during the dielectrophoresis process.

The calculated resistances of the CNTs deposited under different conditions are listed in Table 1. The values are obtained from the electrical characterization of various samples. This table summarizes the results from two materials: SWNTs and MWNTs, two electrode designs: wide electrodes and "teeth"-like electrodes, and six solution concentrations. It can also be used as a reference for future electronics design and experiments.

Our experiments show that the selective deposition of CNTs, including both SWNTs and MWNTs, are highly repeatable. The alignment of CNT thin films, bundles, and individual nanotubes can be achieved using different combinations of solution concentrations and electrode designs. The method presented here can be used in the fabrication of novel CNT-based nanoelectronic devices. Furthermore, we believe that it can also be used in the development of devices beyond electronics, providing a wide range of opportunities. For example, the devices with a controlled amount of CNTs can be used as high-performance sensors for chemical sensing, gas detection, and DNA analysis.

Even though the fabrication and deposition steps presented in this article are still used for small-scale processes, they can be easily extended to large-scale device production. In addition, because the entire process is compatible with the traditional microfabrication technology, it has a high potential to be used in wafer-level fabrication to produce identical devices across the entire surface of the substrate.

| | Resistance (kΩ) | | | |
|---|---|---|---|---|
| | SWNTs | | MWNTs | |
| Concentration (mg/ml) | Wide electrodes | "Teeth"-like electrodes | Wide electrodes | "Teeth"-like electrodes |
| **0.2** | 0.035 | 0.097 | n/a | n/a |
| **0.1** | 0.099 | 0.571 | 0.154 | 0.374 |
| **0.05** | 0.832 | 1.256 | 1.220 | 0.890 |
| **0.025** | 23.641 | 2.806 | 3.926 | 2.356 |
| **0.0125** | 32.015 | 11.055 | 5.576 | 8.938 |
| **0.00625** | n/a | n/a | 20.433 | 14.201 |

Table 1. The calculated resistances of the deposited SWNTs and MWNTs from the electrical characterization.

### 4.3 CNT deposition without an electric field

In order to verify the dielectrophoretic effects on the CNT deposition and alignment, two experiments are conducted for control purposes. Unfortunately, the natural deposition of the CNTs on the substrate is a slow process, resulting in sparsely distributed CNTs across the entire surface. It is difficult to locate the deposited CNTs with the SEM. Therefore, an alternative approach needs to be used.

Because the functionalized CNTs are negatively charged with the covalently attached carboxylic groups, the deposition of CNTs can be enhanced by using substrates with pre-charged surfaces. In this study, we use poly (dimethyldiallylammonium chloride) (PDDA), a positively charged polyelectrolyte, to pre-charge the surface of the silicon substrate (Xue & Cui, 2007). First, the substrate with the 400-μm-wide electrodes is submerged in the PDDA solution for 10 minutes. Next, a drop of CNT solution (with a concentration of 0.1 mg/ml) is placed over the electrodes for 15 min to allow the CNTs to fully deposit on the substrate. Afterward, the substrate is rinsed with DI water and dried with compressed air. Figure 18 shows the SEM images of random networks for both the SWNTs and the MWNTs. In this case, the CNTs cover the entire substrate surface including the gap and the electrodes. There is no CNT alignment observed from these SEM images. Therefore, we can conclude that the alignment of the CNTs in our investigation is solely caused by the dielectrophoresis process.

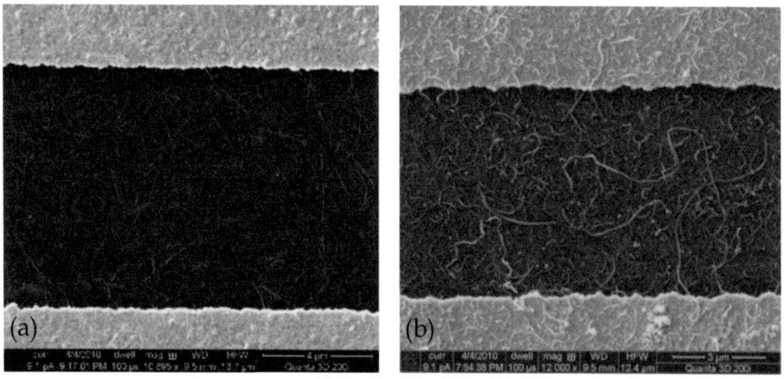


Fig. 18. SEM images (a) a SWNT random network and (b) a MWNT random network. Reprinted with permission from P. Li & Xue, 2010b. @ 2010 American Society of Mechanical Engineers.

## 5. Conclusion

In conclusion, we have investigated the dielectrophoretic deposition of SWNTs and MWNTs to obtain organized nanotubes. The deposited and aligned CNTs exist in three forms: thin films, small bundles, and individual nanotubes. The different alignment results are obtained by using various electrode designs and CNT solutions with various concentrations. In general, electrodes with large widths generate an evenly distributed electric field with parallel field lines; the aligned CNTs are in the form of thin films or sparsely distributed bundles. For electrodes with small width, the electric field is highly concentrated and can induce individual nanotube deposition and alignment. The electrical characterization of the aligned SWNTs shows high linearity between the current and voltage; this means that most deposited SWNTs are metallic. However, the current-voltage curves of the MWNTs are curled; this suggests that more semiconducting MWNTs are mixed in the deposited nanotubes. The measurement data illustrate that the resistance of the aligned CNTs can be adjusted in a wide range. The method presented in this article is inexpensive, reliable, and repeatable. In addition, the fabrication process of the devices is compatible with traditional microfabrication technology; it has a high potential to be used in wafer-level fabrication for the mass production of devices in the future. We believe that our investigation of CNTs can facilitate research activities in this field and enable the development of a wide variety of CNT-based devices such as electronics, sensors, and energy storage and conversion systems.

## 6. Acknowledgment

This work was supported by the Washington State University New Faculty Seed Grant.

## 7. References

Ahmed, W.; Kooij, E. S.; van Silfhout, A. & Poelsema, B. (2009). Quantitative Analysis of Gold Nanorod Alignment after Electric Field-Assisted Deposition. *Nano letters*, Vol. 9, No. 11, (November 2009), pp. 3786-3794, ISSN 1530-6984

Banerjee, S.; Hemraj-Benny, T. & Wong, S. S. (2005). Covalent Surface Chemistry of Single-Walled Carbon Nanotubes. *Advanced Materials*, Vol. 17, No. 1, (January 2005), pp. 17-29, ISSN 0935-9648

Boul, P. J.; Turner, K.; Li, J.; Pulikkathara, M. X.; Dwivedi, R. C.; Sosa, E. D.; Lu, Y.; Kuznetsov, O. V.; Moloney, P.; Wilkins, R.; O'Rourke, M. J.; Khabashesku, V. N.; Arepalli, S. & Yowell, L. (2009). Single Wall Carbon Nanotube Response to Proton Radiation. *The Journal of Physical Chemistry C*, Vol. 113, No. 32, (August 2009), pp. 14467-14473, ISSN 1932-7447

Di Bartolomeo, A.; Rinzan, M.; Boyd, A. K.; Yang, Y.; Guadagno, L.; Giubileo, F. & Barbara, P. (2010). Electrical Properties and Memory Effects of Field-Effect Transistors from Networks of Single- and Double-Walled Carbon Nanotubes. *Nanotechnology*, Vol. 21, No. 11, (March 2010), pp. 115204, ISSN 0957-4484

Dimaki, M. & Bøggild, P. (2004). Dielectrophoresis of Carbon Nanotubes Using Microelectrodes: A Numerical Study. *Nanotechnology*, Vol. 15, No. 8, (August 2004), pp. 1095-1102, ISSN 0957-4484

Gultepe, E.; Nagesha, D.; Casse, B. D. F.; Selvarasah, S.; Busnaina, A. & Sridhar, S. (2008). Large Scale 3D Vertical Assembly of Single-Wall Carbon Nanotubes at Ambient

Temperatures. *Nanotechnology*, Vol. 19, No. 45, (November 2008), pp. 455309, ISSN 0957-4484

Hu, L.; Choi, J. W.; Yang, Y.; Jeong, S.; La Mantia, F.; Cui, L.-F. & Cui, Y. (2009). Highly Conductive Paper for Energy-Storage Devices. *Proceedings of the National Academy of Sciences of the United States of America*, Vol. 106, No. 51, (December 2009), pp. 21490-21494, ISSN 1091-6490

Iijima, S. (1991). Helical Microtubules of Graphitic Carbon. *Nature*, Vol. 354, (November 1991), pp. 56-58, ISSN 0028-0836

Kaempgen, M.; Chan, C. K.; Ma, J.; Cui, Y. & Gruner, G. (2009). Printable Thin Film Supercapacitors Using Single-Walled Carbon Nanotubes. *Nano letters*, Vol. 9, No. 5, (May 2009), pp. 1872-1876, ISSN 1530-6984

Katz, E. & Willner, I. (2004). Biomolecule-Functionalized Carbon Nanotubes: Applications in Nanobioelectronics. *ChemPhysChem*, Vol. 5, No. 8, (August 2004), pp. 1084-1104, ISSN 1439-4235

LeMieux, M. C.; Roberts, M.; Barman, S.; Jin, Y. W.; Kim, J. M. & Bao, Z. (2008). Self-Sorted, Aligned Nanotube Networks for Thin-Film Transistors. *Science*, Vol. 321, No. 5885, (July 2008), pp. 101-104, ISSN 0036-8075

Li, P. & Xue, W. (2010a). Selective Deposition and Alignment of Single-Walled Carbon Nanotubes Assisted by Dielectrophoresis: From Thin Films to Individual Nanotubes. *Nanoscale Research Letters*, Vol. 5, No. 6, (June 2010), pp. 1072-1078, ISSN 1556-276X

Li, P. & Xue, W. (2010b). Dielectrophoretic Assembly of Organized Carbon Nanotubes and Thin Films. *Proceedings of the ASME 2010 International Mechanical Engineering Congress & Exposition*, ISBN 978-0-791-83891-4, Vancouver, British Columbia, Canada, November 12-18, 2010

Li, S.; Liu, N.; Chan-Park, M. B.; Yan, Y. & Zhang, Q. (2007). Aligned Single-Walled Carbon Nanotube Patterns with Nanoscale Width, Micron-Scale Length and Controllable Pitch. *Nanotechnology*, Vol. 18, No. 45, (November 2007), pp. 455302, ISSN 0957-4484

Lim, J.-H.; Phiboolsirichit, N.; Mubeen, S.; Deshusses, M. A.; Mulchandani, A. & Myung, N. V. (2010). Electrical and Gas Sensing Properties of Polyaniline Functionalized Single-Walled Carbon Nanotubes. *Nanotechnology*, Vol. 21, No. 7, (Februar 2010), pp. 75502, ISSN 0957-4484

Liu, H.; Takagi, D.; Chiashi, S. & Homma, Y. (2009). The Controlled Growth of Horizontally Aligned Single-Walled Carbon Nanotube Arrays by a Gas Flow Process. *Nanotechnology*, Vol. 20, No. 34, (August 2009), pp. 345604, ISSN 0957-4484

Monica, A. H.; Papadakis, S. J.; Osiander, R. & Paranjape, M. (2008). Wafer-Level Assembly of Carbon Nanotube Networks Using Dielectrophoresis. *Nanotechnology*, Vol. 19, No. 8, (February 2008), pp. 085303, ISSN 0957-4484

Mureau, N.; Mendoza, E.; Silva, S. R. P.; Hoettges, K. F. & Hughes, M. P. (2006). *In Situ* and Real Time Determination of Metallic and Semiconducting Single-Walled Carbon Nanotubes in Suspension via Dielectrophoresis. *Applied Physics Letters*, Vol. 88, No. 24, (June 2006), pp. 243109, ISSN 0003-6951

Padmaraj, D.; Zagozdzon-Wosik, W.; Xie, L.-M.; Hadjiev, V. G.; Cherukuri, P. & Wosik, J. (2009). Parallel and Orthogonal E-field Alignment of Single-Walled Carbon Nanotubes by AC Dielectrophoresis. *Nanotechnology*, Vol. 20, No. 3, (January 2009), pp. 035201, ISSN 0957-4484

Peng, N.; Zhang, Q.; Li, J. & Liu, N. (2006). Influences of AC Electric Field on the Spatial Distribution of Carbon Nanotubes Formed between Electrodes. *Journal of Applied Physics*, Vol. 100, No. 2, (July 2006), pp. 024309, ISSN 0021-8979

Raychaudhuri, S.; Dayeh, S. A.; Wang, D. & Yu, E. T. (2009). Precise Semiconductor Nanowire Placement through Dielectrophoresis. *Nano letters*, Vol. 9, No. 6, (June 2009), pp. 2260-2266, ISSN 1530-6984

Roberts, M. E.; LeMieux, M. C.; Sokolov, A. N. & Bao, Z. (2009). Self-Sorted Nanotube Networks on Polymer Dielectrics for Low-Voltage Thin-Film Transistors. *Nano letters*, Vol. 9, No. 7, (July 2009), pp. 2526-2531, ISSN 1530-6984

Robertson, J.; Zhong, G.; Telg, H.; Thomsen, C.; Warner, J. H.; Briggs, G. A. D.; Dettlaff-Weglikowska, U. & Roth. S. (2008). Growth and Characterization of High-Density Mats of Single-Walled Carbon Nanotubes for Interconnects. *Applied Physics Letters*, Vol. 93, No. 16, (October 2008), pp. 163111, ISSN 0003-6951

Seo, H.; Han, C.; Choi, D.; Kim, K. & Lee, Y. (2005). Controlled Assembly of Single SWNTs Bundle Using Dielectrophoresis. *Microelectronic Engineering*, Vol. 81, No. 1, (July 2005), pp. 83-89, ISSN 0167-9317

Shim, J. S.; Yun, Y.-H.; Rust, M. J.; Do, J.; Shanov, V., Schulz, M. J. & Ahn, C. H. (2009). The Precise Self-Assembly of Individual Carbon Nanotubes Using Magnetic Capturing and Fluidic Alignment. *Nanotechnology*, Vol. 20, No. 32, (August 2009), pp. 325607, ISSN 0957-4484

Stokes, P. & Khondaker, S. I. (2008). Local-Gated Single-Walled Carbon Nanotube Field Effect Transistors Assembled by AC Dielectrophoresis. *Nanotechnology*, Vol. 19, No. 17, (April 2008), pp. 175202, ISSN 0957-4484

Tsukruk, V.; Ko, H. & Peleshanko, S. (2004). Nanotube Surface Arrays: Weaving, Bending, and Assembling on Patterned Silicon. *Physical Review Letters*, Vol. 92, No. 6, (Feburary 2004), pp. 1-4, ISSN 0031-9007

Wang, Y.; Zhou, Z.; Yang, Z.; Chen, X.; Xu, D. & Zhang, Y. (2009). Gas Sensors based on Deposited Single-Walled Carbon Nanotube Networks for DMMP Detection. *Nanotechnology*, Vol. 20, No. 34, (August 2009), pp. 345502, ISSN 0957-4484

Xiao, Z. & Camino, F. E. (2009). The Fabrication of Carbon Nanotube Field-Effect Transistors with Semiconductors as the Source and Drain Contact Materials. *Nanotechnology*, Vol. 20, No. 13, (April 2009), pp. 135205, ISSN 0957-4484

Xue, W. & Cui, T. (2007). Characterization of Layer-by-Layer Self-Assembled Carbon Nanotube Multilayer Thin Films. *Nanotechnology*, Vol. 18, No. 14, (April 2007), pp. 145709, ISSN 0957-4484

Xue, W. & Cui, T. (2008a). Electrical and Electromechanical Characteristics of Self-Assembled Carbon Nanotube Thin Films on Flexible Substrates. *Sensors And Actuators A: Physical*, Vol. 146, (August 2008), pp. 330-335, ISSN 0924-4247

Xue, W. & Cui, T. (2008b). A Thin-Film Transistor based Acetylcholine Sensor Using Self-Assembled Carbon Nanotubes and $SiO_2$ Nanoparticles. *Sensors and Actuators B: Chemical*, Vol. 134, No. 2, (September 2008), pp. 981-987, ISSN 0925-4005

Xue, W. & Cui, T. (2009). Thin-Film Transistors with Controllable Mobilities based on Layer-by-Layer Self-Assembled Carbon Nanotube Composites. *Solid-State Electronics*, Vol. 53, No. 9, (September 2009), pp. 1050-1055, ISSN 0038-1101

Xue, W.; Liu, Y. & Cui, T. (2006). High-Mobility Transistors based on Nanoassembled Carbon Nanotube Semiconducting Layer and $SiO_2$ Nanoparticle Dielectric Layer. *Applied Physics Letters*, Vol. 89, No. 16, (October 2006), pp. 163512, ISSN 0003-6951

Yuen, F. L.-Y.; Zak, G.; Waldman, S. D. & Docoslis, A. (2008). Morphology of Fibroblasts Grown on Substrates formed by Dielectrophoretically Aligned Carbon Nanotubes. *Cytotechnology*, Vol. 56, No. 1, (January 2008), pp. 9-17, ISSN 0920-9069

Zhou, R.; Wang, P. & Chang, H.-C. (2006). Bacteria Capture, Concentration and Detection by Alternating Current Dielectrophoresis and Self-Assembly of Dispersed Single-Wall Carbon Nanotubes. *Electrophoresis*, Vol. 27, No. 7, (April 2006), pp. 1376-1385, ISSN 0173-0835

# 10

# Aligned Growth of Single-Walled and Double-Walled Carbon Nanotube Films by Control of Catalyst Preparation

Mineo Hiramatsu[1] and Masaru Hori[2]
[1]*Meijo University,*
[2]*Nagoya University*
*Japan*

## 1. Introduction

Carbon nanotubes (CNTs) have attracted significant attention for various potential applications, such as electron field emitter arrays and multi-level interconnections of ultra-large scale next-generation integrated circuits (IC). Many of the proposed applications of CNTs require aligned nanotubes grown on a substrate. Among the various methods used for growing CNTs, plasma-enhanced chemical vapor deposition (CVD) has attracted considerable attention for industrial applications due to its feasibility and potential for large-area production with reasonable growth rates at relatively low temperatures.

For the growth of CNTs on a silicon (Si) substrate or other substrates by CVD, a catalyst metal, such as iron (Fe) or cobalt (Co) is indispensable. The diameters of CNTs almost correspond to the size of the metal catalyst islands or particles on the heated substrate immediately before the supply of a carbon source gas, such as methane ($CH_4$), ethylene ($C_2H_4$), and acetylene ($C_2H_2$). In order to obtain single-walled carbon nanotubes (SWNTs) or double-walled carbon nanotubes (DWNTs) by the CVD method, it is important to control the size of the catalytic particles to approximately several nanometers or less. When using a Si substrate, the formation of metal–silicide as a result of the preheating treatment will complicate the synthesis process. Previously, buffer layers using Al, Ti or $SiO_2$ have been used in order to prevent the formation of metal–silicide (Lee et al., 2003; Kim et al., 2001; Hong et al., 2001). However, it is still difficult to simultaneously control both the catalyst particles' size and the density on the substrate surface. Therefore, control over the preparation of catalytic nanoparticles on the substrate is essential to be able to control the diameters of the nanotubes and to obtain the optimum potential of the CNTs.

In this chapter, growth of aligned CNT films using microwave plasma-enhanced CVD is described, with a particular emphasis on the pretreatment of substrate with catalytic metal nanoparticles. By preparing the Co nanoparticles in a controlled manner, aligned SWNT and DWNT films were fabricated on the Si substrate. Furthermore, area-selective growth of vertical CNTs to form organized SWNT and DWNT microstructures was demonstrated.

## 2. Catalyst particle preparation

When using plasma-enhanced CVD to synthesize CNTs on a substrate, conventionally a catalytic metal thin film is formed on the substrate by sputtering or evaporation methods, and the substrate is typically preheated in order to form catalytic nano-islands in the range of 10–100 nm, prior to the CNT growth. The sizes of the catalyst islands depend on the initial thickness of the catalytic metal thin film. As shown in Fig. 1, thicker catalyst film yields relatively larger catalyst islands, and sometimes multiple CNTs are grown from one catalyst nano-island. On the other hand, thinner film results in the formation of smaller catalyst islands with less density. The control of the size and density of catalyst nanoparticles on the substrate is desirable.

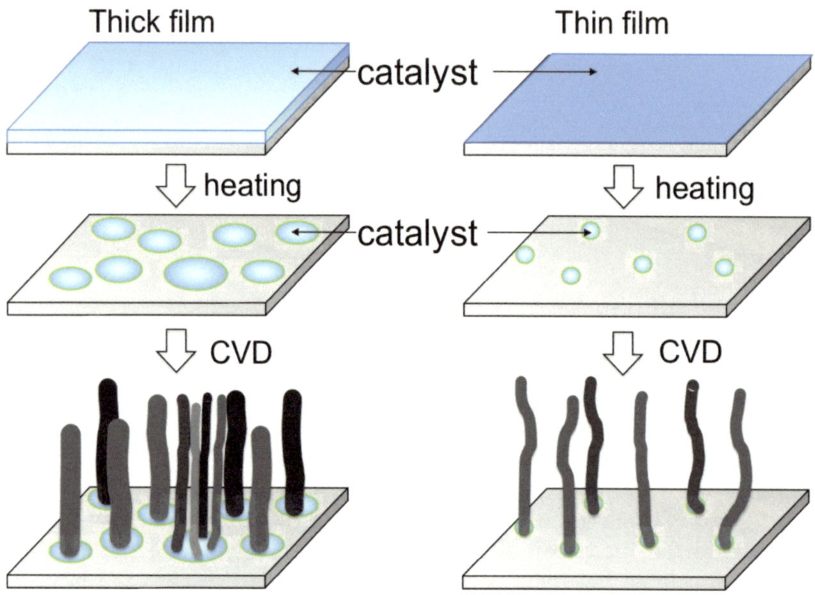


Fig. 1. Schematic of conventional procedure for preparing catalyst nanoparticles or nanoislands and subsequent CVD growth

As a method of controlling the size of catalyst metal islands or particles, zeolite has been used as a support for the catalytic particles. In this method, the size of the catalytic particles is uniformly controlled using the pores of the zeolite as a template, and SWNTs or DWNTs with relatively uniform diameters can be grown (Sugai et al., 2003, 2004). The growth of vertically aligned SWNTs was demonstrated using a catalytic CVD method on a quartz substrate with mono-dispersed Co-Mo catalysts, with 1–2 nm particle size, prepared by a dip-coating method (Murakami et al., 2004). The low temperature growth of aligned SWNTs was attained using a sandwich-like nano-layered structure of $Al_2O_3$ (0.7 nm)/Fe (0.5 nm)/$Al_2O_3$ (5–70 nm) for the preparation of catalytic Fe nanoparticles on a Si substrate (Zhong et al., 2004).

Laser ablation of a metallic target to obtain catalytic particles is effective for preparing high-density catalytic nanoparticles, and the laser ablation technique for catalyst preparation enables the formation of aligned nanotube films with nanotubes diameters of 5 to 10 nm (Hiramatsu et al., 2005). Figure 2 shows a schematic of laser ablation system used for the preparation of catalyst nanoparticles. In this case, KrF excimer laser (wavelength: 248 nm) is used for the ablation of Co target. The excimer laser ablation yields Co nanoparticles with a size of 5–10 nm.

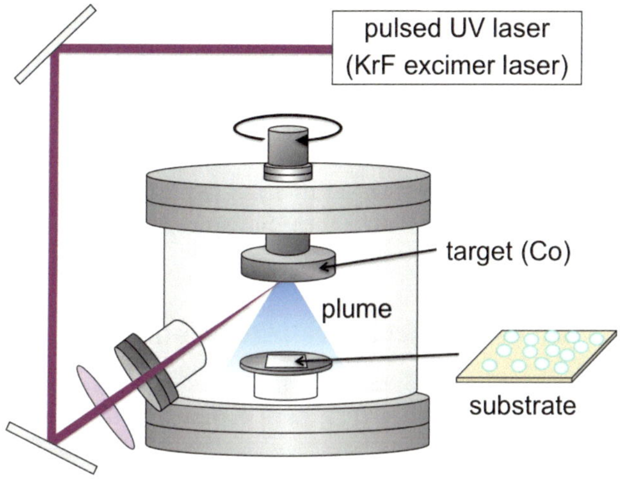


Fig. 2. Schematic of laser ablation system used for the preparation of catalyst nanoparticles

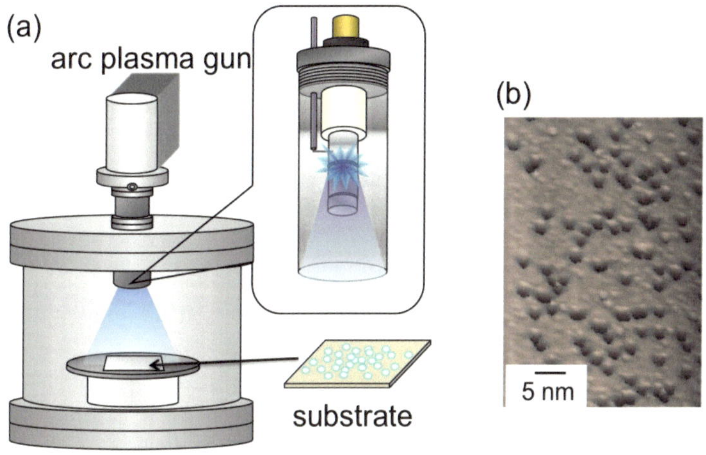


Fig. 3. (a) Schematic of pulsed arc deposition system used for the preparation of catalyst nanoparticles. (b) AFM image of Co-catalyzed Si substrate prepared using pulsed arc discharge. The number of pulses of the arc discharge was 10 shots. (Hiramatsu et al., 2007b) - reproduced with permission from Institute of Pure and Applied Physics

On the other hand, pulsed arc deposition is effective for the preparation of catalyst nanoparticles with a size of 1–2 nm. Figure 3(a) shows a schematic of nanoparticle preparation system using pulsed arc plasma deposition (Arc plasma gun, ULVAC, Inc.) (Agawa et al., 2003; Yamamoto et al., 1998). As a preliminary experiment, Co nanoparticles were directly formed on the Si substrate by pulsed arc discharge. Figure 3(b) shows an atomic force microscopy (AFM) image of the Co-catalyzed Si substrate prepared by pulsed arc discharge with a Co electrode. The number of arc discharge pulses was 10 shots in this case. The cumulative particle density of Co nanoparticles on the substrate was estimated to be $8\times10^{11}$ cm$^{-2}$ after 10 shots of pulsed arc discharges with the Co electrode. It was found that the pulsed arc plasma deposition yielded Co nanoparticles smaller than those formed by laser ablation method, according to the AFM observations shown in Fig. 3(b). The size of Co nanoparticles estimated from the height in AFM is approximately 1-2 nm, which would be suitable for the growth of SWNTs or DWNTs. Actually, in the case of CNT growth on the Si substrate, the formation of metal-silicide as a result of substrate heating process complicates the synthesis process; silicide cannot act as catalyst for CNT growth. Therefore, in our case, Ti or TiN is used as a buffer layer in order to prevent the formation of Co silicide during the rise in temperature, just before introducing carbon source gas into the reaction chamber.

## 3. CNT growth system using microwave plasma-enhanced CVD

Microwave plasma is one of high-density plasmas and is suitable for decomposing $H_2$ molecules to generate H atoms effectively. Figure 4 shows a microwave plasma-enhanced chemical vapor deposition (CVD) system, which has been called ASTeX (Applied Science and Technology, Inc.) type. This system has been extensively used for the growth of diamond films. The ASTeX-type reactor consists of a cylindrical stainless steel chamber. The microwave (2.45 GHz) is coupled from the rectangular waveguide into the cavity via an axial antenna. A discharge called a "plasma ball" is generated above the substrate. The plasma ball provides the substrate heating. The CVD process would be operated at pressures of a few tens of Torr ($10^3$–$10^4$ Pa), and the reactor pressure and microwave power cannot be varied completely independently. At too high pressure or low microwave power, plasma cannot be sustained. On the other hand, if the microwave power is too high for a given pressure, the plasma becomes unstable and tends to jump to the quartz (fused silica) window, occasionally resulting in the destruction of the window by the heat.

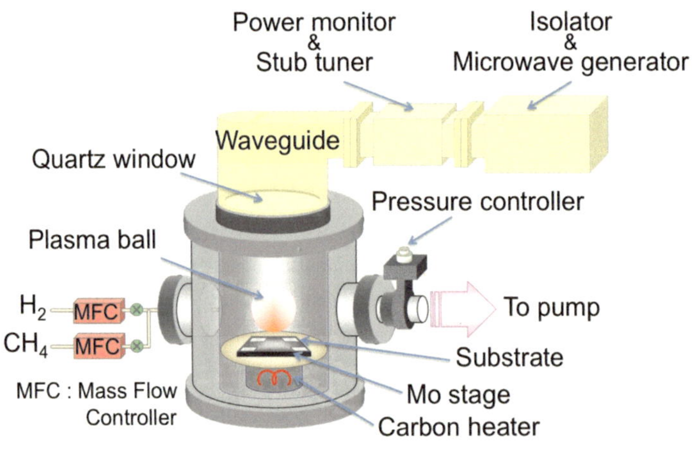


Fig. 4. Schematic of microwave plasma-enhanced CVD system used for CNT growth

In the case of diamond growth, deposition is carried out in a mixture of $CH_4$ or $C_2H_2$ and $H_2$ at substrate temperatures of 700-800 °C. Typical $H_2/CH_4$ flow rate ratio is about 100. The high $H_2/CH_4$ flow rate ratio for diamond growth is essential to remove unwanted non-diamond phase. In the case of the synthesis of CNTs, on the other hand, deposition process is carried out at a low $H_2/CH_4$ flow rate ratio of 1–4. In the present study, CNT films were grown using conventional microwave plasma-enhanced CVD with a 2.45 GHz, 1.5 kW microwave generator. The reactor consists of a cylindrical stainless steel chamber with an inner diameter of 11 cm, and a molybdenum (Mo) substrate stage with a graphite heater that allows for control of the substrate temperature, independent of the microwave input power. The flow rates of $CH_4$ and $H_2$ were 50 and 70 sccm, respectively. The microwave power and total pressure were maintained at 900 W and 70 Torr, respectively. Substrates used for the growth experiments were Si (100). Scanning electron microscopy (SEM) and transmission electron microscopy (TEM) were used to evaluate the morphology of the grown CNTs. Raman spectra for the CNTs were obtained using the 514.5 nm line of an Ar laser and the 632.8 nm line of an He–Ne laser.

## 4. CNT growth experiments

### 4.1 CNT growth from Co nanoparticles prepared using pulsed laser ablation

Growth experiments were carried out using microwave plasma-enhanced CVD on the Co-catalyzed Si substrate at 700 °C. Co nanoparticles of 5–10 nm size as catalyst were prepared by KrF excimer laser (248 nm) ablation using a 99.998% Co target in vacuum at a pressure of $10^{-4}$ Torr at room temperature. The energy density of the excimer laser beam at the target surface was maintained at 5 J/cm$^2$ at a repetition rate of 20 Hz. The distance between the Co target and substrate was approximately 10 cm. The density of Co nanoparticles on the substrate was controlled by varying the ablation period from 60 to 300 min. Furthermore, a TiN thin layer with a thickness of 20 nm was prepared on the Si substrate as a buffer layer to prevent the formation of Co silicide. No heat pretreatment for the catalyst was performed prior to the CNT growth.

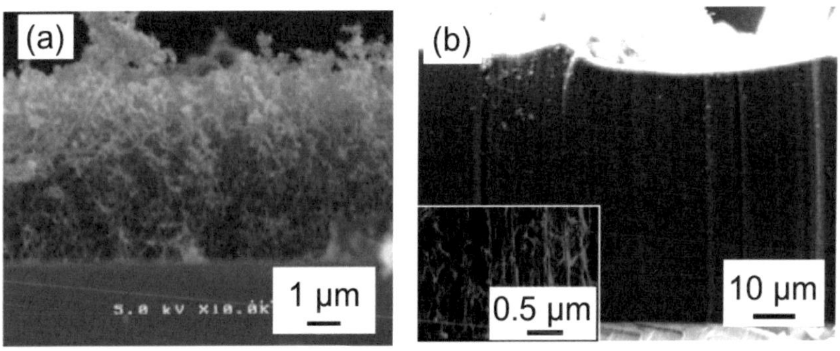


Fig. 5. (a) Cross-sectional SEM image of CNTs with random orientation grown from Co nanoparticles without buffer layer. (b) Cross-sectional SEM image of dense CNT film grown from Co nanoparticles with TiN buffer layer. Inset shows magnified SEM image of dense CNT film. (Hiramatsu et al., 2005) - reproduced with permission from Institute of Pure and Applied Physics

Figure 5(a) shows a cross-sectional SEM image of CNTs grown for 5 min on the Co-catalyzed Si substrate without a buffer layer. As shown in Fig. 5(a), without buffer layer, randomly oriented, curly CNTs with a low density were grown, but the alignment of CNTs was not obtained. Figure 5(b) shows a cross-sectional SEM image of the CNT film grown from Co nanoparticles with TiN buffer layer. In this case, Co nanoparticles were deposited for 180 min, which was nearly optimum condition for dense nucleation. Vertically aligned, dense CNTs were grown, as shown in Fig. 5(b). A magnified SEM image of the dense nanotube film is shown in the inset of Fig. 5(b), indicating that the aligned nanotube forests were composed of nanotube bundles supporting one another. The presence of a TiN buffer layer was found to be effective for CNT growth on a Si substrate.

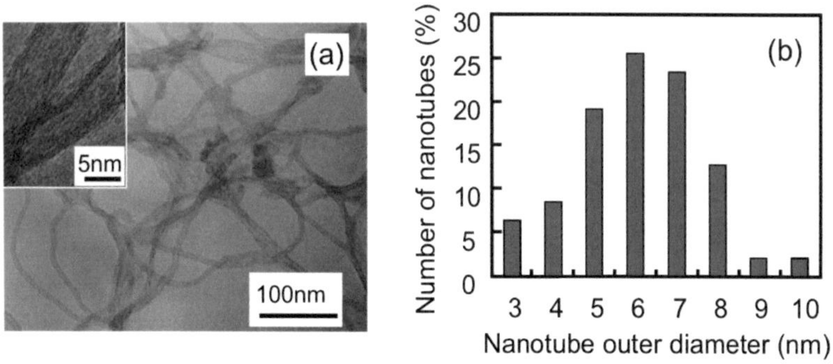


Fig. 6. (a) TEM image of CNTs. (b) Distribution histogram of nanotube outer diameters deduced from TEM observation. (Hiramatsu et al., 2005a) - reproduced with permission from Institute of Pure and Applied Physics

Figure 6(a) shows a typical TEM image of CNTs that were scraped away from the substrate surface and were ultrasonicated in methanol. Note that no Co particles were observed at the nanotube tips with TEM. As shown in the inset of Fig. 6(a), a TEM image reveals the multiwalled structure of the CNTs, containing a clear inner channel with an inner diameter of approximately 4 nm, surrounded by 2 - 5 concentric walls. Figure 6(b) shows a distribution histogram of nanotube outer diameters. CNTs grown using laser-ablated Co nanoparticles had small average diameters and a relatively narrow diameter distribution centered at approximately 6 nm. It was reported that nanotube growth rate was inversely proportional to nanotube diameter (Bower et al., 2000). Namely, the smaller diameter nanotubes would grow at a faster rate in terms of height. In this case, the size of Co islands formed by laser ablation was relatively small and remained unchanged with increasing substrate temperature just before CNT growth, due to the existence of a TiN buffer layer that prevented the formation of Co silicide. Therefore, these Co nanoparticles play an important role as a template for CNT growth, resulting in the determination of nanotube diameters as small as 6 nm and consequently, a high-rate growth of CNTs could be attained. The most significant difference from previous reports is that the catalyst was prepared originally in the form of nanoparticles in this case. Therefore, a nanotube diameter as small as approximately 6 nm was determined by the size of Co particles formed on the TiN buffer layer by laser ablation, resulting in the rapid growth of CNTs at 300 nm/s.

### 4.2 DWNT growth from Co catalysed Si prepared using pulsed arc discharge

As mentioned before, the pulsed arc plasma discharge can yield metal nanoparticles with a size of 1-2 nm, as shown in Fig. 3(b). Meanwhile, it has proven that the presence of a TiN buffer layer is effective for CNT growth on a Si substrate. Accordingly, in order to prepare smaller catalyst nanoparticles for growing CNTs with small diameters, pulsed arc discharge plasma was utilized to prepare Co nanoparticles on the Si substrate with TiN buffer layer. Substrate treatment was performed as follows (Fig. 7). First, a thin TiN layer with thickness of 20 nm was prepared on the Si substrate. The TiN thin layer plays an important role as a buffer layer to prevent the formation of Co silicide during the heating process, just prior to the introduction of $CH_4$ gas into the reaction chamber. Co particles were then deposited on the TiN buffer layer using pulsed arc plasma deposition (Arc plasma gun, ULVAC, Inc.) in vacuum, at a pressure of $10^{-4}$ Torr at room temperature. The arc plasma gun was operated intermittently in a pulsed operation. No heat pre-treatment was performed for the catalyst prior to the CNT growth process. The density of Co nanoparticles on the substrate was controlled by varying the number of pulses in the range from 50 to 250 shots, corresponding to the particle number density of $4 \times 10^{12}$ to $2 \times 10^{13}$ $cm^{-2}$ on the surface. A mixture of $CH_4$ and $H_2$ was used as the source gas. The flow rates of $CH_4$ and $H_2$ were 50 and 70 sccm, respectively. The microwave power and total pressure were maintained at 900 W and 70 Torr, respectively. CNTs were grown on the Co-catalyzed Si substrates in the presence of a TiN buffer layer at a substrate temperature of 700 ˚C.

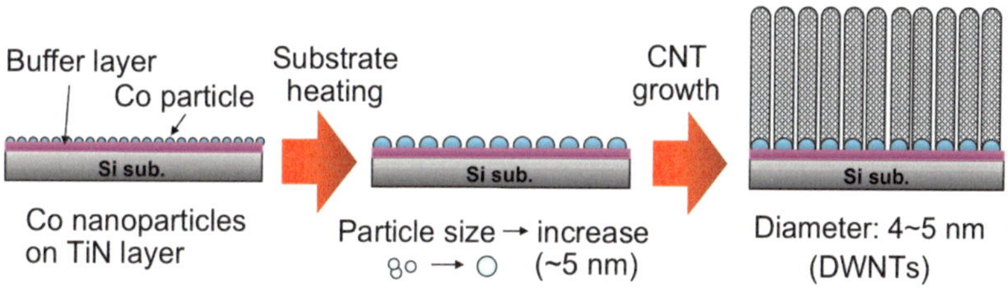


Fig. 7. Schematic of catalyst preparation for growing CNTs with small diameters (DWNTs)

Figure 8 shows a SEM image of Co-catalyzed Si substrate just before introducing $CH_4$ at substrate temperature of 700 ˚C. Co particles were prepared using pulsed arc plasma deposition with 250 pulses, corresponding to the cumulative particle number density of approximately $2 \times 10^{13}$ $cm^{-2}$ on the surface. It was found that Co nanoparticles with size of 4–5 nm were formed on the TiN buffer layer. The pulsed arc plasma deposition using the arc plasma gun yielded Co nanoparticles of about 1–2 nm in size, according to AFM observations of the Co-catalyzed substrate without the TiN buffer layer as shown in Fig. 3(b). In our system, it takes about 10 min to increase the substrate temperature from room temperature to 700 ˚C. During this period, the overlapped or closely adjacent particles would aggregate, presumably resulting in the formation of Co nanoislands of about 3–5 nm in size, which would be suitable for the nucleation of double-walled CNTs.

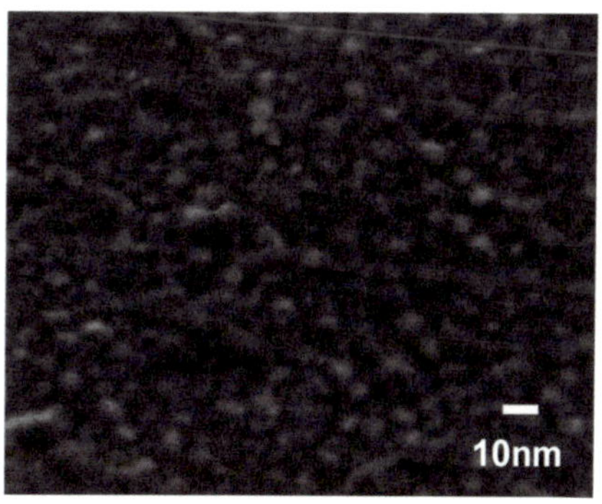


Fig. 8. SEM image of Co-catalyzed Si substrate just before introducing $CH_4$ at substrate temperature of 700 ˚C. Co particles were prepared using pulsed arc plasma deposition with 250 pulses. The cumulative Co particle number density was $2 \times 10^{13}$ $cm^{-2}$ on the surface.

Figures 9(a) and 9(b) show typical cross-sectional and top-view SEM images of the CNT film, respectively. Figures 9(c) and 9(d) show close-up images of the cleaved CNT film. From the morphology observed in Fig. 9(d), it can be seen that individual CNT bundles grew almost vertically via a self-supporting mechanism, due to the extremely high density of the CNTs.

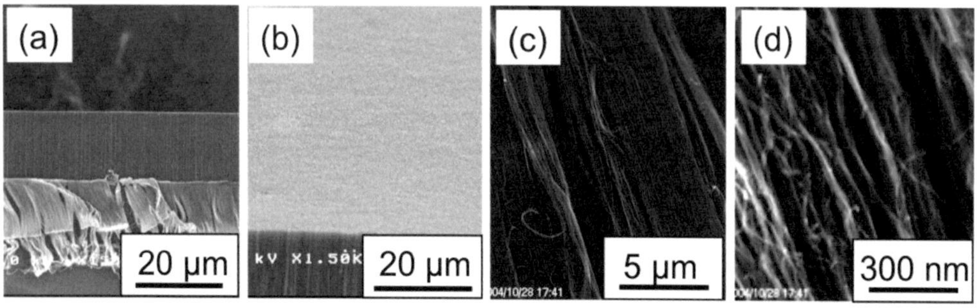


Fig. 9. SEM micrographs of CNT films grown on a Co-catalyzed Si substrate with a TiN buffer layer. The Co particles were prepared using pulsed arc plasma deposition with 250 pulses. (a) Cross-sectional SEM image of the dense CNT film. (b) SEM image showing the surface of the CNT film, (c)(d) Close-up images of cleaved CNT film showing aligned growth of the nanotubes. (Hiramatsu et al., 2005b) - reproduced with permission from Institute of Pure and Applied Physics

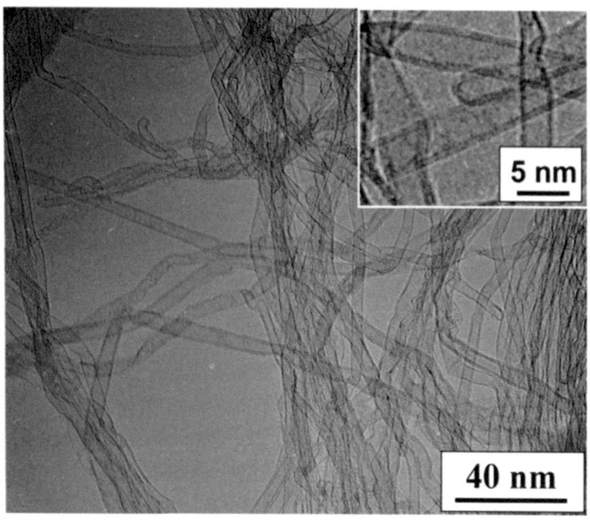


Fig. 10. TEM image of CNTs scraped from the substrate. The CNTs were grown on the Co-catalyzed Si substrates at a substrate temperature of 700 °C. The inset shows a magnified TEM image of typical CNTs. (Hiramatsu et al., 2005b) - reproduced with permission from Institute of Pure and Applied Physics

Figure 10 shows a low-magnification TEM image of typical CNTs. Co nanoarticles were prepared using pulsed arc plasma deposition with 50 pulses. The TEM specimen in Fig. 10 was scraped away from the substrate and was ultrasonicated in methanol. The CNTs are hollow and have a small average diameter of approximately 4.5 nm. Note that no Co particles were observed at the nanotube tips in the TEM micrographs. As shown in the inset of Fig. 10, a magnified TEM image reveals that most of CNTs have a double-walled structure, with a clear inner channel of approximately 4 nm inner diameter. The percentages of single-, double- and triple-walled CNTs were estimated to be 5%, 80%, 15%, respectively. A growth rate curve for the CNT film was obtained by measuring the thickness of the CNT films for different growth periods, according to the observation of cross-sectional SEM images of CNT films. Figure 11 shows the average thickness of the aligned CNT film as a function of the growth period. As shown in Fig. 11, the thickness of the CNT film increased linearly up to 10 min and the CNTs grew at an extremely high rate of 600 nm/s during the first 10 min. Dense DWNT films with a thickness over 500 μm were obtained after 20 min.

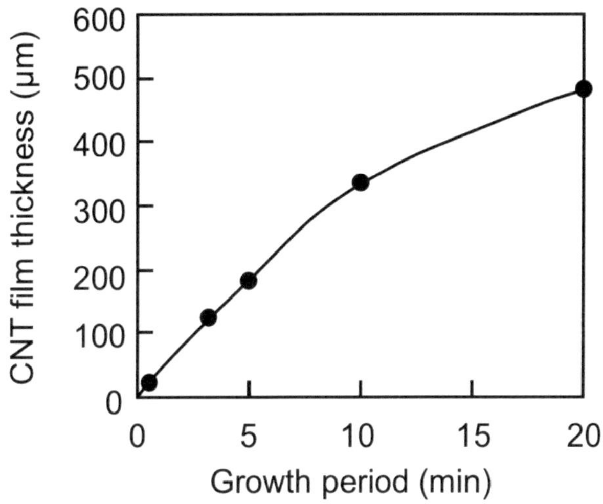


Fig. 11. Average thickness of aligned CNT films as a function of growth period. Co particles were prepared using pulsed arc plasma deposition with 250 pulses. The cumulative Co particle number density was $2 \times 10^{13}$ $cm^{-2}$ on the surface. (Hiramatsu et al., 2005b) - reproduced with permission from Institute of Pure and Applied Physics

Figure 12 shows distributions of the nanotube outer diameters grown from Co nanoparticles of different number densities. Co particles were prepared using pulsed arc plasma deposition with 50 (closed circles) and 250 (open circles) pulses, corresponding to the Co particle number densities of $4 \times 10^{12}$ and $2 \times 10^{13}$ $cm^{-2}$ on the surface, respectively. In the case of growth on Co-catalyzed Si substrates with cumulative Co nanoparticle density of $2 \times 10^{13}$ $cm^{-2}$ (250 pulses), the CNTs had small average diameters of the nanotubes and a narrow diameter distribution centered around 4.5 nm. On one hand, in the case of the CNTs grown with Co nanoparticle density of $4 \times 10^{12}$ $cm^{-2}$ (50 pulses), average outer diameter of CNTs decreased to 3.5–4 nm. The most significant difference from previous reports is that the catalyst was prepared originally in the form of nanoparticles in our study. It has been reported that the nanotube growth rate is inversely proportional to the nanotube diameter (Bower et al., 2000). Namely, the smaller diameter nanotubes would grow at a faster rate in terms of height. In this study, the sizes of the Co particles formed by pulsed arc deposition were relatively small. The resultant Co nanoislands thus clearly play an important role as a template for CNT growth, yielding nanotube diameters as small as 4–5 nm in the case of 250 pulses, for example, and consequently, a fast rate of growth could be attained for the CNTs. The density of nanotubes was roughly estimated to be $10^{12}$ $cm^{-2}$ for the CNTs grown on a Co-catalyzed Si substrate, with a cumulative Co nanoparticle density of $2 \times 10^{13}$ $cm^{-2}$ (250 pulses).

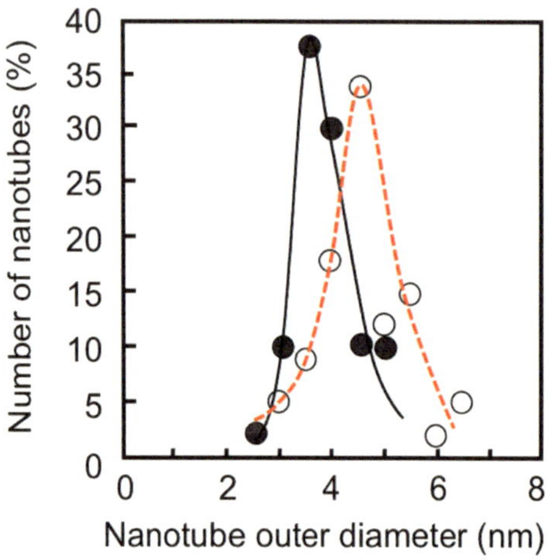


Fig. 12. Distributions of nanotube outer diameters deduced from TEM observations. Co particles were prepared using pulsed arc plasma deposition with 50 (closed circles) and 250 (open circles) pulses, corresponding to the cumulative Co particle number density of $4\times10^{12}$ and $2\times10^{13}$ $cm^{-2}$ on the surface, respectively. (Hiramatsu et al., 2005b)

Figure 13 shows a variation of thickness of DWNT films grown for 3 min and corresponding growth rate as a function of number of pulses of the pulsed arc discharge plasma for the preparation of Co nanoparticles (or estimated cumulative Co nanoparticle number density on the TiN buffer layer). As the cumulative Co particle number density on the substrate increased, the growth rate of the DWNT film increased up to 600 nm/s, as shown in Fig. 13. Too many Co particles on the substrate lead to the increase of the size of Co nanoislands formed during the heating process, resulting in the growth of multiwalled (≥ 3 layers) CNTs with lower growth rate. On the other hand, as the cumulative Co nanoparticle number density on the substrate decreased, average diameter of grown DWNTs decreased slightly as can be seen from Fig. 12, and the growth rate of DWNT films decreased. The reduction of growth rate is attributed to the increase of free space to grow for individual tubes, due to the decrease in the density of Co nanoislands formed by the aggregation of nanoparticles at the nucleation stage of growth. In the case of CNT growth without an electrical field, the CNTs would grow in a curly fashion. If the density of small-sized catalytic islands on the substrate is low, CNTs would grow in random orientations or in a less-aligned manner, resulting in a low growth rate of the CNT film. Figure 14(a) shows a cross-sectional SEM image of the DWNT film grown from Co nanoparticles deposited by 250 shots using pulsed arc plasma (cumulative Co particle number density of $2 \times 10^{13}$ $cm^{-2}$ on the surface). These were nearly optimum conditions for the dense nucleation of DWNTs. As shown in Fig. 14(a), nanotube bundles grew almost straight up due to the high density of DWNTs, corresponding to the high growth rate of 600nm/s for a growth time of 5 min. On the other hand, for growth on a substrate with low-density Co nanoparticles, the growth rate of the DWNT film decreased. Figure 14(b) shows a cross-sectional SEM image of DWNT film grown from Co nanoparticles deposited by 50 shots, corresponding to the Co particle number density of 4 × $10^{12}$ $cm^{-2}$ on the surface. As shown in Fig. 14(b), the DWNTs grew in a curly fashion because individual nanotubes had more free space to grow, resulting in the formation of wavy tubes leading to a reduction of the CNT film growth rate (135 nm/s for growth lasting 5 min) compared to Fig. 14(a).

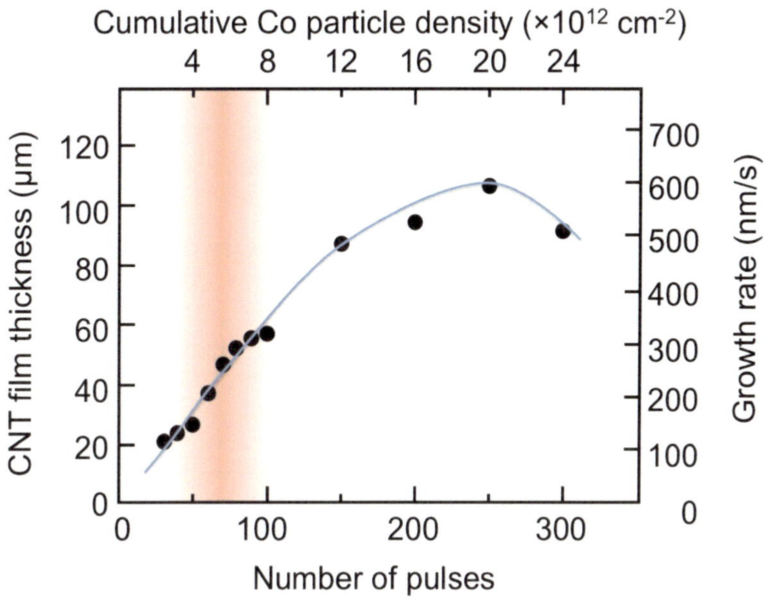


Fig. 13. Variation of thickness of DWNT films grown for 3 min and corresponding growth rate as a function of number of pulses of the pulsed arc discharge plasma for the preparation of Co nanoparticles (or estimated cumulative Co nanoparticle number density on the TiN buffer layer). Masked area corresponds to the condition where self-assembled cone-shaped tips composed of CNT bundles were formed; pulsed arc plasma with 50–100 pulses (see section 4.5).

#### 4.3 SWNT growth from Co and Ti nanoparticles without buffer layer

In the previous section, rapid growth of aligned DWNT films were demonstrated, where catalytic Co nanoparticles were prepared by pulsed arc deposition on a TiN buffer layer. By forming metal nanoparticles employing the pulsed arc discharge with a metal electrode, the density of catalyst nanoparticles with a relatively uniform size can be easily controlled on the substrate. As shown in Fig. 12, by decreasing the number of arc discharge pulses from 250 down to 50 shots, the average diameter of DWNTs decreased from 4.5 nm down to 3.5 nm. However, aligned SWNTs were not grown even at a low density of Co nanoparticles less than 50 pulses, probably because closely adjacent or overlapped Co nanoparticles would easily join together on the TiN surface to increase the size of the nanoislands during the substrate heating, resulting in the formation of less-aligned, low-density DWNTs or randomly oriented, sparse SWNTs after all.

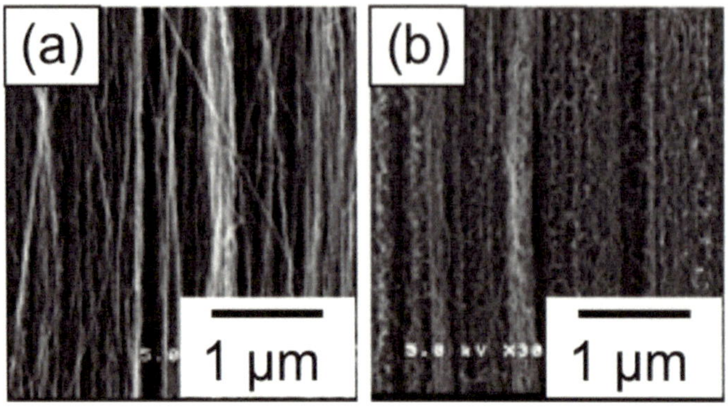


Fig. 14. Cross-sectional SEM images of CNT films grown from Co nanoparticles prepared on TiN buffer layer using pulsed arc plasma with (a) 250 pulses and (b) 50 pulses. (Hiramatsu et al., 2005b) - reproduced with permission from Institute of Pure and Applied Physics

In order to grow SWNT film on a Si substrate, a mixture of Co and Ti nanoparticles was prepared on the Si substrate using pulsed arc plasma deposition with a Co–Ti composite electrode, without a buffer layer. Figure 15 shows a schematic diagram of catalyst preparation for growing CNTs with smaller diameters. Ti is highly reactive with Si to form a silicide, as compared with Co (Murarka, 1983). Accordingly, it is expected that the formation of a Ti–silicide would precede the formation of a Co–silicide, when Ti is mixed with Co. Therefore, Ti prevents the formation of Co–silicide to a certain extent in the substrate heating process, and enables the size of the Co catalyst nanoparticles to be maintained at approximately 1–2 nm. As a result, the fabrication of films composed of vertically aligned SWNTs on Si substrates was attained using microwave plasma-enhanced CVD (Hiramatsu et al., 2007a). Moreover, the controlled preparation of catalyst nanoparticles on the Si substrate was performed by pulsed arc plasma deposition with the alternate use of Co and Ti electrodes. This technique has potential for controlling the size of catalyst nanoislands on the substrate, resulting in the controlled growth of aligned CNTs with single to three walls. By changing the number of cumulative Co nanoparticles, the fabrications of SWNT and DWNT films can be controlled (Hiramatsu et al., 2007b).

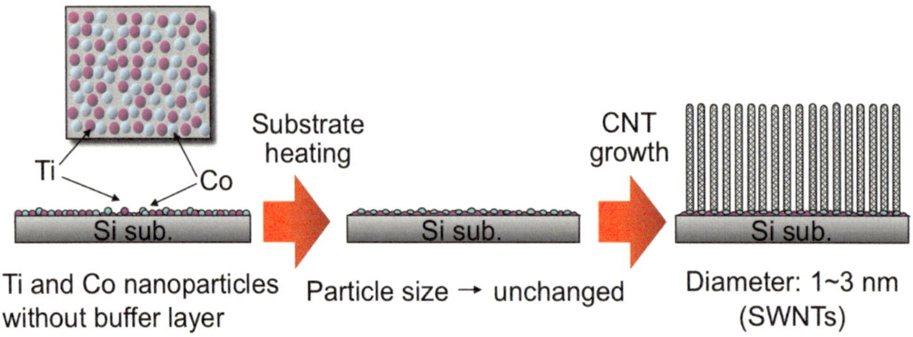


Fig. 15. Schematic of catalyst preparation for growing CNTs with small diameters (SWNTs)

In the case using sintered Ti–Co composite target electrode, the pulsed arc discharge yields a mixture of Co and Ti nanoparticles simultaneously. The sintered Ti–Co composite electrode is commercially available from ULVAC, Inc., and the Co and Ti contents in the composite are 55.2 and 44.8 wt.%, respectively. The pulsed arc discharge was operated with sintered Ti–Co composite electrode at a pressure of $10^{-4}$ Torr at room temperature. Three types of Co-catalyzed Si substrates (types A, B and C) without buffer layer were prepared by setting the number of pulsed arc discharges with the Ti–Co composite electrode at 50, 100 and 250 shots, respectively. Ti nanoparticles mixed with the Co nanoparticles prevent the formation of Co-silicide during the substrate heating process, and enables the size of the Co catalytic nanoparticles to be maintained at approximately 1–2 nm.

CNT films were grown using microwave plasma-enhanced CVD with a 2.45 GHz, 1.5 kW microwave generator. A mixture of $CH_4$ and $H_2$ was used as the source gas. The flow rates of $CH_4$ and $H_2$ were 50 and 70 sccm, respectively. The microwave power and total pressure were maintained at 900 W and 70 Torr, respectively. The growth experiments were carried out for 5 min at a substrate temperature of 700 °C. SEM and TEM were used to evaluate the morphology of the grown CNTs. Raman spectra for the CNTs were obtained using the 514.5 nm line of Ar laser and the 632.8 nm line of He–Ne laser.

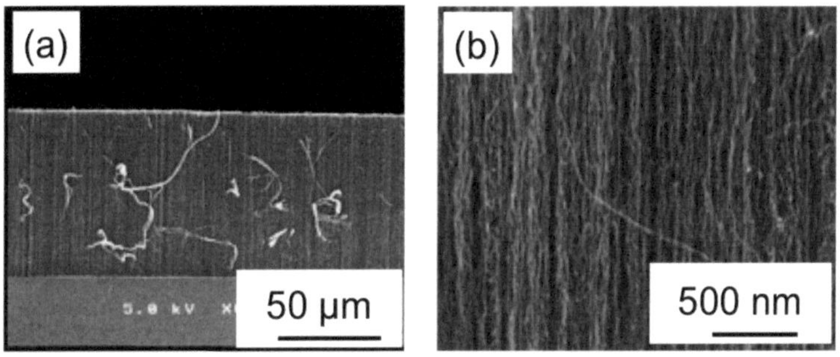


Fig. 16. SEM micrographs of a CNT film grown on a Co-catalyzed Si substrate (type A). The catalytic nanoparticles were prepared by pulsed arc discharge using a sintered Ti–Co composite electrode with 50 pulses. (a) Cross-sectional SEM image of the dense CNT film. (b) Close-up image of the cleaved CNT film showing aligned growth of the nanotubes. (Hiramatsu et al., 2007a) - reproduced with permission from Elsevier

Figure 16(a) shows a typical cross-sectional SEM image of a cleaved CNT film grown for 5 min on a type A substrate. The catalytic nanoparticles were prepared using pulsed arc discharges employing the sintered Ti–Co composite electrode with 50 pulses. A dense and vertically aligned CNT film was observed to grow on the Co-catalyzed Si substrate. Figure 16(b) shows a close-up SEM image of the cleaved CNT film. The morphology in Fig. 16(b) shows that the CNT bundles were not perfectly straight. In this case, the growth rate of the CNT film was approximately 190 nm/s.

Figure 17(a) shows the Raman spectrum at the low frequency band (100–400 $cm^{-1}$) for the CNT sample on the type A substrate measured using the 632.8 nm line. The radial breathing mode (RBM) peaks are clearly observed at 188, 260, 282 and 295 $cm^{-1}$, confirming the existence of SWNTs. Figure 17(b) shows the Raman spectrum at the low frequency band for the same CNT sample measured using the 514.5 nm line. In this case, the RBM peaks are clearly observed at 193, 211, 224, 235, 267 and 290 $cm^{-1}$. The diameter distribution calculated from the frequency of the RBM peaks for the SWNT bundles is in the range between 0.8 and 1.3 nm.

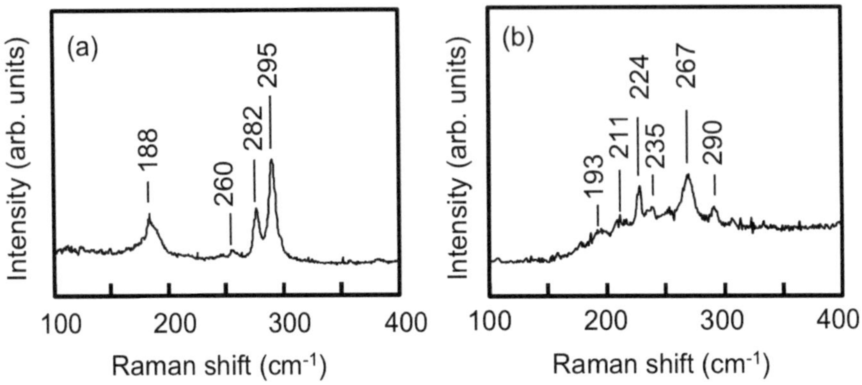


Fig. 17. Raman spectra at the low frequency band (100–400 $cm^{-1}$) for the CNT sample grown on the type 'A' substrate. The Raman spectra (a) and (b) were obtained using the 632.8 nm line of He-Ne laser and the 514.5 nm line of Ar laser, respectively. (Hiramatsu et al., 2007a) - reproduced with permission from Elsevier

In order to evaluate the morphology and diameter distribution of the CNTs, TEM images were taken for the same CNT sample used for the measurement of Raman spectra shown in Figs. 17(a) and 17(b). A distribution histogram of nanotube outer diameters is shown in Fig. 18, and the typical TEM image of the CNTs is shown in the inset in Fig. 18. The TEM specimen was prepared by scraping from the substrate and ultrasonicating in methanol. The inset TEM image in Fig. 18 reveals that most CNTs are SWNTs free of metal particles. The diameters of the SWNTs were measured and the average diameter of the CNTs was determined to be approximately 1 nm, in agreement with the value calculated from the RBM peaks of the Raman spectra in Figs. 17(a) and 17(b). The distribution histogram of nanotube outer diameters shows that the diameters of most CNTs were in the range from 0.5 to 3 nm. The SWNT and DWNT fractions were estimated to be 80 and 20%, respectively.

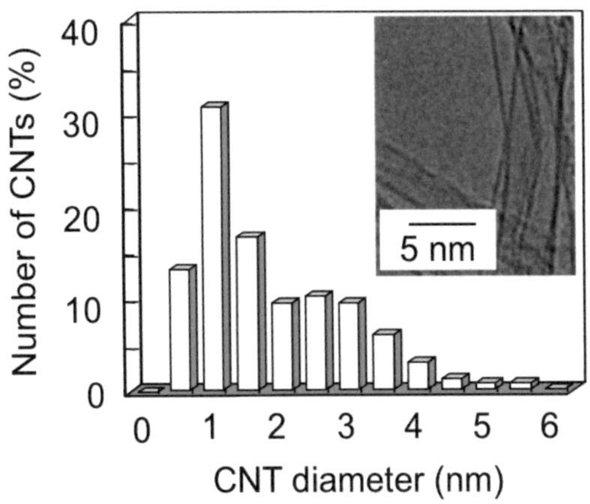


Fig. 18. Distribution histogram of nanotube outer diameters deduced from TEM observations. The CNTs were grown on the type A substrate. The catalytic nanoparticles were prepared by pulsed arc discharges with the sintered Ti-Co composite electrode with 50 pulses. The inset shows a magnified TEM image of typical CNTs. (Hiramatsu et al., 2007a) - reproduced with permission from Elsevier

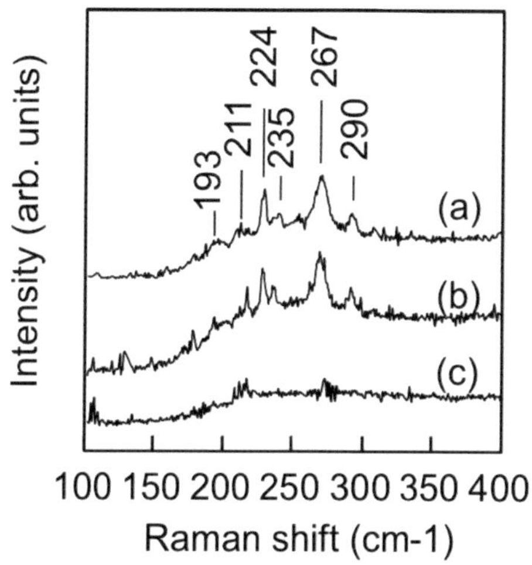


Fig. 19. Raman spectra at the low frequency band (100–400 $cm^{-1}$), measured using the 514.5 nm line of Ar laser, for CNTs grown on Si substrates with different densities of cumulative nanoparticles. Raman spectra (a)–(c) were obtained for CNT films grown on Co-catalyzed Si substrates, types A, B and C, respectively. (Hiramatsu et al., 2007a) - reproduced with permission from Elsevier

CNTs were grown on Si substrates with different densities of cumulative nanoparticles, and the Raman spectra for the CNTs were obtained using the 514.5 nm line of an Ar laser. In Fig. 19, the Raman spectra (a)–(c) in the low frequency band were obtained for CNT films grown on Co-catalyzed Si substrates, types A, B and C, respectively. As mentioned before, in the Raman spectrum (a) for the CNTs grown on the type A substrate, the RBM peaks are clearly observed at 188, 217, 260, 279 and 293 $cm^{-1}$, confirming the existence of SWNTs. In the case of the type B substrate, where the cumulative nanoparticle density was twice that on the type A substrate, the RBM peaks are also clearly observed in the Raman spectrum (b). The Raman spectrum (b) is almost identical to the Raman spectrum (a). On the other hand, the growth rate of the CNT film using the type B substrate was approximately 350 nm/s, which is almost twice that of the type A substrate. When using a substrate with low-density catalytic nanoparticles (type A substrate), the CNT bundles had more space to grow and thereby grew in a curly fashion, as shown in Fig. 16(b). With increasing cumulative density of catalytic nanoparticles, as in the case using the type B substrate, SWNT film with a high growth rate was attained due to the dense nucleation of CNTs from the doubled density of the catalytic nanoparticles as compared to the case using the type A substrate. In contrast, with further increase in the cumulative catalytic nanoparticle density, the RBM peaks disappeared, as seen in the Raman spectrum (c) for the CNTs grown on the type C substrate, resulting in the growth of MWNTs with two walls or more.

Figures 20(a) and 20(b) show SEM images of the surfaces of the type B and C Co-catalyzed Si substrates respectively, just before the introduction of $CH_4$ at a substrate temperature of 700 °C. In Fig. 20(a), nanoparticles with a size of approximately 2 nm were formed on the type B substrate. On the other hand, nanoparticles with a size of approximately 3-4 nm were observed on the type C substrate, as seen in the SEM image in Fig. 20(b). Pulsed arc plasma deposition with the Co electrode yielded Co nanoparticles of about 1-2 nm in size. At an appropriate density of catalytic nanoparticles, the catalytic nanoparticles deposited by pulsed arc discharge play an important role as a template for the growth of SWNTs with diameters as small as 1-2 nm. On the other hand, in the case of growth on a substrate with an excess density of cumulative nanoparticles, MWNTs with 2 or 3 walls were primarily grown. In our system, it takes about 10 min to increase the substrate temperature from room temperature to 700 °C. During this period, overlapping or closely adjacent particles can merge, resulting in the formation of catalytic nanoparticles of approximately 3–4 nm in size.

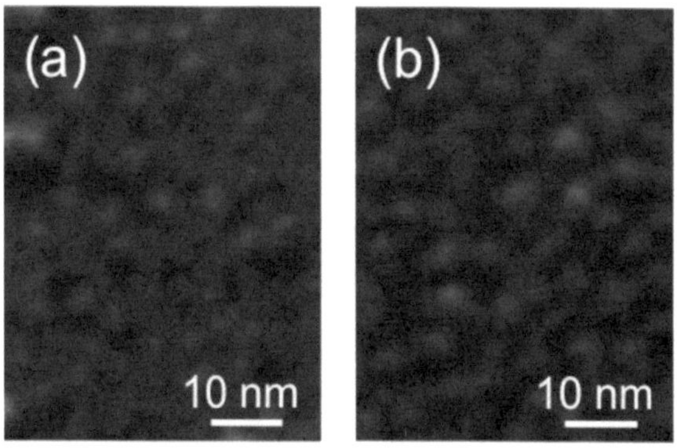


Fig. 20. SEM images of the surface of Co-catalyzed Si substrates: (a) type B and (b) type C, just before introducing $CH_4$ at a substrate temperature of 700 °C. (Hiramatsu et al., 2007a) - reproduced with permission from Elsevier

Further controlled preparation of catalyst nanoparticles on the Si substrate was performed by pulsed arc plasma deposition with the alternate use of Co and Ti electrodes. This technique has potential for controlling the size of catalyst nanoparticles on the substrate, resulting in the controlled growth of aligned CNTs with single to three walls. By changing the number of cumulative Co nanoparticles, the fabrications of SWNT and DWNT films can be controlled.

In this case, Co and Ti nanoparticles are deposited alternately on the Si substrate without a buffer layer, by pulsed arc discharge with the alternate use of Ti and Co electrodes at a pressure of $10^{-4}$ Torr at room temperature. Ti and Co electrodes were used in turn after every 10 shots of pulsed discharge. The number of cumulative pulsed discharges with the Ti electrode was the same as that with the Co electrode for each substrate prepared for the CNT growth experiment hereafter. The density of Co nanoparticles on the substrate was controlled by varying the number of pulsed discharges with the Co electrode in the range from 50 to 250 shots, resulting in cumulative Co particle densities of $4 \times 10^{12}$ to $2 \times 10^{13}$ $cm^{-2}$. From the TEM observation, the pulsed arc discharge with the Ti electrode yields Ti nanoparticles as small as the Co nanoparticles. For Co or Ti film deposition by pulsed arc discharge, the deposition rate of the Co film using the Co electrode was almost the same as that of the Ti film using the Ti electrode. Therefore, it would be assumed that the cumulative densities of Ti and Co nanoparticles are almost the same for each substrate prepared.

Figure 21(a) shows a typical cross-sectional SEM image of a cleaved CNT film grown for 5 min on a Co-catalyzed Si substrate with a cumulative Co nanoparticle density of $4 \times 10^{12}$ $cm^{-2}$. A dense and vertically aligned CNT film was grown on the Co-catalyzed Si substrate. Figure 21(b) shows a magnified SEM image of the cleaved CNT film. From the morphology shown in Fig. 21(b), it can be observed that CNT bundles with a diameter of less than 10 nm grew almost vertically, due to the high density of the CNTs. In this case, the growth rate of the SWNT film was approximately 200 nm/s. Figure 21(c) is a typical top-view SEM image showing the smooth surface of the CNT film. Figure 21(d) shows the TEM image of CNTs, which were scraped away from the substrate and were ultrasonicated in methanol. TEM image shown in Fig. 21(d) reveals that most CNTs are SWNTs without metal particles.

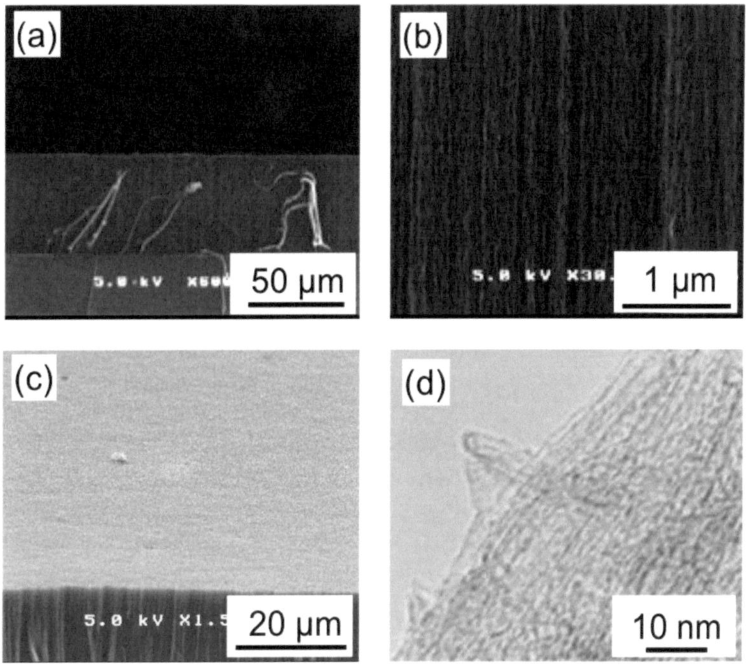


Fig. 21. SEM and TEM micrographs of vertically aligned SWNT film grown for 5 min. (a) Cross-sectional SEM image of cleaved CNT film. (b) Magnified view of the same cleaved CNT film showing aligned growth of nanotubes. (c) SEM image showing the surface of the CNT film. (d) TEM image of CNTs scraped from the substrate. The cumulative Co nanoparticle density was approximately $4 \times 10^{12}$ $cm^{-2}$ for the preparation of the Co-catalyzed Si substrate. (Hiramatsu et al., 2007b) - reproduced with permission from Institute of Pure and Applied Physics

Raman spectra of the CNT samples were measured using the 632.8 nm line of He-Ne laser. In Fig. 22, the Raman spectra (a)–(c) at a low frequency band (100–400 $cm^{-1}$) were obtained for CNT films grown on Co-catalyzed Si substrates with cumulative Co nanoparticle densities of $4 \times 10^{12}$, $1.2 \times 10^{13}$, and $2 \times 10^{13}$ $cm^{-2}$, respectively. In the Raman spectrum shown in Fig. 22(a), for CNTs grown on a Co-catalyzed Si substrate with a cumulative Co nanoparticle density of $4 \times 10^{12}$ $cm^{-2}$, the radial breathing mode (RBM) peaks are clearly observed at 188, 217, 260, 279, and 293 $cm^{-1}$ confirming the existence of SWNTs. The diameter distribution calculated from the frequency of the RBM peaks for SWNT bundles is in the range between 0.8 and 1.3 nm. With an increasing cumulative density of the Co nanoparticles on the Si substrate, the intensities of the RBM peaks in the Raman spectrum (b) decreased, suggesting that the CNTs grown in this case were a mixture of SWNTs and MWNTs with two or three walls. With further increase in the number of cumulative Co nanoparticles up to $2 \times 10^{13}$ $cm^{-2}$, the RBM peaks disappeared as observed in the Raman spectrum (c), resulting in the growth of MWNTs with more than two walls.

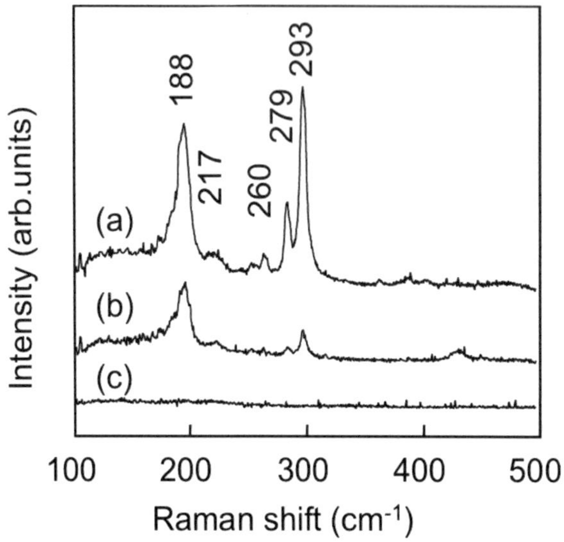


Fig. 22. Raman spectra at low frequency band (100–400 $cm^{-1}$) for CNT samples, measured using 632.8 nm line of He-Ne laser. Raman spectra (a)–(c) were obtained for CNT films grown on Co-catalyzed Si substrates with cumulative Co nanoparticle densities of $4 \times 10^{12}$, $1.2 \times 10^{13}$, and $2 \times 10^{13}$ $cm^{-2}$, respectively. (Hiramatsu et al., 2007b) - reproduced with permission from Institute of Pure and Applied Physics

Figures 23(a)–23(c) show the TEM images of CNTs grown on Co-catalyzed Si substrates with cumulative Co nanoparticle densities of $4 \times 10^{12}$, $1.2 \times 10^{13}$, and $2 \times 10^{13}$ $cm^{-2}$, respectively. Figures 23(d)–23(f) show the distribution histograms of the nanotube outer diameters, corresponding to the CNTs shown in Figs. 23(a)–23(c), respectively. For CNTs grown on a Co-catalyzed Si substrate, with a cumulative Co nanoparticle density of $4 \times 10^{12}$ $cm^{-2}$, the TEM image shown in Fig. 23(a) reveals that most CNTs are SWNTs without metal particles. The diameters of SWNTs were measured and found to be in the range of 1 to 2 nm, in agreement with that calculated from the RBM peaks of the Raman spectrum shown in Fig. 22(a). Note that after an ultrasonic treatment for the preparation of TEM specimens, most SWNTs were still in bundles, due to the van der Waals attraction. The distribution histogram of nanotube outer diameters shown in Fig. 23(d) shows that the CNTs had small average diameters and a narrow diameter distribution centered at approximately 1.5 nm. The percentages of SWNTs and DWNTs were estimated to be 85 and 15%, respectively. On the other hand, for CNTs grown on a Co-catalyzed Si substrate with a cumulative Co nanoparticle density of $1.2 \times 10^{13}$ $cm^{-2}$, majority of the CNTs were DWNTs with an average outer diameter of approximately 4 nm, as observed in the TEM image shown in Fig. 23(b). From the distribution histogram of the nanotube outer diameters shown in Fig. 23(e), the CNTs were a mixture of SWNTs, DWNTs, and MWNTs with three walls. The percentage of DWNTs was estimated to be 60%. With further increase in the cumulative Co nanoparticle density up to $2 \times 10^{13}$ $cm^{-2}$, as shown in Figs. 23(c) and 23(f), the distribution of the grown CNTs changed to a mixture of DWNTs and MWNTs with three or four walls. The diameters of most CNTs were measured and were in the range of 5 to 7 nm.

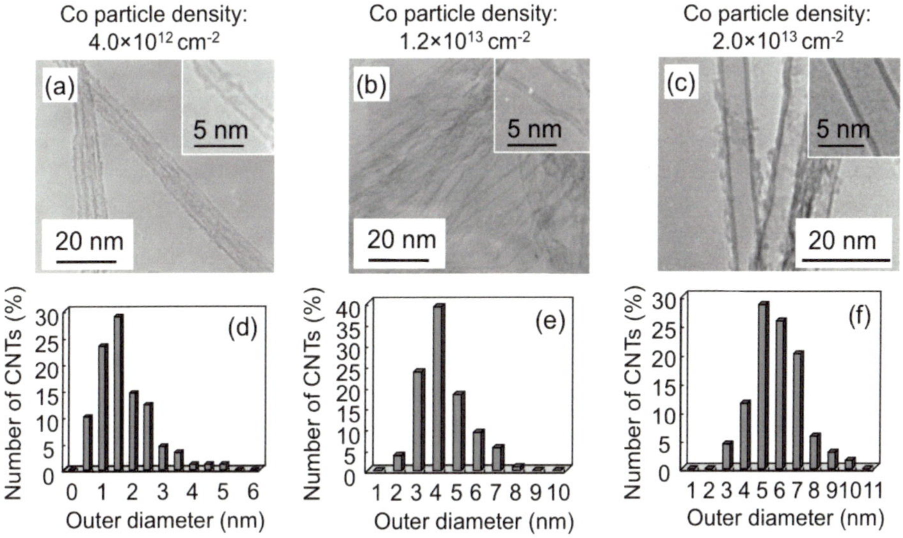


Fig. 23. (a)-(c) TEM images of CNTs grown on Co-catalyzed Si substrates with cumulative Co nanoparticle densities of $4 \times 10^{12}$, $1.2 \times 10^{13}$, and $2 \times 10^{13}$ $cm^{-2}$, respectively. (d)-(f) Corresponding distribution histograms of nanotube outer diameters deduced from TEM observations in (a)-(c), respectively. (Hiramatsu et al., 2007b) - reproduced with permission from Institute of Pure and Applied Physics

### 4.4 Area-selective growth of aligned CNTs

Well-defined, organized CNT structures were thus fabricated. The catalytic nanoparticles were pattered using a lift-off method, and aligned SWNTs were selectively grown on the pattered catalytic particles, which resulted in the fabrication of organized microstructures of aligned SWNTs and DWNTs.

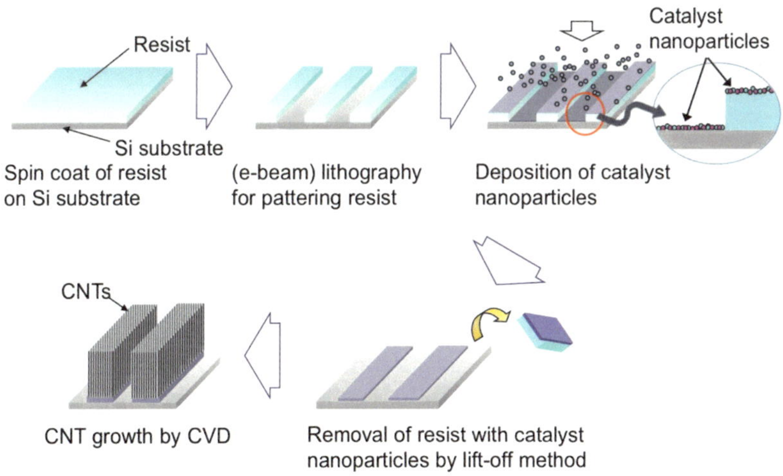


Fig. 24. Schematic diagram for pattering catalyst nanoparticles by lift-off method and area-selective growth of aligned CNTs.

Figure 24 shows a schematic procedure for pattering catalyst nanoparticles by a lift-off method and area-selective growth of aligned CNTs. "Lift-off" is a simple, easy method for patterning films that are deposited. In this case, catalyst nanoparticles are deposited area-selectively on a Si substrate. First, a pattern is defined on the Si substrate using photoresist. By using the pulsed arc discharge, a mixture of Co and Ti nanoparticles are deposited on the photoresist as well as the areas in which the photoresist has been cleared. Then, the photoresist is removed with solvent, taking the metal nanoparticles with it, and leaving only the nanoparticles deposited directly on the substrate. As a result, CNTs are grown area-selectively on the pattered catalyst nanoparticles.

Dense and aligned SWNTs with controlled microstructures were grown from the patterned catalytic nanoparticles, as demonstrated in Figs. 25(a)–25(f). Figure 25(a) shows a SEM top-view image of wavy lines with 1 μm width of vertical SWNTs grown on striped patterns of catalytic nanoparticles, showing area-selective growth of the vertically aligned SWNT film. Furthermore, SWNT pillars with a high aspect ratio were fabricated successfully. Figure 25(b) shows a SEM image of arrays of SWNT cylindrical pillars 100 μm long and 800 nm in diameter, forming a brush structure. A close-up view of the SWNT cylindrical pillars 10 μm long and 800 nm in diameter is shown in Fig. 25(c). Figures 25(d)–25(f) show another examples of organized SWNT microstructures: logos of Nano Factory and Meijo University written in kanji (Chinese character).

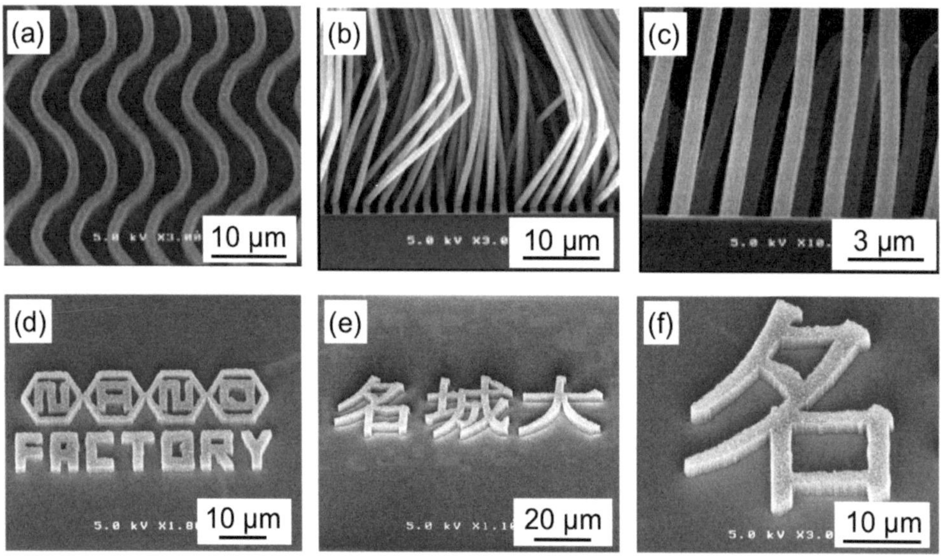


Fig. 25. SEM images of organized SWNT structures grown on patterned catalytic nanoparticles: (a) Wavy lines with 1 μm width, (b) side view of arrays of SWNT cylindrical pillars 100 μm long and 800 nm in diameter, (c) close-up view of SWNT cylindrical pillars 10 μm long and 800 nm in diameter, (d) logo of Nano Factory, (e) logo of Meijo University written in kanji (Chinese character), and (f) close-up view of kanji composed of aligned SWNTs. (Hiramatsu et al., 2007a) - reproduced with permission from Elsevier

CNTs are expected to be used as vertical wiring materials for future large-scale integrated circuits (LSI) interconnects (Horibe et al., 2005; Robertson et al., 2009; Nihei et al., 2010). For this purpose, aligned SWNT pillars were selectively grown on the bottom of the via-holes created in the thick $SiO_2$ film. Figure 26 shows a schematic procedure for growing vertically aligned SWNTs in the via-holes. The 800-nm-thick $SiO_2$ film was etched to form holes with 800 nm in diameter by conventional lithography and dry etching. Before removing the photoresist, catalyst nanoparticles were prepared on the photoresist as well as at the bottom of the holes by pulsed arc plasma deposition with the alternate use of Co and Ti electrodes or the sintered Ti-Co composite electrode. After removing the photoresist, catalyst nanoparticles remained only at the bottom of the holes. Finally, aligned CNTs are grown area-selectively from the catalyst nanoparticles deposited on the bottom of the via-holes.

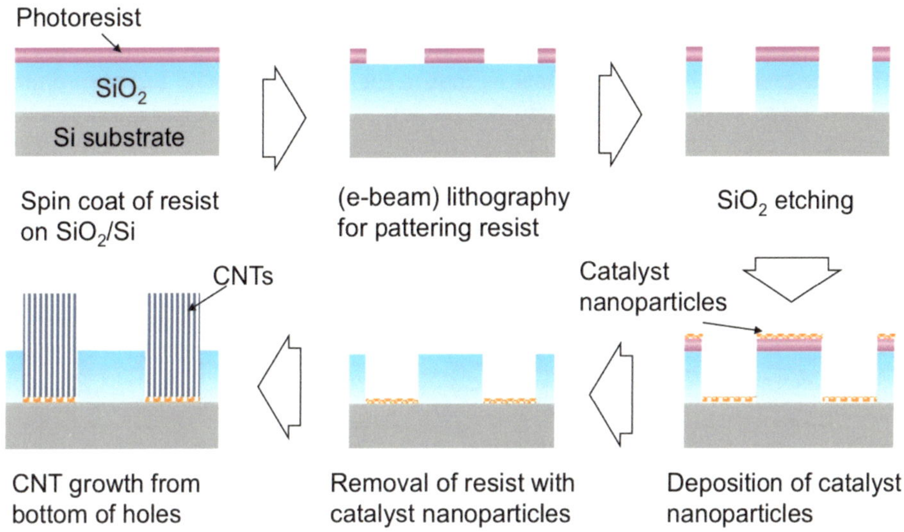


Fig. 26. Schematic diagram for growing vertically aligned SWNTs in the via-holes.

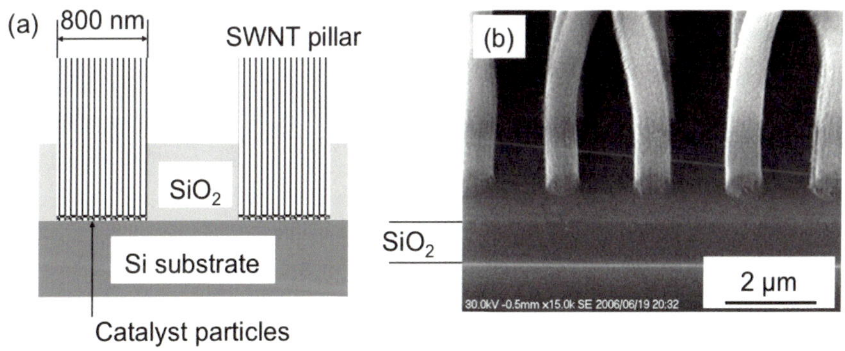


Fig. 27. (a) Schematic of vertical SWNTs embedded in holes drilled in $SiO_2$ film on Si substrate. (b) SEM image of SWNT cylindrical pillars of 800 nm in diameter, grown from the bottom of holes drilled in $SiO_2$ film. (Hiramatsu et al., 2007b) - reproduced with permission from Institute of Pure and Applied Physics

Figure 27(a) shows a schematic of vertically aligned SWNTs embedded in holes drilled in a $SiO_2$ film on the Si substrate. Figure 27(b) shows a SEM image of SWNT cylindrical pillars of

800 nm in diameter, grown from the bottom of the holes drilled in the $SiO_2$ film. In the case of the actual interconnects, planarization process using the chemical mechanical polishing (CMP) should be conducted after the CNT growth. Furthermore, TiN or Ti as the contact layer would exist at the bottom of the via-holes. Therefore, the technique for growing aligned DWNTs shown in Section 4.2 would be suitable for the practical application to the vertical wiring for the future LSI interconnects.

### 4.5 Growth of self-assembled cone-shaped tip arrays for emitter application

When the density of aligned CNTs is high, it is considered that such conditions are effective in the vertical wiring of next-generation integrated circuits. However, from the viewpoint of applications to field electron emission devices, although a threshold electric field that can be expected in the case of DWNTs is as low as that in the case of SWNTs, too high density of the CNTs has an adverse effect on electron emission devices because it reduces the electric fields applied to each nanotube. For this purpose, it is necessary to isolate DWNTs or to create a new shape for bundles. If the density of small-sized catalytic particles on the substrate is significantly low, CNTs would grow in random orientations or in a less-aligned manner, and isolated DWNTs could not be grown.

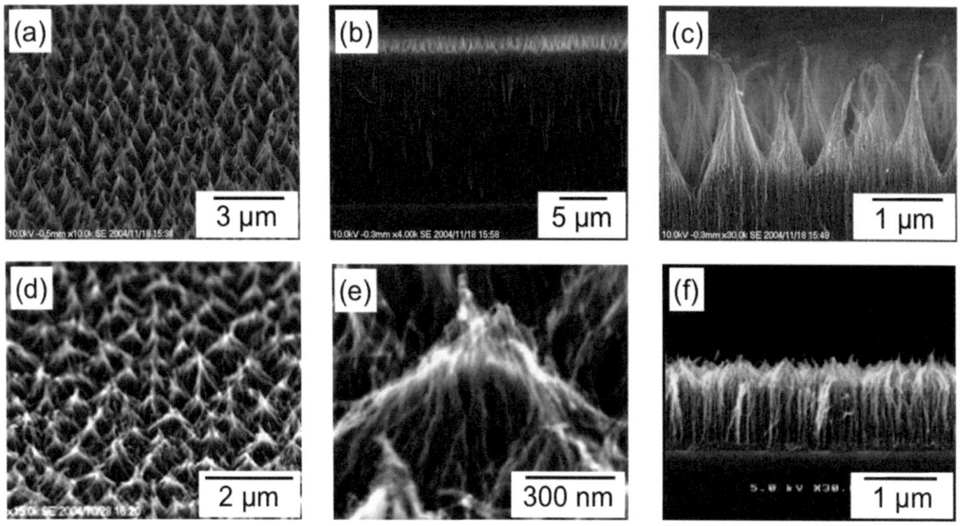


Fig. 28. SEM micrographs of DWNT films with self-assembled conical tip arrays grown on a Co-catalyzed Si substrate with a TiN buffer layer. The Co particles were prepared using pulsed arc plasma deposition with 50 pulses. (a) Tilted-view SEM image of DWNT film grown for 3 min, (b) cross-sectional SEM image of vertically aligned DWNT film, and (c) close-up image of self-assembled conical tips. (d) Tilted-view image, (e) close-up image, and (f) cross-sectional SEM image of DWNT film grown for 5 sec. (Hori et al., 2006)

The density of Co nanoparticles was reduced to some extent in comparison with the case for the growth of dense DWNTs shown in Fig. 9. When the number of shots of pulsed arc plasma was adjusted to approximately 50, the growth rate of CNT films decreased to approximately 200 nm/s. In this case, however, DWNT films possessing self-assembled cone-shaped tips at the top were formed as shown in the SEM photographs of Figs. 28(a)–28(c). A tilted-view SEM image of aligned DWNT film grown on Co-catalyzed Si substrate

with TiN buffer layer is shown in Fig. 28(a). Co particles were prepared using pulsed arc plasma deposition with 50 pulses, corresponding to the cumulative particle number density of approximately 4 × $10^{12}$ $cm^{-2}$ on the surface. As shown in Fig. 28(a), self-assembled conical tips were fabricated on the top of aligned DWNT film. Cross-sectional SEM images of vertically aligned DWNT film with self-assembled conical tip arrays are shown in Figs. 28(b) and 28(c). The height of conical parts was approximately 1 μm, and their distance was approximately 500 nm, as shown in Fig. 28(c). It was found that individual CNTs were not straight, but grown almost vertically, via a self-support mechanism, due to the high density of the CNTs. Figures 28(d)–28(f) show SEM images of DWNT film with conical tips grown for 5 sec. At this moment, the height of conical parts was less than 500 nm and their shape was not clearly defined.
In the case of the growth of single-walled or double-walled thin nanotubes with a certain space, tubes can swing relatively freely in an early stage of the growth. By setting space without causing the falling of the tubes, loosely-packed bundles are easily formed because of van der Waals force to make up cone-shaped tips at the top. It is noted that the self-assembled cone-shaped tips composed of CNT bundles were formed from Co nanoparticles with cumulative particle densities of 4–8 × $10^{12}$ $cm^{-2}$. In Fig. 13, masked area corresponds to the condition where self-assembled cone-shaped tips were formed. The surface morphology of cone-shaped tip arrays does not depend on the growth period of nanotubes, while the CNT film thickness increased linearly with the growth period up to 10 min. In other words, the surface shape of films in the case of 30 seconds growth is the same as that in the case of 10 minutes growth, and the growth process of carbon nanotubes in this experiment is the evidence of base growth.

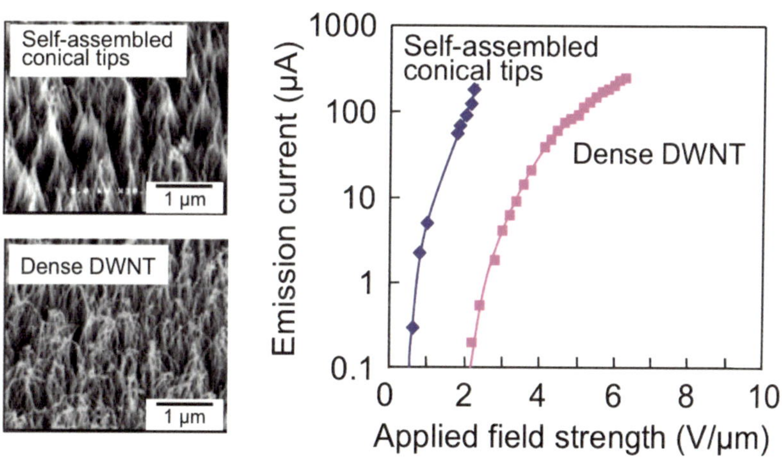


Fig. 29. Field emission characteristics of DWNT film with self-assembled conical tips and dense DWNT film.

Regarding DWNT films with self-assembled cone-shaped tips at the top, the field electron emission characteristics were measured. The field emission properties of as-grown DWNT film with self-assembled conical tips and dense DWNT film were investigated in a homemade parallel-plate diode device with the CNT film as a cathode and a stainless steel rod (1.5 mm in diameter) as an anode in a vacuum chamber ($10^{-6}$ Torr). Negative voltages were applied to the CNT film and corresponding emission currents were recorded. The distance between anode and cathode (CNT film) was kept at 500 μm. The electric field is expressed as the applied voltage divided by the anode to sample distance. The electron field emission properties of DWNT film with self-assembled conical tips and dense DWNT film are shown in Fig. 29. As a result of the measurement, the onset electric field for the electron field emission was below 1 V/μm for the DWNT film with self-assembled conical tips. The above-mentioned "Spindt-type" CNT bundles are thought to be suitable for the electron field emission application.

## 5. Conclusion

We have demonstrated the fabrication of films composed of vertically aligned, SWNTs and DWNTs on Si substrates using microwave plasma-enhanced chemical vapor deposition. As the first step of this study, Co nanoparticles were formed on the substrate using pulsed arc plasma deposition. The pulsed arc discharge with a Co electrode yielded Co nanoparticles approximately 1-2 nm in size on the substrate. A TiN thin film was used as a buffer layer in order to prevent the formation of Co silicide. As a result, a dense, vertically aligned DWNT film was grown rapidly on the Co-catalyzed Si substrate. The CNTs grew at an extremely high rate of 600 nm/s. By forming metal nanoparticles employing the pulsed arc discharge with a metal electrode, the density of catalyst nanoparticles with a relatively uniform size can be easily controlled on the substrate. However, closely adjacent or over-lapped Co nanoparticles would easily join together on the TiN surface to increase the size of the particles during the substrate heating, resulting in the growth of DWNTs.

In order to grow SWNT film on a Si substrate, catalytic nanoparticles were formed on Si substrates using pulsed arc plasma deposition with a Co–Ti composite electrode, without buffer layer. Ti nanoparticles mixed with the Co nanoparticles prevents the formation of Co silicide during the substrate heating process, and enables the size of the Co catalytic nanoparticles to be maintained at approximately 1–2 nm. As a result, the fabrication of films composed of vertically aligned SWNTs on Si substrates was attained. Moreover, the controlled preparation of catalyst nanoparticles on the Si substrate was performed by pulsed arc plasma deposition with the alternate use of Co and Ti electrodes. This technique has potential for controlling the size of catalyst nanoparticles on the substrate, resulting in the controlled growth of aligned carbon nanotubes with single to three walls. By changing the number of cumulative Co nanoparticles, the fabrications of SWNT and DWNT films can be controlled. Furthermore, the catalytic nanoparticles were pattered using a lift-off method, and SWNTs and DWNTs were selectively grown on the pattered catalytic particles, which resulted in the fabrication of organized microstructures of aligned CNTs.

## 6. References

Agawa, Y.; Yamaguchi, K.; Hara, Y.; Amano, S.; Horiuchi, T. & Shen, G. (2003). "Formation of a hafnium nitride film by an arc plasma gun". *ULVAC Technical Journal,* Vol.57E (2003) 1

Bower, C.; Zhou, O. ; Zhu, W.; Werder, D. J. & Jin, S. (2000). "Nucleation and growth of carbon nanotubes by microwave plasma chemical vapor deposition". *Applied Physics Letters,* Vol.77, No.17, pp.2767–2769, DOI:10.1063/1.1319529

Hiramatsu, M.; Taniguchi, M.; Nagao, H.; Ando, Y. & Hori, M. (2005a). "Fabrication of dense carbon nanotube films using microwave plasma-enhanced chemical vapor deposition". *Japanese Journal of Applied Physics,* Vol.44, No.2, pp.1150-1154, DOI: 10.1143/JJAP.44.1150

Hiramatsu, M.; Nagao, H.; Taniguchi, M.; Amano, H. ;Ando, Y. & Hori, M. (2005b). "High-rate growth of films of dense, aligned double-walled carbon nanotubes using microwave plasma- enhanced chemical vapor deposition". *Japanese Journal of Applied Physics,* Vol.44, No.22, pp.L693-L695, DOI: 10.1143/JJAP.44.L693

Hiramatsu, M.; Deguchi, T.; Nagao, H. & Hori, M. (2007a). "Area-selective growth of aligned single-walled carbon nanotube films using microwave plasma-enhanced CVD". *Diamond & Related Materials,* Vol.16, No.4–7, pp.1126–1130, DOI: 10.1016/j.diamond.2006.11.070

Hiramatsu, M.; Deguchi, T.; Nagao, H. & Hori, M. (2007b). "Aligned growth of single-walled and double-walled carbon nanotube films by control of catalyst preparation". *Japanese Journal of Applied Physics,* Vol.46, No.3, pp.L303-L306, DOI: 10.1143/JJAP.46.L303

Hong, W. K.; Chen, K. H.; Chen, L. C.; Tarntair, F. G.; Chen, K. J.; Lin, J. B. & Cheng, H. C. (2001). "Fabrication and characterization of carbon nanotube triodes". *Japanese Journal of Applied Physics,* Vol.40, No.5a, pp.3468-3473, DOI: 10.1143/JJAP.40.3468

Hori, M.; Hiramatsu, M. & Kano, H. (2006). "Carbon nanotube aggregate and method for producing same". WO/2006/120780

Horibe, M.; Nihei, M.; Kondo, D.; Kawabata, A. & Awano, Y. (2005). "Carbon Nanotube Growth Technologies Using Tantalum Barrier Layer for Future ULSIs with Cu/Low-k Interconnect Processes". *Japanese Journal of Applied Physics,* Vol.44, No.7A, pp.5309-5312, DOI: 10.1143/JJAP.44.5309

Kim, U.; Pcionek, R.; Aslam, D. M. & Tomanek, D. (2001). "Synthesis of high-density carbon nanotube films by microwave plasma chemical vapor deposition". *Diamond and Related Materials,* Vol.10, No.11, pp.1947-1951, DOI: 10.1016/S0925-9635(01)00384-3

Lee, S. B.; Teh, A. S.; Teo, K. B.; Chhowalla, K. M.; Hasko, D. G.; Milne, W. I.; Amaratunga, G. A. J. & Ahmed, H. (2003). "Fabrication of carbon nanotube lateral field emitters". *Nanotechnology,* Vol.14, No.2, pp.192-195, DOI: 10.1088/0957-4484/14/2/318

Murakami, Y.; Chiashi, S.; Miyauchi, Y.; Hu, M.; Ogura, M.; Okubo, T. & Maruyama, S. (2004). "Growth of vertically aligned single-walled carbon nanotube films on quartz substrates and their optical anisotropy". *Chemical Physics Letters,* Vol.385, No.3, pp.298-303, DOI: 10.1016/j.cplett.2003.12.095

Murarka, S. P. (1983). *Silicide for VLSI Applications,* Academic Press, New York, p. 72.

Nihei, M.; Kawabata, A.; Sato, M.; Nozue, T.; Hyakushima, T.; Kondo, D.; Ohfuti, M.; Sato, S. & Awano, Y. (2010). "Carbon Nanotube Interconnect Technologies for Future LSIs". In: *Solid State Circuits Technologies,* Jacobus W. Swart, pp.227-238, InTech, ISBN: 978-953-307-045-2

Robertson, J.; Zhong, G.; Hofmann, S.; Bayer, B. C.; Esconjauregui, C. S.; Telg, H. & Thomsen, C. (2009). "Use of carbon nanotubes for VLSI interconnects". *Diamond and Related Materials,* Vol.18, No.5-8, pp.957–962, DOI: 10.1016/j.diamond.2009.02.008

Sugai, T.; Yoshida, H.; Shimada, T.; Okazaki, T.; Bandow, S. & Shinohara, H. (2003). "New Synthesis of High-Quality Double-Walled Carbon Nanotubes by High-Temperature Pulsed Arc Discharge". *Nano Letters,* Vol.3, No.6, pp.769-773, DOI: 10.1021/nl034183+

Sugai, T.; Okazaki, T.; Yoshida, H. & Shinohara, H. (2004). "Syntheses of single- and double-wall carbon nanotubes by the HTPAD and HFCVD methods". *New Journal of Physics,* Vol.6, No.1, (2004), 21, DOI: 10.1088/1367-2630/6/1/021

Yamamoto, Y.; Agawa, Y.; Hara, Y.; Amano, S.; Chayahara, A.; Horono, Y. & Fujii, K. (1998). "Development of a coaxial type vacuum arc evaporation source". *Proceedings of 12th International Conference on Ion Implantation Technology,* pp.1148–1150

Zhong, G.; Iwasaki, T.; Honda, K.; Furukawa, Y.; Ohdomari, I. & Kawarada, H. (2005). "Low temperature synthesis of extremely dense and vertically aligned single-walled carbon nanotubes". *Japanese Journal of Applied Physics,* Vol.44, No.4a, pp.1558-1561, DOI: 10.1143/JJAP.44.1558

# Part 2

# Characterization & Properties of CNTs

# 11

# Study of Carbon Nanotubes Based on Higher Order Cauchy-Born Rule

Jinbao Wang[1,2], Hongwu Zhang[2], Xu Guo[2] and Meiling Tian[1]
[1]*School of Naval Architecture & Civil Engineering, Zhejiang Ocean University,*
[2]*State Key Laboratory of Structural Analysis for Industrial Equipment,*
*Department of Engineering Mechanics, Faculty of Vehicle Engineering and Mechanics,*
*Dalian University of Technology,*
*P.R.China*

## 1. Introduction

Since single-walled carbon nanotube (SWCNT) and multi-walled carbon nanotube (MWCNT) are found by Iijima (1991, 1993), these nanomaterials have stimulated extensive interest in the material research communities in the past decades. It has been found that carbon nanotubes possess many interesting and exceptional mechanical and electronic properties (Ruoff et al., 2003; Popov, 2004). Therefore, it is expected that they can be used as promising materials for applications in nanoengineering. In order to make good use of these nanomaterials, it is important to have a good knowledge of their mechanical properties.

Experimentally, Tracy et al. (1996) estimated that the Young's modulus of 11 MWCNTs vary from 0.4TPa to 4.15TPa with an average of 1.8TPa by measuring the amplitude of their intrinsic thermal vibrations, and it is concluded that carbon nanotubes appear to be much stiffer than their graphite counterpart. Based on the similar experiment method, Krishnan et al. (1998) reported that the Young's modulus is in the range of 0.9TPa to 1.70TPa with an average of 1.25TPa for 27 SWCNTs. Direct tensile loading tests of SWCNTs and MWCNTs have also been performed by Yu et al. (2000) and they reported that the Young's modulus are 0.32-1.47TPa for SWCNTs and 0.27-0.95TPa for MWCNTS, respectively. In the experiment, however, it is very difficult to measure the mechanical properties of carbon nanotues directly due to their very small size.

Based on molecular dynamics simulation and Tersoff-Brenner atomic potential, Yakobson et al. (1996) predicted that the axial modulus of SWCNTs are ranging from 1.4 to 5.5 TPa (Note here that in their study, the wall thickness of SWNT was taken as 0.066nm); Liang & Upmanyu (2006) investigated the axial-strain-induced torsion (ASIT) response of SWCNTs, and Zhang et al. (2008) studied ASIT in multi-walled carbon nanotubes. By employing a non-orthogonal tight binding theory, Goze et al. (1999) investigated the Young's modulus of armchair and zigzag SWNTs with diameters of 0.5-2.0 nm. It was found that the Young's modulus is dependent on the diameter of the tube noticeably as the tube diameter is small. Popov et al. (2000) predicted the mechanical properties of SWCNTs using Born's perturbation technique with a lattice-dynamical model. The results they obtained showed that the Young's modulus and the Poisson's ratio of both armchair and zigzag SWCNTs depend on the tube radius as the tube radius are small. Other atomic modeling studies include first-principles based calculations (Zhou et al., 2001; Van Lier et al., 2000; Sánchez-Portal et al., 1999) and molecular dynamics simulations (Iijima et al., 1996). Although these atomic modeling techniques seem well suited to study problems related to molecular or atomic motions, these calculations are time-consuming and limited to systems with a small number of molecules or atoms.

Comparing with atomic modeling, continuum modeling is known to be more efficient from computational point of view. Therefore, many continuum modeling based approaches have been developed for study of carbon nanotubes. Based on Euler beam theory, Govinjee and Sackman (1999) studied the elastic properties of nanotubes and their size-dependent properties at nanoscale dimensions, which will not occur at continuum scale. Ru (2000a,b) proposed that the effective bending stiffness of SWCNTs should be regarded as an independent material parameter. In his study of the stability of nanotubes under pressure, SWCNT was treated as a single-layer elastic shell with effective bending stiffness. By equating the molecular potential energy of a nano-structured material with the strain energy of the representative truss and continuum models, Odegard et al. (2002) studied the effective bending rigidity of a graphite sheet. Zhang et al. (2002a,b,c, 2004) proposed a nanoscale continuum theory for the study of SWCNTs by directly incorporating the interatomic potentials into the constitutive model of SWCNTs based on the modified Cauchy-Born rule. By employing this approach, the authors also studied the fracture nucleation phenomena in carbon nanotubes. Based on the work of Zhang (2002c), Jiang et al. (2003) proposed an approach to account for the effect of nanotube radius on its mechanical properties. Chang and Gao (2003) studied the elastic modulus and Poisson's ratio of SWCNTs by using molecular mechanics approach. In their work, analytical expressions for the mechanical properties of SWCNT have been derived based on the atomic structure of SWCNT. Li and Chou (2003) presented a structural mechanics approach to model the deformation of carbon nanotubes and obtained parameters by establishing a linkage between structural mechanics and molecular mechanics. Arroyo and Belytschko (2002, 2004a,b) extended the standard Cauchy-Born rule and introduced the so-called exponential map to study the mechanical properties of SWCNT since the classical Cauchy-Born rule cannot describe the deformation of crystalline film accurately. They also established the numerical framework for the analysis of the finite deformation of carbon nanotubes. The results they obtained agree very well with those obtained by molecular mechanics simulations. He et al. (2005a,b) developed a multishell model which takes the van der Waals interaction between any two layers into account and reevaluated the effects of the tube radius and thickness on the critical buckling load of MWCNTs. Gartestein et al. (2003) employed 2D continuum model to describe a stretch-induced torsion (SIT) in CNTs, while this model was restricted to linear response. Using the 2D continuum anharmonic anisotropic elastic model, Mu et al. (2009) also studied the axial-induced torsion of SWCNTs.

In the present work, a nanoscale continuum theory is established based on the higher order Cauchy-Born rule to study mechanical properties of carbon nanotubes (Guo et al., 2006; Wang et al., 2006a,b, 2009a,b). The theory bridges the microscopic and macroscopic length scale by incorporating the second-order deformation gradient into the kinematic description. Our idea is to use a higher-order Cauchy-Born rule to have a better description of the deformation of crystalline films with one or a few atom thickness with less computational efforts. Moreover, the interatomic potential (Tersoff 1988, Brenner 1990) and the atomic structure of carbon nanotube are incorporated into the proposed constitutive model in a consistent way. Therefore SWCNT can be viewed as a macroscopic generalized continuum with microstructure. Based on the present theory, mechanical properties of SWCNT and graphite are predicted and compared with the existing experimental and theoretical data.

The work is organized as follows: Section 2 gives Tersoff-Brenner interatomic potential for carbon. Sections 3 and 4 present the higher order Cauchy-Born rule is constructed and the analytical expressions of the hyper-elastic constitutive model for SWCNT are derived, respectively. With the use of the proposed constitutive model, different mechanical properties of SWCNTs are predicted in Section 5. Finally, some concluding remarks are given in Section 6.

## 2. The interatomic potential for carbon

In this section, Tersoff-Brenner interatomic potential for carbon (Tersoff, 1988; Brenner, 1990), which is widely used in the study of carbon nanotubes, is introduced as follows.

$$V(r_{IJ}) = V_R(r_{IJ}) - B_{IJ} V_A(r_{IJ}) \tag{1}$$

Where

$$V_R(r) = f(r)\frac{D_e}{S-1}e^{-\sqrt{2s}\beta(r-r_e)}, \quad V_A(r) = f(r)\frac{D_e S}{S-1}e^{-\sqrt{2/s}\beta(r-r_e)} \tag{2}$$

$$f(r) = \left\{ \begin{array}{ll} 1 & r < r_1 \\ \frac{1}{2}\left\{1 + cos\left[\frac{\pi(r-r_1)}{(r_2-r_1)}\right]\right\} & r_1 \le r \le r_2 \\ 0 & r > r_2 \end{array} \right\} \tag{3}$$

$$B_{IJ} = \left[1 + \sum_{K(\neq I,J)} G(\theta_{IJK}) f(r_{IK})\right]^{-\delta} \tag{4}$$

$$G(\theta) = a_0\left[1 + \frac{c_0^2}{d_0^2} - \frac{c_0^2}{d_0^2 + (1+cos\theta)^2}\right] \tag{5}$$

with the constants given in the following.

$$D_e = 6.000\ eV, \quad S = 1.22, \quad \beta = 21nm^{-1}, \quad r_e = 0.1390nm$$

$$\delta = 0.50000\,, \quad a_0 = 0.00020813\,, \quad c_0 = 330\,, \quad d_0 = 3.5$$

## 3. The higher order cauchy-born rule

Cauchy-Born rule is a fundamental kinematic assumption for linking the deformation of the lattice vectors of crystal to that of a continuum deformation field. Without consideration of diffusion, phase transitions, lattice defect, slips or other non-homogeneities, it is very suitable for the linkage of 3D multiscale deformations of bulk materials such as space-filling crystals (Tadmor et al., 1996; Arroyo and Belytschko, 2002, 2004a,b). In general, Cauchy-Born rule describes the deformation of the lattice vectors in the following way:

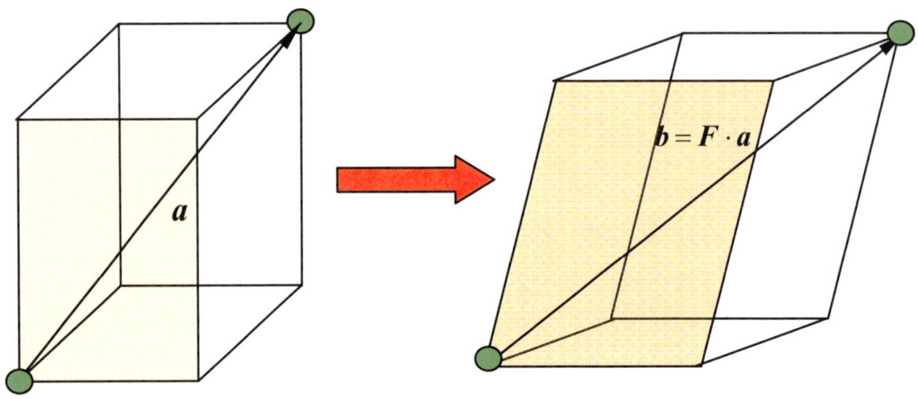


Fig. 1. Illustration of the Cauchy-Born rule

$$\mathbf{b} = \mathbf{F} \bullet \mathbf{a} \tag{6}$$

where $\mathbf{F}$ is the two-point deformation gradient tensor, $\mathbf{a}$ denotes the undeformed lattice vector and $\mathbf{b}$ represents the corresponding deformed lattice vector (see Fig. 1 for reference). In the deformed crystal, the length of the deformed lattice vector and the angle between two neighboring lattice vectors can be expressed by means of the standard continuum mechanics relations:

$$\mathbf{b} = \sqrt{\mathbf{a} \bullet \mathbf{C}\mathbf{a}} \text{ and } \cos\theta = \frac{\mathbf{a}' \bullet \mathbf{C}\mathbf{a}}{||\mathbf{b}'||\,||\mathbf{b}||} \tag{7}$$

where $\mathbf{b}' = \mathbf{F} \bullet \mathbf{a}'$ ($\mathbf{b}'$ and $\mathbf{a}'$ denote the neighboring deformed and undeformed lattice vector, respectively) and $\mathbf{C} = \mathbf{F}^T \bullet \mathbf{F}$ is the Green strain tensor measured from undeformed configuration. $\theta$ represents the angle formed by the deformed lattice vectors $\mathbf{b}$ and $\mathbf{b}'$. Though the use of Cauchy-Born rule is suitable for bulk materials, as was first pointed out by Arroyo and Belytschko (2002; 2004a,b), it is not suitable to apply it directly to the curved crystalline films with one or a few atoms thickness, especially when the curvature effects are dominated. One of the reasons is that if we view SWCNT as a 2D manifold without thickness embedded in 3D Euclidean space, since the deformation gradient tensor $\mathbf{F}$ describes only the change of infinitesimal material vectors emanating from the same point in the tangent spaces of the undeformed and deformed curved manifolds, therefore the deformation gradient tensor $\mathbf{F}$ is not enough to give an accurate description of the length of the deformed lattice vector in the deformed configuration especially when the curvature of the film is relatively large. In this case, the standard Cauchy-Born rule should be modified to give a more accurate description for the deformation of curved crystalline films, such as carbon nanotubes.

In order to alleviate the limitation of Cauchy-Born rule for the description of the deformation of curved atom films, we introduce the higher order deformation gradient into the kinematic relationship of SWCNT. The same idea has also been shown by Leamy et al. (2003).

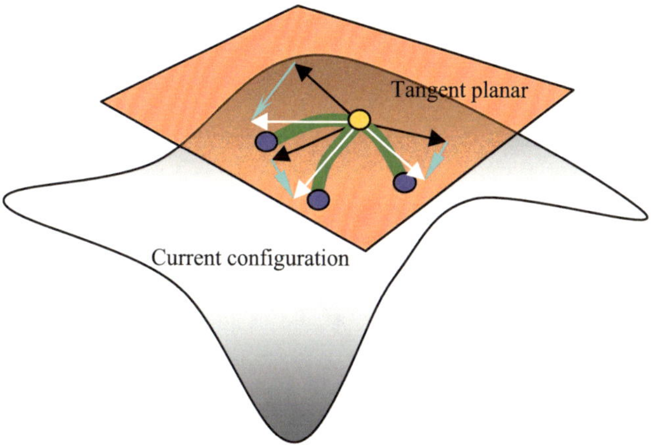


Fig. 2. Schematic illustration of the higher order Cauchy-Born rule

From the classical nonlinear continuum mechanics point of view, the deformation gradient tensor $\mathbf{F}$ is a linear transformation, which only describes the deformation of an *infinitesimal* material line element $d\mathbf{X}$ in the undeformed configuration to an infinitesimal material line element $d\mathbf{x}$ in deformed configuration, i.e.

$$d\mathbf{x} = \mathbf{F} \bullet d\mathbf{X} \tag{8}$$

As in Leamy et al. (2003), by taking the finite length of the initial lattice vector $\mathbf{a}$ into consideration, the corresponding deformed lattice vector should be expressed as:

$$\mathbf{b} = \int_0^{\mathbf{a}} \mathbf{F}(\mathbf{s}) d\mathbf{s} \tag{9}$$

Assuming that the deformation gradient tensor $\mathbf{F}$ is smooth enough, we can make a Taylor's expansion of the deformation field at $\mathbf{s} = \mathbf{0}$, which is corresponding to the starting point of the lattice vector $\mathbf{a}$.

$$\mathbf{F}(\mathbf{s}) = \mathbf{F}(\mathbf{0}) + \nabla\mathbf{F}(\mathbf{0}) \bullet \mathbf{s} + \nabla\nabla\mathbf{F}(\mathbf{0}) : (\mathbf{s} \otimes \mathbf{s}) / 2 + \mathbf{O}(||\mathbf{s}||^3) \tag{10}$$

Retaining up to the second order term of $\mathbf{s}$ in (10) and substituting it into (9), we can get the approximated deformed lattice vector as:

$$\mathbf{b} \approx \mathbf{F}(\mathbf{0}) \bullet \mathbf{a} + \frac{1}{2} \nabla\mathbf{F}(\mathbf{0}) : (\mathbf{a} \otimes \mathbf{a}) \tag{11}$$

Comparing with the standard Cauchy-Born rule, it is obvious that with the use of this higher order term, we can pull the vector $\mathbf{F} \bullet \mathbf{a}$ more close to the deformed configuration (see Fig. 2 for an illustration). By retaining more higher-order terms, the accuracy of approximation can be enhanced. Comparing with the exponent Cauchy-Born rule proposed by Arroyo and Belytschko (2002, 2004a,b), it can improve the standard Cauchy-Born rule for the description of the deformation of crystalline films with less computational effort.

## 4. The hyper-elastic constitutive model for SWCNT

With the use of the above kinematic relation established by the higher order Cauchy-Born rule, a constitutive model for SWCNTs can be established. The key idea for continuum modeling of carbon nanotube is to relate the phenomenological macroscopic strain energy density $W_0$ per unit volume in the material configuration to the corresponding atomistic potential.

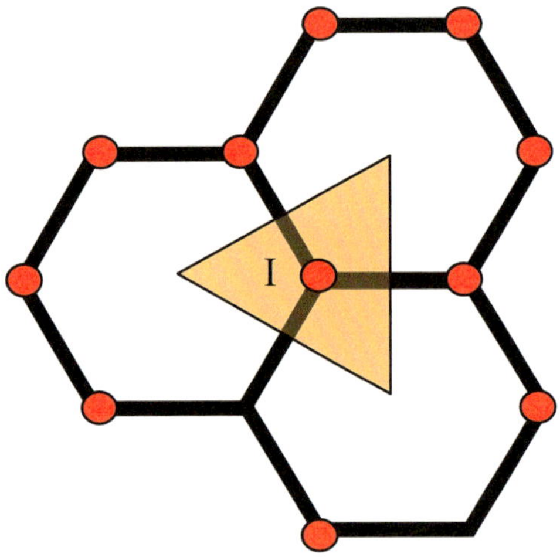


Fig. 3. Representative cell corresponding to an atom $I$

Assuming that the energy associated with an atom $I$ can be homogenized over a representative volume $V_I$ in the undeformed material configuration (i.e. graphite sheet, see Fig. 3 for reference), the strain energy density in this representative volume can be expressed as:

$$W_0 = W_0(|\mathbf{r}_{I1}|,|\mathbf{r}_{I2}|,|\mathbf{r}_{I3}|) = \sum_{J=1}^{3} V_{IJ}(\mathbf{r}_{I1}, \mathbf{r}_{I2}, \mathbf{r}_{I3})/2V_I = W_0(\mathbf{F}, \mathbf{G}) \tag{12}$$

And

$$\mathbf{r}_{IJ} = \mathbf{F} \bullet \mathbf{R}_{IJ} + \mathbf{G} : (\mathbf{R}_{IJ} \otimes \mathbf{R}_{IJ})/2 \tag{13}$$

where $\mathbf{R}_{IJ}$ and $\mathbf{r}_{IJ}$ denote the undeformed and deformed lattice vectors, respectively. $V_I$ is the volume of the representative cell. $\mathbf{F} = F_{ij}\mathbf{e}_i \otimes \mathbf{e}_j$ and $\mathbf{G} = \nabla\mathbf{F} = G_{ijk}\mathbf{e}_i \otimes \mathbf{e}_j \otimes \mathbf{e}_k$ are the first and second order deformation gradient tensors, respectively. Note that here and in the following discussions, a unified Cartesian coordinate system has been used for the description of the positions of material points in both of the initial and deformed configurations.

Based on the strain energy density $W_0$, as shown by Sunyk et al. (2003), the first Piola-Kirchhoff stress tensor $\mathbf{P}$, which is work conjugate to $\mathbf{F}$ and the higher-order stress tensor $\mathbf{Q}$, which is work conjugate to $\mathbf{G}$ can be obtained as:

$$\mathbf{P} = \frac{\partial W_0}{\partial \mathbf{F}} = \frac{1}{2V_I}\sum_{J=1}^{3} \mathbf{f}_{IJ} \otimes \mathbf{R}_{IJ} \tag{14}$$

$$\mathbf{Q} = \frac{\partial W_0}{\partial \mathbf{G}} = \frac{1}{4V_I}\sum_{J=1}^{3} \mathbf{f}_{IJ} \otimes \mathbf{R}_{IJ} \otimes \mathbf{R}_{IJ} \tag{15}$$

where $\mathbf{f}_{IJ}$ is the generalized force associated with the generalized coordinate $\mathbf{r}_{IJ}$, which is defined as:

$$\mathbf{f}_{IJ} = \frac{\partial W}{\partial \mathbf{r}_{IJ}} \tag{16}$$

The corresponding strain energy density can also be rewritten as:

$$W_0 = W / 2V_I \tag{17}$$

Where

$$W = \sum_{J=1}^{3} V_{IJ}(\mathbf{r}_{IJ}\,,\ \mathbf{r}_{IK}\,,\ \theta_{IJK}\ ,\ K \neq I,\ J\ ) \tag{18}$$

denotes the total energy of the representative cell related to atom $I$ caused by atomic interaction. $V_{IJ}$ is the interatomic potential for carbon introduced in Section 2.
We can also define the generalized stiffness $\mathbf{K}_{IJIK}$ associated with the generalized coordinate $\mathbf{r}_{IJ}$ as:

$$\mathbf{K}_{IJIK} = \frac{\partial \mathbf{f}_{IJ}}{\partial \mathbf{r}_{IK}} = \frac{\partial^2 W}{\partial \mathbf{r}_{IJ} \partial \mathbf{r}_{IK}} \tag{19}$$

where the subscripts $I$, $J$ and $K$ in the overstriking letters, such as $\mathbf{f}$, $\mathbf{r}$, $\mathbf{R}$ and $\mathbf{K}$, denote different atoms rather than the indices of the components of tensors. Therefore summation is not implied here by the repetition of these indices.
From (14) and (15), the tangent modulus tensors can be derived as:

$$\mathbf{M}_{\mathbf{FF}} = \frac{\partial^2 W_0}{\partial \mathbf{F} \otimes \partial \mathbf{F}} = \frac{1}{2V_I}\sum_{J=1}^{3}\sum_{K=1}^{3}[\mathbf{K}_{IJIK} \underline{\otimes} (\mathbf{R}_{IJ} \otimes \mathbf{R}_{IK})] \tag{20a}$$

$$\mathbf{M}_{\mathbf{FG}} = \frac{\partial^2 W_0}{\partial \mathbf{F} \otimes \partial \mathbf{G}} = \frac{1}{4V_I}\sum_{J=1}^{3}\sum_{K=1}^{3}[\mathbf{K}_{IJIK} \underline{\otimes} (\mathbf{R}_{IJ} \otimes \mathbf{R}_{IK})] \otimes \mathbf{R}_{IK} \tag{20b}$$

$$\mathbf{M}_{\mathbf{GF}} = \frac{\partial^2 W_0}{\partial \mathbf{G} \otimes \partial \mathbf{F}} = \frac{1}{4V_I} \sum_{J=1}^{3} \sum_{K=1}^{3} [\mathbf{K}_{IJIK} \overline{\otimes} (\mathbf{R}_{IJ} \otimes \mathbf{R}_{IJ})] \otimes \mathbf{R}_{IK} \tag{20c}$$

$$\mathbf{M}_{\mathbf{GG}} = \frac{\partial^2 W_0}{\partial \mathbf{G} \otimes \partial \mathbf{G}} = \frac{1}{8V_I} \sum_{J=1}^{3} \sum_{K=1}^{3} [\mathbf{K}_{IJIK} \overline{\otimes} (\mathbf{R}_{IJ} \otimes \mathbf{R}_{IJ})] \otimes (\mathbf{R}_{IK} \otimes \mathbf{R}_{IK}) \tag{20d}$$

where $[\mathbf{A} \underline{\otimes} \mathbf{B}]_{ijkl} = A_{ik} B_{jl}$, $[\mathbf{A} \overline{\otimes} \mathbf{B}]_{ijkl} = A_{il} B_{jk}$. Compared with the results obtained by Zhang et al. (2002c), four tangent modulus tensors are presented here. This is due to the fact that second order deformation gradient tensor has been introduced here for kinematic description. Therefore, from the macroscopic point of view, we can view the SWNT as a generalized continuum with microstructure.

Just as emphasized by Cousins(1978a,b), Tadmor (1999), Zhang (2002c), Arroyo and Belytschko (2002a), since the atomic structure of carbon nanotube is not centrosymmetric, the standard Cauchy-Born rule can not be used directly since it cannot guarantee the inner equilibrium of the representative cell. An inner shift vector $\boldsymbol{\eta}$ must be introduced to achieve this goal. The inner shift vector can be obtained by minimizing the strain energy density of the unit cell with respect to $\boldsymbol{\eta}$:

$$\hat{\boldsymbol{\eta}}(\mathbf{F}, \mathbf{G}) = \arg(\min_{\eta} W_0(\mathbf{F}, \mathbf{G}, \boldsymbol{\eta})) \Rightarrow \left. \frac{\partial W_0}{\partial \boldsymbol{\eta}} \right|_{\eta = \hat{\eta}} = \mathbf{0} \tag{21}$$

Substituting (21) into $W_0(\mathbf{F}, \mathbf{G}, \boldsymbol{\eta})$, we have:

$$\hat{W}_0(\mathbf{F}, \mathbf{G}) = W_0(\mathbf{F}, \mathbf{G}, \hat{\boldsymbol{\eta}}(\mathbf{F}, \mathbf{G})) \tag{22}$$

Then the modified tangent modulus tensors can be obtained as:

$$\hat{\mathbf{M}}_{\mathbf{FF}} = \frac{\partial^2 \hat{W}_0}{\partial \mathbf{F} \otimes \partial \mathbf{F}} = \mathbf{M}_{FF}\big|_{\eta=\hat{\eta}} - \left[ \frac{\partial^2 W_0}{\partial \mathbf{F} \otimes \partial \boldsymbol{\eta}} \left( \frac{\partial^2 W_0}{\partial \boldsymbol{\eta} \otimes \partial \boldsymbol{\eta}} \right)^{-1} \frac{\partial^2 W_0}{\partial \boldsymbol{\eta} \otimes \partial \mathbf{F}} \right] \Bigg|_{\eta=\hat{\eta}} \tag{23a}$$

$$\hat{\mathbf{M}}_{\mathbf{FG}} = \frac{\partial^2 \hat{W}_0}{\partial \mathbf{F} \otimes \partial \mathbf{G}} = \mathbf{M}_{\mathbf{FG}}\big|_{\eta=\hat{\eta}} - \left[ \frac{\partial^2 W_0}{\partial \mathbf{F} \otimes \partial \boldsymbol{\eta}} \left( \frac{\partial^2 W_0}{\partial \boldsymbol{\eta} \otimes \partial \boldsymbol{\eta}} \right)^{-1} \frac{\partial^2 W_0}{\partial \boldsymbol{\eta} \otimes \partial \mathbf{G}} \right] \Bigg|_{\eta=\hat{\eta}} \tag{23b}$$

$$\hat{\mathbf{M}}_{\mathbf{GF}} = \frac{\partial^2 \hat{W}_0}{\partial \mathbf{G} \otimes \partial \mathbf{F}} = \mathbf{M}_{\mathbf{GF}}\big|_{\eta=\hat{\eta}} - \left[ \frac{\partial^2 W_0}{\partial \mathbf{G} \otimes \partial \boldsymbol{\eta}} \left( \frac{\partial^2 W_0}{\partial \boldsymbol{\eta} \otimes \partial \boldsymbol{\eta}} \right)^{-1} \frac{\partial^2 W_0}{\partial \boldsymbol{\eta} \otimes \partial \mathbf{F}} \right] \Bigg|_{\eta=\hat{\eta}} \tag{23c}$$

$$\hat{\mathbf{M}}_{GG} = \frac{\partial^2 \hat{W}_0}{\partial \mathbf{G} \otimes \partial \mathbf{G}} = \mathbf{M}_{GG}\big|_{\eta=\hat{\eta}} - \left[ \frac{\partial^2 W_0}{\partial \mathbf{G} \otimes \partial \boldsymbol{\eta}} \left( \frac{\partial^2 W_0}{\partial \boldsymbol{\eta} \otimes \partial \boldsymbol{\eta}} \right)^{-1} \frac{\partial^2 W_0}{\partial \boldsymbol{\eta} \otimes \partial \mathbf{G}} \right] \Bigg|_{\eta=\hat{\eta}} \tag{23d}$$

Where

$$\mathbf{M}_{FF}\big|_{\eta \blacklozenge \hat{\eta}} = \frac{1}{2V_I} \sum_{J=1}^{3} \sum_{K=1}^{3} [\hat{\mathbf{K}}_{IJIK} \underline{\otimes} ((\mathbf{R}_{IJ} + \hat{\boldsymbol{\eta}}) \otimes (\mathbf{R}_{IJ} + \hat{\boldsymbol{\eta}}))] \tag{24a}$$

$$\mathbf{M_{FG}}\big|_{\boldsymbol{\eta}=\hat{\boldsymbol{\eta}}} = \frac{1}{4V_I}\sum_{J=1}^{3}\sum_{K=1}^{3}[\hat{\mathbf{K}}_{IJIK}\underline{\otimes}((\mathbf{R}_{IJ}+\hat{\boldsymbol{\eta}})\otimes(\mathbf{R}_{IK}+\hat{\boldsymbol{\eta}})]\otimes(\mathbf{R}_{IK}+\hat{\boldsymbol{\eta}}) \tag{24b}$$

$$\mathbf{M_{GF}}\big|_{\eta=\hat{\eta}} = \frac{1}{4V_I}\sum_{J=1}^{3}\sum_{K=1}^{3}[\hat{\mathbf{K}}_{IJIK}\overline{\otimes}((\mathbf{R}_{IJ}+\hat{\boldsymbol{\eta}})\otimes(\mathbf{R}_{IJ}+\hat{\boldsymbol{\eta}}))]\otimes(\mathbf{R}_{IK}+\hat{\boldsymbol{\eta}}) \tag{24c}$$

$$\mathbf{M_{GG}}\big|_{\eta=\hat{\eta}} = \frac{1}{8V_I}\sum_{J=1}^{3}\sum_{K=1}^{3}[\hat{\mathbf{K}}_{IJIK}\overline{\otimes}((\mathbf{R}_{IJ}+\hat{\boldsymbol{\eta}})\otimes(\mathbf{R}_{IJ}+\hat{\boldsymbol{\eta}}))]\otimes((\mathbf{R}_{IK}+\hat{\boldsymbol{\eta}})\otimes(\mathbf{R}_{IK}+\hat{\boldsymbol{\eta}})) \tag{24d}$$

$$\hat{\mathbf{K}}_{IJIK} = \frac{\partial^2 W}{\partial \mathbf{r}_{IJ}\otimes\partial \mathbf{r}_{IJ}}\bigg|_{\eta=\hat{\eta}\ ,\ \mathbf{r}_{IJ}=\hat{\mathbf{r}}_{IJ}} \tag{25}$$

$$\hat{\mathbf{r}}_{IJ} = \mathbf{F}\bullet(\mathbf{R}_{IJ}+\hat{\boldsymbol{\eta}}) + \mathbf{G}:[(\mathbf{R}_{IJ}+\hat{\boldsymbol{\eta}})\otimes(\mathbf{R}_{IJ}+\hat{\boldsymbol{\eta}})]/2 \tag{26}$$

$$\begin{aligned}\frac{\partial^2 W_0}{\partial \mathbf{F}\otimes\partial\boldsymbol{\eta}}\bigg|_{\eta=\hat{\eta}} = \frac{1}{2V_I}\sum_{J=1}^{3}[\sum_{K=1}^{3}((\hat{\mathbf{K}}_{IJIK}\bullet\mathbf{F})\dot{\otimes}(\mathbf{R}_{IJ}+\hat{\boldsymbol{\eta}})\\ + \underset{\bullet}{\dot{sym}}(\hat{\mathbf{K}}_{IJIK}\bullet\mathbf{G}\bullet((\mathbf{R}_{IK}+\hat{\boldsymbol{\eta}})\otimes(\mathbf{R}_{IJ}+\hat{\boldsymbol{\eta}})))) + \hat{\mathbf{f}}_{IJ}\otimes{}^2\mathbf{1}]\end{aligned} \tag{27}$$

$$\frac{\partial W_0}{\partial\boldsymbol{\eta}}\bigg|_{\eta=\hat{\eta}} = \frac{1}{2V_I}\sum_{J=1}^{3}\hat{\mathbf{f}}_{IJ}\bullet[\mathbf{F}+\underset{\bullet}{\dot{sym}}(\mathbf{G}\bullet(\mathbf{R}_{IJ}+\hat{\boldsymbol{\eta}}))] \tag{28}$$

$$\begin{aligned}\frac{\partial^2 W_0}{\partial\boldsymbol{\eta}\otimes\partial\mathbf{F}}\bigg|_{\eta=\hat{\eta}} = \frac{1}{2V_I}\sum_{J=1}^{3}[\sum_{K=1}^{3}((\mathbf{F}+\underset{\bullet}{\dot{sym}}(\mathbf{G}\bullet(\mathbf{R}_{IJ}+\hat{\boldsymbol{\eta}})))^{\mathrm{T}}\\ \bullet(\hat{\mathbf{K}}_{IJIK}\otimes(\mathbf{R}_{IK}+\hat{\boldsymbol{\eta}}))) + \hat{\mathbf{f}}_{IJ}\underset{\bullet}{\otimes}{}^2\mathbf{1}]\end{aligned} \tag{29}$$

$$\begin{aligned}\frac{\partial^2 W_0}{\partial\boldsymbol{\eta}\otimes\partial\boldsymbol{\eta}}\bigg|_{\eta=\hat{\eta}} = \frac{1}{2V_I}\sum_{J=1}^{3}[\sum_{K=1}^{3}(\mathbf{F}+\underset{\bullet}{\dot{sym}}(\mathbf{G}\bullet(\mathbf{R}_{IJ}+\hat{\boldsymbol{\eta}})))^{\mathrm{T}}\bullet\hat{\mathbf{K}}_{IJIK}\\ \bullet(\mathbf{F}+\underset{\bullet}{\dot{sym}}(\mathbf{G}\bullet(\mathbf{R}_{IK}+\hat{\boldsymbol{\eta}})))) + \hat{\mathbf{f}}_{IJ}\bullet\underset{\bullet}{\dot{sym}}\,\mathbf{G}]\end{aligned} \tag{30}$$

$$\begin{aligned}\frac{\partial^2 W_0}{\partial\mathbf{G}\otimes\partial\boldsymbol{\eta}}\bigg|_{\eta=\hat{\eta}} = \frac{1}{4V_I}\sum_{J=1}^{3}[\sum_{K=1}^{3}(\underset{\bullet}{\dot{sym}}(\hat{\mathbf{K}}_{IJIK}\bullet\mathbf{G}\bullet((\mathbf{R}_{IK}+\hat{\boldsymbol{\eta}})\otimes(\mathbf{R}_{IJ}+\hat{\boldsymbol{\eta}})\otimes(\mathbf{R}_{IJ}+\hat{\boldsymbol{\eta}})))\\ + (\hat{\mathbf{K}}_{IJIK}\bullet\mathbf{F})\dot{\otimes}(\mathbf{R}_{IK}+\hat{\boldsymbol{\eta}})\dot{\otimes}(\mathbf{R}_{IK}+\hat{\boldsymbol{\eta}})) + (\hat{\mathbf{f}}_{IJ}\overline{\otimes}(\mathbf{R}_{IJ}+\hat{\boldsymbol{\eta}})\overline{\otimes}{}^2\mathbf{1})\\ + (\hat{\mathbf{f}}_{IJ}\otimes(\mathbf{R}_{IJ}+\hat{\boldsymbol{\eta}})\otimes{}^2\mathbf{1})]\end{aligned} \tag{31}$$

$$\frac{\partial^2 W_0}{\partial \boldsymbol{\eta}\otimes\partial \mathbf{G}}\Big|_{\eta=\hat{\eta}} = \frac{1}{2V_I}\sum_{J=1}^{3}[\sum_{K=1}^{3}(\frac{1}{2}((\mathbf{F}+\underset{\bullet}{\overset{\bullet}{sym}}(\mathbf{G}\bullet(\mathbf{R}_{IJ}+\hat{\boldsymbol{\eta}})))^{\mathbf{T}} \bullet(\hat{\mathbf{K}}_{IJIK}\otimes(\mathbf{R}_{IK}+\hat{\boldsymbol{\eta}})\otimes(\mathbf{R}_{IK}+\hat{\boldsymbol{\eta}})))) + \underset{\bullet}{\overset{\bullet}{sym}}(\hat{\mathbf{f}}_{IJ}\otimes{}^2\mathbf{1}\otimes(\mathbf{R}_{IJ}+\hat{\boldsymbol{\eta}}))] \tag{32}$$

where ${}^2\mathbf{1}$ is the second order identity tensor. The symbols used in the above expressions are defined as:

$$(\underset{\bullet}{\overset{\bullet}{sym}}[\mathbf{A}\bullet\mathbf{B}\bullet(\mathbf{c}\otimes\mathbf{d})])_{ijk} = (A_{ip}B_{pkn}c_n d_j + A_{ip}B_{pqk}c_q d_j)/2 \tag{33}$$

$$(\mathbf{A}\overset{\bullet}{\otimes}\mathbf{b})_{ijk} = A_{ik}b_j \tag{34}$$

$$[\underset{\bullet}{\overset{\bullet}{sym}}(\mathbf{B}\bullet\mathbf{b})]_{ij} = (B_{ijr}b_r + B_{irj}b_r)/2 \tag{35}$$

$$(\mathbf{a}\underset{\bullet}{\otimes}\mathbf{A})_{ijk} = a_j A_{ik} \tag{36}$$

$$(\underset{\bullet}{\overset{\bullet}{sym}}\mathbf{G})_{ijk} = (G_{ijk}+G_{ikj})/2 \tag{37}$$

$$[(\mathbf{a}\overline{\otimes}\mathbf{b})\overline{\otimes}\mathbf{A}]_{ijkl} = a_i b_k A_{jl} \tag{38}$$

$$(\underset{\bullet}{\overset{\bullet}{sym}}[\mathbf{A}\bullet\mathbf{B}\bullet(\mathbf{c}\otimes\mathbf{d}\otimes\mathbf{d})])_{ijkl} = (A_{ip}B_{plr}c_r d_j d_k + A_{ip}B_{pql}c_q d_j d_k)/2 \tag{39}$$

$$(\mathbf{A}\overset{\bullet}{\otimes}\mathbf{c}\overset{\bullet}{\otimes}\mathbf{c})_{ijkl} = A_{il}c_j c_k \tag{40}$$

$$(\underset{\bullet}{\overset{\bullet}{sym}}(\mathbf{a}\otimes\mathbf{A}\otimes\mathbf{b}))_{ijkl} = (a_j A_{ik} b_l + a_j A_{il} b_k)/2 \tag{41}$$

## 5. Mechanical properties of SWCNTs

It is usually thought that SWCNTs can be formed by rolling a graphite sheet into a hollow cylinder. To predict mechanical properties of SWCNTs, a planar graphite sheet in equilibrium energy state is here defined as the undeformed configuration, and the current configuration of the nanotube can be seen as deformed from the initial configuration by the following mapping:

$$\begin{aligned} x_1 &= \alpha_1 X_1 \\ x_2 &= \alpha_2 R_0 \sin(\frac{X_2}{R_0} + \kappa\alpha_1 X_1) \\ x_3 &= \alpha_2 R_0 (\cos(\frac{X_2}{R_0} + \kappa\alpha_1 X_1) - 1) \end{aligned} \tag{42}$$

where $X_i$, $i=1,2$ is Lagrange coordinate associated with the undeformed configuration (here is a graphite sheet) and $x_i$, $i=1,2,3$ is Eulerian coordinate associated with the deformed configuration. $R$ is the radius of the modeled SWCNT, which is described by a pair of parameters $(n,m)$. The radius $R$ can be evaluated by $R = a\sqrt{m^2+mn+n^2}/2\pi$ with $a = a_0\sqrt{3}$, where $a_0$ is the equilibrium bond length of the atoms in the graphite sheet. $\kappa$ represents the rotation angle per unit length, and parameters $\alpha_1$ and $\alpha_2$ control the uniform axial and circumferential stretch deformation, respectively.

### 5.1 The energy per atom for graphene sheet and SWCNTs

First, based on the present model, the energy per atom of the graphite sheet is calculated and the value of -1.1801 $Kg\,nm^2/s^2$ is obtained. It can be found that the present value agrees well with that of -7.3756 $eV$ ($1eV = 1.6\times10^{-19}$ $Nm$) given by Robertson et al. (1992) with the use of the same interatomic potential.

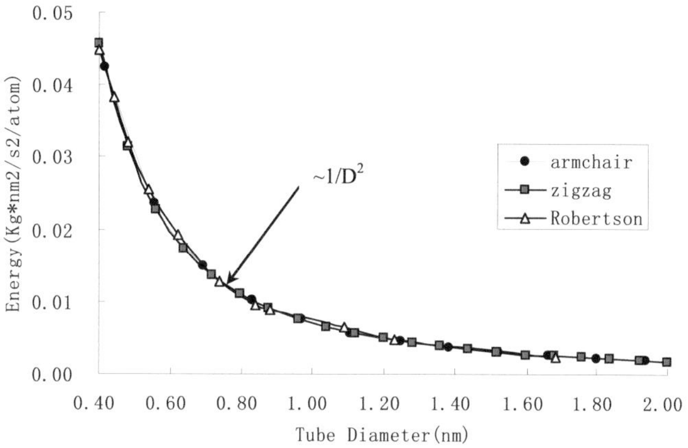


Fig. 4. The energy (relative to graphite) per atom versus tube diameter

The energy per atom as the function of diameters for armchair and zigzag SWNTs relative to that of the graphene sheet is shown in Figure 4. The trend is almost the same for both armchair and zigzag SWNTs. The energy per atom decreases with increase of the tube diameter with $E(D)-E(\infty)=O(1/D^2)$, where $E(\infty)$ represents the energy per atom for graphite sheet.

For larger tube diameter, the energy per atom approaches that of graphite. On the whole, it can be shown that the energy per atom depends obviously on tube diameters, but does not depend on tube chirality. For comparison, the results obtained by Robertson et al. (1992) with the use of both empirical potential and first-principle method based on the same interatomic potential are also shown in Figure 4. It can be found the present results are not only in good agreement with Robertson's results, but also with those obtained by Jiang et al. (2003) based on incorporating the interatomic potential (Tersoff-Brenner potential) into the continuum analysis.

Figure 5 shows the energy per atom for different chiral SWCNTs ((2n, n), (3n, n), (4n, n), (5n, n) and (8n, n)) as a function of tube radius relative to that of the graphene sheet. As is expected, the energy per atom of chiral SWCNTs decreases with increasing tube radius and the limit value of this quantity is -7.3756 eV when the radius of tube is large. From Figure 5, it can be clearly found again that the strain energy per atom depends only on the radius of the tube and is independent of the chirality of SWCNTs, which is similar to armchair and zigzag SWCNTs.

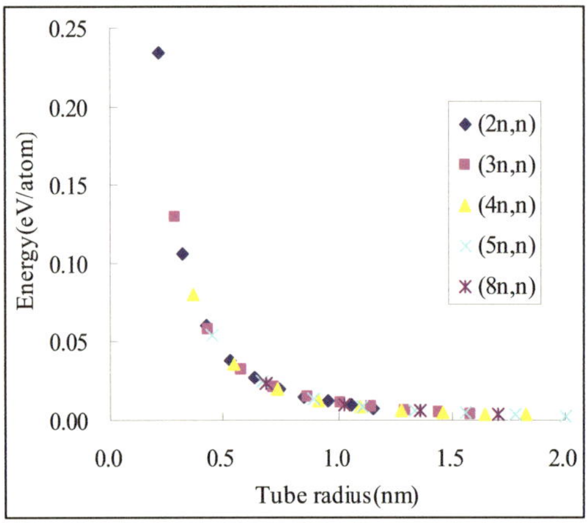


Fig. 5. The strain energy relative to graphite (eV/atom) as a function of tube radius.

### 5.2 Young's modulus and Poisson ratio for graphene sheet and SWCNTs

As shown by Zhang et al. (2002c), the Young's modulus and the Poisson's ratio of planar graphite can be defined from $\hat{\mathbf{M}}_{FF}$ by the following expressions:

$$E = (\hat{\mathbf{M}}_{FF})_{1111} - \frac{(\hat{\mathbf{M}}_{FF})^2_{1122}}{(\hat{\mathbf{M}}_{FF})_{2222}} \tag{43}$$

$$v = \frac{(\hat{\mathbf{M}}_{FF})_{1122}}{(\hat{\mathbf{M}}_{FF})_{1111}} \tag{44}$$

For SWCNTs, we also use the above expressions to estimate their mechanical properties along the axial direction although the corresponding elasticity tensors are no longer isotropic as in planar graphite case. Note that all calculations performed here are based on the Cartesian coordinate system and the Young's modulus $E$ is obtained by dividing the thickness of the wall of SWNT, which is often taken as 0.334nm in the literature.

As for the graphite, the resulting Young's modulus is 0.69TP (see the dashed line in Figure 6a), which agrees well with that suggested by Zhang et al (2002c) and Arroyo and Belytschko (2004b) based on the same interatomic potential (represents by the horizontal solid line in Figure 6a). The Poisson's ratio predicted by the present approach is 0.4295 (see the dashed line shown in Figure 6c), which is also very close to the value of 0.4123 given by Arroyo and Belytschko (2004b) using the same interatomic potential.

As for armchair and zigzag SWCNTs, Figure 6a displays the variations of the Young's modulus with different diameters and chiralities. It can be observed that the trend is similar for both armchair and zigzag SWNTs and the influence of nanotube chirality is not significant. For smaller tubes whose diameters are less than 1.3 nm, the Young's modulus strongly depends on the tube diameter. However, for tubes diameters larger than 1.3 nm, the dependence becomes very weak. As a whole, it can be seen that for both armchair and zigzag SWNTs the Young's modulus increases with increase of tube diameter and a plateau is reached when the diameter is large, which corresponds to the modulus of graphite predicted by the present method. The existing non-orthogonal tight binding results given by Hernández et al.(1998), lattice-dynamics results given by Popov et al. (2000) and the exponential Cauchy-Born rule based results given by Arroyo and Belytschko (2002b) are also shown in Figure 6a for comparison. Comparing with the results given by Hernández et al. (1998) and Popov et al. (2000), it can be seen that although their data are larger than the corresponding ones of the present model, the general tendencies predicted by different methods are in good agreement. From the trend to view, the present predicted trend is also in reasonable agreement with that given by Robertson et al. (1992), Arroyo and Belytschko (2002b), Chang and Gao (2003) and Jiang et al. (2003). As for the differences between the values of different methods, it may be due to the fact that different parameters and atomic potential are used in different theories or algorithms (Chang and Gao, 2003). For example, Yakobson's (1996) result of surface Young's modulus of carbon nanotube based on molecular dynamics simulation with Tersoff-Brenner potential is about 0.36TPa nm, while Overney's (1993) result based on Keating potential is about 0.51 TP nm. Recent *ab initio* calculations by Sánchez-Portal et al.(1999) and Van Lier et al. (2000) showed that Young's modulus of SWNTs may vary from 0.33 to 0.37TPa nm and from 0.24 to 0.40 TPa nm, respectively. Furthermore, it can be found that our computational results agree well with that given by Arroyo and Belytschko (2002b) with their exponential Cauchy-Born rule. They are also in reasonable agreement with the experimental results of $0.8 \pm 0.4$ TP given by Salveta et al. (1999).

Figure 6b depicts the size-dependent Young's moduli of different chiral SWCNTs ((2n, n), (3n, n), (4n, n), (5n, n) and (8n, n)). It can be seen that Young's moduli for different chiral SWCNTs increase with increasing tube radius and approach the limit value of graphite when the tube radius is large. For a given tube radius, the effect of tube chirality can almost be ignored. The Young's modulus of different chiral SWCNTs are consistent in trends with those for armchair and zigzag SWCNTs. For chiral SWCNTs, the trends of the present results are also in accordance with those given by other methods, including lattice dynamics (Popov et al., 2000) and the analytical molecular mechanics approach (Chang & Gao, 2003) .

From Figure 6c, the effect of tube diameter on the Poisson's ratio is also clearly observed. It can be seen that, for both armchair and zigzag SWNTs, the Poisson's ratio is very sensitive to the tube diameters especially when the diameter is less than 1.3 nm. The Poisson's ratio of armchair nanotube decreases with increasing tube diameter but the situation is opposite for that of the zigzag one. However, as the tube diameters are larger than 1.3 nm, the Poisson's ratio of both armchair and zigzag SWNTs reach a limit value i.e. the Poisson's ratio of the planar graphite. For comparison, the corresponding results suggested by Popov et al. (2000) are also shown in Figure 6c. It can be observed that the tendencies are very similar between the results given by Popov et al. (2000) and the present method although the values are different. Moreover, it is worth noting although many investigations on the Poisson's ratio of SWNTs have been conducted, there is no unique opinion that is widely accepted. For instance, Goze et al. (1999) showed that the Poisson's ratio of (10,0), (20,0), (10,0) and (20,0) tubes are 0.275, 0.270, 0.247 and 0.256, respectively. Based on a molecular mechanics approach, Chang and Gao (2003) suggested that the Poisson's ratio for armchair and zigzag SWNTs will decrease with increase of tube diameters from 0.19 to 0.16, and 0.26 to 0.16, respectively. In recent *ab initio* studies of Van Lier et al. (2000), even negative Poisson's ratio is reported.

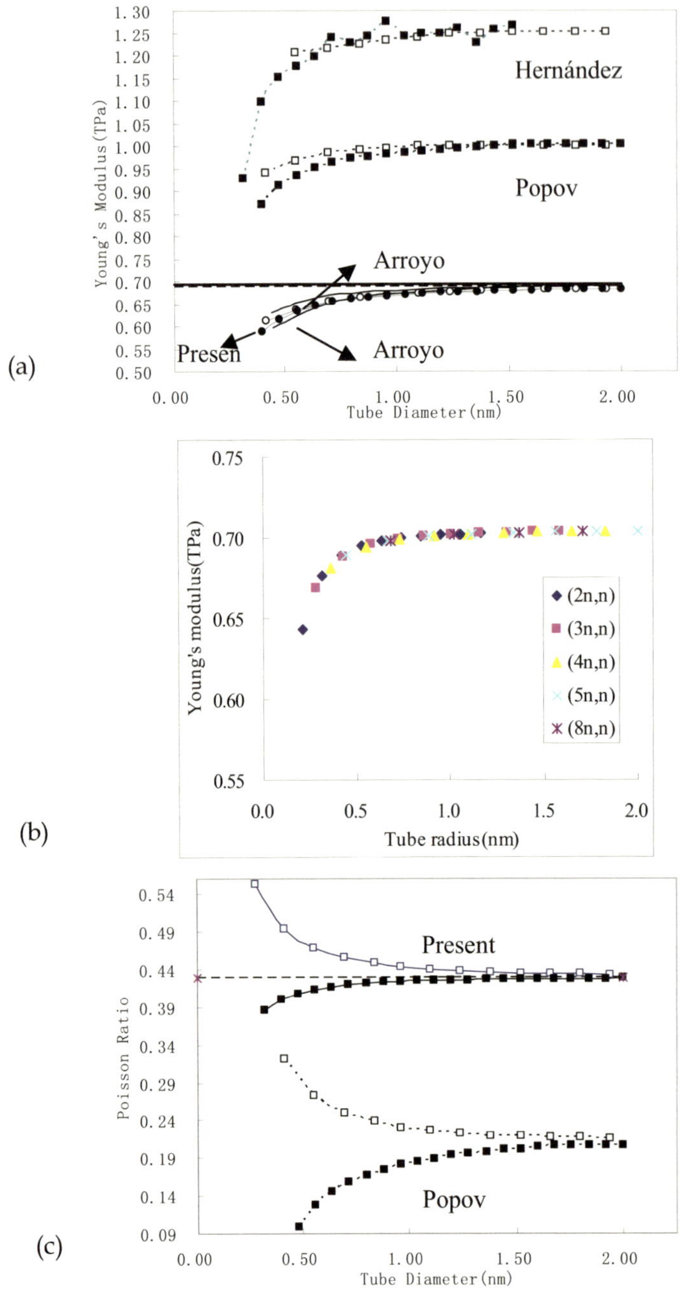


Fig. 6. Comparison between the results obtained with different methods (a) Young's modulus and (b) Young's moduli of chiral SWCNTs versus tube radius. (c)Poisson's ratio. Open symbols denote armchair, solid symbols denote zigzag. Dashed horizontal line denotes the results of graphite obtained with the present approach and the solid horizontal line denotes the results of graphite obtained by Arroyo and Belytschko (2004b) with exponential mapping, respectively.

It also can be seen from Figure 6c that the obtained Poisson's ratio is a little bit high when tube diameter is less than 0.3nm. It may be ascribed to the fact that when tube diameter is less than 0.3nm, because of the higher value of curvature, higher order ($\geq 2$) deformation gradient tensor should be taken into account in order to describe the deformation of the atomic bonds more accurately. Another possible explanation is that for such small values of diameter, more accurate interatomic potential should be used in this extreme case.

### 5.3 Shear modulus for SWCNTs

As for the shear moduli of SWCNTs, to the best of our knowledge, only few works studied this mechanical property systematically since it is difficult to measure them with experiment techniques. Most of these works focus only on the armchair and zigzag SWCNTs.(Popov et al., 2000; Li & Chou, 2003) Thus, the shear moduli of achiral (i.e., armchair and zigzag) SWCNTs are firstly investigated and compared with the existing results (Li & Chou, 2003) for validation of the present model. Then the shear modulus of SWCNTs with different chiralities including (2n, n), (3n, n), (4n, n), (5n, n) and (8n, n) are studied systematically. For determining the shear modulus of SWCNT, it is essential to simulate its pure torsion deformation which can be implemented by incrementally controlling $\kappa$ but relaxing inner displacement $\boldsymbol{\eta}$, parameters $\alpha_1$ and $\alpha_2$ in Equation (42). The shear modulus of SWCNTs can be obtained by the $U$ (strain energy density) and $\kappa$ (twist angle per unit length). Similar to Young's modulus, shear modulus is defined with respect to the initial stress free state.

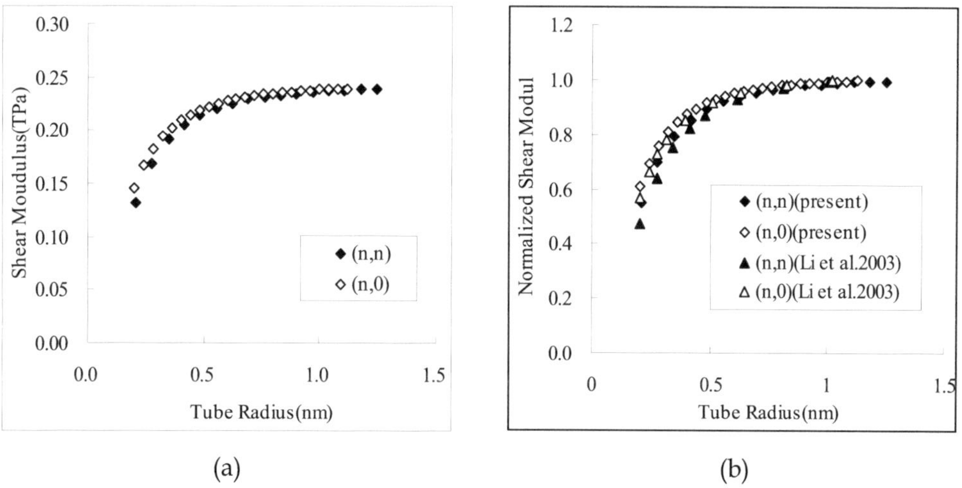


Fig. 7. (a) Shear moduli of armchair and zigzag SWCNTs versus tube radius, (b) Effect of tube radius on normalized shear moduli of armchair and zigzag SWCNTs.

Figure 7a shows the variations of the shear modulus of achiral SWCNTs with respect to the tube radius. It can be found that shear modulus of armchair and zigzag SWCNTs increase with increasing tube radius and approach the limit value 0.24 TPa when the tube radius is large. It is also observed that, similar to the results given by Li & Chou (2003) and Xiao et al. (2005), the present predicted shear moduli of armchair and zigzag SWCNTs hold similar size-dependent trends and the chirality-dependence of shear moduli is not significant.

Figure 7b shows the normalized shear moduli obtained with different methods. The normalization is achieved by using the values of 0.24 TPa and 0.48 TPa which are the limiting values of graphite sheet obtained by the present approach and molecular structural mechanics (Li & Chou, 2003), respectively. Although there is a discrepancy in limit values, it can be found that the size effect obtained by the present study is in good agreement with that of Li and Chou (2003). The difference among the limit values may be attributed to the different atomistic potential and/or force field parameters used in the computation model.
The size-dependent shear modulus of different chiralities SWCNTs are displayed in Figure 8. It is observed that, similar to achiral SWCNTs, the shear moduli of chiral SWCNTs increase with increasing tube radius and a limit value of 0.24 TPa is approaching when the tube radius (also n) is large. For (2n, n) SWCNT, the maximum difference of shear modulus is up to 42%. The dependence of tube chirality is not obvious for chiral SWCNTs. With reference to Figure 7a and Figure 8, it can be found that, at small radius (<1nm), the shear modulus of SWCNTs are sensitive to the tube radius, while at larger radius (>1nm), the size and chirality dependency can be ignored.

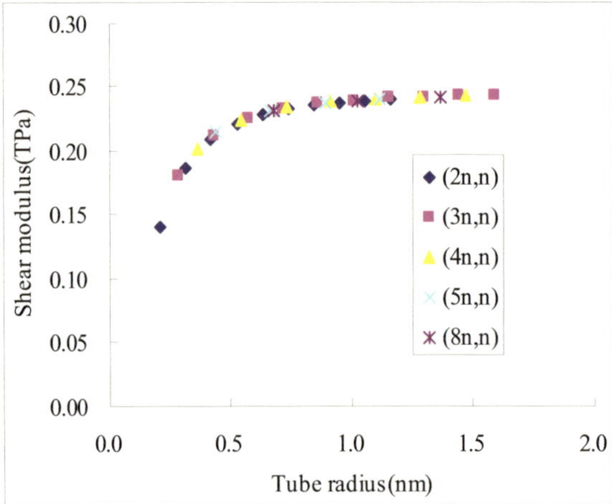


Fig. 8. Shear moduli of chiral SWCNTs versus tube radius.

### 5.4 Bending stiffness for graphene sheet and SWCNTs

In present study, the so-called bending stiffness for graphene sheet refers to the resistance of a flat graphite sheet or the curved wall of CNT with respect to the infinitesimal local bending deformation. The bending stiffness for SWCNTs refers to the bending resistance of the cylindrical tube formed by rolling up graphite sheet with respect to the infinitesimal global bending deformation (see Figure 9 for reference). It should be pointed out that for the first definition, the bending stiffness is an intrinsic material property solely determined by the atomistic structure of the mono-layer crystalline membrane. The second definition, however, is a *structural property* which is determined not only by the bending stiffness of the single atom layer crystalline membrane, but also by the geometry dimensions, such as the diameter of the tube. Unfortunately, these two issues are not well addressed in the past literatures (Kudin et at., 2001; Enomoto et al., 2006).
Based on the higher order Cauchy-Born rule and Equation (42), the strain energy per atom (energy relative to a planar graphite sheet) as a function of the radius of bending curvature can

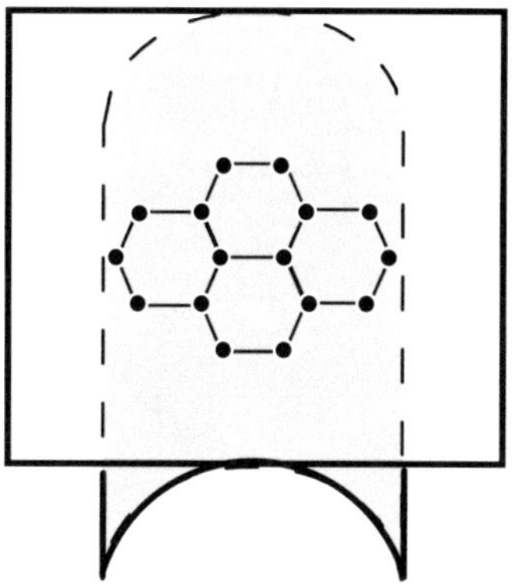

(a)

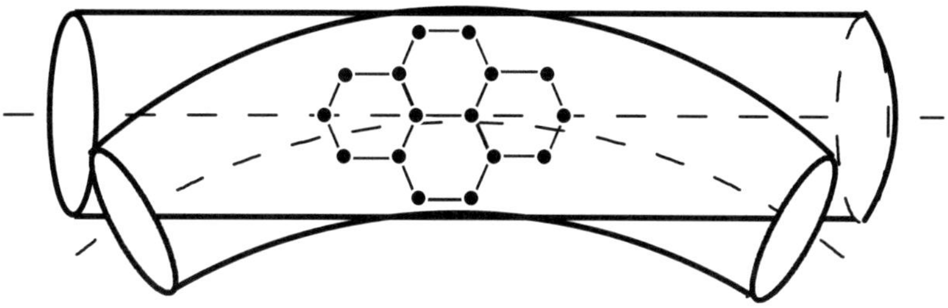

(b)

Fig. 9. (a) Bending of a flat graphite sheet; (b) Bending of a single-walled carbon nanotube

be obtained. By fitting the data of the strain energy and the bending curvature radii with respect to the equation $U = D_{membrane} / 2R_0^2$, one can obtain that the bending stiffness $D_{membrane}$ of the graphite sheet is 2.38 eVÅ$^2$/atom, which is almost independent of its rolling direction. This indicates that the flat graphite sheet is nearly isotropic with regard to bending. The current result agrees well with the effective bending stiffness of graphite sheet 2.20 eVÅ$^2$/atom reported by Arroyo and Belytschko (2004a) with membrane theory and the same interatomic potential under the condition of infinitesimal bending. It is also in good agreement with the result of 2.32 eVÅ$^2$/atom obtained by Robertson et al. (1992) with atomic simulations.

To explore the effective bending stiffness of carbon nanotube based on the higher order Cauchy-Born rule, the following map is used to describe the pure bending deformation of the tube

$$\begin{aligned} x_1 &= \rho\sin(X_1/\rho) - R\sin(X_2/R)\sin\alpha \\ x_2 &= \rho\sin^2(X_1/\rho)/2 + R\sin(X_2/R)\cos\alpha \\ x_3 &= R\cos(X_2/R) \qquad (\alpha = \arctan(X_1/\rho)) \end{aligned} \tag{45}$$

where $R$ is the radius of the modeled SWCNT and $\rho$ is the radius of curvature of the bending tube (curvature of the neutral axis). With the use of this mapping and taking the inner-displacement relaxation into consideration, the strain energy of the bending tube can be computed.

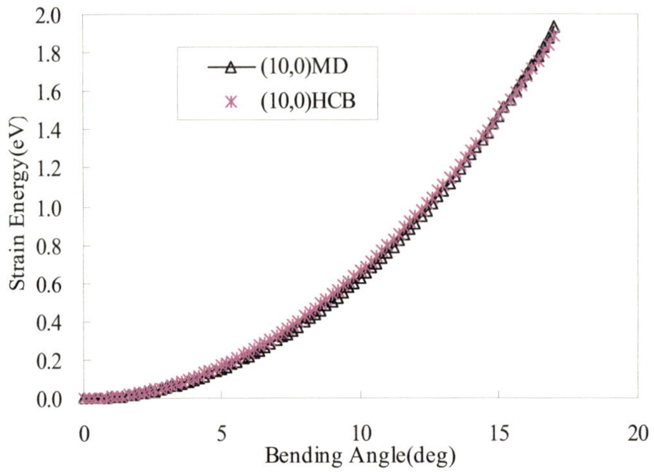


Fig. 10. Comparison of the strain energy of (10,0) SWCNT as a function of the bending angle for HCB(—*—) and MD(—△—) simulation. Herein and after, HCB refers to the continuum theory based on a higher-order Cauchy-Born rule and MD refers to molecular dynamics

Figure 10 show the bending strain energy of zigzag (10,0) SWCNT as a function of bending angle. Here the bending strain energy is defined as the difference between the energy of the deformed tube and that of its straight status. It can be found that the present results obtained with much less computational effort are in good agreement with those of MD simulations.

where $L$ denotes the length of the tube. It can be seen clearly from Equation (46) that the effective bending stiffness of CNTs can be defined as the second derivative of the elastic energy per unit length with respect to the curvature of the neutral axis under pure bending (i.e. constant curvature). Its dimension is $eV \cdot nm$. Figure 11 shows the bending stiffness of different chiral SWCNTs as a function of the tube radius. It can be found that the bending stiffness is almost independent on the chirality of SWCNTs and increases with the increasing of tube radius. Furthermore, using a polynomial fitting procedure, we can approximate the bending stiffness over the considered range of tube radii by the following analytical expression

Once the bending strain energy $U$ is known, the effective bending stiffness of carbon nanotube can be obtained by numerical differentiation based on the following formula

$$U = D_{tube} L \kappa^2 / 2 \tag{46}$$

Just like the derivation of the bending stiffness of the flat graphite sheet, here no representative thickness of the tube is required to obtain the effective bending stiffness of CNTs.

$$\begin{aligned} D_{tube}(eV \cdot nm) = 5583.956(eV/nm^2)R^3(nm^3) + 9.225(eV/nm)R^2(nm^2) \\ - 32.418(eV)R(nm) + 1.517(eV \cdot nm) \end{aligned} \tag{47}$$

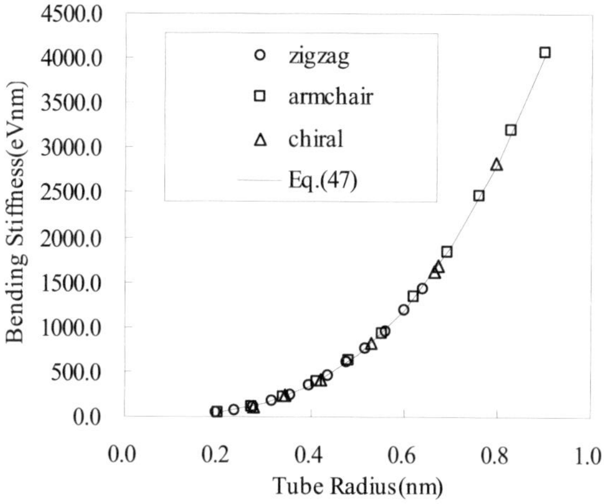


Fig. 11. Variation of bending stiffness with tube radius for different chiral SWCNTs

## 6. Conclusion

In this charpter, a higher order Cauchy-Born rule has been constructed for studying mechanical properties of graphene sheet and carbon nanotubes. In the present model, by including the second order deformation gradient tensor in the kinematic description, we can alleviate the limitation of the standard Cauchy-Born rule for the modeling of nanoscale crystalline films with less computational efforts. Based on the established relationship between the atomic potential and the macroscopic continuum strain energy, analytical expressions for the tangent modulus tensors are derived. From these expressions, the hyper-elastic constitutive law for this generalized continuum can be obtained.

With the use of this constitutive model and the Tersoff-Brenner atomic potential for carbon, the size and chirality dependent mechanical properties (including strain energy, Young's modulus, Poisson's ratio, shear modulus, bending stiffness) of graphene sheet and carbon nanotube are predicted systematically. The present investigation shows that except for Poisson's ratio other mechanical properties (such as Young's modulus, shear modulus, bending stiffness and so on) for graphene sheet and SWCNTs are size-dependent and their chirality-dependence is not significant. With increasing of tube radius, Young's modulus and shear modulus of SWCNTs increase and converge to the corresponding limit values of graphene sheet. As for Poisson's ratio, it can be found that it is very sensitive to the radius and the chirality of SWCNTs when the tube diameter is less than 1.3 nm. The present results agree well with those obtained by other experimental, atomic modeling and continuum concept based studies.

Besides, the present work also discusses some basic problems on the study of the bending stiffness of CNTs. It is pointed out that the bending stiffness of a flat graphite sheet and that of CNTs are two different concepts. The former is an intrinsic material property while the later is a structural one. Since the smeared-out model of CNTs is a generalized continuum with microstructure, the effective bending stiffness of it should be regarded as an independent structural rigidity parameter which can not be determined simply by employing the classic formula in beam theory. It is hoped that the above findings may be helpful to clarify some obscure issues on the study of the mechanical properties of CNTs both theoretically and experimentally.

It should be pointed out that the present method is not limited to a specific interatomic potential and the study of SWCNTs. It can also be applied to calculate the mechanical response of MWCNTs. The proposed model can be further applied to other nano-film materials. The key point is to view them as generalized continuum with microstructures.

## 7. Acknowledgment

This work was supported by the National Natural Science Foundation of China (10802076), the Nature Science Foundation of Zhejiang province (Y6090543), China Postdoctoral Science Foundation (20100470072) and the Scientific Research Foundation of Zhejiang Ocean University.

## 8. References

Arroyo, M. & Belytschko, T. (2002). An atomistic-based finite deformation membrane for single layer crystalline films. Journal of the Mechanics and Physics of Solids, 50, 1941-1977.

Arroyo, M. & Belytschko, T. (2004a). Finite element methods for the non-linear mechanics of crystalline sheets and nanotubes. International Journal for Numerical Methods in Engineering, 59, 419-456.

Arroyo, M. & Belytschko, T. (2004b). Finite crystal elasticity of carbon nanotubes based on the exponential Cauchy-Born rule. Physical Review B, 69, 115415-1-11.

Bhattacharya, K. & James, R.D. (1999). A theory of thin films with applications to microstructures. Journal of the Mechanics and Physics of Solids, 47, 465-502.

Brenner, D.W. (1990). Empirical potential for hydrocarbons for use in simulation the chemical vapor deposition of diamond films. Physical Review B, 42, 9458-9471.

Chang, T.C. & Gao, H.J. (2003) Size-dependent elastic properties of a single-walled carbon nanotube via a molecular mechanics model. Journal of the Mechanics and Physics of Solids, 51, 1059-1074.

Cousins, C.S.G. (1978a). Inner elasticity. Journal of Physics C: Solid State Physics, 11, 4867-4879.

Cousins, C.S.G. (1978b). The symmetry of inner elastic constants. Journal of Physics C: Solid State Physics, 11, 4881-4900.

Enomoto, K.; Kitakata, S.; Yasuhara, T.; Ohtake, N.; Kuzumaki, T. & Mitsuda, Y. (2006) Measurement of Young's modulus of carbon nanotubes by nanoprobe manipulation in a transmission electron microscope. Applied Physics Letters, 88, 153115-1-3.

Garstein, Y.N.; Zakhidov, A.A. & Baughman, R.H. (2003) Mechanical and electromechanical coupling in carbon nanotube distortions. Physical Review B, 68, 115415.

Govindjee, S. & Sackman, J.L. (1999). On the use of continuum mechanics to estimate the properties of nanotubes. Solid State Communication, 110, 227-230.

Goze, C.; Vaccarini, L.; Henrard, L.; Bernier, P.; Hernándz, E. & Rubio, A. (1999). Elastic and mechanical properties of carbon nanotubes. Synthetic Metals, 103, 2500-2501.

Guo, X.; Wang, J.B. & Zhang, H.W. (2006) Mechanical properties of single-walled carbon nanotubes based on higher order Cauchy-Born rule. International Journal of Solids and Structures, 43(5), 1276-1290.

He, X.Q.; Kitipornchai, S. & Liew, K.M. (2005a). Buckling analysis of multi-walled carbon nanotubes: a continuum model accounting for van der Waals interaction. Journal of the Mechanics and Physics of Solids, 53, 303-326.

He, X.Q.; Kitipornchai, S.; Wang, C.M. & Liew K.M. (2005b). Modeling of van der Waals force for infinitesimal deformation of multi-walled carbon nanotubes treated as cylindrical shells. International Journal of Solids and Structures, 42, 6032-6047.

Hernándz, E.; Goze, C.; Bernier, P. & Rubio, A. (1998). Elastic properties of C and BxCyNz composite nanotubes. Physical Review Letters, 80, 4502-4505

Iijima, S. (1991). Helical microtubules of graphitic carbon. Nature, 354, 56-58.

Iijima, S. & Ichlhashi T. (1993). Single-shell carbon nanotubes of 1-nm diameter. Nature, 363, 603-605.

Iijima, S.; Brabec, C.; Maiti, A. & Bernholc, J. (1996). Structural flexibility of carbon nanotubes. Journal of Chemical Physics, 104, 2089-2092.

Jiang, H.; Zhang, P.; Liu, B.; Huang, Y.; Geubelle, P.H.; Gao, H. & Hwang K.C. (2003). The effect of nanotube radius on the constitutive model for carbon nanotubes. Computational Materials Science, 28, 429-442.

Krishnan, A.; Dujardin, E.; Ebbesen, T.W.; Yianilos, P.N. & Treacy, M.M.J. (1998). Young's modulus of single-walled nanotubes. Physical Review B, 58, 14013-14019.

Kudin, D.; Scuseria, G. & Yakobson, B. (2001). C2, BN, and C nanoshell elasticity from ab initio computations. Physical Review B, 64, 235406

Leamy, M.J.; Chung, P.W. & Namburu, R. (2003). On an exact mapping and a higher-order Born rule for use in analyzing graphene carbon nanotubes. Proceedings of the 11th Annual ARL-USMA Technical Symposium, November 5.

Li, C.Y. & Chou, T.W. (2003). A structural mechanics approach for analysis of carbon nanotubes. International Journal of Solids and Structures, 40, 2487-2499.

Liang, H.Y. & Upmanyu, M. (2006) Axial-strain-induced torsion in single-walled carbon nanotubes. Physical Review Letters, 96, 165501.

Mu, W.H.; Li, M.; Wang, W. & Ou-Yang, Z.C. (2009) Study of axial strain-induced torsion of single-wall carbon nanotubes using the 2D continuum anharmonic anisotropic elastic model. New Journal of Physics, 11, 113049.

Odega, G.M.; Gates, T.S.; Nicholson, L.M. & Wise, K.E. (2002). Equivalent-continuum modeling of nano-structured materials. Composites Science and Technology, 62, 1869-1880.

Popov, V.N.; Van Doren, V.E. & Balkanski, M. (2000). Elastic properties of single-walled carbon nanotubes. Physical Review B, 61, 3078-3084.

Popov, V.N. (2004). Carbon nanotubes: properties and application. Materials Science and Engineering R, 43, 61-102

Robertson, D.H.; Brenner, D.W. & Mintmire, J.W. (1992). Energy of nanoscale graphitic tubules. Phyical Review B, 45, 12592-12595.

Ru, C.Q. (2000a). Effective bending stiffness of carbon nanotubes. Physical Review B, 62, 9973-9976.

Ru, C.Q. (2000b). Elastic buckling of single-walled carbon nanotube ropes under high pressure. Physical Review B, 62, 10405-10408

Ruoff, R.S.; Dong, Q. & Liu, W.K. (2003). Mechanical properties of carbon nanotubes: theoretical predictions and experimental measurements. Comptes Rendus Physique, 4, 993-1008.

Sánchez-Portal, D.; Artacho, E. & Soler, J.M. (1999). Ab initio structural, elastic, and vibrational properties of carbon nanotubes. Physical Review B, 59, 12678-12688.

Sunyk, R. & Steinmann, P. (2003). On higher gradients in continuum-atomic modeling. International Journal of Solids and Structures, 40, 6877-6896.

Tadmor, E.; Ortiz, M. & Phillips R. (1996). Quasicontinuum analysis of defects in solids. Philosophy Magazine A, 73, 1529-1563.

Tadmor, E.B.; Smith, G.S.; Bernstein, N. & Kaciras, E. (1999). Mixed finite element and atomistic formulation for complex crystals. Physical Review B, 59, 235-245.

Tersoff, J. (1988). New empirical approach for the structure and energy of covalent systems. Physical Review B, 37, 6991-7000.

Treacy, M.M.J.; Ebbesen, T.W. & Gibson, J.M. (1996). Exceptionally high Young's modulus observed for individual carbon nanotubes. Nature, 381, 678-680.

Van Lier, G.; Van Alsenoy, C.; Van Doren, V. & Geerlings P. (2000). Ab initio study of the elastic properties of single-walled carbon nanotubes and graphene. Chemical Physics Letter, 326, 181-185.

Wang, J.B.; Guo, X.; Zhang, H.W.; Wang, L. & Liao, J.B. (2006a) Energy and mechanical properties of single-walled carbon nanotubes predicted using the higher order Cauchy-Born rule. Physical Review B, 73, 115428.

Wang, J.B.; Guo, X. & Zhang, H.W. (2006b) Nonlinear extension of single-walled carbon nanotubes analyzed by a continuum model based on a higher-order Cauchy-Born rule. Journal of Computational and Theoretical Nanoscience, 3, 798–802.

Wang, J.B.; Guo, X. & Zhang, H.W. (2009a) Higher Order Cauchy-Born Rule Based Study of Chiral Single-walled Carbon Nanotubes. Journal of Computational and Theoretical Nanoscience, 6(7), 1617-1621.

Wang, J.B.; Guo, X. & Zhang, H.W. (2009b) A Revisit of the Bending Stiffness of Graphite Sheet and Single-Walled Carbon Nanotubes. Journal of Computational and Theoretical Nanoscience, 6(10), 2242-2246.

Xiao, J.R.; Gama, B.A. & Gillespie Jr, J.W. (2005) An analytical molecular structural mechanics model for the mechanical properties of carbon nanotubes. International Journal of Solids and Structures, 42, 3075-3092.

Yakobson, B.I.; Brabec, C.J. & Bernholc, J. (1996). Nanomechanics of carbon tubes: instabilities beyond linear response. Physical Review Letters, 76, 2511-2514.

Yu, M.F.; Files; B.S.; Arepalli, S. & Ruoff, R.S. (2000a). Tensile loading of ropes of single wall carbon nanotubes and their mechanical properties. Physical Review Letters, 84, 5552-5555.

Yu, M.F.; Lourie, O.; Dyer, M.J.; Moloni, K.; Kelly, T.F. & Ruoff, R.S. (2000b). Strength and breaking mechanism of multiwalled carbon nanotubes under tensile load. Science, 287, 637-640.

Zhang, H.W.; Wang, L.; Wang, J.B.; Zhang, Z.Q. & Zheng, Y.G. (2008) Torsion induced by axial strain of double-walled carbon nanotubes. Physics Letters A, 372, 3488-3492.

Zhang, P.; Huang, Y.; Gao, H. & Hwang, K.C. (2002a). Fracture nucleation in single-wall carbon nanotubes under tension: a continuum analysis incorporating interatomic potentials. Journal of Applied Mechanics, 69, 454-458.

Zhang, P.; Huang, Y.; Geubelle, P.H. & Hwang, K.C. (2002b). On the continuum modeling of carbon nanotubes. Acta Mechanica Sinica, 18, 528-536.

Zhang, P.; Huang, Y.; Geubelle, P.H.; Klein, P.A. & Hwang, K.C., (2002c). The elastic modulus of single-walled carbon nanotubes: A continuum analysis incorporating interatomic potentials. International Journal of Solids and Structures, 39, 3893-3906.

Zhang, P.; Jiang, H.; Huang, Y.; Geubelle, P.H. & Hwang, K.C., (2004). An atomistic-based continuum theory for carbon nanotubes: analysis of fracture nucleation. Journal of the Mechanics and Physics of Solids, 52, 977-998.

Zhou, G.; Duan, W.H. & Gu, B.L., (2001). First-principles study on morphology and mechanical properties of single-walled carbon nanotube. Chemical Physics Letters, 333, 344-349.

12

# In-Situ Structural Characterization of SWCNTs in Dispersion

Zhiwei Xiao, Sida Luo and Tao Liu
*Florida State University*
*United States*

## 1. Introduction

Owing to its excellent mechanical robustness - high strength, stiffness, toughness (Saito et al., 1998; Baughman et al., 2002), excellent electrical and thermal conductivity and piezoresistivity (Cao et al., 2003; Grow et al., 2005; Skakalova et al., 2006), and versatile spectroscopic and optoelectronic properties (Burghard, 2005; Dresselhaus et al., 2005; Dresselhaus et al., 2007; Avouris et al., 2008), single-walled carbon nanotubes (SWCNTs) offer a great promise as the building blocks for the development of multi-functional nanocomposites (Hussain et al., 2006; Moniruzzaman & Winey, 2006; Green et al., 2009; Chou et al., 2010; Sahoo et al., 2010). To fabricate the SWCNT based multi-functional nanocomposites, one of the most used approaches is through solution or melt processing of SWCNT dispersions in various polymer matrices (Hilding et al., 2003; Moniruzzaman & Winey, 2006; Schaefer & Justice, 2007; Grady, 2009). In addition, the SWCNT dispersions in different liquid media of small molecules, e.g., water or organic solvents, were also proved to be useful for cost-effective processing of SWCNT thin film based novel applications (Cao & Rogers, 2009), e.g., CNT film strain sensors (Li et al., 2004), high mobility CNT thin film transistors (Snow et al., 2005), SWNT thin film field effect electron sources (Bonard et al., 1998) and various CNT film-based transparent electronics (Gruner, 2006). To fully explore the use of SWCNT dispersions for various technologically important applications, it is critical to have a good understanding of the processing-structure relationship of SWCNT dispersions processed by different techniques and methods (Luo et al., 2010).
Regardless of the dispersion processing methods, it has been recognized that, to disperse SWCNTs at a molecular level in either small molecule solvent or polymer solution or melt is extremely difficult (Moniruzzaman & Winey, 2006; Schaefer & Justice, 2007; Mac Kernan & Blau, 2008). The fundamental reasons for such difficulties are threefold. First, the one dimensional tubular structure of SWCNTs imparts this novel species of very high rigidity. When mixed with the solvent of small molecules or flexible chain polymers, the highly rigid nature of SWCNTs as well as its long aspect ratio character (typically >100) results in a competition between the orientational entropy and the packing entropy that drives the mixture towards phase separation (Onsager, 1949; Flory, 1978; Fakhri et al., 2009). The persistence length is a physical measure of the rigidity of a chain-like or worm-like molecule (Tracy & Pecora, 1992; Teraoka, 2002). Depending upon the tube diameter, the theoretically estimated persistence length for an individual SWCNT is as high as of 30 - 1000 μm (Yakobson & Couchman, 2006). This result has been confirmed by the experimental studies

of SWCNT dynamics in aqueous suspension (Duggal & Pasquali, 2006; Fakhri et al., 2009). For comparison, the persistence length of a few widely studied stiff particles/molecules is: 300 nm for the tobacco mosaic virus (TMV), 80 nm for poly (γ-benzyl L-glutamate) (PBLG), and 50 nm for double-stranded DNA (Vroege & Lekkerkerker, 1992). Second, the intertube van der Waals interaction of SWCNTs is very strong. The cohesive energy for a pair of parallel arranged SWCNTs at equilibrium is greater than 2.0 eV/nm (Girifalco et al., 2000). For this reason, one often finds that the SWCNTs organize into a rope or bundle structure in the as-produced materials (Thess et al., 1996; Salvetat et al., 1999). To disperse SWCNTs in a given medium at the molecular level or to exfoliate the SWCNT bundles into individual tubes, the strong intertube cohesive energy has to be overcome. This proved to be a difficult task (O'Connell et al., 2002; Islam et al., 2003; Moore et al., 2003; Zheng et al., 2003; Cotiuga et al., 2006; Giordani et al., 2006; Bergin et al., 2007; Liu et al., 2007; Liu et al., 2009). Lastly, the difficulty to disperse SWCNTs is also attributed to the topological entanglement or enmeshment of long aspect ratio SWCNTs, which could result kinetically quenched fractal structures or aggregates.

Associated with the threefold difficulty to disperse SWCNTs is their hierarchical structures that one may encounter in the dispersion. As schematically shown in Fig. 1, these structures include: 1) the individual tubes with different molecular structure as specified by the rolling or chiral vector (n, m) (Saito et al., 1998); 2) the SWCNT bundles that is composed of multiple individual tubes approximately organized into a 2D hexagonal lattice with their long axis parallel to each other (Thess et al., 1996; Salvetat et al., 1999); 3) the SWCNT aggregates formed by the topological entanglement or enmeshment of individual tubes and/or SWCNT bundles; and 4) the SWCNT networks that span the entire dispersion sample, which may occur as a result of inter-tube, inter-bundle and inter-aggregate connection when the SWCNT loading in the dispersion is high. In a given SWCNT dispersion, the diameter and length of the individual tubes and the SWCNT bundles, the radius of gyration of the SWCNT aggregates, as well as the relative amount of the hierarchical structures of the SWCNTs could be subject to random variations. This brings out the length-scale related polydispersity issues. The length scales of the hierarchical SWCNT structures vary from ~ $10^0$ nm for the diameter of individual tube, ~$10^1$ nm for the diameter of SWCNT bundles, ~ $10^2$ - $10^3$ nm for the length of SWCNT tubes and bundles, ~$10^4$ - $10^5$ nm for the size of SWCNT aggregates, and up to the macroscopic sample size for the SWCNT networks. Given such a broad range of length scales involved in the hierarchical structures of SWCNTs possibly encountered in the dispersion, one can expect that, to quantitatively characterize the structures of SWCNT dispersion and establish the related dispersion processing-structure relationship, a multi-scale characterization approach should be utilized.

The past decades witnessed significant progress being made toward qualitative and quantitative characterization of the SWCNT dispersions by various experimental techniques. Among the different techniques, the microscopy based methods, e.g., optical microscopy (OM), electron microscopy (SEM and TEM) and atomic force microscopy (AFM), have been routinely used for characterizing the SWCNT structures to provide valuable information regarding the diameter, length, and the overall morphology for a given SWCNT sample. However, when applied to characterizing the SWCNT dispersions, the microscopy techniques typically require a sample preparation protocol that converts the dispersion sample from a liquid state to solid state. This may cause the structural changes of the SWCNTs during the sample preparation and thus fail to faithfully provide the desired

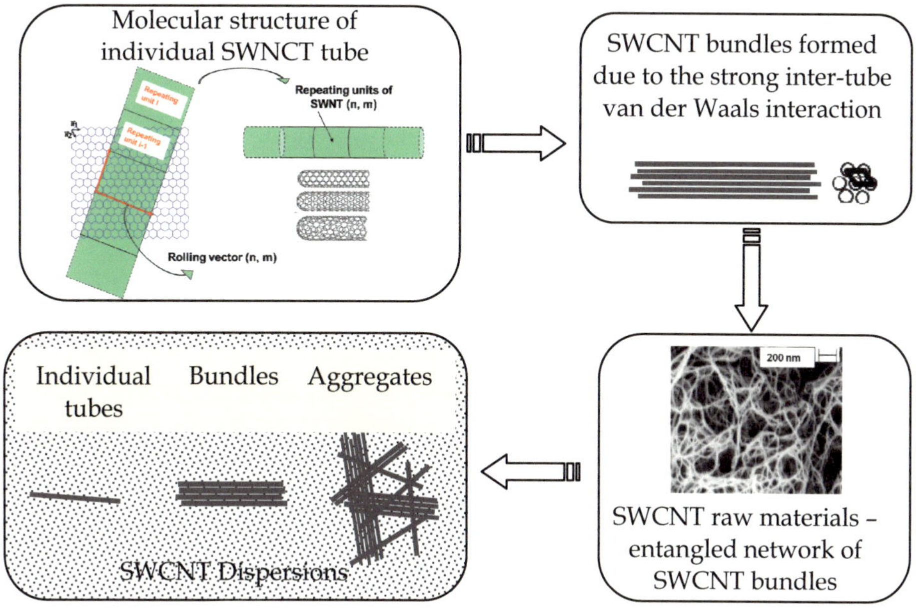


Fig. 1. Schematic SWCNT structures at different length scales

in-situ structural information of the SWCNTs in the dispersion. For this reason, the microscopy technique will not be considered as suitable methods for in-situ structural characterization of SWCNT dispersions. In addition to the microscopy techniques, a few other conventional or non-conventional techniques emerge to show great promise for the in-situ structural characterization of SWCNT dispersions. These emerging techniques include:

1. Viscosity and rheological measurements;
2. Scattering based techniques, e.g., elastic and quasi-elastic light scattering (SLS and DLS), small-angle X-ray and neutron scattering (SAXS and SANS);
3. Sedimentation methods, e.g., analytical and preparative ultracentrifuge method; and
4. Spectroscopic techniques, e.g., simultaneous Raman scattering and photoluminescence spectroscopy.

To facilitate a multi-scale characterization approach for a better understanding of the in-situ SWCNT structures in the dispersion, the above listed experimental techniques and methods will be reviewed in this chapter. For each of the methods, the underlying physical principles and their applications for the in-situ structural characterization of SWCNT dispersions are discussed in the subsequent sections.

## 2. Viscosity and rheological measurements

Suspensions or dispersions, in which the microscopically visible solid particles or fillers are dispersed in a continuous phase like water, organic solvent or polymer solutions, find themselves a great technical importance in many different areas, e.g., biotechnology, cement and concrete technology, ceramic processing, coating and pigment technology, etc. The rheological behavior of a two-phase suspension system has received a great attention and been studied for many years (Batchelor, 1974; Jeffrey & Acrivos, 1976; Russel, 1980; Metzner, 1985; Bicerano et al., 1999; Hornsby, 1999; Larson, 1999; Petrie, 1999). One area concerning the rheological behavior of a suspension system is to understand the shear viscosity of a suspension. To this effect, hundreds of empirical, semi-empirical, and theoretical relationships have been developed for relating the dispersion viscosity, $\eta$, with respect to the volume fraction, the shape of the fillers, and the shear rate under which the viscosity is measured (Bicerano et al., 1999; Hornsby, 1999; Shenoy, 1999). At very low filler volume fraction and zero shear rate, the viscosity $\eta$ of a suspension or dispersion is given by:

$$\frac{\eta}{\eta_0} \approx 1 + [\eta]\varphi \tag{1}$$

where $\eta_0$ is the viscosity of the liquid medium, $\phi$ is the volume fraction of the fillers, and $[\eta]$ is termed as the intrinsic viscosity and it is a dimensionless, scale-invariant functional of the shape of the filler particle. By applying the numerical path integration technique, Douglas et al (Mansfield & Douglas, 2008) presented an accurate expression for the intrinsic viscosity of cylinders applicable to a broad range of aspect ratios ($2.72 < A < \infty$), which is:

$$[\eta] = \frac{8A^2}{45}\left[\ln\left(\frac{4A}{e^{25/12}}\right)\right]^{-1} \times \left[\frac{1 - 1.178t + 1.233t^{1.86} + 1.925t^{6.28} + 0.625t^{12.67}}{1 - 1.094t + 0.757t^{3.83} + 1.344t^{3.83} + 1.978t^{12.07}}\right] \tag{2a}$$

$$A = \frac{L}{d} \tag{2b}$$

$$t = \frac{1}{\ln A} \tag{2c}$$

where $A$ is the aspect ratio of the cylinder and equals to the ratio of the cylinder length $L$ to its diameter $d$. By taking advantage of the rigid rod nature of SWCNTs (Yakobson & Couchman, 2006; Duggal & Pasquali, 2006; Fakhri et al., 2009) and on the basis of Eq. (1) and (2), the aspect ratio of SWCNTs in the dispersion might be determined by the viscosity measurement technique, e.g., the steady-state simple shear experiments.

By following this line of thought, a few studies were carried out for determining the aspect ratio of SWCNT particles in superacid (Davis et al., 2004) and aqueous dispersions (Parra-Vasquez et al., 2007), where the volume fraction of SWCNTs is a few factor of $10^{-5}$. The experimentally determined intrinsic viscosity of SWCNTs in superacid (8300 ± 830) and in aqueous dispersion (7350 ± 750) respectively lead to the estimated aspect ratio of SWCNTs to be 470 ± 30 and 505 ± 35. The similar and very large aspect ratio of the SWCNTs in these two distinctly different dispersion systems indicate the dominant structures of SWCNTs in the dispersion are the individual tubes and/or the SWCNT bundles. It is noted that, the intrinsic viscosity relationship used in these studies is based on a formula given by Batchelor (Batchelor, 1974), which is:

$$[\eta] = \frac{8A^2}{45}\varepsilon \times \left[\frac{1 + 0.64\varepsilon}{1 - 1.5\varepsilon} + 1.659\varepsilon^2\right] \tag{3a}$$

$$\varepsilon = \frac{1}{\ln 2A} \tag{3b}$$

When compared to the accurate expression given by Eq. (2) (Mansfield & Douglas, 2008), the relationship given by Eq. (3) overestimates the intrinsic viscosity by 12% or more for the rods with aspect ratio below 100.

In order to appropriately use Eq. (1) and (2) for estimating the aspect ratio of SWCNTs in the dispersion, the volume fraction of the filler particles has to be kept low. In addition to this requirement, the viscosity measurements also need to be done at a relatively low shear rate. Otherwise, the slender particles with large aspect ratio, e.g., SWCNTs, can be aligned along the flow direction to cause the shear-thinning effect and thus result in a shear-rate dependent intrinsic viscosity, which is not taken into account by Eq. (1) and (2). The Peclet number (Bicerano et al., 1999; Larson, 1999) (*Pe*), defined by the ratio of the characteristic experimental shear rate to the rotational diffusion coefficient of the filler particle,

$$Pe = \frac{\dot{\gamma}}{D_r} \tag{4a}$$

$$D_r = \frac{3k_B T}{\pi \eta_0 L^3}\left(\ln A + 2\ln 2 - \frac{11}{6}\right) \tag{4b}$$

$$D_t = \frac{k_B T}{3\pi \eta_0 L}\left(\ln A + 2\ln 2 - 1\right) \tag{4c}$$

can be used as a criterion to select appropriate experimental conditions to avoid the complication caused by the shear-thinning effect. When *Pe* is smaller than 1, the rotational Brownian motion of the slender particle is able to overcome the shear-field induced alignment and randomize the particle orientation to minimize the shear-thinning effect. In Eq. (4), $k_B$ is the Boltzmann constant and $T$ is the temperature. The rotational and translational diffusion coefficient, $D_r$ and $D_t$, is taken from the work by Bonet Avalos (Avalos et al., 1993) and Yamakawa (Yamakawa, 1975). $D_t$ is given here for completeness and convenience and will be used for a later discussion on the dynamic light scattering technique for characterizing the SWCNT structures.

The steady-state simple shear experiments for the SWCNT dispersion at relatively low particle volume fraction allow one to determine the intrinsic viscosity of SWCNTs and thus infer the particle aspect ratio. In addition to this, the unsteady-state simple shear experiments, e.g., small-amplitude oscillatory flow, also enable one to study the viscoelastic behavior of SWCNT dispersions at relatively high particle volume fraction. Hough et al. (Hough et al., 2004) investigated the dynamic mechanical properties of SWCNT aqueous dispersions with particle volume fraction greater than $10^{-3}$. The observed oscillation frequency independent storage modulus $G'$ and loss modulus $G''$ allow the author to infer the presence of SWCNT network structures in the dispersion. The network structure is formed by the physical association of the SWCNT rods, and the bonding energy responsible for the association is as high as ~ 40 $k_B T$. The similar viscoelastic behavior studies were performed for the SWCNT dispersion in epoxy (Ma et al., 2009) and in unsaturated polyester (Kayatin & Davis, 2009). These polymeric resin based dispersion system presents a strong elastic response at relatively high volume fraction of SWCNTs, which also signifies the formation of SWCNT networks.

In brief, the viscosity and rheological measurements are capable of providing the in-situ structural information of SWCNTs in different dispersing media. The SWCNT structures being probed include the aspect ratio of the individual tubes or SWCNT bundles as well as the network formation of SWCNTs.

## 3. Scattering techniques

For a long time, the elastic scattering techniques, e.g., static light scattering (SLS), small angle X-ray (SAXS) and neutron scattering (SANS) have been widely used for obtaining the structural information of materials of many kinds (Guinier & Fournet, 1955; Glatter & Kratky, 1982; Feigin & Svergun, 1987; Chu, 1991; Higgins & Benoit, 1994). In a typical elastic scattering experiment, a collimated beam of probe particles, e.g., photons in SLS and SAXS, neutrons in SANS, interacts with a sample system that is composed of many scattering units or scatterers. The interaction between the probe beam and the scatterer at position $\vec{r}_i$ produces a spherical scattered wave propagating outwardly from $\vec{r}_i$ toward the detector. The scattering beam intensity recorded by the detector, $I_D$, is a result of the superposition of the multiple spherical scattered waves originated from the many scatterers that are bathed in an illuminated volume $V$ defined by the incident probe beam and the detection optics. $I_D$ is related to the differential scattering cross section $d\sigma/d\Omega$ and given by (Graessley, 2004):

$$\frac{d\sigma}{d\Omega}=\frac{I_D r_D^2}{VI_0}=\sum_{j=1}^{n}\sum_{k=1}^{n} b_j b_k \exp\left[i\vec{q}\cdot\left(\vec{r}_j-\vec{r}_k\right)\right] \tag{5a}$$

$$\vec{q}=\frac{2\pi}{\lambda}\left(\vec{s}_0-\vec{s}_D\right) \tag{5b}$$

$$q=\left|\vec{q}\right|=\frac{4\pi}{\lambda}\sin\frac{\theta}{2} \tag{5c}$$

Normalized by the incident flux of the probe particles, which is the number of the particles impinging on a unit area of the sample per unit time, the differential scattering cross-section $d\sigma/d\Omega$ is defined as the number of scattered particles generated per unit time per unit volume of the sample within a unit solid angle subtended by the detector. In Eq. (5), $b_j$ is the scattering length of the scatterer $j$, a quantity to measure the scattering power of a given species that depends on the details of the probe/scatterer interaction; $I_0$ is the incident beam intensity; $r_D$ is the distance from the scatterer to the detector; $\vec{q}$ is the scattering vector and defined by the difference between the propagation vector of the incident beam ($2\pi\,\vec{s}_0/\lambda$) and that of the scattered beam ($2\pi\,\vec{s}_D/\lambda$); the scattering angle formed by the incident beam and the scattered beam is $\theta$, and $\lambda$ is the wavelength of the incident beam. As noted in Eq. (5), the scattered beam intensity contains the relative spatial position ($\vec{r}_j - \vec{r}_k$) of the scatterers, which forms the basis of using the elastic scattering techniques for characterizing the structures of suspension or dispersions. For a dispersion system of monodispersed particles with random orientation, the generalized differential scattering cross section given by Eq. (5a) can be simplified to (Ballauff et al., 1996; Pedersen, 1997; Peterlik & Fratzl, 2006):

$$\frac{d\sigma}{d\Omega}=n\Delta\rho^2\upsilon^2P(q)S(q) \tag{6}$$

where $n$ is the number density of the particles; $\Delta\rho$ is the difference in scattering length density (scattering length per unit volume of the dispersion particle) between the particles and the dispersing medium; $v$ is the volume of the particle; $P(q)$ is the particle form factor due to the intra-particle contribution to the scattering and characterizes the particle size and shape; and $S(q)$ is the structure factor to reflect the inter-particle contribution to the scattering, which characterizes the relative positions of different particles and contains the interaction information between the particles. Owing to the difficulties of separating the inter- and intra-particle contributions to the dispersion structure, the scattering experiments are usually carried out for dilute dispersion system to minimize the inter-particle contribution. In this case, the structure factor $S(q) = 1$. Without introducing the complication of the inter-particle contribution, the size and shape of the particles in a dilute dispersion can be determined by fitting the scattering intensity with Eq. (6) by applying appropriate form factor $P(q)$. Pedersen (Pedersen, 1997) summarized 27 different form factors, a few of which relevant to the structural characterization of SWCNT dispersions are given below:

1. Form factor for cylinder of length $L$ and radius $R$

$$P_1(q) = \int_0^{\pi/2} \left[ \frac{2J_1(qR\sin\alpha)}{qR\sin\alpha} \frac{\sin[(qL\cos\alpha)/2]}{(qL\cos\alpha)/2} \right]^2 \sin\alpha d\alpha \tag{7}$$

where $J_1(x)$ is the Bessel function of the first kind of order one.

2. Form factor for flexible polymer chain

$$P_2(q) = \frac{2\left[\exp(-q^2R_g{}^2) + q^2R_g{}^2 - 1\right]}{\left(q^2R_g{}^2\right)^2} \tag{8}$$

where $R_g{}^2$ is the mean squared radius of gyration of a Gaussian chain and equals to $(L_c l_k)/6$. $L_c$ is the contour length and $l_k$ is the Kuhn step length of the polymer chain.

3. Form factor for cylinder of length $L$ and radius $R$ with attached $N_c$ Gaussian chains of contour length $L_c$

$$P_3(q) = \frac{1}{(\rho + N_c\rho_c)^2} \left[ \rho^2 P_1(q) + N_c\rho_c^2 P_2(q) + N_c(N_c - 1)\rho_c^2 S_1(q) + 2N_c\rho\rho_c S_2(q) \right] \tag{9a}$$

$$S_1(q) = \left[ \frac{1 - \exp(-q^2R_g{}^2)}{q^2R_g{}^2} \right]^2 \times \int_0^{\pi/2} \{2J_0(qR\sin\alpha)\cos[(qL\cos\alpha)/2]\}^2 \sin\alpha d\alpha \tag{9b}$$

$$S_2(q) = \left[ \frac{1 - \exp(-q^2R_g{}^2)}{q^2R_g{}^2} \right] \times \int_0^{\pi/2} \frac{2J_1(qR\sin\alpha)}{qR\sin\alpha} \frac{\sin[(qL\cos\alpha)/2]}{(qL\cos\alpha)/2} 2J_0(qR\sin\alpha) \cos[(qL\cos\alpha)/2] \sin\alpha d\alpha \tag{9c}$$

where $J_0(x)$ is the Bessel function of the first kind of order zero; $\rho$ and $\rho_c$ is respectively the total excess scattering length of the cylinder and the polymer chains.

Dror (Dror et al., 2005), Yurekli et al. (Yurekli et al., 2004) and Granite et al. (Granite et al., 2010) respectively investigated the structures of styrene-sodium mealeate copolymer and gum arabic wrapped, SDS–stabilized, and pluronic copolymer dispersed SWCNT dispersions by SANS technique. All these studies indicated that the dispersing agents, either the ionic surfactant SDS or the copolymers being used, adsorbed on the SWCNTs to form a core-shell structure, in which the core is formed by thin SWCNT bundles and the shell is attributed to the physical adsorption of the dispersing agents. With the refined cylindrical core-shell form factors, the diameter of the core and the thickness of the shell have been determined by fitting the experimentally determined SANS scattering intensity. It is particularly interesting to note that, for the SDS-stabilized SWCNT dispersions, the SANS experiments indicated that, within the shell, the SDS surfactant molecules do not form any ordered micelle structures but are randomly distributed (Yurekli et al., 2004). One recent molecular dynamic simulation study on the SDS aggregation on SWCNTs (Tummala & Striolo, 2009) supports such a viewpoint. However, another MD simulation study (Xu et al., 2010) reveals a much delicate situation for the SDS structure formation on SWCNTs. Depending upon the diameter of SWCNT as well as the coverage density, the SDS molecules can organize into cylinder-like monolayer structure, hemicylindrical aggregates, and randomly organized structures on the surface of a SWCNT. It is expected that the combined simulation and scattering experiments could ultimately help to have a better understanding of this interesting phenomena.

In addition to the above described form-factor modeling approach, another commonly used method for understanding, analyzing and interpreting the small-angle scattering data is by a much simpler and physically appealing scaling approach (Oh & Sorensen, 1999; Sorensen, 2001). The scaling approach is based on a comparison of the inherent length scale of the scattering, *1/q*, and the length scales in the system of scatterers to qualitatively understand the behaviors of the differential scattering cross section in relation to the structures of the scattering system. Two limiting situations can be used for illustrating the principle of the scaling approach. When the $n$ scatterers are within a *1/q* distance from each other, the phase of the $n$ scattered waves will be in phase and $\vec{q}\cdot\left(\vec{r}_j-\vec{r}_k\right)<1$. In this case, the double sum in Eq. (5a) equals to $n^2$. On the other hand, when the $n$ scatterers are separated from each other by a distance greater than *1/q*, the phase of the $n$ scattered waves will be random and $\vec{q}\cdot\left(\vec{r}_j-\vec{r}_k\right)>1$. In such a case, the double sum in Eq. (5a) equals to $n$. With these results and bear in mind that, for a finite-sized scattering system with uniformly distributed scatterers, the non-zero scattering contribution at a scattering angle other than zero is due to the scatterer density fluctuation on the surface, one can derive a power-law relationship for the scattering intensity of a fractal aggregate with respect to the inherent length scale of *1/q* (Xu et al., 2010). It is stated as:

$$\frac{d\sigma}{d\Omega}\propto I_D\propto n^2(qR_g)^{-D}\quad for\quad a<1/q<R_g \tag{10}$$

where $D$ is the fractal dimension of an aggregate system. For a homogeneous 1D rod, $D = 1$; 2D disk, $D = 2$; and 3D sphere, $D = 3$. Eq. (10) applies to a fractal aggregate system defined by two length scales: $a$ is the size of the scatterer and $R_g$ is the radius gyration of the aggregate. The scaling approach makes the physical significance of the inherent length scale *1/q* more transparent and easier to comprehend.

With the help of Eq. (10), the fractal structures of SWCNTs in the dispersion have been investigated by SAXS (Schaefer et al., 2003a; 2003b), SANS (Zhou et al., 2004; Wang et al.,

2005; Bauer et al., 2006; Hough et al., 2006; Urbina et al., 2008) and SLS (Chen et al., 2004). Depending upon the sample preparation conditions, both the rigid-rod structure of SWCNTs (with $D$ = 1) and the entangled SWCNT fractal networks ($2 < D < 3$) have been observed. It is noted that, among the different scattering techniques being used for characterizing the SWCNT structures in different types of dispersions, the SANS was more popular than the others. This is partially attributed to the relatively high scattering contrast ($\Delta\rho$) of SWCNTs when interact with neutron as compared to X-rays. In addition, the strong optical absorption of SWCNTs in the visible light region could potentially complicate the SLS experiments and make the data interpretation and analysis more difficult. The experimental difficulties related to the SLS technique for the structural characterization of SWCNT dispersions has not been given sufficient attention.

The scattering experiments introduced above rely on measuring the time-averaged scattering intensity as a function of the scattering vector for characterizing the dispersion structures. In addition to this approach, another type of scattering experiments, e.g., dynamic light scattering (DLS) or quasi-elastic light scattering (Chu, 1991; Berne & Pecora, 2000; Teraoka, 2002), is also a valuable technique for in-situ characterizing the dispersion structures. The DLS method takes measurements of the time fluctuation of the scattered beam intensity to determine the time-dependent correlation function of a dynamic system, which provides a concise way for describing the degree to which two dynamic properties are correlated over a period of time. In DLS experiments, the normalized time correlation functions, $g_2(\tau)$, of the scattered light intensity is recorded and given by:

$$g_2(\tau) = \frac{< I_D(t) I_D(t+\tau) >}{< I_D(t) I_D(t) >} \tag{11}$$

which is related to the time correlation function, $g_1(\tau)$, of the scattered electric field ($E_D$)

$$g_1(\tau) = \frac{< E_D(t) E_D(t+\tau) >}{< E_D(t) E_D(t) >} \tag{12}$$

by

$$g_2(\tau) = 1 + \gamma [g_1(\tau)]^2 \tag{13}$$

where $\gamma$ is a constant determined by the specific experiment setup. Both polarized and depolarized DLS experiments can be performed. In the former (latter) experiments, the incident beam is in a vertical polarization direction and the vertically (horizontally) polarized scattered light is detected. Depending upon whether a polarized or depolarized DLS experiment is performed, for a dilute dispersion of rodlike particles, $g_1(\tau)$, is related to the distribution of the diffusion coefficients of the particles by (Chu, 1991; Berne & Pecora, 2000; Lehner et al., 2000; Shetty et al., 2009) :

$$|g_1(\tau)| = \int G(\Gamma) \exp(-\Gamma \tau) d\Gamma \tag{14a}$$

$$\Gamma = D_t q^2 + 6D_r \text{ for depolarized DLS} \tag{14b}$$

$$\Gamma = D_t q^2 \text{ for polarized DLS} \tag{14c}$$

where $G(\Gamma)$ is a distribution function to characterize the polydispersity of the particles; $D_t$ and $D_r$ are respectively the translational and rotational diffusion coefficients of the rods. Upon determination of the rotational and translational diffusion coefficient by the depolarized DLS measurements, one can solve the system equation of Eq. (4b) and (4c) to obtain the length and diameter of the rods. With this approach, Shetty et al (Shetty et al., 2009) and Badaire et al (Badaire et al., 2004) respectively investigated using the polarized DLS technique for in-situ determination of the average length and diameter of functionalized SWCNTs as well as SDS-stabilized SWCNTs in aqueous dispersions. Similar to SLS technique, the strong optical absorption of SWCNTs could also cause the experimental difficulties in using the DLS technique for the structural characterization of SWCNT dispersions.

## 4. Sedimentation characterization techniques

Analytical ultracentrifugation is a powerful and well-known technique in the areas of biochemistry, molecular biology and macromolecular science for characterizing the sedimentation, diffusion behaviors and the molecular weights of both synthetic and natural macromolecules (Fujita, 1975; Laue & Stafford, 1999; Colfen & Volkel, 2004; Brown & Schuck, 2006). The preparative ultracentrifuge also found applications on the characterization of proteins (Shiragami & Kajiuchi, 1990; Shiragami et al., 1990) and macromolecules (Pollet et al., 1979). Fig. 2 schematically shows the operational principle of the ultracentrifugation technique for characterizing the dispersion structures. When the dispersion is subject to centrifugation, the centrifugal force and the thermal agitation respectively cause gravitational drift and Brownian motion of the small particles in the dispersion. As a result, the originally uniformly distributed small particles with concentration of $C_0$ will develop into a certain concentration profile $C(r, t)$ at a given time $t$. The governing equation for describing the particle concentration profile can be derived on the basis of mass balance (Mason & Weaver, 1926; Waugh & Yphantis, 1953; Fujita, 1975; Shiragami & Kajiuchi, 1990) and given by:

$$\frac{\partial C}{\partial t} = D\left(\frac{\partial^2 C}{\partial r^2}\right) - s\omega^2 r_m \frac{\partial C}{\partial r}$$

$$\begin{aligned} & C = C_0 \qquad (t=0) \\ & D\frac{\partial C}{\partial r} = s\omega^2 r_m C \qquad r = r_1, \quad r = r_2; \quad t>0 \\ & r_m = \frac{r_1 + r_2}{2} \end{aligned} \tag{15}$$

where $s$ and $D$ are respectively the sedimentation and translational diffusion coefficient of the particles. For rodlike particles, the relationship between $D$ and its geometric dimension is given by Eq. (4c); and $s$ is given by:

$$s = \frac{m(1-\nu\rho_0)}{3\pi\eta_0 L}(\ln A + 2\ln 2 - 1) \tag{16}$$

In Eq. (16), $m$ is the mass of the particle; $v$ is its partial specific volume and can be approximated by the reciprocal of the particle mass density; and $\rho_0$ is the density of the liquid media.

An approximation is implied in Eq. (15). That is, irrespective of its distance from the center of rotation, the centrifugal field experienced by the particle is uniform and given by $\omega^2 r_m$. With this approximation, Eq. (15) can be solved analytically and the solution can be found in the cited references. With the analytical ultracentrifuge instrument, one can experimentally measure the concentration profile of the dispersion at a given set of centrifugation conditions. Upon fitting the theoretically predicted concentration profile given by Eq. (15), the transport properties, $s$ and $D$, of the particle can be determined, from which the structural information of the particle can be inferred. The analytical ultracentrifuge has recently been reported as a methodology for rapid characterization of the quality of carbon nanotube dispersions (Azoubel & Magdassi, 2010). Nevertheless, no efforts have been pursued for quantitatively extracting the structural information of the carbon nanotube dispersions being studied in this work.

In addition to the analytical ultracentrifuge approach, another sedimentation measurement based characterization technique - preparative ultracentrifuge method (PUM) (Liu et al., 2008) has been recently developed by the authors. The PUM method relies on measuring and analyzing the sedimentation function of a given SWCNT dispersion for quantitative characterizing the transport properties and the structures of SWCNTs. The idea to define the sedimentation function is schematically shown in Fig. 2 and described as follows: when a certain amount of dispersion is subject to centrifugation, the number of particles, $N(V, t=0)$, in a given control volume $V$ before centrifugation will decrease to $N(V, t)$ after time $t$. The sedimentation function is given by the ratio of $N(V, t)$ to $N(V, t=0)$ and related to the particle concentration profile $C(r, t)$ by:

$$F_p(t;s,D,\omega,G,r_1,r_2) = \frac{N(V,t)}{N(V,t=0)} = \frac{\int_{r=r_1}^{r=r_c} C(r,t)A(r)dr}{C_0 \int_{r=r_1}^{r=r_c} A(r)dr} \quad (17)$$

where $A(r)$ is the cross-section area of the centrifuge tube used for performing the PUM experiments. For a given set of centrifugation condition (rotor type, rotation speed and the centrifuge tube geometry), the sedimentation function is uniquely determined by the distributed sedimentation and diffusion coefficients and, therefore, the distributed lengths and diameters of SWCNT particles in a given dispersion. The experimental protocols for measuring the sedimentation function of SWCNT dispersions as well as its theoretical derivation can be found in Liu et al 's work (Liu et al., 2008).

With the analytical solution of Eq. (15) for the concentration profile $C(r, t)$, the experimentally determined sedimentation function can be fitted by Eq. (17) to give the bulk averaged $s$ and $D$ values of a given SWCNT dispersion. It should be noted that, in comparison to the DLS technique, the PUM method intends to have an overestimation of the translational diffusion coefficient $D$. Therefore, to determine the structural information of SWCNTs by the PUM method with Eq. (4c) and Eq. (16), one has to separately measure the diffusion coefficient of the SWCNTs, e.g., by the DLS measurement. The PUM method has been successfully used for studying the processing-structure relationship of SWCNT dispersions processed by sonication and microfluidization techniques (Luo et al., 2010). The comparative studies indicate that, in addition to the energy dissipation rate, the details of the flow field can play a critical role in dispersing and separating the SWCNT bundles into individual tubes.

To examine the PUM method against the commonly used AFM approach for characterizing the SWCNT structures, an individual-tube enriched SWCNT dispersion was prepared. In brief, an SWCNT/SDBS/$H_2O$ dispersion was probe-sonicated for 30 minutes and then subject to ultracentrifugation for ~ 3hrs at 200, 000g. The supernatant, which is concentrated by individual tubes, was collected and examined by both the PUM and the AFM technique for determining the averaged length and diameter of the SWCNT particles. The PUM method was carried out with a fix-angle rotor by the Optima™ MAX-XP ultracentrifuge instrument (Beckman Coulter, Inc.) and the DLS measurement was performed with the Delsa Nano C Particle Size Analyzer (Beckman Coulter, Inc.). The experimentally determined and theoretically fitted sedimentation functions for both the as-sonicated and the individual tube enriched SWCNT dispersions are shown in Fig. 3a. The fitted values of the sedimentation coefficient, *s*, are given in Table 1. In the same table, the diffusion coefficients measured by the polarized DLS method, the bulk averaged length and diameter values calculated with Eq. (4c) and Eq. (16) are also listed. With a spin-coating based sample preparation protocol, the individual tube enriched SWCNT dispersion was also examined by the AFM technique. The representative topography image and the SWCNT length and diameter obtained by AFM are respectively shown in Fig. 3b and listed in Table 1. A reasonable agreement between the AFM measurement and the PUM method has been found for both the length and diameter of the examined individual SWCNTs.

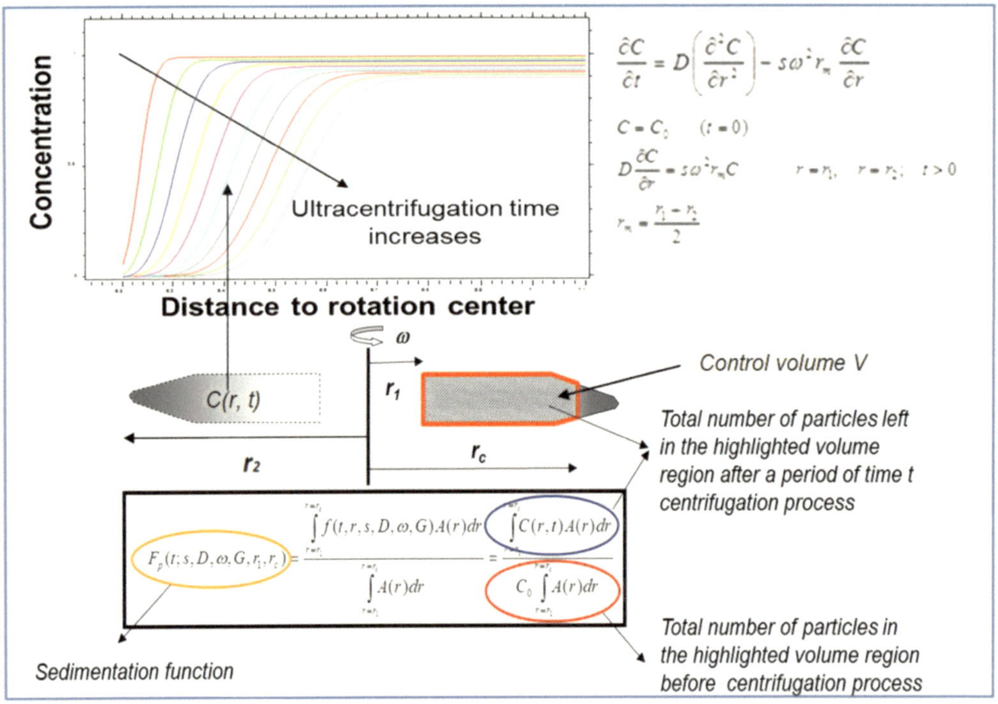


Fig. 2. Operational principle of analytical and preparative ultracentrifuge method for the structural characterization of SWCNT dispersions.

To further validate the PUM method, the sedimentation function for a standard polystyrene (PS) sphere dispersion in water (PS diameter of 100 nm) was determined experimentally and fitted theoretically, and the results are shown in Fig. 3c. Two different types of rotors, fixed-angle and swing-bucket, were used for comparing the effect of rotor geometry. With the sedimentation coefficient determined by the PUM method, the diameter of the PS sphere was accordingly calculated by:

$$d_{ps} = 2\sqrt{\frac{9s\eta_0}{2(\rho_{ps} - \rho_0)}} \tag{18}$$

The results are given in Table 1. The PUM determined PS sphere diameter deviates from the standard value of 100 nm by about 10%. Depending upon whether the fixed-angle rotor or the swing-bucket rotor is used, the PS diameter determined by the PUM method is 89.7 nm and 106.2 nm respectively. The effect of rotor geometry for the PUM method is clear.

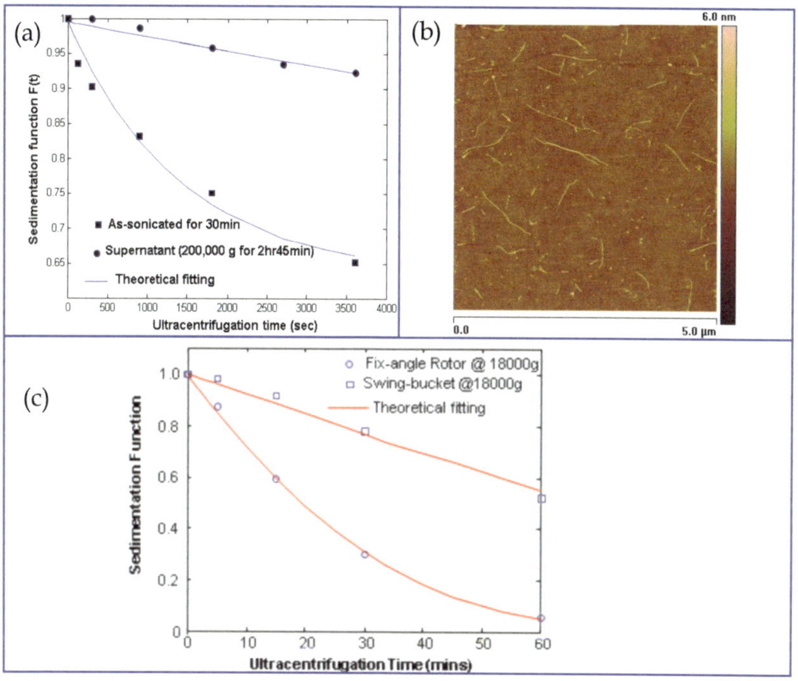


Fig. 3. (a) Experimentally determined and theoretically fitted sedimentation functions for as-sonicated and individual tube enriched SWCNT dispersions; Ultracentrifugation conditions - Fixed-angle rotor, 13,000 g for the as-sonicated dispersion and 65,000 g for the individual tube enriched dispersion; (b) AFM micrograph of the individual tube enriched SWCNT samples. Sample was prepared by spin coating and drying in the air on silicon wafer. (c) Experimentally determined and theoretically fitted sedimentation functions for the standard 100 nm PS sphere dispersion; Ultracentrifugation conditions - Fixed-angle rotor and Swing-bucket, 18, 000 g

Unlike the classical analytical ultracentrifuge approach, in which the concentration profile of the dispersion particles is mapped in the centrifugation process, the PUM method relies on a post-centrifugation process to experimentally determine the sedimentation function. From the instrument perspective, this is a big advantage since there is no complicated real-time detection optics is involved for the PUM method.

| SWCNT/SDBS/$H_2O$ Dispersions | | | | |
|---|---|---|---|---|
| AFM | PUM | | | |
| 200,000 g Centrifuged | As-sonicated | | 200,000 g Centrifuged | |
| L = 603 ± 336 nm | s = $1.76 \times 10^{-11}$ sec | L = 2541 nm | s = $2.40 \times 10^{-13}$ sec | L = 821 nm |
| d=0.94 ± 0.28 nm | D = $1.12 \times 10^{-8}$ $cm^2$/sec | d = 7.6 nm | D = $4.37 \times 10^{-8}$ $cm^2$/sec | d = 0.82 nm |
| Standard 100 nm polystyrene spheres | | | | |
| Standard | PUM | | | |
| 100 nm | Fixed-angle rotor | | Swing-bucket rotor | |
| | s = $2.46 \times 10^{-11}$ sec | d = 89.7 nm | s = $3.45 \times 10^{-11}$ sec | d = 106.2 nm |

Table 1. Comparison of AFM and PUM method for characterizing the SWCNT structures and standard 100 nm PS spheres

## 5. Spectroscopic techniques for charactering the bundling states of SWCNTs

In an as-prepared and well-dispersed SWCNT dispersion, the SWCNTs may either exist as individual tubes or present in a SWNT bundle. The techniques introduced above, including the viscosity and rheological measurements, different scattering techniques, and the sedimentation characterization methods, can hardly provide a reliable estimation on the relative percentage of individual tubes or the exfoliation efficiency of SWCNT bundles in a given dispersion. Given the important roles of bundling states in studying the fundamental photophysics of SWCNT (O'Connell et al., 2002; Torrens et al., 2006; Tan et al., 2007; Tan et al., 2008) and developing high-performance SWCNT-reinforced nanocomposites (Liu & Kumar, 2003; Ajayan & Tour, 2007), it is critical to have the capability for quantitative characterization of the degree of exfoliation for a given SWCNT dispersion.

By observing the broadening and red-shift of the featured absorption peaks of SWCNTs (Hagen & Hertel, 2003), the UV-visible-NIR spectroscopy has been used for qualitatively distinguishing the individual tube enriched SWCNT dispersions from the bundled ones. Moreover, Raman spectroscopy was also intensively used for characterizing the spectral characteristics induced by SWCNT bundling, which includes, e.g., the frequency upshift of the radial breathing mode (RBM) (O'Connell et al., 2004; Izard et al., 2005) and G-band broadening (Cardenas, 2008; Husanu et al., 2008). Using a 785 nm laser as the excitation source, Heller et al. (Heller et al., 2004) demonstrated a positive correlation between the intensity of the 267 $cm^{-1}$ RBM band and the bundling/aggregation states of various SWCNT samples. This valuable observation has been widely used for qualitative determination of the bundling states of SWCNT samples (Graupner, 2007; Kumatani & Warburton, 2008). The authors recently developed a simultaneous Raman scattering and PL spectroscopy technique (SRSPL) (Liu et al., 2009; Luo et al., 2010) to provide a new way for quantitative characterization of the bundling states of SWCNT dispersions.

When a laser interacts with a semi-conductive SWCNT, it can excite both the vibrational and electronic energy transition (Fig. 4a). As a result, one can detect the Raman scattered and the

PL emitted photons to acquire the Raman scattering and photoluminescence spectra (Burghard, 2005; Dresselhaus et al., 2005; Dresselhaus et al., 2007), from which the molecular/atomic and electronic structures of SWCNTs can be inferred.

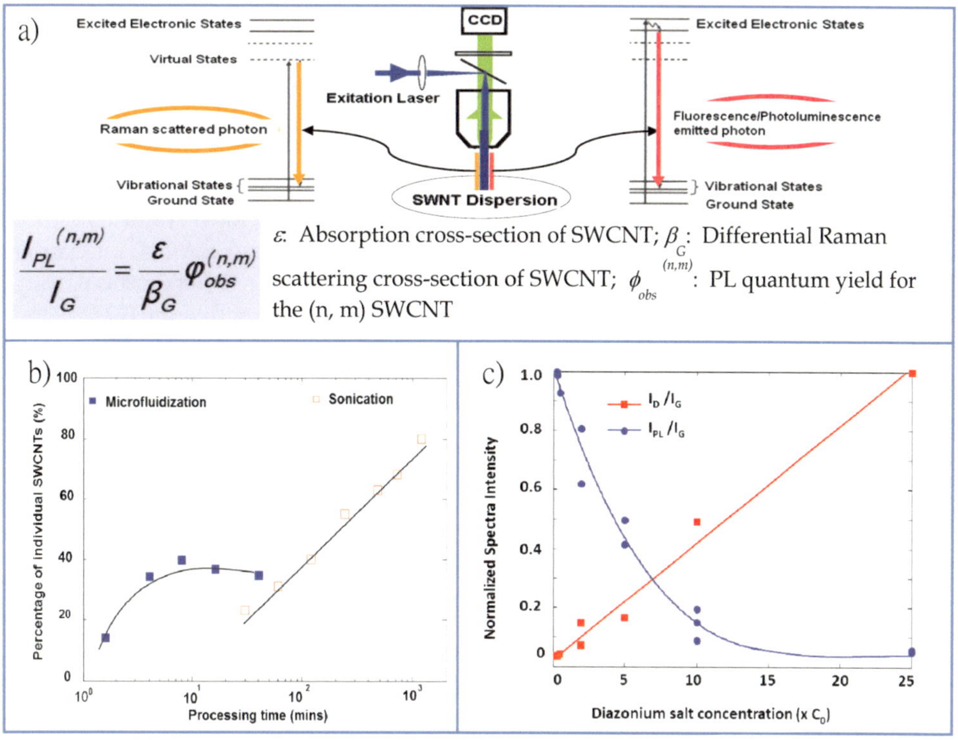


Fig. 4. Simultaneous Raman scattering and photoluminescence spectroscopy (SRSPL) for the degree of exfoliation and the defect density characterization of SWCNTs. a) operation principle of SRSPL method; b) SRSPL determined degree of exfoliation of SWCNTs processed by microfluidization and sonication; c) defect density characterization by SRSPL and Raman D-band for SWCNTs functionalized with diazonium salt.

In general, the Raman and PL spectra are taken separately by two different instruments - Raman spectrometer and fluorometer and analyzed independently. Nevertheless, as demonstrated in Liu et al's work (Liu et al., 2009), there is a significant advantage for acquiring the Raman and PL spectra of SWCNT dispersions simultaneously with the same optics. In this case, without introducing the complicated instrument correction factors, the intensity ratio of a PL band ($I_{PL}$) to a Raman band ($I_{Raman}$) is directly related to the intrinsic optical and spectroscopic properties of SWCNTs by:

$$\frac{I_{PL}}{I_{Raman}} = \frac{\varepsilon}{\beta}\Phi \tag{19}$$

where $\varepsilon$ is the optical absorption cross-section, $\beta$ is the Raman scattering cross-section, and $\Phi$ is the PL quantum yield of the SWCNT. Due to the presence of metallic SWCNTs in its very near neighbor, the PL of a semi-conductive SWCNT can be quenched when it is in a

SWCNT bundle. Using this fact and on the basis of Eq. (19), one can quantitatively determine the percentage of individual tubes or the degree of exfoliation for a given SWCNT dispersion with the SRSPL method (Liu et al., 2009; Luo et al., 2010). Fig. 4b compares the efficiency of the microfluidization and the sonication processes in exfoliating SWCNT bundles as examined by the SRSPL method. Again, it is clear that, the details of the flow field can play a critical role in separating the SWCNT bundles into individual tubes.

In addition to its capability for quantifying the degree of exfoliation, the SRSPL can also be used for characterizing the defect density of chemically functionalized SWCNTs. This is based on that, upon chemical functionalization, the defects introduced on the sidewall of a semi-conductive SWCNT effectively reduced the defect-free segment length, which cause a reduced PL quantum yield (Rajan et al., 2008). Fig. 4c demonstrated the SRSPL method for characterizing the defect density of diazonium salt functionalized SWCNTs (Xiao et al., 2010). In the same figure, the commonly used Raman D-band over G-band ratio (Graupner, 2007) for the same purpose is also shown for comparison. It is clear that the SRSPL and the Raman D-band method complement to each other; the former is suitable for low defect density and the latter is more appropriate for high defect density characterization.

## 6. Conclusion

The hierarchical structures of SWCNTs with a broad range length scales can be found in a dispersion, which may include: 1) individual tubes with different molecular structure as specified by the rolling or chiral vector (n, m); 2) SWCNT bundles that is composed of multiple individual tubes approximately organized into a 2D hexagonal lattice with their long axis parallel to each other; 3) SWCNT aggregates formed by the topological entanglement or enmeshment of individual tubes and/or SWCNT bundles; 4) SWCNT network that spans the overall dispersion sample. In order to establish the processing-structure-property relationship of SWCNT enabled multifunction nanocomposites and SWCNT dispersion related novel applications, an in-situ and quantitative characterization of the hierarchical structures of SWCNTs in the dispersion is necessary. With an emphasis on the underlying physical principles, the recently emerging experimental techniques that enable an in-situ and quantitative structural characterization of SWCNT dispersions are reviewed in this chapter, which include: 1) *Viscosity and rheological measurements; 2) Elastic and quasi-elastic scattering techniques; 3) Sedimentation characterization methods; and 4) Spectroscopic techniques.* Each of these techniques has its own length-scale vantage for the structural characterization of SWCNTs in the dispersion. To fully characterize the hierarchical structures of SWCNTs in the dispersion and understand their roles in controlling the properties and performance of SWCNT enabled multifunction nanocomposites and SWCNT dispersion related novel applications, the best approach is to be able to wisely and coherently utilize the introduced techniques to their advantages. For different reasons, the hierarchical structures of SWCNTs in the dispersion are subject to a certain distribution. This brings out the polydispersity issues, which have not been addressed by the experimental techniques being reviewed here. Future research should be directed toward overcoming this even more challenging issue.

## 7. Reference

Ajayan, P. M. & Tour, J. M. (2007). Materials science - Nanotube composites. *Nature*, Vol. 447, No. 7148, pp. 1066-1068, ISSN 0028-0836

Avalos, J. B.; Rubi, J. M. & Bedeaux, D. (1993). Dynamics of rodlike polymers in dilute solution. *Macromolecules*, Vol. 26, No. 10, pp. 2550-2561, ISSN 0024-9297

Avouris, P.; Freitag, M. & Perebeinos, V. (2008). Carbon-nanotube optoelectronics, In: *Carbon Nanotubes*, pp. 423-454, Springer-Verlag Berlin, ISBN 0303-4216, Berlin, Germany

Azoubel, S. & Magdassi, S. (2010). The formation of carbon nanotube dispersions by high pressure homogenization and their rapid characterization by analytical centrifuge. *Carbon*, Vol. 48, No. 12, pp. 3346-3352, ISSN 0008-6223

Badaire, S.; Poulin, P.; Maugey, M. & Zakri, C. (2004). In situ measurements of nanotube dimensions in suspensions by depolarized dynamic light scattering. *Langmuir*, Vol. 20, No. 24, pp. 10367-10370, ISSN 0743-7463

Ballauff, M.; Bolze, J.; Dingenouts, N.; Hickl, P. & Potschke, D. (1996). Small-angle X-ray scattering on latexes. *Macromolecular Chemistry and Physics*, Vol. 197, No. 10, pp. 359-377

Batchelor, G. K. (1974). Transport properties of two-phase materials with random structure. *Annual Review of Fluid Mechanics*, Vol. 6, pp. 227-255, ISSN 0066-4189

Bauer, B. J.; Hobbie, E. K. & Becker, M. L. (2006). Small-angle neutron scattering from labeled single-wall carbon nanotubes. *Macromolecules*, Vol. 39, No. 7, pp. 2637-2642, ISSN 0024-9297

Baughman, R. H.; Zakhidov, A. A. & de Heer, W. A. (2002). Carbon nanotubes - the route toward applications. *Science*, Vol. 297, No. 5582, pp. 787-792, ISSN 0036-8075

Bergin, S. D.; Nicolosi, V.; Giordani, S.; de Gromard, A.; Carpenter, L.; Blau, W. J. & Coleman, J. N. (2007). Exfoliation in ecstasy: liquid crystal formation and concentration-dependent debundling observed for single-wall nanotubes dispersed in the liquid drug gamma-butyrolactone. *Nanotechnology*, Vol. 18, No. 45, pp. 10, ISSN 0957-4484

Berne, B. J. & Pecora, R. (2000). *Dynamic Light Scattering with Applications to Chemistry, Biology, and Physics* Dover Publications, Inc., ISBN 978-0486411552, New York, USA

Bicerano, J.; Douglas, J. F. & Brune, D. A. (1999). Model for the viscosity of particle dispersions. *Journal of Macromolecular Science-Reviews in Macromolecular Chemistry and Physics*, Vol. C39, No. 4, pp. 561-642, ISSN 0736-6574

Bonard, J. M.; Salvetat, J. P.; Stockli, T.; de Heer, W. A.; Forro, L. & Chatelain, A. (1998). Field emission from single-wall carbon nanotube films. *Applied Physics Letters*, Vol. 73, No. 7, pp. 918-920, ISSN 0003-6951

Brown, P. H. & Schuck, P. (2006). Macromolecular size-and-shape distributions by sedimentation velocity analytical ultracentrifugation. *Biophysical Journal*, Vol. 90, No. 12, pp. 4651-4661, ISSN 0006-3495

Burghard, M. (2005). Electronic and vibrational properties of chemically modified single-wall carbon nanotubes. *Surface Science Reports*, Vol. 58, No. 1-4, pp. 1-109, ISSN 0167-5729

Cao, J.; Wang, Q. & Dai, H. J. (2003). Electromechanical properties of metallic, quasimetallic, and semiconducting carbon nanotubes under stretching. *Physical Review Letters*, Vol. 90, No. 15, pp. 4, ISSN 0031-9007

Cao, Q. & Rogers, J. A. (2009). Ultrathin Films of Single-Walled Carbon Nanotubes for Electronics and Sensors: A Review of Fundamental and Applied Aspects. *Advanced Materials*, Vol. 21, No. 1, pp. 29-53, ISSN 0935-9648

Cardenas, J. F. (2008). Protonation and sonication effects on aggregation sensitive Raman features of single wall carbon nanotubes. *Carbon*, Vol. 46, No. 10, pp. 1327-1330, ISSN 0008-6223

Chen, Q.; Saltiel, C.; Manickavasagam, S.; Schadler, L. S.; Siegel, R. W. & Yang, H. C. (2004). Aggregation behavior of single-walled carbon nanotubes in dilute aqueous suspension. *Journal of Colloid and Interface Science*, Vol. 280, No. 1, pp. 91-97, ISSN 0021-9797

Chou, T. W.; Gao, L. M.; Thostenson, E. T.; Zhang, Z. G. & Byun, J. H. (2010). An assessment of the science and technology of carbon nanotube-based fibers and composites. *Composites Science and Technology*, Vol. 70, No. 1, pp. 1-19, ISSN 0266-3538

Chu, B. (1991). *Laser light scattering: basic principles and practice* (2nd), Academic Press, ISBN 9780486457987 Boston, USA

Colfen, H. & Volkel, A. (2004). Analytical ultracentrifugation in colloid chemistry, In: *Analytical Ultracentrifugation Vii*, Lechner, M. D. and Borger, L., pp. 31-47, Springer-Verlag Berlin, Berlin, Germany

Cotiuga, I.; Picchioni, F.; Agarwal, U. S.; Wouters, D.; Loos, J. & Lemstra, P. J. (2006). Block-copolymer-assisted solubilization of carbon nanotubes and exfoliation monitoring through viscosity. *Macromolecular Rapid Communications*, Vol. 27, No. 13, pp. 1073-1078, ISSN 1022-1336

Davis, V. A.; Ericson, L. M.; Parra-Vasquez, A. N. G.; Fan, H.; Wang, Y. H.; Prieto, V.; Longoria, J. A.; Ramesh, S.; Saini, R. K.; Kittrell, C.; Billups, W. E.; Adams, W. W.; Hauge, R. H.; Smalley, R. E. & Pasquali, M. (2004). Phase Behavior and rheology of SWNTs in superacids. *Macromolecules*, Vol. 37, No. 1, pp. 154-160, ISSN 0024-9297

Dresselhaus, M. S.; Dresselhaus, G.; Saito, R. & Jorio, A. (2005). Raman spectroscopy of carbon nanotubes. *Physics Reports-Review Section of Physics Letters*, Vol. 409, No. 2, pp. 47-99, ISSN 0370-1573

Dresselhaus, M. S.; Dresselhaus, G.; Saito, R. & Jorio, A. (2007). Exciton photophysics of carbon nanotubes. *Annual Review of Physical Chemistry*, Vol. 58, pp. 719-747, ISSN 0066-426X

Dror, Y.; Pyckhout-Hintzen, W. & Cohen, Y. (2005). Conformation of polymers dispersing single-walled carbon nanotubes in water: A small-angle neutron scattering study. *Macromolecules*, Vol. 38, No. 18, pp. 7828-7836, ISSN 0024-9297

Duggal, R. & Pasquali, M. (2006). Dynamics of individual single-walled carbon nanotubes in water by real-time visualization. *Physical Review Letters*, Vol. 96, No. 24, pp. 4, ISSN 0031-9007

Fakhri, N.; Tsyboulski, D. A.; Cognet, L.; Weisman, R. B. & Pasquali, M. (2009). Diameter-dependent bending dynamics of single-walled carbon nanotubes in liquids. *Proceedings of the National Academy of Sciences of the United States of America*, Vol. 106, No. 34, pp. 14219-14223, ISSN 0027-8424

Feigin, L. A. & Svergun, D. I. (1987). *Structure Analysis by Small-Angle X-ray and Neutron Scattering* Plenum Press, ISBN 978-0306426292, New York, USA

Flory, P. J. (1978). Statistical thermodynamics of mixtures of rodlike particles. 5. Mixtures with random coils. *Macromolecules*, Vol. 11, No. 6, pp. 1138-1141, ISSN 0024-9297

Fujita, H. (1975). *Foundations of Ultracentrifugal Analysis* John Wiley & Sons, ISBN 9780471285823, New York, USA

Giordani, S.; Bergin, S.; Nicolosi, V.; Lebedkin, S.; Blau, W. J. & Coleman, J. N. (2006). Fabrication of stable dispersions containing up to 70% individual carbon nanotubes in a common organic solvent. *Physica Status Solidi B-Basic Solid State Physics*, Vol. 243, No. 13, pp. 3058-3062, ISSN 0370-1972

Girifalco, L. A.; Hodak, M. & Lee, R. S. (2000). Carbon nanotubes, buckyballs, ropes, and a universal graphitic potential. *Physical Review B*, Vol. 62, No. 19, pp. 13104-13110, ISSN 0163-1829

Glatter, O. & Kratky, O. (1982). *Small Angle X-ray Scattering* Academic Press, ISBN 978-0122862809, London, UK

Grady, B. P. (2009). Recent Developments Concerning the Dispersion of Carbon Nanotubes in Polymers. *Macromolecular Rapid Communications*, Vol. 31, No. 3, pp. 247-257, ISSN 1022-1336

Graessley, W. W. (2004). *Polymeric Liquids and Networks: Structure and Properties* Garland Science, ISBN 978-0-8153-4169-7, New York and London

Granite, M.; Radulescu, A.; Pyckhout-Hintzen, W. & Cohen, Y. (2010). Interactions between Block Copolymers and Single-Walled Carbon Nanotubes in Aqueous Solutions: A Small-Angle Neutron Scattering Study. *Langmuir*, Vol. 27, No. 2, pp. 751-759, ISSN 0743-7463

Graupner, R. (2007). Raman spectroscopy of covalently functionalized single-wall carbon nanotubes. *Journal of Raman Spectroscopy*, Vol. 38, No. 6, pp. 673-683, ISSN 0377-0486

Green, M. J.; Behabtu, N.; Pasquali, M. & Adams, W. W. (2009). Nanotubes as polymers. *Polymer*, Vol. 50, No. 21, pp. 4979-4997, ISSN 0032-3861

Grow, R. J.; Wang, Q.; Cao, J.; Wang, D. W. & Dai, H. J. (2005). Piezoresistance of carbon nanotubes on deformable thin-film membranes. *Applied Physics Letters*, Vol. 86, No. 9, pp. 3, ISSN 0003-6951

Gruner, G. (2006). Carbon nanotube films for transparent and plastic electronics. *Journal of Materials Chemistry*, Vol. 16, No. 35, pp. 3533-3539, ISSN 0959-9428

Guinier, A. & Fournet, G. (1955). *Small-Angle Scattering of X-rays* (1st), Wiley, New York, USA

Hagen, A. & Hertel, T. (2003). Quantitative analysis of optical spectra from individual single-wall carbon nanotubes. *Nano Letters*, Vol. 3, No. 3, pp. 383-388, ISSN 1530-6984

Heller, D. A.; Barone, P. W.; Swanson, J. P.; Mayrhofer, R. M. & Strano, M. S. (2004). Using Raman spectroscopy to elucidate the aggregation state of single-walled carbon nanotubes. *Journal of Physical Chemistry B*, Vol. 108, No. 22, pp. 6905-6909, ISSN 1520-6106

Higgins, J. S. & Benoit, H. C. (1994). *Polymers and Neutron Scattering* Oxford University Press, ISBN 978-0-19-850063-6, New York, USA

Hilding, J.; Grulke, E. A.; Zhang, Z. G. & Lockwood, F. (2003). Dispersion of carbon nanotubes in liquids. *Journal of Dispersion Science and Technology*, Vol. 24, No. 1, pp. 1-41, ISSN 0193-2691

Hornsby, P. R. (1999). Rheology, compounding and processing of filled thermoplastics, In: *Mineral Fillers in Thermoplastics I*, pp. 155-217, Springer-Verlag Berlin, ISBN 0065-3195, Berlin, Germany

Hough, L. A.; Islam, M. F.; Hammouda, B.; Yodh, A. G. & Heiney, P. A. (2006). Structure of semidilute single-wall carbon nanotube suspensions and gels. *Nano Letters*, Vol. 6, No. 2, pp. 313-317, ISSN 1530-6984

Hough, L. A.; Islam, M. F.; Janmey, P. A. & Yodh, A. G. (2004). Viscoelasticity of single wall carbon nanotube suspensions. *Physical Review Letters*, Vol. 93, No. 16, pp. 4, ISSN 0031-9007

Husanu, M.; Baibarac, M. & Baltog, I. (2008). Non-covalent functionalization of carbon nanotubes: Experimental evidence for isolated and bundled tubes. *Physica E-Low-Dimensional Systems & Nanostructures*, Vol. 41, No. 1, pp. 66-69, ISSN 1386-9477

Hussain, F.; Hojjati, M.; Okamoto, M. & Gorga, R. E. (2006). Review article: Polymer-matrix nanocomposites, processing, manufacturing, and application: An overview. *Journal of Composite Materials*, Vol. 40, No. 17, pp. 1511-1575, ISSN 0021-9983

Islam, M. F.; Rojas, E.; Bergey, D. M.; Johnson, A. T. & Yodh, A. G. (2003). High weight fraction surfactant solubilization of single-wall carbon nanotubes in water. *Nano Letters*, Vol. 3, No. 2, pp. 269-273, ISSN 1530-6984

Izard, N.; Riehl, D. & Anglaret, E. (2005). Exfoliation of single-wall carbon nanotubes in aqueous surfactant suspensions: A Raman study. *Physical Review B*, Vol. 71, No. 19, pp. 7, ISSN 1098-0121

Jeffrey, D. J. & Acrivos, A. (1976). The rheological properties of suspensions of rigid particles. *Aiche Journal*, Vol. 22, No. 3, pp. 417-432, ISSN 0001-1541

Kayatin, M. J. & Davis, V. A. (2009). Viscoelasticity and Shear Stability of Single-Walled Carbon Nanotube/Unsaturated Polyester Resin Dispersions. *Macromolecules*, Vol. 42, No. 17, pp. 6624-6632, ISSN 0024-9297

Kumatani, A. & Warburton, P. A. (2008). Characterization of the disaggregation state of single-walled carbon nanotube bundles by dielectrophoresis and Raman spectroscopy. *Applied Physics Letters*, Vol. 92, No. 24, pp. 3, ISSN 0003-6951

Larson, R. G. (1999). *The Structure and Rheology of Complex Fluids* Oxford University Press, ISBN 978-019-5121-97-1, New York, USA

Laue, T. M. & Stafford, W. F. (1999). Modern applications of analytical ultracentrifugation. *Annual Review of Biophysics and Biomolecular Structure*, Vol. 28, pp. 75-100, ISSN 1056-8700

Lehner, D.; Lindner, H. & Glatter, O. (2000). Determination of the translational and rotational diffusion coefficients of rodlike particles using depolarized dynamic light scattering. *Langmuir*, Vol. 16, No. 4, pp. 1689-1695, ISSN 0743-7463

Li, Z. L.; Dharap, P.; Nagarajaiah, S.; Barrera, E. V. & Kim, J. D. (2004). Carbon nanotube film sensors. *Advanced Materials*, Vol. 16, No. 7, pp. 640-+, ISSN 0935-9648

Liu, T. & Kumar, S. (2003). Effect of orientation on the modulus of SWNT films and fibers. *Nano Letters*, Vol. 3, No. 5, pp. 647-650, ISSN 1530-6984

Liu, T.; Luo, S. D.; Xiao, Z. W.; Zhang, C. & Wang, B. (2008). Preparative Ultracentrifuge Method for Characterization of Carbon Nanotube Dispersions. *Journal of Physical Chemistry C*, Vol. 112, No. 49, pp. 19193-19202, ISSN 1932-7447

Liu, T.; Xiao, Z. W. & Wang, B. (2009). The exfoliation of SWCNT bundles examined by simultaneous Raman scattering and photoluminescence spectroscopy. *Carbon*, Vol. 47, No. 15, pp. 3529-3537, ISSN 0008-6223

Liu, Y. Q.; Gao, L.; Zheng, S.; Wang, Y.; Sun, J.; Kajiura, H.; Li, Y. & Noda, K. (2007). Debundling of single-walled carbon nanotubes by using natural polyelectrolytes. *Nanotechnology*, Vol. 18, No. 36, pp. 6, ISSN 0957-4484

Luo, S. D.; Liu, T. & Wang, B. (2010). Comparison of ultrasonication and microfluidization for high throughput and large-scale processing of SWCNT dispersions. *Carbon*, Vol. 48, No. 10, pp. 2992-2994, ISSN 0008-6223

Ma, A. W. K.; Chinesta, F. & Mackley, M. R. (2009). The rheology and modeling of chemically treated carbon nanotubes suspensions. *Journal of Rheology*, Vol. 53, No. 3, pp. 547-573, ISSN 0148-6055

Mac Kernan, D. & Blau, W. J. (2008). Exploring the mechanisms of carbon-nanotube dispersion aggregation in a highly polar solvent. *Epl*, Vol. 83, No. 6, pp. 6, ISSN 0295-5075

Mansfield, M. L. & Douglas, J. F. (2008). Transport properties of rodlike particles. *Macromolecules*, Vol. 41, No. 14, pp. 5422-5432, ISSN 0024-9297

Mason, M. & Weaver, W. (1926). The settling of small particles in a fluid. *Physical Review*, Vol. 23, pp. 412-426

Metzner, A. B. (1985). Rheology of suspensions in polymeric liquids. *Journal of Rheology*, Vol. 29, No. 6, pp. 739-775, ISSN 0148-6055

Moniruzzaman, M. & Winey, K. I. (2006). Polymer nanocomposites containing carbon nanotubes. *Macromolecules*, Vol. 39, No. 16, pp. 5194-5205, ISSN 0024-9297

Moore, V. C.; Strano, M. S.; Haroz, E. H.; Hauge, R. H.; Smalley, R. E.; Schmidt, J. & Talmon, Y. (2003). Individually suspended single-walled carbon nanotubes in various surfactants. *Nano Letters*, Vol. 3, No. 10, pp. 1379-1382, ISSN 1530-6984

O'Connell, M. J.; Bachilo, S. M.; Huffman, C. B.; Moore, V. C.; Strano, M. S.; Haroz, E. H.; Rialon, K. L.; Boul, P. J.; Noon, W. H.; Kittrell, C.; Ma, J. P.; Hauge, R. H.; Weisman, R. B. & Smalley, R. E. (2002). Band gap fluorescence from individual single-walled carbon nanotubes. *Science*, Vol. 297, No. 5581, pp. 593-596, ISSN 0036-8075

O'Connell, M. J.; Sivaram, S. & Doorn, S. K. (2004). Near-infrared resonance Raman excitation profile studies of single-walled carbon nanotube intertube interactions: A direct comparison of bundled and individually dispersed HiPco nanotubes. *Physical Review B*, Vol. 69, No. 23, pp. 15, ISSN 1098-0121

Oh, C. & Sorensen, C. M. (1999). Scaling approach for the structure factor of a generalized system of scatterers. *Journal of Nanoparticle Research*, Vol. 1, pp. 369-377.

Onsager, L. (1949). The effects of shape on the interaction of colloidal particles. *Annals of the New York Academy of Sciences*, Vol. 51, No. 4, pp. 627-659, ISSN 0077-8923

Parra-Vasquez, A. N. G.; Stepanek, I.; Davis, V. A.; Moore, V. C.; Haroz, E. H.; Shaver, J.; Hauge, R. H.; Smalley, R. E. & Pasquali, M. (2007). Simple length determination of single-walled carbon nanotubes by viscosity measurements in dilute suspensions. *Macromolecules*, Vol. 40, No. 11, pp. 4043-4047, ISSN 0024-9297

Pedersen, J. S. (1997). Analysis of small-angle scattering data from colloids and polymer solutions: modeling and least-squares fitting. *Advances in Colloid and Interface Science*, Vol. 70, pp. 171-210, ISSN 0001-8686

Peterlik, H. & Fratzl, P. (2006). Small-angle X-ray scattering to characterize nanostructures in inorganic and hybrid materials chemistry. *Monatshefte Fur Chemie*, Vol. 137, No. 5, pp. 529-543, ISSN 0026-9247

Petrie, C. J. S. (1999). The rheology of fibre suspensions. *Journal of Non-Newtonian Fluid Mechanics*, Vol. 87, No. 2-3, pp. 369-402, ISSN 0377-0257

Pollet, R. J.; Haase, B. A. & Standaert, M. L. (1979). Macromolecular characterization by sedimentation equilibrium in the preparative ultra-centrifuge. *Journal of Biological Chemistry*, Vol. 254, No. 1, pp. 30-33, ISSN 0021-9258

Rajan, A.; Strano, M. S.; Heller, D. A.; Hertel, T. & Schulten, K. (2008). Length-dependent optical effects in single walled carbon nanotubes. *Journal of Physical Chemistry B*, Vol. 112, No. 19, pp. 6211-6213, ISSN 1520-6106

Russel, W. B. (1980). Review of the role of colloidal forces in the rheology of suspensions. *Journal of Rheology*, Vol. 24, No. 3, pp. 287-317, ISSN 0148-6055

Sahoo, N. G.; Rana, S.; Cho, J. W.; Li, L. & Chan, S. H. (2010). Polymer nanocomposites based on functionalized carbon nanotubes. *Progress in Polymer Science*, Vol. 35, No. 7, pp. 837-867, ISSN 0079-6700

Saito, R.; Dresselhaus, G. & Dresselhaus, M. S. (1998). *Physical Properties of Carbon Nanotubes* Imperial College Press, ISBN 1-86094-223-7, London, UK

Salvetat, J. P.; Briggs, G. A. D.; Bonard, J. M.; Bacsa, R. R.; Kulik, A. J.; Stockli, T.; Burnham, N. A. & Forro, L. (1999). Elastic and shear moduli of single-walled carbon nanotube ropes. *Physical Review Letters*, Vol. 82, No. 5, pp. 944-947, ISSN 0031-9007

Schaefer, D.; Brown, J. M.; Anderson, D. P.; Zhao, J.; Chokalingam, K.; Tomlin, D. & Ilavsky, J. (2003). Structure and dispersion of carbon nanotubes. *Journal of Applied Crystallography*, Vol. 36, pp. 553-557, ISSN 0021-8898

Schaefer, D. W. & Justice, R. S. (2007). How nano are nanocomposites? *Macromolecules*, Vol. 40, No. 24, pp. 8501-8517, ISSN 0024-9297

Schaefer, D. W.; Zhao, J.; Brown, J. M.; Anderson, D. P. & Tomlin, D. W. (2003). Morphology of dispersed carbon single-walled nanotubes. *Chemical Physics Letters*, Vol. 375, No. 3-4, pp. 369-375, ISSN 0009-2614

Shenoy, A. V. (1999). *Rheology of filled polymer system* (1st), Springer, ISBN 978-041-2831-0-03, New York, USA

Shetty, A. M.; Wilkins, G. M. H.; Nanda, J. & Solomon, M. J. (2009). Multiangle Depolarized Dynamic Light Scattering of Short Functionalized Single-Walled Carbon Nanotubes. *Journal of Physical Chemistry C*, Vol. 113, No. 17, pp. 7129-7133, ISSN 1932-7447

Shiragami, N. & Kajiuchi, T. (1990). Precipitation of protein by ultracentrifuge with angle rotor. 1. Model for sedimentation process. *Bioprocess Engineering*, Vol. 5, No. 2, pp. 85-88, ISSN 0178-515X

Shiragami, N.; Kajiuchi, T. & Matsuda, A. (1990). Precipitation of protein by ultracentrifuge with angle rotor. 2. Experimental. *Bioprocess Engineering*, Vol. 5, No. 3, pp. 103-105, ISSN 0178-515X

Skakalova, V.; Kaiser, A. B.; Woo, Y. S. & Roth, S. (2006). Electronic transport in carbon nanotubes: From individual nanotubes to thin and thick networks. *Physical Review B*, Vol. 74, No. 8, pp. 10, ISSN 1098-0121

Snow, E. S.; Campbell, P. M.; Ancona, M. G. & Novak, J. P. (2005). High-mobility carbon-nanotube thin-film transistors on a polymeric substrate. *Applied Physics Letters*, Vol. 86, No. 3, pp. 3, ISSN 0003-6951

Sorensen, C. M. (2001). Light scattering by fractal aggregates: A review. *Aerosol Science and Technology*, Vol. 35, No. 2, pp. 648-687, ISSN 0278-6826

Tan, P. H.; Hasan, T.; Bonaccorso, F.; Scardaci, V.; Rozhin, A. G.; Milne, W. I. & Ferrari, A. C. (2008). Optical properties of nanotube bundles by photoluminescence excitation and absorption spectroscopy. *Physica E-Low-Dimensional Systems & Nanostructures*, Vol. 40, No. 7, pp. 2352-2359, ISSN 1386-9477

Tan, P. H.; Rozhin, A. G.; Hasan, T.; Hu, P.; Scardaci, V.; Milne, W. I. & Ferrari, A. C. (2007). Photoluminescence spectroscopy of carbon nanotube bundles: Evidence for exciton energy transfer. *Physical Review Letters*, Vol. 99, No. 13, pp. 4, ISSN 0031-9007

Teraoka, I. (2002). *Polymer Solutions: An Introduction to Physical Properties* John Wiley & Sons, Inc., ISBN 978-0471389293, New York, USA

Thess, A.; Lee, R.; Nikolaev, P.; Dai, H. J.; Petit, P.; Robert, J.; Xu, C. H.; Lee, Y. H.; Kim, S. G.; Rinzler, A. G.; Colbert, D. T.; Scuseria, G. E.; Tomanek, D.; Fischer, J. E. & Smalley, R. E. (1996). Crystalline ropes of metallic carbon nanotubes. *Science*, Vol. 273, No. 5274, pp. 483-487, ISSN 0036-8075

Torrens, O. N.; Milkie, D. E.; Zheng, M. & Kikkawa, J. M. (2006). Photoluminescence from intertube carrier migration in single-walled carbon nanotube bundles. *Nano Letters*, Vol. 6, No. 12, pp. 2864-2867, ISSN 1530-6984

Tracy, M. A. & Pecora, R. (1992). Dynamics of rigid and semirigid rodlike polymers. *Annual Review of Physical Chemistry*, Vol. 43, pp. 525-557, ISSN 0066-426X

Tummala, N. R. & Striolo, A. (2009). SDS Surfactants on Carbon Nanotubes: Aggregate Morphology. *Acs Nano*, Vol. 3, No. 3, pp. 595-602, ISSN 1936-0851

Urbina, A.; Miguel, C.; Delgado, J. L.; Langa, F.; Diaz-Paniagua, C. & Batallan, F. (2008). Isolated rigid rod behavior of functionalized single-wall carbon nanotubes in solution determined via small-angle neutron scattering. *Physical Review B*, Vol. 78, No. 4, pp. 5, ISSN 1098-0121

Vroege, G. J. & Lekkerkerker, H. N. W. (1992). Phase-transitions in lyotropic colloidal and polymer liquid-crystals. *Reports on Progress in Physics*, Vol. 55, No. 8, pp. 1241-1309, ISSN 0034-4885

Wang, H.; Christopherson, G. T.; Xu, Z. Y.; Porcar, L.; Ho, D. L.; Fry, D. & Hobbie, E. K. (2005). Shear-SANS study of single-walled carbon nanotube suspensions. *Chemical Physics Letters*, Vol. 416, No. 1-3, pp. 182-186, ISSN 0009-2614

Waugh, D. F. & Yphantis, D. A. (1953). Transient solute distributions from the basic equation of the ultracentrifuge. *Journal of Physical Chemistry*, Vol. 57, No. 3, pp. 312-318, ISSN 0022-3654

Xiao, Z. W.; Wang, B. & Liu, T. (2010). Simultaneous Raman scattering and photoluminescence spectroscopy for quantifying the chemically induced defects of single-walled carbon nanotubes, *Proceedings of American Society for Composites Twenty-Fifth Technical Conference*, Dayton, OH, USA, September 2010

Xu, Z. J.; Yang, X. N. & Yang, Z. (2010). A Molecular Simulation Probing of Structure and Interaction for Supramolecular Sodium Dodecyl Sulfate/Single-Wall Carbon Nanotube Assemblies. *Nano Letters*, Vol. 10, No. 3, pp. 985-991, ISSN 1530-6984

Yakobson, B. I. & Couchman, L. S. (2006). Persistence length and nanomechanics of random bundles of nanotubes. *Journal of Nanoparticle Research*, Vol. 8, No. 1, pp. 105-110, ISSN 1388-0764

Yamakawa, H. (1975). Viscoelastic properties of straight cylindrical macromolecules in dilute solution. *Macromolecules*, Vol. 8, No. 3, pp. 339-342, ISSN 0024-9297

Yurekli, K.; Mitchell, C. A. & Krishnamoorti, R. (2004). Small-angle neutron scattering from surfactant-assisted aqueous dispersions of carbon nanotubes. *Journal of the American Chemical Society*, Vol. 126, No. 32, pp. 9902-9903, ISSN 0002-7863

Zheng, M.; Jagota, A.; Semke, E. D.; Diner, B. A.; McLean, R. S.; Lustig, S. R.; Richardson, R. E. & Tassi, N. G. (2003). DNA-assisted dispersion and separation of carbon nanotubes. *Nature Materials*, Vol. 2, No. 5, pp. 338-342, ISSN 1476-1122

Zhou, W.; Islam, M. F.; Wang, H.; Ho, D. L.; Yodh, A. G.; Winey, K. I. & Fischer, J. E. (2004). Small angle neutron scattering from single-wall carbon nanotube suspensions: evidence for isolated rigid rods and rod networks. *Chemical Physics Letters*, Vol. 384, No. 1-3, pp. 185-189, ISSN 0009-2614

# 13

# Microwave Absorption Characteristics of Carbon Nanotubes

Xiao-Gang Sun, Ming Gao, Cheng Li and Yiqiang Wu
*Nanchang University, Sun nanotech Co Ltd*
*China*

## 1. Introduction

Radar-absorption materials (RAM) have a growing and widespread applications in broadcasting and television, radar technology and microwave dark-room. In particular, in military applications, stealth weapons exhibit crucial effect in war. The research of RAM has been accelerating with the stealth technology and significant progresses have been achieved The investigations of RAM began before War II. It has experienced a process from traditional RAM to advanced RAM represented by nano-RAM. These advanced RAM mainly include nano-metals and alloys, nano-oxides, nano-SiC, nano-ferrite, nano-graphite, nano-SiC, nano-SiN, nano conductive polymers and carbon nanotubes (CNTs) [1]. Nano RAM has light-weight, excellent compatibility and broad bandwidth, and considered as important potential RAM for industrial application. CNTs are a very important RAM. They possess high strength and toughness [2,3], excellent electrical and thermal conductivity [4-7].The investigations of microwave absorption of CNTs had gained a big momentum due to its potential applications in the stealth technology of aircrafts, military equipment and microwave dark-room..

Cao[8] investigated microwave absorption properties of CNTs-polyester and achieved promising results with the absorbing peak value of -14 dB and bandwidth of 10.50 GHz(R<-5dB). Sheng [9] tested the radar wave absorption properties of CNTs coated with nickel. The results show the increase of bandwidth, but the absorption peak decreases. The peak and bandwidth of the RAM of CNTs coated with nickel reach -11.85dB and 2.23 GHz(R<-10dB). Lin [10,11] investigated the microwave-absorption properties of Co-filled and Fe-fillled carbon nanotubes respectively. The results indicate that the microwave-absorbing properties of Co-filled CNTs was improved. The maximum reflection loss is about -39.32 dB and the bandwidth corresponding to the reflection loss below -10 dB is 3.47 GHz. With increasing thickness, the maximum reflection loss shifts to lower frequency. The fe-filled CNTs also show excellent microwave absorption properties.Che[12] investigated the properties of microwave absorption and electromagnetic interference shielding of Fe-filled CNTs. The results shows the microwave absorption and shielding effect was enhanced substantially. The shape of Fe filled into CNTs is largely related to the shielding effect and microwave absorption. Zhao[13] investigate the Microwave absorbing property Ni-coated and Ag nanowires filled carbon nanotubes. CNTs were coated with nickel by an electroless plating technique, CNTs were filled with Ag nanowires via a wet chemical method and epoxy was used as substrate materials.

The microwave absorption properties of the filled CNTs were improved compared with the unmodified CNTs/epoxy composites. Microwave absorption peaks of Ni-coated CNTs/epoxy composites moved to the higher frequencies. For the Ag nanowires filled CNTs, the absorption curve illustrates that reflection loss of the corresponding composites is below -10 dB in the range of 7.2–9.0 GHz, and the minimum value is-19.19 dB at 7.8 GHz. The microwave absorbing peak of the composites moves to the low frequency by filling the Ag nanowires into the CNTs. This shows that the absorption peak frequency of the CNTs/epoxy composites can be manipulated easily by plating different nickel coatings onto the surface of CNTs or filling Ag nanowires into CNTs. It was considered that the microwave absorptions of composites containing CNTs or Ag nanowires filled CNTs result mainly from dielectric loss rather than magnetic loss. In contrast, the microwave absorption of Ni-coated CNTs/epoxy composites was attributed to both dielectric and magnetic losses. Li[14] used multiwalled carbon nanotubes(MWCNTs) and carbonyl iron (CI) as composite microwave absorption agents to investigate the microwave absorption properties of the composite materials. The results show the microwave absorption of the composites was substantially enhanced by adding as little as 1wt% of MWCNTs.The absorption peak of composite reached -22.2dB . the R (dB) values less than-20 dB was broadened from 1.2 (from 14.2 to13.0GHz for pure CI composites) to 2.4GHz (from 13.6 to 16.0GHz for the MWCNTs added CI composites) at the whole 2–18GHz band. The enhancement of the microwave absorption was considered from originating from the combination of MWCNTs and pure CI particles. Zhao[14] employ γ-ray to radiate the composite microwave absorption agent to improve electromagnetic properties of Ni-coated CNTs and achieve a positive results. The prepared MWCNTs/Ni composite was demonstrated to be a ferromagnetic material, and the relative complex permeability and permittivity were studied. The reflection loss of MWCNTs were greatly improved in the frequency range between 3.8GHz and 18 GHz, with a maximum absorption of −7.2 dB at 6.4 GHz. Zhan[15] employed Typical hydrothermal process to synthesize CNTs/Fe3O4 inorganic hybrid material for microwave absorption agent and measured the complex permittivity and permeability of the hybrid materials. The electromagnetic properties were greatly enhanced.Wang[16] prepared a new three element hybrid materials for microwave absorption. The iron-filled carbon nanotubes were coated with FeCo by an electroless plating method to prepare the microwave absorption materials and value the microwave absorption characteristics of them. The results demonstrate that the soft magnetic characteristics of iron-filled carbon nanotubes can be improved after being coated with FeCo alloy nanoparticles which results in more effective microwave absorption. Peng[17] used number simulation method to value the microwave characteristics of MWCNTs. The study focuses on the dielectric properties of CNTs/polymer composite with low CNTs concentration at low-frequency range of 50 MHz–3 GHz. Kim[18] investigated the microwave absorption characteristics of the sandwich constructions composed of CNTs/epoxy and PVC foam. The results show the reflection loss of the hybrid composite structure obtained from the free space measurement system in the X-band frequency range with the absorbing bandwidth of-10 dB was 3.3 GHz (8.2–11.5 GHz) and the maximum and minimum EM absorption rate were 97% and 84%, respectively. Zhang[19] investigated the dielectric, magnetic, and microwave absorbing properties of $Sm_2O_3$-filled multiwalled carbon nanotubes (MWCNTs) .The complex permittivity and permeability were measured at a microwave frequency range of 2–18 GHz. $Sm_2O_3$ nanoparticles encapsulated in the cavities of MWCNTs enhance the magnetic loss of MWCNTs. The calculated results indicate that the bandwith of the modified MWCNTs is much broader than that of unfilled MWCNTs. The maximum reflectivity (R) is about −12.22 dB at 13.40 GHz and corresponding bandwidth below −5 dB is more than 5.11 GHz. With the increase of thickness, the peak of R shifts to lower frequency. The multiple absorbing peaks appeared in the $Sm_2O_3$-filled CNT composites , which helps to broaden microwave absorbing bandwidth. Rosa[20] investigated thr multiphase composite materials filled with multiwall carbon nanotubes (MWCNTs), short nickel-coated carbon fibers and millimeter-long carbon fibers with various weight fractions . The effective complex permittivity of several composite samples is measured in the frequency range from 8 GHz to 18 GHz. The obtained results show that the addition of the MWCNTs into the mixture allows tuning the EM properties of the composite filled with the short nickel-coated fibers. Numerical simulations are also performed. The best performing screens in the Ku-band have thicknesses of about 2.13 mm and 1.57 mm, minimum reflection of about -73 dB and -45 dB and bandwidth of 6 GHz and 5 GHz, respectively. Fan[21] prepared CNT/polymer composites and measured the electromagnetic characteristics and microwave absorption properties of them in a frequency band of 20-18GHz. The test results demonstrated that the maximum absorbing value reached 17.61 dB and 24.27 dB with a loading of 4wt% and 8wt% of CNTs and the correspoding absorbing peak at 7.6 GHz, 15.3 GHz respectively. The dielectric loss was considered as main attribution to The microwave absorption of CNTs composites rather than magnetic loss.

Although many progresses have been achieved, the bandwidth and peak of RAM need to be improved further and enhanced for military and civil applications. In this paper, the microwave absortion characteristics of CNTs were further investigated. The raw MWCNTs, doped MWCNTs and aligned MMCNTs were used as microwave absorption agents and epoxy resin was used as matrix to pepare the composites to measure the properties of microwave absorption of CNTs.

## 2. Experiment

### 2.1 Synthesis of carbon nanotubes

#### 2.1.1 Growth of carbon nanotubes

Carbon nanotubes were grown by the catalytic decomposition of hydrocarbon precursor gas such as acetylene,ethylene, ethanol or methane.In this investigations, liquified petrolum gas (LPG) was used as carbon sources. Metal catalysts mainly include nickel, cobalt, iron, or a combination of them. The metal catalyst nanoparticles can be fabricated by sol-geo, co-deposition processes. The size of catalyst particles has big impact on the diameters of the nanotubes that are to be grown. The smaller particle grow smaller diameter of carbon nanotubes. In the paper, iron was employed as catalyst which play a crucial role in the nucleation and growth of carbon nanotubes in the thermal chemical vapor deposition (CVD) process. Nanotubes grow at the sites of the metal catalyst. The carbon-containing gas is broken apart at the surface of the catalyst particle, and the carbon is transported to the edges of the particle, where it forms the nanotubes. The substrates of catalyst include diatomite, MgO or $Al_2O_3$ to increase the surface area for higher yield of the catalytic reaction of the carbon feedstock with the metal particles. The removal of the catalyst support commonly employed an acid treatment method, which sometimes could destroy the original structure of the carbon nanotubes. In this investigation, carbon nanotubes themslef were used as substrates which result in higher purity of as-produced carbon nanotubes. The details of the MWNTs preparation method [ 22 ] were described elsewhere.

### 2.1.2 Growth of aligned carbon nanotubes

Ferrocene was dissolved in solution of xylene used also as the carbon source. The flat quartz glass substrates were put on the middle of the electrical furnace. The furnace was heated to about 850 ℃. Argon gas was introduced to the furnace to eliminate air and then a mixture gas of argon and hydrogen was introduced as the carrier gas. The solution of ferrocene and xylene was injected into furnace and was vapored. The vapor then went into the reactive area where the ACNTs grew on surface of quartz glasses.

### 2.1.3 Preparation of the composites

#### 2.1.3.1 Preparation of MWCNT/epoxy composites

Carbon nanotubes with different loading weight of CNTs were respectively added into epoxy resin and are were sufficiently mixed by high-speed stirring dispersion and ultrasonication. The mixture was smeared onto an aluminum plate layer by layer until the thickness of the composites reached 4mm.

#### 2.1.3.2 Preparation of ACNTs/epoxy composites

The films of ACNTs were peeled off from the quartz glasses. Then the films were put on the surface of a aluminum plate with size of 180X180mm.and were fixed by epoxy resin to prepare the samples. After curled, the microwave absorption characteristics of the composites were measured.

### 2.2 CNTs characterization

In this work, a FEI Quanta 200 scanning electronic microscopy (SEM) and a Hitachi H-600 transmission electronic microscopy (TEM) were used to observe the structure and morphology of CNTs. A TG/DTA Pyris diamond was employed for thermogravimetric analysis to ascertain the purity of carbon nanotubes.

### 2.3 Microwave absorption characteristics of CNTs

Microwave absorption characteristics of CNT/epoxy composites are measured through radar absorption materials (RAM) measuring system of arch method reflectivity.

## 3. Results and discusion

Fig.1 is a TEM image of CNTs and Fig.2 is a SEM image of CNTs. The images demonstrate the CNTs are multiwalled carbon nanotubes(MWCNTs) and have a diameter of 10-30nm with a length of about several micrometers. The images also show that the MWCNTs were entangled with each other to become clusters that are the the most main challenge to meet for effective industrial applications for most fileds. Specially for composite materials. Fig.3 shows the TEM image of dispersed ACNTs. Fig.4 and Fig.5 show SEM images of Low-magnifying and high- magnifying SEM image of ACNTs. and Fig.6 is micro-image of ACNTs shot by common camera. Fig.7 demostrates the top morphology of ACNTs.

Fig.4 and Fig.5 show that ACNTs were vertically grown on the substrates. The growth oriention are contained each other in growth and they only can vertically grown on the surface of substrates. So the carbon nanotube arrays were shaped. Fig.6 shows the top of ACNT film looks like the brushwood. Fig.3 shows that CNTs in ACNTs are multi-walled carbon nanotubes (MWCNTs). The MWCNTs have a diameter of 30~50 nm.

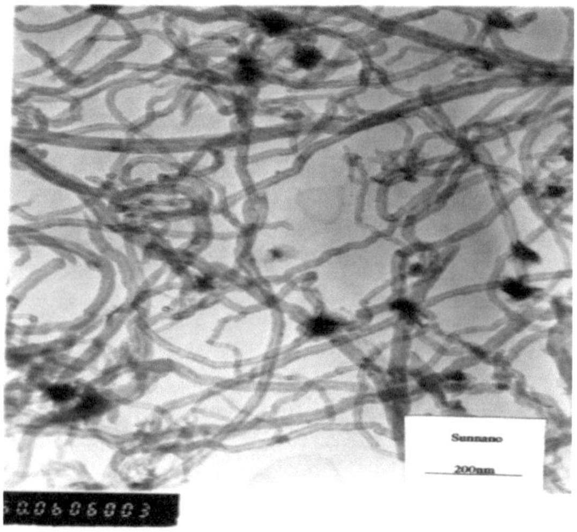


Fig. 1. TEM image of MCWNTs

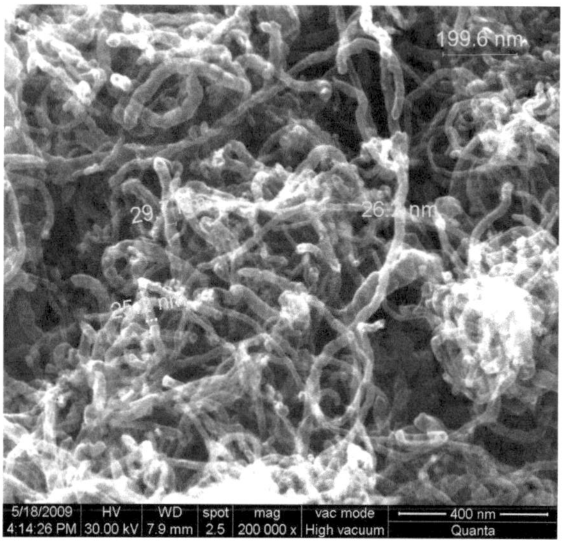


Fig. 2. SEM image of MCWNTs

The purity of CNTs reaches over 95%.The thermogravimetric analysis (TGA) of MWCNTs was shown in Fig.8. The MWCNTs have high thermal stability. It can be found that weight was lost slowly from 50 to 550°C, corresponding to the loss of little water and a few amorphous carbon. At the temperature range from 550 °Cto 700, the weight decreased sharply to 2.20 wt%, indicating that the combustion of the MWCNTs started at 550°C. Also a thing to note is that the curve slope maintained almost the same value from 550 to 700°C. It is illustrated that the MWCNTs were combusted at a constant speed, suggesting that the MWCNTs reached to a high purity at 550°C. After 700°C, the weight of the sample remains unchanged. The remainders may be assigned to catalyst including 2.20 % iron and aluminium metal oxides. Therefore, 550°C can be considered as an optimum temperature to burn out amorphous carbon particles for CNT purification. The figure demonstrate the purity of measured CNTs reached 97.80%.

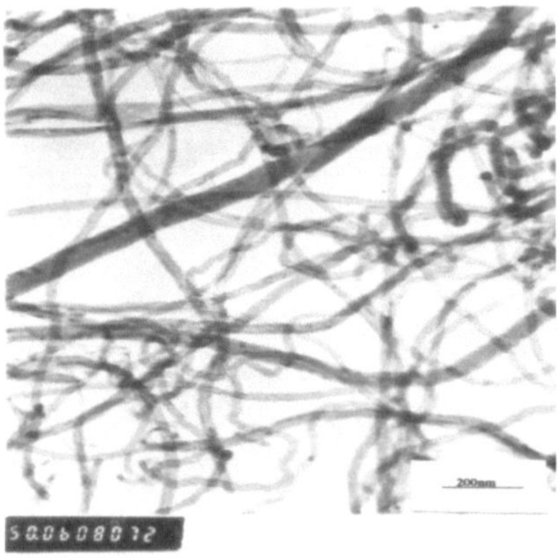

Fig. 3. TEM image of dispersed ACNTs

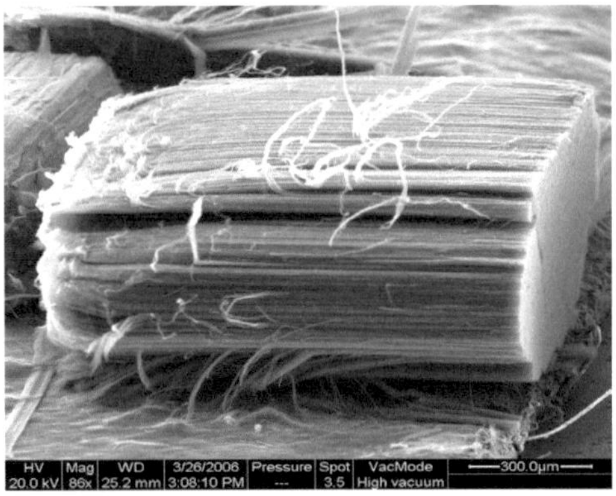


Fig. 4. Low-magnifying SEM image of ACNTs

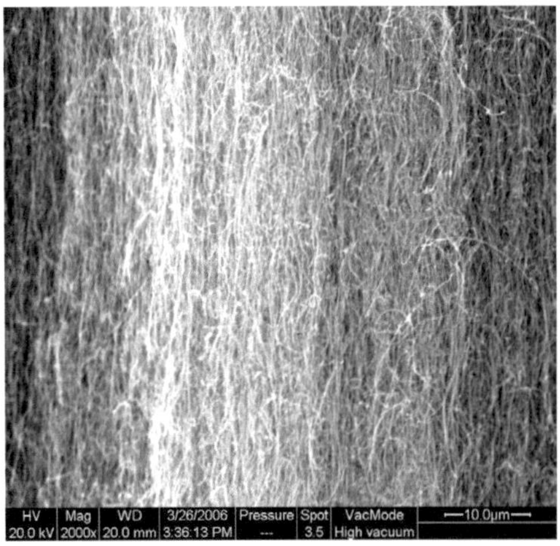


Fig. 5. SEM image of ACNTs

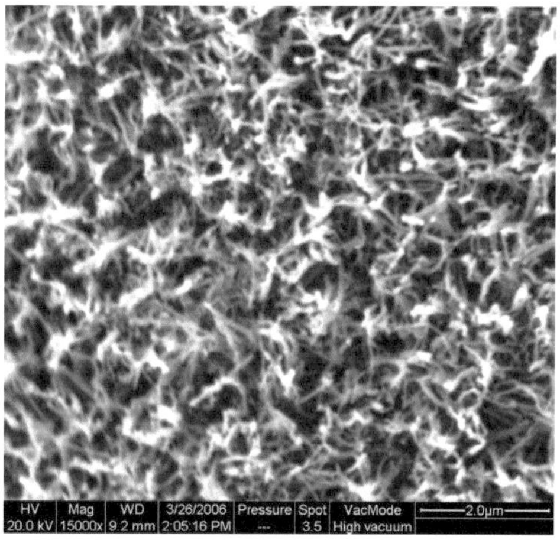


Fig. 6. Top SEM image of ACNTs

Fig. 7. Macro image of ACNTs

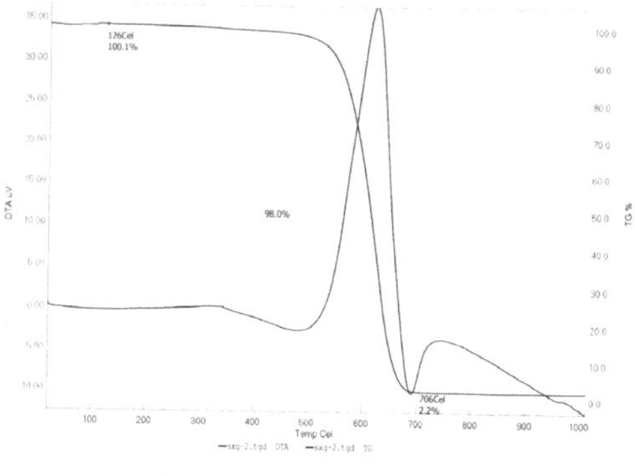


Fig. 8. Thermogravimetric curves of MWCNTs

The Fig.9 shows SEM image of cross-section of CNT/epoxy composites. The figures demostrate the MWCNTs were well dispersed in epoxy resin matrix but some entangled MWCNTs can still be observed in the epoxy resim matrix.

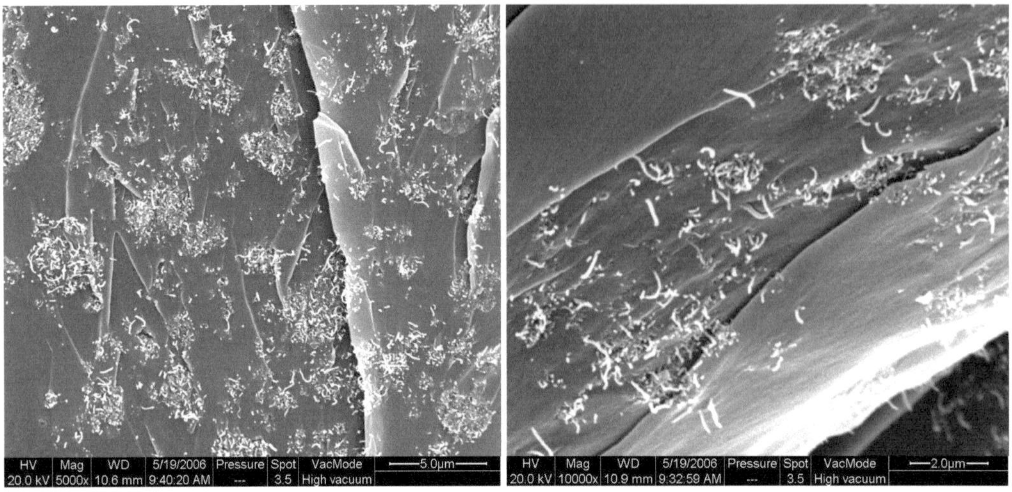


Fig. 9. SEM image of cross-fracture of MWCNT/epoxy composites

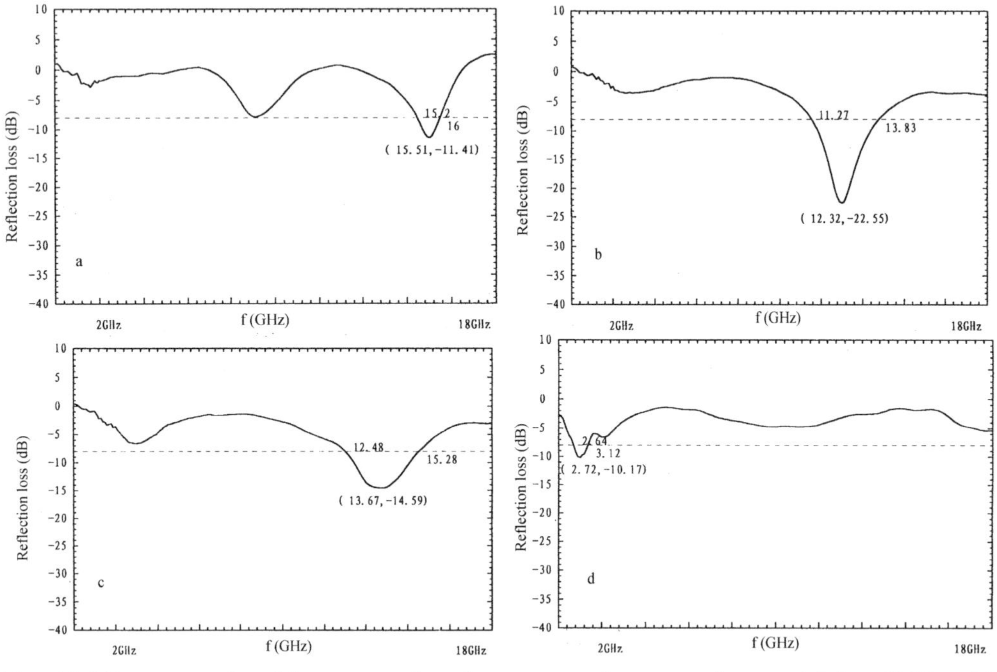


Fig. 10. Microwave absorbing properties of epoxy matrix composites containing CNTs with different loading of (a)5wt%, (b) 8wt%, (c) 10wt% and (d) 20wt%.

Fig.10 shows the reflection loss curves of different MWCNT/epoxy composites. The figures of reflecton loss show carbon nanotubes can effectively absorb radar waves in frenquency band of 2-18GHz.The absorbing peak, absorbing peak value and bandwidth change with different loading of carbon nanotubes. The wave absorption paprameters were showed in table.1. The sample b obtained the largest absorbing peak value of -22.55 dB at 12.32 GHz with a bandwidth of 2.56 GHz (R<-10 dB) and 4.16 GHz (R<-5 dB) respectively.

The sample c obtained the largest bandwidth of of 2.80 GHz（R<-10 dB）and 6.24 GHz（R<-5 dB）respectively. Another important characteristic, as demonstrated in Fig10a,c,d is double absorbing peaks, The second largest peak value reach -7.73 dB at 9.2 GHz，-6.62 at 4.40 Ghz and -5.39 dB at 17.80 GHz for sample a, c,d. The duble absorbing peaks is a peculiar properties of carbon nanotubes which can improve wave absorbing properties and enhance the bandwidth.

| Samples No. | CNT (%) | Depth (mm) | peak value (R/dB) | absorbing peak (f/GHz) | bandwidth (f/GHz) R<-10dB | R<-5dB |
|---|---|---|---|---|---|---|
| a | 5 | 8 | -11.41 | 15.51 | 0.79 | 3.2 |
| b | 8 | 8 | -22.55 | 12.32 | 2.56 | 4.16 |
| c | 10 | 8 | -14.59 | 13.67 | 2.8 | 6.24 |
| d | 20 | 8 | -10.17 | 2.72 | 0.47 | 2.56 |

Table 1. Radar waves absorbing properties of CNT/EP composites

Fig.11 shows microwave absorption characteristics of rare earth doped MWCNT composites (a) 5wt% CNT and 1% CeO,and (b) 8wt% CNT and 1% CeO. The Fig.11 indicated the radar waves absorbing properties of carbon nanotubes were substantially improved after modification by rare earth oxides. The highest absorbing peak of the sample reachs -29.10 dB at 10.88 GHz and the bandwidth reaches 7.68 GHz（R<-10dB）loaded with 1wt% rare earth and 8% carbon nanotubes compared with undoped MWCNT with the absorbing peak value of 12.32dB and bandwidth of 2.80 GHz（R<-10 dB）. The peak value increased by 16.87 dB and bandwidth increased by 4.88GHz.

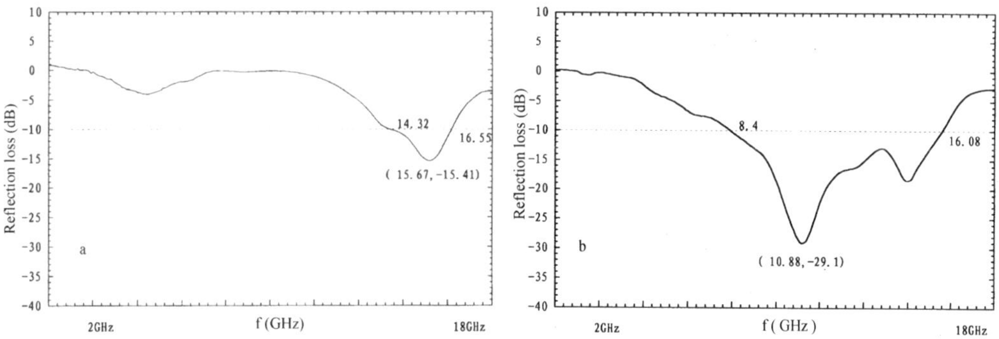


Fig. 11. Microwave absorption characteristics of rare earth dopped MWCNT composites.

Rare earth elements are of great importance in magnetic, electronic, and optical materials because of the number of unpaired electrons in their 4f shells. The unique electronic, optical and chemical properties make them useful in radar wave absorption. The rare earth 4f shells were not fully filled. This results in the magnetic moment. The magnetic moment of rare

earth atoms or ions are related with not only the factor $g_l$ but although angular momentum J. These unique characteristics enhance largely the microwave absorption of CNTs.
Fig.12 demostrates microwave absorption characteristics of aligned carbon nanotubes with different film thickness of (a) 20μm, (b) 2mm and (c) 1mm. The microwave absorption parameters were listed in table2.

| Samples | Depth (mm) | Absorbing peak (R/dB) | Absorbing peak (f / GHz) | Frequency (f / GHz) R<-10dB | Width R<-5dB |
|---|---|---|---|---|---|
| a | 0.2 | -15.87 | 17.83 | 4.25 | 6.40 |
| b | 1 | -10.02 | 17.83 | 0.16 | 4 |
| c | 2 | -7.54 | 13.80 | 0 | 6.40 |

Table 2. Microwave absorption parameters of ACNTs

As showed in Fig.4 and Fig.5, the aligned carbon nanotubes were vertically grown on the substrates in oriention. they are restricted in growth oriention and only can vertically be grown on the surface of substrates and so the array of carbon nanotubes was shaped. The carbon nanotubes in array parallels each other and resulted in anisotropy of aligned carbon nanotubes. Fig.6 shows the top of ACNT films looking like the hassock. This morphology is similar to jungle structure of radar wave absorption disguise and is beneficial to microwave absorption[23]. When electromagnetic wave reaches the top surface of ACNTs, the part was reflected back into air and other enters into ACNT jungle. The electromagnetic wave was hold in the hassock of ACNTs and was absorbed and exhausted in repeat reflection.
The microwave absorption curves and absorption parameters of ACNTs show ANTs have excellent microwave absorption properties in higher band in 2~18GHz. The absorbing peak and bandwidth change with different thickness of films of ACNTs. The absorption peak of the sample A reaches -15.87dB at 17.83 GHz and has a bandwidth of 4.25 GHz（R<-10dB） and 6.40GHz （R<-5 dB）. The absorption peak of the sample B reaches -10.02 dB at 17.83 GHz and has a bandwidth of 0.16 GHz（R<-10dB） and 4.00GHz （R<-5 dB）. The absorption peak of the sample C reaches -7.54 dB at13.80 GHz and has a bandwidth of 5.40 GHz （R<-5 dB）.
The test results indicate the absorption peak of ACNTs decreases with the increase of thickness of film of ACNTs. The peak decreases from -15.87 dB (sample A) to -7.54 dB (Fig.c)。The peak is situated at higher band of 2～18 GHz. This may be resulted from the increase reflection rate of microwave on thicker film of ACNTs. The microwave absorption bandwidth of ACNTs also changes with thickness of films of ANTs. The bandwidth of reflection loss less than -10dB decreases from 4.24 GHz (sample A) to zero (Fig.c). The bandwidth of reflection loss less than -5dB first decreases from6.40 GHz (Fig.a) to 4.00 (Fig.b) and then again increase 6.40 GHz (Fig.c). These do not exhibits obvious trend.
Nanomaterials have a new wave absorption mechanism from the effects of small size, surface and quanta size. The quanta size effect brings out the breakage of periodic boundaries and make the characteristics of voice, light, electron, magnet and energetics

changed , so they substantially enhance the properties of microwave absorption of nanomaterials. CNTs have vast surface, high rate atoms of surface and substantive hanging-bonds. These effect of interface polarization and mulriple dispersion improve the electromagnetic wave absorption of CNTs. The quanta size effect makes energy level of electrons to be splitted. The gaps of energy just contain microwave energy level （$10^{-2}\sim10^{-5}$ eV）,and so produces a new way of wave absorption. Further more, CNT has two kinds of structures including the symmetry and chirality. The chiral structures enhance the radar wave absorption performance of CNTs. CNTs were dispersed into polymer matrix to form electrically conductive network in polymer. The energy of electromagnetic waves was also attenuated in network resistors, which is similar to the resistive type of wave absorption materials.

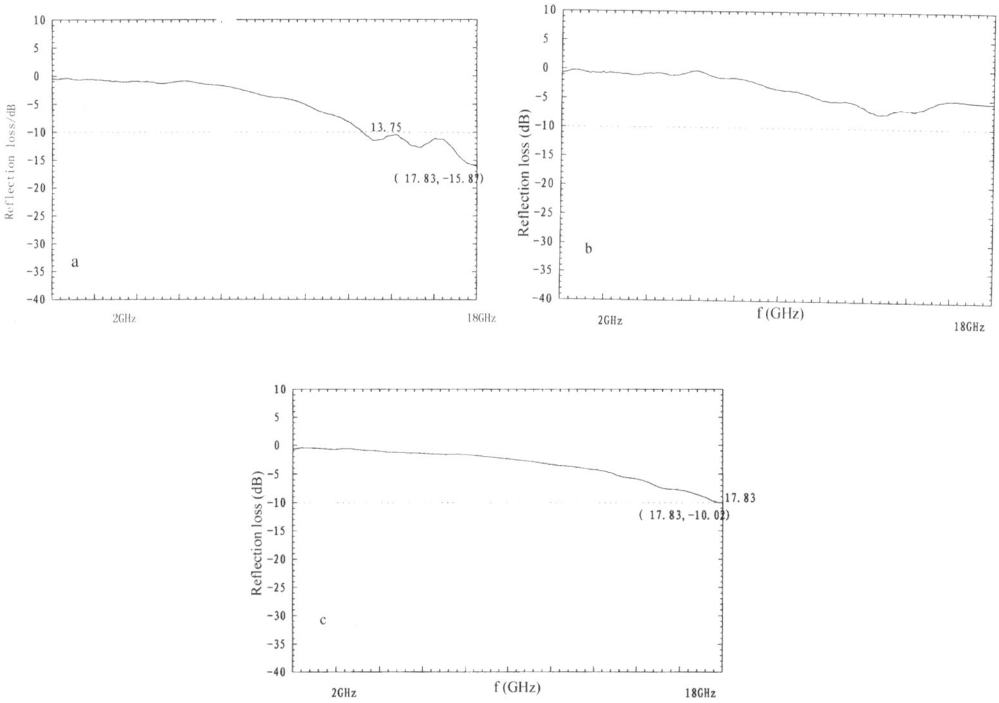


Fig. 12. Microwave absorption characteristics of aligned carbon nanotubes with different film thickness of (a) 20μm, (b) 2mm and (c) 1mm.

Rare earth elements are of great importance in magnetic, electronic, and optical materials because of the number of unpaired electrons in their 4f shells. The unique electronic, optical and chemical properties make them useful in radar wave absorption. The rare earth 4f shells were not fully filled. This results in the magnetic moment. The magnetic moment of rare earth atoms or ions are related with not only the factor $g_1$ but although angular momentum J. These unique characteristics enhance largely the microwave absorption performance of CNTs.

## 4. Conclusion

Summary, carbon nanotubes including multiwalled carbon nanotubes, rare earth doped carbon nanotubes and aligned carbon nanotubes were investigated for stealth applications. The reaserch demostrated that carbon nanotubes have good microwave absorption properties. The rare earth doped carbon nanotubes showed better microwave absorption properties. CNTs doped with rare earth demonstrate good improvement of microwave absorption properties. The absorbing peak of the CNTs doped with 1wt% rare earth reached -29.10 dB at 10.88 GHz and the bandwidth reached 7.68 GHz（R<-10dB）loading with 8% carbon nanotubes compared with undopped MWCNT with the absorbing peak value of 12.32dB and bandwidth of 2.80 GHz（R<-10 dB）. The peak value increased by 16.87 dB and bandwidth increased by 4.88GHz.

Alinged carbon nanotubes showed good microwave absorption prperties in higher band of 2~18 GHz. The microwave absorption peak value of ACNTs decreased with the increase of thickness of film of ACNTs. The peak value decreased from -15.87 dB to -7.54 dB. The peak is situated at higher band of 2～18 GHz. The microwave absorption bandwidth of ACNTs also changes with the thickness of films of ANTs. The bandwidth of reflection loss less than -10dB reached 4.24 GHz.

Carbon nanotbes demonstrate big potential in industrial applications for microwave absorption. The futher investigation was needed to devolop the applications, specailly for stealth arms.

## 5. References

[1] Sumio Iijima. Helical Microtubules of Graphitic Carbon. Nature. 1991, 354:56-58

[2] M.R.Falvo, G. J. Clary, R M Taylor, et al. Bending and Buckling of Carbon Nanotubes Under Large Strain.Nature. 1997, 389:582.

[3] J P Salvetat, J M Bonard, N H Thomson, et al. Mechanical Properties of Carbon Nanotubes. Appl. Phys. A. 1999, 69: 255-260.

[4] B G Demczyk, Y M Wang, J Cumings, et al. Direct Mechanical Measurement of the Tensile Strength and Elastic Modulus of Multiwalled Carbon Nanotubes. Material Science and Engineering .2002, A334:173-178.

[5] Savas Berber, Young-Kyum Kwon, David Tomanek. Unusually High Thermal Conductivity of Carbon Nanotubes. Phys.Rev.Lett. 2000,84:4613-4616.

[6] Ebbesen T W, Lezec H J, Hiura H, et al. Electrical Conductivity of Individual Carbon. Nature. 1996, 382:54-56.

[7] Z K Tang, Lingyun Zhang, N Wang, et al. Superconductivity in 4 Angstrom Single-walled Carbon Nanotubes. Science. 2001, 292:3462-3465.

[8] SHEN Zeng-min, ZHAO Dong-lin. Study on the Microwave Absorbing Poperty of Composite Material Containing Carbon Nanotubes with Ni Coating. New Carbon Materials, 2001,16(1):1-3.

[9] CAO Mao-sheng, GAO Zheng-juan, ZHU Jing. Materials Engineering, 2003,2:34-36.

[10] Haiyan Lin, Hong Zhu *, Hongfan Guo, Liufang Yu. Microwave-absorbing properties of Co-filled carbon nanotubes. Materials Research Bulletin 43 (2008) 2697–2702.

[11] Haiyan Lin, Hong Zhu , Hongfan Guo, Liufang Yu. Investigation of the microwave-absorbing properties of Fe-filled carbon nanotubes. Materials Letters 61 (2007) 3547–3550.

[12] Renchao Che, Lian-Mao Peng,Xiao Feng Duan, et al. microwave absorption enhancement and conplex permitivity and permeability of Fe encapsulated within carbon nanotubes. Adv.Mater2004,16, No.5:401-404.

[13] Yong Li , Changxin Chen , XiaoyanPan , Yuwei Ni, Song Zhang , Jie Huang , Da Chen , Yafei Zhang. Multiband microwave absorption films based on defective multiwalled carbon nanotubes added carbony liron/acrylicresin. PhysicaB,(2009):1-4.

[14] Hongtao Zhao, Xijiang Hana, Miaofei Han, Lifang Zhang, Ping Xu. Preparation and electromagnetic properties of multiwalled carbon nanotubes/Ni composites by γ-irradiation technique. Materials Science and Engineering B 167 (2010) 1–5.

[15]YingqingZhan,RuiZhao,YajieLei,FanbinMeng,JiachunZhong,Xiaobo Liu. A novelcarbonnanotubes/Fe3O4 inorganichybridmaterial:Synthesis,Characterizationandmicrowaveelectromagnetic properties. .Magn.Magn.Mater.(2010):1-5.

[16] Chen Wang, Ruitao Lu, Feiyu Kang, Jialin Gu, Xuchun Gui, DehaiWu. Synthesis and application of iron-filled carbonnanotubes coated with FeCo alloynanoparticles. Journal ofMagnetismand Magnetic Materials 321 (2009) 1924–1927.

[17] Peng Zhihua, Peng Jingcui, Peng Yanfeng, Ou Yangyu, Ning Yantao. Complex permittivity and microwave absorption properties of carbon nanotubes/polymer composite: A numerical study. Physics Letters A 372 (2008) 3714–3718.

[18] Po Chul Kim, Dai Gil Lee. Composite sandwich constructions for absorbing the electromagnetic waves. Composite Structures 87 (2009) 161–167.

[19] Lan Zhang a, Hong Zhu. Dielectric, magnetic, and microwave absorbing properties of multi-walled carbon nanotubes filled with Sm2O3 nanoparticles. Materials Letters 63 (2009) 272–274.

[20] Igor Maria De Rosa, Adrian Dinescu , Fabrizio Sarasini, Maria Sabrina Sarto, Alessio Tamburrano. Effect of short carbon fibers and MWCNTs on microwave absorbing properties of polyester composites containing nickel-coated carbon fibers. Composites Science and Technology 70 (2010) 102–109.

[21] Zhuangjun Fan, Guohua Luo, Zengfu Zhang, Li Zhou, Fei Wei. Electromagnetic and microwave absorbing properties of multi-walled carbon nanotubes/polymer composites. Materials Science and Engineering B 132 (2006) 85–89.

[22] Xiaoshu Zeng ,Xiaogang Sun, Guoan Cheng, Xiaosong Yan,Xianliang Xu. Production of multi-wall carbon nanotubes on a large scale. Physica B 323 (2002) 330–332.

[23] C.X. Hu, Z.H. Wang, W.Z. Li, Stealth coating technology, first ed., Chemical Industrial Publishing, Bei Jing, 2004.

# 14

# Structural Instability of Carbon Nanotube

I-Ling Chang
*National Cheng Kung University*
*Taiwan*

## 1. Introduction

Since Iijima reported MWCNTs in 1991, CNTs have captured the intensive attention of researchers worldwide due to the combination of their expected structural perfection, small size, low density, high stiffness, high strength, and excellent electronic properties. CNTs have been widely adopted as microscopic probing tips (Dai et al., 1996; Hafner et al., 2001), nanocomposites reinforcements (Bower et al., 1998; Jin et al., 1998), nanotweezers (Kim & Lieber, 1999), and nanoactuators (Baughman et al., 1999; Fennimore et al., 2003) due to their slender and high aspect ratio structures. Meanwhile, nanotubes are also highly susceptible to buckling under compression, which is a structural instability. Once the buckling of CNTs occurs, the load-carrying capability would suddenly reduce and lead to possible catastrophic failure of the nanotubes, which significantly limit the loading strengths of the probing tips and compressive strengths of nanocomposite structures. Even the physical properties such as conductance of carbon nanotube can be influenced by the occurrence of buckling (Postma et al., 2001). Hence, it is crucial to understand the mechanism of nanotube buckling and even predict the onset of buckling in order to improve the nanotube applications.

A review of the relevant literature shows that significant studies have employed both experimental (Falvo et al., 1997; Iijima et al., 1996; Thostenson & Chou, 2004; Waters et al., 2004) and theoretical (Ru, 2000; Yakobson & Avouris, 2001) approaches to investigate the bucking behaviors of CNTs. However due to the difficulties encountered at nanoscale, the experimental investigation of the buckling behaviors of CNTs remains a challenging problem and individual factors that affect buckling could not be easily identified. In theoretical study, the CNTs are commonly treated as beams or thin-shell tubes with certain wall thickness and elastic constants and, thus, it is difficult to consider the chirality and size effects on buckling behavior of CNTs because the continuum assumption disregards the discrete nature of atomic structures (Ru, 2000; Yakobson & Avouris, 2001). Some researchers attempted to introduce the atomic-continuum method combining the atomic detail in the continuum description and examine the various properties of CNTs (Chang, 2004; Guo et al., 2008; Li & Chou, 2003a, 2003b). The atomic-continuum method could shorten the computational time in larger atomic system.

As the fast development and rapid advancement of computers, molecular approaches have become important tools and are widely applied to study the factors that would influence the buckling of CNTs (Buehler et al., 2004; Cao & Chen, 2006a, 2006b; Huh & Huh, 2008; Liew et al., 2004; Ozaki et al., 2000). Although some researchers already discussed various aspects of the CNT buckling behavior, systematic analysis on the effect of geometry (i.e. radius, length and length-to-radius ratio) and chirality (i.e. armchair, zigzag and chiral) on the buckling mechanism is still lacking to the best of our knowledge. Consequently, the present study employs MD simulations based on the Tersoff many-body potential function (Tersoff, 1986, 1988, 1989) to perform a systematic and comprehensive investigation into the buckling behaviors of single-walled CNTs under uniaxial compressive displacement loading. Besides, the applicability of the continuum buckling theory, which has been well developed for thin tubes, on predicting the buckling behavior of the CNT will also be examined.

## 2. Methodology

Atomic model of the single-walled CNT with radius $r$ and axial length $L$ are illustrated in Fig. 1. All simulations are performed at room temperatures, 300$K$ using a rescaling method and Newton's equations of motion are solved using a fifth order Gear's predictor-corrector algorithm. The empirical Tersoff many-body potential (Tersoff, 1986, 1988, 1989), which is commonly adopted in CNTs molecular simulation studies to provide quick estimation and significant insight into the thermo-mechanical behavior, is employed to describe the interatomic interaction between the carbon atoms. The force acting on an individual atom is obtained by summing the forces contributed by the surrounding atoms. The initial atomic models of CNT will be relaxed under NTP ensemble for 10000 time steps with a 1 *fs* step size to make sure the nanotubes reach their equilibrium states.

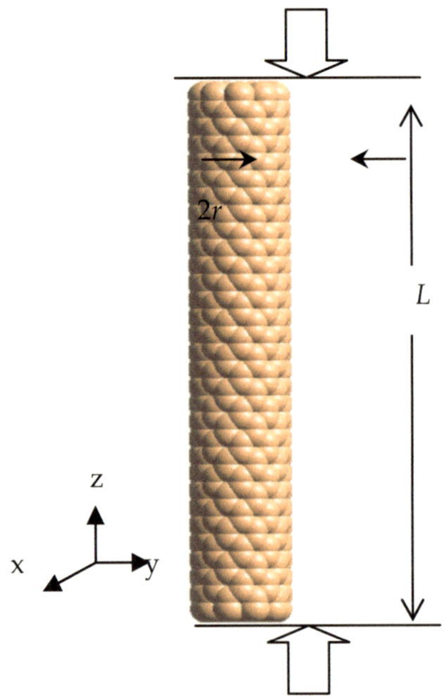


Fig. 1. The schematic presentation of the CNT under uniaxial compression.

During the simulations, the nanotube is compressed uniaxially and incrementally along $z$ direction by a uniform strain under NTV ensemble. The periodic boundary condition (PBC) is applied in the axial direction, which served as a loading mechanism to apply uniform displacements. Minimum image criterion is adopted to implement the periodic boundary condition. In order to reduce the loading strain rate, additional MD steps are applied for the relaxation of the CNT after each stage of the compression. The equilibrated configuration will be used as the initial state for the next loading step. In this study, the applied strain increment is 0.5%, then the CNT atomic system is relaxed for the interval of 10 *ps* and the corresponding strain rate is 0.05% $ps^{-1}$. The total energy, pressure and atomic configuration of the relaxed structure are monitored at each loading step to determine whether the buckling of the CNT occurs. Once a sudden jump in the total energy, as illustrated in Fig. 2, is observed, the atomic configuration of the nanotube would be inspected. Buckling strains (or called critical compressive strains) can be indicated from the jumps in the total energy and pressure.

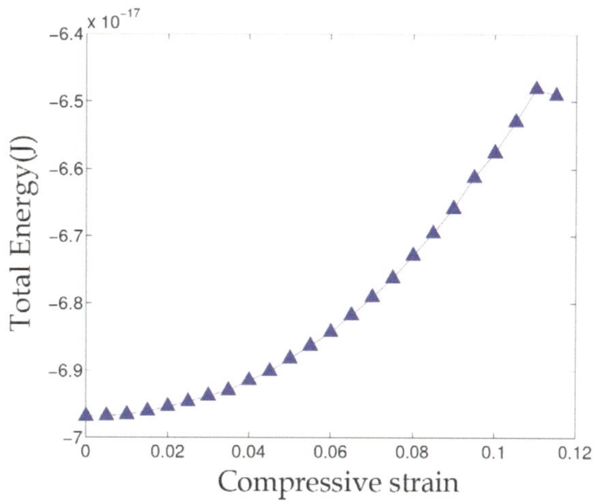


Fig. 2. The relation between the system energies and compressive strains for (5,5) CNT with length of 7.38Å.

CNTs can be considered as a graphene sheet rolled into a cylindrical shell and three distinct types of nanotubes could be classified as armchair, zigzag and chiral according to the way of rolling-up. In order to assess the influence of geometry and chirality on the buckling behavior of the single-walled CNTs, various radii, lengths, length-to-radius ratios and helical types of CNTs would be simulated and elucidate the dependence of buckling strain.

The length-to-radius ratio (or called slenderness ratio) is defined as $S.R. = \frac{L}{r}$. The radius and translation length, which is the smallest periodic axial distance, of $(m,n)$ CNTs can be calculated as

$$r = \frac{a\sqrt{3(n^2 + m^2 + mn)}}{2\pi}$$

$$T_{(m,n)} = 2\pi r \frac{\sqrt{3}}{d_R}$$

where $a$ is the interatomic C–C bond length, 1.42Å and $d_R$ is the greatest common divisor of $2n + m$ and $2m + n$. Since the periodic boundary condition is employed as a loading mechanism, the simulated CNT length must be multiple of the translation length so that the complete six-membered ring structure of the CNT could be preserved. The geometric parameters, radius *r* and length *L*, for different chiralities of CNTs are listed in Table 1-3, respectively. Similar radius, length and slenderness ratio for armchair, zigzag and chiral CNTs are chosen, if possible, for the simulation. Since the translation lengths for (6,4), (9.6) and (12,8) chiral CNTs are 18.569 Å, the choices of slenderness ratios for chiral CNTs are more limited.

| (*m*,*n*) | (5,5) | (8,8) | (10,10) |
|---|---|---|---|
| Radius(Å) | 3.39 | 5.424 | 6.78 |
| Length(Å) / *S.R.* | 7.38 / 2.18 | 12.3 / 2.27 | 14.76 / 2.18 |
| | 12.30 / 3.63 | 19.68 / 3.63 | 24.595 / 3.63 |
| | 36.89 / 10.88 | 59.03 / 10.88 | 73.79 / 10.88 |
| | 61.49 / 18.14 | 98.38 / 18.14 | 122.98 / 18.14 |
| | 86.08 / 25.39 | 137.73 / 25.39 | 172.165 / 25.39 |
| | 147.57 / 43.53 | 236.11 / 43.53 | 295.14 / 43.53 |

Table 1. The radii and lengths of the modeled armchair CNTs

| (*m*,*n*) | (9,0) | (14,0) | (17,0) |
|---|---|---|---|
| Radius(Å) | 3.523 | 5.48 | 6.655 |
| Length(Å) / *S.R.* | 8.52 / 2.42 | 12.78 / 2.33 | 17.04 / 2.56 |
| | 12.78 / 3.63 | 21.3 / 3.89 | 25.56 / 3.84 |
| | 38.34 / 10.88 | 59.64 / 10.88 | 72.42 / 10.88 |
| | 63.9 / 18.14 | 97.98 / 17.88 | 119.28 / 17.92 |
| | 89.46 / 25.39 | 140.58 / 25.65 | 170.4 / 25.60 |
| | 153.36 / 43.53 | 238.56 / 43.53 | 289.68 / 43.53 |

Table 2. The radii and lengths of the modeled zigzag CNTs

| (*m*,*n*) | (6,4) | (9,6) | (12,8) |
|---|---|---|---|
| Radius(Å) | 3.413 | 5.119 | 6.825 |
| Length(Å) / *S.R.* | 18.569 / 5.44 | 18.569 / 3.62 | 18.569 / 2.72 |
| | 37.138 / 10.88 | 55.707 / 10.88 | 74.276 / 10.88 |
| | 55.707 / 16.32 | 92.845 / 18.14 | 129.982 / 18.92 |
| | 92.845 / 27.20 | 129.982 / 25.39 | 167.712 / 24.57 |
| | 148.55 / 43.52 | 222.83 / 43.53 | 297.10 / 43.53 |

Table 3. The radii and lengths of the modeled chiral CNTs

## 3. Results

MD approaches are utilized to simulate the uniaxial compression test of CNTs with different geometries and chiralities. The size and chirality effects on buckling behaviors are studied and systematically compared through the critical compressive strains. It is noticed that two distinct types of buckling configurations are commonly observed as shown in Fig. 3 (a) and (b). One is shell wall buckling with kinks on the wall with the tube's centerline remaining straight, which is a local instability and the other is column buckling, which buckle as a whole. Moreover, there is another intriguing type of initial buckling configuration, which starts with a bump on the wall as shown in Fig. 3(c), and the CNT will eventually turn into column or shell wall buckling. It is also observed that the buckling kinks or bends appear mostly near the middle of the nanotubes, which indicates the loading mechanism does not impose any extra constraint on the two ends. Unlike the velocity controlled loading (Jeng et al, 2004), the drastic deformation starts from the imminent places near the ends due to Poisson's effect. Hence, it is difficult to separate the boundary constraint effect from the other investigating factors. It is also noted that the radial distribution function before and after buckling is quite similar as shown in Fig. 4, which implies that the bond structures do not change due to buckling.

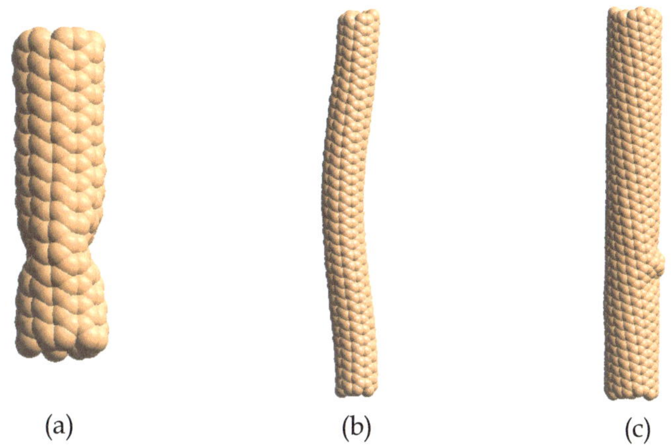


Fig. 3. The buckled shapes of CNTs. (a) shell wall, (b) column and (c) bump on the wall.

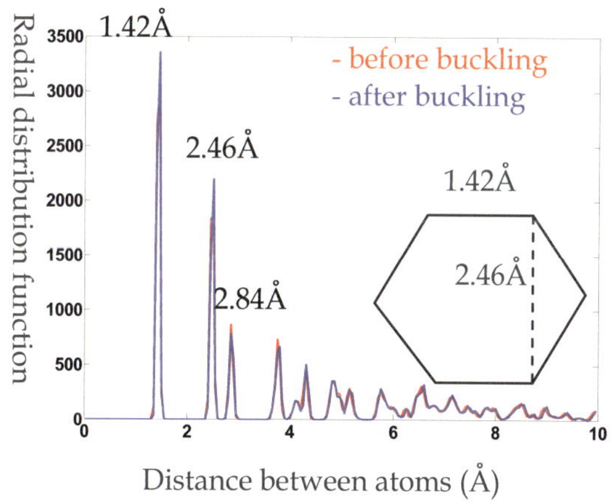


Fig. 4. The radial distribution functions before and after buckling.

The buckling strains for different chirality and slenderness ratio of CNTs are shown in Fig. 5, 6 and 7, respectively. The filled marks indicate column buckling and the empty one indicate shell wall buckling. In general, the nanotube's buckling behavior transits from shell wall type for a short tube to column type for a long tube of the same radius irrespective to the chirality of the CNTs. For slender nanotubes ( $S.R. \geq 20$ ), axial compression results in "global buckling" while the nanotubes undergo "local buckling" for stout nanotubes ( $S.R. \leq 10$ ). It is noticed that the slenderness ratio has strong influence on the buckling strains and the buckling strains decrease rapidly with the increase of slenderness ratio particularly for CNTs with smaller radii. Under similar length-to-radius ratio, it is noted that the buckling strain decreases as the radius of the CNT increases especially for CNTs with smaller slenderness ratios. It is observed that the chirality of the CNTs does not affect the buckling behaviors and buckling strains significantly.

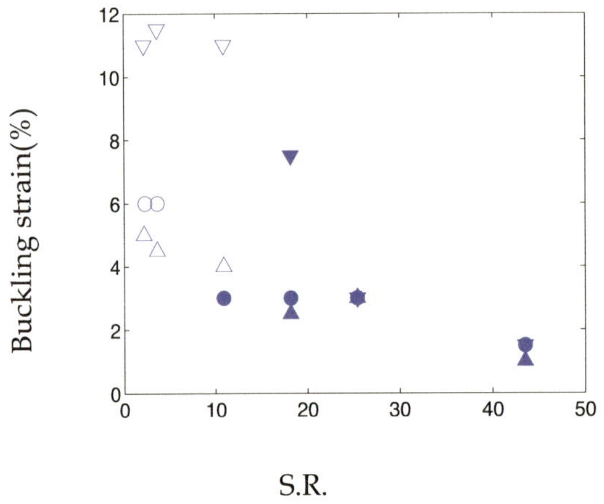


Fig. 5. The relationship between the buckling strain and slenderness ratio for armchair CNTs. ▽▼: (5,5), ○●: (8,8), △▲: (10,10) CNT.

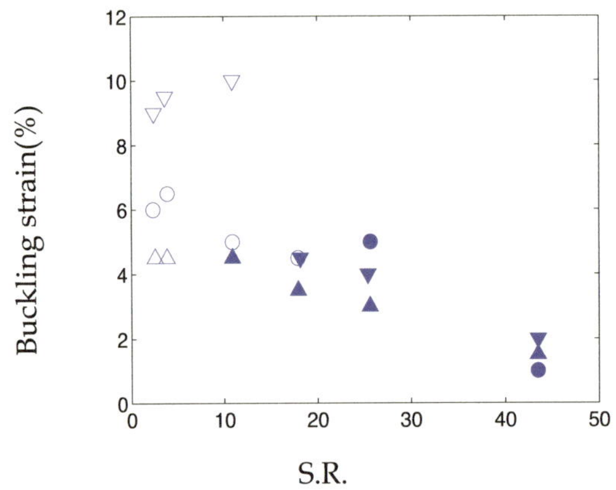


Fig. 6. The relationship between the buckling strain and slenderness ratio for zigzag CNTs. ▽▼: (9,0), ○●: (14,0), △▲: (17,0) CNT.

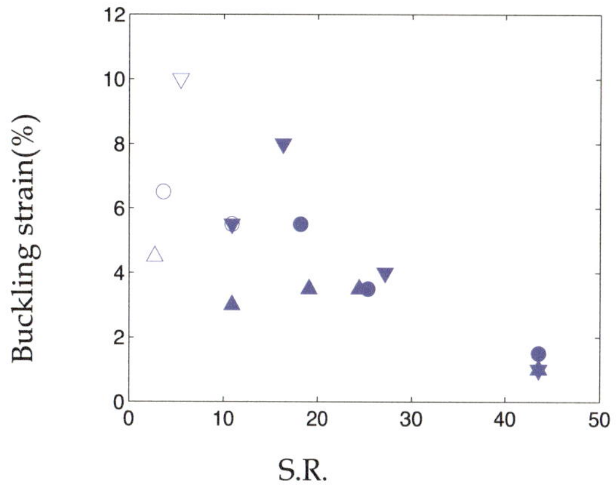


Fig. 7. The relationship between the buckling strain and slenderness ratio for chiral CNTs. ▽▼: (6,4), ○●: (9,6), △▲: (12,8) CNT.

## 4. Analysis and discussion

From current simulation results, it is very difficult to deduce a clear relationship between the critical compressive buckling strain and the CNT geometry and helical type, not to mention predicting the buckling type. In continuum mechanics, there already exist theories to calculate the buckling strain of thin shell tube depending on the buckling type in macroscopic scale. However, the applicability of continuum buckling theory in nanoscale is still an unsolved question. Hence, the continuum predictions on the buckling strains and the simulation results will be compared in order to examine the applicability.

By using Euler-Bernoulli beam theory, the critical buckling strain of a tube with both ends clamped displaying the column-like buckling behavior can be described as (Landau et al., 1986)

$$\varepsilon_{cr}^{column} = \frac{\pi^2\left[\left(r+\frac{t}{2}\right)^2-\left(r-\frac{t}{2}\right)^2\right]}{L^2}. \quad (1)$$

where $t$ is the thickness of the tube. If $t \ll r$, the equation could be rewritten as

$$\varepsilon_{cr}^{column} = 2\left(\frac{\pi r}{L}\right)^2 \quad (2)$$

It is noticed that the critical compressive buckling strain decreases as the inverse square of the slenderness ratio for column-type buckling. On the other hand, the critical compressive buckling strain of the tube displaying the shell-like buckling behavior is (Libai & Simmonds, 1998; Timoshenko & Gere, 1961)

$$\varepsilon_{cr}^{shell} = \frac{1}{\sqrt{3\left(1-\nu^2\right)}}\frac{t}{r} \approx 0.588\frac{t}{r} \quad (3)$$

The Poisson's ratio, v, of the CNTs is chosen as 0.19 (Yakobson et al. 1996). It is noted that the critical compressive strain of shell-like buckling depends only on the tube thickness and radius, but is independent of the tube length.

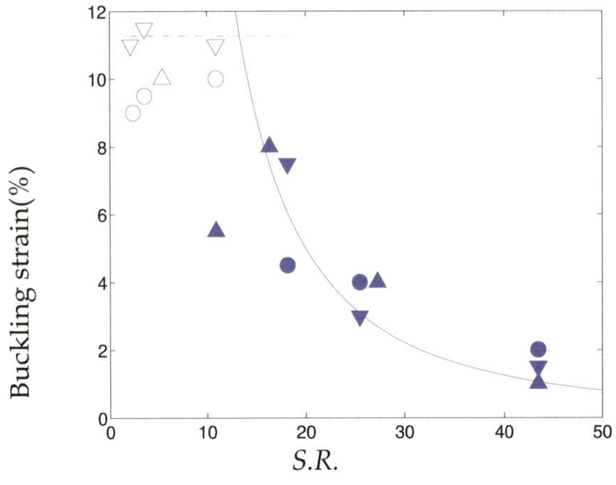


Fig. 8. The comparisons between the continuum predictions and molecular simulation results for CNTs. ▽▼:(5,5), ○●:(9,0) and △▲:(6,4)

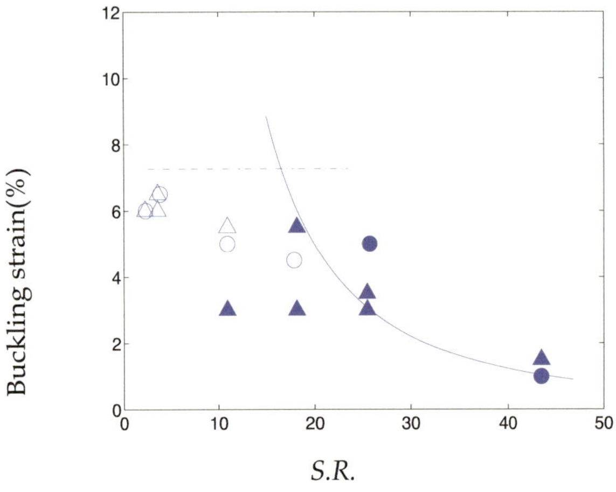


Fig. 9. The comparisons between the continuum predictions and molecular simulation results for CNTs. ▽▼:(8,8), ○●:(14,0) and △▲:(9,6)

Fig. 8, 9 and 10 illustrate the comparisons between the continuum predictions on the buckling strains and the simulation results for respective set of similar radii. One of the uncertainties of the continuum model is the effective nanotube thickness. Here in this research, the effective nanotube thickness is chosen as 0.66Å (Yakobson et al. 1996), which is roughly the atomic radius of carbon, instead of the commonly used CNT wall thickness, 3.4 Å. Since the nanotube thickness is not negligible as compared to the radius, Eq. (1) is adopted in the comparison. The dash lines illustrate the continuum prediction of shell-like buckling while the solid lines depict the prediction based on Euler-Bernoulli beam theory. It is observed that the buckling resistance does not show obvious dependence on the CNT chirality under similar radii. The continuum prediction of shell-like buckling can serve as an upper bound for predicting the shell wall buckling but significantly overestimate the buckling strains for nanotubes with higher length-to-radius ratio irrespective to the nanotube chirality. As the slenderness ratio becomes higher, the continuum prediction based on Euler-Bernoulli beam theory could capture the trend of column type buckling but also overestimate the buckling strains for nanotubes with smaller length-to-radius ratio. It is observed that those buckling strain which deviate significantly from the continuum prediction are at the intersection region of the two continuum theories. It is speculated that the difference in buckling strain between the continuum theories and simulation results at the transition region could be due to the competition between two buckling mechanisms.

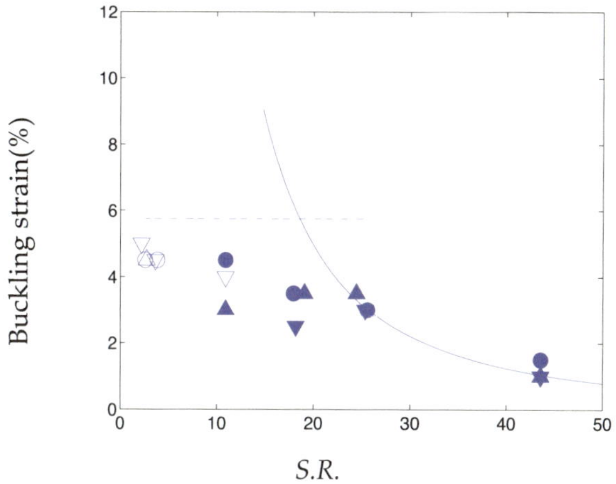


Fig. 10. The comparisons between the continuum predictions and molecular simulation results for CNTs. ▽▼: (10,10), ○●: (17,0) and △▲: (12,8)

## 5. Conclusion

In this work, molecular dynamics approach is employed to study the buckling behaviors of single-walled carbon nanotubes with different geometric sizes and chiralities under room temperature. Based on the MD simulation results, it is observed that the nanotube's buckling behavior transits from shell wall type for a short tube to column type for a long tube of the same radius irrespective to the chirality of the CNTs. Moreover, the buckling strain is getting smaller as the CNT becomes slender for most nanotubes, which implies that the slender nanotubes have lower buckling resistance. Under similar length-to-radius ratio, it is noted that the buckling strain decreases as the radius of the CNT increases especially for CNTs with smaller slenderness ratios. From the comparison with the prediction made by continuum buckling theories, it is concluded that the corresponding buckling strain and buckling type predicted by the continuum theory could agree reasonably well with MD simulations of the CNTs under compression except at the transition region. From the findings of this paper, it is suggested that the continuum buckling theory with proper choice of parameters, i.e., wall thickness and Poisson's ratio, could capture the trend of the buckling strain on the length-to-radius ratio disregarding to the helical types (i.e., armchair, zigzag and chiral) of the nanotubes and, hence, could serve as a primitive guideline in predicting the buckling strain of the CNTs.

## 6. Acknowledgment

This research work is supported by National Science Council of Taiwan under the grants NSC97-2221-E-194-015 and NSC98-2221-E-194-012-MY2. The support of AFOSR under Contract No. FA2386-09-1-4152 AOARD 094152 is also acknowledged. We are also grateful to the Taiwan National Center for High-performance Computing for computer time and facilities.

## 7. Reference

Baughman, R. H.; Cui, C. X.; Zakhidov, A. A.; Iqbal, Z.; Barisci, J. N.; Spinks, G. M.; Wallace, G. G.; Mazzoldi, A.; De Rossi, D.; Rinzler, A. G.; Jaschinski, O.; Roth, S. & Kertesz, M. (1999). Carbon Nanotube Actuators, *Science,* Vol.284, No.5418, (May 1999), pp.1340-1344, ISSN 0036-8075

Bower, C.; Rosen, R.; Jin, L.; Han J. & Zhou, O. (1999). Deformation of Carbon Nanotubes in Nanotube-Polymer Composites, *Applied Physics Letters,* Vol.74, No. 22, (May 1999), pp.3317-3319, ISSN 0003-6951

Buehler, M.; Kong, Y. & Gao, H. (2004). Deformation Mechanisms of Very Long Single-Wall Carbon Nanotubes Subject to Compressive Loading, *Journal of Engineering Materials and Technology*, Vol.126, No.3, (July 2004), pp.245-249, ISSN 0094-4289

Cao, G. & Chen, X. (2006a). Mechanisms of Nanoindentation on Single-Walled Carbon Nanotubes: The Effect of Nanotube Length, *Journal of Materials Research*, Vo.21, No. 4, (April 2006), pp.1048-1070, ISSN 0884-2914

Cao, G. & Chen, X. (2006b). Buckling Behavior of Single-walled Carbon Nanotubes and a Targeted Molecular Mechanics Approach, *Physical Review B*, Vol.73, No.16, (October 2006), pp.155435, ISSN 1098-0121

Chang, T. (2004). Buckling of Single-Walled Carbon Nanotubes via a Hybrid Atomic/Continuum Approach, *Acta Mechanica Sinica*, Vol.36, No.6, (November 2004), pp.744-748,

Dai, H.; Hafner, J. H.; Rinzler, A. G.; Colbert, D. T. & Smalley, R. E. (1996). Nanotubes as Nanoprobes in Scanning Probe Microscopy, *Nature*, Vol.384, No.6605, (November 1996), pp.147-150, ISSN 0028-0836

Falvo, M. R.; Clary, G. J.; Taylor, R. M.; Chi, V.; Brooks, F. P.; Washburn, S. & Superfine, R. (1997). Bending and Buckling of Carbon Nanotubes under Large Strain, *Nature,* Vol.389, No. 6651,( October 1997 ), pp.582-584, ISSN 0028-0836

Fennimore, A. M.; Yuzvinsky, T. D.; Han, W. Q.; Fuhrer, M. S.; Cumings, J. & Zettl, A. (2003). Rotational Actuators based on Carbon Nanotubes, *Nature*, Vol.424, No.6947, (July 2003), pp. 408-410, ISSN 0028-0836

Guo, X.; Leung, A.Y.T.; He, X.Q.; Jiang H. & Huang Y. (2008). Bending Buckling of Single-Walled Carbon Nanotubes by Atomic-Scale Finite Element, *Composites Part B-Engineering*, Vol.39, No.1, pp.202-208, ISSN 1359-8368

Hafner, J. H.; Cheung, C. L.; Woolley A. T. & Lieber, C. M. (2001). Structural and Functional Imaging with Carbon Nanotube AFM Probes, *Progress in Biophysics & Molecular Biology*, Vol.77, No.1, pp.73-110, ISSN 0079-6107

Huh, J. & Huh H. (2008). Effect of Helicity on the Buckling Behavior of Single-Wall Carbon Nanotubes, *International Journal of Modern Physics B*, Vol.22, No.31-32, (December 2008), pp.5872-5877, ISSN 0217-9792

Iijima, S.; Brabec, C.; Maiti, A. & Bernholc, J. (1996). Structural Flexibility of Carbon Nanotubes, *Journal of Chemical Physics*, Vol.104, No.5, (February 1996), pp.2089-2092, ISSN 0021-9606

Jeng, Y.-R.; Tsai, P.-C. & Fang, T.-H. (2004). Effects of Temperature and Vacancy Defects on Tensile Deformation of Single-Walled Carbon Nanotubes, *Journal of Physics and Chemistry of Solids*, Vol.65, No.11, (November 2004), pp.1849-1856, ISSN 0022-3697

Jin, L.; Bower, C. & Zhou, O. (1998). Alignment of Carbon Nanotubes in a Polymer Matrix by Mechanical Stretching, *Applied Physics Letters*, Vol.73, No.9,( August 1998), pp.1197-1199, ISSN 0003-6951

Kim, P. & Lieber, C. M. (1999). Nanotube Nanotweezers, *Science*, Vol.286, No.5447, (December 1999), pp.2148-2150, ISSN 0036-8075

Landau, L. D.; Pitaevskii, L. P.; Lifshitz, E. M. & Kosevich, A. M. (1986). *Theory of Elasticity*, Butterworth-Heinemann, ISBN 978-075-0626-33-0, Oxford, UK

Li, C & Chou, T.-W. (2003a). A Structural Mechanics Approach for the Analysis of Carbon Nanotubes, *International Journal of Solids and Structures*, Vol.40, No.10, (May 2003), pp. 2487-2499, ISSN 0020-7683

Li, C & Chou, T.-W. (2003b). Single-Walled Carbon Nanotubes as Ultrahigh Frequency Nanomechanical Resonators, *Physical Review B*, Vol.68, No.7, (August 2003), pp.073405, ISSN 1098-0121

Libai, A. & Simmonds, J. G. (2005). *The Nonlinear Theory of Elastic Shells*, Cambridge University Press, ISBN 978-052-1019-76-7, Cambridge, UK

Liew, K. M.; Wong, C. H.; He, X. Q.; Tan, M. J. & Meguid, S. A. (2004). Nanomechanics of Single and Multiwalled Carbon Nanotubes, *Physical Review B*, Vol.69, No.11, (March 2004), pp.115429, ISSN 1098-0121

Ozaki, T.; Iwasa, Y. & Mitani, T. (2000). Stiffness of Single-Walled Carbon Nanotubes under Large Strain, *Physical Review Letters*, Vol.84, No.8, (February 2000), pp.1712-1715, ISSN 0031-9007

Postma, H. W. C.; Teepen, T.; Yao, Z.; Grigoni, M. & Dekker, C. (2001). Carbon Nanotube Single-Electron Transistors at Room Temperature, *Science*, Vol.293, No.5527, (July 2001), pp.76-79, ISSN 0036-8075

Ru, C. Q. (2000). Effective Bending Stiffness of Carbon Nanotubes, *Physical Review B*, Vol.62, No.15, (October 2000), pp.9973-9976, ISSN 0163-1829

Tersoff, J. (1986). New Empirical-Model for the Structural-Properties of Silicon, *Physical Review Letters*, Vol.56, No.6, (February 1986), pp.632-635, ISSN 0031-9007

Tersoff, J. (1988). New Empirical-Approach for the Structure and Energy of Covalent Systems, *Physical Review B*, Vol.37, No.12, (April 1988), pp.6991-7000, ISSN 0163-1829

Tersoff, J. (1989). Modeling Solid-State Chemistry-Interatomic Potentials for Multicomponent Systems, *Physical Review B*, Vol.39, No.8, (March 1989 ), pp.5566-5568, ISSN 1098-0121

Thostenson, E. T. & Chou, T.-W. (2004). Nanotube Buckling in Aligned Multi-Wall Carbon Nanotube Composites, *Carbon*, Vol.42, No.14, pp.3015-3018, ISSN 0008-6223

Timoshenko, S. P. & Gere, J. M. (1961). *Theory of Elastic Stability*, McGraw-Hill, ISBN 978-048-6472-07-2, New York, USA

Waters, J. F.; Riester, L.; Jouzi, M.; Guduru, P. R. & Xu, J. M. (2004). Buckling Instabilities in Multiwalled Carbon Nanotubes under Uniaxial Compression, *Applied Physics Letters*, Vol.85, No.10, (September 2004), pp.1787-1789, ISSN 0003-6951

Yakobson, B. I. & Avouris, P. (2001). Mechanical Properties of Carbon Nanotubes, *Topics in Applied Physics*, Vol.80, pp.287-327, ISSN 0303-4216

Yakobson, B. I.; Brabec, C. J. & Bernholc, J. (1996). Nanomechanics of Carbon Tubes: Instabilities beyond Linear Response, *Physical Review Letters*, Vol.76, No.14, (April 1996), pp.2511-2514, ISSN 0031-9007

# 15

# Molecular Dynamics Simulation Study on the Mechanical Properties and Fracture Behavior of Single-Wall Carbon Nanotubes

Keka Talukdar and Apurba Krishna Mitra
*Department of Physics, National Institute of Technology, Durgapur*
*India*

## 1. Introduction

Tremendous research motivation has been observed in the composite science and technology since the discovery of carbon nanotubes by Sumio Iijima (Iijima, 1991). Carbon nanotubes are highly crystallized tubular allotropes of carbon with hexagonal pattern repeating itself in space. The in-plane C-C bond is strong covalent σ bond. In contrary, there exists a weak π bond out of plane which acts in between the shells of a multi-wall carbon nanotube or in between different single-wall carbon nanotubes in a bundle. The combination of the high strength and high elastic modulus along the axial direction and the low density with high aspect ratio of the tubes has imparted in them excellent mechanical properties such that they may be used as reinforcing fibers in a polymer matrix to prepare low weight, high strength structural composites. As a component of a fiber-filled composite, the exact knowledge of the mechanical characteristics of the carbon nanotubes is necessary to tailor them for specific use. However, the proper strength and failure behavior of the nano materials can only perfectly be understood by atomistic simulation. In the nano regime, the continuum mechanics is inadequate and an atomistic description of the system is necessary. Moreover the carbon nanotubes are found to consist of various types of defects. Defects present inside their structure often result in some very complex phenomena in their failure process in the atomic scale which can be handled only by atomistic simulation. Here the power of molecular dynamics simulation technique is exploited in investigating the mechanical characteristics of various types of defect-free as well as defective tubes with a varying number of Stone-Wales defects with different combinations. The effect of interlayer interaction between different single-wall tubes in a bundle and the role of potential functions in the mechanical behavior of different carbon nanotubes are also investigated here.

## 2. Theoretical predictions and experimental observations

Nanoelectromechanical systems or super strong composite materials, which are of great interest in the present days, can be realized in practice by fabrication of materials using carbon nanotubes (CNTs). The choice of CNTs in these fields is found to be very much beneficial for their tremendous high strength and low density. Young's modulus of a solid depends on the nature of the chemical bonding of the constituent atoms. Due to the presence of strong covalent σ bond the axial Young's modulus should be equal to the in plane elastic modulus of graphite (1.04 TPa). Measurements with HRTEM or AFM reveal that CNTs have high Young's modulus which is close to 1 TPa, i.e. 100 times that of steel. All experimental observations (Krishnan et al., 1998; Salvetat et al., 1999a, 1999b; Treacy et al., 1996;) predict such high stiffness of the CNTs or their bundles. Very high Young's modulus of 2.8–3.6 TPa, for SWCNT and 1.7–2.4 TPa for MWCNT have also been observed in some studies (Lourie et al., 1998). Y value of 1.28 ± 0.59 TPa was found in one investigation (Wong et al., 1997). Scanning electron microscopy was used by Yu et al. (Yu et al., 1999a, 2000b, 2000c) for direct measurement of the tensile properties. Young's modulus obtained ranges from 0.32-1.47 TPa with a mean of 1.002 TPa for single- wall carbon nanotubes (SWCNTs) and 0.27-0.95 TPa for multi-wall carbon nanotubes (MWCNTs).

But the prediction of their mechanical properties (Batra & Sears, 2007; Belytschko et al., 2002; Chou, et al., 2010; Coluci et al., 2007; Dereli & Özdoğan, 2003; Liew et al., 2004; Yakobson et al., 1996) often ends with some uncertainties especially due to some unavoidable defects produced in them during their production, purification or functionalization. Stone-Wales (SW) defects (Stone & Wales, 1986), vacancies, pentagons, heptagons, lattice-trapped states, ad-dimers etc. are many types of defects that can appear in the CNT structure. Influence of defects can be observed in the mechanical properties of CNTs as well as in their electronic or magnetic properties. However, the effect of Stone-Wales defects has been investigated by many researchers (Belytschko et al., 2002; Chandra et al., 2004; Nardelli et al., 1998; Pozrikidis, 2009; Song et al., 2006; Troya et al., 2003; Tunvir et al., 2008) to obtain widely varying results. Such type of defect is produced by $90^0$ rotation of a C-C bond and thus producing two pentagons and two heptagons by conversion from four hexagons of carbon atoms. No matter what the process or potential adopted, reduction of failure strength and ductility has been reported by most of the authors. Some authors observed that chemical reactiivities were enhanced for a defective zigzag SWCNT compared to and planar graphene ( Picozzi et al., 2004). The role of vacancy defects or holes on the mechanical properties of CNTs was studied in many theoretical investigations such as that of Mielke et al.(Mielke et al., 2004), Xiao & Hou (Hou & Xiao, 2007) and Wang et al. (Wang et al., 2007). Decreasing effects on the failure stress and failure strain along with the scattering of stress values were reported by Troya et al.(Troya et al., 2003) for the introduction of 1, 2 and 5 adjacent and diagonal defects. Semiempirical quantum mechanical calculations of Troya included MSINDO and PM3 methods. They used 2nd generation Brenner potential for simulation. Mielke et al. (Mielke et al., 2004) performed MD simulation in three different approaches- DFT on the plane wave basis, PM3 approach and MTBG2. They took pristine as well as defective CNTs with one and two atom vacancies.

The role of defects on carbon nanotube elastic moduli and failure behavior was studied by tight binding molecular dynamics simulation by Haskins et al. (Haskins et al., 2007). Molecular simulation was used by Yang et al. (Yang et al., 2007) to determine the fracture strength of single-wall carbon nanotubes containing different concentrations of randomly distributed point defects. Another attempt to explain the discrepancies between the theoretical and the experimental results in predicting the mechanical properties of CNTs was made by Huq et al. (Huq et al., 2008). Shtogun and Woods (Shtogun & Woods, 2010) has calculated the mechanical properties of radially deformed defective single-wall carbon nanotube by first principle density functional theory calculations. Two different modes of formation of SW defects (mode 1 and mode 2, also referred as SW-A and SW-B) were considered by molecular mechanics model. Pozrikidis (Pozrikidis, 2009) has shown that inclined, axial and circumferential defect orientations have a strong influence on the mechanical response of zigzag and armchair SWCNTs. An attempt to explain the reason of scattering in data of the mechanical properties of CNTs was made by Tunvir et al (Tunvir et al., 2008) with Morse potential. To investigate the interference effect of spatial arrangements between two neighboring vacancy and SW defects with respect to the loading direction, their relative distances as well as their local orientations ware varied and the results were compared. But they considered the defects in the middle of a (10, 10) SWCNT and the defects were considered in only one side of the tube.

In 1995 Ebbensen & Takada (Ebbesen & Takada, 1995) got an experimental evidence of topological defects such as Stone-Wales defects in CNTs. Miyamoto et al. (Miyamoto et al., 2004) made an experimental identification of SW defects in CNT samples. Thus the experiments established the facts that the lower values of Young's modulus and tensile strength of CNTs originated because of the presence of defects in them, in particular the SW defects.

The enhancement in the mechanical performance of materials is reported in the literature (Dalton et al., 2003; Koizol et al, 2007; Mora et al., 2009; ) while preparing composites using CNTs as reinforcing agents. Based on finite element method, MD simulation study (Kuronuma et al., 2010) has been done to explore the fracture behavior of cracked carbon nanotube based polymer composites. Seo et al. (Seo et al., 2010) concluded that the MWCNTs were the good and appropriate materials to improve elastic behavior of the composites with a large increase in their flexural strength and Young's modulus in the tune of about 60%. In spite of so much research attempts discussed above, there still exists enough scope of investigation about the mechanical properties of the CNTs and an elaborate study can only lead to better design of high-strength CNT-composites.

## 3. The aim of this study

All types of carbon nanotubes can be used for low weight structural composites. NASA is developing materials using CNTs for space applications, where weight driven cost is the major concern (Despres et al., 1995, Iijima et al., 1996). Sufficient improvement in the research work in this field has been achieved to enhance the adhesion of the CNTs with the host to build composite materials for various purposes (Andrews et al., 1999; Chae et al., 2006; Kearns and Shambaugh, 2002). During processing or purification, different chemical groups are attached with the CNTs either via "end and defect-site" or by "sidewall functionalization" process. The chemical attachment of the less curved sidewalls usually requires the presence of sidewall defects such as Stone–Wales defects (Balasubramanian & Burghard, 2005; Banerjee et al., 2005). For better bonding of carbon nanotubes with polymer matrix in a CNT-polymer composite, suitable functionalization of the CNTs is essential. Different types of chemical functional groups can be attached (Kang et al., 2010; Qi et al., 2003) in these processes without hampering nanotube properties. So the presence of defects in the CNT structure is not always degrading.

Using monochromatic light or electron beam, suitable number of defects can be produced in the CNT structure for specific applications. A review (Krasheninnikov & Banhart, 2007) on the engineering of CNTs and other carbon materials with electron ion beams discus the recent advancements in this respect. Defects can serve the purpose of grafting functional groups directly at the defect site (Canto et al., 2011). Addition reaction is most favorable in the position of Stone-Wales defects at the carbon-carbon double bond. Again attaching functional groups to the CNTs is necessary to achieve proper adhesion of them with the polymer matrix while forming composites. Electron beam exposure at high temperatures induces structural defects in CNTs which promote the joining of tubes via cross-linking of dangling bonds and thus producing Y or T junctions. This fact is both experimentally and theoretically proved by Terrones et al. (Terrones et al., 2002). Defect-induced junctions between single- wall, double-wall carbon nanotubes and also between metal crystals are possible (Manzo et al., 2010) as electron irradiation produces structural transformation in the CNTs and metal atoms (Banhart, 1999). Fusion of two (5, 5) SWCNTs into a (10, 10) tube via a zipper mechanism by SW transformation is also reported by Yoon et al. under HRTEM (Yoon et al., 2004). Under electron irradiation, the experimental and theoretical evidence of the change of CNT's diameter is also another consequence of the artificial production of defects (Ajayan et al., 1998). So production of defects in certain amount and in specific positions of a CNT is not a difficult task with the newly developed experimental techniques.

Defects were previously thought to be only unwanted, but now it is shown that they can be controllably produced for preparing novel nanodevices. So the need to identify and quantify them is also required. Measurements done by polarization dependent scanning transmission X-ray microscopy (Felten et al., 2010) or Raman spectroscopy (Miyata et al, 2011) can now identify and quantify defects. Defect spacing can further be varied by varying the synthesis technique (Fan et al., 2005).

Observing the new trends in the progress in the defect-controlled manipulation of the nanotube properties in various fields of science and technology, we have critically investigated their mechanical properties involving two major defects, mainly SW defects. As the SWCNTs have cylindrical geometry, defects may arise on its surface anywhere. Thus defects may appear not only in one side but they may be situated opposite to each other. Keeping this in mind, more than one defect are considered at different positions of a zigzag (10, 0), chiral (5, 3) and armchair (5, 5) SWNTs with varying separating distances and with different separating angles to simulate them in atomic scale. Also 2$^{nd}$ generation Tersoff-Brenner (Brenner et al., 2002) potential is incorporated in our study which can explain nicely the creation and destruction of bonds in a CNT structure. Defects are distributed in all parts of the tube and moreover, diagonal, overlapping and neighboring defects are taken along with many other possible arrangements. Fracture modes are modeled and compared with a detailed study of the mechanical response of the tubes viz, young's modulus, failure strength and ductility.

Experimental data (Yu et al., 2000c) for a single-wall carbon nanotube bundle under tensile load show that the Young's modulus can be as high as 1.47 TPa and tensile strength may range from 13-53 GPa. In spite of so many attempts to explain the discrepancies between the theoretical and experimental data regarding the mechanical properties of SWCNTs, some observed discrepancies between theory and experiment are still unanswered. SWCNTs have a natural tendency to form bundles and thus in the most of the experiments undertaken by various investigators, SWCNT bundles have been used as samples. So we must take into account the influence of bundle formation in our calculations of mechanical properties of SWCNTs. In this work, we have compared the observed mechanical properties of three different types of SWCNTs separately and also the bundles of each type and a bundle of their mixture. The results are explained with the overlapping of density of states (DOS) of

the CNTs in a bundle. So this is an attempt to carry out more realistic theoretical investigations to facilitate comparison with experimental data. 2nd generation reactive empirical bond order potential (Brenner et al., 2002) is adopted for simulation. For this study we have restricted the potential to be smoothed-off at the cut-off region. Lennard-Jones 6-12 potential (Lennard-Jones, 1924) function is adopted for interlayer interaction.
To explain the above discrepancies, different types of SWCNTs are simulated using three different potential functions and the results are compared.

## 4. Atomistic modeling and simulation

By atomistic modeling one can understand the interaction between the constituent molecules of a material and thus the behavior of the material with various external constraints, their deformation and failure process can also be understood. Starting from the simple laws of Physics to describe the position and momentum of each atom in a material and solving the equations of motion of the system of atoms or molecules, one can obtain the dynamic behavior of all the particles of the system. Thermodynamical behavior of gases and liquids (Alder & Wainwright, 1957; Alder & Wainwright, 1959; Allen & Tildesly, 1989; Rahman, 1964) was the first, where molecular dynamics simulation was performed. Later, this process has been used to find the mechanical behavior of solids. The failure process of solids is associated with so many computational complexities that modeling and simulation have now become a very exciting area of research. At first, a mathematical model is to be developed for a physical problem. Then the equations resulting from that model building (Ashby, 1996) are to be solved. The model should be such that it should be able to capture the physical features of the problem. Sidney Yip of MIT (Yip, 2005) has stated that modeling is the physicalization of a concept, simulation its computational realization. Modeling requires the knowledge about the physics of the system i.e. about its constituents or the behavior of the particles. Simulation requires the technique to solve enormous numerical equations related to the complicated systems. The behavior of cracks, dislocations, grain boundary processes can be very successfully investigated by atomistic modeling (Buehler, 2008). In modern materials modeling this process has gained immense importance. Any complex problem can be solved by knowing Newton's laws and the nature of interaction of the atoms. In the present work mechanical properties of the CNTs are investigated by modeling and simulation. Nowadays, researchers have considered the computer as a tool to do experiment, similar as experimentalists do in their laboratory. Computational experiments thus need to build a suitable model of a physical problem, set up some equations to represent the problem, run simulation and to interpret the results of the simulation process.

### 4.1 Molecular dynamics simulation

Molecular dynamics simulation is a form of computer simulation in which atoms and molecules are allowed to interact for a period of time by approximations of known physics, giving a view of the motion of the particles. Position and momentum of each particle is updated using a suitable algorithm. In materials science, this is a very effective process. We call molecular dynamics simulation a computer simulation technique where the time evolution of a set of interacting atoms is followed by integrating their equations of motion. Material behavior and their deformation are described in atomistic modeling. The purpose of molecular dynamics simulation is to understand the properties of assemblies of molecules in terms of their structure and the microscopic interactions between them. Something new, which cannot be found in other ways, can be learnt by simulation.

### 4.2 Present method of simulation

We have performed MD simulation using the software 'Brenner code' which is an open source code and which has been customized and modified as necessitated by our scheme of work. The coordinates of a carbon nanotube generated in the program can be used for simulation. The programs are all written in FORTRAN language. All programs are to be compiled separately to run the main program. Microsoft developer studio is used as the platform of manipulating programs. Time step of 0.5 fs is chosen for the molecular dynamics simulation presented here. We have tested that slight changes in the time step affect the simulation result negligibly. Also The evolution of the system energy with respect to time is noticed and it is observed that 30000-50000 time steps are required to achieve convergence in energy minimization and hence to get equilibrium condition. However, simulation can be run using other software also, like TINKER, LAMMPS etc.

A thermostat is required to lead a system to a desired temperature i.e. to enable a NVT or NPT ensemble. To modify the equations of motion to obtain a specific thermodynamical ensemble, the velocities of the atoms are rescaled by a scaling factor. We have used Berendsen thermostat (Berendsen et al., 1984) in this study to maintain a fixed temperature.

In this thermostat, velocity is rescaled by a factor $\lambda=\sqrt{(T_0/T(t)}$

In each step velocity is rescaled so that the rate of increase of temperature is proportional to the difference in temperature; $dT(t)/dt= 1/\tau\ (T_0- T(t))$

So the scale factor is given by, $\lambda = [1+ \Delta t/\tau\ (T/T_0-1)]^{1/2}$

when $\Delta t$ is the time step and $\tau$ is a coupling constant. It is called rise time which gives the strength of the coupling of the system with the hypothetical heat bath at temperature $T_0$.

## 5. Verification of the dependence of mechanical response of single walled carbon nanotubes with the position and arrangement of stone- wales defects

### 5.1 Theoretical method

The two pentagons and two heptagons that are produced by the $90^0$rotation of a C-C bond in the honeycomb structure of a SWCNT may be oriented in different manner. The tube symmetry and the inclination of the axis joining the pentagons or heptagons with the tube axis decide whether the excess strain of formation of defect is released or not while straining the tube. Three different SWCNTs are taken and analyzed in our study. The zigzag, chiral and armchair tubes are 42.5, 59.6 and 49.19 Å long respectively with aspect ratios 10.9, 21.75 and 14.5. The zigzag and armchair tubes contain 400 atoms and the chiral tube contains 392 atoms as a whole. Berendsen thermostat is used to allow small changes in the velocities of the atoms such that temperature of the system reaches a value close to 300K. The tubes are stretched in the axial direction keeping other end fixed. By stretching the tubes in small strain increments, the equilibrium potential energy were calculated by simulation at first in absence of any defect and then with 1, 2, 3 and 4 defects at different positions of the tubes. Stretch is applied along the axis of the tubes. No other constraint is there in other directions and force is zero in any other direction.

Stress is calculated from the energy-strain curve as $\sigma = 1/V\ (dE/d\varepsilon)$ where $\sigma$ is the longitudinal stress, V the volume of the tube, $\varepsilon$ the strain and E the strain energy of the tube. Volume of the tube is found as $V=2\pi r\delta rl$ where r is the inner radius of the tube, $\delta r$ its wall thickness and l the length of the tube. We have taken δr as 0.34 nm, which has been the standard value, used by most of the authors. To calculate stress from the energy-strain curve we have used a linear relationship for the elastic region and appropriate non-linear equations for the segments of high strain deformation regions. Young's modulus is found from the slope of the linear portion of the stress-strain curve. By changing the (z, r, θ) values, defects are created at different positions of the same tube. θ is changed by $90^0$ from its first value to form two oppositely directed defects. To produced two diagonal defects θ is changed by $45^0$ and z is adjusted to the desired value.

### 5.2 Characteristics of defect free tubes

With Brenner's 2nd generation REBO potential an armchair SWCNT shows remarkable ductility of 32% and tensile strength of 196.3 GPa [Fig 1]. A chiral tube breaks at 30% strain while its failure stress is calculated to be 149.88 GPa. The ductility and failure stress of a zigzag tube, in contrary, are much less and that are 18% and 115.4 GPa respectively. Linear elastic region extends upto 7% strain, beyond which the nature of the curve changes to non-linearity. Young's modulus is found for the pristine tube as 1.06 TPa for a zigzag tube, 0.891 for a chiral tube and 0.814 for an armchair tube. Our calculation matches with the experimental values of failure stress by Demczyk (Demczyk et al., 2002) which is 150±45 GPa. Young's modulus value is also close to the experimental value of 1.28 TPa by Wong et al. (Wong, 1997) and 1.25 TPa by Krishan et al. (Krishnan et al., 1998). Maximum strain can be compared with the results of quantum mechanical calculations (Ozaki et al., 2000; Troya et al., 2003) though 10-13% maximum strain and failure stress between 13-52 GPa were observed by Yu et al. (Yu et al., 2000c) experimentally. Fracture of a zigzag SWNT [Fig 2(a)] is found to be brittle in our calculation. As shown in Fig 2(b) and 2(c), failure patterns are not so sharp for a chiral and an armchair tube.

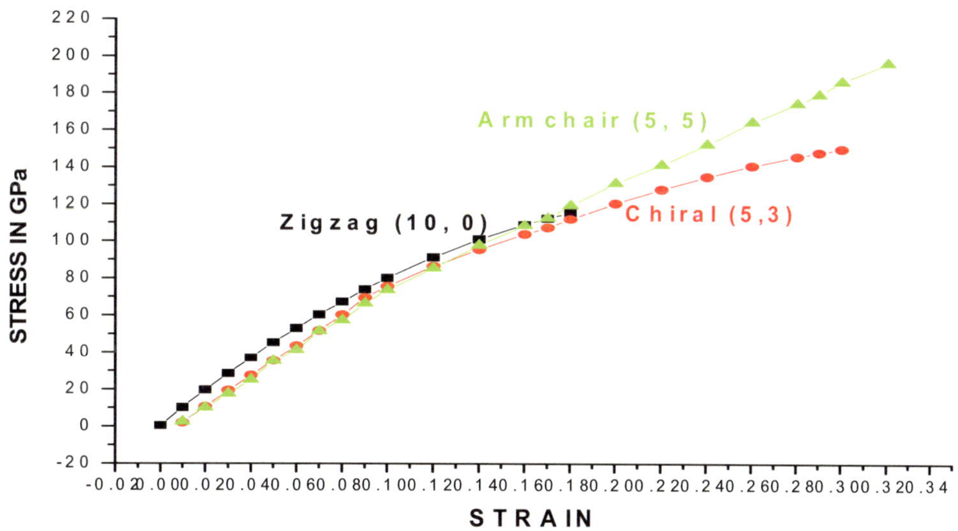


Fig. 1. Stress-Strain curves for defect free SWCNTs

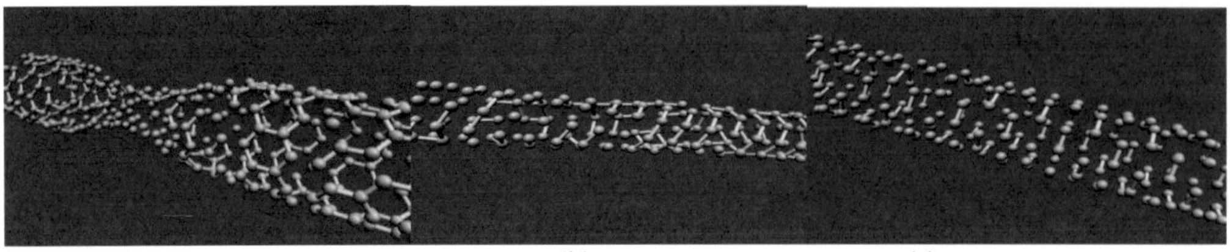

(a) (b) (c)

Fig. 2. Fracture patterns of a defect free (a) zigzag (10, 0) (b) Chiral (5, 3) and (c) Armchair (5, 5) SWCNTs

### 5.3 Modeling of defects

A 5/7/7/5 defect is realised theoretically by rotation of a C-C bond by $90^0$ and thus converting two pair of hexagons to two pentagons and two heptagons. Single defect is created in any desired place by adjusting (z, r, θ) values. For more than one defect z values are changed to produce separated defects and orientation between two defects or oppositely situated defects are produced by changing θ. Symmetric defects mean that they are situated symmetrically on both sides of z=0 position. Same θ for a number of defects produces some defects on the same line. Where θ varies by $180^0$, oppositely situated defects are produced. Thus, for four different combinations of z, r and θ as (-4.26,3.91,90), (-12.78,3.91,90), (4.26,3.91,-90), (12.78,3.91,-90), four symmetrical but oppositely situated defects are produced. Fig 3 shows a (8, 8) SWCNT with two defects at different positions.

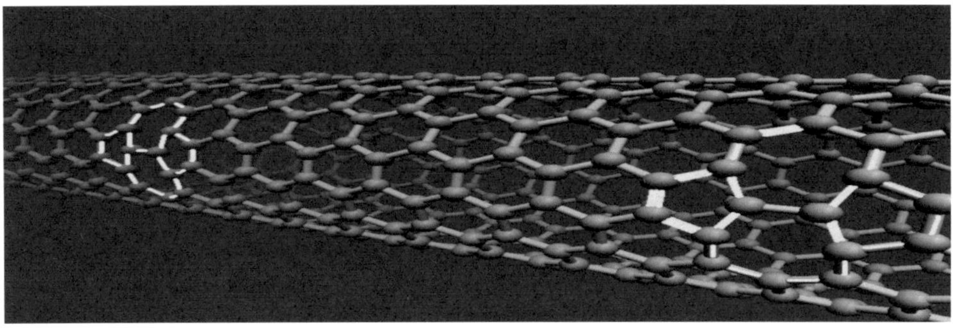

Fig. 3. Two SW-defects shown in a (8, 8) SWCNT

### 5.4 Variation of tube properties with defects

#### 5.4.1 Effect of a single defect

With only one defect an SWCNT shows a scattering of data for all mechanical characteristics for different position of its formation. The variation is most pronounced for an armchair tube. Fig 4 shows the stress-strain curves for a (5, 5) SWCNT with a single defect at different positions. The curves for z=-15 and z=-20 show completely different nature. A defect at the edge which is opposite to the loading side and is oriented parallel to the axis of the tube is shown by the pink line in Fig 4. The different nature of the curve proves such defect does not affect the mechanical behavior in a considerable amount. The defect energy ($E_d$) for a defect near that edge (z=15.98 or 19.17) is greater than that of other defects for small strains but lower for large strains. While straining the tube with such defects at the far end, excess strain due to the formation of the defect is released irrespective of its orientation. However, $E_d$ is always lower for a pristine tube compared to others at small strains but higher for large strains. Tabulated results for the variation of the mechanical properties of the three types of tubes are given in Table 2, 3 and 4. Mechanical responses of the chiral and zigzag tubes show some variation. Maximum reduction of Young's modulus is observed with a single defect for a (5,5) tube with a defect situated near the edge and the reduction is 29%. However, the maximum reductions of Y values are 24% and 16% for a (10,0) and a (5,3) tube respectively. Ductility of a zigzag tube is not influenced by any defect present in its structure. But whatever is the position of the defect, breaking starts from the defect site. Fig 5 reveals that breaking of a chiral or armchair tube is brittle in nature. Necking is observed in case of a zigzag tube For a pristine tube we have observed that the applied force results in many SW rotations after 10% strain as reported by Zhang et al. (Zhang et al., 1998). Stress-strain curves [Fig 4] show the variation of the mechanical behavior of the tube mainly in the non-linear portion after inclusion of a single defect at different places.

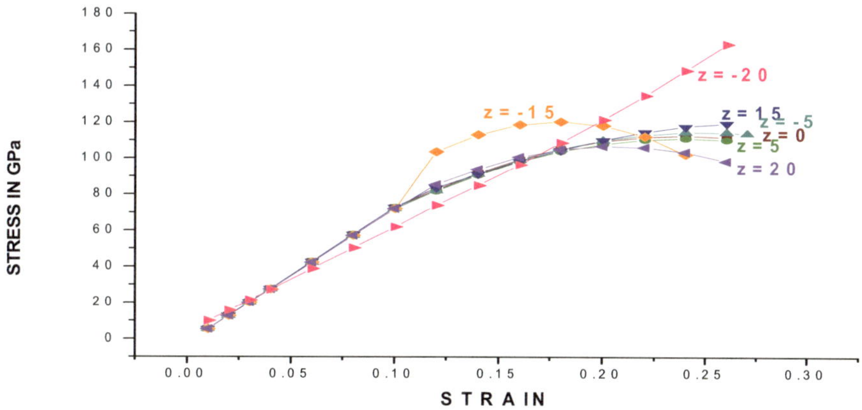


Fig. 4. Stress-strain curves for a (5, 5) SWCNT with a single defect at different positions

For a chiral tube tensile strength reduces by 34% and strain by 23% [Table 3] and the reductions are 45.6% and 25% respectively for an armchair tube [Table 4]. $E_d$ for defects situated near the edge in case of the chiral and zigzag tubes are higher than that situated in the mid position like a zigzag tube. As the bond between two pentagons is very week, the tube with such SW defect where pentagons oriented along the loading direction breaks easily. Where pentagons are inclined with the axis, the tube offers more resistance before breaking.

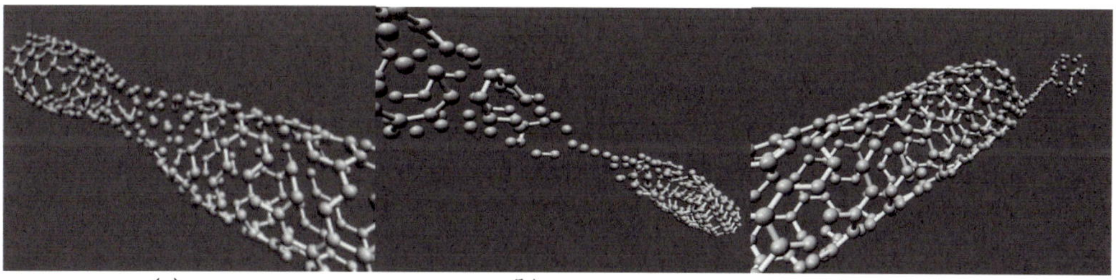

(a) (b) (c)

Fig. 5. Breaking patterns of a (a) (10,0) SWNT with 1 defect at (z, r, θ) position (15.98, 3.87, 9) (b) Chiral tube with a defect at (0.61, 2.65, 92) (c) armchair tube with a defect at (15.37, 3.37,90)

### 5.4.2 Effect of a couple of defects

More variation of result is obtained after the inclusion of 2 defects [Fig 6]. But tensile strength and Y value are higher than the result of inclusion of one defect except for two diagonally situated defects and when the defects are situated on the same line separated by a distance. Thus for the combinations (-1.06, 3.87, -9), (1.06, 3.87, 27) and (-3.19, 3.87, 27), (3.19, 3.87, 27) the reduction of tensile strength is 41.9% and 39.4% for the (10, 0) tube. When the two defects are not on the same line, rather situated opposite to each other with z position 0 and 8.52, the defects can not reduce the strength. In each case the nature of defect-defect interference is different. The stress-strain curves are shown in Fig 6 which distinguishes between different defect separation and orientation. A defect can be represented by an edge dislocation which produces gliding motion .The glide is accompanied with attachment or breaking of bonds when atoms move together and the changes are observed for the atoms in the line of glide motion. The atoms and bonds below and above the defect change. So attraction or repulsion between defects is possible according to their distance of separation and orientation (Samsonidze et al., 2002). If attraction occurs, the moving dislocations pile up such that the energy of the system slowly decreases due to the application of the external force. On the other hand any external force should bring the system to the minimum energy state easily if repulsive interaction acts between the defects. So repulsive interference occur for the above two combinations and for the defects at (0, 3.91, 90) and (8.52, 3.91, -90) attraction takes place. Here, the two glide motion moves opposite to each other. The orientation angle i.e. the angle between the lines joining the two pentagons of the two defects takes a major part in defect- defect correlation.

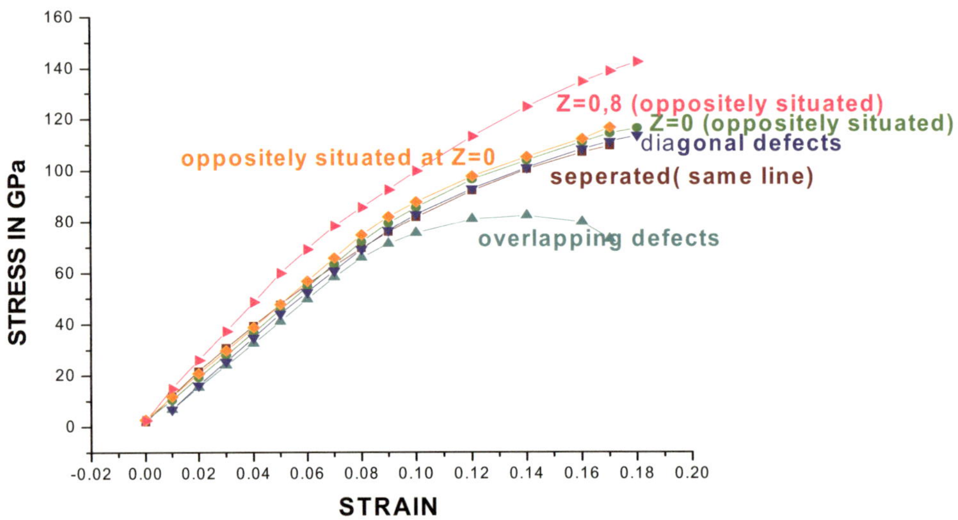


Fig. 6. Stress-strain curves for a (10,0) SWCNT with 2 defects at different positions

For a chiral tube, two diagonal defects and two oppositely situated defects at the mid position reduce its strength by considerable amounts. The striking fact is that, the two overlapping defects situated side by side have no such remarkable influence on the strength. The stress-strain curves of a chiral and armchair SWCNT with 2 defects at different positions are depicted in Fig 7 and Fig 8.

Remarkable differences are observed for an armchair tube. Diagonal defects no longer change the stress-strain curve [Fig 7] compared to the defect free tube. The armchair symmetry helps attractive correlation between two diagonal defects while evidence of repulsive correlation is clear in case of the chiral tube. Repulsive correlation is much weaker for a zigzag tube in case of diagonal defects.

Breaking of a (10, 0) tube with 2 defects is accompanied with deformation of the tube in different places as shown in Fig 9(a). Other two tubes sometimes get twisted before fracture, especially for diagonal defects [Fig 9(c)]. It is surprising to note that breaking of a chiral tube [Fig 9(b)] with two defects at (4.56, 2.65, 59) and (-4.72, 2.70, 53) does not initiates from any defect site. Ruther, the tube breaks at a position near the edge of the tube. The interaction of the chiral symmetry with the orientations of the defects as well as the angle between the pentagons of the defects serves a crucial role in the fracture of an SWCNT.

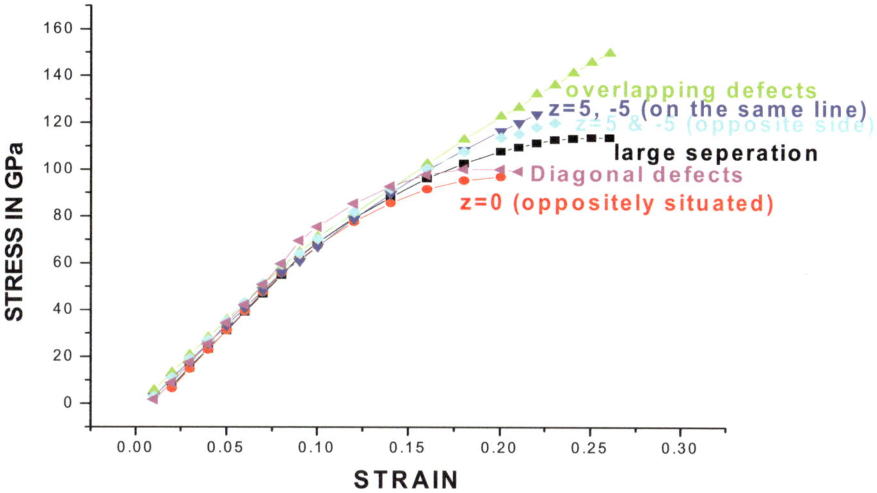


Fig. 7. Stress-strain curves of a (5, 3) SWNT with 2 defects at different positions

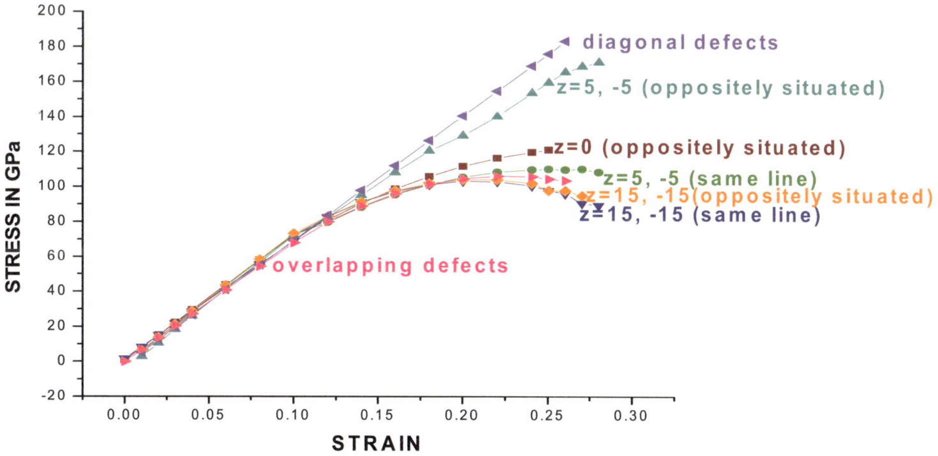


Fig. 8. Stress-strain curves of a (5, 5) SWNT with 2 defects at different positions

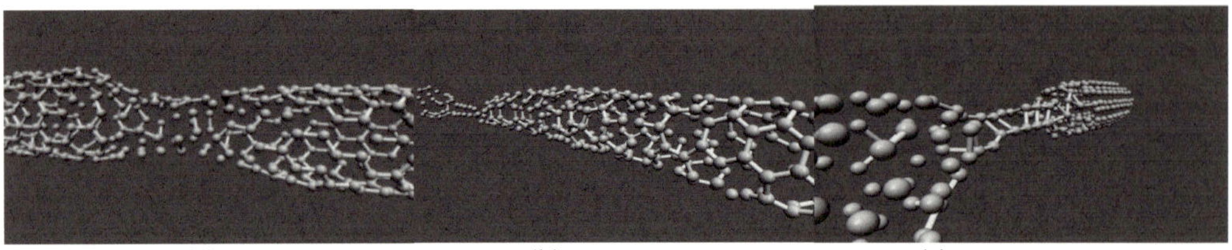

(a) (b) (c)

Fig. 9. Breaking of a **(a)** (10,0) SWCNT with two defects at (-1.06,3.87,-9) and (1.06,3.87,27) **(b)** (5,3) SWCNT with two defects at (4.56, 2.65, 59) and (-4.72, 2.70, 53) **(c)** (5, 5) SWCNT with two defects at (-0.61,3.37,90) and (0.61, 3.37,126)

### 5.4.3 Influence of three defects

Different combinations of three defects are taken to study the effect of odd number of defects. An aggregation of three diagonal defects in a zigzag tube reduces the strength of the tube to 89.58 GPa[Fig 10]. Reduction of strength is also remarkable when the defects are close and almost on the same line (orientation of one defect differs only by $9^0$). Like two defects on the same line, three such defects exactly on the same line have no such pronounced effect on the tensile strength of the tube. The middle defect here is influenced complicatedly by the other two defects. In each case different results are obtained. For the defects that are oriented by some angle with respect to each other, curvature plays a significant role in decreasing the strength. For defects on the same line, mainly the separation between the defects is important. In our study maximum strain is less hampered by the inclusion of defects for a zigzag tube.

It is revealed from our study that accumulation of a number of defects has no extra influence on the mechanical response of a zigzag SWCNT. $E_d$ increases with the increase of number of defect. But ultimately interference between defects leads to adjustment of equilibrium energy leaving some changes in stiffness and maximum tensile capacity but ultimate strain reduces only from 18% to 17% in some cases and sometimes it is fixed to 18%. The defects

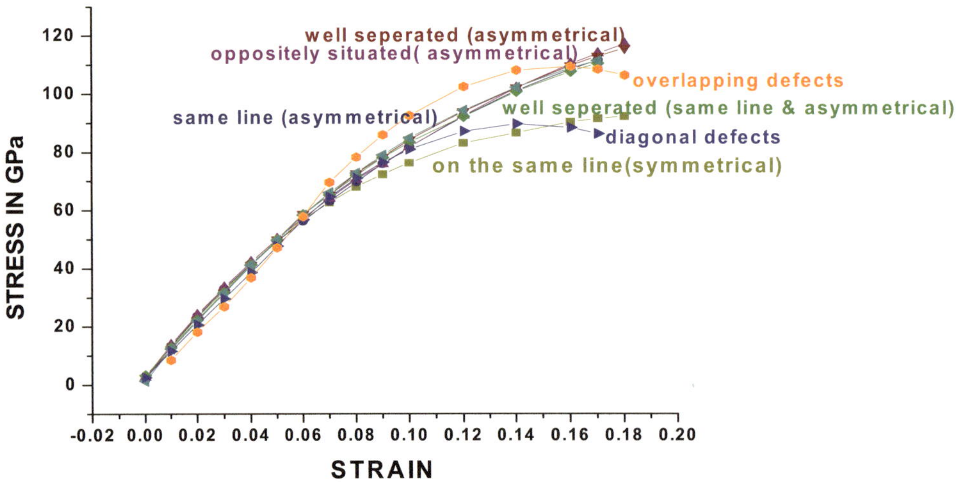


Fig. 10. Stress-strain curves for a (10, 0) SWCNT with 3 defects at different positions

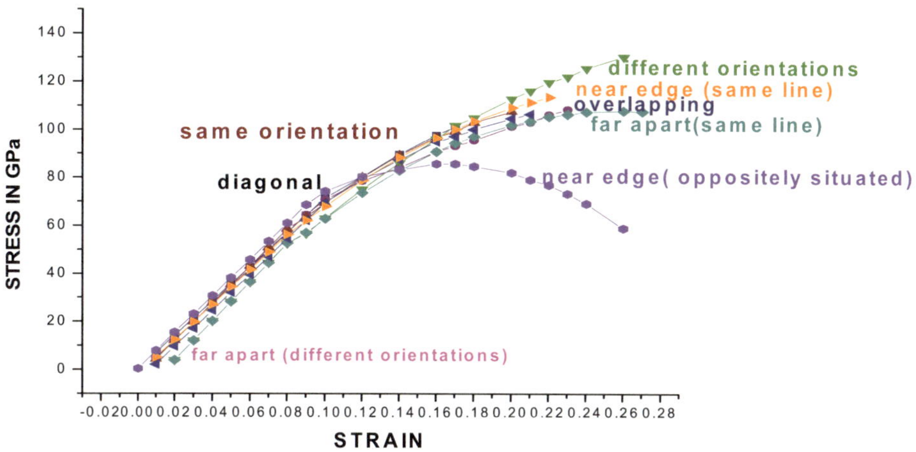


Fig. 11. Stress-strain curves for a (5, 3) SWNT with 3 defects at different positions

created on equal distances and symmetrically on either side of the Z=0 position interact among themselves in such a manner that the stiffness increases than the value when the defects are asymmetric but on the same line. The observation seems to resemble with the observations of Tunvir et al. (Tunvir et al., 2008). The result of inclusion of 3 defects for a (5, 3) and (5, 5) tubes are shown in Fig 11 and 12 below. Three overlapping defects in case of a (5, 5) SWCNT change the pattern of the stress-strain curve [Fig 12] significantly. Failure stress is dropped to 85.79 GPa from 196.3 GPa and maximum strain from 32% to 26% in this case. But maximum strain is reduced mostly for 3 oppositely situated asymmetrical defects [Table 4]. In contrary, maximum stress increased to 148.2 GPa.

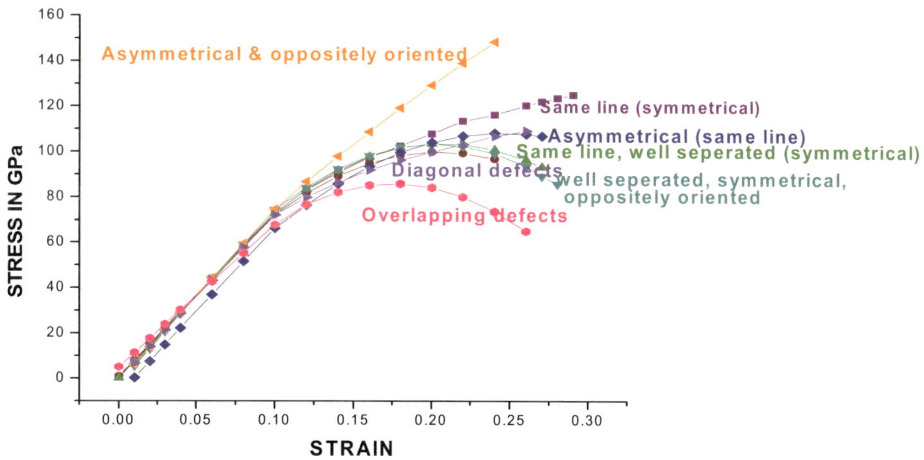


Fig. 12. Stress-strain curves for a (5, 5) SWNT with 3 defects at different positions

Breaking of a tube with three symmetrical defects on the same line often originates from the middle defect. But sometimes, for example for 3 overlapping defects in a zigzag tube breaking [Fig 13(a)] is not so sharp. It includes maximum bond breaking with necking. Fig 13(b) shows the fracture of a chiral tube with 3 defects while Fig 13(c) is the same picture for an armchair tube. As stated earlier, twisting of the tube in Fig 13(b) is prominent.

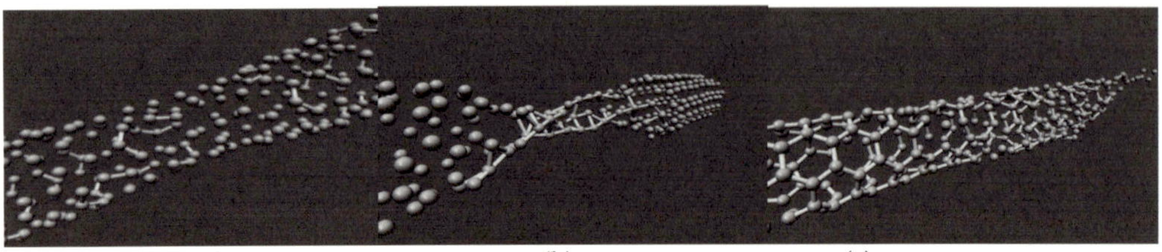

(a) (b) (c)

Fig. 13. Breaking of a (a) (10, 0) SWCNT with 3 defects at (0.00,3.91,90), (4.26,3.91,126), (3.91, 3.87, 171), (b) chiral SWCNT with 3 defects at (0.00,3.91,90), (-8.52,3.91,-90), (-17.04.3.89,90) and (c) an armchair SWCNT with 3 defects at .(-5.53,3.37,-90),(0.61,3.37,90), (5.53,3.37,-90)

### 5.4.4 Results of inclusion of 4 defects at different positions

When four SW defects are introduced in the three different SWCNTs, an armchair tube is found to be influenced greatly by it [Table 4]. Tensile strain of an armchair SWCNT is dropped by 53%. The maximum strain of a chiral SWCNT [Table 3]is greatly influenced by the inclusion of 4 defects while such change is not observed for a zigzag tube [Fig 14, Table 2]. Two pairs of diagonal defects reduce the ductility of a chiral tube by 46%. 16% failure strain in this case is close to the experimental result of M. F. Yu (Yu et al, 2000c). Sharp breakings are observed for a chiral [Fig 15(b)] and an armchair [Fig 15(c)]tubes where breaking of a zigzag tube [Fig 15(a)] is not always so sharp.

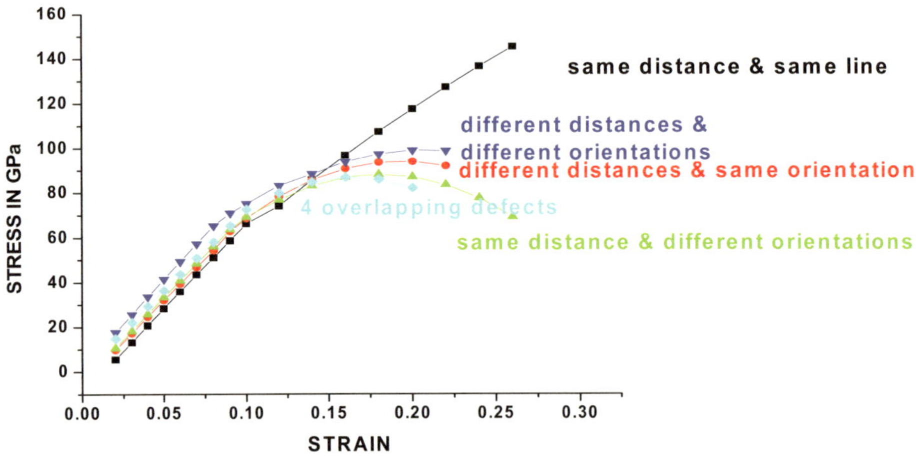


Fig. 14. Stress-Strain curves of a (10, 0) SWNT with 4 defects at different positions

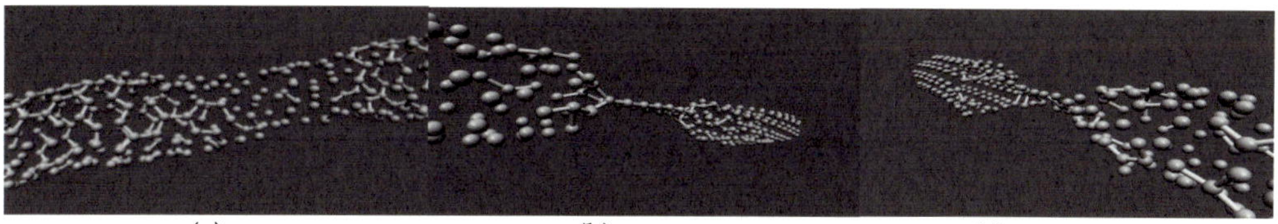


Fig. 15. Breaking of **(a)** a (10,0) SWCNT with 4 defects at (z, r, θ) positions ( -4.26,3.91,90), (12.78,3.91,90), ( 4.26,3.91,90), (-12.78,3.91,90) **(b)** (5,3) SWCNT with 4 defects at (-5.02, 2.73,97), (0.61, 2.65, 92), (3.04, 2.65, 99), (7.45, 2.70,90) **(c)** (5,5) SWCNT with 4 defects at (-4.3,3.37,90), (-1.84,3.37,126), (5.53,3.37,90) and (6.76,3.37,126)

### 5.4.5 Comparison between the three tubes considering the same defects inside

The difference between the mechanical responses of the three types of SWCNTs is studied and they are tabulated in Table 2, 3 and 4. When the three tubes are considered with the same defective condition, large variations are observed. For example, Fig 16 and Fig 17 are the stress strain curves of three tubes with 2 diagonal defects [Fig 16] and four symmetrical defects on the same line [Fig 17] respectively. Nature of the curves is completely changed in the two cases. That proves the symmetry dependent characteristics of single-wall carbon nanotubes when their mechanical properties are concerned. From Table 1, a summary of the results can be obtained.

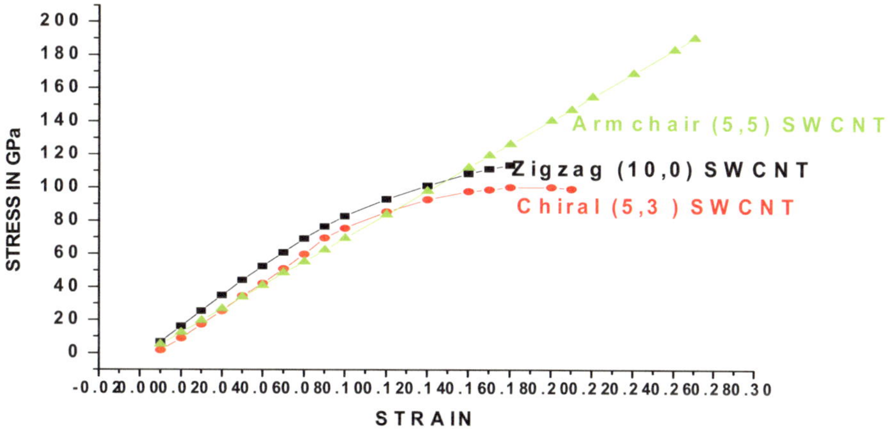


Fig. 16. Comparison of the strain-strain curves of the three tubes for 2 diagonal defects

The above study reveals that the structure of a CNT has special effect on its mechanical behavior. Zigzag symmetry has less influence on the ductility, where chiral and zigzag symmetry affect the ductility greatly. So the ductility of a zigzag SWCNT is not influenced by the presence of SW defects in its structure. A chiral tube shows less stability with increasing number of defects. Neighboring defects, especially overlapping defects reduce the strength of a zigzag and armchair tubes mostly but for chiral tube effect of overlapping defects is not so pronounced. For a cylindrical geometry of a CNT, their mechanical response is thus can be influenced by the position, arrangement and orientation of the defects as well as by the symmetry of the tube. Table 1 gives the comparison in tabulated form.

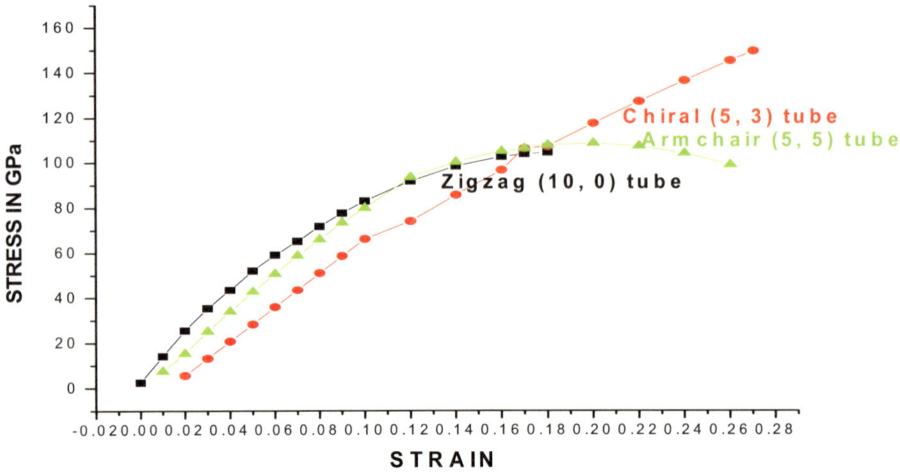


Fig. 17. Comparison of the stress-strain curves of the three tubes for 4 symmetric defects on the same line

| SWCNT | Maximum reduction of Y value and (z, r, θ) position | Maximum reduction of failure stress and (z, r , θ) position | Maximum reduction of ductility and (z, r , θ) position |
|---|---|---|---|
| Zigzag: (10, 0) | 23% for 4 defects at (0.00,3.91,90),(4.26,3.91,90) (-4.26,3.91,90),(7.45,3.87,99) | 41.9% for 2 defects at (-1.06,3.87,-9), (1.06,3.87,27) | No significant reduction |
| Chiral: (5, 3) | 10.4% for 3 defects at (-21.15,2.73,94), (0.61,2.65,92), (21.15,2.7,86) | 42.9% for three defects at (-19.93,2.7,97), (0.46,2.73,24), (19.93,2.73,83) | 46.7% for 4 defects at (-6.24, 2.7, 94),(-2.59, 2.73,105) (0.61, 2.65, 92), (3.8, 2.73, 79) |
| Armchair: (5, 5) | 29.7% for 1 defect at (-19.06, 3.37, 36) | 56.3% for three defects at (-1.84,3.37,90),(0.61,3.37,90), (3.07,3.37,90) | 31.25% for 4 defects at (-4.3,3.37,90), (1.84,3.37,126) (5.53,3.37,90),(6.76,3.37,126) |

Table 1. Maximum reduction of Young's modulus, failure stress and ductility for the zigzag, chiral and armchair SWCNTs and the corresponding (z, r, θ) values

Unlike the bulk materials, defects of various kinds are not always degrading the material properties, but they may also be beneficial for fixing up the point of chemical functionalization, charge injection and symmetry breaking effects and and also may facilitate spectroscopic characterization process. Defect controlled future applications of CNTs have now been paid attention and efforts are made to successfully create suitable type and quantity of defects in their structure for specific purposes without compromising their other excellent properties. Production of Y or T junction for producing electronic devices or functionalization with different chemical groups at the defect sites to enhance cohesion of the CNT fibers with the matrix element are some of the common uses of defect sites.

The calculated result showing the interaction between SW defects however is still to be proved experimentally. But the sites of SW defects are used for functionalisation of chemical

| | Position of defect/defects (Z, r, θ) | Young's modulus (TPa) | Failure Stress(GPa) | Maximum Strain |
|---|---|---|---|---|
| Single defect: | 1. (0, 3.91, 18) | 0.832 | 105.34 | 17% |
| | 2. (9.58, 3.87, 99) | 0.859 | 104.66 | 18% |
| | 3. (-9.58, 3.87, 99) | 0.847 | 103.59 | 17% |
| | 4. (15.98, 3.87, 9) | 0.824 | 116.67 | 18% |
| | 5. ( -15.98, 3.87, 9) | 0.846 | 104.14 | 17% |
| | 6. (19.17, 3.91, 36) | 0.821 | 127.71 | 18% |
| | 7. (-19.17, 3.91, 36) | 0.829 | 106.67 | 17% |
| 2 defects : | 1. (-4.26,3.91,54), (4.26,3.91,54) | 0.891 | 109.44 | 17% |
| | 2. (-4.26,3.91,90), (4.26,3.91,-90) | 0.873 | 116.38 | 18% |
| | 3. (0,3.91,90), (8.52,3.91,-90) | 1.00 | 141.7 | 18% |
| | 4. (-1.06,3.87,-9), (1.06,3.87,27) | 0.849 | 82.31 | 18% |
| | 5. (2.13.3.91,0), (0.00,3.91,90) | 0.893 | 113.5 | 18% |
| | 6. (0.00,3.91,-18), (0.00,3.91,162) | 0.900 | 116.58 | 18% |
| | 7. (-3.19,3.87,27), (3.19,3.87,27) | 0.832 | 85.87 | 18% |
| 3 defects : | 1. (-4.26,3.91,99), (0.00,3.91,90), (4.26,3.91,90) | 0.904 | 92.19 | 18% |
| | 2.(0.00,3.91,90), (-4.26,3.91,90), (8.52,3.91,90) | 0.865 | 111.41 | 17% |
| | 3. (-4.26,3.91,-90), (4.26,3.91,90), (17.04,3.91,90) | 0.95 | 117.02 | 18% |
| | 4. (0.00,3.91,90), (-8.52,3.91,-90), (-17.04.3.89,90) | 0.94 | 115.6 | 18% |
| | 5. (0.00,3.91,90), (-8.52,3.91,90), (-17.04,3.91,90) | 0.924 | 110.36 | 17% |
| | 6. (0.00,3.91,90), (8.52,3.91,90), (17.04,3.91,90) | 0.946 | 111.28 | 17% |
| | 7. (0.00,3.91,90), (4.26,3.91,126), (3.91,3.87,171) | 0.892 | 89.58 | 17% |
| | 8. (0.00,3.91,90), (3.91,3.87,99), (7.45,3.87,99) | 0.898 | 109.28 | 18% |
| 4 defects : | 1. ( -4.26,3.91,90), (12.78,3.91,90), ( 4.26,3.91,90), (-12.78,3.91,90) | 0.941 | 104.5 | 18% |
| | 2. (-4.26,3.91,90), (-12.78,3.91,90), (4.26,3.91,-90), (12.78,3.91,-90) | 0.859 | 109.88 | 17% |
| | 3. (-4.26,3.91,90), (12.78,3.91,90), (4.26,3.91,-90), (-12.78,3.91,-90) | 0.868 | 115.5 | 18% |
| | 4. (-8.52,3.91,90), (-4.26,3.91,90), (0.00,3.91,90), (4.26,3.91,90) | 0.871 | 89.6 | 18% |
| | 5. (0.00,3.91,90), (4.26,3.91,90) (-4.26,3.91,90), (7.45,3.87,99) | 0.814 | 112.17 | 18% |

Table 2. Mechanical behavior of a (10,0) SWCNT with one, two, three and four defects at different positions

groups, mainly carboxylic group which in turn, is exploited for better adhesion between CNT and matrix in the composite materials. Composites made up of CNT as reinforcements have many potential applications, e.g. for making of spacecrafts, sports goods etc. Y or T junctions may be prepared utilizing the presence of defects which can be used for preparing nano transistor. The functionining of those devices may be influenced by the presence of odd or even number of defects. Also, several such junctions can be produced and linked up for building up of mesh like architectural construction. The interaction between the odd and even number of defects may be exploited to strengthen such structure. The extraordinary mechanical, electrical and thermal properties of the CNTs have been exploited in biological applications also, basically in drug delivery system.

| | Position of defect/defects (Z, r, θ) | Young's modulus (TPa) | Failure Stress(GPa) | Maximum Strain |
|---|---|---|---|---|
| Single defect: | 1. (0.61, 2.65, 92) | 0.808 | 133.51 | 24% |
| | 2. (9.89, 2.7, 97) | 0.752 | 115.25 | 23% |
| | 3. (-9.89, 2.73, 83) | 0.774 | 98.907 | 21% |
| | 4. (14.91, 2.65, 0) | 0.743 | 111.09 | 23% |
| | 5. ( -14.91, 2.65, 0) | 0.81 | 121.41 | 22% |
| | 6. (20.24, 2.7, 39) | 0.78 | 114.57 | 21% |
| | 7. (-19.63, 2.73, 53) | 0.809 | 111.47 | 22% |
| 2 defects | : 1. (-1.83 ,2.65,84), (0.61,2.65,92) | 0.75 | 149.72 | 26% |
| | 2. (4.56,2.65,59), (-4.72,2.7,-53) | 0.757 | 129.95 | 24% |
| | 3. (5.02,2.7,83), (-5.02,2.7,-83) | 0.803 | 119.93 | 23% |
| | 4. (-1.22,2.65,-4), (1.22,2.65,4) | 0.832 | 100.1 | 21% |
| | 5. (-19.93,2.7,97), (19.93,2.73,83) | 0.791 | 113.73 | 26% |
| | 6. (0.61,2.65,92),(-0.61,2.65,-92) | 0.821 | 96.76 | 20% |
| 3 defects | : 1.(0.46,2.73,24), (3.04,2.65,119), (5.63,2.73,174) | 0.715 | 101.14 | 17% |
| | 2.(-0.76,2.7,20), (2.89,2.73,31), (6.09,2.65,18)) | 0.742 | 108.26 | 23% |
| | 3. (-5.02,2.73,97), (0.61,2.65,92), (5.02,2.7,83) | 0.775 | 107.43 | 20% |
| | 4. (-15.52,2.65,88), (0.61,2.65,92), (14.3,2.65,88) | 0.809 | 130.23 | 26% |
| | 5. (-15.52,2.65,88), (0.46,2.73,24), (15.52,2.65,-88) | 0.748 | 106.35 | 21% |
| | 6. (-21.15,2.73,94), (0.61,2.65,92), (21.15,2.7,86) | 0.729 | 113.408 | 22% |
| | 7. (-19.93,2.7,97), (0.46,2.73,24), (19.93,2.73,83) | 0.758 | 85.57 | 26% |
| 4 defects | : 1. ( -5.02, 2.73, 97), (0.61, 2.65, 92), ( 3.04, 2.65, 99), (7.45, 2.7, 90) | 0.74 | 93.81 | 20% |
| | 2. (-15.52, 2.65, 88), (-7.45, 2.73, 90), (7.45, 2.7, 90), (14.30, 2.65, 88) | 0.758 | 145.22 | 26% |
| | 3. (-15.52, 2.65, 88), (-5.02, 2.73, 97), (5.02, 2.73, -97), (15.52, 2.65, -88) | 0.756 | 88.09 | 18% |
| | 4. (-19.93, 2.7, 97), (-9.89,2.73, 83), (0.00,3.91,90), (4.26,3.91,90) | 0.792 | 98.92 | 20% |
| | 5. (-6.24, 2.7, 94), (-2.59, 2.73, 105) (0.61, 2.65, 92), (3.8, 2.73, 79) | 0.814 | 86.81 | 16% |

Table 3. Mechanical behavior of a (5, 3) SWCNT with one, two, three and four defects at different positions

For hydrogen storage, the defect sites are responsible for better absorption. The interaction between defects should be studied for the full exploitation of CNT properties, particularly in composite materials for achieving the target of their possessing very high strength. While functionalizing a CNT for making composite materials, the reduction of the strength of the CNT due to defect-defect interaction may be controlled by functionalizing the CNTs at proper defect positions.

Not only the number of defects, but their position and orientation are also important while analyzing their effects on the mechanical properties of the carbon nanotubes. This dependence varies with the symmetry of the tubes. Young's modulus of a zigzag tube with four defects placed symmetrically almost on the same line is reduced largely compared to a defect free tube. Two diagonal and overlapping defects are found to have maximum reducing effect on failure stress for the tube. The ductility of a zigzag tube is almost insensitive to the presence of defects in its structure.

For a chiral tube, the maximum reduction of Y value is noticed for three well separated symmetrically situated defects. The failure stress of a chiral tube is mostly influenced by three well separated defects (symmetrically situated around z=0 position). 46.7% reduction of failure strain is calculated for a chiral tube with four defects for some specific combinations. For an armchair tube reduction of Y value is maximum for 1 defect near the edge. 56.3% reduction of strength is reported for the tube with three defects on the same line but asymmetrically placed about z=0. 31.25% reduction of failure strain is reported for an armchair tube with four defects for some specific combinations.

There are many possibilities of the arrangement of different number of defects due to which the strength and stiffness of the CNTs may vary differently. In our investigation, we have observed the variation of these mechanical properties (strength and stiffness) for several combinations of separating distances as well as different orientations of the defects in the structure of the CNTs. In some cases, the reduction of these major properties is significant whereas in some other cases it is less. Now, if we can functionalize the CNTs at those defect sites where, the reduction of their strength or stiffness is negligible, then it may be possible to fabricate composite materials with such functionalized CNTs without losing their strength or stiffness much. For example, for two SW defects at (z, r, θ) positions of (0, 3.91, 90) and (8.52, 3.91,-90), the reduction of both the strength and stiffness of the zigzag SWCNT (10, 0) is minimum. So, functionalizing the tube at those defect sites will not reduce much of the strength of the composite material reinforced by them. Development of better experimental technique to identify the suitable sites of defects for functionalization will help the composite builders in this respect in near future.

## 6. Variation of mechanical properties and fracture behavior of single wall carbon nanotubes on bundle formation

Using the 2$^{nd}$ generation REBO potential with smoothing at cut-off region, MD simulation is carried out for three different types of tubes. Then bundle of each type is taken containing three of them in each bundle and then one bundle of the mixture of the three types is taken and simulated separately. Lengths of the single zigzag, armchair and chiral tubes are 85.18 Å 98.58 Å, 119.26 Å respectively. As before, room temperature is maintained by Berendsen thermostat. By stretching the tubes in small strain increments, the equilibrium potential energy is calculated by simulation. Keeping one end fixed the other end of the bundle is stretched gradually from the unstretched condition in such a way that all the tubes are

| | Position of defect/defects (Z, r, θ) | Young's modulus (TPa) | Failure Stress(GPa) | Maximum Strain |
|---|---|---|---|---|
| Single defect: | 1. (0.61, 23.37, 90) | 0.749 | 112.56 | 26% |
| | 2. (5.53, 3.37, 90) | 0.749 | 111.05 | 26% |
| | 3. (-5.53, 3.37, 90) | 0.741 | 141.42 | 27% |
| | 4. (15.37, 3.37,90) | 0.749 | 119.44 | 26% |
| | 5. ( -15.37, 3.37,90) | 0.741 | 120.26 | 24% |
| | 6. (19.06, 3.37, 90) | 0.741 | 106.72 | 28% |
| | 7. (-1906, 3.37, 36) | 0.572 | 178.16 | 28% |
| 2 defects | : 1. (0.61, 3.37,90), (0.61,3.37, -90) | 0.717 | 120.97 | 25% |
| | 2. (-5.53, 3.37, 90),(5.53,3.37, 90) | 0.74 | 109.67 | 28% |
| | 3. (-5.53, 3.37, 90), (5.53, 3.337, -90) | 0.77 | 190.38 | 30% |
| | 4. (-15.37, 3.37, 90), (15.37,3.37,90) | 0.672 | 103.09 | 20% |
| | 5. (-15.37,3.37,90),(15.37,3.37,-90)) | 0.744 | 103.97 | 27% |
| | 6. (-0.61,3.37,90), (0.61, 3.37,126) | 0.711 | 190.32 | 27% |
| | 7. (-0.61,3.37, 90), (1.84,3.37,90) | 0.678 | 106.01 | 26% |
| 3 defects | : 1.(-5.53,3.37,90), (0.61, 3.37,90), (5.53, 3.37,90) | 0.726 | 124.83 | 29% |
| | 2.(-5.53,3.37,-90),(0.61,3.37,90), (5.53,3.37,-90) | 0.714 | 99.78 | 24% |
| | 3. (-20.29,3.37,90),(0.61,3.37,90) (20.29,3.37,90) | 0.726 | 103.42 | 27% |
| | 4. (-20.29,3.37,-90)(0.61,3.37,90), (20.29,3.37,-90) | 0.766 | 103.18 | 28% |
| | 5. (0.61,3.37,90),(6.76,3.37,90) (15.37,3.37,90) | 0.732 | 108 | 27% |
| | 6. (0.61,3.37,90), (6.67,3.37,-90) (15.37,3.37,-90) | 0.76 | 148.3 | 24% |
| | 7. (-1.84,3.37,90),(-0.61,3.37,126) (1.84,3.37,162) | 0.724 | 109.02 | 26% |
| | 8. (-1.84,3.37,90),(0.61,3.37,90) (3.07,3.37,90 | 0.625 | 85.79 | 26% |
| 4 defects | : 1. ( -11.68,337,90), (-4.3,3.37,90) ( 4.3,3.37,90),(11.68,3.37,90 | 0.813 | 108.57 | 26% |
| | 2. (-11.68,3.37,-90),(-4.3,3.37,-90), (4.3,3.37,90), (11.68,3.37,90) | 0.764 | 98.54 | 26% |
| | 3. (-11.68,3.37,-90),(4.3,3.37,-90) (-4.3,3.37,90),(11.68,3.37,90) | 0.679 | 92.82 | 26% |
| | 4. (-4.3,3.37,90),(-1.84,3.37,126) (5.53,3.37,90),(6.76,3.37,126) | 0.706 | 91.09 | 22% |
| | 5. (-1.84,3.37,90),(0.61,3.37,90) (4.3,3.37,90),( 6.76,3.37,90) | 0.773 | 113.12 | 26% |

Table 4. Mechanical behavior of a (5, 5) SWCNT with one, two, three and four defects at different positions

stretched equally. Intertube interaction is modeled with Lennard-Jones potential. For a SWCNT bundle, volume of each tube is calculated separately and then added to get the total volume.

### 6.1 Mechanical characteristics of isolated tubes

A zigzag SWCNT exhibits a high Young's modulus value of 1.47 TPa with 16% ductility. Calculated failure stress is 76.77 GPa. Stress strain curves of the three tubes are shown in Fig 8.1. For the (5, 0) tube, stress increases slowly upto 8% more or less linearly and then flattens giving a yield point near 9% strain. On straining the tube beyond 12% strain for a zigzag SWCNT, as reported by Zhang et al. (Zhang et al., 1998), we have observed SW rotation that ultimately leads to rupture of the tube. After 18% strain, the tube breaks totally with the formation of a series of SW rotations. The stress strain curves of isolated (5, 5) and (5, 3) tubes [Fig 18] are somewhat different in nature. Also their failure stress and failure strain values are different. The single (5, 5) tube has the highest fracture strain of 26% compared to the other two. Table 5 gives the detailed picture of the failure stress and strain values. However, the chiral SWCNT shows the maximum breaking stress of 115.65 GPa.

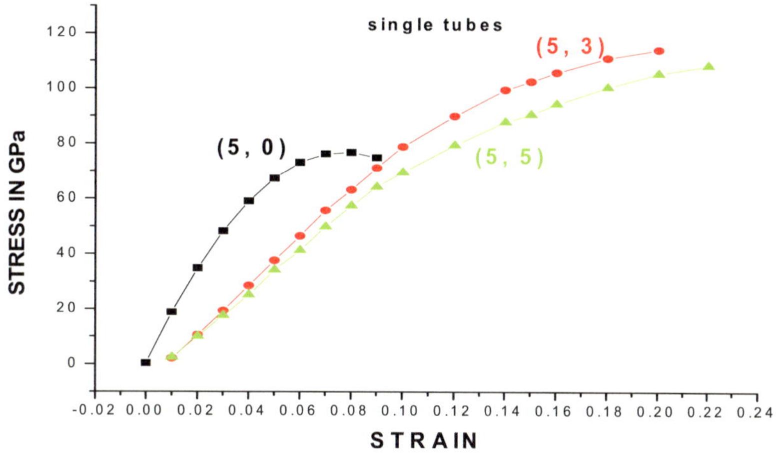


Fig. 18. Stress-strain curves of three different types of isolated tubes.

Young's modulus values agree very well with the experimental data of Krishnan et al. (Krishnan et al. 1998) Wong et al. (Wong et al., 1997), and Treacy et al. (Treacy et al., 1996). Failure stress or strain of the single tubes does not match with the experimental findings of Yu et al. (Yu et al., 2000c) which, we will show in the next part of this section, match with experiment when mixture of bundles of nanotubes are formed. Yu et al. (Yu et al., 2000c) observed 10-13% maximum strain for SWCNT bundle with failure stress varying in between 13- 52 GPa. For single tubes, our calculated values of failure stress matches with Mielke et al (Mielke, 2004) and Troya et al. (Troya, 2003) with the same potential.

Failure of these tubes [Fig 19] does not show complete rupture of the tubes. Rather they exhibit breaking of bonds all over the tubes at the breaking point.

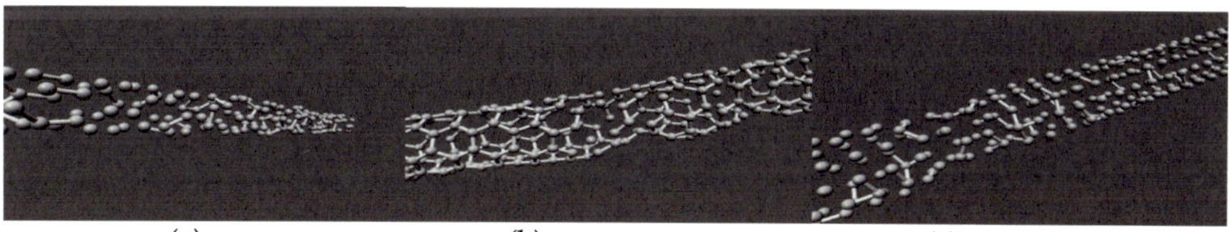

(a) (b) (c)

Fig. 19. Breaking patterns of an isolated (a) zigzag (b) chiral (c) armchair SWCNT

### 6.2 Characteristics of a bundle of three different types of SWCNTs

Single- wall carbon nanotubes that are grown by laser ablation, (Eklund et al., 2002; Guo et al., 1995; Thess et al., 1996;) arc discharge (Bethune et al., 1993; Ebbesen & Ajayan, 1992; Iijima, 1991) or the most recent HiPCO method always occur in bundles. These bundles are held together by weak interactions between the tubes. The bundles of SWCNTs are different from the isolated tubes in two respects. A bundle contains nanotubes of different chirality and slightly different diameters (Henrard et al., 2000). Bundling also changes their properties by tube interaction. Surprisingly, Young's modulus is noticeably increased to 1.60 TPa again and failure strain is reduced to 9% [Fig 20]. Failure stress is also reduced much giving a value of 68.50 GPa. Fracture of a SWCNT bundle with three (5, 3) chiral tubes is depicted in Fig 21 which are very much different in nature from the stress-strain curves of isolated tubes [Fig 19]. Table 5 shows the results in tabulated form.

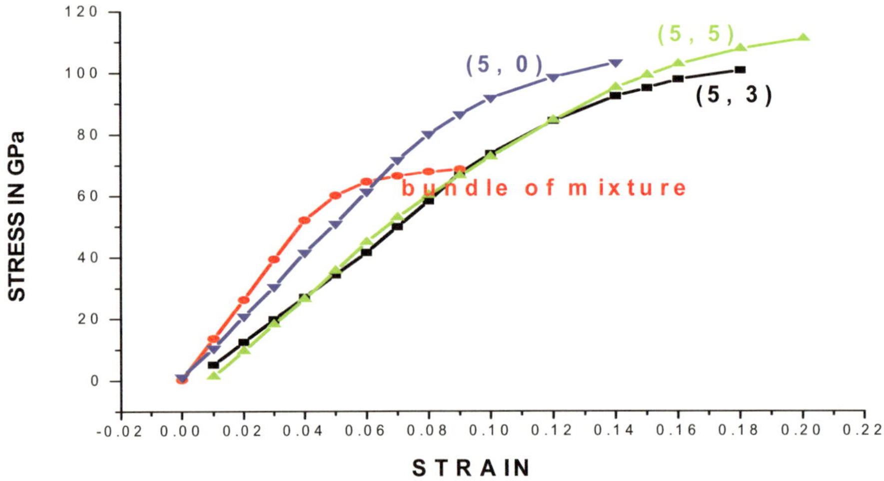


Fig. 20. Stress-strain curves of bundles of different SWCNTs

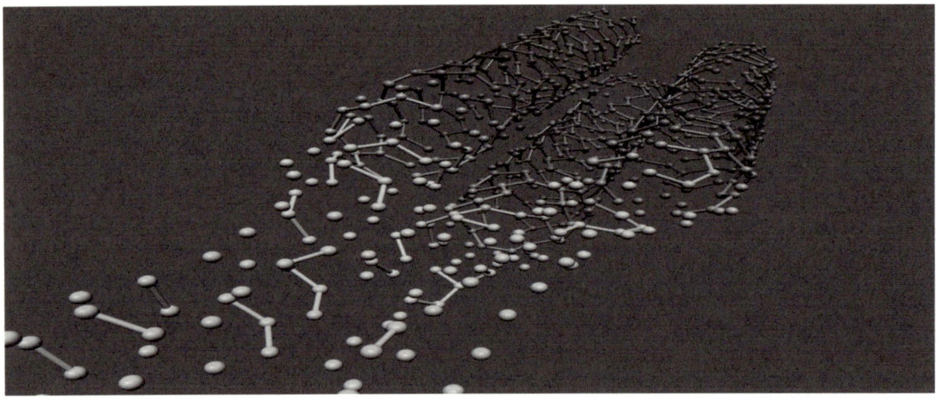

Fig. 21. Failure of a bundle of three (5, 3) SWCNTs

This changed behavior may be explained by overlapping of energy bands in nanotube bundles. Scanning tunneling spectroscopy (Wildoer et al., 1998), resonant Raman scattering (Jorio et al., 2001) and optical absorption or emission measurements (Connell et al., 2002) confirmed the electronic DOS in carbon nanotubes. Ouyang et al (Ouyang et al., 2001) experimentally found a pronounced dip in the electronic DOS of a (8, 8) SWCNT at Fermi level inside a bundle, compared to an isolated tube. This is due to symmetry breaking by other tubes in proximity. Not always a gap is produced, but gap may also be closed for such a tube. Closing of band gap is observed for a (10, 0) tube in a bundle in (Reich et al., 2002). Due to this type of overlapping of electronic bands, attraction or repulsion arises inside the SWCNT bundle. All intermolecular/van der Waals forces are anisotropic which means that they depend on the relative orientation of the molecules. The induction and dispersion interactions are always attractive, irrespective of orientation, but the electrostatic interaction changes sign upon rotation of the molecules. That is, the electrostatic force can be attractive or repulsive, depending on the mutual orientation of the molecules giving rise to different interaction between the tubes of a bundle.

In the low energy part of the band structure the bundling of the nanotubes changes the electronic properties by symmetry breaking and by the intratube dispersion perpendicular to $K_z$.. That also holds good for larger electronic energies also. Again, when isolated tubes of different symmetries are present, energy bands are strongly split. Reich et al. (Reich et al., 2002) showed that in a high symmetry packing of (6, 6) nanotube the degenerate bands of isolated tube remained degenerate by symmetry in the crystal. In their study, it was also revealed that the dispersion of the electronic bands perpendicular to $k_z$ is less in a zigzag (10, 0) tubes than a (6, 6) tubes. The first two valance states at the Γ point of the brillouin zone results in a strong dispersion in the corresponding states perpendicular to $k_z$ for armchair tubes. Chiral tubes are likely to be more complicatedly influenced by symmetry breaking and band splitting. The change of behavior of the curves for an isolated tube and their bundles can thus be explained by the change of DOS which must be taken into account while comparing a theoretical result with the experimental findings.

| SWCNT | Young's Modulus (TPa) | Failure stress (GPa) | Failure strain |
|---|---|---|---|
| Zigzag (5,0) | | | |
| Single: | 1.468 | 76.770 | 14% |
| Bundle: | 1.013 | 105.008 | 15% |
| Armchair (5,5) | | | |
| Single: | 0.792 | 107.913 | 26% |
| Bundle : | 0.867 | 112.032 | 21% |
| Chiral (5,3) | | | |
| Single: | 0.831 | 115.652 | 22% |
| Bundle: | 0.729 | 101.017 | 18% |
| Mixture of three Different tubes | 1.602 | 68.500 | 9% |

Table 5. Young's modulus, tensile strength and ductility of different SWCNT types and their bundles

### 6.3 Effect of the number of tubes in the SWCNT bundle on their mechanical behavior

We have investigated the influence of an increasing number of tubes in an armchair (5, 5) single-wall carbon nanotube bundle on their mechanical properties under tensile and compressive loading by molecular dynamics simulation. 2nd generation Brenner bond order potential is adopted for energy minimization in the simulation process.

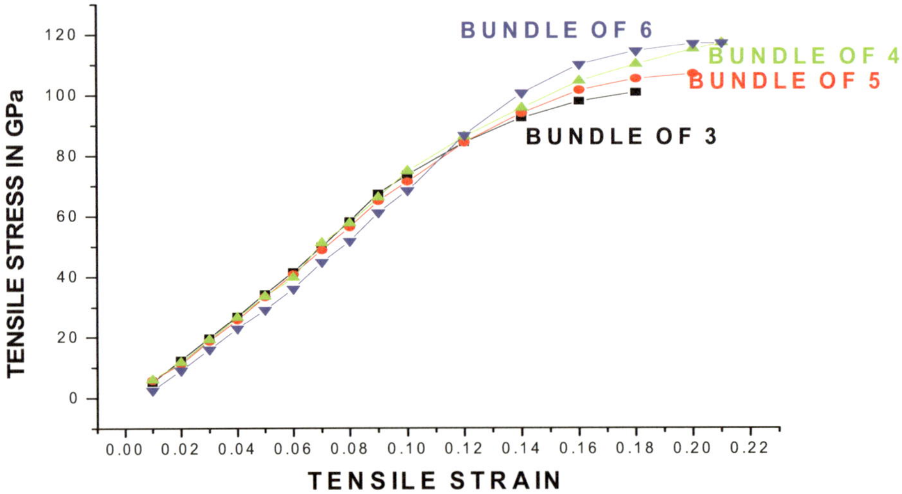


Fig. 22. Tensile stress vs. strain curves of bundles of (5, 5) SWCNTs with different number of tubes

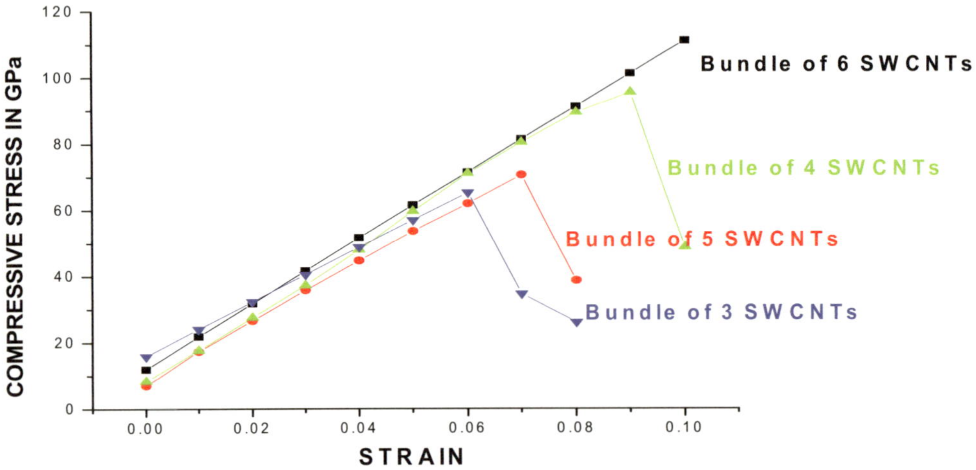


Fig. 23. Compressive stress vs. strain curves of bundles of (5, 5) SWCNTs with different number of tubes

From the stress-strain curves of the bundles [Fig 22 and 23] it can be inferred that with the increase of the number of tubes in the bundle, the bundle becomes more stiff and strong. Resemblance of the curves is noticed for 4 and 6 membered bundles in Fig 8.9. The same Figure also shows the resemblance of the curves for 3 and 5 membered bundles. The Young's modulus decreases while the compressive modulus increases with the number of tubes in a bundle [Fig 24]. But, overall increase of tensile and compressive stress [Fig 25] is reported in this study. The maximum Young's modulus of 0.73 TPa and maximum compressive modulus of 1.00 TPa are obtained in our calculation for a bundle of three tubes. These are within the range of the experimental findings of Yu et al. (Yu et al., 2000c) and Treacy et al. (Treacy, 1996) Failure strength of about 100 GPa can be matched with the observations by Demczyk et al. (Demczyk et al., 2002) Ductility, tensile and compressive strengths agree well with the experimental results as well as with other MD simulation calculations.

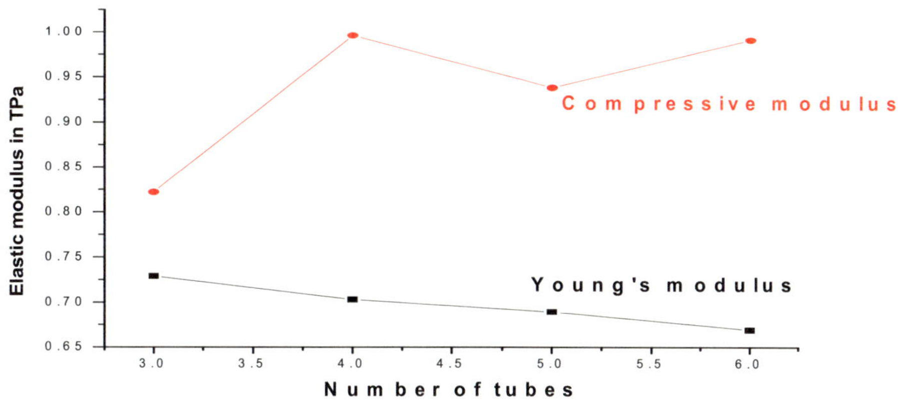


Fig. 24. The variation of elastic moduli with the number of tubes in a bundle

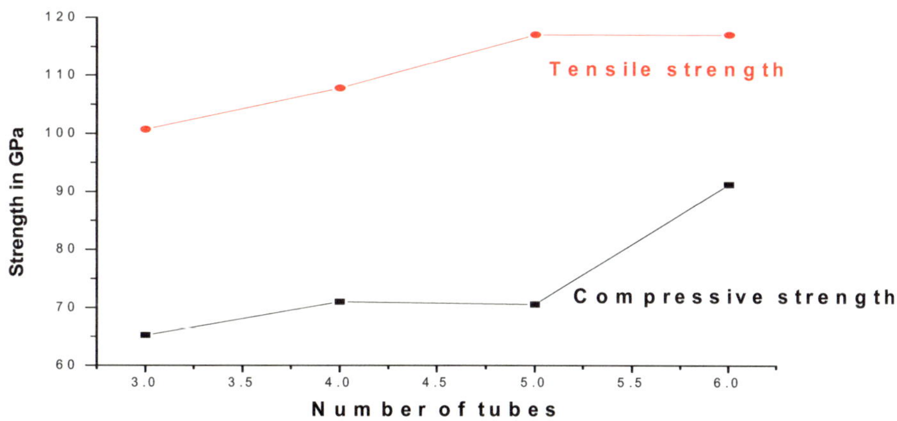


Fig. 25. The variation of strength with the number of tubes in a bundle

When the fracture of these bundles is modeled, it is observed that the bundle consisting of a fewer number of tubes breaks completely at the maximum stain value where all of its members break equally. On the other hand, only one or two tubes of a bundle with a larger number of tubes break completely at the same strain value. Compression produces kinks in the structure of the SWCNT bundles. They show complete collapse from the position of kink formation at a certain strain. The cross section of the tubes in a bundle of the carbon nanotube changes on the application of force from circular to oblate and more deformed shape appears on the increase of strain. The breaking and buckling of the bundles are depicted in Fig 26 (a-j).

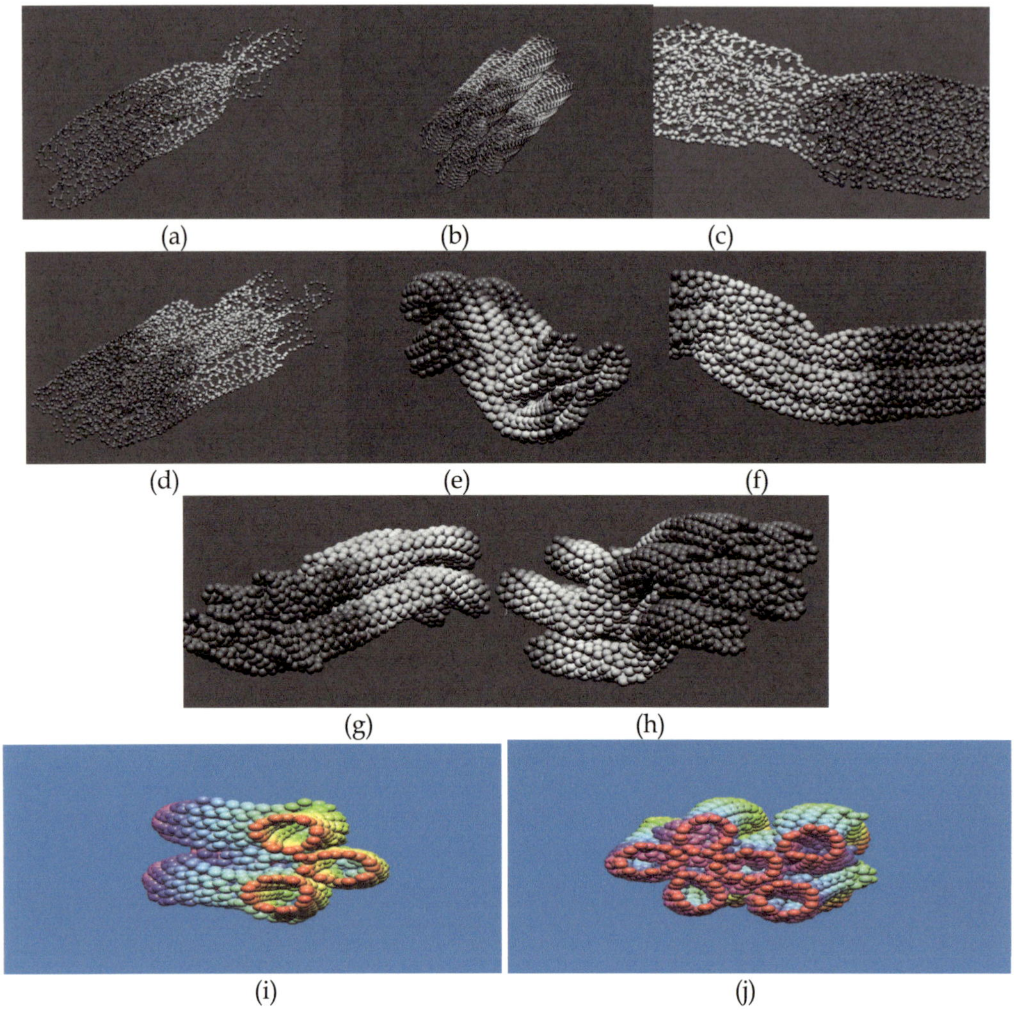

Fig. 26. **(a)** Breaking of a 3 membered bundle **(b)** A bundle of 4 tubes **(c)** breaking of a 5 membered bundle. It breaks from the middle portion. **(d)** A six membered bundle breaks near the end (not all the tubes are broken at a time) **(e)** buckling of a bundle of 3 tubes **(f)** buckling with 4 tubes **(g)** buckling with 5 tubes (tubes are distorted near the loading edge **(h)** buckling of a six membered bundle **(i)** Distortion of the cross section of the tubes in a bundle of three SWCNTs under pressure. **(j)** Distortion of the cross section is more for a bundle of five SWCNTs. Different cross-sectional changes are observed for different tubes in the same bundle

Comparing the MD simulation results of different SWCNTs and their bundles, it can be concluded that a major reason of non-compatibility of the theoretical results with the experimentally obtained low values of failure strain and failure stress is the existence of the SWCNTs in the form of bundles where different types of tubes may be present. While forming the bundles, the zigzag tube resists the external force strongly and the armchair tube breaks symmetrically on the application of the tensile force. 25% reduction of maximum strain is observed for a bundle of three chiral SWCNTs compared to a single tube. The reduction is maximum for a bundle of the mixture of three types and we get 43.75% decrease of the strain from the lowest strain of 16% for a single zigzag tube. Overlapping of density of states is responsible for the changed behavior of the bundle of nanotubes. Also the number of tubes in a SWCNT bundle affects their mechanical properties in a great extent.

## 7. Dependence of the mechanical properties of SWCNTs on the choice of interatomic potential functions

The choice of potential function in explaining the mechanical properties of CNTs is a vital factor. In Fig 1, we have already shown the stress-strain curves of three different types of defect-free SWCNTs with 2nd generation REBO potential. An armchair (5, 5), a chiral (5, 3) and a zigzag (5, 0) SWCNT exhibit stress-strain curves as shown in Fig 27 with tight binding potential. In a (5, 0) SWCNT the plastic flow region contains a plateau after which, the stress increases to 161.00 GPa. The armchair and the chiral tubes show lesser tensile strength of 141.52 GPa and 125.61 GPa respectively. But their failure strain values are same as 26%. In tight binding approximation, the (5, 5) and the (5, 3) tubes show some ductility before failure.

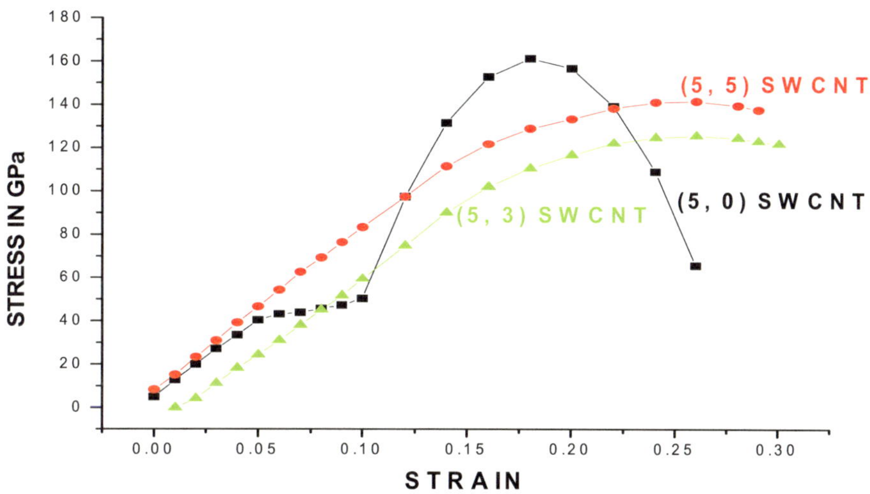


Fig. 27. Stress-strain curves of three different types of SWCNTs with tight binding potential

For three different types of defect-free tubes, the mechanical characteristics are tabulated below (Table 6). Results for the smoothed off REBO potential are taken from Fig 18.

From Table 5, we can conclude that reasonable results are obtained for REBO potential with smoothed off cut-off function. For the (5, 0) tube failure strength of 76.77 GPa and failure strain of 14% are much less than the values with other potentials. Other tubes also show much lower values of the mechanical characteristics with the former potential with respect to the others. So, the 2nd generation REBO potential with smoothed off cut-off function gives more reasonable values with respect to the experimental values (Yu et al., 2000c). The discrepancies, still present may be due to the availability of the experimental samples in the form of bundles or due to the presence of defects in the samples.

| SWCNT | Tight Binding | | | 2nd generation REBO | | | Smoothed off REBO | | |
|---|---|---|---|---|---|---|---|---|---|
| | Y.M. (TPa) | F.S. (GPa) | F.Sr. | Y.M. (TPa) | F.S. (GPa) | F.Sr. | Y.M. (TPa) | F.S. (GPa) | F.Sr. |
| (5, 0) | 1.13 | 161.00 | 18% | 1.44 | 144.70 | 20% | 1.47 | 76.77 | 14% |
| (5, 5) | 0.81 | 141.52 | 26% | 0.81 | 196.30 | 32% | 0.79 | 107.91 | 26% |
| (5, 3) | 1.00 | 125.61 | 26% | 0.89 | 149.88 | 30% | 0.83 | 115.65 | 22% |

Table 6. Comparison of the results obtained for three different potentials for defect-free tubes

## 8. Conclusions

The dependence of mechanical properties of SWCNTs on various factors is discussed in this chapter. So by the quantitative investigation it can be concluded that for engineering interest, the mechanical properties of the CNTs can be tailored by introducing suitable number of defects in suitable positions and orientations. In the present situation, the designing of composites with desired properties is a challenging job which depends on more understanding of the properties of these reinforcing agents.

A striking change in the mechanical properties of a SWCNT is noticed when some of them form a bundle. Though we have obtained the Young's modulus and failure stress values in the experimentally specified range, the failure strain is higher than that of the experimental values for a single tube. An armchair, a zigzag and a chiral SWCNTs show higher failure strain than their bundles. Due to the overlapping of density of states a bundle of their mixture shows a drastic reduction of the failure strain (from 18% for a defect-free tube to 9% on bundle formation). Repulsive interaction between these tubes is observed in this study. Breaking of symmetry of a tube in proximity of the others is the possible reason of such changes. When we increased the number of tubes in a bundle of an armchair SWCNT, the bundle becomes stronger and stiffer. With more tubes, a bundle possesses high buckling strength also. The cross section of the tubes in a bundle deform under pressure. The cross section changes from circular to oval under the application of pressure. More deformation is observed with the increase of compressive force.

We have tried to increase the acceptability of the results obtained in the simulation procedure by choosing different potentials, in particular the Tersoff-Brenner potential and tight binding potential. Tight binding potential is used as it is.

But we have carried out simulation with 2nd generation REBO potential and also with smoothed off 2nd generation REBO potential to avoid the sudden hike of energy near the cut-off region. More reasonable results are obtained for a smoothed-off REBO potential at the cut-off region.

## 9. References


Ajayan, P. M.; Ravikumar, V. & Charlier, J.-C. (1998). Surface Reconstructions and Dimensional Changes in Single-Walled Carbon Nanotubes. *Physical Review Letters*, vol. 81, No. 7, pp. 1437-1440, ISSN 0031-9007

Alder, B.J. & Wainwright, T.E. (1957). Phase Transition for a Hard Sphere System. *Journal of Chemical Physics*, vol. 27, No. 5, pp. 1208-1209, ISSN 0021-9606

Alder, B.J. & Wainwright, T.E. (1959). Studies in Molecular Dynamics. I. General Method. *Journal of Chemical Physics*, vol. 31, No. 2, pp. 459-466, ISSN 0021-9606

Allen, M.P. & Tildesly, D.J. (1989). *Computer Simulation of Liquids*. Oxford University press, ISBN 0 19 855645, Oxford, New York

Andrews, R.; Jacques, D.; Rao, A.M.; Rantell, T.; Derbyshire, F. & Chen, Y. et al. (1999). Nanotube composite carbon fibers. *Applied Physics Letters*, vol. 75, No. 9, pp 1329-3, ISSN 0003-6951

Ashby, M.F. (1996). MOdelling of materials problem. Journal of Computer-Aided Materials Design. Vol. 3, No. 1-3, pp.95-99, ISSN 0928-1045

Balasubramanian, K. & Burghard, M. (2005). Chemically functionalized carbon nanotubes. *Small*, vol. 1, No. 2, pp. 180-192, ISSN 1613-6829

Banerjee, S.; Hemraj-Benny, T. & Wong, S.S. (2005). Covalent Surface Chemistry of Single-Walled Carbon Nanotubes. *Advanced Materials*, vol. 17, No. 1, pp. 17-29, ISSN 0935-9648

Banhart, F. (1999). Irradiation effects in carbon nanostructures. *Reports on Progress in Physics*, vol. 62, No. 8, pp. 1181, ISSN 0034-4885

Batra, R.C. & Sears, A. (2007). Uniform radial expansion/contraction of carbon nanotubes and their transverse elastic moduli. *Modelling and Simulation in Material Science and Engineering*, vol. 15, No. 8, 835, ISSN 0965-0393

Belytschko, T.; Xiao, S.P.; Schatz, G.C. & Ruoff, R.S. (2002). Atomistic simulations of nanotube fracture. *Physical Review B*, vol. 65, No. 23, pp. 235430-8, ISSN 1050-2947

Berendsen, H.J.C.; Postma, J.P.M.; van Gunsteren, W.F.; DiNola, A. & Haak, J. R. (1984). Molecular dynamics with coupling to an external bath. *Journal of Chemical Physics*, vol. 81, No. 8, pp. 3684-3690, ISSN 0021-9606

Bethune, D.S.; Kiang, C. H.; De Vries, M. S.; Gorman, G.; Savoy, R. & Vazquez, J. et al. (1993). Cobalt-catalyzed growth of carbon nanotubes with single-atomic-layer walls. *Nature*, vol. 363, pp. 605-607, ISSN 0028-0836

Brenner, D. W.; Shenderova, O.A.; Harrison, J.A.; Stuart, S.J.; Ni, B. & Sinnot, S.B. (2002). A second-generation reactive empirical bond order (REBO) potential energy expression for hydrocarbons. *Journal of Physics: Condensed Matter*, vol. 14, No. 4, 783-802, ISSN 1098-0121

Buehler, M.J. (2008). *Atomistic Modeling of Material Failure*. Springer Science and Business Media, ISBN 978-0-387-76425-2, New York, USA

Canto, E. D.; Flavin, K.; Movia, D.; Navio, C.; Bittencourt, C. & Giordani, S. (2011). A Critical Investigation of Defect Site Functionalization on Single Walled Carbon Nanotubes. *Chemistry of Materials*, vol. 23, No. 1, pp. 67-74, ISSN 0897-4756

Chae, H.G.; Minus, M. L. & Kumar, S. (2006). Oriented and Exfoliated Single Wall Carbon Nanotubes in Polyacrylonitrile. *Polymer*, vol. 47, No. 10, pp. 3494-3504, ISSN 0032-3861

Chandra, N.; Namilae, S. & Shet, C. (2004). Local elastic properties of carbon nanotubes in the presence of Stone-Wales defects. *Physical Review B*, vol. 69, No. 9, pp. 094101-12

Chou, T. W.; Gao, L.; Thostenson, E.T.; Zhang, Z. & Byun, J.H. (2010). An assessment of the science and technology of carbon nanotube-based fibers and composites. *Composites Science and Technology*, vol. 70, No.1, pp. 1-19, ISSN 0266-3538

Coluci, V.R.; Pugno, N.M.; Dantas, S.O.; Galvao, D.S. & Jorio, A. (2007). Atomistic simulations of the mechanical properties of 'super' carbon nanotubes. *Nanotechnology*, vol 18, No. 33, pp. 335702, ISSN 1550-7033

Dalton, A.B.; Collins, S.; Muñoz, E.; Razal, J.M.; Ebron, V.H. & Ferraris, J.P. et al. (2003). Super-tough carbon nanotube fiber. *Nature*, vol. 423, No. 6941, pp. 703-703, ISSN 0028-0836

Demczyk, B.G.; Wang, Y.M.; Cumings, J.; Hetman, M.; Han, W. & Zettl, A. et al. (2002). Direct mechanical measurement of the tensile strength and elastic Modulus of multiwalled carbon nanotubes. *Materials Science and Engineering A*, vol. 334, pp. 173-178, ISSN 0921-5093

Dereli, G. & Özdoğan, C. (2003). Structural stability and energetics of single walled carbon nanotubes under uniaxial strain. *Physical Review B*, vol. 67, pp. 035416-11

Despres, J. F.; Daguerre, E. & Lafdi, K. (1995). Flexibility of graphene layers in carbon nanotubes, *Carbon*, vol. 33, No. 1, pp. 87-92, ISSN 0008-6228

Ebbesen, T. W. & Ajayan, P. M. (1992). Large scale synthesis of carbon nanotubes. *Nature*, vol. 358, pp. 220-222, ISSN 0028-0836,

Ebbesen, T. W. & Takada, T. (1995). Topological and $SP^3$ defect structures in nanotubes. *Carbon*, vol. 33, No. 7, pp. 973-978, ISSN 0008-6223

Eklund, P.C.; Pradhan, B.K.; Kim, U.J.; Xiong, Q.; Fischer, J.E. & Friedman, A.D. et al. (2002). Large scale production of

single-walled carbon nanotubes using ultrafast pulses from a free electron laser. *Nano Letters*, vol. 2, No. 6, pp. 561-566, ISSN 1530-6984

Fan, Y.; Goldsmith, B. R. & Collins, P. G. (2005). Identifying and counting point defects in carbon nanotubes. *Nature Materials*, vol. 4 , pp. 906-8911, ISSN 1476-1122

Felten, A.; Gillon, X.; Gulas, M.; Pireaux, J. J. ; Ke, X. & Tendeloo, G.V. et al. (2010). Measuring Point Defect Density in Individual Carbon Nanotubes Using Polarization-Dependent X-ray Microscopy. *ACS Nano*, vol. 4, No. 8, pp. 4431-6, ISSN 1936-0851

Guo, T.; Nikolaev, P.; Thess, A.; Colbert, D. T. & Smalley, R. E. (1995). Catalytic growth of single-walled manotubes by laser vaporization. *Chemical Physics Letter*, vol. 243, No.s 1-2, pp. 49-54, ISSN 0009-2614

Haskins, R.W.; Maier, R.S.; Ebeling, R.M.; Marsh, C.P.; Majure, D.L. & Bednar, A.J. et al. (2007). Tight-binding molecular dynamics study of the role of defects on carbon nanotube moduli and failure. *Journal of Chemical Physics*, vol. 127, No. 7, pp. 074708, ISSN 0021-9606

Henrard, L.; Loiseau, A.; Journet, C. & Bernier, P. (2000). *European Physical Journal B*. vol. 13, No. 4, pp. 661-669, ISSN 1434-6028

Hou, W. &. Xiao, S. (2007). Mechanical Behaviors of Carbon Nanotubes with Randomly Located Vacancy Defects *Journal of Nanoscience and Nanotechnology*, Vol. 7, No. 12, pp. 4478-4485, ISSN 1550-7033

Huq, A.M.A.; Goh, K.L.; Zhou, Z.R. & Loao, K. (2008). On defect interactions in axially loaded single-walled carbon nanotubes. *Journal of Applied Physics*, vol. 103, No. 5, pp. 054306-7, ISSN: 0021-8979

Iijima, S. (1991). Helical microtubules of graphitic carbon. *Nature*, vol. 354, pp. 56-58, ISSN 0028-0836

Iijima, S.; Brabec, C.; Maiti, A. & Bernholc, J, (1996). Structural flexibility of carbon nanotubes, *Journal of Chemical Physics*, vol. 104 , No. 5, pp. 2089-2092, ISSN 0021-9606

Jorio, A.; Filho, A.G.S.; Dresselhaus, G.; Dresselhaus, M.S.; Saito, R. & Hafner, J. H. et al. (2001). Joint density of electronic states for one isolated single-wall carbon nanotube studied by resonant Raman scattering. *Physical Review B*. vol. 63, No. 24, pp. 245416-4, ISSN 0163-1829

Kang, H.; Lim, S.; Park, N.; Chun, K.Y. & Baik, S. (2010). Improving the sensitivity of carbon nanotube sensors by benzene functionalization. *Sensors and Actuatuators B: Chemical*, vol. 147, No. 1, pp. 316-321, ISSN 0925-4005

Kearns, J.C. & Shambaugh, R.L. (2002). Polypropylene fibers reinforced with carbon nanotubes. *Journal of Applied Polymer Science*, vol. 86, No. 8, pp.2079-2084, ISSN 0021-8995

Koziol, K.; Vilatela, J.; Moisala, A.; Motta, M.; Cunniff, P.; Sennett, M. & Windle, A. (2007). High-Performance Carbon Nanotube Fiber. *Science*, vol. 318, No. 5858, pp. 1892-1895, ISSN 0965-0393

Krasheninnikov, A.V. & Banhart, F. (2007). Engineering of nanostructured carbon materials with electron or ion beams, *Nature Materials*, vol. 6, pp. 723-733, ISSN 1476-1122

Krishnan, A.; Dujardin, E.; Ebbesen, T.W., Yianilos, P.N. & Treacy, M.M.J. (1998). Young's modulus of single-walled nanotubes. *Physical Review B*, vol. 58, No. 20, pp. 14013-14019, ISSN 0163-1829

Kuronuma, Y.; Shindo, Y.; Takeda, T. & Narita, F. (2010). Fracture behaviour of cracked carbon nanotube-based polymer composites: Experiments and finite element simulations. *Fatigue and Fracture of Engineering Materials and Structures*, vol. 33, No. 2, pp. 87-93, ISSN 8756-758X

Lennard-Jones, J.E. (1924). On the Determination of Molecular Fields-II, Proceedings of Royal Society of London A, vol. 106, No. 738, pp. 463-477, ISSN 0950–1207

Liew, K.M.; Wong, C.H.; He, X.Q.; Tan, M.J. & Meguid, S.A. (2004). Nanomechanics of single and multiwalled carbon nanotubes. *Physical Review B*, vol. 69, No. 11, pp. 115429-8, ISSN 0163-1829

Lourie, O. & Wagner, H D. (998). Transmission electron microscopy observations of fracture of single-wall carbon nanotubes under axial tension, *Applied Physics Letters*, vol. 73 No. 24, pp. 3527

Mielke, S. L.; Troya, D.; Zhang, S.; Li, J.L.; Xiao, R.C. & Ruoff, R.S. et al.. (2004). The role of vacancy defects and holes in the fracture of carbon nanotubes. *Chemical Physics Letters*, vol. 390, pp 413-420 ISSN 0009-2614

Miyamoto, Y.; Rubio, A.; Berber, S.; Yoon, M. & Tomanek, D. (2004). Theoretical identification of Stone-Wales defects in nanotubes. *Physical Review B*, vol. 69, pp. 121413-4, 0163-1829

Miyata, Y.; Mizuno, K. & Kataura, H. (2011). Purity and Defect Characterization of Single-Wall Carbon Nanotubes Using Raman Spectroscopy. *Journal of Nanomaterials*, doi:10.1155/2011/786763, ISSN 1687-4129

Mora, R.J.; Vilatela, J.J. & Windle, A. (2009). Properties of composites of carbon nanotube fibres. *Composites Science and Technology*, vol. 69, No 10, pp. 1558-1563, ISSN 0266-3538

Nardelli,M.B.; Yakobson, B.I. & Bernholc, J. (1998). Mechanism of strain release in carbon nanotubes. *Physical Review B*, vol. 57, No. 8, pp. R4277-R4280, ISSN 0163-1829

O'Connell, M.J.; Bachilo, S.M.; Huffman, C.B.; Moore, V.C.; Strano, M.S. & Haroz, E.H. et al. (2002). Band Gap

Fluorescence from Individual Single-Walled Carbon Nanotubes. *Science*, vol. 297, No. 5581, pp. 593-596, ISSN 0036-8075

Ouyang, M.; Huang, J.-L.; Cheung, C.L. & Lieber, C.M. (2001). Energy Gaps in "Metallic" Single-Walled Carbon Nanotubes. *Science*, vol. 292 No. 5517, pp. 702-705, ISSN 0036-8075

Ozaki, T.; Iwasa, Y. & Mitani, T. (2000). Stiffness of single-walled carbon nanotubes under large strain. *Physical Review Letters*, vol. 84, No. 8, pp. 1712-1715, ISSN 0031-9007

Picozzi, S.; Santucci, S.; Lozzi, L.; Valentini, L. & Delley, D. (2004). Ozone adsorption on carbon nanotubes: the role of Stone-Wales defects. *Journal of Chemical Physics*, vol. 120, No. 5, 7147-7152, ISSN 0021-9606

Pozrikidis, C. (2009). Effect of the Stone-Wales defect on the structure and mechanical properties of single-wall carbon nanotubes in axial stretch and twist. *Archive of Applied Mechanics*, vol. 79, No. 2, pp.113-123, ISSN 0939-1533

Qi, P.; Vermesh, O.; Grecu, M.; Javey, M.; Wang, Q. & Dai, H. et al. (2003). Toward Large Arrays of Multiplex Functionalized Carbon Nanotube Sensors for Highly Sensitive and Selective Molecular Detection. *Nano Letters*, vol. 3, No. 3, pp. 347-351, ISSN 1530-6984

Rahman, A. (1964). Correlations in the Motion of Atoms in Liquid Argon. *Physical Review A*, vol. 136, No. 2A, pp. A405-A411, ISSN 1050-2947

Reich, S.; Thomsen, C. & Ordejon, P. (2002). Electronic band structure of isolated and bundled carbon nanotubes. *Physical Review B*, vol. 65, No. 15, pp. 155411-11

Rodríguez-Manzo, J. A.; Tolvanen, A. ; Krasheninnikov, A. V.; Nordlund, K.; Demortière, A. & Banhart, F. (2010). Defect-induced junctions between single- or double-wall carbon nanotubes and metal crystals. *Nanoscale*, vol. 2, No. 6, pp. 901-905. ISSN 2040-3364

Salvetat, J.-P.; Kulik, A.J.; Bonard, J.-M.; Briggs, G.A.D.; Stöckli, T. & M´et´enier, K. et al. (1999a). Elastic modulus of ordered and disordered multiwalled carbon nanotubes. *Advanced Materials*, vol. 11, No. 2, pp. 161-165, ISSN 0935-9648, ISSN 0935-9648

Salvetat, J.-P. ; Briggs, G.A.D.; Bonard, J.-M.; Bacsa, R.R.; Kulik, A.J. & Stöckli, T. et al. (1999b). Elastic and shear moduli of single-walled carbon nanotube ropes. *Physical Review Letters*, vol. 82, No. 5, pp. 944-947, ISSN 0031-9007

Samsonidze, Ge. G.; Samsonidze, G.G. & Yakobson, B.I. (2002). Energetics of Stone-Wales defects in deformations of
monoatomic hexagonal layers. *Computational Materials Science*, vol. 23, pp. 62-72, ISSN 0927-0256

Seo, M.K.; Byun, J.H. & Park, S.J. (2010). Studies on Morphologies and Mechanical Properties of Multi-walled Carbon Nanotubes/Epoxy Matrix Composites. *Bull. Korean Chem. Soc*, vol. 31, No. 5, 1237-1240, ISSN 0253-2964

Shtogun, Y.V. & Woods, Y.V. (2010). Mechanical properties of defective single wall carbon nanotubes. *Journal of Applied Physics*, vol. 107, No. 6, pp. 061803-6, ISSN 0021-8979

Song, J.; Jinag, H. & Shi, D.L. (2006). Stone–Wales transformation: Precursor of fracture in carbon nanotubes. International, Journal. of Mechanical *Science*, vol. 48, No. 12, pp. 1464-1470, ISSN 0965-0393

Stone, A.J. & Wales, D.J. (1986). Theoretical studies of icosahedral $C_{60}$ and some related species. *Chemical Physics Letters*, vol. 128, No. 5-6, 501-503, ISSN 0009-2614

Terrones, M.; Banhart, F.; Grobert, N.; Charlier, J.-C..; Terrones, H. & Ajayan, P. M. (2002). Molecular junctions by joining single-walled carbon nanotubes. *Physical Review Letters*, vol. 89, No. 7, pp. 075505-4, ISSN 0031-9007

Thess, A.; Lee, R.; Nikolaev, P.; Dai, H.; Petit, P. & Robert, J. et al. (1996). Crystalline ropes of metallic carbon nanotubes. *Science*, vol. 273, No. 5274, pp. 483-487, ISSN 0036-8075

Treacy, M M J.; Ebbesen, T.W.; Gibson, J.M. (1996). Exceptional high Young's modulus. observed for individual carbon nanotubes, *Nature*, vol. 381, No. 6584, pp. 678-680, ISSN 0028-0836

Troya, D.; Mielke, S.L. & Schatz, G.C. (2003). Carbon nanotube fracture - differences between quantum mechanical mechanisms and those of empirical potentials. *Chemical Physica Letters*. Vol. 382, No. 1-2, pp. 133-141

Tunvir, K.; Kim, A. & Nahm, S.H. (2008). The effect of two neighboring defects on the mechanical properties of carbon nanotubes. *Nanotechnology*, vol. 19, No. 6, pp. 065703, ISSN 1550-7033

Yoon, M.; Han, S.; Kim, G.; Lee, S.B.; Berber, S.; Osawa, I.E. & Ihm, J. et al. (2004). Zipper Mechanism of Nanotube Fusion: Theory and Experiment. *Physical Review Letters*, vol. 92, No. 7, pp. 075504-4, ISSN 0031-9007

Wang, Q.; Duan, W.H.; Richards, N.L. & Liew, K.M. (2007). Modeling of fracture of carbon nanotubes with vacancy defect. *Physical Review B*, vol. 75, No.-20, pp. 201405(R)-4, ISSN 0163-1829

Wildoer, J.W.G.; Venema, L.C.; Rinzler, A.G.; Smalley, R.E. & Dekker, C. (1998). Electronic structure of atomically resolved carbon nanotubes. *Nature*, vol. 391, pp. 59-62, ISSN 0028-083

Wong, E.W.; Sheehan, P.E. & Lieber, C.M. (1997). Nanobeam mechanics : Elasticity, strength and toughness of nanorods and nanotubes. *Science*, vol. 277, No. 5334, pp. 1971-1975, ISSN 0036-8075

Yakobson, B.I.; Brabec, C.J. & Bernholc, J. (1996). Nanomechanics of Carbon Tubes: Instabilities beyond Linear Response. *Physical Review Letters*, vol. 76, No. 14, 2511-2514, ISSN 0031-9007

Yang, M. Koutsos, V. & Zaiser, M. (2007). Size effect in the tensile fracture of single-walled carbon nanotubes with defects. *Nanotechnology*, vol. 18, No. 15, 155708, ISSN 1550-7033

Yip, S. (2005) Lecture notes, MIT, URL- http://ocw.mit.edu/courses/nuclear-engineering/22-a09-career-options-for-biomedical-research-fall-2006/lecture-notes/advisorsem05.pdf, Last accessed on 13-4-2011

Yu, M.F.; Dyer, M.J.; Skidmore, G.D.; Rohrs, H.W.; Lu, X.K. & Ausman, K.D. et al. (1999a). Three-dimensional manipulation of carbon nanotubes under a scanning electron microscope. *Nanotechnology*, vol. 10, No. 3, pp. 244, ISSN 1550-7033

Yu, M.F.; Lourie, O.; Dyer, M.J.; Molony, K.; Kelly, T.F. & Ruoff, R.S. (2000b). Strength and Breaking Mechanism of Multiwalled Carbon Nanotubes Under Tensile Load. *Science*, vol. 287, No. 5453, pp. 637-640, ISSN 0036-8075

Yu, M.F.; Files, B.S.; Arepally, S. & Ruoff, R.S. (2000c). Tensile Loading of Ropes of Single Wall Carbon Nanotubes and their Mechanical Properties. *Physical Review Letters*, vol. 84, No. 24, pp. 5552-5555, ISSN 0031-9007

Zhang, P.; Lamert, P.E. & Crespi, V.H. (1998). Plastic deformations of carbon nanotubes. *Physical Review Letters*, vol. 81, No. 24, pp. 5346-5349, ISSN 0031-9007

# 16

# Microscopic Structure and Dynamics of Molecular Liquids and Electrolyte Solutions Confined by Carbon Nanotubes: Molecular Dynamics Simulations

Oleg N. Kalugin[1], Vitaly V. Chaban[1,2] and Oleg V. Prezhdo[2]
*[1]V. N. Karazin Kharkiv National University,*
*[2]University of Rochester,*
*[1]Ukraine,*
*[2]USA*

## 1. Introduction

Carbon nanotubes (CNT) are a completely new carbon material that are expected to become typical raw material for nanotechnology, applied to such broad fields as composite materials, electronic devices, drug delivery nanocapsules, etc. (Abrahamson & Nair, 2008; Abrikosov et al., 2005; Ahmad et al., 2006; Ajayan & Zhou, 2001; Avouris, 2002; Back & Shim, 2006; Bordjiba et al., 2008; Danilov et al., 2005; Eletskii, 1997; Hilder & Hill, 2008; Z. Liu et al., 2008; F. Yang et al., 2008; X. Yang et al., 2008; Zhang et al., 2007).

One of the most interesting and promising is application of CNTs as electrode materials (Centeno et al., 2007; Janes et al., 2007; Huang et al., 2008; R. Lin et al., 2009). Nanoporous carbon exhibits excellent charge-discharge properties and a stable cyclic life. Moreover, activated composite carbon films generate high specific capacitance, laying the foundation for a new generation of double-layer super-capacitors (SC) (Endo et al., 2008; Huang, et al., 2008; Wu & Xu, 2006). CNT provides an ideal model for investigating the microscopic details of fluid transport in these nanoporous carbon structures. SC design requires polar, but aprotic solvents such as acetonitrile (AN). In spite of the great fundamental and practical importance of AN, its structural and dynamical properties inside CNTs have never been investigated yet.

The fuel cells (Li et al., 2004; Maclean & Lave, 2003), which directly transform the chemical reaction energy between hydrogen and oxygen into electric energy, are seen as the energy source of the next-generation. With their environmentally friendly and high efficiency characteristics, the cells are being researched and developed as the future energy for automobiles and as energy generation for the houses. Since the CNTs have the possibility of clearly surpassing raw materials used so far, the aspects of applying it to fuel cell electrodes is under consideration (M.L. Lin et al., 2008). The most promising fuel cells are based on methanol (MeOH, $CH_3OH$) which is renewable and easily storable (Convert et al., 2001; Gu & Wong, 2006; Hsieh & J.Y. Lin, 2009; Hsieh et al., 2009; H.S. Liu et al, 2006; Qi et al.; 2006;

Schultz et al., 2001; Suffredini et al., 2009; C.H. Wang et al., 2007; Z. Wang et al., 2008). In view of the present large interest on the methanol behavior inside nanopores, it is highly informative to carry out MD simulation on liquid methanol confined by CNTs to elucidate an influence of CNT internal diameter on microscopic structure and dynamic (transport) properties of this alcohol.

Dimethyl sulphoxide (DMSO) is an important polar aprotic solvent, widely used in the chemical industry, biology and medicine, that dissolves both polar and nonpolar compounds (Martin & Hanthal, 1975; Yu & Quinn, 1994, 1998). Due to its distinctive property of penetrating the skin very readily, DMSO is an imprescriptible agent in medicine used as a carrier for transporting remedies into a human body. In this context, transport properties of liquid DMSO inside the biological nanoporous materials are of potential interest. From this point of view, internal space of CNTs can be considered as ideal model for investigation of DMSO behavior in the molecular-scale confined space.

From the fundamental point of view, the comparison of microscopic properties of confined molecular liquids with significantly differing kinds of molecular structure is of potential interest. The possibility to estimate and predict transport properties of these liquids inside carbon nanoporous structures forms a basis for their future nanotechnological and pharmaceutical applications together with carbon nanotubes. Speaking in a more general case, CNTs can be substituted with nanoporous carbon. Unfortunately, investigation of the confined liquids by means of direct experimental techniques is still quite a tricky task, so atomistic computer simulations are of ultimate importance.

In the present paper, the influence of spatial confinements caused by internal space of Single Walled Carbon Nanotubes (SWCNTs) and Multi Walled Carbon Nanotubes (MWCNTs) on microscopic structure and particle dynamics of the confined non-aqueous molecular liquids acetonitrile, methanol, dimethyl sulphoxide (AN, MeOH, DMSO) and infinitively diluted solutions of $Li^+$ in MeOH and solutions of $Et_4NBF_4$ of finite concentrations in AN are investigated conducting molecular dynamics (MD) simulations on them.

## 2. Details of molecular dynamics simulations

### 2.1 SWCNTs–based systems

A series of MD simulations of non-capped armchair SWCNTs with liquid AN, MeOH, MeOH+$Li^+$ and DMSO located both inside and outside them have been performed. The simulated systems (Fig. 1) were implemented as square parallelepipeds with a ratio of side lengths approximately equal to the ratio of the length to the diameter of the corresponding SWCNT. The SWCNT centre-of-mass coincided with a geometrical centre of the molecular dynamics cells. The SWCNTs were surrounded by a few layers of solvent molecules (outside solvent) allowing the solvent particles to migrate both inside and outside the nanotube during the simulation. As an example, in Fig. 1 and 2 the snapshots of MD simulation cells from simulations of AN confined inside SWCNT(15,15) and the sketch of the MD cell along with SWCNT(22,22) and MeOH molecules are shown. Table 1 summarizes the designation, parameters, and some simulation details of the modeled systems. For reference purpose, the corresponding properties of bulk systems are discussed as well. For all the modeled systems, the values dielectric constant and density of the liquids were put equal to the experimental ones (Poltoratchkij, 1984).

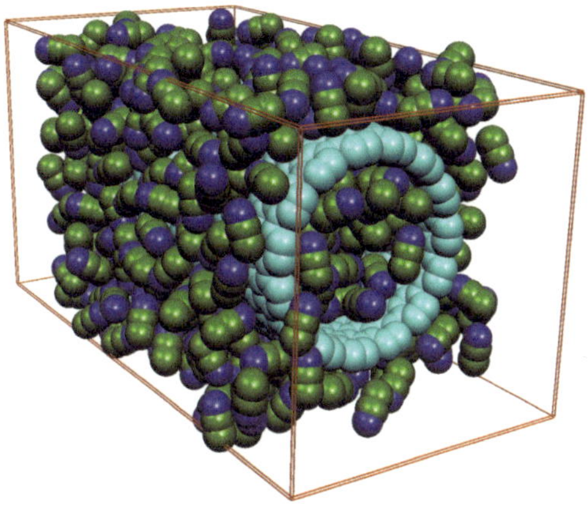

Fig. 1. Snapshots of MD simulation cells with AN confined inside SWCNT(15,15).

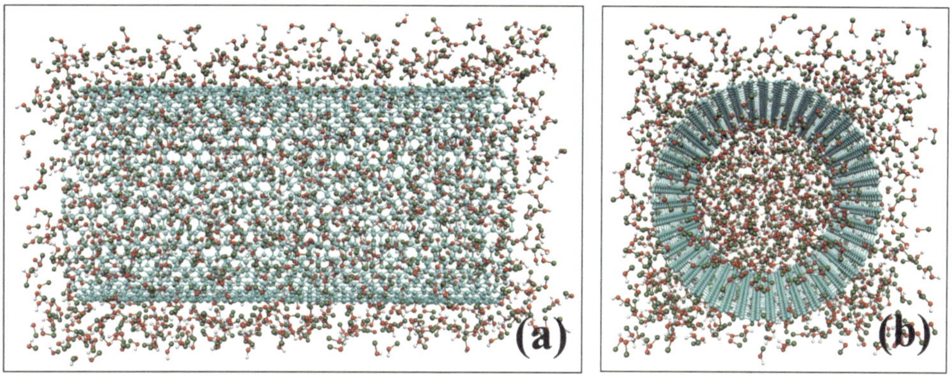


Fig. 2. Sketch of the MD cell along with SWCNT (22,22) and MeOH molecules: (a) side view, (b) cross-sectional view.

## 2.2 MWCNTs – based systems

In contrast to the previous systems, the structural and dynamic properties of the solutions of $Et_4NBF_4$ of finite concentration in acetonitrile were modeled inside of MWCNTs without any ionic or molecular species outside. The length of the greatest edge MD cell (square parallelepiped) was chosen equal to the tube length. This, along with periodic boundary conditions in an axial direction, allows one to create a model equivalent to infinitely long nano-sized carbon channel of the cylindrical form filled by electrolyte solution. Such systems have a great practical interest for development of modern double-layer super-capacitors. For maintenance of realistic interparticle interaction potential of a solution inside of inner volume of MWCNT, a number of walls of MWCNTs was chosen equal to three in order to the total thickness of MWCNT was bigger than cut-off radius of van-der-Waals interactions in the MD cell.

The composition of the modelled systems consisting of AN molecules, $Et_4N^+$ and $BF_4^-$ ions ($Et_4NBF_4$) and MWCNTs is presented in Table 2. Dielectric constant of the solution was taken equal to its value for pure acetonitrile, and density at 298 K was measured in our laboratory.

| System | Solvent | SWCNT | SWCNT diameter, nm | SWCNT length, nm | Number of molecules | Number of ions |
|---|---|---|---|---|---|---|
| IA | AN | Bulk | – | – | 216 | 0 |
| IIA | | (26,26) | 3.526 | 7.010 | 1530 | 0 |
| IIIA | | (22,22) | 2.984 | 6.026 | 887 | 0 |
| IVA | | (19,19) | 2.577 | 5.042 | 608 | 0 |
| VA | | (15,15) | 2.035 | 4.058 | 432 | 0 |
| VIA | | (11,11) | 1.493 | 3.074 | 281 | 0 |
| VIIA | | (8,8) | 1.087 | 2.091 | 216 | 0 |
| IM | MeOH | Bulk | – | – | 324 | 0 |
| IIM | | (22,22) | 2.984 | 6.030 | 428 | 0 |
| IIIM | | (15,15) | 2.035 | 6.030 | 890 | 0 |
| IVM | | (8,8) | 1.087 | 6.030 | 1530 | 0 |
| VM | | Bulk | | | 323 | 1 $Li^+$ |
| VIM | | (22,22) | 2.984 | 6.030 | 427 | 1 $Li^+$ |
| VIIM | | (15,15) | 2.035 | 6.030 | 889 | 1 $Li^+$ |
| VIIIM | | (8,8) | 1.087 | 6.030 | 1529 | 1 $Li^+$ |
| ID | DMSO | Bulk | – | – | 572 | 0 |
| IID | | (22,22) | 2.984 | 6.030 | 572 | 0 |
| IIID | | (15,15) | 2.035 | 6.030 | 918 | 0 |
| IVD | | (8,8) | 1.087 | 6.030 | 1000 | 0 |

Table 1. Designation and some parameters of the modelled systems based on SWCNTs with AN, MeOH and DMSO.

| System | MWCNT | MWCNT inner diameter, nm | Number of AN molecules | Number of electrolytes molecules | Density, kg/m$^3$ | Molarity, mol/l |
|---|---|---|---|---|---|---|
| IE | (15,15)<br>(20,20)<br>(25,25) | 1.655 | 133 | 7 $Et_4NBF_4$ | 842.8 | 0.8428 |
| IIE | (19,19)<br>(24,24)<br>(29,29) | 2.197 | 236 | 12 $Et_4NBF_4$ | 842.8 | 0.8428 |
| IIIE | (22,22)<br>(27,27)<br>(32,32) | 2.604 | 333 | 17 $Et_4NBF_4$ | 842.8 | 0.8428 |
| IVE | - | Bulk | 333 | 17 $Et_4NBF_4$ | 842.8 | 0.8428 |

Table 2. Some parameters of the modelled systems based on the solutions of $Et_4NBF_4$ in AN confined by MWCTNs.

As an example, the snapshot of MD simulation cell representing the system IIE is shown in Fig. 3.

### 2.3 General approach

The MD simulations of all systems were performed with a 1 fs time-step, in the NVT ensemble, with periodic boundary conditions in all directions at 298 K using Berendsen thermostat with a characteristic thermostat time equal to 100 fs. System equilibrations have been performed over 200 ps for pure solvents and 2000 ps for electrolyte solutions, and the data were collected over at least 5 runs of 500 ps and 5000 ps for molecular liquids and $Et_4NBF_4$–AN systems, respectively. The simulations have been carried out using the proprietary software package MDCNT (Molecular Dynamics inside Carbon NanoTubes) (Kalugin et al., 2006) developed by us at the Department of Inorganic Chemistry of V.N. Karazin Kharkiv National University.

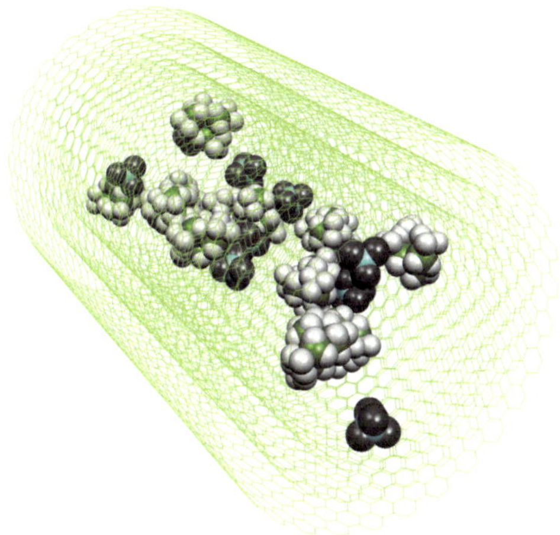

Fig. 3. Snapshots of MD simulation cell representing the system IIE. The AN molecules are not sown.

The site-site interactions between all atom pairs in the system are given by the sum of Lennard-Jones (LJ) 12-6 and Coulomb potentials,

$$U(r_{ij}) = 4\varepsilon_{ij}\left[\left(\frac{\sigma_{ij}}{r_{ij}}\right)^{12} - \left(\frac{\sigma_{ij}}{r_{ij}}\right)^{6}\right] + \frac{q_i q_j e^2}{4\pi\varepsilon_0 r_{ij}} \tag{1}$$

where $\varepsilon_{ij}$ and $\sigma_{ij}$ are the LJ parameters between sites $i$ and $j$ of distinct molecules, $q_i$ is the partial charge on site $i$, and $r$ is the site-site separation. Cross interactions were obtained from Lorentz-Berthelot combining rules, $\varepsilon_{ij} = \sqrt{\varepsilon_{ii}\varepsilon_{jj}}$ and $\sigma_{ij} = (\sigma_{ii} + \sigma_{jj})/2$. Shifted force potential was employed for the LJ part of the potential, whereas the reaction field method was used to calculate the long-range Coulombic part. In this work we used the well-tested force field models for solvent (the three-site rigid A3 (Mountain, 1997) for AN, the three-site rigid H1 (Haughney et al. 1987) for MeOH, and the four-site rigid VG (H. Liu et al., 1995) for DMSO) which imply the intermolecular interactions to be a sum of Coulomb and Lennard-Jones (LJ) (12, 6) potentials, rigid bonds and fixed angles in the solvent molecules. Usage of the force field model with rigid bond and angle values is justified by the different time scales of intra- and intermolecular motions in the case of these liquids. The techniques of reaction field and shifted force were applied to Coulomb and LJ (12, 6) interactions, respectively.

LJ-parameters for $Li^+$ were restored from the van-der-Waals parameters (Peng et al., 1997) according to the procedure described earlier (Kalugin et al., 2003) and already applied in our previous works.

The force-field of the CNT carbon atoms was taken to be purely LJ (Van Gunsteren et al., 1996). The geometrical parameters of ideal armchair CNTs were generated by the proprietary algorithm (Kalugin et al., 2006) with the length of carbon-carbon bond equal to 0.1421 nm. The carbon atoms of the CNTs were held fixed during the MD simulations. The CNT was assumed to be rigid with a fixed carbon-carbon bond length equal to 0.1421 nm. Our previous test simulations have shown that the effect of carbon nanotube flexibility is not critical for the discussed properties of the confined molecules (Chaban et al., 2010a). All the potential parameters used in the present MD simulations are enumerated in Table 3.

| Particle | Geometry | Site | $\sigma_{ii}$, nm | $\varepsilon_{ii}$, kJ/mol | $q_i$ (e) |
|---|---|---|---|---|---|
| CNT | $r_{CC}$ = 0.1421 nm | C | 0. 33611 | 0.405868 | 0 |
| AN | $r_{CN}$ = 0.117 nm, $r_{CMe}$ = 0.146 nm, ∠MeCN = 180° | Me | 0.36 | 1.588 | +0.269 |
| | | C | 0.34 | 0.416 | +0.129 |
| | | N | 0.33 | 0.129 | -0.398 |
| MeOH | $r_{OHo}$ = 0.095 nm, $r_{OMe}$ = 0.142 nm, ∠$H_o$OMe = 108.53° | $H_o$ | 0 | 0 | +0.431 |
| | | O | 0.3083 | 0.7312 | -0.728 |
| | | Me | 0.3861 | 0.7579 | +0.297 |
| DMSO | $r_{SO}$ = 0.153 nm, $r_{SMe}$ = 0.180 nm, ∠OSMe = 106.75°, ∠MeSMe = 97.4° | S | 0.356 | 1.29699 | +0.139 |
| | | O | 0.263 | 1.7154 | -0.459 |
| | | Me | 0.366 | 0.9414 | +0.160 |
| $Li^+$ | - | Li | 0.0826 | 26.158 | +1 |
| $BF_4^-$ | tetrahedron: $r_{BF}$ =0.139 nm, $r_{FF}$=2.27 nm | B | 0 | 0 | +0.9756 |
| | | F | 0.30 | 0.2845 | -0.4939 |
| $Et_4N^+$ ($D_{2d}$ symmetry) | (Kalugin et al., 2005) | $C_1$ | 0.391992 | 0.48959 | +0.03423 |
| | | $C_2$ | 0.3875 | 0.73227 | -0.28586 |
| | | $H_1$ | 0 | 0 | +0.08854 |
| | | $H_2$ | 0 | 0 | +0.10958 |
| | | N | 0 | 0 | -0.01667 |

Table 3. The potential parameters for interacting sites of the AN, MeOH and DMSO molecules, $Li^+$, $BF_4^-$, $Et_4N^+$ ions and CNT. "Me" stands for the methyl group as united atom (site).

### 2.4 Simulated properties

Keeping in mind that there should be specific structure patterns in the cylindrical confined area, we proposed the single-atom distribution functions, $P_\alpha(r,z)$, of cylindrical symmetry

(Fig. 4). The cylindrical distribution functions (CDF) is defined as (Chaban et al., 2009, Chaban & Kalugin, 2010)

$$P_{\alpha}(r,z) = \rho_{\alpha}(r,z) / < \rho_{\alpha} >^{\text{inside CNT}} \quad (2)$$

where $\rho_{\alpha}(r,z)$ and $< \rho_{\alpha} >^{\text{inside CNT}}$ are the local and mean atomic density of α species, respectively, inside of the SWCNT (Fig. 4).

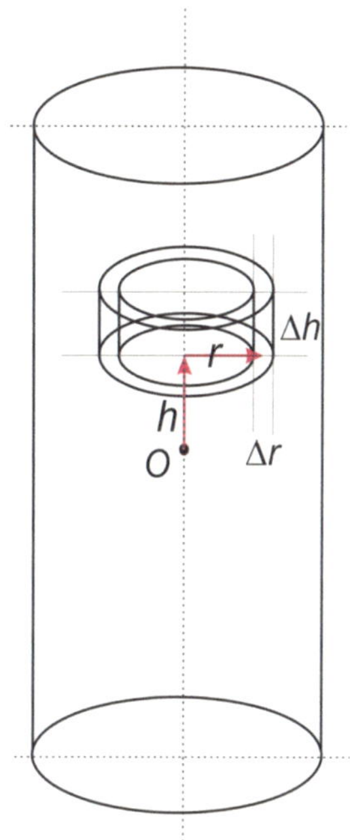


Fig. 4. The definition of the cylindrical distribution functions $P_{\alpha}(r,z)$ (CDF).

The local densities, $\rho_{\alpha}(r,z)$, were calculated by dividing the confined (by the SWCNT) space into a number of slices in axial (*z*) direction along the axis of the SWCNT (with a step of 0.02 nm) and a number of cylindrical shells in radial (*r*) direction perpendicular to axial one (with the same step) and then taking the statistical average for the local density for each slice or shell.

To examine the re-orientational dynamics of solvent molecules inside CNTs, we evaluate the re-orientational autocorrelation functions (ACFs) of the unit vector $\mathbf{u}$ along the molecular dipole $\boldsymbol{\mu}$,

$$C_{\mu\mu}(t) = \langle \mathbf{u}(0) \cdot \mathbf{u}(t) \rangle / \langle \mathbf{u}(0) \cdot \mathbf{u}(0) \rangle \quad (3)$$

The translational self-diffusion coefficients (SDC) were derived from velocity autocorrelation functions via the Green-Kubo equation,

$$D = \lim_{t \to \infty} \frac{1}{3} \int_0^{\infty} \langle \mathbf{v}(0) \mathbf{v}(t) \rangle dt \quad (4)$$

## 3. Results and discussion

### 3.1 AN based systems in SWCNTs

The solvent structure of AN inside SWCNTs was analyzed by computing the cylindrical distribution function $P_{\alpha}(r,z)$ of the atomic density along the CNT axial ($z$) and radial ($r$) directions. Figure 5 shows two examples of the distributions of nitrogen atomic density of AN molecules confined inside the (15,15) and (19,19) armchair CNTs with the internal diameters of 1.655 nm and 2.197 nm, respectively. The same oscillatory behavior of the atomic density along the radial direction was observed for all CNTs and all AN interacting sites, including N, C and $CH_3$. The atomic density is maximal near the CNT wall, where molecular correlations are reinforced by the space confinement. The second maximum in the atomic density is seen about 0.35 nm after the first maximum. The height of the second maximum is ~1.5 times smaller than the height of the first maximum. This trend was seen in all cases. At distances larger than 0.7 nm from the CNT wall the confinement effects can be neglected.

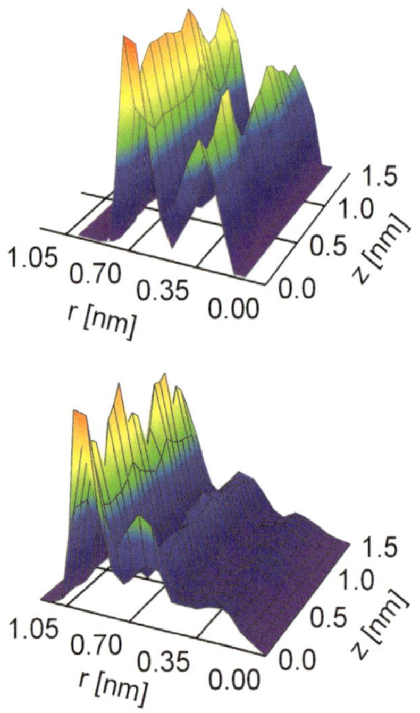


Fig. 5. Cylindrical distribution functions of nitrogen atomic density inside of CNTs (15,15) (top) (19,19) (bottom). The point (0, 0) on the each graph corresponds to the geometrical centre of the SWCNT.

One can expect that the spatial confinement of the AN molecules at distances shorter than 0.7 nm from the CNT wall should have an important impact on the solvent dynamical and transport properties. The confinement effect is seen with both the re-orientation dynamics and the diffusion coefficient. These types of motion are slowed down inside CNTs. The diffusion shows a uniform dependence on the CNT diameter, while the re-orientation exhibits strong anomalies in the CNTs whose radii are less than 0.7 nm, as elucidated below.

In order to examine the orientation dynamics of the AN molecules inside CNTs, we evaluated the orientation autocorrelation function (ACF) of the unit vector $\mathbf{u}$ along the direction of the molecular dipole $\boldsymbol{\mu}$, $C_{\mu\mu}(t)$. The long-time behavior of $C_{\mu\mu}(t)$ extending beyond 2 ps is well described by a single exponential decay: $\ln C_{\mu\mu}(t) = \text{const} - t/\tau_{\mu}$. The corresponding orientation relaxation times, $\tau_{\mu}$, calculated using the least squares method from the slopes of $\ln C_{\mu\mu}(t)$ at times between 2 and 10 ps are summarized in Table 4. The re-orientation dynamics of AN molecules inside CNTs is drastically slower than in bulk liquid. The orientation relaxation times significantly exceed the bulk value and increase with decreasing CNT diameter. A uniform behavior is seen with the nanotubes from (26,26) to (11,11), whose diameters are more than twice larger than the 0.7 nm confinement distance, discussed above. The orientation relaxation time of AN inside the (11,11) CNT, whose diameter is 1.1 nm, jumps to an extremely large value of 102 ps. Surprisingly, the relaxation time for the (8,8) tube with diameter of only 0.7 nm is quite small $\tau_{\mu}$ = 18.1 ps, and is much closer to that of the (15,15) tube, $\tau_{\mu}$ = 11.9 ps, than the (11,11) tube, even though the spatial confinement effects should be strongest in the (8,8) CNT.

| System | CNT inner diameter, nm | AN self-diffusion coefficient, $D{\cdot}10^9$, m$^2$s$^{-1}$ | AN orientation relaxation time, $\tau_{\mu}$, ps |
|---|---|---|---|
| IA | – | $3.240 \pm 0.004$ | 3.9 |
| IIA | 3.526 | $2.52 \pm 0.05$ | 7.9 |
| IIIA | 2.604 | $2.27 \pm 0.05$ | 9.0 |
| IVA | 2.197 | $2.03 \pm 0.12$ | 9.1 |
| VA | 1.655 | $1.69 \pm 0.12$ | 11.9 |
| VIA | 1.113 | $1.09 \pm 0.22$ | 102 |
| VIIA | 0.707 | $0.76 \pm 0.07$ | 18.1 |

Table 4. System parameters and dynamic properties of AN molecules confined by SWCNTs.

Translational diffusion of AN inside CNTs is of great importance to a variety of applications. The self-diffusion coefficient $D$ was calculated by the Green-Kubo formula. In order to avoid the open-end boundary effects, only AN molecules located more than one molecule diameter (0.6 nm) away from the nanotube ends were used to calculate $D$. The values reported in Table 2 clearly show that the diffusion coefficient of AN inside CNTs decreases with decreasing CNT diameter. The change between bulk and the 1nm (8,8) CNT is a factor of 4. The behavior of the translational diffusion coefficient is uniform, in contrast to the corresponding variation in the orientation relaxation time, Table 1. This result is very important for such practical applications as double-layer SC, which require steady solvent diffusion inside nanoporous carbon of varying pore-diameter distributions. The spatial confinement influences the translation motion to a lesser extent than the rotational motion, as follows from data reported in Table 4.

Optimization and development of electrochemical devices based on nanoporous carbon requires an analytic expression for the self-diffusion coefficient of a liquid inside the nanopores of arbitrary diameter and length. In the absence of a general theory of fluid diffusion in porous materials, we extended the recently proposed description of liquid transport under steric confinement of a solid matrix (Sevriugin et al., 2003) and obtained a

simple analytic expression for the observed trend in the diffusion coefficient (Kalugin et al., 2008), as described below.
Reference (Sevriugin et al., 2003) shows that self-diffusion of a liquid in a heterogeneous system is decreased relative to the pure liquid according to

$$D = D_0 \exp(-P_{st}) \tag{5}$$

where $D_0$ is the bulk self-diffusion coefficient, and $P_{st}$ is the probability of steric restrictions imposed on a particle of a fluid by the surrounding matrix. The probability $P_{st}$ is determined by the space distribution of these steric constrains or, in the simplest case, by the confinement geometry.
Consider a liquid molecule that diffuses distance $\Lambda$ and collides with a CNT wall, see insert in Fig. 6. Collisions occur when the molecule is close to the CNT wall and moves towards the wall. The thickness of the solvent layer that is sufficiently close to the wall to produce a collision can be estimated by the mean-free path of diffusion $d_m$. The directionality of the molecular motion is accounted for by the following average

$$\langle \Lambda \rangle = \int \Lambda d\Omega \Big/ \int d\Omega \ \langle \Lambda \rangle = \int \Lambda d\Omega \Big/ \int d\Omega \tag{6}$$

in which the integration is performed over the solid angle that is directed towards the CNT wall.

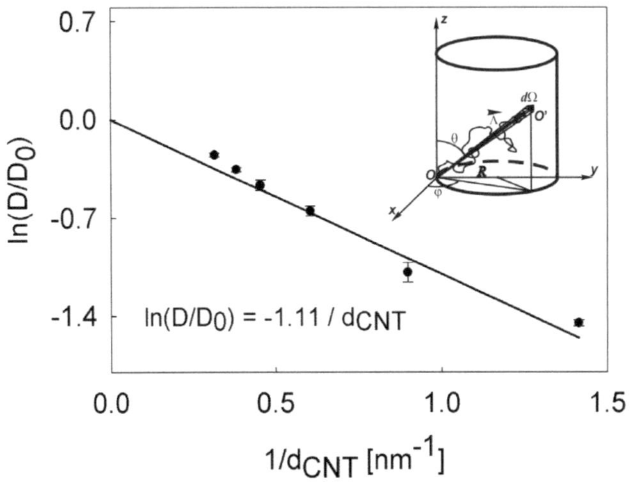


Fig. 6. Relative AN self-diffusion coefficient as a function of CNT inverse diameter. Inset: Diffusion path of a particle inside a CNT.

Geometric considerations lead to

$$\langle \Lambda \rangle \approx 0.64d \tag{7}$$

where $d$ is the internal diameter of the CNT. The probability that the motion of the diffusing particle will be impeded by a collision with the wall is given by the ratio $d_m / \langle \Lambda \rangle$. Then, the expression (2) for self-diffusion coefficient in the confined geometry becomes

$$D = D_0 \exp(-d_m / 0.64d) \tag{8}$$

or, explicitly inserting the value of the mean-free diffusion path for bulk AN, $d_m$=0.65 nm:

$$D = D_0 \exp(-1.01 / d) \tag{9}$$

Figure 6 demonstrates good agreement between this simple theoretical expression and the results of the MD simulation, given by the filled circles.

### 3.2 MeOH based systems in SWCNTs

To examine a long-range structure of a liquid methanol inside the SWCNTs, we have calculated the cylindrical distribution function $P_\alpha(r,z)$ of the atomic density along the CNT axial ($z$) and radial ($r$) directions (α - any site of MeOH or $Li^+$) of cylindrical symmetry. Contour plots of cylindrical distribution functions $P_\alpha(r,z)$ for the hydrogen (α =H) and oxygen (α =O) atoms of MeOH molecules inside the SWCNT (8,8), SWCNT (15,15) and SWCNT (22,22) are plotted in Fig. 7. Cylindrical atomic density distribution (Fig. 7) of MeOH confined in the CNTs demonstrates the series of intertransient maxima oriented at an angle of ~30° with respect to the CNT axes. It is interesting to note a similar character of $P_\mathrm{H}(r,z)$ and $P_\mathrm{O}(r,z)$ distributions, that evidently indicates the chain-like hydrogen bond network in confined methanol. It should be also stressed, that CNT diameter influences the

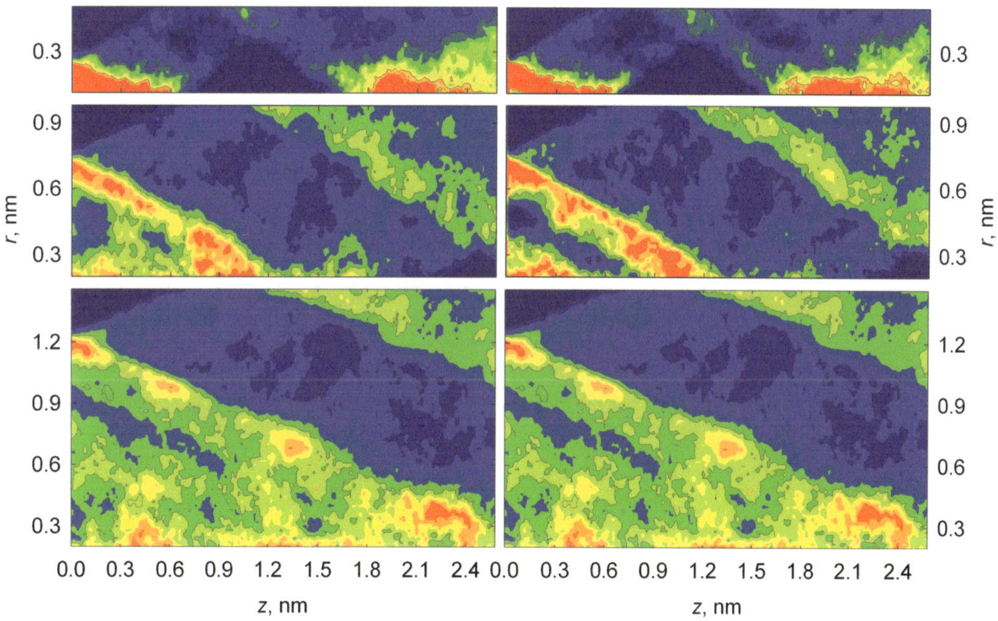


Fig. 7. Contour plots of cylindrical distribution functions $P_\alpha(r,z)$ for the hydrogen (α=H) (left) and oxygen (α=O) (right) atoms of MeOH molecules inside the SWCNT (8,8) (top), SWCNT (15,15) (middle), and SWCNT (22,22) (bottom) derived from MD simulations on systems IIM-IVM. The point (0, 0) on the each graph corresponds to the geometrical centre of the SWCNT.

intensity of density distribution functions, but in general form of density anisotropy. These observations allow us to make a conclusion about layered long-range structure of methanol with helix-like distribution of H-bonds inside the CNTs.

The mentioned above conclusion about layered structure in liquid methanol inside the CNT is completely confirmed by instantaneous configuration of the methanol molecules inside the SWCNT (15,15) (Fig. 8). The molecular helix-like chains formed by the hydrogen-bonded MeOH molecules inside the CNTs are clearly observed.

Bearing in mind the discussed above changes in long-range structure of liquid MeOH driven by CNTs, it is interesting to identify the CNT influence on infinitely diluted solution of $Li^+$. In the present study, we simulated the configurations when $Li^+$ has already gone into the CNT at maximum depth (3 nm far from any end of the CNT) and is initially equidistant from the CNT walls. Cylindrical distribution function for $Li^+$, $P_{Li}(r,z)$, shows the permanent location of the cation around its initial position ($r$ = 0, $z$ = 0). The mentioned deviation from the centre point is the same inside all the investigated CNTs and does not depend on the CNT diameter.

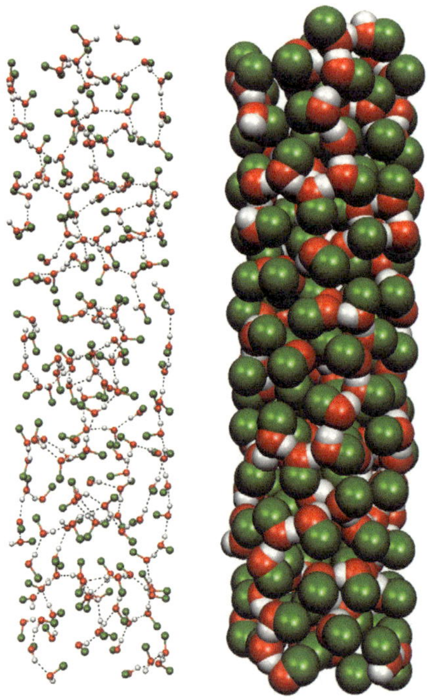

Fig. 8. Instantaneous configuration of the methanol molecules inside the SWCNT (15,15) ($d_{CNT}$=2.984 nm) at 298 K from MD simulation on system IIM in "balls and sticks" (left) and "spacefill" (right) formats. H-bonds between MeOH molecules are indicated by dashed lines. The helix nature of the H-bond network is clearly seen.

To examine the re-orientational dynamics of the MeOH molecules inside CNTs, we evaluate the re-orientational autocorrelation functions (ACFs) of the unit vector u along the molecular dipole μ, $C_{\mu\mu}(t)$. The long time (t > 4 ps) behavior of $C_{\mu\mu}(t)$ is well described by single exponent $\ln C_{\mu\mu}(t) = \text{const} - t/\tau_{\mu}$. The corresponding re-orientational relaxation times,

$\tau_\mu$, calculated from the slopes of $\ln C_{\mu\mu}(t)$ (by using least squares method) at long times (4 ≤ t ≤ 10 ps) are summarized in Table 5. It is interesting to note that the re-orientational relaxation times of MeOH molecules inside the CNTs significantly exceed the corresponding value for the bulk solvent and increase with the CNT diameter decrease. For example, inside the SWCNT (8,8), $\tau_\mu$ increases more than two times in comparison with bulk.
Self-diffusion coefficients, $D$, of the centre of mass of MeOH molecules and lithium-ion were calculated from the velocity autocorrelation functions via the Green-Kubo relation.
The resulting $D_{MeOH}$ and $D_{Li+}$ values for all the simulated systems are summarized in Table 5. The diffusion coefficients of MeOH molecules confined by CNTs are noticeably lower as compared with bulk and for the SWCNT (8,8) $D_{MeOH}$ value is about two times lower than that without space restrictions. At the same time, diffusion coefficients of MeOH molecules depend slightly on CNT diameter, $d_{CNT}$. Only $0.35 \cdot 10^{-9}$ m$^2$ s$^{-1}$ grows in $D_{MeOH}$ value occurs when $d_{CNT}$ increases by a factor of three (from 1.087 nm to 2.984 nm). It allows us to conclude that the average diffusion coefficient of MeOH inside the CNT of a given diameter is sharply decreased by the first layer of parietal solvent molecules which lose one of their degrees of freedom and the structure of this layer is reinforced in the maximally possible way inside the CNT. This conclusion is in complete agreement with the H-bond structure of confined MeOH already discussed. Diffusion coefficients slowdown inside the CNTs is an important feature of the direct methanol fuel cells.
In contrast to the methanol diffusion coefficient, the decrease in $D_{Li}{}^+$ values with decreasing $d_{CNT}$ is insignificant (Table 5). This is probably caused by a location of lithium ion along with its stable solvation shell far from the internal CNT wall where the solvent dynamics is not affected sufficiently by the space restrictions.

| System | $D_{MeOH}\,10^9$, m$^2$ s$^{-1}$ | $\tau_\mu$, ps | System | $D_{Li^+}\,10^9$, m$^2$ s$^{-1}$ |
|---|---|---|---|---|
| IM | 2.16 ± 0.02 | 12 | VM | 0.53 ± 0.07 |
| IIM | 1.30 ± 0.02 | 18 | VIM | 0.50 ± 0.06 |
| IIIM | 1.25 ± 0.06 | 23 | VIIM | 0.46 ± 0.07 |
| IVM | 0.95 ± 0.06 | 28 | VIIIM | 0.41 ± 0.07 |

Table 5. The self-diffusion coefficients and microscopic dipole relaxation times, $\tau_\mu$, of the methanol molecules from MD simulations of the systems IM-IVM and self-diffusion coefficients of the lithium ion from MD simulations of the systems VM-VIIIM.

### 3.3 DMSO–based systems in SWCNTs

In order to clarify a long-range structure of the confined DMSO we have also analyzed the cylindrical distribution functions $P_\alpha(r,z)$. The CDFs $P_O(r,z)$ of the oxygen atoms of DMSO inside all the investigated SWCNTs are shown in Fig. 9.
The highest atomic density is observed near the inner wall of SWCNT. It indicates the particular reinforcements of molecular interactions and the increase of molecules ordering at the distances ~0.5 nm and less from the carbon atoms. That distances correspond to the first liquid layer from the wall. Inside the smallest SWCNT (8, 8), the above tendency can not be observed because all the confined DMSO molecules are located near the SWCNT inner wall (see Fig. 9).

On contrary to the acetonitrile and methanol confined by SWCNTs, no special long-range pattern in the case of DMSO is observed. Inside the SWCNT (22,22) one can identify low grade second layer of the solvent molecules at the distances ca. 0.7-0.8 nm from the SWCNT inner wall. Low sensitivity of DMSO to the presence of the SWCNT can be explained by the strong spatial correlations between the DMSO molecules due to anti-parallel alignment of their dipole moments. It appears that anisotropic dipole-dipole intermolecular interactions, that specify the structure of DMSO in liquid phase, are more powerful as compare with solvophobic interactions of the solvent molecules with carbon atoms of the SWCNT. Nevertheless, the space confinements given by the SWCNT inner wall promote these dipole-dipole correlations, but such reinforcement can not cover more then one molecular layer due to the short-range character of the dipole correlations in the liquid DMSO.

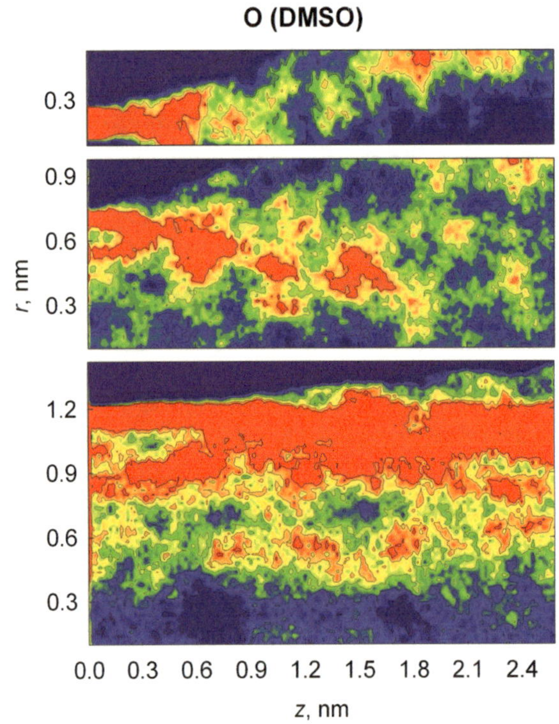


Fig. 9. Contour plots of cylindrical distribution functions, $P_{\alpha}(r,z)$, for Oxygen atoms of DMSO inside the SWCNTs (8, 8), (15, 15) and (22, 22) in the top, middle and bottom, respectively derived from MD simulations. The point (0, 0) on the each graph corresponds to the geometrical centre of the SWCNT.

The self diffusion coefficients (SDCs) derived from VACFs are 0.4±0.1, 0.56±0.05, 0.62±0.05, 1.06±0.05 ($\times 10^{-9}$ $m^2/s$) for SWCNTs (8, 8), (15, 15), (22, 22) and bulk, respectively. The simulated SDCs are well correlated with a value of SWCNT diameter. Performing a complex analysis of the SDCs of DMSO inside the SWCNTs against bulk value together with CDFs, $P_{\alpha}(r,z)$, one can assume that the general slowdown of DMSO SDCs in the case of SWCNTs is defined by the fraction of DMSO molecules in the first layer near the inner wall of the nanotube.

### 3.4 Acetonitrile solutions of $Et_4NBF_4$ in MWCNTs

In the $Et_4NBF_4$ solutions in AN both in the bulk phase and in confinements the wide distribution of the ionic clusters size is found (Fig. 10). The average cluster size (a number of cations $Et_4N^+$ and anions $BF_4^-$ consisting a cluster) slightly increase from 8.4 to 12.2 as the inner diameter of MWCNTs decreases from the infinity (bulk solvent) to 1.655 nm.

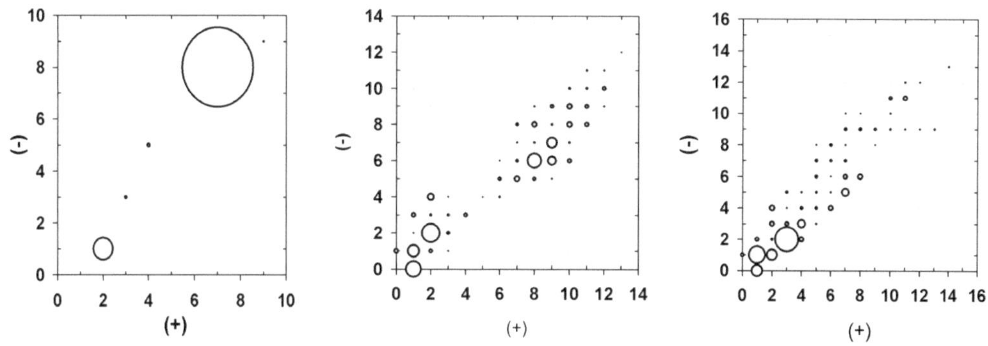


Fig. 10. Distribution of the cluster sizes for the systems IE (left), IIE (middle) and IVE (right). (+) and (-) stand for the $Et_4N^+$ and $BF_4^-$ ions, respectively. Diameter of the circles reflects a relative probability of the formation of the corresponding cluster.

It is clarified that dynamical properties of electrolyte solutions confined by carbon nanotube are defined by two general factors: geometry of confinements and ionic subsystem structure of confined solution. Translational diffusion of AN and ions of of $Et_4NBF_4$ inside MWCNTs is of great importance for improvement of super capacitors. The self-diffusion coefficients $D$ were calculated by the Green-Kubo formula and listed in Table 6.

| System | MWCNT inner diameter, nm | $D$(AN) $10^9$, $m^2 s^{-1}$ | $D(Et_4N^+)10^9$, $m^2 s^{-1}$ | $D(BF_4^-)$ $10^9$, $m^2 s^{-1}$ |
|---|---|---|---|---|
| IE | 1.655 | 0.63 | 0.25 | 0.29 |
| IIE | 2.197 | 0.74 | 0.28 | 0.35 |
| IIIE | 2.604 | 0.87 | 0.34 | 0.40 |
| IVE | bulk | 2.20 | 0.81 | 0.93 |

Table 6. Self-diffusion coefficients of the acetonitrile and ions of $Et_4NBF_4$ solutions in AN from MD simulations of the systems IE-IVE.

The values reported in Table 6 clearly show that the diffusion coefficient of AN and ions inside MWCNTs decreases with decreasing internal diameter of CNT. It should be note that slowdown of ions SDCs inside of MWCNTs is not as drastic as one would expect. This is very important for optimization of pore size of the carbon nanomaterials used for the development of the modern double-layer super-capacitors.

## 4. Conclusion

To recapitulate, we elucidated the details of structure, re-orientational dynamics and translational diffusion of AN confined inside CNTs with diameters ranging from 1 nm to 3.5

nm. The geometric confinement creates a strong periodic pattern in the AN structure near the CNT wall, with the persistence length of 0.7 nm. The orientation relaxation time increases with decreasing CNT diameter and shows a highly non-uniform behavior for small CNTs, associated with specific solvent structures in the tightly confined spaces. The translational diffusion coefficient changes continuously with decreasing CNT diameter, even for the smallest CNTs. The observed dependence of the diffusion coefficient on the CNT size was described analytically with a simple model, which can be applied for optimization of electrochemical devices based on nanoporous carbon.

It was revealed that although local order of MeOH does not differ from bulk, long-range structure is helix-like near the inner CNT walls. Whereas the mechanism of hindered translations of the MeOH molecules and $Li^+$ confined in the CNTs does not differ from bulk, translational mobility of MeOH molecules is noticeably lower than in bulk ones due to the translation self-diffusion slowdown within the first layers of the parietal solvent molecules. In contrast to MeOH, decrease of translational self-diffusion coefficients of $Li^+$, located along the CNT axis, is insignificant.

As it follows from the results of the performed molecular dynamics simulations on liquid dimethyl sulphoxide confined by single-walled carbon nanotubes at 298 K, the local order of DMSO analyzed in terms of site-site intermolecular radial distribution functions is similar to that in the bulk except the case with the smallest SWCNT (8, 8). Meanwhile, the microscopic structure of the confined DMSO expressed in terms of cylindrical distribution functions demonstrates well pronounced changes in a local atomic density in the vicinity of the inner wall of the SWCNT.

The translational self-diffusion coefficients of the centre-of-mass of the DMSO molecules are lower by a factor of 2-3 then bulk ones and are evidently correlated with a value of the SWCNT diameter. It is shown that the slowdown of self-diffusion coefficient of DMSO confined in the SWCNTs is reduced by the first layer of DMSO molecules close to the SWCNT wall, and this is caused by the reinforcement of the dipole-dipole correlations among DMSO molecules. The SDCs of DMSO inside of the SWCNTs are of interest for its transportation through SWCNT-based capsules for drug delivery. Based on the re-orientational dynamics of liquid DMSO in confinements analysis, it was shown that cryoprotective properties of DMSO are stipulated by the considerable growth of microscopic relaxation times in the presence of spatial confinements.

The adequacy of analytical model of rectilinear dependence of reduced diffusion coefficient logarithm on the inverse diameter of CNT for AN, the solvent without specific intermolecular interactions, was demonstrated.

On the base of self-diffusion coefficients of $Et_4N^+$ and $BF_4^-$ inside the CNT considerable decrease carbon nanomaterials with effective diameters of up to 3 nm are suggested as an electrode material for modern electrochemical double-layer supercapacitors.

## 5. Acknowledgements

The authors acknowledge computational support of the Ukrainian-American Laboratory in Computational Chemistry established between Kharkiv, Ukraine and Jackson, MS, U.S.A. The funding was provided in part by grants from the National Science Foundation, CHE-0701517, and Petroleum Research Fund of the American Chemical Society, 46772-AC6. V. V. C. acknowledges the funding from Fund for Fundamental Studies of V. N. Karazin Kharkiv

National University (grants # 0107V000666 and #0109U001426). O. N. K. acknowledges Yury Sapronov's Kharkiv City charitable fund for the financial support of this investigation.

## 6. References

Abrahamson, J.T., Nair, N. (2008). Modelling the increase in anisotropic reaction rates in metal nanoparticle oxidation using carbon nanotubes as thermal conduits. *Nanotechnology*, Vol. 19, No. 19, (April 2008), pp. (195701.1-195701.8), ISSN 0957-4484.

Abrikosov, A.A., Livanov, D.V., & Varlamov, A.A. (2005). Electronic spectrum and tunneling properties of multiwall carbon nanotubes. *Physical Review B*, Vol. 71, No. 16, (April 2005), pp. (165423.1-165423.8), ISSN 1089-5647.

Ahmad, K., Pan, W., & Shi, S.L. (2006). Electrical conductivity and dielectric properties of multiwalled carbon nanotube and alumina composites. *Applied Physics Letters*, Vol. 89, No. 17, (September 2006), pp. (133122.1-133122.3), ISSN 0003-6951.

Ajayan, P.M., Zhou, O.Z. (2001). Applications of carbon nanotubes, In: *Carbon Nanotubes, Topics in Applied Physics, Vol. 80,* M. S. Dresselhaus, G. Dresselhaus & Ph. Avouris (Eds.), pp. (391-425), Springer, ISBN 978-3-540-41086-7, Berlin, Heidelberg.

Avouris, P. (2002). Carbon nanotube electronics. *Chemical Physics*, Vol. 281, No. 2-3, (March 2002), pp. (429-445), ISSN 0301-0104.

Back, J.H., Shim, M. (2006). pH-dependent electron-transport properties of carbon nanotubes. *Journal of Physical Chemistry B*, Vol. 110, (October 2006), pp. (23736-23741), ISSN 1089-5647.

Bordjiba, T., Mohamedi, M., & Dao, L.H. (2008). Charge storage mechanism of binderless nanocomposite electrodes formed by dispersion of CNTs and carbon aerogels. *Journal of the Electrochemical Society*, Vol. 155, No. 2, (December 2007), pp. (A115-A124), ISSN 0013-4651.

Centeno, T.A., Hahn, M., Fernández, J.A., Kötz, R., & Stoeckli, F. (2007). Correlation between capacitances of porous carbons in acidic and aprotic EDLC electrolytes. *Electrochemistry Communications*, Vol. 9, No. 6, (June 2007), pp. (1242-1246), ISSN 1388-2481.

Chaban, V.V., Kalugin, O.N. (2009). Structure and Dynamics in Methanol and its Lithium Ion Solution Confined by Carbon Nanotubes. *Jounal of Molecular Liquids*, Vol. 145, No. 3, (May 2009), pp. (145–151), ISSN 0167-7322.

Chaban, V.V., Kalugin, O.N. (2010). Liquid dimethyl sulphoxide confined by carbon nanotubes. *Jounal of Molecular Liquids*, Vol. 151, No. 2-3, (February 2010), pp. (113–116), ISSN 0167-7322.

Chaban, V.V., Kalugin, O.N., Habeninicht, B.F., & Prezhdo, O.V. (2010). The Influence of the Rigidity of a Carbon Nanotube on the Structure and Dynamics of Confined Methanol. *Journal of the Physical Society of Japan*, Vol. 79, No. 6, (June 2010), pp. (064608.1-064608.5), ISSN 1347-4073.

Convert, P., Coutanceau, C., Crouigneau, P., Gloaguen, F., & Lamy, C. (2001). Electrodes modified by electrodeposition of CoTAA complexes as selective oxygen cathodes in a direct methanol fuel cell. *Journal of Applied Electrochemistry*, Vol. 31, No. 9, (September 2001), pp. (945-952), ISSN 0021-891X.

Danilov, M.O., Melezhik, A.V., & Danilenko, N.I. (2005). Carbon nanotubes as catalyst supports for oxygen electrodes. *Russian Journal of Applied Chemistry*, Vol. 78, No. 11, (November 2005), pp. (1849-1853), ISSN 1070-4272.

Eletskii, A.V. (1997). Carbon nanotubes. *Uspekhi Fizicheskikh Nauk*, Vol. 167, No. 9, (September 2005), pp. (945-972), ISSN 1996-6652.

Endo, M., Strano, M.S., & Ajayan, P.M. (2008). Potential applications of carbon nanotubes, In: *Carbon Nanotubes, Topics in Applied Physics, Vol. 111*, A. Jorio, G. Dresselhaus, M. S. Dresselhaus (Eds.), pp. (13-62), Springer, ISBN 978-3-540-72864-1, Berlin, Heidelberg.

Gu, Y.J., Wong, W.T. (2006). Nanostructure PtRu/MWNTs as anode catalysts prepared in a vacuum for direct methanol oxidation. *Langmuir*, Vol. 22, No. 26, (November 2006), pp. (11447-11452), ISSN 0743-7463.

Haughney, M., Ferrario, M., & McDonald, I.R. (1987). Molecular-dynamics simulation of liquid methanol. *Journal of Physical Chemistry*, Vol. 91, No. 19, (September 1987), pp. (4934-4940), ISSN 0022-3654.

Hilder, T.A., Hill, J.M. (2008). Carbon nanotubes as drug delivery nanocapsules. *Current Applied Physics*, Vol. 8, No. 3-4, (May 2008), pp. (258-261), ISSN 1567-1739.

Hsieh, C.T., Lin, J.Y. (2009). Fabrication of bimetallic Pt-M (M = Fe, Co, and Ni) nanoparticle/carbon nanotube electrocatalysts for direct methanol fuel cells. *Journal of Power Sources*, Vol. 188, No. 2, (March 2009), pp. (347-352), ISSN 0378-7753.

Hsieh, C.T., Lin, J.Y., & Yang, S.Y. (2009). Carbon nanotubes embedded with PtRu nanoparticles as methanol fuel cell electrocatalysts. *Physica E-Low-Dimensional Systems & Nanostructures*, Vol. 41, No. 3, (January 2009), pp. (373-378), ISSN 1386-9477.

Huang, J.S., Sumpter, B.G., & Meunier, V. (2008). A universal model for nanoporous carbon supercapacitors applicable to diverse pore regimes, carbon materials, and electrolytes. *Chemistry-A European Journal*, Vol. 14, No. 22, (July 2008), pp. (6614-6626), ISSN 0947-6539.

Huang, C.W., Chuang, C.M., Ting, J.M., & Teng, H.S. (2008). Significantly enhanced charge conduction in electric double layer capacitors using carbon nanotube-grafted activated carbon electrodes. *Journal of Power Sources*, Vol. 183, No. 1, (August 2008), pp. (406-410), ISSN 0378-7753.

Janes, A., Kurig, H., & Lust, E. (2007). Characterisation of activated nanoporous carbon for supercapacitor electrode materials. *Carbon*, Vol. 45, No. 6, (February 2007), pp. (1226-1233), ISSN 0008-6223.

Kalugin, O.N., Adya, A.K., Volobuev, M.N., & Kolesnik, Ya.V. (2003). Solvation of solvophilic and solvophobic ions in dimethyl sulphoxide: microscopic structure by molecular dynamics simulations. *Physical Chemistry Chemical Physics*, Vol. 5, No. 8, (March 2003), pp. (1536-1546), ISSN 1463-9076.

Kalugin, O.N., Pazura, Yu.I., & Kolesnik, Ya.V. (2005). Internal structure of tetraalkylammonium ions in infinitely diluted solutions in acetonitrile, dymethyl sulphoxide and methanol. *Kharkiv University Bulletin*, Vol. 669, No. 13(36), pp. (162-168), ISSN 0453-8048.

Kalugin, O.N., Chaban, V.V., & Kolesnik, Y.V. (2006). Molecular dynamics simulation of liquid acenotitrile and solution of $Li^+$ in it inside carbon nanotubes by using

MDCNT package. *Kharkov University Bulletin*, Vol. 731. Chemical Series, No. 14(37), (December 2006), pp. (41-58), ISSN 0453-8048.

Kalugin, O.N., Chaban, V.V., Loskutov, V.V., & Prezhdo, O.V. (2008). Uniform Diffusion of Acetonitrile inside Carbon Nanotubes Favors Supercapacitor Performance. *Nano Letters*, Vol. 8, No. 8, (July 2008), pp. (2126-2130), ISSN 1530-6984.

Li, W.Z., Liang, C.H., & Xin, Q. (2004). Application of novel carbon nanomaterials in low-temperature fuel cell catalysts. *Chinese Journal of Catalysis*, Vol. 25, No. 10, (October 2004), pp. (839-843), ISSN 1872-2067.

Lin, M.L., Huang, C.C., Lo, M.Y., & Mou, C.Y. (2008). Well-ordered mesoporous carbon thin film with perpendicular channels: Application to direct methanol fuel cell. *Journal of Physical Chemistry C*, Vol. 112, (January 2008), pp. (867-873), ISSN 1932-7447.

Lin, R., Taberna, P.L., Chmiola, J., Guay, D., Goqotsi, Y., & Simon, P. (2009). Microelectrode Study of Pore Size, Ion Size, and Solvent Effects on the Charge/Discharge Behavior of Microporous Carbons for Electrical Double-Layer Capacitors. *Journal of the Electrochemical Society*, Vol. 156, No. 1, (October 2008), pp. (A7-A12), ISSN 0013-4651.

Liu, H., Muller-Plathe, F., van Gunsteren, W.F. (1995). A force field for liquid dimethyl sulfoxide and physical properties of liquid dimethyl sulfoxide calculated using molecular dynamics simulation. *Journal of American Chemical Society*, Vol. 117, No. 15, (April 1995), pp. (4363-4366), ISSN 0002-7863.

Liu, H.S., Song, C.J., Zhang, L., Zhang, J., Wang, H., & Wilkinson, D.P. (2006). A review of anode catalysis in the direct methanol fuel cell. Journal of Power Sources, Vol. 155, No. 2, (April 2006), pp. (95-110), ISSN 0378-7753.

Liu, Z., Chen, K., Davis, C., Sherlock, S., Cao, Q., Chen, X., & Dai, H. (2008). Drug delivery with carbon nanotubes for in vivo cancer treatment. *Cancer Research*, Vol. 68, No. 16, (August 2008), pp. (6652-6660), ISSN 0008-5472.

Maclean, H.L., Lave, L.B. (2003). Evaluating automobile fuel/propulsion system technologies. *Progress in Energy and Combustion Science*, Vol. 29, No. 1, (February 2003), pp. (1-69), ISSN 0360-1285.

Martin, D., Hanthal, H.G. (1975). *Dimethyl Sulfoxide*, Wiley, New York.

Mountain, R.D. (1997). Shear viscosity and dielectric constant of liquid acetonitrile: A computer simulation study. *Journal of Chemical Physics*, Vol. 107, No. 10, (September 1997), pp. (3921-3923), ISSN 0021-9606.

Peng, Z, Ewig, C.S., Hwang, H.-J., Waldman, M., & Hagler, A.T. (1997). Derivation of Class II Force Fields. 4. van der Waals Parameters of Alkali Metal Cations and Halide Anions. *Journal of Physical Chemistry A*, Vol. 107, No. 39, (September 1997), pp. (7243-7252), ISSN 1089-5639.

Poltoratchkij, G.M. (1984). *Thermodynamic characteristics of non-aqueous electrolyte solutions. Handbook*, Chemistry, Leningrad, USSR

Qi, L., Xie, X.F., Xu, J.M., & Zhou, Q.F. (2006). Transport phenomena and numerical simulation related to water in direct methanol fuel cell. *Progress in Chemistry*, Vol. 18, No. 012, (December 2006), pp. (1725-1734), ISSN 1005281X.

Schultz, T., Zhou, S., & Sundmacher, K. (2001). Current status of and recent developments in the direct methanol fuel cell. *Chemical Engineering & Technology*, Vol. 24, No. 12, (December 2001), pp. (1223-1233), ISSN 0930-7516.

Sevriugin, V.A., Loskutov, V.V., & Skirda, V.D. (2003). Dependence of the self-diffusion coefficient of liquid molecules in a porous medium on its geometric parameters. *Colloid Journal*, Vol. 65, No. 5, (September 2003), pp. (602-605), ISSN 1061-933X.

Suffredini, H.B., Salazar-Banda, G.R., & Avaca, L.A. (2009). Carbon supported electrocatalysts prepared by the sol-gel method and their utilization for the oxidation of methanol in acid media. *Journal of Sol-Gel Science and Technology*, Vol. 49, No. 2, (November 2008), pp. (131-136), ISSN 0928-0707.

Van Gunsteren, W.F., Billeter, S.R., Eising, A.A., Hünenberger, P.H., Kürger, P., Mark, A.E., Scott, W.R.P. & Tironi, I.G. (1996). *Biomolecular Simulation: The GROMOS 96 Manual and User Guide.* Zürich Biomos b.v. ISBN 3 7281 2422 2, Zürich, Groningen.

Wang, C.H., Dub, H.Y., Tsai, Y.T., Chen, C.-P., Huang, C.-J., Chen, L.C., Chen, K.H., & Shin, H.-C. (2007). High performance of low electrocatalysts loading on CNT directly grown on carbon cloth for DMFC. *Journal of Power Sources*, Vol. 171, No. 1, (September 2007), pp. (55-62), ISSN 0378-7753.

Wang, Z., Zhu, Z.Z., Li, Y.X. , & Li, H.L. (2008). Highly dispersed palladium nanoparticles on functional MWNT surfaces for methanol oxidation in alkaline solutions. *Chinese Journal of Chemistry*, Vol. 26, No. 4, (April 2008), pp. (666-670), ISSN 1614-7065.

Wu, F., Xu, B. (2006). Progress on the application of carbon nanotubes in supercapacitors. *New Carbon Materials*, Vol. 21, No. 2, (June 2006), pp. (176-184), ISSN 1872-5805.

Yang, F., Fu, D., Long, J., & Ni, Q.X. (2008). Magnetic lymphatic targeting drug delivery system using carbon nanotubes. *Medical Hypotheses*, Vol. 70, No. 4, (October 2007), pp. (765-767), ISSN 0306-9877.

Yang, X., Zhang, Z., Liu, Z., Ma, Y., Yang, R., & Chen, Y. (2008). Multi-functionalized single-walled carbon nanotubes as tumor cell targeting biological transporters. *Journal of Nanoparticle Research*, Vol. 10, No. 5, (November 2007), pp. (815-822), ISSN 1388-0764.

Yu, Z.W., Quinn, P.J. (1994). Dimethyl-Sulfoxide - a Review of Its Applications in Cell Biology. *Bioscience Reports*, Vol. 14, No. 6, (December 1994), pp. (259-281), ISSN 0144-8463.

Yu, Z.W., Quinn, P.J. (1998). The modulation of membrane structure and stability by dimethyl sulphoxide (Review). *Molecular Membrane Biology*, Vol. 15, No. 1, (January 1998), pp. (59-68), ISSN 0968-7688.

Zhang, R., Baxendale, M., & Peijs, T. (2007). Universal resistivity-strain dependence of carbon nanotube/polymer composites. *Physical Review B*, Vol. 76, No. 19, (November, 2007), pp. (195433.1-195433.5), ISSN 1098-0121.

# 17

# Comparison of NQR of $O_2$, $N_2$ and CO on Surface of Single-Walled Carbon Nanotubes and Chemisorption of Oxygen-Doped on the Surface of Single-Walled Carbon Nanotubes: A DFT and NMR Computational Study

S. A. Babanejad[1], F. Ashrafi[1], A. Ghasemi[1], N. Salarzadeh[1], M. Rahimova[2], G. H. Babanejad[3], G. Babanejad[3] and N. Babanejad[3]
*[1]Payam e Noor University, Sari, Mazandaran, I.R.*
*[2]National University of Tajikistan,*
*[3]Multi Media University,*
*[1]Iran*
*[2]Tajikistan*
*[3]Malaysia*

## 1. Introduction

The discovery of carbon nanotubes produced by graphite, first reported by Iijima in 1991(S. Iijima & T. Ichihasi 1993), ushered in a new and very amazing research field in compacting gases by physisorption methods. The changes in electrical resistance, by adsorption of certain gas molecules are considerable, for example by adsorption of $O_2$, $N_2$, $NH_3$ and $H_2$ (H. Chang et al. 2001). Figure 1 depicts a $C_{72}H_{16}$ tube modeling an armchair (4, 4) SWCNTs which demonstrates the stated effect on the electronic structure of single-walled carbon nanotubes (SWCNTs).

Comparing the adsorption of gases on the surface, using computational methods substantially reduces costs and thus NQR were used in related investigations. Even at low concentration, due to the charge transfer between gases and tube, gas physisorption can change the conductivity of CNTs. Electronic properties of SWCNT have been studied in a number of theoretical works (H. van C. Houten et al., 1992; H. Aijki & T. Ando, 1993; J. W. G. Wildoer et al., 1998; A. Bezryadin, 1998) and optimized forms of nanotube can be designed by a precise positioning of various gases on considered carbon atoms. Theoretical studies have found that this single-walled carbon nanotube has novel electronic properties, which can be semiconducting, depending on their radius or chiralities (T. Hertel et al., 1998; J. W. Mintmire et al., 1992; N. Hamada et al., 1992; R. Saito et al., 1992; M. Rao et al., 1997; H. Kataura et al., 1999). According to the electron-transmission mechanism on the surface of CNTs, the detected gas can be classified into reducing and oxidizing gaseous species. The electrical resistance of CNTs was found to increase when exposed to reduce gaseous species as $N_2$ (J. Zhao et al., 2002), whereas being exposed to oxidizing ones as $O_2$ (P. G. Collins et al., 2000) decreased. A new SWCNT gas sensor would be fulfilled by utilizing such electrical characteristics. In this study, $N_2$, $O_2$, and CO adsorption mechanism on carbon nanotubes was investigated by the surface CNT gas sensor. The physisorption of $N_2$, $O_2$, and CO at the open ended single-walled carbon nanotube (SWCNT) has been investigated. It was found that $N_2$, $O_2$, and CO can be physisorbed at the surface site of armchair SWCNT that makes the N–N, O-O, and C-O bonds active. This can be attributed to the decisive effect of the local end carbon atoms arrangement of the open-ended SWCNT surface. In addition, comparison of the adsorption amount on CNTs forcefully increases this viewpoint that present theoretical study on open-ended SWCNTs shows larger adsorption capacity. In this paper, a computational study of the adsorption of $N_2$, $O_2$, and CO on the surface the open-ended SWCNT is reported. The isosteric adsorption and the binding energy were compared with adsorption predicted on surface SWCNT. Nuclear experimental techniques such as nuclear quadrupolar resonance (NQR) (G. K. Semin et al., 1975) are widely used to study the geometry and electronic structure of molecules. For non-magnetic dielectrics, this response gives information about coordination and geometry around each nucleus with spin $I > 0$. It is known that when nuclei with spin$>1/2$ are put in an electric field gradient (EFG) (A. R. Kessel & V. L. Ermakov 1999.; A.K. Khitrin et al., 2001), decayed spin energy levels are created. NQR methods are applied to produce high external magnetic fields and some kind of internal interaction in order to form a non decayed energy spectrum. However, the field has recently started to produce good products and an increasing amount of experimental and theoretical data is becoming available. There are two naturally occurring isotopes for nitrogen-14 with natural abundance of 99.635% and nuclear spin I=1and N-15 with natural abundance of 0.365% and I=1/2. Despite its rich natural plentifulness, present N-14 applied an electric quadrupole moment (NQR) in our study. Dependence of these parameters on length and diameter of CNTs are also considered. The NQR measurable parameters are quadrupole coupling constant ($C_Q$) and asymmetry parameter ($\eta_Q$) which both are also reproduced by quantum chemical calculations of the electric field gradient (EFG) tensors. Nuclei with spin angular momentum $I > 1/2$ have the nuclear electric quadrupole moment, which interacts with the electric field gradient (EFG) tensor originated at the site of quadrupole nuclei. For O-17, N-14, and C-13 spin angular momentums are $\frac{5}{2}$, $\frac{3}{2}$ and $\frac{1}{2}$, respectively. Therefore O-17 and N-14 are very sensitive to the electronic density at the sites of nuclei and feel changes by any disturbance. Calculation by computational methods has a long history in the study of materials used in energy technologies. As the number of articles in this field suggest, these metodes continue to play a substantial role which is, moreover, likely to grow in the future (C. R. A. Catlow, et al., 2010). Carbon nano-tubes (CNT) are nano structures derived from rolled grapheme planes (fig. 2) (S. Iijima & T. Ichihasi, 1993) whose electronic properties can be controlled. Zhao et al., (J. Zhao, 2002) studied the adsorption of various gas molecules ($NO_2$, $O_2$, $NH_3$, $N_2$, $CO_2$, $CH_4$, $H_2O$, $H_2$, Ar) on both single SWNT and SWNT bundles using first principles method. The self-consistent field (SCF) electronic structure calculations are performed based on density functional theory (DFT) (T. Schimizu & M. Tsukada, 1993; M. Lynch & P. A. Hu, 2000; N. D. McClenaghan et al., 2000; C. Noguera, 2001). Jhi et al., (S.H. Jhi, 2000), theoretically studied the effect of oxygenation on the electronic and magnetic properties of SWNT their calculation for the density of states shows that weak coupling between carbon and oxygen leads to conducting states near the band gap. One possible way to modify the electronic and vibronic properties is a charge transfer during their intercalation and fictionalization (E.B. Barros et al., 2007; A.G. Souza Filho et al., 2006). Depending on their diameter and felicity it was predicted that they can be semiconductors or metals (R. Saito et al., 1992; J. W. Mintmire et al., 1992). They can also sustain large current densities (Ph. Avouris, 2000), and their electrical properties can be modified by doping (R.S. Lee, 1997). The conductance of a single oxygen doped (6, 6) nano-tube decreases by about 30% with respect to that of the perfect nano-tube (N.D. Lang, 2000 & 1998). Ulbricht et al (H. Ulbricht et al., 2002) concluded that no evidence for a more strongly bound chemisorbed species or for dissociative oxygen adsorption was found. The effects of oxygen chemisorption on a nano-tube based field effect transistor have been controversial as to whether it induces oxygen-doping of the nano-tube body or the work function increase in the semiconductor electrode. The doping effect could be more influential in devices with longer nano-tubes (S. A. Babanejad et al., 2010; F. Ashrafi et al., 2010). In this study, two contributions to the resistance of nano-tubes were investigated. First, we calculate the contact from oxygen chemisorption on a nano-tube with model semiconductor (5, 0) zigzag and (4, 4) arm chair single-walled carbon nano-tubes (fig 2). Then we concentrate on the resistance produced by substitutional defects. We show the chemical-shielding ($\sigma_{ii}$) tensors were converted to isotropic chemical-shielding (iso) and anisotropic chemical-shielding ($\Delta\sigma$) and asymmetric ($\mu_j$) parameters of $^{17}O$ and $^{13}C$ atoms for the optimized structures (tables 3 and 4). The study of electronic and structural properties of oxygen-doped single wall nano-tubes have been performed (M. Mirzaei & N. L. Hadipour, 2006). The calculation of NMR (M. J. Duer, 2002) parameters using DFT techniques have become a major and powerful tool in the investigation of molecular structure. The calculations showed consistent results with the Computational ones. The tensors originating at the sites ($A_1$, $A_2$, $A_3$ and $A_4$) of half-spin magnetic nuclei make available important trends about the electronic properties at the sites of these nuclei. The tensors were computed in the optimized structures by high-level quantum chemical calculations (F. Ashrafi et al., 2010; G. Wu, 2002). In this computational evaluation, the influence of oxygen-doping on the electrostatic properties of zigzag (5, 0) and arm chair (4, 4) CNTs are studied via the tensors calculations at the sites of $^{17}O$ nuclei in two case representative O-doped models (A. S. Ghasemi et al., 2010). The length of 7.1 Å and 4.8 Å were obtained for (5,0) and (4,4) single-wall nanotube including oxygen-doped (O-doped), respectively. The forms indicated in figures 2 and 3 are considered in calculations (tables 2 to5).

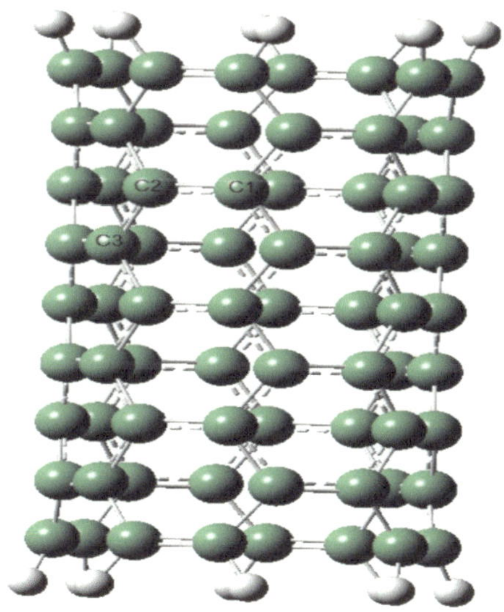


Fig. 1. CNTs(4,4)

| Model (configuration) | $R_{C-C}$ (A°) | $R_{C-X}$ (A°) (X=O, N, C) | $R_{O-O}$ (A°) | $R_{C-O}$ (A°) | $R_{N-N}$ (A°) | Δ $E_{ab}$ (ev) | Dipole momentum (Debye) |
|---|---|---|---|---|---|---|---|
| CNT | $(C-C)_1$=1.424<br>$(C-C)_2$=1.419<br>$(C-C)_3$=1.438<br>$(C-C)_4$=1.405<br>$(C-C)_5$=1.437<br>$(C-C)_6$=1.437 | - | - | - | - | - | 0.4358 |
| $N_2$-CNTs-A1 | $(C-C)_1$=1.48<br>$(C-C)_2$=1.481<br>$(C-C)_3$=1.481<br>$(C-C)_4$=1.481 | $(C-N)_1$=1.515<br>$(C-N)_2$=1.515 | - | -<br>- | 1.250 | -77910.48 | 3.5448 |
| $N_2$-CNTs-A2 | $(C-C)_1$=1.504<br>$(C-C)_2$=1.516<br>$(C-C)_3$=1.517<br>$(C-C)_4$=1.500 | $(C-N)_1$=1.521<br>$(C-N)_2$=1.525 | - | | 1.255 | -77909.73 | 1.6172 |
| $O_2$-CNTs-A1 | $(C-C)_1$=1.487<br>$(C-C)_2$=1.487<br>$(C-C)_3$=1.487<br>$(C-C)_4$=1.487 | $(C-O)_1$=1.465<br>$(C-O)_2$=1.465 | 1.485 | -<br>- | -<br>- | --79023.84 | 3.1747 |
| $O_2$-CNTs-A2 | $(C-C)_1$=1.500<br>$(C-C)_2$=1.507<br>$(C-C)_3$=1.563<br>$(C-C)_4$=1.459 | $(C-O)_1$=1.436<br>$(C-O)_2$=1.408 | 2.66 | - | - | -79024.74 | 3.4475 |
| CO-CNTs-A1 | $(C-C)_1$=1.481<br>$(C-C)_2$=1.48<br>$(C-C)_3$=1.48<br>$(C-C)_4$=1.48 | $(C-C)_1$=1.584<br>$(C-O)_2$=1.501 | - | 1.315 | - | -78013.55 | 3.4764 |
| CO-CNTs-A2 | $(C-C)_1$=1.480<br>$(C-C)_1$=1.480<br>$(C-C)_1$=1.461<br>$(C-C)_1$=1.582 | $(C-C)_1$=1.477<br>$(C-O)_2$=1.562 | - | 1.374 | | -78012.27 | 2.6888 |

Table 1. Calculated adsorption energies $E_{ab}$ (eV), bond energies(A°) and dipole momentum (Debye) of the $O_2$ and $N_2$ and CO adsorbed on surface armchair (n, n), n=4 nanotube.

## 2. Computational details

In this study $O_2$ and $N_2$ molecules adsorption behaviors on the end and surface of single-walled nanotube is taken in to consideration. A (4, 4) CNT containing 72 carbon atoms with length of 9.8A° and a diameter of 5.6 A° is selected for this purpose. Saturating carbon dangling bonds with 16 hydrogen atoms is necessary because there is no periodic boundary conditions in molecular calculations and also due to limitation of nanotube length and lack of homogeneity for ending atoms, symmetry breaks down and some changes in geometrical properties are proved for ending atoms during optimization processes. Optimization of a sample system includes relaxation of atoms to lower forces from other constituents on each atom. Calculations were carried out with Gaussian98 suite of programs at all-electron level (M. J. Frisch et al., 1998). It has been established that DFT is able to accurately treat such systems due to incorporation of the exchange-correlation effects (V. Barone et al., 2004; W. L. Yim, & Z. F. Liu, 2004; X. Lu et al., 2005). In quadrupolar spin system, the electric field gradient (EFG) tensor at nitrogen-14 and oxygen-17 nuclear sites has axial symmetry (asymmetry parameter $\eta = 0$ ). The existence of the zero asymmetry parameter was one of the reasons why this compound is considered to present such interest (S. A. Babanejad et al., 2010; F. Ashrafi et al., 2010; E. A. Hill & , J. P. Yesinowski, 1997; A. Abragam, 1961). Geometry optimizations and EFG calculations were performed using 6-311G* basis set with B3LYP functional (H. S. Kang, 2006; S. Hou, 2004). The interaction between nuclear electric quadrupole moment and EFG at quadrupole nucleus is described with Hamiltonian (A. Abragam, 1961) $\hat{H} = \frac{e^2 Q q_{ZZ}}{4I(2I-1)}[3\hat{I}_Z^2 - \hat{I}^2) + \eta_Q(\hat{I}_X^2 - \hat{I}_Y^2)]$ where eQ is the nuclear electric quadrupole moment, I is the nuclear spin and $q_{zz}$ is the largest component of EFG tensor. The principal components of the EFG tensor, $q_{ii}$, are computed in atomic unit $(1 \text{ au} = 9.717365 \times 10^{21} \text{ V m}^{-2})$, with $|q_{xx}| \le |q_{yy}| \le |q_{zz}|$ and $q_{xx} + q_{yy} + q_{zz} = 0$. These diagonal elements are related by a symmetry parameter $\eta_Q = |(q_{yy} - q_{xx})/q_{zz}|$ and $0 \le \eta_Q \le 1$, that measures the deviation of EFG tensor from axial symmetry (E. A. C. Lucken, 1992). Cluster model is proved to be valid for nanotubes (G. E. Froudakis et al., 2003; D. C. Sorescu et al., 2001). The computed $q_{zz}$ component of EFG tensor is used to obtain nuclear quadrupole coupling constant from the equation $C_Q = e^2 Q q_{ZZ} / h$ (E. A. C. Lucken, 1992).

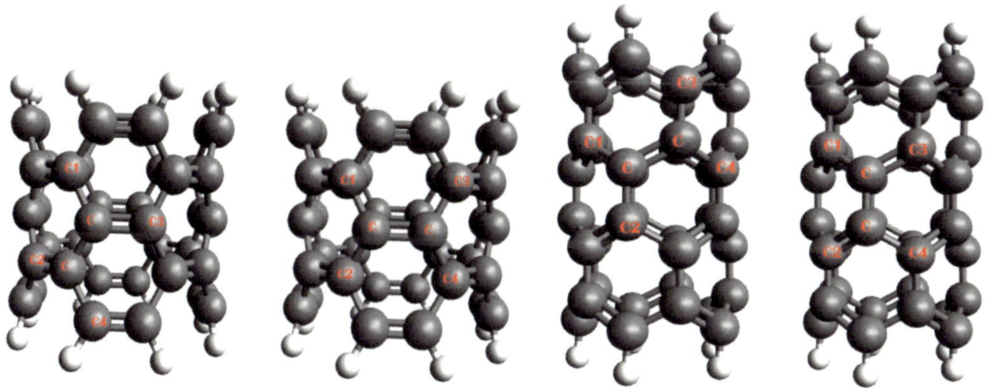

CNT(4,4)($A_1$) CNT(4,4)($A_2$) CNT(5,0)($A_1$) CNT(5,0)($A_2$)

Fig. 2. ($A_1$ & $A_2$) The (5, 0) and (4, 4) SWCNT

In the present study, the effects of oxygen ($O_2$) molecules chemisorption on SWCNTs of models (4, 4) and (5, 0) was investigated. In order to investigate the electronic structure in semiconductor nanotube contacts of $O_2$ molecules, the computations were fully implemented by Gaussian 98 Software package (E.B. Barros et al., 2007; A.G. Souza Filho et al., 2006; M. J. Frisch et al., 1998). Geometry optimizations were performed using 6-31G* basis set with DFT/B3LYP functional (R. G. Parr & W. Yang, 1994; A. D. Becke, 1993). NMR $^{17}O$ and $^{13}C$ chemical shielding calculations were computed at B3LYP/6-311G* level of theory using gauge including atomic orbitals (GIAO) approach (K. Wolinski et al., 1990). The undoped models (4, 4) and(5, 0) consisted of 40 C atom with length of 4.8 Å and 7.1 Å are chosen for the purpose, respectively. In absence of periodic boundary conditions in molecular calculations, it is necessary to saturate the carbon dangling bonds with hydrogen atoms. Curvature of small tubes is a crucial feature responsible for intense interaction of atoms in tubes. Quantum chemical calculated tensors at the principal axes system (PAS) ($\sigma_{11} \leq \sigma_{22} \leq \sigma_{33}$) is converted to a diagonal matrix with $\sigma_{11}$ , $\sigma_{22}$ and $\sigma_{33}$ components , measurable NMR parameters, chemical shielding isotropic ($\sigma_{iso}$), chemical shielding anisotropic ($\Delta\sigma$) and asymmetric ($\mu_j$) are used, respectively (M. J. Duer, 2002). This shows a second-order change in the molecular energy.

| Model | $r_{C-C}$ | $r_{C-O}$ | $R_{O-O}$ | $\Delta E_{abs}$ - DFT |
|---|---|---|---|---|
| CNT(4,4)($A_1$) | $(C-C)_1$=1.421<br>$(C-C)_2$=1.422<br>$(C-C)_3$=1.421<br>$(C-C)_4$=1.422 | - | - | - |
| CNT(4,4)($A_2$) | $(C-C)_1$=1.451<br>$(C-C)_2$=1.419<br>$(C-C)_3$=1.422<br>$(C-C)_4$=1.452 | - | - | - |
| CNT(4,4)-$O_2$($A_3$) | - | $(C-O)_1$=1.333<br>$(C-O)_2$=1.334<br>$(C-O)_3$=1.334<br>$(C-O)_4$=1.333 | 2.563 | -43762.03 |
| CNT(4,4)-$O_2$($A_4$) | - | $(C-O)_1$=1.378<br>$(C-O)_2$=1.395<br>$(C-O)_3$=1.386<br>$(C-O)_4$=1.384 | 2.547 | -43763.24 |

[a] All calculated distances are in Å. All calculated binding energies are in electron volt (eV).

Table 2. Calculated structural parameters and binding energies of $O_2$ Chemisorption on the (4 , 4) SWCNT [a].

$$E = E_0 + B_0 \chi B_0 + \sum_{i=1}^{N} \mu_i \sigma B_0 + \cdots \tag{1}$$

The summation is taken over the O nuclei in the system. We are not interested in the magnetic susceptibility, $\chi$, but only in the bilinear response property.

| Model | $r_{C-C}$ | $r_{C-O}$ | $r_{O-O}$ | $\Delta E_{ads}$ - DFT |
|---|---|---|---|---|
| CNT(5,0)($A_1$) | $(C-C)_1$=1.437<br>$(C-C)_2$=1.451<br>$(C-C)_3$=1.437<br>$(C-C)_4$=1.451 | - | - | - |
| CNT(5,0)($A_2$) | $(C-C)_1$=1.426<br>$(C-C)_2$=1.451<br>$(C-C)_3$=1.408<br>$(C-C)_4$=1.437 | - | - | - |
| CNT(5,0)-$O_2$($A_3$) | - | $(C-O)_1$=1.408<br>$(C-O)_2$=1.408<br>$(C-O)_3$=1.408<br>$(C-O)_4$=1.409 | 2.421 | -43652.95 |
| CNT(5,0)-$O_2$($A_4$) | - | $(C-O)_1$=1.375<br>$(C-O)_2$=1.373<br>$(C-O)_3$=1.335<br>$(C-O)_4$=1.399 | 2.563 | -43653.63 |

[a] All calculated distances are in Å. All calculated binding energies are in electron volt (eV).

Table 3. Calculated structural parameters and Chemisorption energies of $O_2$ adsorbed on the (5, 0) SWCNT [a].

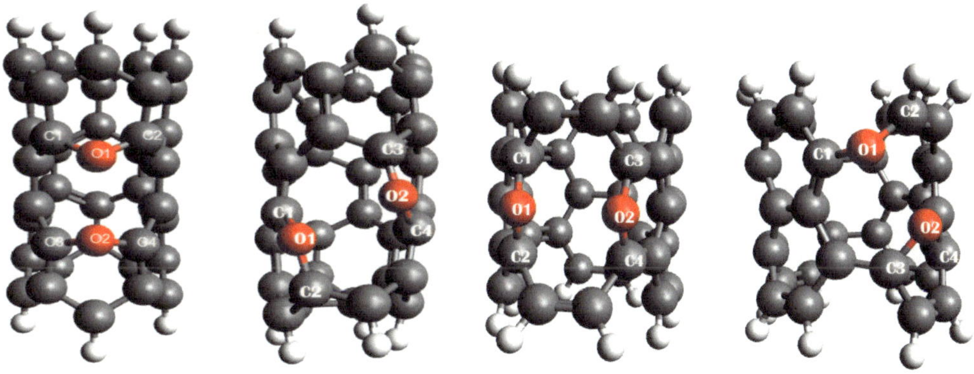


CNT(4,4)-$O_2$($A_3$) CNT(4,4)-$O_2$($A_4$) CNT(5,0)-$O_2$($A_3$) CNT(5,0)-$O_2$($A_4$)

Fig. 3. Chemisorption configurations of an $O_2$ molecule (The sites A3, A4 of (4, 4) and(5, 0) SWCNT-$O_2$, respectively)

$$\sigma_{ij} = \left( \frac{\partial^2 E}{\partial B_i \partial \mu_j} \right)_{B_i = \mu_j \to 0} \tag{2}$$

Where $\mu_j$ is the components of magnetic moment and $B_i$ is external magnetic field. The principal components for specification of shielding are defined by this coordinate system as following equation (C.M. Marian, & M. Gastreich, 2001).

$$\Delta\sigma = \frac{3}{2}(\sigma_{33} - \sigma_{iso}) \quad , \quad \sigma_{iso} = \frac{(\sigma_{11} + \sigma_{22} + \sigma_{33})}{3} \quad , \quad \eta_\sigma = \frac{3}{2}\left(\frac{\sigma_{22} - \sigma_{11}}{\Delta\sigma}\right) \tag{3}$$

In which $\sigma_{iso}$, $\Delta\sigma$ and are isotropic, anisotropic and asymmetric parts of tensor, respectively and in certain cases vanishes.

## 3. Results and discussion

Geometries, binding energies and NQR (4, 4) SWCNT interacted with $O_2$ ,$N_2$ and CO molecule species have studied in this work. The calculated geometry parameters and binding energies, dipole momentum and EFG tensors have shown in tables 1 and 5 in the following sections, molecular geometries and binding energies, $E_{ab}$, $C_Q$, EFG tensors and the data obtained from $O_2$, $N_2$, and CO molecules adsorptions are discussed, separately.
In the present work, two models of zigzag (5,0)and armchair (4, 4) SWCNTs with specefied tube lengths are studied using quantum chemical calculations (figs. 2 and 3). Chemical shielding tensors of H-capped (5, 0) and (4, 4) SWCNTs interacted with oxygen molecules are obtained. The calculated geometry parameters and binding energies and $^{17}O$ and $^{13}C$ chemical shielding tensors are presented in tables 2 to 5. The molecular geometries and binding energies and NMR chemical shielding tensors resulted from oxygen molecular chemisorptions are discussed in following sections, sepa.

### 3.1 Geometries properties and adsorption and binding energies

In this study, the use of electronic properties of nano tubes has been established to appear field of spin-electronics, a field that influences the electron's spin degree of freedom for transfer and storage of information and communication. The optimized geometries of calculated configurations of $O_2$, $N_2$, and CO molecules adsorbed on (4, 4) SWCNT are schematically displayed in fig. 4. Geometrical parameters, adsorption energies and dipole moment are summarized in table1. The nature of stationary points are confirmed by vibrational frequency calculations at the B3LYP/6-311G* level. For nitrogen, oxygen and CO molecules we have considered distinct adsorption sites, marked as CNT, CNT–$O_2$, CNT–$N_2$ and CNT–CO adsorption energies, $\Delta E_{ab}$, (Table1) are calculated using:

$$E_{ad} = E_{tot}(\text{molecule}O_2 + \text{CNT}_S) - E_{tot}(\text{CNT}_S) - E_{tot}(\text{molecule}O_2) \tag{4}$$

$$E_{ad} = E_{tot}(\text{molecule}N_2 + \text{CNT}_S) - E_{tot}(\text{CNT}_S) - E_{tot}(\text{molecule}N_2) \tag{5}$$

$$E_{ad} = E_{tot}(\text{moleculeCO} + \text{CNT}_S) - E_{tot}(\text{CNT}_S) - E_{tot}(\text{moleculeCO}) \tag{6}$$

Where, $E_{tot}$(CNT), $E_{tot}$ ($O_2$), $E_{tot}$ (CNT+$O_2$), $E_{tot}$ (N2), $E_{tot}$ (CNT+$N_2$), $E_{tot}$ (CO) and $E_{tot}$ (CNT+CO) are the energies of the optimized tubes, which are adsorption systems, respectively. By this explanation, $E_{ad} < 0$ corresponds to exothermic adsorption which leads to local minima stable for adsorption of gas molecules on the surface of nanotube. Armchair (4, 4) nanotube has two different C–C bonds ((C1–C2) =1.405A° and (C2–C3) =1.438A°) thus suggests two distinct adsorption sites. A diagrammatic view of this form is showed in fig. 4.

CNT, N2-CNT-A1&2, O2-CNT-A1&2 and CO2-CNT-A1&2. Such a structure has also been observed for other SWCNTs (S. Dag et al., 2003; H. He et al., 1998; S. P. Walch, 2003). For the molecular $O_2$–CNTs, $N_2$-CNTs, and CO–CNTs systems, O2, N2 and CO seemed to place parallel to the outer surface of the tube. Geometry calculations of distortion caused by the oxygen and nitrogen and carbon monoxide molecules on the (C1–C2) bond are changed partly. Placing the oxygen molecule in CNT-A1, CNT-A2 sites doesn't change the bridge distance of (C2–C3) considerably. Two different types of adsorbed $O_2$, $N_2$ and CO molecules were recognized (Fig. 4. CNT, $N_2$ -CNT-A1, $N_2$,- CNT-A2, $O_2$ -CNT-A1, $O_2$,- CNT-A2, CO-CNT-A1 and CO-CNT-A2). The calculated adsorption energies were predicted to be -77910.48 and -77909.73 eV for $N_2$ and -79023.84 and -79024.74 eV for $O_2$ and -78013.55 and -78012.27 eV for CO, respectively. The length of nanotube have selected with regard to the length of unit cell of nanotube. Such adsorptions of $O_2$ molecule are known as cycloaddition which is very similar to those found for larger diameter tubes (M. J. Duer, 2002; Y. F. Zhang & Z.F. Liu, (2004). Nitrogen molecules adsorbed with a comparatively lower rate and almost never formed a chemical binding with the carbon nanotube. The geometry of (4, 4) tube is considerably modified when such oxidation occurs and physisorbed product is formed. The electron configuration of $O_2$ is KK $(\sigma_{2s})^2$ $(\sigma_{2s}^*)^2$ $(\sigma_{2pz})^2$ $(\pi_{2px})^2$ $(\pi_{2py})^2$ $(\pi_{2px}^*)^1$$(\pi_{2py}^*)^1$. The electron configuration of $N_2$ is KK $(\sigma_{2s})^2$ $(\sigma_{2s}^*)^2$ $(\pi_{2px})^2$ $(\pi_{2py})^2$ $(\pi_{2pz})^2$, and the transferred electron is placed in the half-filled anti-bonding orbital of $O_2$, thus weakens the O–O bond. The electron can't enter into $N_2$ molecule binding orbital because the binding orbital is filled. This arrives to either $sp^3$ hybridization for two carbon atoms or breaking of one C–C bond. Two different types of adsorbed $O_2$, $N_2$, and CO species were identified (Fig. 4. and Table 1). Also, the dipole moments were calculated by Gaussian software and have shown in table1. Obtained values demonstrate that as the dipole moment becomes bigger, the absolute value of bond energy increases. We can explain this reality as following: the big dipole moment relies to the large distance between electron clouds, then, as the distance becomes bigger the absolute value of bond energy will become higher. By comparing the obtained results with Jordan's one (D. C. Sorescu et al., 2001). It is well known that the tendency for $sp^2$–$sp^3$ re hybridization upon $O_2$ adsorption is strong for thin nanotubes, because highly bent $sp^2$ bonding of thin nanotubes is favored for the transition to $sp^3$ bonding. According to adsorption energy and dipole moment parameters in table1, $O_2$ molecule shows the highest adsorption rate.

This is a general reason for the binding performed studies, which shows that nitrogen molecules energy values of adsorption on armchair model with determined diameter and length have about twice differences in grandeur. Based on performed calculations, we approach that the adsorption accomplishes over open ends of nanotubes has more advantages. In addition, all these energies are positive which demonstrate the reaction is improbable. Based on these results, we can conclude that the physical adsorption over the surface area of nanotube occurs very hard and so this is an appropriate case.

Also, In this section, stable configurations of oxygen molecule chemisorption at the surface of SWCNT are discussed. After optimized structures were obtained, geometrical parameters and binding energies of the models structure of these oxygen molecule attached to the zigzag (5, 0)and armchair (4, 4) SWCNTs were calculated as shown in Figures (2) and (3). The results at the level of the B3LYP DFT method and the 6-311G* standard basis set are summarized in tables 1 and 2. Upon chemisorption of a $O_2$ molecule on the C-C bond at the surface, the molecule $O_2$ dissociates toward the O-O bond lengths. Chemisorption on nano-tube increases from 1.21 Å and 2.528 Å to 2.563 Å for (4, 4) and (5, 0) SWCNT, respectively.

We have considered two distinct chemisorption sites, marked as $A_1$, $A_2$, $A_3$ and $A_4$ (table 1 and 2). CNT and CNT–$O_2$ binding energies, $E_{ad}$, are calculated using, $E_{ad} = E_{tot}$(molecule$O_2$ + $CNT_S$) -$E_{tot}$($CNT_S$) - $E_{tot}$(molecule $O_2$) Where, $E_{tot}$ (CNT), $E_{tot}$ ($O_2$) and $E_{tot}$ (CNT+$O_2$) are the energies of the optimized tubes, that are chemisorption and tube–adsorb ate systems, respectively. Armchair (4, 4) and zigzag (5, 0) tube has different C–C bonds thus offers two distinct chemisorption sites (table 1and 2) before and after the doping of O atoms, the bond length of in SWNT-$A_1$(4, 4) from.
$(C-C)_{1,3}$=1.421 A° and $(C-C)_{2,4}$=1.422 A° decreased to 1.333A°-1334 A° and bond length of in SWNT-$A_2$(4, 4) ) from $(C-C)_1$=1.451A° , $(C-C)_2$=1.419A° , $(C-C)_3$=1.422A° and $(C-C)_4$=1.452A° decreased to 1.378 -1.395 before and after the doping of O atoms, the bond length of in SWNT-$A_3$(5,0) from $(C-C)_{1,3}$=1.437A° and $(C-C)_{2,4}$=1.451A° decreased to 1.335 A° - 1.399 A° bond length of in SWNT-$A_4$(5, 0) ) from $(C-C)_1$=1.426 A°, $(C-C)_2$=1.451A° , $(C-C)_3$=1.408A° and $(C-C)_4$=1.437A° increased to 1.454-1.475 Density functional calculations of SWNT, efficient process of charge transfer between the oxygen molecule and the nano-tube is found to substantially reduce the susceptibility of the π-electrons of the nano-tube to modification by oxygen while maintaining stable doping. Oxygen chemisorption can be achieved with $O_2^+$ ion implantation 28668ce95cc(T. Kamimura et al., 2005).

### 3.2 The $N_2$, $O_2$ and CO NQR parameters and $^{17}O$ NMR parameters (4, 4) and (5,0)

Semiconducting SWCNTs are ballistic conductors with two and one spin degenerate conducting channel(s), respectively (A. Bachtold et al., 2000; V. Krstić et al., 2000). The channels belong to the first $\pi$ and $\pi^*$-band of the delocalized $\pi$-electron system. The N-14, O-17, C-13 NQR parameters ($C_Q$ and $\eta$) in the geometrically optimized SWCNTs model armchair (4, 4) were estimated by EFG tensors calculations at the B3LYP level of the DFT method and the 6-311G* standard basis set.
Tables 6 shows the calculated NQR and EFG tensors for SWCNTs $\eta$ parameter of $O_2$, $N_2$ and CO adsorption on the CNTs surface has a remarkable effect on EFG tensors. A glimpse to $\eta$ values presented in table 2 reveals that the for N-14 and O-17 changes in EFG tensor for molecular adsorptions are quite significant which is in complete agreement with calculations. The B3LYP/6-311G* calculations indicate that all three principal components of the EFG tensor($q_{ii}$) and associated asymmetry parameter are affected due to adsorption of oxygen, nitrogen and CO molecules. For the ($O_2$–CNT) and ($N_2$–CNT) systems, the EFG tensors of CNT (4, 4)- $O_2$(A2) and CNT(4, 4)-$N_2$(A1) are more significantly affected compared to CNT(4, 4)- $O_2$ (A1) and CNT(4,4)-$N_2$(A2), respectively. As previously mentioned, oxygen molecules adsorption at the CNT (4, 4)-$O_2$(A2) leads to the $O_1$–$O_2$ bond cleavage and $N_2$ molecules adsorption at the CNT(4, 4)-$N_2$(A1) breaks C1-C2 bond. Therefore, a noticeable change in the field gradient, especially at the $C_1$ and C2 is detected. $O_2$ adsorptions produce more EFG change at CNT (4, 4)- $O_2$(A2 which can be attributed to their hybridization effect (from $sp^2$ to $sp^3$). This is consistent with the bond angle distortion from 120 to 109, induced by oxygen and nitrogen adsorption. The principle components of EFG tensor change significantly after CO adsorption at C1 and C2 atoms in CNT(4, 4)-CO (A2).
New data and presentation of results are given here for O-doping computational NMR parameters of oxygen nuclei for two models (4,4) and (5, 0) of CNTs (Table 3and 4). Oxygen molecule chemisorptions of SWCNTs have remarkable influence on NMR tensors, which is in complete accordance with the facts mentioned above. Consequently, it has been $^{17}O$

indicated that for the H-capped SWCNTs, the calculated $^{17}O$ chemical shielding values at the ends are smaller than in the tube's center if the carbon is directly bound to hydrogen; otherwise it is larger (H.J. Liu, 2007). It is also depicted that.
chemical shielding components converge in a way similar to that of the chemical shifts when increasing the tube length albeit not as smoothly as the isotropic shielding.

| Model | $^{17}O$ | atoms | $\sigma_{ii}$ $(\sigma_{11}, \sigma_{22}, \sigma_{33})^b$ | $\sigma_{\iota o o}$ | $\Delta\sigma$ | $\eta_\sigma$ |
|---|---|---|---|---|---|---|
| CNT(4, 4)<br>($A_1$) | - | $C_1$<br>$C_2$<br>$C_3$<br>$C_4$ | (-1.16; -1.16; 163.87)<br>(-1.16; -1.16; 163.87)<br>(-0. 73; -0. 73; 163.79)<br>(-0. 75; -0. 75; 163.74) | 53.8495<br>53.8495<br>54.1090<br>54.0800 | 165.0308<br>165.0308<br>164.5215<br>164.4900 | 0.0000<br>0.0000<br>0.0000<br>0.0000 |
| CNT(4, 4)<br>($A_2$) | - | $C_1$<br>$C_2$<br>$C_3$<br>$C_4$ | (-0.71; -0.71; 163.77)<br>(-40.9778; 52.7796; 159.7877)<br>(-35.3100; 13.8404; 175.9600)<br>(-1.17; -1.17; 163.85) | 54.1157<br>57.1965<br>51.4968<br>53.8354 | 164.4815<br>153.8868<br>186.6948<br>165.0219 | 0.0000<br>2.4588<br>1.4317<br>0.0000 |
| CNT(4, 4)-<br>$O_2$<br>($A_3$) | $O_1$ | $C_1$<br>$C_2$ | (-66.8441; 54.0846; 83.9711)<br>(-66.4764; 53.8616; 84.4001) | 23.7372<br>23.9284 | 90.3508<br>90.7075 | 2.0077<br>1.9900 |
| | $O_2$ | $C_3$<br>$C_4$ | (-66.7714; 54.0488; 83.9700)<br>(-66.3669; 54.0373; 84.0975) | 23.7491<br>23.9226 | 90.3313<br>90.2623 | 2.0063<br>2.0009 |
| CNT(4, 4)-<br>$O_2$<br>($A_4$) | $O_1$ | $C_1$<br>$C_2$ | (-53.8794; 39.8243; 95.9593)<br>(-48.5238; 12.7128; 100.4479) | 27.3014<br>21.5456 | 102.9869<br>118.3535 | 1.3648<br>0.7761 |
| | $O_2$ | $C_3$<br>$C_4$ | (-29.8179; 7.3599; 103.6236)<br>(-66.1822; 56.8643; 97.6708) | 27.0552<br>29.4509 | 114.8526<br>102.3298 | 0.4856<br>1.8037 |

[a] Calculated $\sigma_{ii}$, $\sigma_{iso}$ and $\Delta\sigma$ values are in ppm
[b] In each raw, the first number is for $\sigma_{11}$, the second number is for $\sigma_{22}$ and the third number is for $\sigma_{33}$.

Table 4. Calculated $^{17}O$ NMR parameters for CNT, $O_2$–CNT (4, 4) systems [a]

| Model | $^{17}O$ | atoms | $\sigma_{ii}$ ($\sigma_{11}$, $\sigma_{22}$, $\sigma_{33}$)[b] | $\sigma_{iso}$ | $\Delta\sigma$ | $\eta_\sigma$ |
|---|---|---|---|---|---|---|
| CNT(5, 0) ($A_1$) | - | $C_1$ | (-118.0860; 12.8629; 163.87) | 18.0597 | 212.0138 | 10.8763 |
| | | $C_2$ | (-116.9704; 13.8079; 159.2077) | 18.6817 | 210.7890 | 10.5005 |
| | | $C_3$ | (-118.1198; 12.8584; 159.4017) | 18.0467 | 212.0325 | 10.8866 |
| | | $C_4$ | (-116.9332; 13.8073; 159.2050) | 18.6930 | 210.7680 | 10.4911 |
| CNT(5, 0) ($A_2$) | - | $C_1$ | (-45.2746; 155.1751; 325.7105) | 145.2037 | 270.7602 | 2.0707 |
| | | $C_2$ | (-118.1198; 12.8584; 159.4017) | 18.0467 | 212.0325 | 10.8866 |
| | | $C_3$ | (-27.4737; 155.3875; 319.5851) | 149.1663 | 255.6282 | 1.8388 |
| | | $C_4$ | (-0.0454; 120.08; 120.08) | 78.5398 | 62.3103 | 0.0000 |
| CNT(5, 0)-$O_2$ ($A_3$) | $O_1$ | $C_1$ | (-63.2476; 63.5826; 90.1893) | 30.1748 | 90.0217 | 6.3048 |
| | | $C_2$ | (-69.3226; 65.3214 - 8.8003i; 65.3214 + 8.8003i) | 20.4401 | 67.3219 +13.2005i | 9.8809 - 0.6458i |
| | $O_2$ | $C_3$ | (-31.5986; 74.3457; 104.6805) | 49.1425 | 83.3070 | 3.2338 |
| | | $C_4$ | (-72.6502; 74.9677 -19.9352i; 74.9677 +19.9352i) | 25.7617 | 73.8090 +29.9028i | 8.5952 - 1.1607i |
| CNT(5, 0)-$O_2$ ($A_4$) | $O_1$ | $C_1$ | (-48.8614; 22.4174; 75.9510) | 16.5023 | 89.1730 | 1.1990 |
| | | $C_2$ | (-72.6368; 16.2112; 92.0007) | 11.8584 | 120.2134 | 1.1086 |
| | $O_2$ | $C_3$ | (-48.4295; 2.6526; 96.4404) | 16.8878 | 119.3289 | 0.6421 |
| | | $C_4$ | (-128.7315; 13.2045; 82.9991) | -10.8427 | 140.7627 | 1.5125 |

[a] Calculated $\sigma_{ii}$, $\sigma_{iso}$ and $\Delta\sigma$ values are in ppm
[b] In each raw, the first number is for $\sigma_{11}$, the second number is for $\sigma_{22}$ and the third number is for $\sigma_{33}$.

Table 5. Calculated $^{17}O$ NMR parameters for CNT, $O_2$–CNT (5, 0) systems [a].

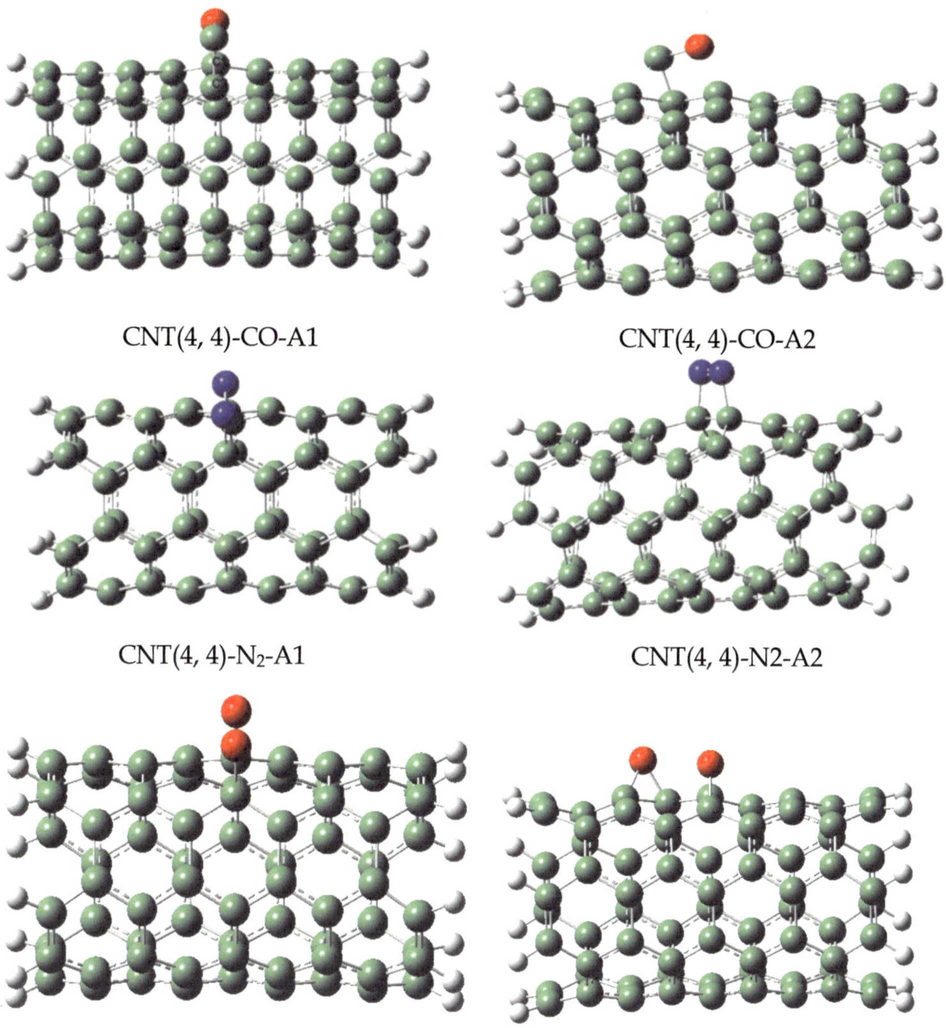


Fig. 4. ( A1) and (A2) adsorption configurations of an Oxygen. Nitrogen and Carbon monoxide molecule.

Chemical shielding tensors and chemical shifts are efficient parameters for characterization of carbon nanotubes. Calculation of these shielding tensors for oxygen nuclei reveals that increasing length and diameter of SWCNTs $-A_1$ (5, 0) chemical shielding will cause O nuclei converge on nanotube surface. Results are consistent with strong interaction between the tube and $O_2$ molecule in SWCNTs $-A_1$(5, 0) . This is consistent with previous results derived from band structure calculations (A. Rubio et al., 1994; X. Balase, 1994). On the other hand, the calculated $^{17}O$ chemical shielding values in the middle of the CNT (4, 4) and CNT (5, 0) seem to approach values 53.8495 ppm, 54.1090ppm and 78.5398 ppm, 149.1663 ppm, respectively (table 3and 4). The NMR chemical shielding of finite SWCNTs were found to converge very slowly, if at all, to the infinite limit, indicating that hydrogen capped tube fragments are not necessarily good models of infinite systems. As the length of the fragment increases, these orbitals do not yield a contribution to the electron density along the tube (except at the ends) and must therefore be regarded as artifacts due to treating the finite-sized systems. More recently, this group indicated that (E.Zurek et al., 2008) also, the introduction of oxygen atoms is theoretically predicted to give rise to chiral current flow along the nanotube due to symmetry breaking (Y. Liu, & H. Guo, 2004; Y. Miyamoto, (1996). Due to $O_2$ chemisorptions the calculated $^{17}O$ NMR parameters of those interacted carbon atoms are also modified. As understood by comparison of sites ($A_1$, $A_2$, $A_3$ and $A_4$), the carbon atoms included in $O_2$ chemisorptions become more shielded. Among the four NMR principal components, intermediate shielding component, $\sigma_{22}$, shows more change from nanotube to the $O_2$–CNT system. The discrepancy between the $^{17}O$ chemical shielding tensor for the sites ($A_1$, $A_2$, $A_3$ and $A_4$) systems must be attributed to the different nature of the frontier orbital's which will have an influence on the $^{17}O$ chemical shielding. However, this theoretical considerations and predictions are undermined by recent experimentally investigations where chiral currents have been observed in undoped single-walled carbon nanotubes (V. Krstić et al., 2002). The interest in oxygen-doped CNTs in terms of application is the control of the type of charge carriers within the carbon nanotubes. This control is one key-issue for a successful implementation of CNTs in nanotubes and molecular electronics. $O_2$-CNTs should show significant advantages over nanotubes for gas sensor applications, due to their reactive tube surfaces, and the sensitivity of their transport characteristics to the

| Model | Atom | $q_{xx}$ | $q_{yy}$ | $\eta_Q$ | $q_{zz}$ |
|---|---|---|---|---|---|
| CNT(4, 4) | C1 | 0.095540 | 0.137945 | -0.233485 | 0.18 |
| | C2 | 0.115713 | 0.118493 | -0.234206 | 0.01 |
| | C3 | 0.115853 | 0.118548 | -0.234401 | 0.01 |
| CNT(4, 4)-O2 (A1) | C1 | 0.114913 | 0.217092 | -0.332005 | 0.30 |
| | C2 | 0.115191 | 0.217377 | -0.332568 | 0.30 |
| CNT(4, 4)-O2(A2) | C1 | 0.056708 | 0.235503 | -0.292211 | 0.61 |
| | C2 | 0.134179 | 0.172331 | -0.306510 | 0.12 |
| CNT(4, 4)-CO (A1) | C1 | 0.131771 | 0.279074 | -0.410845 | 0.36 |
| | C2 | 0.026205 | 0.139259 | -0.165464 | 0.68 |
| CNT(4, 4)-CO (A2) | C1 | 0.002444 | 0.226654 | -0.229098 | 0.98 |
| | C2 | 0.201458 | 0.223643 | -0.425101 | 0.05 |
| CNT(4, 4)-N2(A1) | C1 | 0.062574 | 0.133920 | -0.196494 | 0.36 |
| | C2 | 0.007669 | 0.136100 | -0.143769 | 0.89 |
| CNT(4, 4)-N2(A2) | C1 | 0.046359 | 0.090398 | -0.136757 | 0.32 |
| | C2 | 0.037639 | 0.158783 | -0.196422 | 0.62 |

Table 6. Calculated Carbon-13 EFG parameters for the CNTs, $N_2$–CNTs and $O_2$–CNTs systems.

presence, distribution and chemistry of oxygen. Peng and Cho first suggested $O_2$-CNT for use in gas sensors, due to the ability of oxygen dopants to bind to incoming gas species (S. Peng & K. Cho, 2003). The oxygen in the nanotubes can be seen as regular defects which change the chemical behavior of tubes.

### 3.3 Electronic properties

Collins et al., (P. G. Collins et al., 2000) found that oxygen gas has dramatic effects on conductivity, thermoelectric power, and the local density of states of nanotube. Kang et al., (D. Kang wt al., 2005) reported the characteristic behavior of afield effect transistor based on an individual SWCNT upon exposure to $O_2$ at various pressures, and attributed the device behaviors to oxygen induced p-type doping and thus the Fermi level pinning near the top of the valence band.

The total density of states (TDOS) of these carbon nanotubes quantum structure states in the valence band shown in fig. 5. We study the efficacy on electronic properties of SWCNT by $O_2$ chemisorptions. The calculated band gap of nanotube model zigzag (5,0) CNT($A_{1,2}$) is about 0.77eV and of nanotube model armchair (4, 4) CNT($A_{1,2}$) is about (2.79) eV, which is very close to DFT simulated for the other small diameter nanotubes (W. L. Yim, & Z. F. Liu, 2004; M. A. Rantner et al., 1998). The existence of this small band gap suggests that the system can be converted into a Narrow-gap semiconductor material. Such a big decrease of the band gap of the zigzag (5, 0) and armchair (4, 4) nanotubes upon $O_2$ chemisorption on seat models may originates from the changes of the highest occupied molecular orbital (HOMO) and the lowest unoccupied molecular orbital (LUMO) of these systems. The electron conduction mechanism is expected to be tunneling when the Fermi levels of contacts lie within the HOMO-LUMO gap of a short length molecule, as for the case of these SWCNT (V. Derycke, 2002). As shown in Figure3, the HOMO and LUMO densities of pure zigzag (5,0) and armchair (4,4) SWCNT are mainly positioned at nanotube wall, while those of the open-ended SWCNT with $O_2$ chemisorption at the seat models are mainly localized within the C-C bonds (Fig. 3). Indeed, Kang et al. (D. Kang et al., 2005) have used the LUMO of $O_2$ to allege that this charge trapping transfer should be interpreted as p-type doping regardless of the actual strength of the process. In this study we find much more evidence for both effects. This would occur when electrons accumulate on the nanotube, preventing whole conduction while the Fermi level is pinned at the valence band maximum, creating a barrier to electron introducing (D. Kang et al., 2005). All of these, suggest that different electronic properties of an open zigzag (5, 0) and armchair (4, 4) SWCNT can be achieved through chemisorption of $O_2$ on nanotubes by the same chemisorption on different sites. When molecular oxygen is chemisorbed on the SWCNT, the interaction of them is very strong which cause changes in electronic properties of these nanotubes. Thus, the presence of molecular $O_2$ increases the bond gap energy of previous zigzag (5, 0) and armchair (4, 4) SWCNT. With chemisorption of $O_2$, the band gap is calculated for model CNT(4,4)-$O_2$($A_3$) 1.47 eV, for model CNT(4, 4)-$O_2$($A_4$) 2.9 eV, for model CNT(5,0)-$O_2$($A_3$) 1.44 eV and for model CNT(5,0)-$O_2$($A_4$) 1.37eV. The effect of $O_2$ chemisorption on zigzag (5, 0) and armchair (4, 4) SWCNT increases the bond gap energy.

The effect is more obvious for CNT (4, 4)-$O_2$ ($A_4$). Also the TDOS plots for the sites ($A_1$, A2, A3 and A4) are significantly differ from the zigzag (5, 0) and armchair (4, 4) nanotube near an electric filed, which may result in a conductance change of nanotube up to chemisorption of $O_2$.

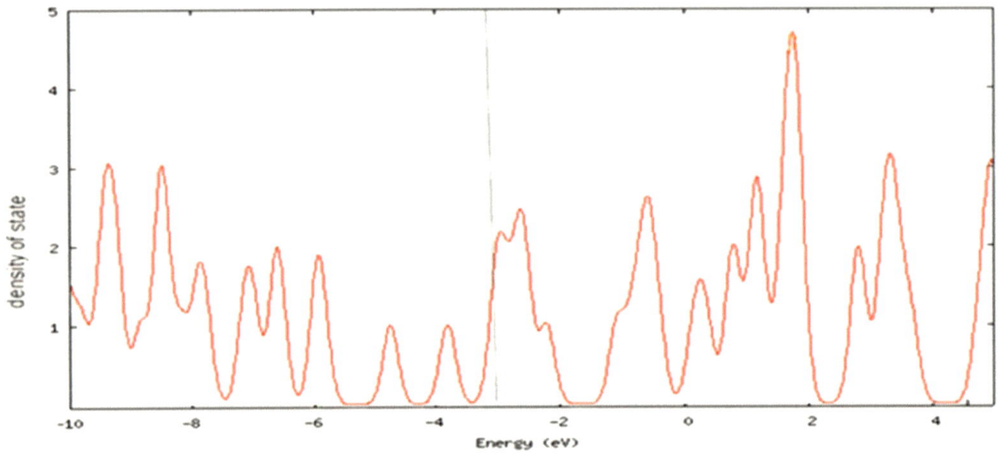


(a) CNTs (5,0) ($A_1$)

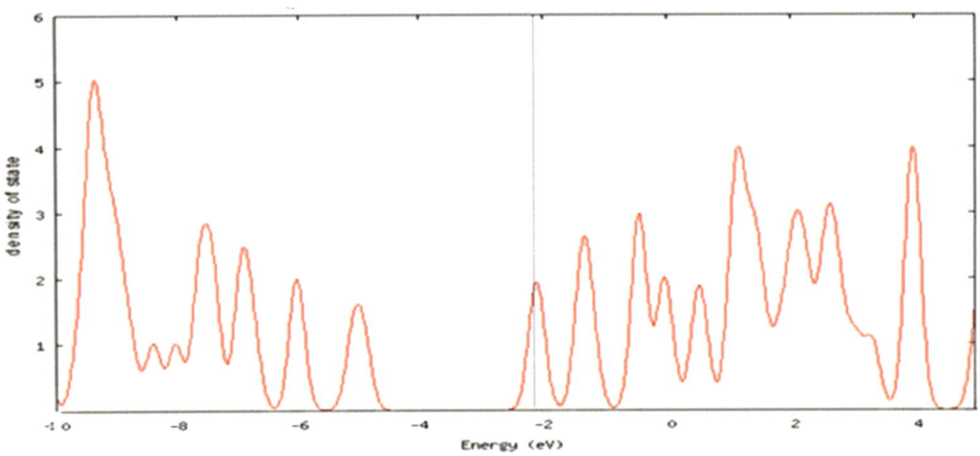


(b) CNTs (4,4) ($A_2$)

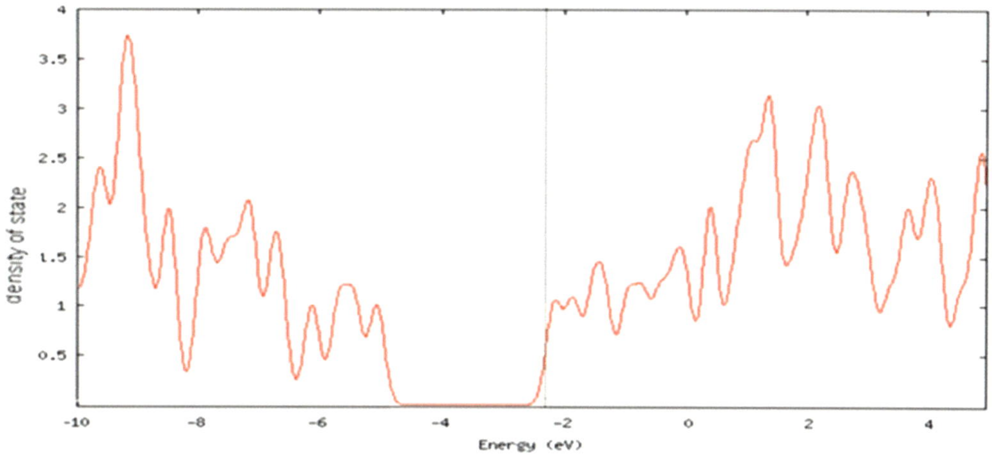


(c) CNT (4,4)-$O_2$ ($A_3$)

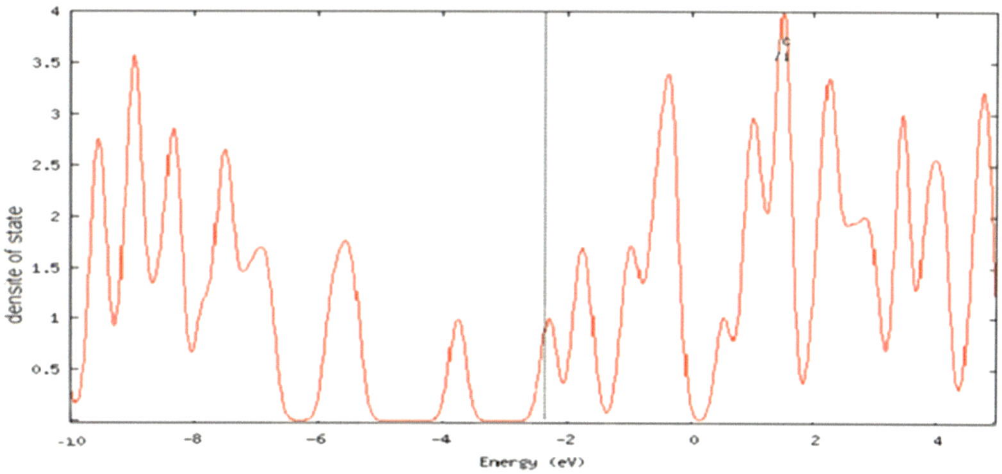


(d) CNT (4,4)-$O_2$ ($A_4$)

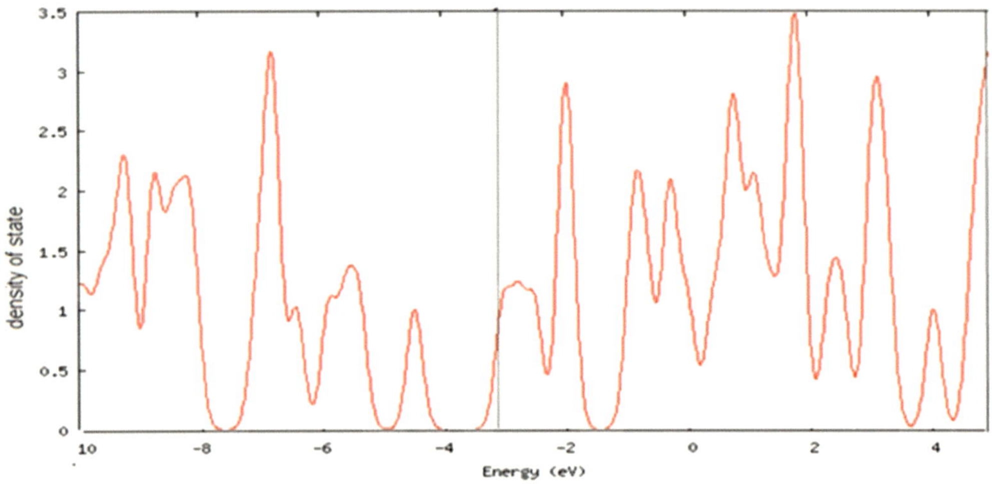


(e) CNT (5,0)-$O_2$ ($A_3$)

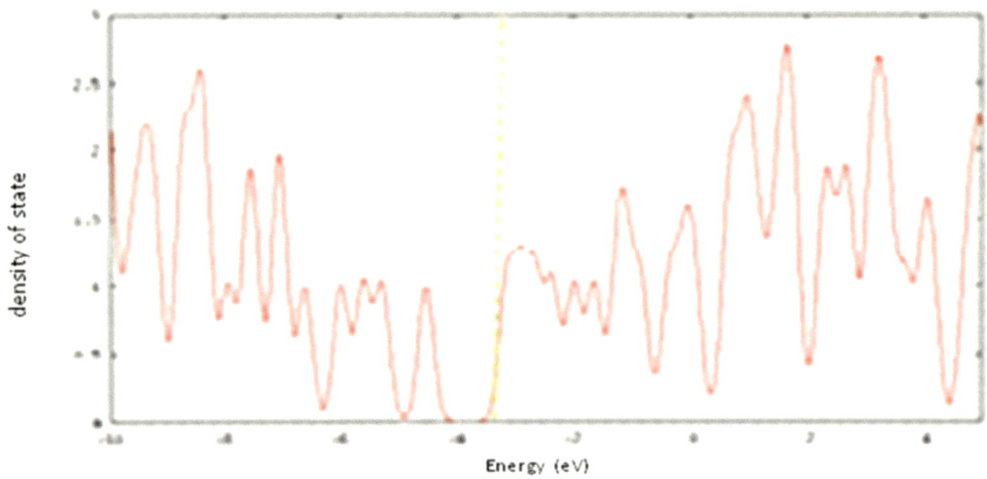


(f) CNT ( 5,0)-$O_2$ ($A_4$)

Fig. 5. Electronic density of states for(a) CNT (5,0) ($A_1$), (b) CNT (4,4) ($A_2$), , (c) CNT(4,4)-$O_2$($A_3$), (d) CNT(4,4)-$O_2$($A_4$),(e) CNT(5,0)-$O_2$($A_3$), and (f) CNT(5,0)-$O_2$($A_4$) SWCNT systems. Dashed line denotes the Fermi level.

For chemisorptions model CNTs (5, 0) ($A_1$), we found that band gaps below electric filed become narrow and new local energy levels occur near the Fermi level, which result in the nearly continuous local density of state (DOS) peaks below electric filed. Furthermore, bonding the $O_2$ on model CNTs (4, 4) ($A_2$) appears a peak near electric filed with a band gap about 2.79 eV. The TDOS for ($O_2$–CNT) system is presented in CNT (4, 4)-$O_2$ ($A_4$). For this system significant changes in the DOS are observed near the Fermi level, similar to the $O_2$ chemisorptions. However, the chemisorption of oxygen molecule is further increases band gap (2.9eV) and reduces the electrical conductance of the zigzag (5, 0) and armchair (4, 4) SWCNT.

This situation is naturally different from the oxygen molecule chemisorptions on the wall of nanotubes. In CNT (5,0) – $O_2$ ($A_3$) model, oxygen molecule with a minor change in length of O-O band would be doped on carbon nanotube zigzag model.

## 4. Conclusion

In summary, our theoretical studies show that CNTs can be used as mechanical sensors. We found the electronic properties of CNTs are sensitive to the adsorptions of oxygen, nitrogen and carbon monoxide gases on the surface. According to DFT theory and hybrid functional B3LYP are applied to study NQR of $N_2$, $O_2$ and CO calculations the electronic structure properties of (4, 4) SWCNT.

We calculated Oxygen-17, Nitrogen-14 and Carbon-13 EFG tensors in the various structures of thymine an optimized isolated gas-phase, gas monomer and the target molecule in X-CNTs (X=$O_2$, $N_2$ and CO). The calculated results have different effects on the EFG tensor at Oxygen-17, Nitrogen-14 and Carbon-13 nuclei. Theoretical calculations are performed to characterize the behavior of $N_2$ and $O_2$ molecules adsorption on external surface and chemisorption on the end of armchair (4, 4), SWCNTs.

The results show that as the diameter of armchair tubes increases, the binding energy of $N_2$ and $O_2$ molecules decreases. The equilibrium $N_2$-CNT and $O_2$-CNT on the surface distance exhibits considerable sensitivity to the type of tube. The calculated $N_2$-CNT, $O_2$-CNT and CO-CNT bond lengths are (1.515-1.525 ˚A, 1.408-1.465 ˚A )and( 1.477-1.584 ˚A ) on the surface of armchair(4, 4), respectively.

Adsorptions are also dependent upon the nanotube family and radius. Moreover, CQ (Nitrogen-14 and Oxygen-17) of armchair nanotube increases as the diameter of tube decreases while in our study, the diameter of tube increased. The obtained CQ parameters for Nitrogen-14(A1, A2) and Oxygen-17(A1, A2) are(4.148, 4.017)MHZ and(4.466,1.833) MHZ ,respectively. Due to the physisorption, NQR parameters of nitrogen and oxygen molecular are also alerted.

In summary, we studied the influence of substitutional $O_2$ on the single-walled carbon nanotubes conformation and a quantum-chemical calculation was performed. The calculations show that the combination of hexagons and $O_2$ molecule concentration produces kinks that include the regular shaped nanotubes.

The GIAO calculations at the B3LYP/6-311G* level using DFT optimized geometries provided isotropic shielding tensors that correlated well with the observed chemical shift data.

The calculated values provided the unambiguous definite assignment of the observed $^{17}O$ - NMR calculative data and can be used in the prediction of the chemical shifts of known SWCNT molecules. The present calculations can also be used to predict chemical shift data for species the formation of which has not yet been observed. For four $O_2$ chemisorption model, we found band gaps above Fermi level become narrower and new local energy levels occur near the Fermi level, which result in the nearly continuous DOS peaks below Fermi level.

In overall of our studies, it is worthwhile to replace the pure nanotubes by chemically doped nanotubes and exploit the new phenomena.

## 5. References


Abragam, A. (1961). *Principles of Nuclear Magnetism*, Clarendon Press,Oxford, p. 166, ISBN 019852014X(PBK).

Aijki, H. & Ando, T. (1993). Energy Bands of Carbon Nanotubes in Magnetic Fields, *J. Phys. Soc. Japan*, Vol.62, 1255, ISSN 1347-4073.

Ashrafi, F.; Ghasemi, A.S.; Babanejad, S.A. & Rahimof, M. (2010). Optimization of Carbon Nanotubes for Nitrogen Gas Adsorption, *Research J. of Appl.Sci. Eng. Tech*, Vol. 2 (6), pp. 547-551, ISSN 2040-7467.

Avouris, Ph.; Martel, R.; Ikeda, H.; Hersam, M.; Shea, Tomanek D., Enbody, R.J., Eds.; H.R. & Rochefort, A. (2000).*Science and Applications of Nanotubes*, Kluwer Academic/Plenum Publishers: New York, ISBN 0306466872.

Babanejad, S.A.; Ashrafi, F. & Ghasemi, A. S. (2010). Optimization of adsorption of oxygen gas on Carbon nanotubes surface, *Archives of Applied Science Research*, Vol. 2 (5), , pp. 438 - 443, ISSN 0975-508x .

Balase, X.; Rubio, A.; Louie, S.G. & Cohen, M.L. (1994). StabilityandBandGapConstancy of Boron-Nitride Nanotubes, *Europhys. Lett.* Vol.28, 335, ISSN 1286-4854(Online), 0295-5075(Print).

Bachtold, A.; Fuhrer, M. S.; Plyasunov, S.; Forero, M.; Anderson, E.H. & Zettl, A. (2000). Scanned prob microscopy of electronic transport in carbon nanotubes, *Phys. Rev. Lett.*, Vol.84, 6082-5, Issue 26, ISSN 1079-7114(Online), 0031- 9007(Print).

Barone,V.; Heyd, J. & Scuseria, G.E. (2004). Effect of oxygen chemisorption on the energy gap of chiral single-walled carbon nanotubes, *Chem. Phys. Lett.* Vol.389, 289, ISSN 0009-2614.

Barros, E.B. et al. (2007). Raman spectroscopy of double-walled carbon nanotubes treated with $H_2SO_4$, *Phys. Rev. B* 76, 045425, ISSN 1550-235x(Online), 1098-0121(Print).

Bezryadin, A.; Verschuren, A. R. M.; Tans, S. J. & Dekker, C. (1998). Multiprobe transport experiments on individual single-wall carbon nanotubes, *Phys. Rev. Lett.* Vol.80, 4036, Issue 18, ISSN 1079-7114(Online), 0031-9007(Prit).

Becke, A.D. (1993). Density-functional thermochemistry. III. The role of exact exchange, *J. Chem. Phys.* Vol.98, 5648, ISSN 0009-2014.

Catlow, C. R. A. et al. (2010). Advances in computational studies of energy materials, doi: 10.1098/rsta. *0111 Phil. Trans. R. Soc. A.*, ISSN 1471-2970.

Chang, H.; Lee, J.; Lee, D. S. M. & Lee, d Y. H. (2001). Adsorption of NH3 and NO2 Molecules on Carbon Nanotubes,*Appl. Phys. Lett.* Vol.79, id. 8363-3865, Issue 23. (Received 18 April 2001; accepted 27 September 2001), ISSN 1077-3118 (Online), 0003-6951(Print)

Collins, P. G.; Bradley, Keith.; Ishigami, Masa, & Extreme, A. Z. (2000). Oxygen Sensitivity of Electronic Properties of Carbon Nanotubes, *Science*. Vol.287,5459- 1801. ISSN 1095-9203 (Online), 0036-8075 (Print).

Collins, P.G.; Bradley, k.; Ishigami, M. & Zettl, A. (2000). Extreme Oxygen Sensitivity of Electronic Properties of Carbon Nanotubes, *Science*, Vol.287, 1801, 1095-9203(Online), 003608075(Print).

Dag, S.; lseren, O.Gu¨ & Ciraci, S. (2003). Effect of the adsorption of oxygen on electronic structures and geometrical parameters of armchair single-wall carbon nano-tubes: A density functional study", Chem. Phys. Lett. Vol.380, 1, Issues 1-2, ISSN 0009-2014.

Derycke, V.; Martel, R,; Appenzeller, J. & Avouris, P. (2002). Controlling doping and carrier injection in carbon nanotube transistors, *Appl. Phys. Lett*, Vol.80(15), 2773–5, ISSN 1882-0786(Online), 1882-0778(Print).

Duer, M.J. (2002). *Solid State NMR Spectroscopy*, Blackwell Science Ltd., London, DOI: 10.1002/9780470999394.ch2, ISBN 0-632-05351-8.

Froudakis, G.E. et al. (2003). Density functional theory study of atomic oxygen, $O_2$ and $O_3$ adsorptions on the H- capped (5,0) single-walled carbon nanotube, Phys. Rev. B., Vol.68, 115435, Issue 8, ISSN 1550-235x(Online), 1098- 0121(Print).

Frisch, M.J. et al. (1998). *Examination of* Gaussian-*Type Basis Sets on Alkali Metal Iodides*, GAUSSIAN 98. Gaussian 98, Gaussian Inc., Pittsburgh PA, , *J. Phys. Chem. A.*, ISSN 1520-5215(Online), 1089-5632(Print).

Ghasemi, A. S.; Ashrafi, F.; Babanejad, S. A. & Rahimof M. (2010). A Computational NMR Study of Chemisorption of Nitrogen-Doped on the surface of Single-Walled Carbon Nanotubes, *Archives of Applied Science Research*, Vol.2 (4), 262, ISSN 0975-508X.

Hamada, N.; Sawada, S.; & Oshiyama, A. (1992). New one-dimensional conductors: Graphitic microtubules, *Phys. Rev. Lett.* Vol.68, 1579, Issue 10, ISSN 1079-7114(Online), 0031-9007(Prit).

He , H.; Klinowski, J.; Forster, M. & Lerf ,A. (1998). Density functional theory study of atomic oxygen, $O_2$ and $O_3$ adsorptions on the H-capped (5,0) single-walled carbon nano-tube, *Chem. Phys. Lett. B*. 287, 53, Issue 8, ISSN 0009- 2014.

Hertel, T.; Walkup, R. E.; & Avouris, Ph. (1998). Deformation of carbon nanotubes by surface van der Waals forces,*Phys. Rev. B*, Vol.58, 13870, Issue 20, ISSN 1550-235x (Online), 1098-0121 (Prit) .

Hill, E.A. & Yesinowski, J.P. (1997). Solid-state N nuclear magnetic resonance techniques for studying slow molecular motions, *J. Chem. Phys*. Lett. 107,. 346, ISSN 0009-2014 .

Hou, S.; Shen, Z.; Zhang, J.; Zhao, X. & Xue, Z. (2004). Ab initio calculations on the open end of single-walled BN nanotubes, *Chem. Phys. Lett.* Vol.393, 179. Issues 1-3, ISSN 0009-2014.

Houten, H. van C.; Beenakker, W. J. & van Wees, B. J. (1992), *In Semiconductors and Semimetals*, Vol.35, M. A. Reed, ed. (Academic Press, New York, 1992). ISBN: 0-12-752135-6.

Iijima, S. & Ichihasi, T. (1993). Single-shell carbon nanotubes of 1-nm diameter, *Nature*,. Vol. 363 (1993) pp.603-605, ISSN. 0028-0836.

Jhi, S.H.; Louie, S.G. & Cohen, M.L. (2000). A comparative study of $O_2$ adsorbed carbon nanotubes. *Phys. Rev. Lett.*Vol. 85, 1710, ISSN 1079-7114(Online), 0031-9007(Print).

Kamimura, T; .Yamamoto, K.; Kawai, T. & Matsumoto, K. (2005). Oxygen interaction with single-walled carbon nanotubes Japan. *J. Appl. Phys.* Vol.44, p. 8237, 1882-0786(Online), 1882-0778(Print).

Kang, H.S. (2006) Theoretical study of boron nitride nanotubes with defects in nitrogen-rich synthesis ,*J. Phys. Chem. B.,* 110, 4621, Issue 7, ISSN 1520-5207(Online), 1520-6106(Print).

Kang, D.; Park, N.; Ko, J.; Bae, E. & Park, W. (2005). *Oxygen-induced p-type doping of a long individual single walled carbon nanotube.* Nanotechnology, Vol.16(8), 1048–52, ISBN195S0957-448(02)30254-X.

Kataura, H. et al. (1999). Optical Properties of Single-Wall Carbon Nanotubes, *Synth. Met.* Vol.103, 2555-2558,ISSN 0379-6779.

Kessel, A.R. & Ermakov, V.L. (1999). Multiqubit spin, *JETP Lett.*, Vol.70, 61, ISSN 0021-3640.

Khitrin, A.K.; Song, H. & Fung,, B.M. (2001). NMR simulation of an eight-state quantum system, *Phys. Rev.* A. Vol. 631, Issue 3, 1050-2947(Online), 1094-1622(Print).

Krstić, V.; Roche, S. & Burghard, M. (2000). Phase breaking in three-terminal contacted single –walled carbon nanotube bundles, *Phys. Rev. B*, Vol.62, R16353, ISSN 1550-235x(Online), 1098-0121(Print).

Krstić, V.; Roche, S.; Burghard, M.; Kern, K. & Rikken, G.L.J.A. (2002). Single and Multiwall Nitrogen doped Nanotubes, *J. Chem. Phys.*, Vol.117, 1315, ISSN 0009-2014.

Lang, N.D. (2000 & 1998). Electron Interference Effects on the Conductance of Doped Carbon Nanotubes. *Phys. Rev. Lett.,* Vol.81, 3515. & *Phys. Rev. Lett.*, Vol.84, 358, ISSN1079-7114(Online), 0031-9007(Print).

Lee, R.S.; Kim, H.J.; Fisher, J.E.; Thess, A. & Smalley, R.E. (1997). Nanotube Bundles Doped with K and Br, *Nature* (London), Vol.388, 255, ISSN. 0028-0836.

Liu, Y. & Guo, H. (2004). Current distribution in B- and N-doped carbon nanotubes, *Phys. Rev. B.*, Vol.69, 115401, ISSN 1550-235x(Online), 1098-0121(Print).

Liu, H.J.; Zhai, J.P.; Chan, C.T. & Tang, Z.K. (2007). *Density functional theory study of atomic oxygen, O2 and O3 adsorptions on the H-capped (5,0) single-walled carbon nanotube,* Nanotechnology, Vol.18, 65704, ISBN 195S0957- 448(02)30254-X.

Lu, X.; Chen, Z.F. & Schleyer, P.V. (2005). Are Stone-Wales Defect Sites Always More Reactive Than Perfect Sites inthe Sidewalls of Single-Wall Carbon Nanotubes?, *J. Am. Chem. Soc.* Vol.127, 20 ,Georgia 30602-2525,1520-5120(Online), 0002-7863(Print).

Lucken, E. A. C. (1992). *Nuclear Quadrupole Coupling Constants*", Academic Press, London, ISBN 13-978-0124584501.

Lynch, M. & Hu, P. (2000). A Density functional theory study of CO and atomic oxygen chemisorption on, Pt(111). *Surf Sci*; Vol.458, ISSN 00396028.

Marian, C.M. & Gastreich, M. (2001). *Structure-property relationships in boron nitrides: The $^{15}N$- and $^{11}B$ chemical shifts*, Solid State Nucl. Mag. Vol.19, 29-44, ISSN 0926-2040.

McClenaghan, N. D.; Hu, P. & Hardacre, C. A. (2000). Density functional theory study of the surface relaxation and reactivity of $Cu_2O$, *Surf Sci.,* Vol.100, 464(2–3), 223-232, ISSN 00396028.

Miyamoto, Y. (1996). Mechanically stretched carbon nanotubes: Induction of chiral current, *Phys. Rev. B.*, Vol.54, R11149, ISSN 1550-235x (Online), 1098-0121(Print).

Mintmire, J.W.; Dunlap, B.I.; & White, C.T. (1992). Are fullerene tubules metallic?, *Phys. Rev. Lett.* Vol.68, 631, Issue 5, ISSN 1079-7114(Online), 0031-9007(Prit).

Mintmire, J.W.; Dunlap, B.I.; White, C.T. (1992). Are fullerene tubules metallic? *Phys. Rev. Lett.*, Vol.68, 631, ISSN1079-7114(Online), 0031-9007(Print).

Mirzaei, M. & Hadipour, N.L. (2006). An investigation of hydrogen-bonding effects on the nitrogen and hydrogen electric field gradient and chemical shielding tensors in the 9-methyladenine real crystalline structure: a density functional theory study, *J. Phys. Chem. A.* Vol.110, 4833, ISSN 1520-5215(Online), 1089-5639(Print).

Noguera, C. (2001). Insulating oxides in lowdi mensionality: a theoretical review, *Surf Rev Lett.*, Vol. 8, 121–67, ISSN 0218-625x.

Parr, R.G. & Yang, W. (1994). *Density Functional Theory of Atoms and Molecules*, Oxford University Press, ISBN 13: 9780195092769.

Peng, S. & Cho, K. (2003). Ab Initio Study of Doped Carbon Nanotube Sensors, *Nano Lett.*, 3(4), 513-517, ISSN 1530- 6992(Online), 1530-6984(Print).

Rantner, M.A. et al. (1998). Molecular wires: charge transport, mechanisms, and control In Molecular Electronics, Science and Technology, 2–37. Ann. N.Y. Acad. Sci. 852: The New York Academy of Sciences, NewYork, doi: 10.1196/annals.1292.001, ISSN 1095-9203(Online), 0036-8075(Print).

Rao, M. et al. (1997). Diameter-Selective Raman Scattering from Vibrational Modes in Carbon Nanotubes Science, Vol.275, 187, *Science.* ISSN 1095-9203 (online), 0036-8075 (print).

Rubio, A.; Corkill, J.L. & Cohen, M.L. (1994). Theory of graphitic boron nitride nanotubesm, *Phys. Rev. B.*, Vol.49, 5081, ISSN 1550-235x(Online), 1098-0121(Print).

Saito, R.; Fujita, M.; Dresselhaus,G. & Dresselhaus. M.S. (1992). New one-dimensional conductors: Graphitic microtubules, *Phys. Rev. B*, Vol.*46*, 1804, Issue 3, ISSN 1550-235x (Online), 1098-0121 (Prit).

Saito, R.; Fujita, M.; Dresselhaus, G. & Dresselhaus, M.S. (1992). Electronic structure of chiral graphene tubules, *Appl. Phys. Lett.*, Vol.60, 2204, ISSN 1882-0786(Online), 1882-0778(Print).

Semin, G.K.; Babushkina, T.A. & Yakobson, G.G. (1975). *Nuclear Quadrupole Resonance in Chemistry*, Wiley, New York, ISBN 0470775807.

Sorescu, D. C. K.; Jordan, D. & Avouris, Ph. (2001). Theoretical Study of Oxygen Adsorption on Graphite and the (8,0) Single-walled Carbon Nano-tube P, J. phys. Chem. B., 105, 11227-11232, Issue 7, ISSN1520-5207(Online), 1520- 6106(Print).

Sorescu, D.C.; Jordan, K.D. & Avouris, P. (2001). Theoretical Study of Oxygen Adsorption on Graphite and the (8,0) Single-walled Carbon Nanotube, *J. Phys. Chem. Lett. B.* Vol.105(45), 11227-11232, ISSN 0009-2014.

Souza Filho, A.G. et al. (2006). Resonance Raman scattering studies in $Br_2$-adsorbed double-wall carbon nanotubes, *Phys. Rev. B*, Vol.73, 235413, ISSN 1550-235x(Online), 1098-0121(Print).

Schimizu, T. & Tsukada, M. (1993). Origin of the different formation modes of the oxygen added rowoverlayer on, Ag(110) and Cu(110) surfaces. *Surf Sci.;* Vol.295(1-2):L1017–22, ISSN 00396028.

Ulbricht, H.; Moos, G. & Hertel, T. (2002). Interaction of C60 with carbon nanotubes and ... of large molecules from solid surfaces, *Phys. Rev. Lett. B.*, Vol.66, 75404, ISSN1550-235x(Online), 1098-0121(Print).

Walch, S. P. (2003). A theorytical study of fundamental single-wall carbon nanotubes, ONION calculation", *Chem. Phys. Lett.* Vol.374 , 501, ISSN 0009-2014.

Wildoer, J. W. G.; Venema, L. C.; Rinzler, A. G.; Smalley, R. E. & Dekker, C. (1998). Electronic structure of atomically resolved carbon nano- tubes, *Nature*, vol. 391, pp. 59, May 1998 *Houston*, Texas *77251, USA*, ISSN. 0028- 0836.

Wolinski, K.; Hilton, J.F. & Pulay, P. (1990). Efficient implementation of the gauge-independent atomic orbital method for NMR chemical shift calculations, *J. Am. Chem. Soc.*, Vol.112, 8251, ISSN 1520-5126(Online), 0002- 7863(Print).

Wu, G.; Dong, S.; Ida, R. & Reen, N. (2002). A solid-state $^{17}O$ nuclear magnetic resonance study of nucleic acid bases, *J. Am. Chem. Soc.* Vol.124, 1768, ISSN 1520-5126(Onliner), 0002-7863(Print).

Yim, W.L. & Liu, Z.F. (2004). Increased reactivity of single wall carbon nanotubes at carbon ad-dimer defect sites, *Chem. Phys. Lett.* Vol.398, 297, ISSN 0009-2014.

Yim, W.L. & Liu, Z.F. (2004). A reexamination of the chemisorption and desorption of ozone on the exterior of a (5,5) single walled carbon nanotube, *Chem. Phys. Lett.* Vol.398, 297, Issues 4-6, ISSN 0009-2614.

Zhang, Y.F. & Liu, Z.F. (2004). Oxidation of Zigzag Carbon Nanotubes by Singlet O2: Dependence on the Tube Diameter and the Electronic Structure, *J. Phys. Chem. B.*, Vol.*108* (31), pp 11435–11441, Issue 7, ISSN 1520- 5207(Online), 1520- 6106(Print).

Zhang, CJ. & Hu, P. (2000). Why oxygen atoms must be activated from hollow sites to bridge sites in catalytic, CO oxidation? *J Am Chem Soc* ., Vol.122, 2134–5, ISSN 1520-5126(Online), 0002-7863(Print).

Zhao, J.; Buldum, A.; Han, J. & Ping Lu, J. (2002). *Gas molecule adsorption in carbon nanotubes and nanotube bundles*, Nanotechnology Vol.13, no. 2, pp. 195–200, ISBN 195S0957-448(02)30254-X.

Zurek, E.; Pickard, C.J. & Autschbach, B.J. (2008). Determining the Diameter of Functionalized Single-Walled Carbon Nanotubes with $^{13}C$ NMR: A Theoretical Study, *J. Phys. Chem. C.*, Vol.112, 9267-9271, ISSN 1932-7455(Online), 1932-7447(Print).

# Part 3

## Applications

18

# Smart Materials and Structures Based on Carbon Nanotube Composites

Sang-ha Hwang[1], Young-Bin Park[1*], Kwan Han Yoon[2] and Dae Suk Bang[2]
[1]*School of Mechanical and Advanced Materials Engineering*
*Ulsan National Institute of Science and Technology (UNIST)*
[2]*Department of Polymer Science and Engineering*
*Kumoh National Institute of Technology*
*Korea*

## 1. Introduction

Since the first discovery of carbon nanotubes (CNTs) in 1991, CNTs have generated enormous research activities in many areas of science and engineering due to their combined exceptional mechanical, thermal and electronic properties. These properties make nanotubes ideal, not only for a wide range of applications but also as a test-bed for fundamental scientific studies (Baughman *et al.*, 2002). They can be described as a graphite sheet rolled up into a nanoscale tube. Two structural forms of CNTs exist: single-walled (SWCNTs) and multi-walled (MWCNTs) nanotubes. CNT lengths can be as short as a few hundred nanometers or as long as several micrometers. SWCNT have diameters between 1 and 10 nm and normally capped ends. In contrast, MWCNT diameters range from 5 to a few hundred nanometers because their structure consists of many concentric cylinders held together by van der Waals forces. CNTs are synthesized in a variety of ways, such as arc discharge, laser ablation, high pressure carbon monoxide (HiPCO), and chemical vapor deposition (CVD) (Dresselhaus, 1997). CNTs exhibit excellent mechanical, electrical, thermal and magnetic properties. The exact magnitudes of these properties depend on the diameter and chirality of the nanotubes and whether their structure is single- or multi-walled. Fig. 1 shows a segment of a single graphene plane that can be transformed into a carbon nanotube by rolling up into a cylinder. To describe this structure, a chiral vector is defined as $OA = na_1 + ma_2$, where $a_1$ and $a_2$ are unit vectors for the hexagonal lattice of the graphene sheet, $n$ and $m$ are integers, along with a chiral angle $\theta$, which is the angle of the chiral vector with respect to the $x$ direction. Using this $(n, m)$ scheme, the three types of nanotubes are characterized. If $n = m$, the nanotubes are called "armchair". If $m = 0$, the nanotubes are called "zigzag". Otherwise, they are called "chiral". The chirality of nanotubes has significant impact on their transport properties, particularly the electronic properties. For a given $(n, m)$ nanotube, if $(2n + m)$ is a multiple of 3, then the nanotube is metallic, otherwise the nanotube is a semiconductor. Each MWCNT contains a multi-layer of graphene, and each layer can have different chiralities, so the prediction of its physical properties is more complicated than that of SWCNT (Jin & Yuan, 2003).

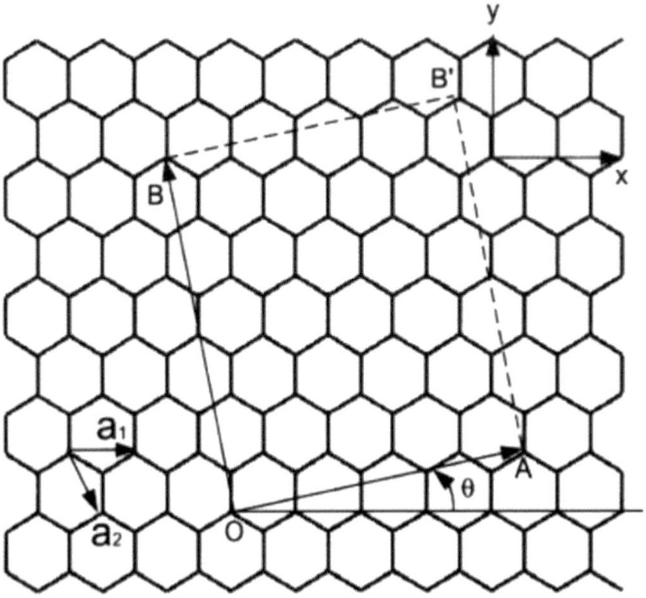


Fig. 1. The graphite plane of nanotube surface coordinates (Jin & Yuan, 2003).

The basic structure of CNTs is comprised of *sp*2 carbons. This *sp*2 structure provides CNTs with higher mechanical properties compared to any materials, even diamonds. It is well known that the mechanical properties of CNTs exceed those of any existing materials. Although there is no consensus on the exact mechanical properties of CNTs, theoretical and experimental results have shown exceptional mechanical properties of CNTs with Young's modulus of 1.2 TPa and tensile strength of 50–200 GPa (Coleman *et al.*, 2006).

Other excellent physical properties of CNTs have also attracted much attention. The properties are summarized and compared with other carbon allotropes in Table 1. Because of their unique properties, many promising applications and potential practical applications have been reported, such as field emission materials, catalyst support, electronic devices, nanotweezers, reinforcements in high performance composites, supercapacitors, hydrogen storage and high sensitivity sensors and actuators. These are just a few possibilities that are currently being explored. As research continues, new applications will also develop (Dresselhaus *et al.*, 2004).

| Property | Graphite | Diamond | Fullerene | CNTs | |
|---|---|---|---|---|---|
| | | | | SWCNT | MWCNT |
| Specific gravity (g $cm^{-3}$) | 1.9-2.3 | 3.5 | 1.7 | 0.8 | 1.8 |
| Electrical conductivity (S $cm^{-1}$) | 4000 | $10^{-2}$-$10^{-15}$ | $10^{-5}$ | $10^{2}$-$10^{6}$ | $10^{3}$-$10^{5}$ |
| Electron mobility ($cm^2 V^{-1} s^{-1}$) | 2.0 | 1800 | 0.5 | ~$10^5$ | $10^4$-$10^5$ |
| Thermal conductivity (W $m^{-1}K^{-1}$) | 298 | 900-2320 | 0.4 | 6000 | 2000 |
| Coefficient of thermal expansion ($K^{-1}$) | $-1 \times 10^6$ | $(1\text{-}3) \times 10^6$ | $6.2 \times 10^{-5}$ | ~0 | ~0 |
| Thermal stability (in air) (K) | 450-650 | <600 | 600 | >600 | >600 |

Table 1. Different physical properties of carbon allotropes (Ma *et al.*, 2010)

## 2. Processing of carbon nanotube composites

CNT-based polymer composite materials are being utilized in an increasing number of applications including automotive, aerospace, defence, sporting goods and infrastructure sectors. This is due to their high durability, high strength, light weight, design and process flexibility, *etc.* Thermosets such as epoxy, unsaturated polyester, gels, as well as thermoplastics have been used as the matrix. The conductivity, strength, elasticity, toughness, and durability of formed composites may all be substantially improved by the addition of nanotubes.

Especially electrically conductive CNT-polymer composites are used in anti-static packaging applications, as well as in specialized components in the electronics, automotive, and aerospace sectors. The incorporation of conductive filler particles into an insulating polymer matrix leads to bulk conductivities at least exceeding the anti-static limit of $10^{-6}$ S/m. Common conductive fillers are metallic or graphitic particles in any shape (spherical, platelet-like or fibrous) and size. However, the incorporation of CNTs allows lower percolation threshold compared to other conductive fillers (Fig. 2). The use of CNTs as a conductive filler in polymers is their biggest current application (Bal & Samal, 2007).

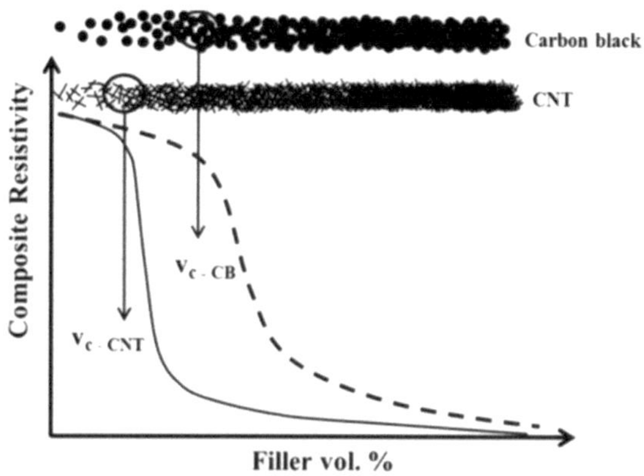


Fig. 2. Illustration of CNT network percolation ($v_c$) compared with carbon black

The effective utilization of CNTs in composite applications depends strongly on the ability to homogeneously disperse them throughout the matrix without destroying their integrity. Therefore, it has become clear that the issues of dispersion, alignment, and stress transfer are crucial, and often problematic at nanoscale. However, in order to be able to utilize CNTs and their properties in real-world applications, CNT-based nanocompsites provide a pathway to realize the properties of these fascinating nanostructures at macroscopic levels by bridging over a range of length scales.

### 2.1 Carbon nanotube dispersion

The potential of using nanotubes as a constituent of polymer composites has not been presently realized mainly because of the difficulties associated with dispersion and processing. High aspect ratio, combined with high flexibility, increase the possibility of nanotube entanglement and close packing. The low dispersity comes from the tendency of pristine nanotubes to assemble into bundles or ropes like shown in Fig. 3 (Thess *et al.*, 1996). Thus, a significant challenge in developing high-performance CNT-polymer composites is to introduce the individual CNTs in a polymer matrix in order to achieve better dispersion and alignment and strong interfacial interactions, to improve the load and electron transfer across the CNT-polymer matrix interfaces.

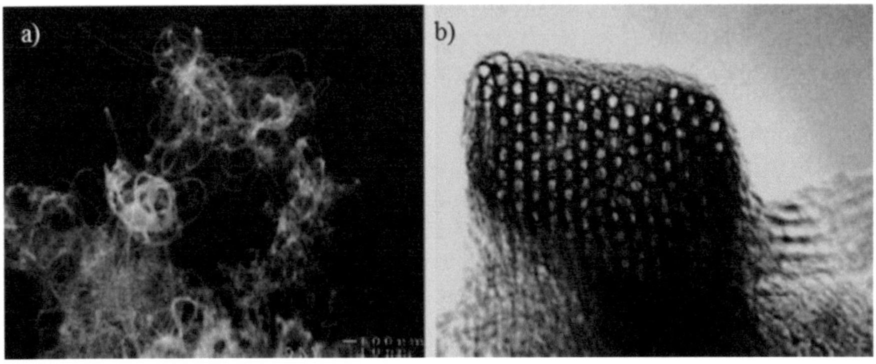


Fig. 3. (a) SEM image of entangled SWCNT agglomerates and (b) TEM image of a SWCNT bundle (Thess *et al.*, 1996).

### 2.1.1 Mechanical dispersion of carbon nanotubes

**Ultrasonication**

Although in most cases, it is very difficult to get a homogeneous dispersion of the CNTs in the polymeric matrix, ultrasonication is a very effective method of dispersion and de-agglomeration of CNTs, as ultrasonic waves of high-intensity ultrasound generates cavitation in liquids. There are two major methods for delivering ultrasonic energy into liquids, the ultrasonic bath (Fig. 4. (a)) and the ultrasonic horn (Fig. 4(b)). Ultrasonication disperses solids primarily through a microbubble nucleation and collapse sequence. The ultrasonication bath has a higher frequency (40–50 kHz) than cell dismembrator horns (25 kHz). Ultrasonication of fluids leads to three physical mechanisms: cavitation of the fluid, localized heating, and the formation of free radicals. Cavitation, the formation and implosion of bubbles, can cause dispersion (Lu *et al.*, 1996).

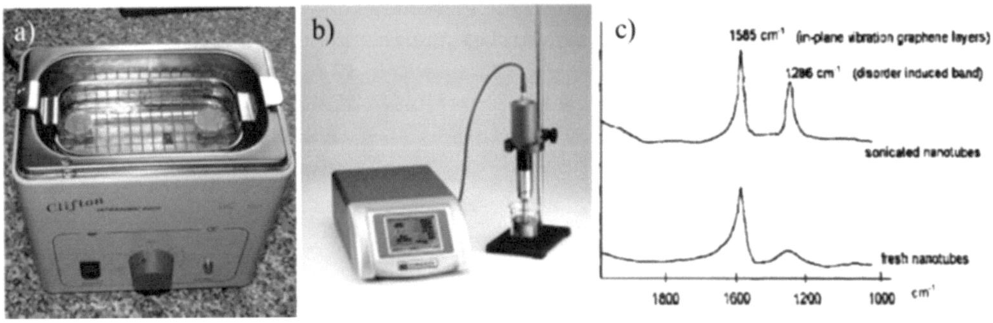


Fig. 4. (a) Bath type, (b) horn type sonicator and (c) Raman spectra of CNTs before and after sonication

However, ultrasonication affects not only CNT dispersion but also its length and diameter (Fig. 4 (c)). After reducing their lengths during ultrasonication, SWCNTs rearrange into superropes. These superropes have diameters more than 20 times the initial bundle diameter (Shelimova *et al.*, 1998). In MWCNTS, ultrasonication creates expansion and peeling or fractionation of MWCNT graphene layers. The destruction of MWCNTs seems to initiate on the external layers and travel towards the center. It has been reported that the nanotube layers seem quite independent, so MWCNTs would not only get shorter, but actually thinner with time (Lu et al., 1996). There have been attempts to develop less destructive ultrasonication methods. One example is ultrasonication with diamond crystals, a method that reportedly destroys the SWCNT bundles but not the tubes. Raman spectra showed typical SWCNT peaks even after 10 hours of treatment with this method (Haluska *et al.*, 2001).

**Ball Milling**

Ball milling is a method that is usually used to grind bulk materials into fine powder. During milling, a high pressure is generated locally due to the collision between the rigid balls in a sealed container (Fig. 5). Cascading effect of balls reduces the size of material to fine powder. Balls are usually made by ceramic, flint pebbles and stainless steel.

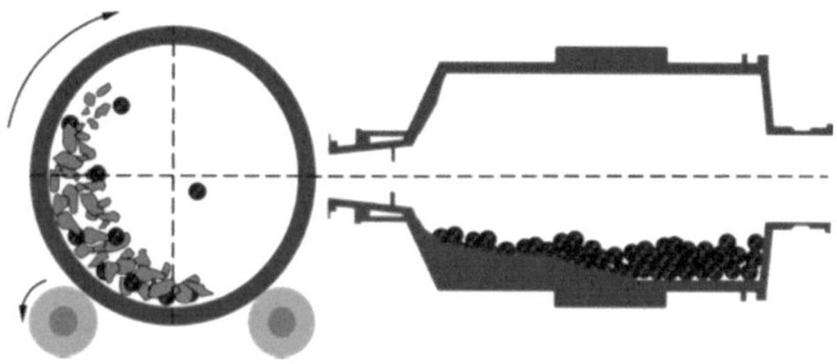

Fig. 5. Schematics of ball milling technique

Ball milling has been successfully applied to CNT dispersion into polymer matrices. To obtain narrow length and diameter distributions of CNTs and to open the nanotubes for improved sorption capacity for gases, ball-milling is a very useful method (Awasthi *et al.*, 2002). However, it has also been observed that a large amount of amorphous carbon is created which clearly indicates that the tubes are damaged in different ways and that ball-milling is a destructive method (Jia *et al.*, 1999).

**Calendering (Three-Roll mill)**

Calendering, also commonly known as three-roll-milling (Fig. 5 (a)) is a dispersion technique that employs both shear flow and extensional flow created by rotating rolls of different speed to mix and disperse CNTs or other nanoscale fillers into polymers or other viscous matrixes. The first and third rollers (usually called the feed and apron rolls, respectively) in Fig. 5 (b) rotate in the same direction while the center roller rotates in the opposite direction. In order to create high shear rates, angular velocity of the center roll must be higher than that of feed roll ($\omega 2 > \omega 1$). As the resin suspension is fed into the narrow gap ($\delta$) between feed and center rolls, the liquid mixture flows down covering (essentially coating) the adjacent rolls through its surface tension under intensive shear forces. At the end of each subsequent intended dwell time, the processed resin suspension is collected by using a scraper blade in contact with the apron roll. This milling cycle can be repeated several times to maximize dispersion.

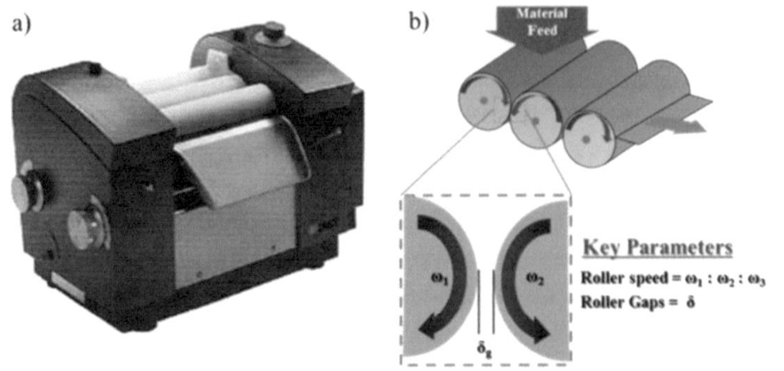


Fig. 6. (a) Calendering machine (three-roll mill) used for CNT dispersion into a polymer matrix and (b) general schematic diagram of its mechanism.

One of the unique advantages of this technique is that the gap width between the rollers can be mechanically or hydraulically adjusted and maintained, thus it is easy to obtain a controllable and narrow size distribution of particles in viscous materials. In some operations, the width of gaps can be decreased gradually to achieve the desired level of particle dispersion (Viswanathan *et al.*, 2006). A typical calendering machine and its principle are shown schematically in Fig. 5. The employment of a calender to disperse CNTs in a polymer matrix has become a very promising approach to achieve relatively good CNT dispersion according to some recent reports (Thostenson & Chou, 2006a).

However, the fed material should be in a viscous state when mixed with CNTs, thus this tool may not be applied to disperse CNTs into thermoplastic matrices, such as polyethylene, polypropylene and polystyrene. In contrast, CNTs can be conveniently dispersed into the liquid monomer or oligomer of thermosetting matrices, and nanocomposites can be obtained via *in situ* polymerization.

**Extrusion (Melt Compounding)**

Extrusion is a popular technique used to disperse CNTs into solid polymers, including most thermoplastics, where thermoplastic pellets mixed with CNTs are fed into the extruder hopper. In particular, twin-screw extruders (Fig. 7. (a)) are used extensively for CNT-polymer mixing and compounding. The modular design of twin-screw extruder allows this operation to be designed specifically for the formulation being processed (Fig 7. (b)) (Bauhofer & Kovacs, 2009). For example, the two screws may be co-rotating or counter-rotating, intermeshing or non-intermeshing. In addition, the configurations of twin-screw extruders themselves may be varied using forward conveying elements, reverse conveying elements, kneading blocks, and other designs in order to achieve each CNT-polymer mixing characteristics. This technique is particularly useful in producing CNT-polymer composites with high filler contents. However, care must be taken to prevent CNT damages due to excessive shear stresses imposed during the extrusion process.

Polymer melt compounding is useful, especially in industry because it does not demand additional processes. However, the melt compounding studied and optimized so far has been mostly focused on micro-compounders at lab-scale. Scale-up of these techniques are not just a matter of size but also a matter of different rheological and thermodynamical issues (Oh & Hong, 2010).

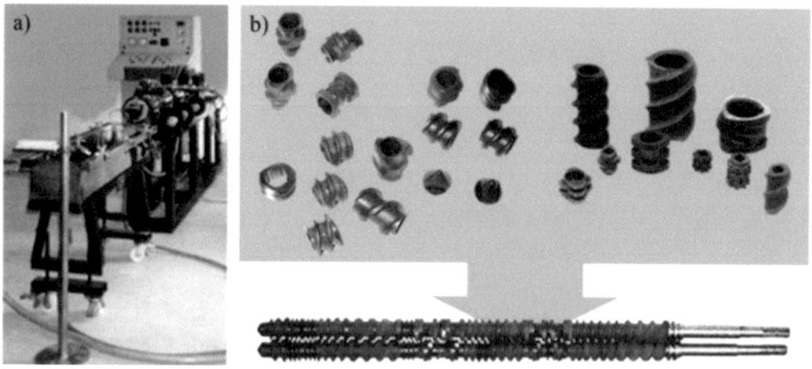


Fig. 7. (a) Lab scale twin screw extruder and (b) screw modular design.

### 2.1.2 Dispersion of carbon nanotube based on functionalization

**Covalent Functionalization**

Functionalization of CNTs is an effective way to minimize nanotube interaction, which helps to better disperse and stabilize the CNTs within a solvent or polymer matrix. There are several approaches for functionalization of CNTs, including covalent and non-covalent functionalizations.

In the case of covalent functionalization, the structure of CNTs is disrupted by changing $sp^2$ carbon atoms to $sp^3$ carbon atoms, and the physical properties of CNTs, such as electrical and thermal conductivities, are influenced. However, functionalization of CNTs with covalent bonding can improve dispersity in solvents and polymers. Generally, surface modification starts from acid treatment, which create -COOH and -OH functional groups on the CNT during oxidation by oxygen, reactive gas, sulfuric acid, nitric acid and other concentrated acids or their mixtures. The quantitative amounts of -COOH and -OH functional groups depend on oxidation conditions and oxidizing agent. Nanotube ends can be opened and residual catalyst and amorphous carbons are removed during the oxidation process (Spitalskya *et al.*, 2010).

Carboxylic functionalized CNT surfaces can be further used to chemically attach other small molecules or macromolecules through the reaction of the oxidation-induced functional groups. One of the common chemical reactions with acid functionalized CNT is the amidization, in which amide bond between amine group moieties is formed. One example is the use of amino-functionalized MWCNTs in epoxy systems to yield improved mechanical properties (Stevens *et al.*, 2003). The improved mechanical performance in these functionalized systems may reflect both the enhanced dispersion and an improved surface interaction between CNT and polymer matrix. Further improvements in solubility can be achieved by fluorination, again leading to improvements in both the stiffness and strength, with the addition of 1 wt.% of oxidized and fluorinated SWCNTs. Also, it has been established that the electrical properties of MWCNTs change after fluorination, leading to a wide range of electrical structures, from insulating over to semiconducting and metallic-like behavior (Seifert *et al.*, 2000).

The chemically functionalized CNTs can produce strong interfacial bonds with many polymers, allowing CNT-based nanocomposites to possess high mechanical and functional properties.

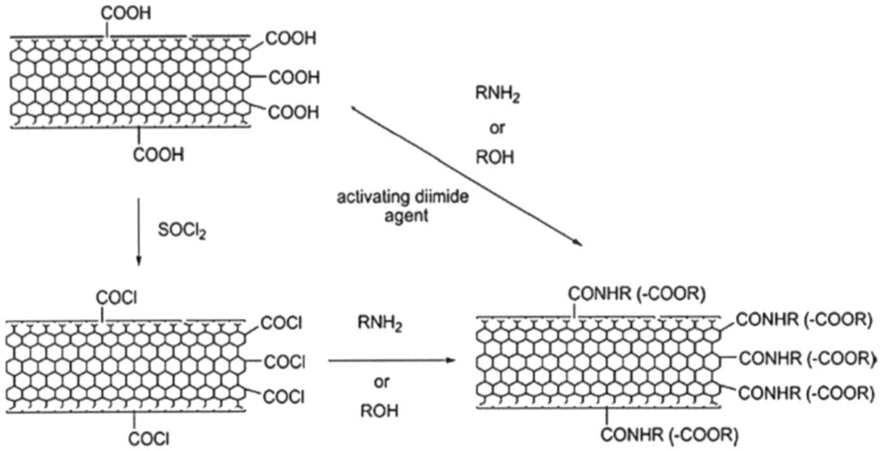


Fig. 8. Schematic representation of amidization process which starts from oxidized CNTs (Spitalsky *et al.*, 2010).

**Non-covalent Functionalization**

A non-covalent method used to modify CNT surface is popular functionalization method since it does not compromise the physical properties of CNTs. The electrostatic repulsion provided by adsorbed surfactants stabilizes the nanotubes against the strong van der Waals interactions between the nanotubes, hence preventing agglomeration. This repulsive and attractive force balance creates a thermodynamically stable dispersion, which results in separation of CNTs from the bundles into individual nanotubes. Anionic surfactants, such as sodium dodecylsulfate (SDS) and sodium dodecylbenzene sulfonate (NaDDBS), are commonly used to disperse CNT aggregation in polar media. The interaction between the surfactants and the CNTs depends on the nature of the surfactants, such as its alkyl chain length, headgroup size, and charge (Fig. 9) (Ma *et al.*, 2010).

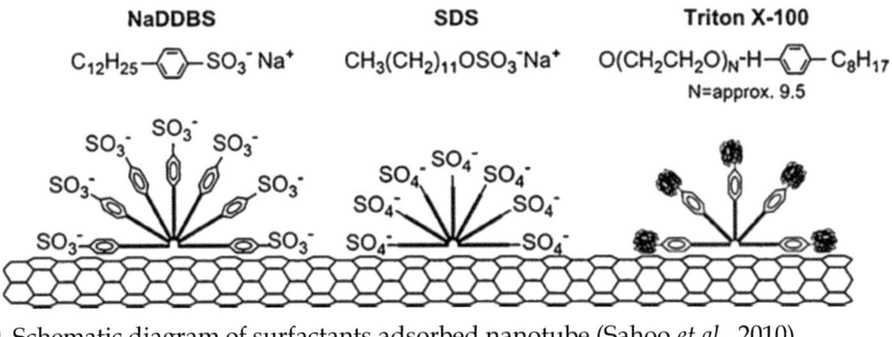


Fig. 9. Schematic diagram of surfactants adsorbed nanotube (Sahoo *et al.*, 2010).

The physical interaction of polymers with CNTs to make specific formation can be explained by the 'wrapping' mechanism which is π-stacking interactions between the polymer and the nanotube surface. Usually, wrapping polymer consists of aromatic groups on main chain or substitutional groups. For example, polyvinyl pyrrolidone (PVP) or polystyrene sulfonate (PSS) wrapped CNT shows improved dispersity and electrical properties compared to those of the individual components (Cheng *et al.*, 2008).

Small angle neutron scattering studies demonstrated a non-wrapping conformation of polymers in CNT dispersions. In these cases, different structures and compositions of copolymers efficiently act as stabilizers. The suggested mechanism of non-wrapping is that one of the blocks in block copolymers adsorbed to the nanotubes surfaces and another solvophilic blocks act as a steric barrier that leads to the formation of stable dispersions of individual CNTs above a threshold concentration of the polymer. A study of the stabilization effect produced by different diblock or multiblock copolymers led to the conclusion that selective interaction of the different blocks with solvent is essential in order to obtain stable colloidal dispersions of CNTs (Nativ-Roth *et al.*, 2007).

### 2.2 Control of carbon nanotube orientation

Similar to conventional fiber-reinforced composites, both mechanical properties and functional properties, such as electrical, thermal and optical properties of CNT-polymer composites are directly related to the alignment direction of CNTs in the matrix. Recently, this topic has drawn much attention due to the advance in nanocomposite processing techniques and the limitaions of randomly oriented, discontinuous nanotube composites.

#### 2.2.1 Orientation of carbon nanotube by yarn formation

Recent advances in fabrication of CNTs allow to grow up to several millimeters in length, and these CNTs are possibly aligned to continuous macroscopic SWCNT fibers (Fig. 10). This provides an opportunity for fabricating continuous nanotube reinforced composites.

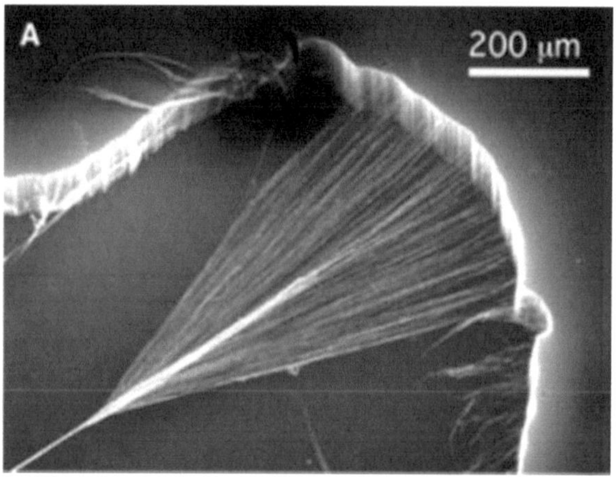


Fig. 10. SEM image of direct yarn formation from MWCNT forest (Zhang et al., 2004).

It has been reported that free-standing arrays of millimeter long, vertically aligned multiwalled nanotubes exhibit supercompressibility, outstanding fatigue resistance, and viscoelastic characteristics. Continuously aligned nanotube reinforced polydimethylsiloxane (PDMS) composite shows remarkably enhanced compressive modulus and strength, anisotropic characteristics, and damping capability (Ci *et al.*, 2008).

#### 2.2.2 Force field orientation

The first method developed to fabricate aligned CNTs in polymer matrix was by "cutting" an CNT-epoxy nanocomposite. This process is simply explained by the nature of rheology in composite media on nanometer scales and flow-induced anisotropy produced by the "cutting" process. The fact that CNTs do not break and are aligned after cutting also suggests that they have excellent mechanical properties along the nanotube direction. However, the orientation of CNTs in CNT-epoxy composite is affected by the cut slice thickness, since the alignment effect is only effective near the slice surface (Ajayan *et al.*, 1994).

A solution approach involved a SWCNT-dispersed surfactant solution (sodium dodecyl sulfate, SDS) injected through a syringe needle into a polyvinyl alcohol (PVA) solution. Because the PVA solution is more viscous than the SWCNT suspension, there is a shear contribution in the flow at the tip of the syringe needle, the flow-induced alignment is maintained by the PVA solution, and SWCNTs are rapidly stuck together as they are injected out from the syringe. By pumping the polymer solution from the bottom, meter-long ribbons are easily drawn, and well-oriented PVA-CNT composite fibers and ribbons are formed by a simple process. It offers a method to align CNTs by a flow field (Vigolo *et al.*, 2000).

The more effective and convinient method in CNT orientation is uniaxially stretching of polymer-CNT composite films. CNT-polymer composite films and fibers produced by any process can be drawn uniaxially showing higher conductivity along the stretched direction than the direction perpendicular to it. Also, the mechanical properties such as elastic modulus and yield strength of composite fibers increased with draw ratio, and CNTs in the composite fibers were better aligned. It is also possible to prepare aligned CNT composite films by extruding the composite melt through a rectangular die and drawing the film prior to cooling. For example, as compared to the drawn polystyrene (PS) film, the tensile strength and modulus of the PS-MWCNT composite films were greater (Thostenson & Chou, 2002).

However, PS-MWCNT composites prepared by spin casting at high speed showed that MWCNTs were aligned in specific angles relative to the radial direction: 45° and 135 ° on average. The presence of 2.5 vol.% MWCNTs doubles the tensile modulus and transforms the film from insulating to conducting. It is also noted that the CNTs have higher orientation than the polymer matrix during melt-drawing of the polymer-CNT composites (Bhattachacharyya *et al.*, 2003).

### 2.2.3 Electric or magnetic field induced orientation

Studies of SWCNT alignment using electric or magnetic fields have usually involved epoxies or polyesters as matrices because of their low viscosity before cure. Under the electric field, it was shown that both AC and DC electric fields can be used to induce the formation of aligned CNT networks spanning the gap between electrodes in contact with the dispersion. With increasing field strength, the quality of these networks and the resulting bulk conductivity of the composite material can be enhanced(Martin *et al.*, 2004). However, at high CNT content, thus high viscosity of molten resin system, the magnetic field-induced alignment of polymeric materials is more effective in CNT alignment. This technique has been the focus of several research efforts, initiated by the first use of high magnetic field to align MWCNTs in a polyester matrix to produce electrically conductive and mechanically anisotropic composites. A high magnetic field is an efficient and direct means to align CNTs. For example, to align MWCNT dispersed in methanol suspension, a magnetic field greater than 7 T is demanded. For the CNT alignment in a polymer, even higher magnetic field would be demanded because of high viscosity. Under a high magnetic field of 10 T, it has been shown that MWCNTs were aligned in the monomer solution during their polymerization and MWCNTs were aligned parallel to the magnetic field inside the polymer matrix (Camponeschi *et al.*, 2007). Recently, magnetic field aligned polycarbonate (PC) and CNT-epoxy composites hae been reported and they suggested that aligning CNTs in polymer matrices can improve mass transport property as well as electrical conduction. It is also viewed that CNTs are better aligned in a PC matrix using magnetic field as compared to an electric field (Abdalla *et al.*, 2010).

### 2.2.4 Electrospinning induced orientation of carbon nanotube in polymeric nanofiber

Among several approaches to align nanotubes, the electrospinning technique has recently ben used to incorporate CNTs in a polymeric matrix to form composite nanofibers, combining the benefits of nanofibers with the merits of CNTs. Due to the sink flow and the high extension of the electrospun jet, it is expected to align the nanotubes during the electrospinning process, as was also predicted by a mathematical model. However, the distribution and alignment of the nanotubes in the nanofibers are strongly associated with the quality of the nanotube dispersion prepared before addition of the spinnable polymer solution. Generally, well-dispersed MWCNTs were incorporated as individual elements mostly aligned along the nanofiber axis. Conversely, irregular nanotubes were poorly aligned and appeared curled, twisted, and entangled. It is also suggested that the nanofiber diameter, the interaction between the spun polymer and the nanotubes and wetting ability are important factors affecting the alignment and distribution of the nanotubes. This was demonstrated by the difference in the alignment of SWCNTs in polyacrylonitrile (PAN) and polylactic acid (PLA) nanofibers (Ko *et al.*, 2003). More recent work to incorporate SWCNTs into PEO nanofibers by the electrospinning process showed SWCNTs were embedded in PEO in a more regular form since SWCNTs are much smaller and uniform in shape and size, as compared to MWCNTs. On the other hand, their stronger tendency to bundle up into coiled aggregates introduces a pronounced difficulty. Therefore, special attention is given to the dispersion process, which is essential for successful alignment of the nanotubes by the electrospinning process. Structural analysis of the composite nanofibers in terms of the distribution and orientation of both the nanotubes and the polymer matrix has been studied (Salalha *et al.*, 2004).

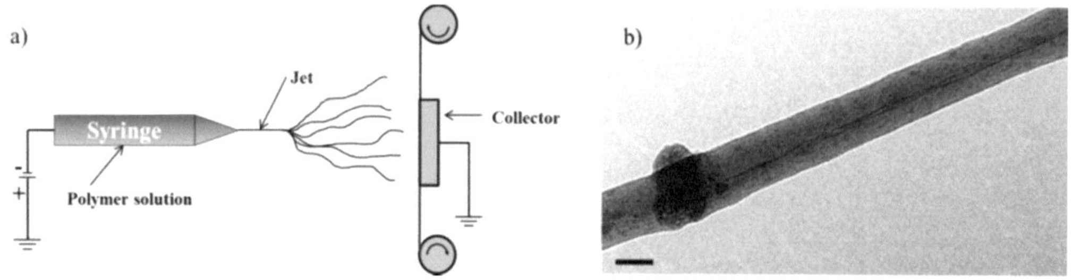


Fig. 11. (a) Simple schematic presentation of electrospinning and (b) TEM image of SWCNT aligned PEO nanofiber. Scale bar = 50 nm (Salalha *et al.*, 2004)

## 3. Electrical properties of carbon nanotube composites

CNTs have clearly demonstrated their capability as fillers in conductive polymer composites. Percolation theory predicts that there is a critical concentration at which composites of insulating polymers become electrically conductive. According to the percolation theory, conductivity of composite ($\sigma_c$) can be estimated from the following equation.

$$\sigma_c = A(V - V_c)^{\beta} \tag{1}$$

Where $V$ is the CNT volume fraction, $V_c$ is the CNT volume fraction at the percolation threshold, and $A$ and $\beta$ are fitted constant. The percolation threshold has been reported to ranging from 0.0025 wt.% to several wt.%. The percolation threshold for the electrical conductivity in CNT-polymer composites depends on degree of surface modification dispersion, alignment, CNTs aspect ratio, polymer types and processing methods. The electrical conductivity and percolation threshold of CNT-polymer composites are summarized in Table 2.

| Polymer matrix | CNT type | Maximum filler content (wt.%) | Processing or dispersion method | Maximum electrical Conductivity (S/m) | Percolation threshold (wt/%) | Reference |
|---|---|---|---|---|---|---|
| PS | SWCNTs | 2 | Solution mixing | $10^{-3}$ | 0.27 | Chang et al., 2006 |
| PS | Aligned CNT array | | Drop Casting | 1330 | | Peng et al., 2009 |
| HDPE | Acid-SWCNTs | 6 | Extrusion | $10^{-1}$ | ~4 | Zhang et. al., 2006 |
| LDPE | Acid - MWCNTs | 10 | Ball mill | ~2 | ~1-3 | Gorrasi et al., 2007 |
| PP | MWCNTs | 10.7 | Melt mixing | 4.6 | 1.1 | Miˇcuˇsík et al., 2009 |
| PMMA | SWCNTs | 25 | Coagulation | $10^{-1}$ | ~1 | Narayan et al., 2009 |
| PMMA | MWCNTs | 0.4 | Solution mixing | $3\times10^{3}$ | 0.003 | Kim et al., 2004 |
| PMMA | Aligned CNT | - | Drop casting | 1250 | - | Peng et al., 2009 |
| PC | MWCNT | 15 | Extrusion | 20 | 1-2 | Potschke et al., 2002 |
| PC | PPE-SWCNTs | 10 | Solution mixing | $4.8\times10^{2}$ | 0.11 | Ramasubramaniam et al., 2003 |
| Nylon 6 | MWCNTs | 10 | Melt mixing | 0.1 | 2-2.5 | Krause et al., 2009 |
| Nylon 6,6 | MWCNTs | 10 | Melt mixing | 0.1 | 0.5-1 | Krause et al., 2009 |
| PDMS | MWCNTs | 2.5 | Ultrasonica-tion | 0.02 | 1.5 | Khosla A & Gray BL, 2009 |

| | | | | | | |
|---|---|---|---|---|---|---|
| PI | MWCNTs | 1.5 | | $3.83\times10^{-4}$ | - | Xiaowen et al., 2005 |
| PI | Acid-MWCNTs | 7 | Solution mixing | $3.8\times10^{-6}$ | - | Yuen et al., 2007 |
| PU | MWCNTs | 27 | Solution casting | $2\times10^{3}$ | 0.009 | Koerner et al., 2005 |
| PET | SWCNTs | 5 | Melt mixing | ~1 | 0.024 | Hernandez et al., 2009 |
| Epoxy | Aligned SWCNT | 5 | Solution casting | $\sim10^{-5}$ | 0.5 | Qing et al., 2008 |
| Epoxy | Silane-MWCNT | 1 | Solution mixing | $1.67\times10^{-2}$ | - | Lee et al., 2011 |
| Epoxy | MWCNT | 8 | 3-roll mill | $2.3\times10^{4}$ | 0.0117 | Iosif et al., 2009 |
| Epoxy | SDS-MWCNTs | 0.5 | Bulk mixing | $2.5\times10^{-7}$ | - | Santos et al., 2008 |
| Nafion | SWCNTs | 18 | Solution mixing | $3.2\times10^{3}$ | - | Landi et al., 2002 |

Table 2. Electrical properties of CNT-polymer composites

## 4. Smart, multifunctional applications of carbon nanotube composites

CNT-based polymer composites have found numerous multifunctional applications owing to their capability to serve as reinforcing, lightweighting agents and a material platform for electrostatic discharging, electromagnetic interference shielding, radar absorbing, mechanical/chemical sensing, energy harvesting, and flame retardation. Smart applications can be categorized into sensing and actuation, and this chapter will primary focus on the review of research on electromechanical sensing using CNT-based polymer composites. The studies on sensors and actuators based on CNTs and their composites up to 2007 are well summarized by Li *et al.* (Li *et al.*, 2008), and this chapter primarily presents more recent studies.

### 4.1 CNT Nanocomposites for electromechanical sensing

Electromechanical sensing and structural health monitoring typically utilize the piezoresistive behavior of the electrically conductive network formed by CNTs in polymer matrices, that is, the behavior characterized by a change in resistivity with respect to the structural deformation incurred by an external load. For example, when a CNT nanocomposite is subjected to a tensile load, the percolated CNT network is disrupted, resulting in an increase in resistivity. The variation in resistivity under a load is attributed to the variation in contact configurations and tunnelling distances among the contacting CNTs upon nanocomposite deformation.

Initial studies on piezoresistivity of conductive CNT network involved free-standing CNT films or sheets, also known as "buckypapers." CNT buckypapers are typically made by filtration, similar to the papermaking process, where the CNTs are uniformly dispersed in a solvent, usually with the aid of surfactants, and subsequently passed through a filtering paper on which the CNTs are eventually deposited, dried, and detached. The CNT sheets were bonded to the surfaces of various substrates, including brass (Li *et al.*, 2004; Vemuru et al., 2009), aluminium (Li *et al.*, 2004), and fiberglass (Kang *et al.*, 2006). While these substrates were loaded under tension or flexure, the resistance between two electrodes attached to the CNT sheet was measured *in situ*. Most of these studies employed isotropic, randomly oriented CNT networks, and showed that the resistance increase linearly under tension and decreased linearly under compression. The isotropy allows multi-directional, multi-location strain measurements.

It has been reported that CNTs can be added to various materials, including general-purpose thermoplastics and thermosets, specialty polymers, such as polyvinylidene fluoride (PVDF) and shape memory polymers, elastomers, and concrete, and utilize the piezoresistivity of the nanocomposites for strain or pressure sensing.

#### 4.1.1 Thermoplastic-based nanocomposites

The research group at the University of Cincinnati (Kang *et al.*, 2006) reported comprehensive research work on strain sensing using buckypapers and SWCNT-polymethylmethacrylate(PMMA) composites. Fig. 12(a) shows the strain response of a SWCNT buckypaper sensor, which shows higher sensitivity in the linear bending range. However, it shows saturated strain behavior above 500 microstrains, which is probably attributed to the slippage among CNT bundles due to the weak van der Waals interactions at nanotube interfaces. When the sensor is compressed, the individual CNTs do not slip as much as compared to the tension case, resulting in the lack of saturation. Fig. 12(b) shows the strain response of composite sensor at varying CNT loadings. Although the composite strain sensors show lower sensitivities than buckypaper, they show linear symmetric strain response trends in both compression and tension. The interfacial bonding between CNTs and the polymer reduces slip and effectively increases the strain in the sensor.

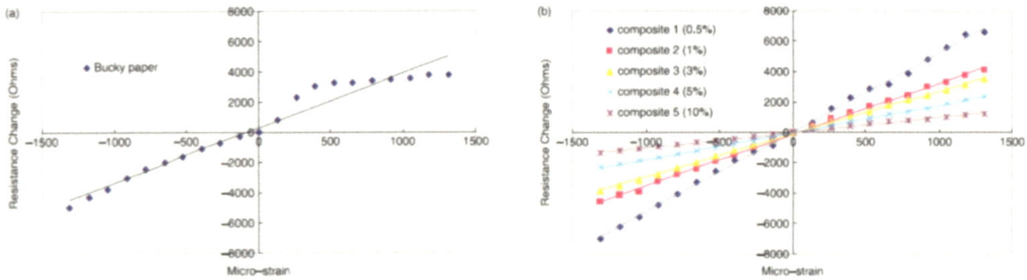


(a) Buckypaper (b) SWCNT-PMMA composite

Fig. 12. Piezoresistive response of: (a) buckypaper sensor and (b) SWCNT-PMMA composite sensor (Kang *et al.*, 2006a, 2006b)

Pham et al. (Pham *et al.*, 2008) reported the development of conductive, MWCNT-filled, polymer composite films that can be used as strain sensors with tailored sensitivity. The electrical resistance of MWCNT-PMMA composite films subjected to tensile strains was measured, and the potential applications of the films as strain sensors with a broad range of tunable sensitivity were investigated. The surface resistivity of the films was observed to increase with increasing tensile strain, which is due to the reduction in conductive network density and increase in inter-tube distances induced by applied strains. The highest sensitivity achieved in this study was almost an order of magnitude greater than conventional resistance strain gages (Fig. 13). A semi-empirical model, based on the percolation theory, was developed to identify the relationship between applied strain and sensitivity factor (Fig. 14). Not only can the sensitivity be tailored over a broad, but also it can be increased significantly be having the conductive filler content approach the percolation threshold.

Zhang et al. (Zhang *et al.,* 2006) presented a study on MWCNT-polycarbonate(PC) composites as multifunctional strain sensors, where a 5 wt.% composite showed instantaneous electrical resistance response to linear and sinusoidal dynamic strain inputs and a sensitivity of ~3.5 times that of a typical strain gage. Billoti *et al.* (Billoti *et al.*, 2010) presented a study on thermoplastic polyurethane (TPU) fibers containing MWCNTs, fabricated via an extrusion process, which demonstrated a tuneable level of electrical conductivity. The observation of Arrhenius dependence of zero-shear viscosity and the assumption of simple inverse proportionality between the variation of conductivity, due to network formation, and viscosity allow a universal plot of time variation of conductivity to be composed, which is able to predict the conductivity of the extruded fibers. The same nanocomposite fibers also demonstrated good strain sensing abilities, which were shown to be tunable by controlling the extrusion temperature.

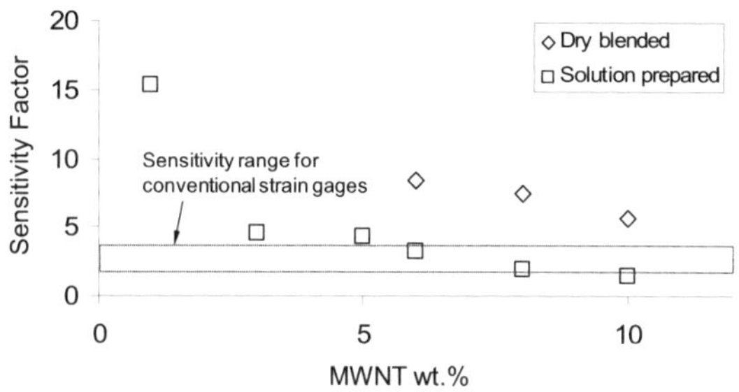


Fig. 13. Comparison of sensitivity factors between MWCNT-PMMA films and conventional resistance strain gages

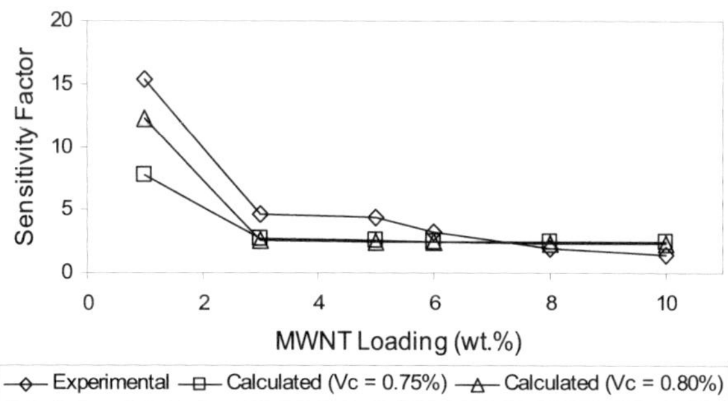


Fig. 14. Calculated and experimental sensitivity factors of MWCNT-PMMA films

Abraham et al. (Abraham *et al.,* 2008) reported the development and characterization of a CNT-PMMA nanocomposite flexible strain sensor for wearable health monitoring applications. These strain sensors can be used to measure the respiration rhythm which is a vital signal required in health monitoring. A number of strain sensor prototypes with different CNT compositions have been fabricated and their characteristics for both static as well as dynamic strain have been measured. Bautista-Quijanoa *et al.* (Bautista-Quijanoa *et al.*, 2010) reported the electrical and piezoresistive responses of thin polymer films made of polysulfone (PSF) modified with 0.05–1% w/w MWCNTs. Gage factors were measured for films with 0.2–1% CNT weight loadings. The films were then bonded to macroscopic aluminum specimens and evaluated as strain sensing elements during quasi-static and cycling tensile loading. Excellent piezoresistive capabilities were found for films with MWCNT loadings as low as 0.5% w/w.

CNTs were added to a piezoelectric polymer, PVDF, to for various smart applications, including strain sensing. Deshmukh *et al.* (Deshmukh *et al.*, 2009) presented an experimental evidence of the creation of an electrostrictive response in PVDF by addition of small quantities of CNTs. It was demonstrated that the piezoelectric response of nanocomposites can be dramatically enhanced through addition of conductive nanoparticles such as CNTs without additional weight penalties. Most importantly, these improvements were achieved at much lower actuation voltages and were accompanied by an increase in both mechanical and dielectric properties. In the work by Kim *et al.* (Kim *et al.*, 2008), CNTs were included in a PVDF matrix to enhance the properties of PVDF. The CNT-PVDF composite was fabricated by solvent evaporation and melt pressing. The inclusion of CNT allowed the dielectric properties of PVDF to be adjusted such that lower poling voltages can be used to induce a permanent piezoelectric effect in the composite. The CNT-PVDF composites were mounted on the surface of a cantilever beam to compare the voltage generation of the composite against homogeneous PVDF thin films.

#### 4.1.2 Thermoset-based nanocomposites

The primary types of thermosets used as the matrices for strain sensing nanocomposites include epoxy, vinyl ester, and polyimide, among which epoxies are most popular. In the work by Wichmann *et al.* (Wichmann *et al.*, 2009), electrically conductive epoxy based nanocomposites based on MWCNTs and carbon black were investigated concerning their potential for strain sensing applications with electrical conductivity methods. It was found that the nanocomposites exhibited a distinct resistance vs. strain behavior in the regime of elastic deformation, which is in good agreement with prevalent theories about charge carrier transport mechanisms in isolator/conductor composites. Applying an analytical model, it was shown that the piezoresistivity of nanocomposites may contribute valuable information about the conductive network structure and charge carrier transport mechanisms occurring in the nanocomposites. The authors also developed a direction-sensitive bending strain sensor consisting of a single block of MWCNT-epoxy composite by generating a gradient in electrical conductivity throughout the material (Wichmann *et al.*, 2008).

Zhang *et al.* (Zhang *et al.*, 2007) demonstrated a simple, effective and real-time diagnostic, and repair technique featuring MWCNTs that are infiltrated into epoxy. It was shown that by monitoring volume and through-thickness resistances, one can determine the extent and propagation of fatigue-induced damage such as crack and delamination growth in the vicinity of stress concentrations (Fig. 15). The conductive nanotube network also provides opportunities to repair damage by enabling fast heating of the crack interfaces; the authors show up to 70% recovery of the strength of the undamaged composite.

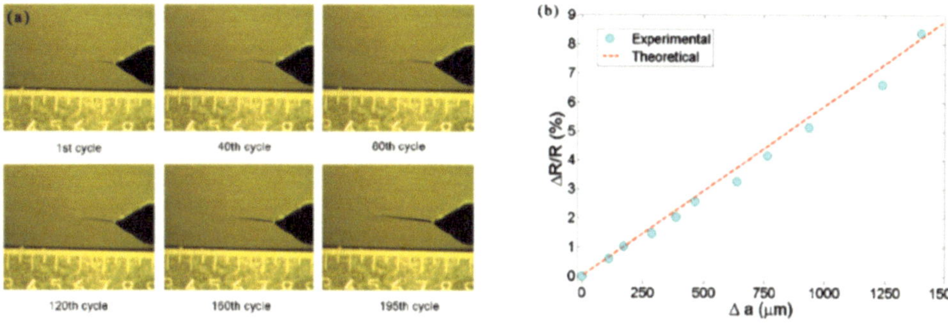


Fig. 15. Detection of real-time fatigue crack growth: (a) snapshots of fatigue crack growth; (b) the change in electrical resistance across the crack interface

Quasi-static and dynamic strain sensing of CNT-epoxy composites was studied by Anand and Mahapatra (Anand & Mahapatra, 2009), and de la Vega, *et al.* (de la Vega *et al.*, 2011) characterized the local and global stress response of SWCNT–epoxy composites by simultaneous Raman spectroscopic and electrical measurements on nanocomposite specimens subjected to various levels of surface strain. Both the Raman G-band resonance frequency and the electrical resistance of the composite are found to change monotonically with strain until an inflection point is reached at 1.5% strain.

Thostenson *et al.* (Thostenson *et al.*, 2009) synthesized vinyl ester monomer from the epoxy resin to overcome processing challenges associated with volatility of the styrene monomer in vinyl ester resin. Calendering was employed for MWCNT dispersion in vinyl ester monomer and the subsequent processing of nanotube/vinyl ester composites. The high aspect ratios of the carbon nanotubes were preserved during processing, and an electrical percolation threshold below 0.1 wt.% carbon nanotubes in vinyl ester was observed. A systematic study of the effect of SWCNTs on the enhanced piezoresistive sensitivity of polyimide nanocomposites from below to above percolation was reported by Kang *et al.* (Kang *et al.*, 2009). The maximum piezoresistive stress coefficient obtained just above the percolation threshold concentration (0.05 wt.%) exceeded those of metallic piezoresistive materials by two orders of magnitude.

### 4.1.3 Elastomer-based nanocomposites

Hwang *et al.* (Hwang *et al.*, 2011) fabricated a piezoresistive composite using MWCNTs as a conductive filler and polydimethylsiloxane (PDMS) as a polymer matrix, which operated in the extremely small pressure range required for finger-sensing. To achieve a homogeneous dispersion of MWCNTs in PDMS, the MWCNTs were modified by a polymer wrapping method using poly(3-hexylthiophene) (P3HT). The percolation threshold of the composites was significantly lowered by the presence of P3HT. The electrical conductivity and piezoresistive sensitivity of the composite were found to strongly depend on the P3HT concentration. The well-dispersed P3HT-MWCNT/PDMS composite showed good piezoresistive characteristics in the 0–0.12 MPa pressure range.

Wang *et al.* (Wang *et al.*, 2010) studied the piezoresistivity of a multi-walled carbon nanotube filled silicone rubber composite under uniaxial pressure. The experimental results showed that the active carboxyl radical on multi-walled carbon nanotubes can effectively improve the homogeneous distribution and alignment of conductive paths in the composite. As a result, the composite presented positive piezoresistivity with improved sensitivity and sensing linearity for pressure, both of which are key parameters for sensor applications. Elastomeric composites based on ethylene-propylene-diene-monomer (EPDM) filled with MWCNTs showed improved mechanical properties as compared to the pure EPDM matrix (Ciselli *et al.*, 2010). A linear relation was found between conductivity and deformations up to 10% strain, which means that such materials could be used for applications such as strain or pressure sensors. Cyclic experiments were conducted to establish whether the linear relation was reversible, which is an important requirement for sensor materials.

High-elasticity CNT-methylvinyl silicone rubber (VMQ) nanocomposites with a high sensitivity and linear piezoresistive behavior was fabricated by dispersing conductive MWCNTs with different aspect ratios, AR = 50 and 500, into rubber matrix homogeneously (Dang *et al.*, 2008). It was found that the percolation threshold of the nanocomposites containing AR = 50 MWCNTs was significantly lower than those containing AR = 500 MWCNTs. Extremely sensitive positive-pressure coefficient effect of the resistance and excellent cyclic compression under low pressure were also observed in the MWCNT-VMQ nanocomposites with AR = 50 MWCNTs at relatively low loadings.

### 4.2 CNT-based multiscale hybrid composite for electromechanical sensing

Multiscale hybrid composites (MHCs) are defined as composites consisting of at least three constituents having more than two different length scales. The most common type consists of the resin (macro), continuous (unidirectional or woven) fiber fabric (micro), and nanoparticles (nano) (Fig. 16). Conventional continuous fiber-reinforced plastics (FRPs) are characterized by extremely high in-plane modulus and stiffness (fiber-dominated) and poor through-thickness properties (matrix-dominated). In MHCs, high-performance nanomaterials are added to improve the through-thickness properties and, at the same time, to impart multifunctionalities to the composites.

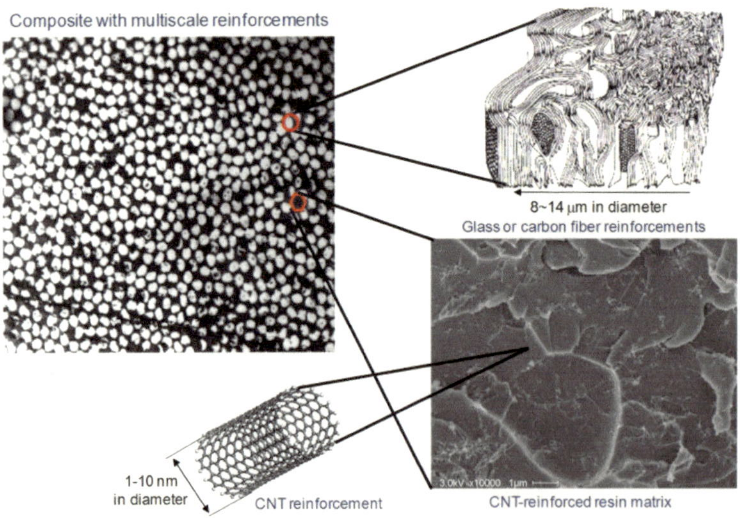


Fig. 16. Concept of multiscale hybrid composites

Research on structural health monitoring of MHCs using the piezoresistivity of the percolated network of CNTs has been pioneered by the University of Delaware. A typical MHC manufacturing process involved dispersion of CNTs in the resin using a three-roll mill, followed by composite fabrication using vacuum-assisted resin transfer molding. The MHCs thus obtained were tested under various loading conditions to *in situ* monitor and detect various failure modes, including delamination, matrix damage and fiber breakage as shown in Fig. 17 (Thostenson & Chou, 2006b), and crack growth under fatigue (Gao *et al.*, 2009). Similar research was performed in parallel by Technische Universitat Hamburg-Harburg (Boger *et al.*, 2008). Kim *et al.* (Kim *et al.*, 2010) applied 3D braided textile as reinforcement and used CNTs as the sensing components for structural health monitoring of 3D braided composites.

An alternative way to incorporate CNTs in FRPs is to surface treat or coat the fibers with CNTs, instead of dispersing them in the resin. Specific methods include dipping the fibers into a CNT solution (Gao *et al.*, 2010), aid of coupling agents (Sureeyatanapas & Young, 2009), lay-by-layer deposition (Loh *et al.*, 2009), and direct growth of CNTs on the fibers using electrophoresis (Bekyarova *et al.*, 2007). Another unique method is to embed continuous CNT fibers (Alexopoulos *et al.*, 2010) or threads (Abot *et al.*, 2010) in FRPs for large-area strain sensing.

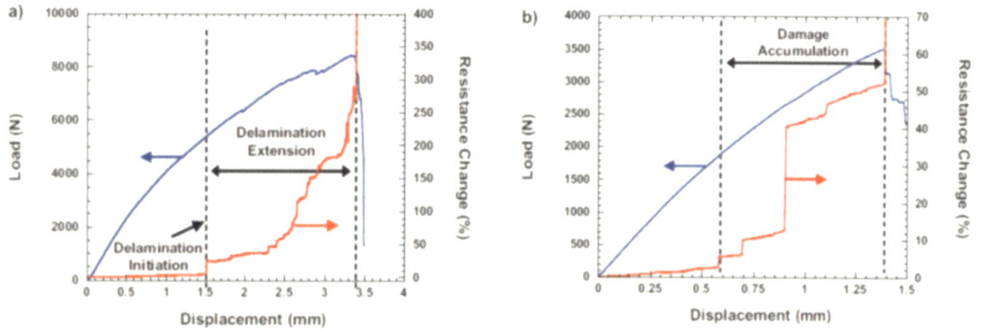


Fig. 17. Load-displacement and resistance response of: (a) a five-ply unidirectional composite with the center ply intentionally cut to initiate delamination; (b) a $(0/90)_s$ cross-ply composite showing accumulation of damage due to microcracks (Thostenson & Chou, 2006b)

## 5. Conclusion

CNTs have made inroads into multifunctional, smart applications, particularly strain sensing for structural health monitoring. A vast number of studies have focused on tailoring the mechanical and electrical properties of CNT-based nanocomposites by controlling CNT dispersion, orientation, and CNT-matrix interface at the nanoscale. The insights gained from the electromechanical behavior of CNT nanocomposites have open up a new field in structural health monitoring of multiscale hybrid composites. Although fundamental studies on processing-structure-property relationship in CNT nanocomposites need to be continued, allied efforts will need to be devoted to large-area strain mapping, cumulative stress/strain tracking, damage detection and life prediction algorithms, and data acquisition and analysis to fully utilize the smart sensing and actuation capabilities of CNT nanocomposites.

## 6. Acknowledgment

This research was supported by Basic Science Research Program through the National Research Foundation of Korea (NRF) funded by the Ministry of Education, Science and Technology (Grant No. 2009-0070548).

## 7. References

Abdalla M.; Dean D., Theodore M., Fielding J., Nyairo E., Price G. (2010). Magnetically processed carbon nanotube/epoxy nanocomposites: Morphology, thermal, and mechanical properties. *Polymer*, Vol.51, Issue.7, (March 2010), pp. 1614-1620, ISSN 0032-3861

Abot, J. L.; Song, Y., Vatsavaya, M. S., Medikonda, S., Kier, Z., Jayasinghe, C., Rooy, N., Shanov, V. N. & Schulz, M. J. (2010). Delamination detection with carbon nanotube thread in self-sensing composite materials. *Compos. Sci. Technol.*, Vol. 70, (March 2010), pp. 1113–1119, ISSN 0266-3538

Abraham, J. K.; Aryasomayajula, L., Whitchurch, A., & Varadan, V. K. (2008). Carbon nanotube strain sensors for wearable patient monitoring applications, *Proc. of SPIE*, Vol. 6931, ISSN 0277-786X

Ajayan P.; Stephen O., Colliex C., Trauth D.(1994). Aligned Carbon Nanotube Arrays Formed by Cutting a Polymer Resin—Nanotube Composite, *Science*, Vol.265, Issue.5176 (August 1994), pp. 1212-1214, ISSN 0036-8075

Alexopoulos, N. D.; Bartholome, C., Poulin, P., & Marioli-Riga, Z. (2010). Structural health monitoring of glass fiber reinforced composites using embedded carbon nanotube (CNT) fibers. *Compos. Sci. Technol.*, Vol. 70, pp. 260–271, ISSN 0266-3538

Allaoui A.; Bai S., Cheng H., Bai J. (2002). Mechanical and electrical properties of a MWNT/epoxy composite. *Compos. Sci. Technol.*, Vol.62, Issue.15, (November 2002), PP. 1993–1998, ISSN 0266-3538

Anand, S. V. & Mahapatra, D. R. (2009). Quasi-static and dynamic strain sensing using carbon nanotube/epoxy nanocomposite thin films. *Smart. Mater. Struct.*, Vol. 18, (March 2009), 45013, ISSN 0964-1726

Awasthi K.; Kamalakaran R., Singh A., Srivastava O. (2004). Ball-milled carbon and hydrogen storage. *Int. J. Hydrog. Energy*, Vol.27, Issue.4, pp. 425–32, ISSN 0360-3199

Bal S. & Samal S. (2007). Carbon Nanotube Reinforced Polymer Composite-A State of The Art. *Bull. Mat. Sci.*, Vol.30, Issue.4, (August 2007), pp. 379–386, ISSN 0250-4707

Bautista-Quijanoa, J. R.; Avilésa, F., Aguilara, J. O., & Tapia, A. (2010). Strain sensing capabilities of a piezoresistive MWCNT-polysulfone film. *Sensors and Actuators A*, Vol. 159, (March 2010), pp. 135–140, ISSN 0924-4247

Bekyarova, E.; Thostenson, E. T., Yu, A., Kim, H., Gao, J., Tang, J., Hahn, H. T., Chou, T.-W., Itkis, M. E., & Haddon, R. C. (2007). Multiscale carbon nanotube-carbon fiber reinforcement for advanced epoxy composites. *Langmuir*, Vol. 23, pp. 3970-3974, ISSN 0743-7463

Bhattachacharyya A.; Sreekumar T., Liu T., Kumar S., Ericson L., Hauge H., Smalley R. (2003) Crystallization and orientation studies in polypropylene/single wall carbon nanotube composite. *Polymer*, Vol.44, Issue.8, Issue. (April 2003), pp. 2373-2377, ISSN 0032-3861

Baughman, R.; Zakhidov, A. & de Heer, W. (2002). Carbon Nanotubes, The Route Toward Applications. *Science*, Vol.297, No.5582, (August 2002), pp. 787–792, ISSN 0036-8075

Bauhofer W. & Kovacs J. (2009). A review and analysis of electrical percolation in carbon nanotube polymer composites, *Compos. Sci. Technol.*, Vol. 69, Issue.10, (August 2009) pp. 1486-1498 ISSN 0266-3538

Bilotti, E.; Zhang, R., Deng, H., Baxendale, M., & Peijs, T. (2010). Fabrication and property prediction of conductive and strain sensing TPU/CNT nanocomposite fibres, *J. Mater. Chem.*, Vol. 20, pp. 9449–9455, ISSN 0959-9428

Boger, L.; Wichmann, M. H. G., Meyer, L. O., & Schulte, K. (2008). Load and health monitoring in glass fibre reinforced composites with an electrically conductive nanocomposite epoxy matrix. *Compos. Sci. Technol.*, Vol. 68, (January 2008), pp. 1886–1894, ISSN 0266-3538

Camponeschi E.; Vance R., Al-Haik M., Garmestani H., Tannenbaum R. (2007). Properties of carbon nanotube–polymer composites aligned in a magnetic field, *Polymer*, Vol. 45, Issue.10, (September 2007), pp.2037-2046, ISSN 0032-3861

Chang T.; Kisliuk A., Rhodes S., Brittain W., Sokolov A. (2006) Conductivity and mechanical properties of well-dispersed single-wall carbon nanotube/polystyrene composite. *Polymer*, Vol.47, Issue.22, (October 2006) pp. 7740–7746, ISSN 0032-3861

Cheng F.; Imin P., Maunders C., Botton G., Adronov A. (2008). Soluble, discrete supramolecular complexes of single-walled carbon nanotubes with fluorene-based conjugated polymers. *Macromolecules*, Vol.41, Issue.7, (March 2008) pp. 2304–2308, ISSN 0024-9297

Ci L.; Suhr J., Pushparaj V., Zhang X., Ajayan P. (2008). Continuous Carbon Nanotube Reinforced Composites. *Nano Lett.*, Vol.8, No.9, (August 2008), pp. 2762-2766 ISSN 1530-6984

Ciselli, P.; Lu, L., Busfield, J. J. C., & Peijs, T. (2010). Piezoresistive polymer composites based on EPDM and MWNTs for strain sensing applications. *e-Polymers*, Vol. 14, (February 2010), ISSN 1618-7229

Coleman J.; Khan U., Blau W and Gun'ko Y. (2006). Small but Strong: A Review of The Mechanical Properties of Carbon Nanotube–Polymer Composites. *Carbon*, Vol. 44, Issue.9 (August 2006) pp. 1624–1652, ISSN 0008-6223

Dang, Z.-M.; Jiang, M.-J., Xie, D., Yao, S.-H., Zhang, L.-Q., & Bai, J. (2008). Supersensitive linear piezoresistive property in carbon nanotubes/silicone rubber nanocomposites. *J. Appl. Phys.*, Vol. 104, (July 2008), 24114, ISSN 0021-8979

de la Vega, A.; Kinloch, I. A., Young, R. J., Bauhofer, W., & Schulte, K. (2011). Simultaneous global and local strain sensing in SWCNT–epoxy composites by Raman and impedance spectroscopy. *Compos. Sci. Technol.*, Vol. 71, pp. 160–166, ISSN 0266-3538

Deshmukh, S.; Ounaies, Z., & Krishnamoorti, R. (2009). Polymer nanocomposites as electrostrictive materials, *Proc. of SPIE*, Vol. 7289, ISSN 0277-786X

Dresselhaus M. (1997). Future Directions in Carbon Science. *Ann. Rev. Mater. Res.*, Vol.27, Issue.1 (August 1997) pp. 1–34, ISSN 1531-7331

Dresselhaus M.; Dresselhaus G. Charlier J. & Hernández E. (2004). Electronic, thermal and mechanical properties of carbon nanotubes. *Philos T R Soc A*, Vol.15, Issue.362, (October 2004) pp. 2065-2098, ISSN 1364-503X

Gao, L.; Thostenson, E. T., Zhang, Z., & Chou, T.-W. (2009). Sensing of damage mechanisms in fiber-reinforced composites under cyclic loading using carbon nanotubes. *Adv. Funct. Mater.*, Vol. 19, pp. 123–130, ISSN 1616-301X

Gao, S.-L.; Zhuang, R.-C., Zhang, J., Liu, J.-W., & Mader, E. (2010). Glass fibers with carbon nanotube networks as multifunctional sensors. *Adv. Funct. Mater.*, Vol. 20, (May 2010), pp. 1885–1893, ISSN 1616-301X

Gorrasi J.; Sarno M., Di Bartolomeo A., Sannino D., Ciambelli P., Vittoria V. (2007) Incorporation of carbon nanotubes into polyethylene by high energy ball milling: morphology and physical properties. *J. Polym. Sci. Pt. B-Polym. Phys.*, Vol.45, Issue.5, (January 2007) pp. 597–606, ISSN 0887-6266

Haluska, M.; Hulman, M., Hirscher, M., Becher, M., Roth, S., Stepanek I. and Bernier P. (2001) Hydrogen storage in mechanically treated single wall carbon nanotubes. *AIP Conf. Proc. Electronic Properties of Molecular Nanostructures*, Vol. 591, pp. 603–608.

Hernandez J.; Garcia-Gutierrez M., Nogalas A., Rueda D., Kwiatkowska M., Szymczyk A., Roslaniec Z., Concheso A., Guinea I., Ezquerra T. (2009). Influence of preparation procedure on the conductivity and transparency of SWCNT–polymer composites. *Compos. Sci. Technol.*, Vol.69, Issue.11-12 (September 2009), pp. 1867–1872, ISSN 0266-3538

Hwang, J.; Jang, J., Hong, K., Kim, K. N., Han, J. H., Shin, K., & Park, C. E. (2011). Poly(3-hexylthiophene) wrapped carbon nanotube/poly(dimethylsiloxane) composites for use in finger-sensing piezoresistive pressure sensors. *Carbon*, Vol. 49, pp. 106–110, ISSN 0008-6223

Iosif D.; Rosca, Suong V., Hoa. (2009). Highly conductive multiwall carbon nanotube and epoxy composites produced by three-roll milling. *Carbon*, Vol.47, Issue.8, (July 2009), pp. 1958–1968, ISSN 0008-6223

Jia, Z.; Wang Z., Liang J., Wei B. and Wu D. Production of short multi-walled carbon nanotubes. *Carbon*, Vol.37 Issue.6, (1999) pp. 903–906. ISSN 0008-6223

Jiang X.; Bin Y., Matsuo M. (2005). Electrical and mechanical properties of polyimide–carbon nanotubes composites fabricated by in situ polymerization. *Polymer*, Vol.46, Issue.18, (August 2005) pp. 7418–7424, ISSN 0032-3861

Jin Y. & Yuan F. (2003). Simulation of Elastic Properties of Single-Walled Carbon Nanotubes. *Compos. Sci. Technol.*, Vol.63, Issue.11, (August 2003) pp. 1507-1515, ISSN 0266-3538

Kang, I.; Yun, Y. H., Kim, J. H., Lee, J. W., Gollapudi, R., Subramaniam, S., Narasimhadevara, S., Hurd, D., Kirikera, G. R., Shanov, V., Schulz, M. J., Shi, D., Boerio, J., Mall, S., & Ruggles-Wren, M. (2006). Introduction to carbon nanotube and nanofiber smart materials. *Compos. Pt. B*, Vol. 37, (March 2006), pp. 382–394, ISSN 1359-8368

Kang, J. H.; Park, C., Scholl, J. A., Brazin, A. H., Holloway, N. M., High, J. W., Lowther, S. E., & Harrison, J. S. (2009). Piezoresistive characteristics of single wall carbon nanotube/polyimide nanocomposites. *J. Polym. Sci.: Part B: Polym. Phys.*, Vol. 47, pp. 994–1003, ISSN 0887-6266

Kim H.; Kim K., Lee S., Joo J., Yoon H., Cho S., Lyu S., Lee C. (2004) Charge transport properties of composites of multiwalled carbon nanotube with metal catalyst and polymer: application to electromagnetic interference shielding. *Curr. Appl. Phys.*, Vol.4, Issue.6, (November 2004) pp. 577–580, ISSN 1567-1739

Kim, J.; Loh, K. J., & Lynch, J. P. (2008). Piezoelectric polymeric thin films tuned by carbon nanotube fillers, *Proc. of SPIE*, Vol. 6932, ISSN 0277-786X

Kim, K. J.; Yu, W.-R., Lee, J. S., Gao, L., Thostenson, E. T., Chou, T.-W., & Byun, J.-H. (2010). Damage characterization of 3D braided composites using carbon nanotube-based in situ sensing, *Compos. Pt. A*, Vol. 41, pp. 1531–1537, ISSN 1359-835X

Ko F., Gogotsi Y.; Ali A., Naguib N., Ye H., Yang G., Li C., Willis P. (2003). Electrospinning of Continuous Carbon Nanofiber Yarns. *Adv. Mater.*, Vol.15, Issue.14 (July 2003) pp. 1161–1165, ISSN 0935-9648

Koerner H.; Liu W., Alexander M., Mirau P., Dowty H., Vaia R. (2005). Deformation-morphology correlations in electrically conductive CNT–thermoplastic polyurethane nanocomposites. *Polymer*, Vol.46, Issue.12, (May 2005), pp. 4405–4420 ISSN 0032-3861

Krause B.; Pötschke P., Häuser L. (2009). Influence of small scale melt mixing conditions on electrical resistivity of carbon nanotube–polyamide composites. *Compos. Sci. Technol.*,Vol.69, Issue.10, (August 2009) pp. 1505–1515, ISSN 0266-3538

Landi B.; Raffaelle R., Heben M., Alleman J., Van Derveer W., Gennett T. (2002). Single wall carbon nanotube–nafion composite actuators. *Nano Lett.*, Vol.2, No.11, (October 2002), pp. 1329–32, ISSN 1530-6984

Lee J.; Rhee K., Park S. Silane modification of carbon nanotubes and its effects on the material properties of carbon/CNT/epoxy three-phase composites. *Compos. Pt. A-Appl. Sci. Manuf.*, Vol.42, Issue.5 (May 2011) 478–483, ISSN 1359-835X

Li, C.; Thostenson, E. T., & Chou, T.-W. Chou (2008). Sensors and actuators based on carbon nanotubes and their composites: A review. *Compos. Sci. Technol.*, Vol. 68 (January 2008), pp. 1227–1249, ISSN 0266-3538

Li, X.; Levy, C., & Elaadil, L. (2008). Multiwalled carbon nanotube film for strain sensing. *Nanotechnology*, Vol. 19, (January 2008), 45501, ISSN 0957-4484

Li, Z.; Dharap, P., Nagarajaiah, S., Barrera, E. V., and Kim, J. D. (2008). Carbon nanotube film sensors. *Adv. Mater.*, Vol. 16, Issue 7, (April 2008), pp. 640-643, ISSN 0935-9648

Loh, K. J.; Hou, T.-C., Lynch, J. P., & Kotov, N. A. (2009). Carbon nanotube sensing skins for spatial strain and impact damage identification. *J. Nondestruct. Eval.*, Vol. 28, (March 2009), pp. 9–25, ISSN 0195-9298

Lu K.; Lago R., Chen Y., Green M, Harris P., Tsang S. (1996) Mechanical damage of carbon nanotubes by ultrasound. *Carbon*, Vol.34, Issue. 6, pp. 814–816 ISSN 0008-6223

Ma P.; Siddiquia N., Maromb G & Kim J. (2010). Dispersion snd Functionalization of Carbon Nanotubes for Polymer-Based Nanocomposites: A Review. *Compos. Pt. A-Appl. Sci. Manuf.*, Vol.41, Issue.10, (October 2010), pp. 1345-1367, ISSN 1359-835X

Martin C.; Sandler J., Windle A., Schwarz M., Bauhofer W., Schulte K., Shaffer M. (2005). Electric field-induced aligned multi-wall carbon nanotube networks in epoxy composites. *Polymer*, Vol.46, Issue.3 (January 2005) pp. 877–886 ISSN 0032-3861

Mičušík M.; Omastová M., Krupa I, Prokeš J., Pissis P., Logakis Pandis C., Pötschke P., Piontek J. (2009). A comparative study on the electrical and mechanical behaviour of multi-walled carbon nanotube composites prepared by diluting a masterbatch with various types of polypropylene. *J. Appl. Polym. Sci.*, Vol.113, No.4, (April 2009) pp. 2536–2551, ISBN 0021-8995

Narayan C.; Yayong L., Kaikun Y., Weiqun P., Spandan M., Howard W. (2009) Single-Walled Carbon Nanotube/Poly(methyl methacrylate) Composites for

Electromagnetic Interference Shielding, *Polym. Eng. Sci.*, Vol.49, Issue.8, (May 2009), pp. 1627–1634 ISSN 0032-3888

Nativ-Roth E.; Shvartzman-Cohen R., Bounioux C., Florent M., Zhang D., Szleifer I., Yerushalmi-Rozen R. (2007). Physical adsorption of block copolymers to SWNT and MWNT: a nonwraping mechanism. *Macromolecules*, Vol.40, Issue.10 (April 2007) pp. 3676–3685, ISSN 0024-9297

Oh J.; Ahn K. & Hong J. (2010). Dispersion of entangled carbon nanotube by melt extrusion. *Korea-Aust. Rheol. J.*, Vol.22, Issue.2, (June 2010), pp. 89-94 ISSN 1226-119X

Peng H. & Sun X. (2009). Highly aligned carbon nanotube/polymer composites with much improved electrical conductivities. *Chem. Phys. Lett.*, Vol.471, Issue.1-3, (March 2009), pp. 103–105, ISSN 0009-2614

Pham, G. T.; Park, Y.-B., Liang, Z., Zhang, C, & Wang, B. (2008). Processing and modeling of conductive thermoplastic/carbon nanotube films for strain sensing. *Compos. Pt. B*, Vol. 39, pp. 209–216, ISSN 1359-8368

Pötschke P.; Fornes T., Paul D. (2002) Rheological behaviour of multiwalled carbon nanotubes/polycarbonate composites. *Polymer*, Vol.43, Issue.11, (May 2002), pp. 3247–3255, ISSN 0032-3861

Ramasubramaniam R.; Chen J., Liu H. (2003). Homogeneous carbon nanotube/ polymer composites for electrical applications. *Appl. Phys. Lett.*, Vol.83, Issue.14, (October 2003) pp. 2928–2930, ISSN 0003-6951

Sahoo N.; Ranab S., Cho J., Li L., Chan S. (2010) Polymer nanocomposites based on functionalized carbon nanotubes. *Prog. Polym. Sci.*, Vol.35, Issue.7, (July 2010) pp.837-867, ISSN 0079-6700

Salalha W.; Dror Y., Khalfin R., Cohen Y, Yarin A., Zussman E. (2004). Single-Walled Carbon Nanotubes Embedded in Oriented Polymeric Nanofibers by Electrospinning. *Langmuir*, Vol.20 No.20, (September 2004), pp. 9852–9855, ISSN0743-7463

Sandler J., Kirk J., Kinloch I., Shaffer M., Windle A. (2003). Ultra-low electrical percolation threshold in carbon-nanotube-epoxy composites. *Polymer*, Vol.44, Issue.19, (September 2003), pp. 5893–5899, ISSN 0032-3861

Santos A.; Leite T., Furtado C., Welter C., Pardini L., Silva G. (2008). Morphology, thermal expansion, and electrical conductivity of multiwalled carbon nanotube/epoxy composites. *J. Appl. Polym. Sci.*, Vol.108, No.2, (January 2008) pp. 979–986, ISSN 0021-8995

Seifert G.; Kohler T., Frauenheim T. (2000) Molecular wires, solenoids, and capacitors by sidewall functionalization of carbon nanotubes. *Appl. Phys. Lett.*, Vol.77, Issue.9, (June 2000) pp. 1313–1315 ISSN 0003-6951

Shelimova K.; Esenalieva R., Rinzlera A., Huffman C. and Smalley R. (1998) Purification of single-wall carbon nanotubes by ultrasonically assisted filtration. *Chem. Phys. Lett.*, Vol.282, Issue.5-6 (January 1998), pp. 429-434, ISSN 0009-2614

Spitalskya Z.; Tasisb D., Papagelis K, Galiotis C. (2010). Carbon nanotube–polymer composites: Chemistry, processing, mechanical and electrical properties. *Prog. Polym. Sci.*, Vol.35, Issue.3, (March 2010), pp. 357-401, ISSN 0079-6700

Stevens J.; Huang A., Peng H., Chiang I., Khabashesku V, Margrave J. (2003) Sidewall amino-functionalization of SWNTs through fluorination and subsequent reactions with terminal diamines. *Nano Lett.*, Vol.3, No.3, (January 2003), pp. 331–336, ISSN 1530-6984

Sureeyatanapas, P. & Young, R. J. (2009). SWNT composite coatings as a strain sensor on glass fibres in model epoxy composites. *Compos. Sci. Technol.*, Vol. 69, pp. 1547–1552, ISSN 0266-3538

Thess A.; Lee R., Nikolaev P., Dai H., Petit P., Robert J., Xu C., Lee H., Kim S., Rinzler A., Colbert D., Scuseria G., Tomanek D, Fischer J., Smalley R. (1996). Crystalline ropes of metallic carbon nanotubes. *Science*, Vol.273, No.5274, (July 1996), pp. 483–488, ISSN 0036-8075

Thostenson E. & Chou T. (2002). Aligned multi-walled carbon nanotube-reinforced composites: processing and mechanical characterization *J. Phys. D-Appl. Phys.*, Vol.35, Issue.16 (August 2002) pp.L77-L80 ISSN 0022-3727

Thostenson E. T. & Chou T.-W. (2006). Processing-structure-multi-functional property relationship in carbon nanotube/epoxy composites. *Carbon*, Vol.44, Issue.14 (November 2006) PP. 3022–3029. ISSN 0008-6223

Thostenson, E. T. & Chou, T.-W. (2006). Carbon nanotube networks: sensing of distributed strain and damage for life prediction and self healing. *Adv. Mater.*, Vol. 18, (October 2006), pp. 2837–2841, ISSN 0935-9648

Thostenson, E. T.; Ziaee, S., & Chou, T.-W. (2009). Processing and electrical properties of carbon nanotube/vinyl ester nanocomposites. *Compos. Sci. Technol.*, Vol. 69, pp. 801–804, ISSN 0266-3538

Vemuru, S. M.; Wahi, R., Nagarajaiah, S., & Ajayan, P. M. (2009). Strain sensing using a multiwalled carbon nanotube film. *J. Strain Anal. Eng.*, Vol. 44, pp. 555-562, ISSN 0309-3247

Vigolo B., Pe´nicaud A., Coulon C., Sauder C., Pailler R., Journet C., Bernier P., Poulin P. (2000). Macroscopic Fibers and Ribbons of Oriented Carbon Nanotubes. *Science*, Vol.290 No.5495, (November 2000), pp. 1331-1334, ISSN 0036-8075

Viswanathan V.; Laha T., Balani K., Agarwal A., Seal S. (2006). Challenges and advances in nanocomposite processing techniques. *Mater. Sci. Eng. R-Rep.*, Vol.54, Issue.5-6, (November 2006), pp. 121-285, ISSN 0927-796X

Wang Q.; Dai J., Li W., Wei Z., Jiang J. (2008). The effects of CNT alignment on electrical conductivity and mechanical properties of SWNT/epoxy nanocomposites. *Compos. Sci. Technol.*, Vol.68, Issue.7-8 (June 2008) pp. 2208–2213, ISSN 0266-3538

Wang, P.; Geng, S., & Ding, T. (2010). Effects of carboxyl radical on electrical resistance of multi-walled carbon nanotube filled silicone rubber composite under pressure. *Compos. Sci. Technol.*, Vol. 70, (May 2010), pp. 1571–1573, ISSN 0266-3538

Wichmann, M. H. G.; Buschhorn, S. T., Boger, L., Adelung, R., & Schulte, K. (2008). Direction sensitive bending sensors based on multi-wall carbon nanotube/epoxy nanocomposites. *Nanotechnology*, Vol. 19, (October 2008), 475503, ISSN 0957-4484

Wichmann, M. H. G.; Buschhorn, S. T., Gehrmann, J., & Schulte, K. (2009). Piezoresistive response of epoxy composites with carbon nanoparticles under tensile load. *Phys. Rev. B*, Vol. 80, (December 2009), 245437, ISSN 1098-0121

Yuen S.; Ma C., Lin Y., Kuan H. (2007). Preparation, morphology and properties of acid and amine modified multiwalled carbon nanotube/polyimide composite. *Compos. Sci. Technol.*, Vol.67, Issue.11-12, (September 2007) pp. 2564–2573, ISSN 0266-3538

Zhang M.; Atkinson K., Baughman R. (2004). Multifunctional Carbon Nanotube Yarns by Downsizing an Ancient Technology. *Science*, Vol.306, (April 2004), pp. 1358-1361 ISSN 0036-8075.

Zhang Q.; Rastogi S., Chen D., Lippits D., Lemstra P. (2006). Low percolation threshold in single-walled carbon nanotube/high density polyethylene composites prepared by melt processing technique. *Carbon*, Vol.44, Issue.4, (April 2006) pp. 778–785, ISSN 0008-6223

Zhang, W.; Suhr, J., & Koratkar, N. (2006). Carbon nanotube/polycarbonate composites as multifunctional strain sensors. *J. Nanosci. Nanotechnol.*, Vol. 6, pp. 960-964, ISSN 1533-4880

Zhang, W.; Sakalkar, V., & Koratkar, N. (2007). In situ health monitoring and repair in composites using carbon nanotube additives. *Appl. Phys. Lett.*, Vol. 91, (September 2007), 133102, ISSN 0003-6951

19

# Nonlinear Optical Properties of Graphene and Carbon Nanotube Composites

Jun Wang[1a], Yu Chen[1b], Rihong Li[1a], Hongxing Dong[1a], Long Zhang[1a], Mustafa Lotya[2], Jonathan N. Coleman[2] and Werner J. Blau[2]
[1a]*Key Laboratory of Materials for High-Power Laser, Shanghai Institute of Optics and Fine Mechanics, Chinese Academy of Sciences*
[1b]*Key Laboratory for Advanced Materials, Department of Chemistry, East China University of Science and Technology*
[2]*School of Physics and the Centre for Research on Adaptive Nanostructures and Nanodevices (CRANN), Trinity College Dublin*
[1]*China*
[2]*Ireland*

## 1. Introduction

The rapid development of nanoscience and nanotechnology provides lots of new opportunities for nonlinear optics. A growing number of nanomaterials have been shown to possess remarkable nonlinear optical (NLO) properties, which promotes the design and fabrication of nano and nano-scale optoelectronic and photonic devices (Xia et al. 2003; Avouris et al. 2008; Hasan et al. 2009; Bonaccorso et al. 2010; Loh et al. 2010; Coleman et al. 2011). The wonderful carbon allotropes discovered in recent decades are the most representative products of nanotechnology: from 3D carbon nanoparticles (graphite), to 0D fullerenes, to 1D carbon nanotubes (CNTs), and then to 2D graphenes discovered most recently. Interestingly, all of these nano-carbons exhibit diverse NLO properties. For instance, carbon black suspensions show strong thermally-induced nonlinear scattering (NLS) effect and hence optical limiting (OL) for intense ns laser pulses (Mansour et al. 1992); fullerenes show large third-order optical nonlinearity and reverse saturable absorption (RSA) at certain wavelength band (Tutt et al. 1992); CNTs show ultrafast second- and third-order nonlinearities and saturable absorption (SA) in the near infrared (NIR) region (Hasan et al. 2009); and graphenes show ultrafast carrier relaxation time and ultra-broad-band resonate NLO response (Bonaccorso et al. 2010).

Optical limiting is an important NLO phenomenon, which can be utilized to protect delicate optical instruments, especially the human eye, from intense laser beams (Tutt et al. 1993). As shown in Fig. 1, ideally an optical limiter should strongly attenuate intense, potentially dangerous laser beams, while exhibiting high transmittance for low intensity ambient light. Generally speaking, there are two main mechanisms for passive OL: nonlinear absorption (NLA) and NLS. The former can be further divided into multi-photon absorption (MPA), RSA and free-carrier absorption (FCA). Up to date, numerous inorganic and organic materials, such as phthalocyanines (O'Flaherty et al. 2003; de la Torre et al. 2004), porphyrins

(Blau et al. 1985; Senge et al. 2007), organic dyes (He et al. 1995; He et al. 2008), metal nanoclusters (Sun et al. 1999; Wang et al. 2009), quantum dots (He et al. 2007), etc. have been found to possess OL response. Carbon-related nanomaterials are actually a main branch in the field of OL materials (Chen et al. 2007; Wang et al. 2009). It has been confirmed that fullerene shows RSA induced OL, and nanotubes and graphene show NLS induced limiting. However, the most important point is that the advantage of these carbon nanomaterials manifests themselves in tailorable chemical properties by binding functional materials, e.g., polymers, organic molecules and metal nanoparticles, forming versatile OL composites (Chen et al. 2007; Wang et al. 2009; Bottari et al. 2010).

The large surface energy of nanotubes imposes restrictions on the formation of individual nanotubes in most inorganic and organic solvents. For solubilized nanotubes, one can employ polymers or organic molecules to functionalize, covalently or noncovalently, the surface of nanotubes. In the same manner, pristine single- or few-layer graphene is also difficult to exist stably in many organic solvents. It is thus very significant to design and synthesize nanotube- and graphene-based solution-processed organic/polymeric materials, which is a key step for the development of viable nano-carbon OL devices (Chen et al. 2007; Wang et al. 2009).

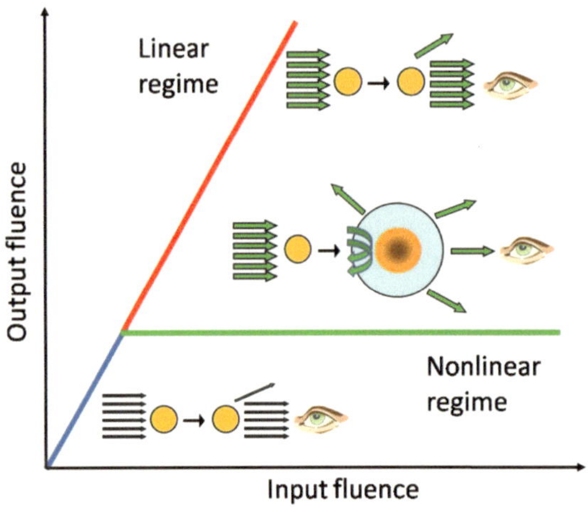


Fig. 1. The response of an ideal optical limiter.

## 2. Mechanisms

### 2.1 Nonlinear scattering (NLS)

Thermally induced NLS may be the most common nonlinear phenomenon for various nanomaterial systems, such as nanotubes, nanorods, nanowires, nanosheets, nanoribbons, nanospheres, nanodots etc. (Wang et al. 2009). An effective scattering process can disperse the highly intense beam into a larger spatial dimension and hence reduce the intensity of the direct incident beam. According to Mie scattering theory, the nanoscale particles alone cannot scatter a light beam effectively. The effective scattering arises from the formation of scattering centres with size of the order of the wavelength of the incident laser beam. The formation of scattering centres, initiating from nanoparticles, has three possible origins.

The induced scattering centres consist of two origins: the formation and growth of solvent bubbles, which is due to the thermal energy transfer from the nanotubes to the solvent; and the formation and expansion of carbon microplasmas, which is due to the ionization of nanotubes. The former takes place at the lower incident energy fluence, while the latter takes place at higher fluences. Belousova et al. developed a theoretical model to explain the OL of carbon nanoparticles (Belousova et al. 2003; Belousova et al. 2004). In this model, the whole limiting process is described theoretically by three steps: the dynamics of the formation and expansion of solvent vapour bubbles; the Mie scattering of the expanding bubbles; and the nonlinear propagation through the scattering medium. Although the objects of modeling are quasi-spherical carbon nanoparticles, the Mie theory-based prediction works qualitatively for nanotubes and is helpful for understanding bubble growth dynamics and thus the OL process in CNT suspensions. As an example, Fig. 2a shows the variations of absorption and scattering cross sections as radius of gas bubbles in carbon nanoparticle suspensions, and Fig. 2b illustrates the inside pressure, expansion rate and radius of a gas bubble as functions of illumination time (Belousova et al. 2004). Moreover, Belousova's simulation indicates that the scattering cross section increases significantly with the increasing size of vapor bubbles, meanwhile the absorption cross section decreases until it is negligible when the bubbles grow, effectively limiting the incident power.

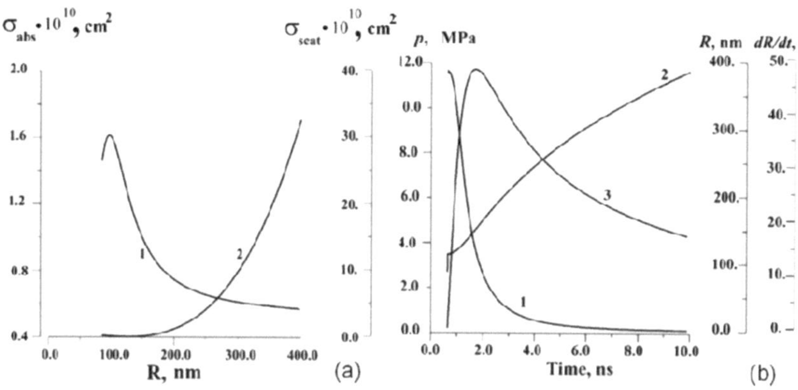


Fig. 2. Variations of (1) absorption and (2) scattering cross sections as radius of gas bubbles in carbon nanoparticle suspensions (a), and the inside pressure (1), expansion rate (2) and radius (3) of a gas bubble as functions of illumination time (b) (Belousova et al. 2004).

### 2.2 Reverse saturable absorption (RSA)

The process of RSA involves multi-step, excited state absorption (ESA) from the singlet ground state to the first excited triplet state via the first excited singlet state. The most representative materials include phthalocyanines, porphyrins, fullerenes, etc. A general five-level model, as shown in Fig. 3, has been considered to simulate the RSA process in the phthalocyanine system (O'Flaherty et al. 2003; O'Flaherty et al. 2004). The vibrational levels of the electronic states are ignored. Generally, for this five-level system after initial excitation, the first excited singlet state $S_1$ is populated, from here the electrons may be subsequently excited into $S_2$ within the pulse width of the laser. Once in $S_2$, they rapidly relax to $S_1$ again. From $S_1$, the population may undergo an intersystem crossing to the first excited triplet $T_1$ with a time constant $\tau_{isc}$ and thereafter undergo excitations and relaxations to and from $T_2$. Thus, the population is exchanged cyclically between $S_1$ and $T_1$, as the lifetime of $T_1$ ($\tau_{ph}$) is very long in comparison to $\tau_{isc}$. With further simplify matters, it was assumed that relaxation out of states $S_2$ and $T_2$ is very rapid so that the population of these two levels may be neglected. Furthermore, stimulated emission from $S_1$ is excluded due to the small fluorescence quantum yield.

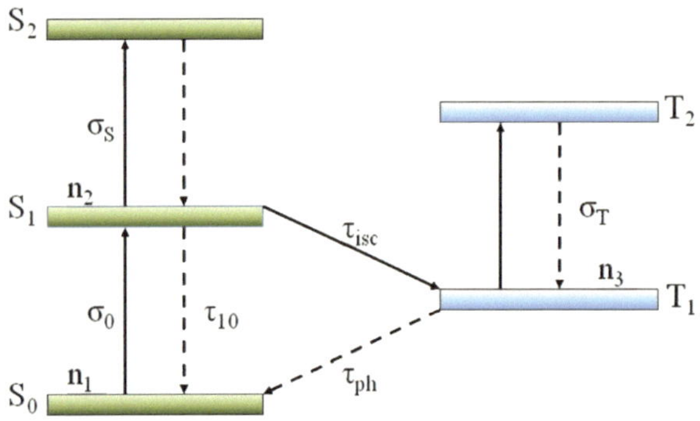


Fig. 3. Illustration of a five-level RSA process. Si represents singlet levels, and Ti represents triplet levels. Solid arrows imply an excitation resulting from photon absorption and dashed arrows represent relaxations.

The extinction of incident beam is governed by the propagation equation

$$\partial I/\partial z=-\alpha_{NL}I=-(\sigma_0 N_1+\sigma_s N_2+\sigma_T N_3)I \quad (1)$$

where the nonlinear absorption coefficient $\alpha_{NL}$ is composed of the ground state absorption $\sigma_0 N_1$, the first excited singlet state absorption $\sigma_s N_2$ and the first excited triplet state absorption $\sigma_T N_3$. N and σ refer to the population and absorption cross section of specific energy levels. Under the steady state approximation $\alpha_{NL}$ can be derived in the form,

$$\alpha_{NL}(F,F_{sat},\kappa)=\alpha_L(1+F/F_{sat})^{-1}(1+\kappa F/F_{sat}) \quad (2)$$

where $\alpha_L$ is the linear absorption coefficient, κ is the ratio of excited state cross section ($\sigma_{ex}$) to ground state cross section ($\sigma_0$), $\kappa=\sigma_{ex}/\sigma_0\approx\sigma_T/\sigma_0$, F represents the energy density and $F_{sat}$ is the energy density at which the ground state absorption saturates (O'Flaherty et al. 2003; O'Flaherty et al. 2004). This model reproduces the RSA effects and highlights the crucial role that the ESA plays in the overall absorption coefficient. Considering this expression for the nonlinear absorption coefficient, one can state that higher κ values combined with lower $F_{sat}$ values define more efficient OL ability.

### 2.3 Multi-photon absorption (MPA)

A multi-photon process is one which occurs through the simultaneous absorption of two or more photons via virtual states in a medium, as shown in Fig. 4. Many metals, semiconductor nanomaterials, quantum dots, organic chromophores and conjugated polymers possess multi-photon absorption induced OL effects (He et al. 2008). For two-photon absorption (TPA), the process can be described by a propagation equation with "Beer-Lambert" format

$$\partial I/\partial z=-(\alpha+\beta I)I \quad (3)$$

where α in unit of $m^{-1}$ is the linear absorption coefficient and β in unit of m/W is the TPA coefficient. Provided that the linear absorption is very small at lower intensity, we obtain the solution for the transmission intensity

$$I(L)=I_0/(1+I_0\beta L). \tag{4}$$

It is clearly seen from the solution that the transmission intensity decreases as the incident intensity increases, resulting in OL phenomenon. The ability of TPA induced OL is strongly dependent on the TPA coefficient, the incident intensity, as well as the propagation length L. The TPA coefficient is related to the TPA cross section, a function of the exciting wavelength. The OL of TPA materials is more effective for shorter incident pulses, since the intensity of shorter pulses (ps or fs) is much higher than that of longer pulses (ns). The three-photon absorption process exhibits very similar characteristics.

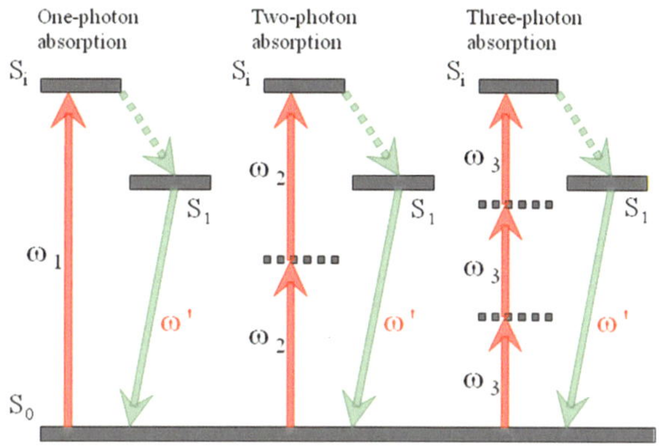


Fig. 4. Multi-photon absorption process.

In addition, it is worth discussing the difference of RSA and TPA processes under high intensity approximation. For RSA process, the nonlinear transmittance originates completely from the non-saturable ESA at very high intensities, hence tends to converge to a minimum transmittance $T_{RSA}$, which had been observed by Blau et al. in tetraphenylporphyrins in 1985 (Blau et al. 1985). The authors deduced analytically the expression for the minimum transmittance $T_{RSA}$, given by

$$T_{RSA}=T_0^{\kappa} \tag{5}$$

where $\kappa=\sigma_{ex}/\sigma_0$ is the ratio of excited state cross section ($\sigma_{ex}$) to ground state cross section ($\sigma_0$). Obviously, the minimum transmission for RSA is a non-zero quantity, which is dependent on κ, as well as the low intensity linear transmittance $T_0$. For TPA process, the transmitted intensity I(L) approaches a constant $1/\beta L$ at very high intensity $I_0\rightarrow+\infty$, and hence the final transmittance $T_{TPA}=I(L)/I_0$ can reach to zero, resulting in a complete optical limitation. On the contrary, the RSA process theoretically cannot realize the complete limiting operation. Such difference between RSA and TPA is important for designing practical optical limiters.

### 2.4 Free-carrier absorption (FCA)

In semiconductors, carriers can be generated by one-photon or two-photon exciting. As shown in Fig. 5, these electron/hole pairs, by absorbing additional photons, can be excited to states higher/lower in the conduction/valence band. The process is named 'free-carrier absorption', which is similar to ESA in molecular system (Boggess et al. 1986). It should be pointed out that there are four possible processes in a FCA medium – linear absorption, TPA, one-photon induced FCA and two-photon induced FCA. For the simplest case, the linearly excited one-photon induced FCA in Fig. 5 can be described by the propagation equation

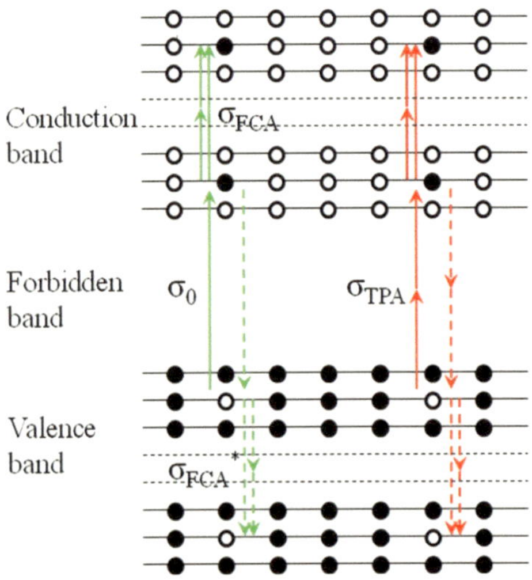


Fig. 5. Free-carrier absorption in semiconductor.

$$\partial I/\partial z=-(\alpha+\sigma_{FCA}N)I, \tag{6}$$

where $\sigma_{FCA}$ is the FCA cross section. With the carrier density N given by $\partial N/\partial t=\alpha I/h\nu$, one can get an approximate solution for the propagation equation

$$T=T_0/[1+(1-T_0)(F_0\sigma_{FCA}/4h\nu)], \tag{7}$$

where $T_0$ is the linear transmission. When the peak incident fluence $F_0$ increases, the total transmission T decreases, resulting in an OL effect. For the most complicated case, all four processes take place in a FCA medium, then we have (Boggess et al. 1986; Tutt et al. 1993)

$$\partial I/\partial z=-(\alpha+\beta I^2+\sigma_{FCA}N)I \tag{8}$$

and

$$\partial N/\partial t=\alpha I/h\nu+\beta I^2/2h\nu. \tag{9}$$

A range of semiconductor nanoparticles, metal nanocomposites and quantum dotes exhibit FCA-induced OL effects. The FCA-induced NLO response is independent on the incident pulse duration, provided that the duration is shorter than the diffusion and recombination processes of free carriers. FCA is also insensitive to the particle size and geometry. It can work in both solid state films and suspensions, covering broad temporal and wavelength ranges. In many nanomaterials, FCA can coexist with NLS and TPA since the generation of free carriers can arise from a TPA process.

## 3. Graphene composites

Doubts about the stability of 2D crystals were finally dispelled by the discovery of graphene, a hexagonally symmetric, covalently bonded 2D carbon monolayer (Novoselov et al. 2004; Novoselov et al. 2005; Geim et al. 2007). Possessing excellent electronic properties, graphene provides a route to study fundamental quantum phenomena, such as the quantum hall effect in condensed-matter materials (Geim et al. 2007). Up to $10^5$ $cm^2/V \cdot s$ mobility of charge carriers, which behave asmassless Dirac fermions in graphene, motivates the development of graphene-based electronic devices, challenging traditional silicon-based electronics (Geim et al. 2007).

In addition to the outstanding electronic, mechanical and thermal properties, graphene has been discovered to possess unique optical and photonic properties, which are summarized as follows.

1. The Dirac electrons in graphene have a linear dispersion between energy and momentum near the Dirac point, resulting in a continuously resonate optical response in a broadband spectral region from the visible to the near infrared (> 2.5 μm) (Geim et al. 2007).
2. Monolayer graphene shows wavelength independent linear optical absorption. For any low intensity light wave, the absorbance rigorously follows $\pi \cdot \alpha \approx 2.3\%$ per layer, where $\alpha$ is the fine-structure constant. As a result, the absorbance of multilayer graphene is proportional to the number of layers (Nair et al. 2008).
3. Graphene possesses ultrafast carrier dynamics due to the ultrafast carrier-carrier scattering and carrier-phonon scattering. Under the fs pulse excitation, the intraband equilibrium time is as short as ~100 fs and the intreband relaxation time is on a ps timescale (Dawlaty et al. 2008).
4. Graphene has significant NLO properties. Depending on the different experimental conditions, graphene and graphene oxide show NLS (Wang et al. 2009), ESA, TPA (Liu et al. 2009) or saturable absorption (SA) (Bao et al. 2009; Sun et al. 2010). Four-wave mixing experiment confirmed that the effective nonlinear susceptibility $|\chi^{(3)}|$ is as large as $10^{-7}$ esu in graphene flakes (Hendry et al. 2010). The second harmonic generation was also observed from a multi-layer graphene film (Dean et al. 2009).
5. Graphene oxide (GO) is a 2D network of mixed $sp^2$ and $sp^3$ carbon bondings. The isolated nanoscale $sp^2$ domains in the $sp^3$ matrix leads to a bandgap in GO. The width of the bandgap can be controlled by the size, shape and fraction of the $sp^2$ clusters, achieving a tunable photoluminescence and electroluminescence (Eda et al. 2010; Loh et al. 2010).

Before 2008, the study of photonic and optoelectronic properties of graphene have remained theoretical. With the help of development of the low-cost, high-yield method for mass production of graphene, the experimental study of NLO properties of graphene and graphene derivatives has developed very rapidly since 2009. Hereinafter, we introduce the NLO properties of graphene and its functionalized derivatives.

### 3.1 Graphene and graphene oxide

In contrast to micromechanical cleavage (Novoselov et al. 2005) and epitaxial growth (de Heer et al. 2007), a recently developed liquid-phase exfoliation technique provides a low-cost, high-yield method for mass production of unoxidized, defect-free graphene (Hernandez et al. 2008; Lotya et al. 2009). In this method, the sieved graphite powder was dispersed in a range of organic solvents. After the low power sonication treatment and subsequent mild centrifugation to remove macroscopic aggregates, the homogeneous dark dispersions were obtained. All dispersions were stable against sedimentation and with only minimal aggregation occurring over a period of weeks. Experimental and theoretical analyses reveal that the surface energies of the selected solvents, e.g., N-methyl-2-pyrrolidone (NMP), N,N-dimethylacetamide (DMA), g-butyrolactone (GBL) etc., match very well that of graphite (~70–80 mJ m$^{-2}$), resulting in a minimal energy cost of overcoming the van der Waals forces between two graphene sheets, hence the effective exfoliation to graphene single or few layers (Bergin et al. 2008; Coleman 2009).

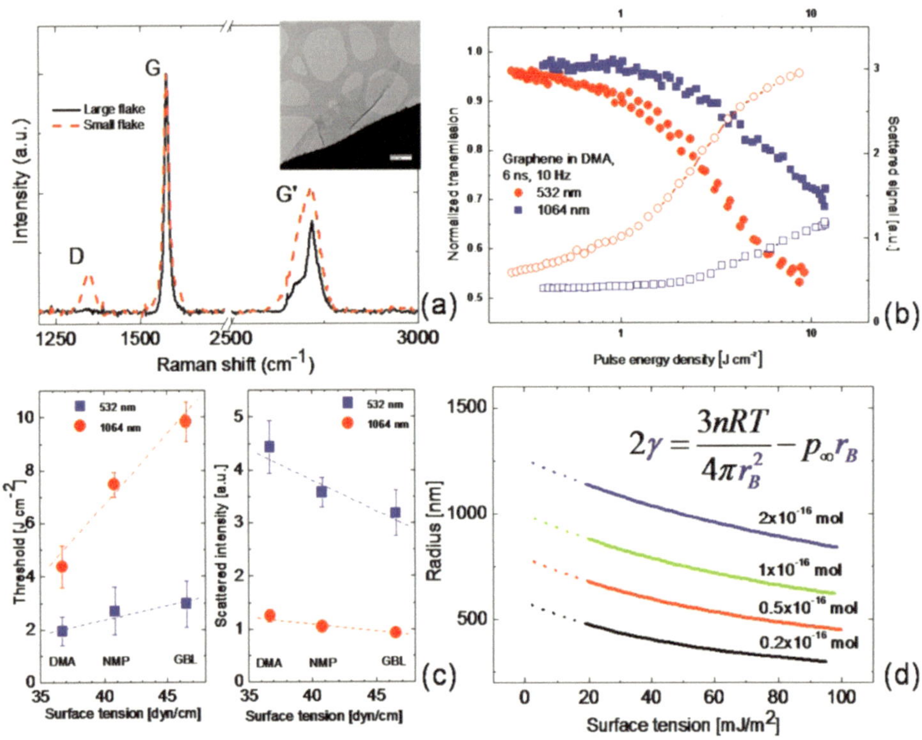


Fig. 6. TEM image, Raman spectra (a) and broadband OL (b) of the graphene dispersions. Limiting threshold and scattered intensity as functions of surface tension of solvents (c). Radius of the bubbles as a function of surface tension of solvents(d) (Wang et al. 2009).

Figure 6(a) show the TEM image and the Raman spectrum of graphene flakes prepared in γ-butyrolactone by the liquid-phase exfoliation technique. In addition to the clear TEM graph of the single-layer graphene flakes, the invisible D peak, as well as the clear G line and characterized 2D band, in the Raman spectrum witness the existence of defect-free monolayer and few-layer graphenes. We recently demonstrated that the liquid-phase exfoliated graphene dispersions exhibit broadband OL for ns pulses at 532 and 1064 nm, as shown in Fig. 6(b) (Wang et al. 2009). NLS, originating from the thermally induced solvent bubbles and microplasmas, is responsible for this nonlinear behaviour. The surface tension of the solvents has a strong influence on the OL performance of the graphene dispersions. As shown in Fig. 6(c), it is clear seen that the lower the surface tension, the smaller the limiting threshold and the larger the scattered intensity. We established a simple model to estimate the radius of the gas bubbles as a function of the surface tension of the dispersant. The result in Fig. 6(d) reveals that the lower surface tension results in the larger bubble size, hence more effective scattering and OL. In addition, the graphene flakes exhibit a similar OL response to that of $C_{60}$ and SWNTs.

Zhou et al. prepared a stable graphene solution by reducing GO using a simple and clean hydrothermal dehydration method, which can effectively remove oxygen-containing groups in GO and restore the aromatic rings (Zhou et al. 2009). The NLO properties of the reduced GO were measured by adsorbing the graphene on the end of an optical fiber, which guides a 1560 nm cw or 5 ns pulses laser beam for irradiation. The graphene exhibits a tunable NLA as well as OL response for the NIR light by changing the preparation conditions, i.e., temperature and pressure, and hence the oxygen functional groups and structural defects in graphene, which was confirmed by XPS, NMR and Raman spectroscopy.

The NLO properties of GO were studied by Liu et al. (Liu et al. 2009). Synthesized using the modified Hummers method, the GO was dispersed in DMF for the linear optical and NLO characterizations. UV-Vis spectrum of the GO dispersions shows an absorption peak at 268 nm, followed by a monotonously decreasing towards long wavelength region. Individual GO sheets were observed in AFM graph. The pulse open aperture Z-scan study verified that the RSA and TPA are mainly responsible for the NLO response of the GO solutions under ns and ps pulses at 532 nm, respectively. However, the contribution from NLS was not reported in Liu's paper. Feng et al. investigated the NLO and OL properties of a range of graphene derivatives, namely, graphene nanosheets, GO nanosheets, graphene nanoribbons and GO nanoribbons (Feng et al. 2010). Broadband NLO responses at 532 and 1064 nm were demonstrated in these graphene derivatives. Whereas the four derivatives exhibit different OL behavior, the NLS dominates the NLO response at 1064 nm while both the NLS and NLA contribute at 532 nm. Overall, the reduced graphenes possess better OL performance than the corresponding GO precursors due to the increased conjugation and crystallinity. The similar phenomenon was observed by Zhao et al., who found that the limiting response of graphene nanosheets is better than that of the GO nanosheets owing to the extended π conjugation in graphene (Zhao et al. 2010). In addition to the solvent dependent limiting properties studied, broadband limiting effect was realized as well using graphene nanosheets, which exhibit promising limiting at 532, 730, 800 and 1300 nm.

As with CNTs, the demonstration of graphene for OL renders graphene and related materials as a new class of nanomaterial for photonic and optoelectronic nanodevices (Bonaccorso et al. 2010). In the same way that nanotubes serve not only as nonlinear scatters but also as host material for functional counterparts, which we introduce below, this unique 2D nanomaterial could be a promising host for an optical limiter as well as for other photonic devices. Benefiting from the rich oxygen-containing groups, such as carboxyl and carbonyl groups on the edge and hydroxyl and epoxy groups on the basal plane, GO sheets can be decorated readily with a range of functional organic and inorganic materials by covalent or noncovalent combination, forming diverse nanohybrids with certain function (Loh et al. 2010).

### 3.2 Organic molecule functionalized graphene composites

For the NLO and OL applications, Xu et al. synthesized the first graphene hybrid by functionalizing with a metal-free porphyrin - TPP-$NH_2$. As shown in Fig. 7, the soluble graphene nanohybrid exhibits an improved OL performance compared with $C_{60}$, GO, TPP-NH2 and the mixture of the TPP-NH2 and GO (Xu et al. 2009). A more detailed NLO study reveals that the combination of multiple nonlinear mechanisms, i.e. RSA, TPA, NLS, as well as photo-induced electron transfer results in the superior OL performance of the nanohybrid (Liu et al. 2009). The similar accumulation effect resulting in improved OL was confirmed in oligothiophene-graphene (Liu et al. 2009; Zhang et al. 2009) and fullerene-graphene (Liu et al. 2009; Zhang et al. 2009) nanohybrid systems as well. Very recently, the NLO properties of covalently linked graphene-metal porphyrins composite materials, namely, graphene-zinc porphyrin and graphene-copper porphyrins, were reported by Krishna et al. (Krishna et al. 2011). Effective combination of the different OL mechanisms, say, NLA, TPA, NLS and energy transfer in the graphene-porphyrin composites results in the improved OL effect for ns pulses at 532 nm. In the hybrid system, the existence of NLS, arising from the graphene moiety, can largely increase the damage threshold of the nano-composites. An energy transfer model based on the graphene-porphyrin hybrids was developed and verified that the energy transfer from porphyrin to graphene enhances the TPA of the system.

The role of energy transfer in the graphene based NLO materials was investigated by Mamidala et al., who blended the electron acceptor GO with positively charged porphyrin and negatively charged porphyrin, respectively (Mamidala et al. 2010). The NLO response of the positively charged porphyrin-GO system is much larger than that of the negatively charged porphyrin-GO system, confirming the important role of the energy transfer in such donor-acceptor complexes. While NLS dominates the OL effect, the energy transfer facilitates the deactivation of the hybrids, resulting in energy dissipation via the non-radiative decay and hence the effective heat accumulation in the hybrids or heat transfer from GO to the adjacent solvent. More pronounced energy transfer effect was seen in the porphyrin-Au nanoparticle complex, probably due to the better electron accepting ability of Au in comparison with the GO.

The analogous energy/electron transfer enhanced NLS was observed from a GO-dye ionic complex ($PNP^+GO^-$) (Balapanuru et al. 2010). Compared with the pristine GO and the dye PNPB, the charge-transfer composite exhibits much larger light scattering signal as well as nonlinear transmission and OL for ns pulses at both 532 and 1064 nm. The organic dye can effectively absorb the incident laser energy and transfer to the GO, resulting in the ionization of the GO or further transfer to solvent, forming microplasmas or vapor bubbles for NLS. From the above works, it should be pointed out that the energy transfer effect may inspire deeply the design and synthesis of the new OL hybrid materials.

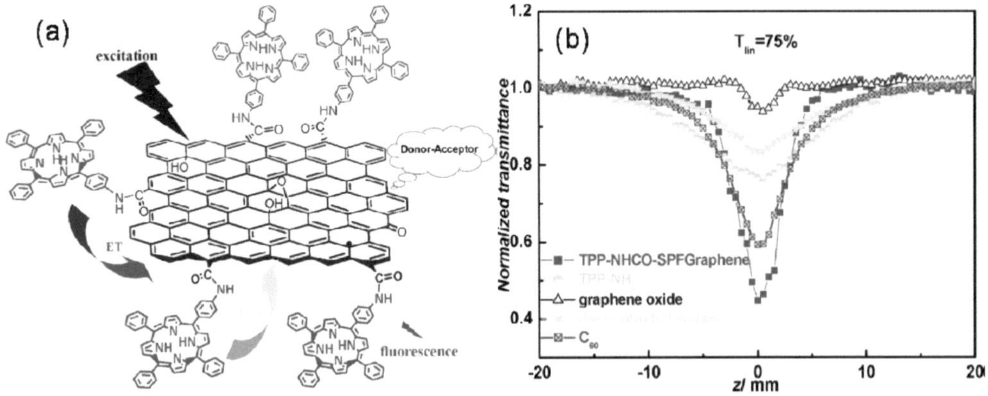


Fig. 7. The structure of the $TPP-NH_2$ functionalized GO (a) and the NLO response of the $TPP-NH_2$-GO compared with $C_{60}$, GO, TPP-NH2 and the mixture of the TPP-NH2 and GO (b) (Xu et al. 2009).

Very recently, we synthesized a soluble GO covalently functionalized with zinc phthalocyanine (PcZn), by an amidation reaction (Zhu et al. 2011). As shown in Fig. 8(a), the formation of an amido bond between PcZn and GO was confirmed by X-ray photoelectron and Fourier transform infrared spectroscopy. Fig. 8(b) presents the OL behavior of the GO-PcZn, GO and PcZn. It can be clearly seen that at the same level of linear transmission, GO-PcZn dispersions present much better OL performance than both GO and PcZn. As a result of the covalent link between GO and PcZn, The enhanced OL response at 532 nm can be attributed to the effective combination of the different NLO mechanisms, i.e., RSA of PcZn, and NLS and TPA of GO. It is likely that the significant scattering signal from the pure PcZn solution results from the formation of PcZn nanoparticles, as reported in []. Although PcZn did not make any significant contribution to the OL at 1064 nm [], it is surprising that the GO-PcZn dispersions have much greater OL response than GO. Coincidently, as shown in Fig. 8(b), the scattered curve from the GO-PcZn dispersions is steeper than that from GO as well. Whereas the origin of such large improvement of the OL at 1064 nm is not clear yet, it is possible that the energy transfer plays some role for the enhanced OL. After all, the GO-PcZn hybrid material has much better broadband NLO and OL performance than the GO alone.

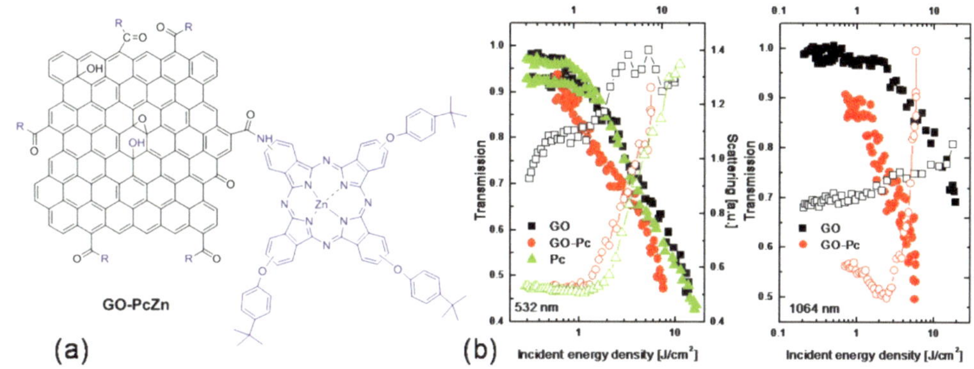


Fig. 8. The structure of the GO-PcZn composite (a) and the OL response of the GO-PcZn (b) (Zhu et al. 2011).

### 3.3 Polymer functionalized graphene composites

As mentioned above, graphene is insoluble in many organic solvents. To obtain solution-processed graphene polymer composites, the thermally-reduced graphene oxides (RGOs) were functionalized with poly(N-vinylcarbazole) (PVK) through generation of anions along the PVK backbone by using sodium hydride, followed by subsequent nucleophilic addition of these anionic species into the π-conjugated structure of the RGO platelets (Li et al. 2011). The structure of the RGO-PVK is depicted in Fig. 9(a). The wt% of RGO in the resulting polymer was estimated as 11.21%. Sonicated for 10 min in THF, the RGO-PVK dispersions are stable for at least one month (see Fig. 9(b)). Typical open aperture Z-scan results are depicted in Figs. 9(c) and 9(d). In contrast to PVK, which does not show any OL effect, the resulting hybrid material RGO-PVK displayed very good broadband NLO and OL responses at 532 and 1064 nm due to the effective combination of different NLO mechanisms, say, NLS and TPA.

Midya et al. synthesized a polymer functionalized RGO composite. The polymer used to covalently link with RGO is based on fluorene-thiophene-benzothiadazole as a donor-spacer-acceptor triad (Midya et al. 2010). With the good solubility in a range of common used organic solvents, the composite solution exhibits excellent OL performance for 532 nm ns pulses. With the help of the donor-acceptor electron transfer structure, the polymer-RGO hybrids show more effective NLS and hence OL than that of carbon nanotubes, RGO, or the polymer alone. However, the TPA from the polymer triads of the hybrids cannot be ruled out.

Aiming to the solid state NLO devices, Zhao et al. studied the OL response of graphene and GO nanosheets in a polymer gel matrix polyvinyl alcohol (PVA) (Zhao et al. 2010). The graphene-PVA composites exhibit a transparent and solid-like structure and possess remarkable OL effect for ns pulses at 532 nm. Operated at 10 Hz pulses, the graphene-PVA matrix emerge bleaching and degradation of the limiting performance after the first a few shots. This issue can be fixed by melting the PVA at 60-80 °C to rehomogenize the graphene in gel.

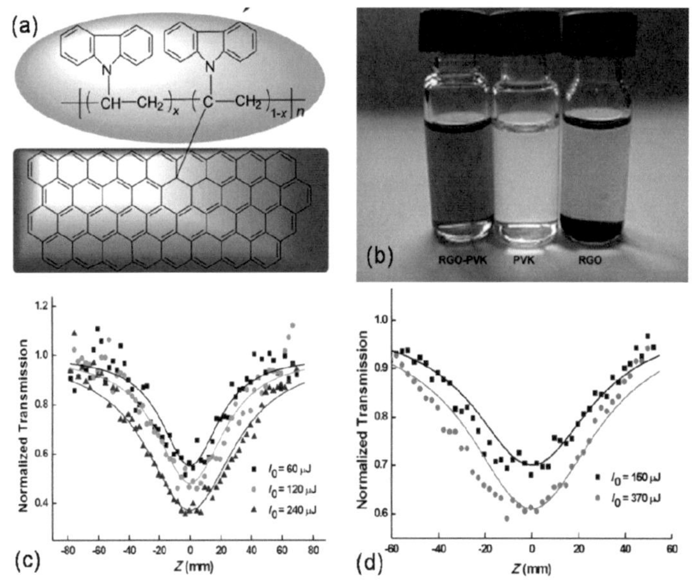


Fig. 9. The structure (a) and the solubility (b) of the RGO-PVK composite. The NLO responses of the RGO-PVK at 532 nm (c) and 1064 nm (d) (Li et al. 2011).

### 3.4 Nanostructure functionalized graphene composites

The linking of inorganic nanostructures on graphene nanosheets can result in the breakage of the electronic and molecular structures and the extended π conjugation of the graphene, and hence lower the device performance. Recently, Feng et al. developed a facile approach to preserve the lossless formation of graphene composite, in which the graphene was decorated with CdS quantum dots (QDs) by using benzyl mercaptan (BM) as the interlinker (see Fig. 10(a)) (Feng et al. 2010). TEM image reveals that the ~3 nm diameter CdS QDs are distributed uniformly on the surface of graphene nanosheets. As shown in Fig. 10(b), the CdS-graphene composite possesses outstanding broadband OL properties, mainly due to

NLS and FCA, for 532 and 1064 nm ns pulses. However, the energy transfer from the QDs to graphene cannot be ruled out. In addition, a $Fe_3O_4$ nanoparticles functionalized GO composite for OL was reported by Zhang et al. (Zhang et al. 2010).

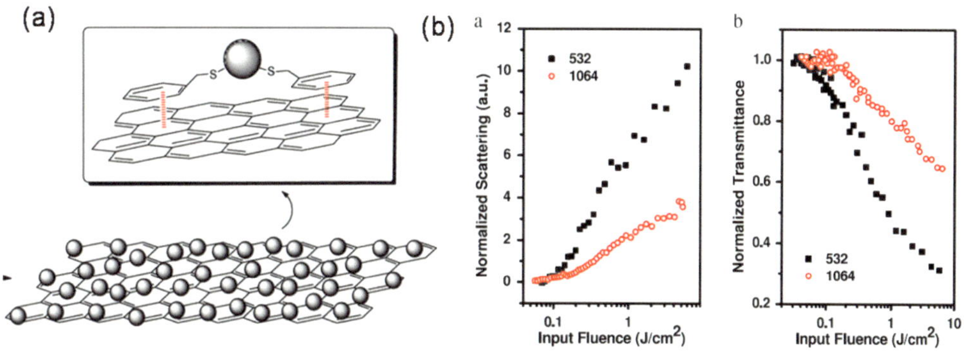


Fig. 10. The structure (a) and broadband OL (b) of the CdS-graphene composite (Feng et al. 2010).

## 4. Carbon nanotube composites

As 1D nanostructured materials, CNTs have attractive mechanical, electrical, and thermal properties, which have found many potential applications in the field of nanoscience and nanotechnology. In the past decade, CNTs have been extensively studied as an OL material (Chen et al. 2007; Wang et al. 2009). It is appealing that the nanotubes combine the advantages of the other two allotropes - carbon black has broadband OL and the fullerene acts as a favourable counterpart for functional materials. CNTs exhibit a significant OL effect covering a broad wavelength range from the visible to the NIR. Most importantly, the tailorable chemical properties of CNTs promote the synthesis of versatile nanotube composites by binding functional materials, e.g. metal nanoparticles, organic molecules and polymers.

### 4.1 Carbon nanotubes

Following the investigation of carbon black suspensions for OL, people started to realize that the CNT could be a new class of carbon nanomaterial for OL in 1998. Sun et al. and Chen et al. reported for the first time the OL property of nanotube suspensions (Sun et al. 1998; Chen et al. 1999). The broadband OL response was demonstrated using ns laser pulses and NLS was proposed as the primary mechanism for OL. In addition, the wavelength, solvent and bundle size effects were considered in their works. Vivien et al. studied systematically the OL performance, dynamics and mechanism of CNT suspensions by employing a series of experimental methods, e.g. Z-scan, the time-resolved pump-probe technique, white light emission measurement, the nonlinear transmittance experiment and the shadowgraphic imaging technique (Vivien et al. 1999; Vivien et al. 2000; Vivien et al. 2002; Vivien et al. 2002). Solvent bubble growth and the phase transition of CNTs at a range of incident fluences were observed, which confirmed that NLS, arising from solvent bubble and carbon vapour bubble formation, dominates the NLO properties of CNT suspensions. The impact of the incident beam wavelength and pulse duration on the OL performance has been studied as well. As described in subsection 2.1, one can simulate the growth dynamics of these bubbles in suspensions.

CNTs tend to aggregate into large bundles due to the high surface energy, which is a serious obstacle when it comes to real-life applications. People have found that CNTs can exist stably as individual nanotubes or small bundles in a range of amide solvents for reasonable periods of time. A typical example is the demonstration of large-scale debundling of single-walled nanotubes (SWNTs) by diluting nanotube dispersions with the solvent Nmethyl-2-pyrrolidinone (NMP) (Giordani et al. 2006). Experimental and theoretical analyses reveal that the surface energies of NMP and some other solvents, i.e. N,N-dimethylacetamide (DMA) and N,N-dimethylformamide (DMF) match very well with that of the nanotube. This results in a minimal energy cost to overcome the van der Waals forces between two nanotubes, and hence the effective debundling (Coleman 2009).

In recent years, we carried out a series of fundamental research on the OL mechanism, performance and its influence factor of the SWNT dispersions. The NLO properties of individual nanotubes were investigated in NMP, where the population of individual nanotubes was observed to increase as the concentration is decreased, with up to ~70% of all dispersed objects being individual nanotubes at a concentration of $4.0\times10^{-3}$ mg ml$^{-1}$ (Wang et al. 2008). AFM measurements reveal that the root-mean-square diameter of nanotubes decreases to less than 2 nm at $8.0\times10^{-3}$ mg ml$^{-1}$ before saturating at this level. Figure 11(a) shows the linear and NLO coefficients, deduced by open aperture Z-scan, as functions of the concentration of the SWNT dispersions in NMP. As the concentration of SWNTs is increased, the nonlinear extinction and OL effects improve significantly, while the limiting thresholds decrease gradually. Even with smaller sizes, the individual nanotubes still exhibit superior OL performance for 532 nm ns pulses than phthalocyanine nanoparticles and $Mo_6S_{4.5}I_{4.5}$ nanowires. The inset of Fig. 11(a) shows the difference between NLS-dominated nanotubes and RSA-dominated phthalocyanines. The nonlinear transmission of the SWNT dispersions has a distinct discontinuity, corresponding to a limiting threshold. The transmission is roughly constant when the energy fluence is below the threshold. When the incident fluence exceeds the threshold, the transmission decreases significantly. The limiting threshold implies that the nanotubes transfer enough heat energy to the surrounding solvent to cause the solvent to vaporize and grow to the critical size, in order to effectively scatter the incident beam. In contrast, the transmission of the phthalocyanines decreases with increasing incident energy. There is no evidence of the limiting threshold for phthalocyanines in the figure. Moreover, improved OL performance was found from the same nanotubes in DMF (Wang et al. 2008). As shown in Fig. 11(b), the DMF dispersions show superior nonlinear extinction effects and lower limiting thresholds. The static light scattering results in the inset of Fig. 11(b) proved that the DMF dispersions have the larger average bundle size, which in combination with the lower boiling point and surface tension of DMF, results in the superior optical limiting performance.

On the other hand, we showed that the OL performances of SWNT dispersions in NMP were enhanced significantly by blending a range of organic solvents or by increasing the temperature of the dispersions up to 100 °C (see Fig. 11(c) and 11(d)). While both nanotube bundle size and various solvent parameters have an influence on the OL responses, we verified experimentally that the surface tension of the solvent plays a more important role than the viscosity or boiling point; the appropriate solvent properties contribute to the NLS dominated OL phenomenon more than the bundle size (Wang et al. 2010). As the appropriate thermodynamic properties of the solvents are much more important for improving the OL performance, the solvent parameters were controlled by either changing the temperature of the dispersions or blending a secondary solvent (Wang et al. 2010). While

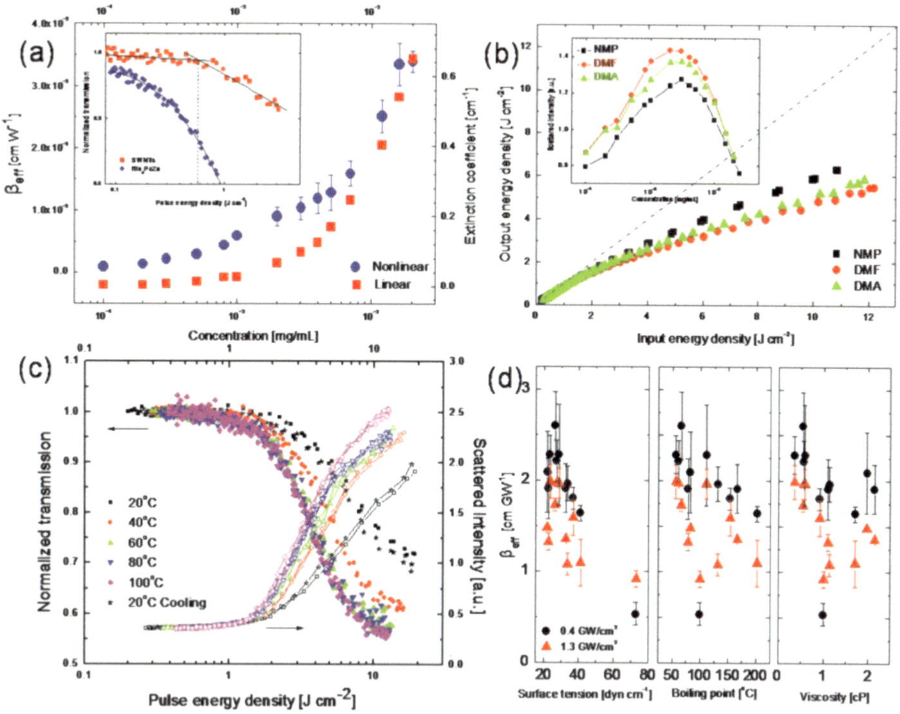


Fig. 11. The linear and NLO coefficients of the individual nanotube dispersions in NMP (a) (Wang et al. 2008). The OL of the nanotubes in different solvents (b) (Wang et al. 2008). The OL of the nanotube dispersions as a function of temperature (c) (Wang et al. 2010). Nonlinear extinction coefficient of the nanotube dispersions as a function of surface tension, boiling point and viscosity of the binary solvent mixtures (d) (Wang et al. 2010).

| Effects on optical limiting | | Optical limiting response |
|---|---|---|
| Structure of CNTs | SWNT, MWNT | SWNT ≈ MWNT |
| | Bundle diameter | The larger > The smaller |
| | Length | The longer ≥ The shorter |
| | Aspect ratio | The larger > The smaller |
| | Number density | The denser > The sparser |
| Physical properties of dispersant | Boiling point | The higher < The lower |
| | Surface tension | The larger < The smaller |
| | Viscosity | The higher < The lower |
| Laser source | Wavelength | The longer < The shorter |
| | Pulse duration | The longer > The shorter |
| | Repetition rate | The higher < The lower |

Table 1. Summary of the factors that influence the OL responses of CNT dispersions. The signs of inequality indicate the contrast of OL responses.

the OL performance can be varied freely by increasing or decreasing the temperature from room temperature to 100 °C, the reduction of temperature below the freezing point of NMP and then down as far as -80 °C has little influence on the limiting performance. As a result of adding a small amount of organic solvent into the NMP dispersions, the NLO responses were enhanced significantly due to the reduction of surface tension and other parameters, as shown in Fig. 11(d). By contrast, the addition of water leads to a decrease in the optical limiting response. Nanotube dispersions in water/surfactant exhibit a similar limiting performance to the nanotubes in NMP. Our results reveal that the OL performance of the nanotube dispersions can be engineered by adjusting the solvent properties. Because the CNT dispersions are typical of the thermally induced light scattering dominated OL materials, we believe the conclusions fit not only the nanotubes but also other nanomaterials with the similar limiting mechanism.

### 4.2 Organic molecule functionalized nanotube composites

Most of the OL studies on pristine nanotubes concentrate on the physical mechanism and its influencing factors as summarized in Table 1. Although pristine nanotubes possess broadband limiting effects, the nanotubes alone could not satisfy all requirements for laser protection. The development of complex CNT composites is expected to enable practical OL devices. Whereas a lot of organic dyes exhibit NLA at certain wavelength bands, the optical limiting effect in nanotubes covers a broad wavelength range from the visible to the NIR. Nonlinear absorbers, i.e. phthalocyanines, have a quick response time in the ps regime, while nanotubes generally respond at best in the ns regime. Merging the complementary temporal and spatial nonlinear characteristics of NLA compounds and nanotubes has resulted in the development of nonlinear absorber-CNT hybrids by covalent or noncovalent link.

A TPA chromophore, Stilbene-3, and a SWNT mixture was prepared by Izard et al. (Izard et al. 2004). The cumulative OL effect was observed when the two moieties have comparable OL responses. If one moiety dominates, the whole limiting performance is close to that of the moiety. The composites, which exhibit both NLS and TPA, are expected to work in a broad temporal and spectral range. Webster et al. blended a RSA dye, 1,10,3,3,30,30-hexamethylindotricarbocyanine iodide (HITCI), with functionalized nitrogen-doped multi-walled nanotubes (MWNTs) to enhance the nonlinear transmittance of the whole system (Webster et al. 2005). The blended composite exhibits an improvement in the OL performance in comparison with the two individual materials. At the low intensity regime, the nonlinear response is dominated by the RSA dye HITCI before the NLS becomes significant. After the onset of NLS at the high intensity regime, nanotubes dominate the optical limiting. Blau and co-workers demonstrated the superior optical limiting effect from a noncovalently linked tetraphenylporphyrin–nanotube composite (Ni Mhuircheartaigh et al. 2006). The transmission electron microscope (TEM) image in Fig. 12(a) shows clearly the adhesion of porphyrin molecules to the outside of double-walled nanotubes by van der Waals interaction. The photo-induced electron transfer effects from covalently or noncovalently linked RSA dye-nanotube composites have been widely studied, which may help to improve the NLO response of such complex material systems. Recently, we reported the linear and NLO properties of a range of phthalocyanine-nanotube blends (see the inset of Fig. 12(b)) (Wang et al. 2008). The addition of nanotubes did not change the linear UV-visible absorption characteristics of phthalocyanines but resulted in significant fluorescence quenching. Due to the solvent effect, the phthalocyanine-nanotube composites in DMF exhibit a larger nonlinear response than those in NMP. As shown in Fig. 12(b), the blends enhanced the OL performance in the higher energy density region when compared to the phthalocyanine solutions. In agreement with Webster et al.'s result, phthalocyanines influenced the OL effect in the lower energy density region, while the nanotubes played a more critical role in the attenuation of incident laser light in the higher energy density region. Overall, the OL behavior of the composites was increased with further addition of nanotubes.

Apart from the noncovalently-linked dye-nanotube composites, de la Torre et al. described the synthesis and characteristics of covalently functionalized single-walled nanotubes with metallophthalocyanines (de la Torre et al. 2003). Liu et al. synthesized covalently linked porphyrin-SWNT composites (Liu et al. 2008). The structures of the porphyrin-functionalized nanotubes are illustrated Fig. 12(c). Compared with $C_{60}$, individual nanotubes and porphyrins, the composite solutions show outstanding optical limiting responses for ns laser pulses at 532 nm. The authors attributed the superior performance to the effective combination of the NLO mechanism and the photo-induced electron transfer between porphyrins and nanotubes.

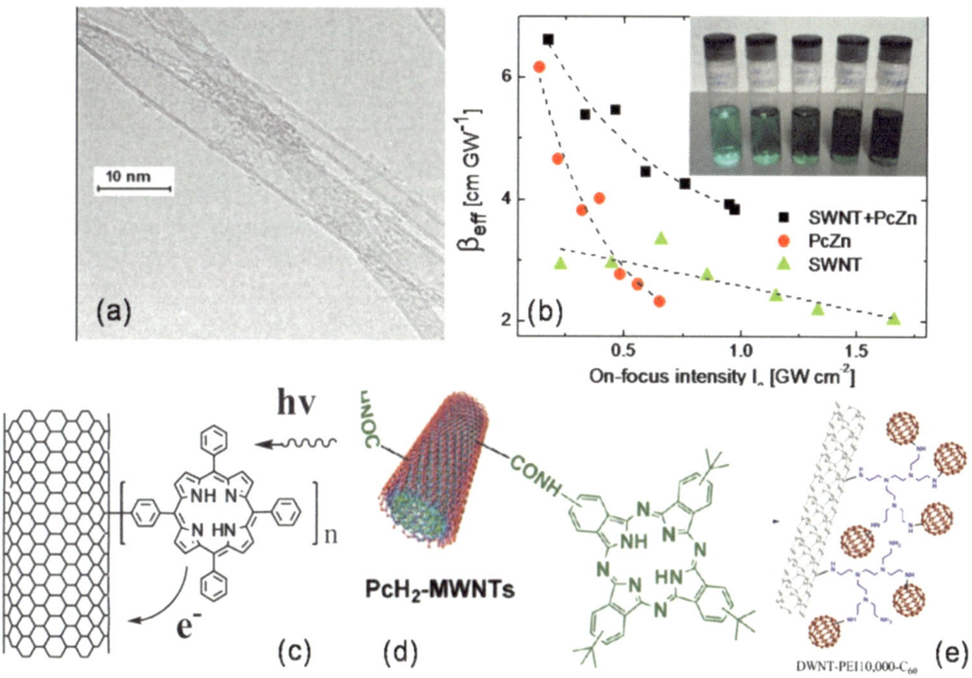


Fig. 12. TEM image showing the adhesion of organic porphyrin molecules to the outside of DWNT (a) (Ni Mhuircheartaigh et al. 2006). The nonlinear extinction coefficient as a function of on-focus intensity for various phthalocyanine–nanotube composites in DMF (b) (Wang et al. 2008). The structures of the porphyrin-SWNT (c) (Liu et al. 2008), $PcH_2$-MWNT (d) (He et al. 2009) and DWNT-$C_{60}$ (e) (Liao et al. 2010).

Chen and his coworkers synthesized an unsymmetrically substituted metal-free phthalocyanine-covalently functionalized MWNT ($PcH_2$-MWNT) hybrid composite, in which the wt % of MWNTs in the resulting product was found to be 35% (He et al. 2009). The molecular structure is given in Fig. 12(d). A considerably quenching of the fluorescence intensity was found in the photoluminescence spectrum of $PcH_2$-MWNTs. This observation suggests a quenching of the singlet excited $PcH_2$ by the covalently linked MWNTs. This material exhibits strong scattering at higher intensities, which evidently comes from the MWNT counterpart. The nonlinear response of $PcH_2$ is due to RSA, while that of $PcH_2$-MWNTs is due to both RSA and NLS, which could be two conflicted mechanisms for OL, giving rise to suppression of the whole nonlinear response of $PcH_2$-MWNTs.

Liao et al. synthesized a double-walled nanotube-fullerene (DWNT-$C_{60}$) hybrid by covalently linking DWNT and $C_{60}$ by amination reaction with polyethylenimine (see Fig. 12(e)) (Liao et al. 2010). The nanohybrid can be dispersed in poly(m-phenylenevinylene-co-2,5-dioctoxy-p-phenylenevinylene) (PmPV) toluene solutions via 20 min sonication treatment. Both the hybrid dispersions and the polymer composites exhibit promising limiting effect, while the former works better due to the solvent effect discussed above. When dispersed in PmPV or chlorobenzene, the nanohybrid is expected to merge complementary temporal and spatial NLO characteristics of fullerene and CNTs, resulting in an enhanced OL. The OL performance of the DWNT-$C_{60}$ hybrids is superior to those of $C_{60}$ and SWNTs at the same level of transmission (~80%). Whereas NLS is an evident mechanism, RSA from $C_{60}$ moieties has significant contribution. Photo-induced charge transfer between the DWNT and $C_{60}$ moieties may also play an important role on the enhanced OL.

### 4.3 Polymer functionalized nanotube composites

As we mentioned above, nanotubes tend to aggregate into large bundles in most inorganic and organic solvents because of their relatively high surface energy, which is a serious obstacle when it comes to real-life applications. It is thus of great interest to design and prepare soluble nanotubes, which allows the easy manufacture of large-area thin film optoelectronic devices by spin coating or screen-printing technologies. Covalently or noncovalently functionalizing the surface of nanotubes by polymers is a simple and low-cost method to produce soluble nanotube and graphene composites.

A breakthrough in exploring the noncovalent interaction of the nanotube and polymer was made by Curran et al. who adopted a conjugated polymer, PmPV (see Fig. 13(a)), to disperse and purify the nanotubes, resulting in property modified nanocomposites (Curran et al. 1998). The coiled polymer conformation allows it to surround the layers of the nanotubes, permitting sufficiently close intermolecular proximity for π-π interaction to occur. The PmPV has a bright yellow color while the PmPV-nanotube composite possesses a deep green color, implying the strong interaction between the polymer chains and the nanotubes. As shown in Fig. 13(b), a clear wrapping effect of individual nanotubes by the PmPV matrix was observed by TEM. PmPV is an appropriate polymer to disperse CNTs while retaining the superior optical response from the nanotubes. O'Flaherty et al. prepared two kinds of polymer-nanotube composite by dispersing nanotubes into PmPV and poly(9,9-di-n-octylfluorenyl-2,7'-diyl) (PFO), respectively (O'Flaherty et al. 2003; O'Flaherty et al. 2003). Both of these composite systems showed an excellent OL effect on ns laser pulses at 532 nm. The strong back and front scattered light signals, with characteristics of Mie scattering, indicate evidence of the NLS origin of OL.

For soluble nanotube polymer composites, the preparation procedure usually involves mixing nanotube dispersions with solutions of the polymer and then evaporating the solvents in a controlled way. The solution mixing approach is limited to polymers that freely dissolve in common solvents. An alternative method for producing a homogeneous dispersion of nanotubes is to incorporate nanotubes into thermoplastic polymers at the temperature higher than the melting point of these polymers or, to in situ polymerize the suitable monomers, such as styrene, aniline, phenylacetylene, and other monomers in the presence of nanotubes. Hereinafter, we introduce several covalently functionalized nanotube polymer composites for optical limiting.

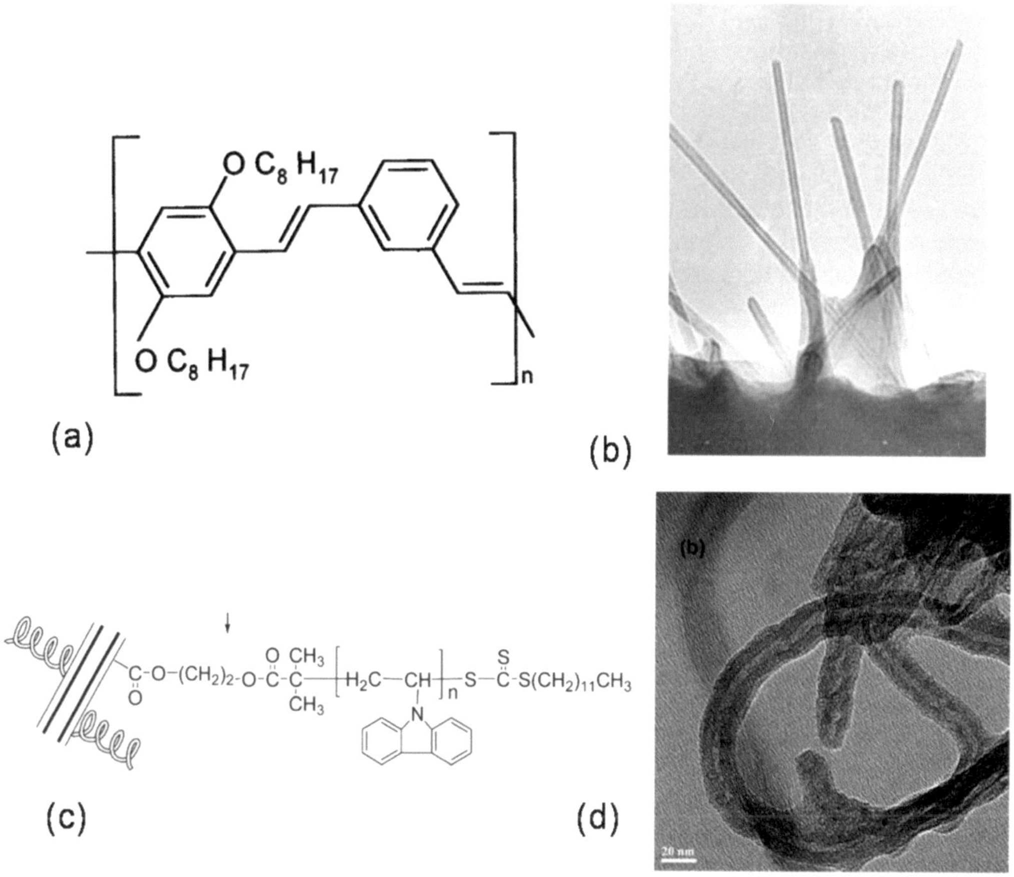


Fig. 13. The molecular structure of PmPV (a) and TEM image of nanotubes in PmPV (b) (Curran et al. 1998). The structure (c) and TEM image (d) of the MWNT-PVK hybrid (Zhang et al. 2010).

A series of poly(N-vinylcarbazole)-grafted MWNT (MWNT-PVK) hybrid materials were synthesized in the presence of S-1-Dodecyl-S'-(α, α'-dimethyl-α''-acetic acid) trithiocarbonate (DDAT)-covalently functionalized MWNTs (MWNT-DDAT) as reversible addition-fragmentation chain transfer (RAFT) agent (Zhang et al. 2010). In that work, we used a new RAFT agent, DDAT-covalently functionalized MWNTs, first, and then grafted the PVK chains onto the surface of MWNTs to produce the soluble MWNT-PVK hybrid materials by RAFT polymerization, as shown in Fig. 13(c). High-resolution TEM graphs reveal that the MWNTs were coated by a layer of organic species whose thickness depends on the molecular size and the quantity covalently attached onto the surface of MWNTs. The average diameter of MWNT-COOH is about 14 nm, while that of MWNT-PVK increases to 23-25 nm, as shown in Fig. 13(d). Incorporation of the PVK moieties onto the nanotube surface can considerably improve the solubility and processability of the nanotubes. For all MWNT-PVK hybrid materials, they are soluble in some common organic solvents such as toluene, THF, chloroform, DMF and others. At the same level of linear transmission, the MWNT-PVK with 79.2% PVK moieties in the material structure possesses best optical limiting performance for the ns pulses at 532 nm in comparison with the other MWNT-PVK composites, MWNTs and $C_{60}$. Light scattering, originating from the thermal-induced microplasmas and/or microbubbles, is responsible for the optical limiting. Subsequently, a new PVK-covalently grafted SWNT (SWNT-PVK) hybrid material was synthesized via an in situ anionic polymerization reaction of N-vinylcarbazole and the negatively charged SWNTs (Li et al. 2011). Same as the MWNT-PVK, appearance of the PVK moieties onto the surface of nanotubes significantly improves the solubility and processability of the SWNTs. At the same level of linear transmission, the SWNT-PVK dispersions show better optical limiting performance than the pristine SWNT dispersions. Micro-plasma and/or micro-bubble induced NLS is considered as the main mechanism for the OL.

In addition to the non-conjugated polymer, i.e., PVK, we also adopt conjugated polymer to functionalized covalently nanotubes. A new conjugated polymer PCBF with pendent amino groups in the polymer side chains was synthesized by the Suzuki coupling reaction (Niu et al. 2011). Then, this polymer was used to react with MWNTs with surface-bonded acryl chloride moieties to give a soluble donor-acceptor type MWNT-PCBF hybrid material, in which PCBF was chosen as electron donor, whereas the MWNT itself may serve as the electron acceptor. The TEM graph implies that the average thickness of PCBF covalently grafted onto the MWNTs is around 10.4 nm. After the low power sonication treatment, the MWNT-PCBF in tetrahydrofuran (THF) is stable for at least one month at a concentration as high as 5 g/L. It can be clearly seen that MWNT-PCBF exhibited excellent optical limiting performance. The MWNT-PCBF manifests the remarkable broadband OL with a comparable limiting performance for both 532 and 1064 nm pulses. The strong scattering signals indicate that the thermally induced NLS is responsible for the OL.

### 4.4 Nanostructure functionalized nanotube composites

The optical properties of CNTs can be modified by coating functional composites. Chin et al. successfully improved the transmission of nanotubes in the near UV region by coating silicon carbide or silicon nitride on the surface (Chin et al. 2004). The high transmission nanotube composites incorporated with good OL performances are appropriate for the development of laser protection devices. The same authors further employed polycrystalline Au or Ag nanoparticles as coatings deposited on the outside of multi-walled nanotubes (Chin et al. 2005). Broadband OL effects for ns pulses at 532 nm and 1064 nm were demonstrated in the functionalized nanotube composites. Enhanced limiting performance for 532 nm pulses was observed from the composites when compared with pristine nanotubes. The surface plasmon absorption (SPA) of Au and Ag coatings at 532 nm is attributed to the enhancement of the NLS as well as the optical limiting effect in the nanotube composites. However, polycrystalline Ni- and Ti-coated nanotubes did not show significant improvement for optical limiting since Ni and Ti nanoparticles do not exhibit SPA around 532 nm. Moreover, it should be mentioned that the CNT and carbon nanoparticle mixtures were studied as a class of optical limiting nanomaterial as well (O'Flaherty et al. 2003).

Recently, Zhan and her coworkers synthesized a MWNT composite by functionalizing the sidewalls of nanotubes with CdS QDs using a two-step approach, with in situ polymerized thiophene as interlinker (Feng et al. 2010). TEM, XRD and TGA analyses verified that the thiophene coating formed on the surface of the MWCNTs by means of π electron interactions and the subsequent coupling of CdS QDs. As a consequence of interparticle coupling and the low percentage of CdS in the MWNT-PTh-CdS, the absorption of CdS becomes weaker and broader. Strong PL quenching of the CdS was observed after bonding to the nanotubes due to electron/energy transfer from the excited CdS QDs to the nanotubes. The MWNT-PTh-CdS exhibit a remarkable OL enhancement in comparison with the pristine MWNTs, especially at 1064 nm, owing to the presence of CdS QDs linked by conducting PTh to the MWCNTs and the subsequent electron/energy transfer facilitated NLS.

The same authors further prepared a series of functionalized MWNT composites by coating different conducting, semiconducting, and insulating materials, i.e., crystalline Au nanoparticles, $TiO_2$ nanoclusters, and amorphous $SiO_2$ nanoshells, on the sidewalls of the nanotubes (Zheng et al. 2010). The synthesis employed a combination of self-assembly and sol-gel technique. The structures and the TEM images of the three composites are illustrated in Fig. 14. The composites with Au-, $TiO_2$-, and $SiO_2$-coatings exhibit respectively the superior, equivalent, and inferior OL performance in comparison with the pristine nanotubes. As discussed above, the distinct OL response is likely due to the different electron/energy transfer strength, which largely influences the NLS process. In the three coatings, the conducting Au nanoparticles show the most effective electron transfer to the metallic nanotubes, resulting in the best NLS and OL.

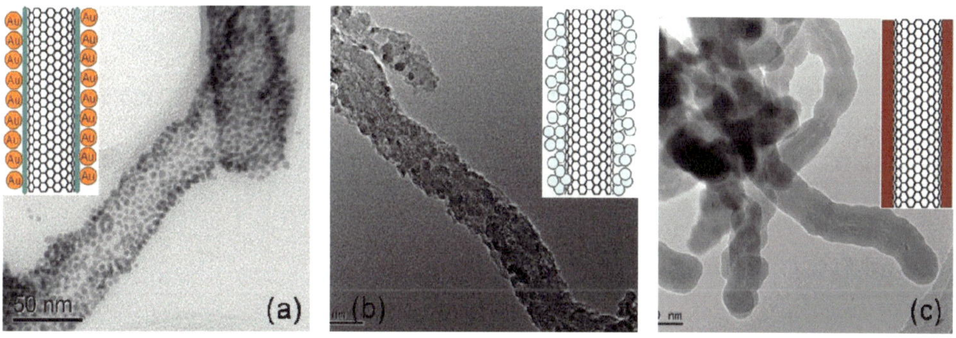


Fig. 14. The structures and TEM images of MWNTs functionalized with crystalline Au nanoparticles (a), $TiO_2$ nanoclusters (b) and amorphous $SiO_2$ nanoshells (c) (Zheng et al. 2010).

## 5. Summary and remarks

1. It is seen from the literature statistics in Fig. 15 using the *ISI Web of Science* that the development of OL keeps vigorously in recent decade. Especially, the involvement of nanotube and graphene invigorates this tendency. As we mentioned above, the excellent chemical activity of graphene and nanotubes provides a broad platform for various functional counterparts, forming multi-component, multi-functional hybrid composites with wider spatial and temporal responses for OL.

2. The derivatives of graphene and nanotube represent a key branch in the field of OL. In most of such nanohybrids, it is being attached importance to the electron/energy transfer from functional moiety to graphene or nanotube, which is considered playing an influential role on improving OL performance.
3. While the chemical synthesis and characterization of the OL materials develops rapidly, the corresponding NLO testing technique and theoretical analysis seems have reached a plateau. Merely a few papers report new measurement method or theoretical modelling for the OL materials (Belousova et al. 2003; Belousova et al. 2004; Venkatram et al. 2005; Gu et al. 2008; Rayfield et al. 2010). It is short of the NLO theory specific to the multi-component, multi-mechanism nanohybrids, which is probably the bottleneck restricts a ultimately improvement of the OL performance. The research of OL briefly consists of materials, mechanisms, the design of OL device. Aiming to industry capable OL devices, a balanced development of the three aspects is undoubtedly urgent.

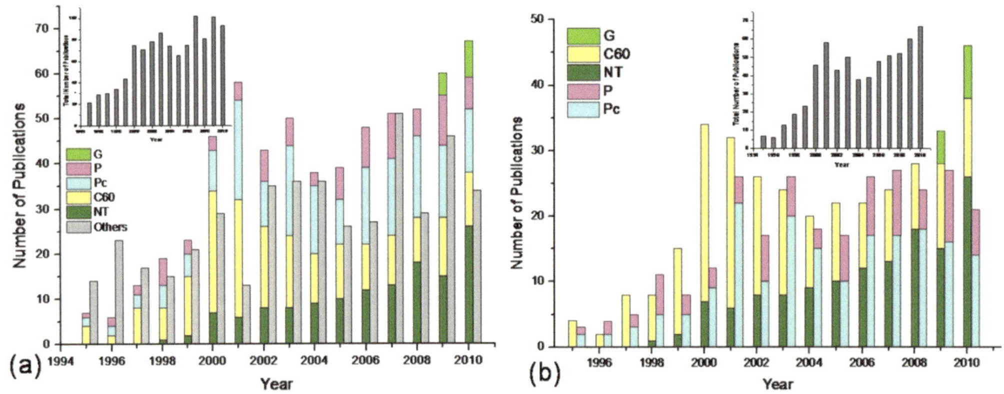


Fig. 15. A literature statistics using the *ISI Web of Science* restricted to the most plausible candidates for OL, namely, CNTs, graphene, $C_{60}$, phthalocyanines and porphyrins, shows an increasing trend in publications on CNTs, graphene, and their derivatives compared with other materials. Only original articles were included, review articles were excluded.

## 6. Acknowledgments

This work was supported by the starting grant of the 100-Talent Program of SIOM, Chinese Academy of Sciences (1108221-JR0) and the National Natural Science Foundation of China (50802103 and 51072207). Y.C. thanks for the financial supports of the National Natural Science Foundation of China (20676034 and 20876046), the Ministry of Education of China (309013), the Fundamental Research Funds for the Central Universities, the Shanghai Municipal Educational Commission for the Shuguang fellowship (08GG10) and the Shanghai Eastern Scholarship.

## 7. References

Avouris, P., M. Freitag and V. Perebeinos (2008). Carbon-nanotube photonics and optoelectronics. *Nature Photonics*, Vol. 2, No. 6, pp. 341-350.

Balapanuru, J., J. X. Yang, S. Xiao, Q. L. Bao, M. Jahan, L. Polavarapu, J. Wei, Q. H. Xu and K. P. Loh (2010). A Graphene Oxide-Organic Dye Ionic Complex with DNA-Sensing and Optical-Limiting Properties. *Angewandte Chemie-International Edition*, Vol. 49, No. 37, pp. 6549-6553.

Bao, Q., H. Zhang, Y. Wang, Z. Ni, Y. Yan, Z. X. Shen, K. P. Loh and D. Y. Tang (2009). Atomic-Layer Graphene as a Saturable Absorber for Ultrafast Pulsed Lasers. *Advanced Functional Materials*, Vol. 19, No.19, pp. 3077–3083.

Belousova, I. M., N. G. Mironova, A. G. Scobelev and M. S. Yur'ev (2004). The investigation of nonlinear optical limiting by aqueous suspensions of carbon nanoparticles. *Optics Communications*, Vol. 235, No. 4-6, pp. 445-452.

Belousova, I. M., N. G. Mironova and M. S. Yur'ev (2003). Theoretical investigation of nonlinear limiting of laser radiation power by suspensions of carbon particles. *Optics and Spectroscopy*, Vol. 94, No. 1, pp. 86-91.

Bergin, S. D., V. Nicolosi, P. V. Streich, S. Giordani, Z. Y. Sun, A. H. Windle, P. Ryan, N. P. Niraj, Z. T. Wang, L. Carpenter, W. J. Blau, J. Boland, J. P. Hamilton and J. N. Coleman (2008). Towards Solutions of Single-Walled Carbon Nanotubes in Common Solvents. *Advanced Materials*, Vol. 20, No. 10, pp. 1876-1881.

Blau, W., H. Byrne, W. M. Dennis and J. M. Kelly (1985). Reverse saturable absorption in tetraphenylporphyrins. *Optics Communications*, Vol. 56, No. 1, pp. 25-29.

Boggess, T. F., K. M. Bohnert, K. Mansour, S. C. Moss, I. W. Boyd and A. L. Smirl (1986). SIMULTANEOUS MEASUREMENT OF THE 2-PHOTON COEFFICIENT AND FREE-CARRIER CROSS-SECTION ABOVE THE BANDGAP OF CRYSTALLINE SILICON. *Ieee Journal of Quantum Electronics*, Vol. 22, No. 2, pp. 360-368.

Bonaccorso, F., Z. Sun, T. Hasan and A. C. Ferrari (2010). Graphene photonics and optoelectronics. *Nature Photonics*, Vol. 4, No.9, pp. 611-622.

Bottari, G., G. de la Torre, D. M. Guldi and T. Torres (2010). Covalent and Noncovalent Phthalocyanine-Carbon Nanostructure Systems: Synthesis, Photoinduced Electron Transfer, and Application to Molecular Photovoltaics. *Chemical Reviews*, Vol. 110, No. 11, pp. 6768-6816.

Chen, P., X. Wu, X. Sun, J. Lin, W. Ji and K. L. Tan (1999). Electronic structure and optical limiting behavior of carbon nanotubes. *Physical Review Letters*, Vol. 82, No. 12, pp. 2548-2551.

Chen, Y., Y. Lin, Y. Liu, J. Doyle, N. He, X. D. Zhuang, J. R. Bai and W. J. Blau (2007). Carbon nanotube-based functional materials for optical limiting. *Journal of Nanoscience and Nanotechnology*, Vol. 7, No. 4-5, pp. 1268-1283.

Chin, K. C., A. Gohel, W. Z. Chen, H. I. Elim, W. Ji, G. L. Chong, C. H. Sow and A. T. S. Wee (2005). Gold and silver coated carbon nanotubes: An improved broad-band optical limiter. *Chemical Physics Letters*, Vol. 409, No. 1-3, pp. 85-88.

Chin, K. C., A. Gohel, H. I. Elim, W. Ji, G. L. Chong, K. Y. Lim, C. H. Sow and A. T. S. Wee (2004). Optical limiting properties of amorphous SixNy and SiC coated carbon nanotubes. *Chemical Physics Letters*, Vol. 383, No. 1-2, pp. 72-75.

Coleman, J. N. (2009). Liquid-Phase Exfoliation of Nanotubes and Graphene. *Advanced Functional Materials*, Vol. 19, No. 23, pp. 3680-3695.

Coleman, J. N., M. Lotya, A. O'Neill, S. D. Bergin, P. J. King, U. Khan, K. Young, A. Gaucher, S. De, R. J. Smith, I. V. Shvets, S. K. Arora, G. Stanton, H. Y. Kim, K. Lee, G. T. Kim, G. S. Duesberg, T. Hallam, J. J. Boland, J. J. Wang, J. F. Donegan, J. C. Grunlan, G. Moriarty, A. Shmeliov, R. J. Nicholls, J. M. Perkins, E. M. Grieveson, K. Theuwissen, D. W. McComb, P. D. Nellist and V. Nicolosi (2011). Two-Dimensional

Nanosheets Produced by Liquid Exfoliation of Layered Materials. *Science*, Vol. 331, No. 6017, pp. 568-571.

Curran, S. A., P. M. Ajayan, W. J. Blau, D. L. Carroll, J. N. Coleman, A. B. Dalton, A. P. Davey, A. Drury, B. McCarthy, S. Maier and A. Strevens (1998). A composite from poly(m-phenylenevinylene-co-2,5-dioctoxy-p-phenylenevinylene) and carbon nanotubes: A novel material for molecular optoelectronics. *Advanced Materials*, Vol. 10, No. 14, pp. 1091-1093.

Dawlaty, J. M., S. Shivaraman, M. Chandrashekhar, F. Rana and M. G. Spencer (2008). Measurement of ultrafast carrier dynamics in epitaxial graphene. *Applied Physics Letters*, Vol. 92, No. 4, pp. 042116.

de Heer, W. A., C. Berger, X. S. Wu, P. N. First, E. H. Conrad, X. B. Li, T. B. Li, M. Sprinkle, J. Hass, M. L. Sadowski, M. Potemski and G. Martinez (2007). Epitaxial graphene. *Solid State Communications*, Vol. 143, No. 1-2, pp. 92-100.

de la Torre, G., W. J. Blau and T. Torres (2003). A survey on the functionalization of single-walled nanotubes: The chemical attachment of phthalocyanine moieties. *Nanotechnology*, Vol. 14, No.7, pp. 765-771.

de la Torre, G., P. Vaquez, F. Agullo-Lopez and T. Torres (2004). Role of structural factors in the nonlinear optical properties of phthalocyanines and related compounds. *Chemical Reviews*, Vol. 104, No. 9, pp. 3723-3750.

Dean, J. J. and H. M. v. Driel (2009). Second harmonic generation from graphene and graphitic films. *Applied Physics Letters*, Vol. 95, No. 26, pp. 261910.

Eda, G., Y. Y. Lin, C. Mattevi, H. Yamaguchi, H. A. Chen, I. S. Chen, C. W. Chen and M. Chhowalla (2010). Blue Photoluminescence from Chemically Derived Graphene Oxide. *Advanced Materials*, Vol. 22, No. 4, pp. 505–509.

Feng, M., R. Q. Sun, H. B. Zhan and Y. Chen (2010). Decoration of carbon nanotubes with CdS nanoparticles by polythiophene interlinking for optical limiting enhancement. *Carbon*, Vol. 48, No. 4, pp. 1177-1185.

Feng, M., R. Q. Sun, H. B. Zhan and Y. Chen (2010). Lossless synthesis of graphene nanosheets decorated with tiny cadmium sulfide quantum dots with excellent nonlinear optical properties. *Nanotechnology*, Vol. 21, No. 7, pp. 075601.

Feng, M., H. B. Zhan and Y. Chen (2010). Nonlinear optical and optical limiting properties of graphene families. *Applied Physics Letters*, Vol. 96, No. 3, pp. 033107.

Geim, A. K. and K. S. Novoselov (2007). The rise of graphene. *Nature Materials*, Vol. 6, No. 3, pp. 183-191.

Giordani, S., S. D. Bergin, V. Nicolosi, S. Lebedkin, M. M. Kappes, W. J. Blau and J. N. Coleman (2006). Debundling of single-walled nanotubes by dilution: Observation of large populations of individual nanotubes in amide solvent dispersions. *Journal of Physical Chemistry B*, Vol. 110, No. 32, pp. 15708-15718.

Gu, B., W. Ji, P. S. Patil, S. M. Dharmaprakash and H. T. Wang (2008). Two-photon-induced excited-state absorption: Theory and experiment. *Applied Physics Letters*, Vol. 92, No. 9, pp 091118.

Hasan, T., Z. Sun and A. C. Ferrari (2009). Nanotube–polymer composites for ultrafast photonics. *Advance Materials*, Vol. 21, No. 38-39, pp. 3874-3899.

He, G. S., J. D. Bhawalkar, C. F. Zhao and P. N. Prasad (1995). Optical limiting effect in a two-photon absorption dye doped solid matrix. *Applied Physics Letters*, Vol. 67, No. 17, pp. 2433-2435.

He, G. S., L. S. Tan, Q. D. Zheng and P. N. Prasad (2008). Multiphoton Absorbing Materials: Molecular Designs, Characterizations, and Applications. *Chemical Reviews*, Vol. 108, No. 4, pp. 1245-1330.

He, G. S., K. T. Yong, Q. D. Zheng, Y. Sahoo, A. Baev, A. I. Ryasnyanskiy and P. N. Prasad (2007). Multi-photon excitation properties of CdSe quantum dots solutions and optical limiting behavior in infrared range. *Optics Express*, Vol. 15, No. 20, pp. 12818-12833.

He, N., Y. Chen, J. Bai, J. Wang, W. J. Blau and J. Zhu (2009). Preparation and Optical Limiting Properties of Multiwalled Carbon Nanotubes with π-Conjugated Metal-Free Phthalocyanine Moieties. *Journal of Physical Chemistry C*, Vol. 113, No. 30, pp. 13029-13035.

Hendry, E., P. J. Hale, J. Moger, A. K. Savchenko and S. A. Mikhailov (2010). Coherent Nonlinear Optical Response of Graphene. *Physical Review Letters*, Vol. 105, No. 9, pp. 097401.

Hernandez, Y., V. Nicolosi, M. Lotya, F. M. Blighe, Z. Sun, S. De, I. T. McGovern, B. Holland, M. Byrne, Y. K. Gun'Ko, J. J. Boland, P. Niraj, G. Duesberg, S. Krishnamurthy, R. Goodhue, J. Hutchison, V. Scardaci, A. C. Ferrari and J. N. Coleman (2008). High-yield production of graphene by liquid-phase exfoliation of graphite. *Nature Nanotechnology*, Vol. 3, No. 9, pp. 563-568.

Izard, N., C. Menard, D. Riehl, E. Doris, C. Mioskowski and E. Anglaret (2004). Combination of carbon nanotubes and two-photon absorbers for broadband optical limiting. *Chemical Physics Letters*, Vol. 391, No. 1-3, pp. 124-128.

Krishna, M. B. M., V. P. Kumar, N. Venkatramaiah, R. Venkatesan and D. N. Rao (2011). Nonlinear optical properties of covalently linked graphene-metal porphyrin composite materials. *Applied Physics Letters*, Vol. 98, No. 8, pp. 081106.

Li, P.-P., Y. Chen, J. Zhu, M. Feng, X. Zhuang, Y. Lin and H. Zhan (2011). Charm-Bracelet-Type Poly(N-vinylcarbazole) Functionalized with Reduced Graphene Oxide for Broadband Optical Limiting. *Chemistry - A European Journal*, Vol. 17, No. 3, pp. 780-785.

Li, P. P., L. J. Niu, Y. Chen, J. Wang, Y. Liu, J. J. Zhang and W. J. Blau (2011). In situ synthesis and optical limiting response of poly(N-vinylcarbazole) functionalized single-walled carbon nanotubes. *Nanotechnology*, Vol. 22, No. 1, pp. 015204.

Liao, K.-S., J. Wang, D. Früchtl, N. J. Alley, E. Andreoli, E. P. Dillon, A. R. Barron, H. Kim, H. J. Byrne, W. J. Blau and S. A. Curran (2010). Optical limiting study of double wall carbon nanotube-fullerene hybrids. *Chemical Physics Letters*, Vol. 489, No. 4-6, pp. 207-211.

Liu, Y. S., J. Y. Zhou, X. L. Zhang, Z. B. Liu, X. J. Wan, J. G. Tian, T. Wang and Y. S. Chen (2009). Synthesis, characterization and optical limiting property of covalently oligothiophene-functionalized graphene material. *Carbon*, Vol. 47, No. 13, pp. 3113-3121.

Liu, Z. B., J. G. Tian, Z. Guo, D. M. Ren, F. Du, J. Y. Zheng and Y. S. Chen (2008). Enhanced Optical Limiting Effects in Porphyrin-Covalently Functionalized Single-Walled Carbon Nanotubes. *Advanced Materials*, Vol. 20, No. 3, pp. 511-515.

Liu, Z. B., Y. Wang, X. L. Zhang, Y. F. Xu, Y. S. Chen and J. G. Tian (2009). Nonlinear optical properties of graphene oxide in nanosecond and picosecond regimes. *Applied Physics Letters*, Vol. 94, No. 2, pp. 021902.

Liu, Z. B., Y. F. Xu, X. Y. Zhang, X. L. Zhang, Y. S. Chen and J. G. Tian (2009). Porphyrin and Fullerene Covalently Functionalized Graphene Hybrid Materials with Large Nonlinear Optical Properties. *Journal of Physical Chemistry B*, Vol. 113, No. 29, pp. 9681-9686.

Loh, K. P., Q. L. Bao, G. Eda and M. Chhowalla (2010). Graphene oxide as a chemically tunable platform for optical applications. *Nature Chemistry*, Vol. 2, No. 12, pp. 1015-1024.

Lotya, M., Y. Hernandez, P. J. King, R. J. Smith, V. Nicolosi, L. S. Karlsson, F. M. Blighe, S. De, Z. M. Wang, I. T. McGovern, G. S. Duesberg and J. N. Coleman (2009). Liquid Phase Production of Graphene by Exfoliation of Graphite in Surfactant/Water Solutions. *Journal of the American Chemical Society*, Vol. 131, No. 10, pp. 3611-3620.

Mamidala, V., L. Polavarapu, J. Balapanuru, K. P. Loh, Q.-H. Xu and W. Ji (2010). Enhanced nonlinear optical responses in donor-acceptor ionic complexes via photo induced energy transfer. *Optics Express*, Vol. 18, No. 25, pp. 25928-25935.

Mansour, K., M. J. Soileau and E. W. Van Stryland (1992). Nonlinear optical properties of carbon-black suspensions (ink). *Journal of the Optical Society of America B-Optical Physics*, Vol. 9, No. 7, pp. 1100-1109.

Midya, A., V. Mamidala, J. X. Yang, P. K. L. Ang, Z. K. Chen, W. Ji and K. P. Loh (2010). Synthesis and Superior Optical-Limiting Properties of Fluorene-Thiophene-Benzothiadazole Polymer-Functionalized Graphene Sheets. *Small*, Vol. 6, No. 20, pp. 2292-2300.

Nair, R. R., P. Blake, A. N. Grigorenko, K. S. Novoselov, T. J. Booth, T. Stauber, N. M. R. Peres and A. K. Geim (2008). Fine structure constant defines visual transparency of graphene. *Science*, Vol. 320, No. 5881, pp. 1308-1308.

Ni Mhuircheartaigh, E. M., S. Giordani and W. J. Blau (2006). Linear and nonlinear optical characterization of a tetraphenylporphyrin-carbon nanotube composite system. *Journal of Physical Chemistry B*, Vol. 110, No. 46, pp. 23136-23141.

Niu, L., P. Li, Y. Chen, J. Wang, J. Zhang, B. Zhang and W. J. Blau (2011). Conjugated polymer covalently modified multiwalled carbon nanotubes for optical limiting. *Journal of Polymer Science Part A: Polymer Chemistry*, Vol. 49, No. 1, pp. 101-109.

Novoselov, K. S., A. K. Geim, S. V. Morozov, D. Jiang, M. I. Katsnelson, I. V. Grigorieva, S. V. Dubonos and A. A. Firsov (2005). Two-dimensional gas of massless Dirac fermions in graphene. *Nature*, Vol. 438, No. 7065, pp. 197-200.

Novoselov, K. S., A. K. Geim, S. V. Morozov, D. Jiang, Y. Zhang, S. V. Dubonos, I. V. Grigorieva and A. A. Firsov (2004). Electric field effect in atomically thin carbon films. *Science*, Vol. 306, No. 5696, pp. 666-669.

Novoselov, K. S., D. Jiang, F. Schedin, T. J. Booth, V. V. Khotkevich, S. V. Morozov and A. K. Geim (2005). Two-dimensional atomic crystals. *Proceedings of the National Academy of Sciences of the United States of America*, Vol. 102, No. 30, pp. 10451-10453.

O'Flaherty, S. A., R. Murphy, S. V. Hold, M. Cadek, J. N. Coleman and W. J. Blau (2003). Material investigation and optical limiting properties of carbon nanotube and nanoparticle dispersions. *Journal of Physical Chemistry B*, Vol. 107, No. 4, pp. 958-964.

O'Flaherty, S. M., J. J. Doyle and W. J. Blau (2004). Numerical approach for optically limited pulse transmission in polymer-phthalocyanine composite systems. *Journal of Physical Chemistry B*, Vol. 108, No. 45, pp. 17313-17319.

O'Flaherty, S. M., S. V. Hold, M. E. Brennan, M. Cadek, A. Drury, J. N. Coleman and W. J. Blau (2003). Nonlinear optical response of multiwalled carbon-nanotube dispersions. *Journal of the Optical Society of America B-Optical Physics*, Vol. 20, No. 1, pp. 49-58.

O'Flaherty, S. M., S. V. Hold, M. J. Cook, T. Torres, Y. Chen, M. Hanack and W. J. Blau (2003). Molecular engineering of peripherally and axially modified phthalocyanines for optical limiting and nonlinear optics. *Advanced Materials*, Vol. 15, No. 1, pp. 19-32.

Rayfield, G. W., A. Sarkar, S. Rahman, J. P. Godschalx and E. W. Taylor (2010). Mechanistic studies for optical switching materials for space environments. *Nanophotonics and Macrophotonics for Space Environments Iv*. E. W. Taylor and D. A. Cardimona. Bellingham, Spie-Int Soc Optical Engineering. 7817.

Senge, M. O., M. Fazekas, E. G. A. Notaras, W. J. Blau, M. Zawadzka, O. B. Locos and E. M. N. Mhuircheartaigh (2007). Nonlinear optical properties of porphyrins. *Advanced Materials*, Vol. 19, No. 19, pp. 2737-2774.

Sun, X., R. Q. Yu, G. Q. Xu, T. S. A. Hor and W. Ji (1998). Broadband optical limiting with multiwalled carbon nanotubes. *Applied Physics Letters*, Vol. 73, No. 25, pp. 3632-3634.

Sun, Y. P. and J. E. Riggs (1999). Organic and inorganic optical limiting materials. From fullerenes to nanoparticles. *International Reviews in Physical Chemistry*, Vol. 18, No. 1, pp. 43-90.

Sun, Z. P., T. Hasan, F. Torrisi, D. Popa, G. Privitera, F. Q. Wang, F. Bonaccorso, D. M. Basko and A. C. Ferrari (2010). Graphene Mode-Locked Ultrafast Laser. *Acs Nano*, Vol. 4, No. 2, pp. 803-810.

Tutt, L. W. and T. F. Boggess (1993). A review of optical limiting mechanisms and devices using organics, fullerenes, semiconductors and other materials. *Progress in Quantum Electronics*, Vol. 17, No. 4, pp. 299-338.

Tutt, L. W. and A. Kost (1992). Optical Limiting Performance of C60 and C70 Solutions. *Nature*, Vol. 356, No. 6366, pp. 225-226.

Venkatram, N., D. N. Rao and M. A. Akundi (2005). Nonlinear absorption, scattering and optical limiting studies of CdS nanoparticles. *Optics Express*, Vol. 13, No. 3, pp. 867-872.

Vivien, L., E. Anglaret, D. Riehl, F. Bacou, C. Journet, C. Goze, M. Andrieux, M. Brunet, F. Lafonta, P. Bernier and F. Hache (1999). Single-wall carbon nanotubes for optical limiting. *Chemical Physics Letters*, Vol. 307, No. 5-6, pp. 317-319.

Vivien, L., P. Lancon, D. Riehl, F. Hache and E. Anglaret (2002). Carbon nanotubes for optical limiting. *Carbon*, Vol. 40, No. 10, pp. 1789-1797.

Vivien, L., J. Moreau, D. Riehl, P. A. Alloncle, M. Autric, F. Hache and E. Anglaret (2002). Shadowgraphic imaging of carbon nanotube suspensions in water and in chloroform. *Journal of the Optical Society of America B-Optical Physics*, Vol. 19, No. 11, pp. 2665-2672.

Vivien, L., D. Riehl, F. Hache and E. Anglaret (2000). Nonlinear scattering origin in carbon nanotube suspensions. *Journal of Nonlinear Optical Physics & Materials*, Vol. 9, No. 3, pp. 297-307.

Wang, J. and W. J. Blau (2008). Linear and nonlinear spectroscopic studies of phthalocyanine-carbon nanotube blends. *Chemical Physics Letters*, Vol. 465, No. 4-6, pp. 265-271.

Wang, J. and W. J. Blau (2008). Nonlinear optical and optical limiting properties of individual single-walled carbon nanotubes. *Applied Physics B-Lasers and Optics*, Vol. 91, No. 3-4, pp. 521-524.

Wang, J. and W. J. Blau (2008). Solvent effect on optical limiting properties of single-walled carbon nanotube dispersions. *Journal of Physical Chemistry C*, Vol. 112, No. 7, pp. 2298-2303.

Wang, J. and W. J. Blau (2009). Inorganic and Hybrid Nanostructures for Optical Limiting. *Journal of Optics A – Pure and Applied Optics*, Vol. 11, No. 2, pp. 024001.

Wang, J., Y. Chen and W. J. Blau (2009). Carbon Nanotubes and Nanotube Composites for Nonlinear Optical Devices. *Journal of Materials Chemistry*, Vol. 19, No. 40, pp. 7425-7443.

Wang, J., D. Früchtl and W. J. Blau (2010). The importance of solvent properties for optical limiting of carbon nanotube dispersions. *Optics Communications*, Vol. 283, No. 3, pp. 464-468.

Wang, J., D. Früchtl, Z. Sun, J. N. Coleman and W. J. Blau (2010). Control of Optical Limiting of Carbon Nanotube Dispersions by Changing Solvent Parameters. *The Journal of Physical Chemistry C*, Vol. 114, No. 13, pp. 6148-6156.

Wang, J., Y. Hernandez, M. Lotya, J. N. Coleman and W. J. Blau (2009). Broadband Nonlinear Optical Response of Graphene Dispersions. *Advanced Materials*, Vol. 21, No. 23, pp. 2430-2435.

Webster, S., M. Reyes-Reyes, X. Pedron, R. López-Sandoval, M. Terrones and D. L. Carroll (2005). Enhanced nonlinear transmittance by complementary nonlinear mechanisms: a reverse-saturable absorbing dye blended with nonlinear-scattering carbon nanotubes. *Advanced Materials*, Vol. 17, No. 10, pp. 1239-1243.

Xia, Y., P. Yang, Y. Sun, Y. Wu, B. Mayers, B. Gates, Y. Yin, F. Kim and H. Yan (2003). One-Dimensional Nanostructures: Synthesis, Characterization, and Applications. *Advanced Materials*, Vol. 15, No. 5, pp. 353-389.

Xu, Y., Z. Liu, X. Zhang, Y. Wang, J. Tian, Y. Huang, Y. Ma, X. Zhang and Y. Chen (2009). A Graphene Hybrid Material Covalently Functionalized with Porphyrin: Synthesis and Optical Limiting Property. *Advanced Materials*, Vol. 21, No. 12, pp. 1275-1279.

Zhang, B., J. Wang, Y. Chen, D. Fruchtl, B. Yu, X. D. Zhuang, N. He and W. J. Blau (2010). Multiwalled Carbon Nanotubes Covalently Functionalized with Poly(N-vinylcarbazole) via RAFT Polymerization: Synthesis and Nonlinear Optical Properties. *Journal of Polymer Science Part a-Polymer Chemistry*, Vol. 48, No. 14, pp. 3161-3168.

Zhang, X. L., X. Zhao, Z. B. Liu, Y. S. Liu, Y. S. Chen and J. G. Tian (2009). Enhanced nonlinear optical properties of graphene-oligothiophene hybrid material. *Optics Express*, Vol. 17, No. 26, pp. 23959-23964.

Zhang, X. Y., Z. B. Liu, Y. Huang, X. J. Wan, J. G. Tian, Y. F. Ma and Y. S. Chen (2009). Synthesis, Characterization and Nonlinear Optical Property of Graphene-C-60 Hybrid. *Journal of Nanoscience and Nanotechnology*, Vol. 9, No. 10, pp. 5752-5756.

Zhang, X. Y., X. Y. Yang, Y. F. Ma, Y. Huang and Y. S. Chen (2010). Coordination of Graphene Oxide with Fe3O4 Nanoparticles and Its Enhanced Optical Limiting Property. *Journal of Nanoscience and Nanotechnology*, Vol. 10, No. 5, pp. 2984-2987.

Zhao, B. S., B. B. Cao, W. L. Zhou, D. Li and W. Zhao (2010). Nonlinear Optical Transmission of Nanographene and Its Composites. *Journal of Physical Chemistry C*, Vol. 114, No. 29, pp. 12517-12523.

Zheng, C., M. Feng and H. B. Zhan (2010). The synthesis of carbon nanotube based composites with conducting, semiconducting, and insulating coatings and their optical limiting properties. *Carbon*, Vol. 48, No. 13, pp. 3750-3759.

Zhou, Y., Q. L. Bao, L. A. L. Tang, Y. L. Zhong and K. P. Loh (2009). Hydrothermal Dehydration for the "Green" Reduction of Exfoliated Graphene Oxide to Graphene and Demonstration of Tunable Optical Limiting Properties. *Chemistry of Materials*, Vol. 21, No. 13, pp. 2950-2956.

Zhu, J., Y. Li, Y. Chen, J. Wang, B. Zhang, J. Zhang and W. J. Blau (2011). Graphene oxide covalently functionalized with zinc phthalocyanine for broadband optical limiting. *Carbon*, Vol. 49, No. 6, pp. 1900–1905.

# 20

# Design and Demonstration of Carbon Nanotubes (CNTs)-Based Field Emission Device

Tian Jin-shou[1], Li Ji[2], Xu Xiang-yan[1] and Wang Jun-feng[1]
[1]*State Key Laboratory of Transient Optics and Photonics,Xi'an Institute of Optic and Precision Mechanics of CAS*
[2]*Institute of Optoelectronics, Shenzhen University, Shenzhen, China*

## 1. Introduction

Since its discovery, carbon nanotube (CNT), possessing a series of particularly electrical and mechanical property as well chemical stability, has been considered as one of the most advanced electronics materials. A lot of research has been extensively carrying through on CNTs' potential applications for instance, gas storage, quanta lead, electron device, catalyst carrier, and etc. Among its various applications, more talked is its application as field emission cathode (FEC) material to make large-area, full-colored and high efficiency displays or lighting devices (FEC-LED) because of its excellent field emission property. Compared with the other material of field emission cathode, CNT-FEC devices have a series of unique performances, such as higher field emission efficiency, lower power consume, lower cost, non pollute problem in its production processes, and so on. Those advantages come from the following physical and chemical mechanism:

1. CNT possesses so large aspect ratio in structure that the field enhancement factor CNT can reach 30000 to 50000 for single CNT, 800 to 3000 for CNT film, no mater what state , standing or lying they are. Therefore CNT is such an excellent field emitter that CNT-FEC can easily provide necessary current under a lower drive voltage with much lower power consume.
2. CNT-FED workmanship is simply, the material cost is low. When the manufacture is enlarged, it can compete with other technologies of display or lighting devices in price.

On the other hand there are still some theoretical and technical problems need to be solved before CNT-FED being get more competitive applications on the market, for example, its feasibility of a large scale production, full colored field, optimal structure design, spatial homogeneity, stability, lifespan, as well as its low cost fabrication technology. this chapter mainly concerns these problems mentioned and gives some elementary discussion in for our further understand of them, including :

- The research on the field emission properties of CNT and computer simulation based on Fowler-Nordheims theory;

- The research on influence of relative height between cathode and gate on electron transmission efficiency.
- Structural Analysis on a Field Emission Display Panel Based on CNTs.

## 2. Research on the field emission properties of CNTs

### 2.1 The main points and technical background

The theory on CNT's field emission is a base for FED theoretical analysis and optimal design, which deals closely with device efficiency, power consume, operation stability, life time, and so on. There are a lot of different viewpoints on the mechanism of the CNTs field emitters. Some researchers approved an idea that the CNTs field emission accords with the Fowler-Nordheims tunneling theory and the electrons are emitted from the top of CNTs, whose work function approaches the value of graphites. On the other hand, Collins, Zettle and Bonard[2] considered that: the CNTs field emission is more complex than the one expected by F-N tunneling theory. They proposed a new CNT field emission prototype. Gulyaev[3] believed that: the closely relationship of field emission with temperature for both single-walled and multi-walled CNTs shows that the CNTs are low work function field emitters. But Rinzler[4] held that the CNTs are very sharply high work function emitter. The reason why the CNTs field emission increasing with a high temperature is due to the carbon atom's reconstruction on the tip of CNTs, resulting in the increasing field enhancement factor. Dean *et al*[5], divided the field emission of carbon into three processes: adsorption state, clean state and high current state. We have also observed in our experiment that the properties curve of CNT field emission is not a strict straight line, which changes with the electric field intensity. So we established a simple prototype to explain the observed CNT emission characteristics. The main research result is that the F-N curve's non-linear in experiment can be explained as the disappearance of adsorbate (mainly $H_2O$) , which changes the effect work function of CNTs. When they are bombarded by the remained gas particles in device, the tube caps are flattened and their lengths shortened, resulting in its electron emission contribution of short CNTs to total currents enhancing with the increase of electric field intensity.

### 2.2 Experiment description

The adopted CNTs are produced by chemical vapor deposition (CVD), and observed by electron microscopy. The CNT material purity is very high and having a lot of CNTs to aggregate and tangle together generally, so they need to be dispersed with ultrasonic process. We take a proper amount of CNTs powder and pour it into acetone liquor and then ultrasonic process for more than half an hour. Sucking the suspension of CNTs through filter paper and air dried thoroughly,then mix the dry CNTs with silver paste. Because of the highly viscousity of silver paste, it must be mixed adequately with special tool, and then repeated it like this several times until the content of CNTs is appropriate. Thus enough emission current can be not only got, but also the field shielding effect between adjacent CNTs can be avoided. The mixture of silver paste and CNTs is printed on the clean glass substrate by silk-screen printing technology, put in muffle furnace, and then heated at 300°C for half an hour, making silver paste solidified. On one hand the silver paste can solidify the CNTs, on the other hand; it makes the CNTs emission electrons supply continually. Resulting from the different melting point between CNTs and silver paste, a trench is etched on the silver paste. By adjusting the energy and scanning speed of laser beam, not only the silver paste should be etched,forming electrical

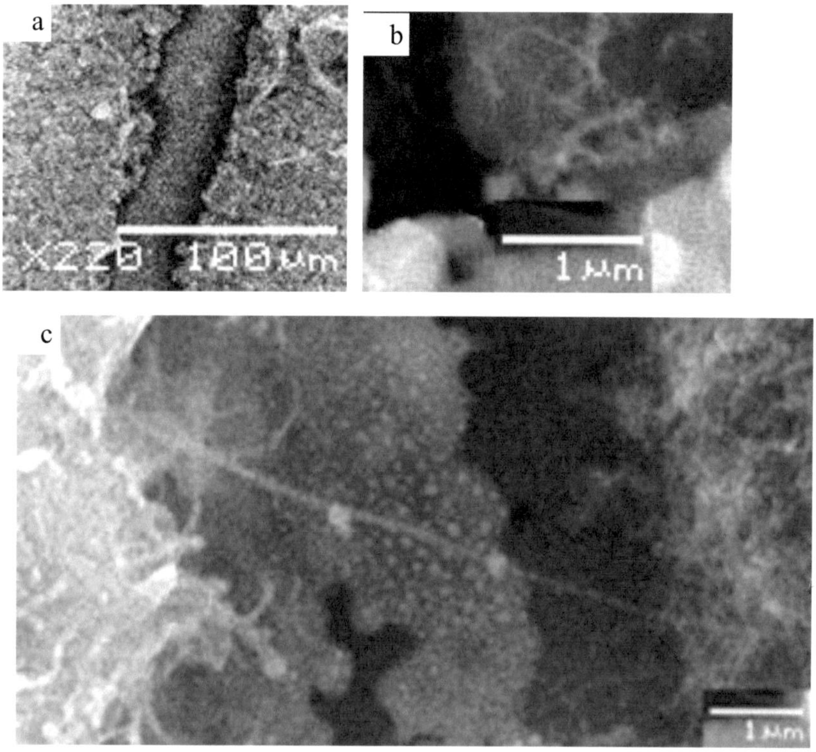


Fig. 1. (a) a trench etched with laser; (b) carbon nanotube on the edge of trench; (c) a carbon nanotube with the length of 5µm

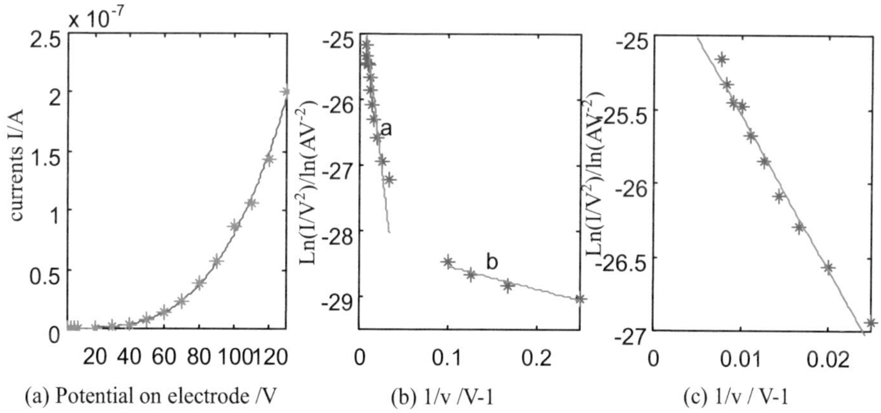


Fig. 2. (a). Emitting current vs. Voltage, (b). F-N plot（Potential ranged from 4V~130V）(c). F-N plot（Potential ranged from 40V~130V）

insulation between the cathode and gate (the resistance between the cathode and gate shoud be more than 50MΩ at least), but also ensuring CNTs left at the edge of the trench. Then the field emission character of CNTs on both side of trench can be measured in vacuums with a low current testing instrument (such as Cathely 6517A). Fig.1(a) shows that a trench is etched on the silver layer mixed with CNTs by laser beam, in the middle of which there are some little silver particles left, whose width is about 35um. There are some CNTs on both side of the trench, rarely seen them in the middle of trench, which is likely to be sputtered out with the silver particles. In the trench some longer CNTs can be observed occasionally. Seeing Fig. 1(c), the length of a CNTs outside of the silver layer in trench is about 5um. The edge of the trench is not very even and the length of CNTs is not the same.

The character of CNTs field emission is shown as Fig. 2, which shows that the curve of CNTs field emission matches very well with F-N curve when applied voltage between the electrodes exceeds 40V ; while flaten a lot, when less than 40V. (shown as curve b in Fig. 2 (b)).

### 2.3 Theoretical prototype and analysis

In field emission experiments, the measured F-N curve is not a strict straight line which has a little difference compared with the result deduced from the F-N Theory. As curve b in Fig. 2(b), this is a relatively common phenomenon, but people trend to show the emission characteristic curve in specific voltage range, such as a curve in Fig. 2c. We have seldom seen somebody explain the phenomenon shown as in curve a) and b) in Fig. 2b. From the F-N formula,

$$\frac{i(L)}{A}=(1.54\times10^{-6})\frac{\beta^2(L)V^2}{\varphi}\exp(-\frac{6.83\times10^7\varphi^{3/2}}{\beta(L)V}) \tag{1}$$

In equation (1), $L$ is the length of CNTs and $A$ is the effective emitting area of CNTs. It can be observed that only two parameters that infulence the F-N curve slope, namely, the field enhancement factor β and work function φ. For CNTs, β has something with the length, radii, and the shape of the CNTs, while φ relates with the CNTs adsorption. In order to explain the reason why the F-N curve slope in experiment changed with the voltage changing, let us to suppose that the length distribution of CNTs at the edge of the trench obeying with normal distribution, and CNTs shape at the up end and their radii are the same, and each CNT field emission is independent of different field enhancement factor β($L$). So the total field emission current I is:

$$I=\sum_{l}G(L)i(L) \tag{2}$$

In equation (2), G (L) is the number of the CNTs with the length distributed in the range of L± 0.2μm, if some of the CNTs is shortend at the higher emission current because of heat or bormbardment of the charged particles(in fact, the disappearance of some kind protroding parts on the top of CNTs can be considered in the same way as that of shortening the length of CNTs), so at different voltage, the length of CNTs have different cutoff values, which impact the CNTs field enhancement factor. Fig. 3 show the F-N curves at two different work function and five different cutoff lengths. It can be seen that the shorter cutoff length is, the greater the work function is, and the greater the slope of curves is. To be simple, we supposed that the expected value of the CNTs length is 3μm, whose mean square deviation is 1μm, and all the radii of CNTs is about 3nm.There are several methods about the calculation of the field enhancement factor, for different length ranges[6,7]. We make use of the calculation method β（L）=$KL/r$ commonly used for CNTs. Here K is the statistic of the previous field enhancement factor and a correctable coefficient given when we choose CNTs length ranges. Now the experimental phenomena can be explained below: At the low voltage, the emission current is mainly emitted from longer CNTs, but little from shorter ones. The longer the CNTs are, the greater the field enhancement factor is, and the smaller the slope of F-N curve is. During this stage, the field emission process is dominated by the CNTs in absorbate formed by water vapor, which can be formed into the C-H-O-H bond on the surface of CNTs, and reduces the effective work function of the CNTs surface[8]. These two factors result in a smaller F-N curve slope ( $\varphi^{1.5}/\beta$) at the beginning of the CNTs field emission.

As the electric field intensity increases further, the CNTs field emission current increases and the tiny protuberances on the tip of CNTs are melted or smoothened under bombardment by residual gas particles. Meanwhile, the absorbate on the top of CNTs is also removed, and then the effective work function increase to 5 eV, which is equal to that of net carbon. It can be seen that the shorter CNTs and the greater work function will lead to the greater slope of F-N curve, and the contribution of shorter CNTs to field emission is growing with the increasing of electric strength on the top of CNT. Fig. 4 show the values of CNTs field emission current at different voltage and different cut-off lengths ranged from 1μm to 4.7μm, to a certain cut-off length, the lengths of CNTs follows the Guassian distribution. It can be seen that: When the applied voltage is lower (such as 6V) , the emission current is mainly contributed by CNTs of length 4.6μm or more; When the applied voltage increases to 130V, the current emitted by CNTs with length of 2.8μm or more will contribute a large proportion to the total. So we can get a conclusion: with the reduction of CNTs length and increasing of its work function according to improvement of the voltage, the slope of the F-N curve becomes more flat, When the applied voltage is more than 40V, the field emission is mainly adopted by clean CNTs with work function of about 5eV, and the F-N curve strictly follows a straight line ( shown as Fig. 2(c))

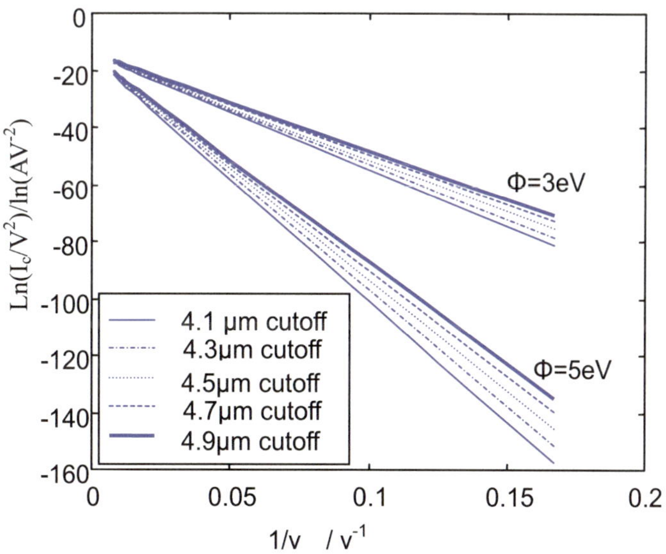


Fig. 3. F-N curves of different cutoff length and work function

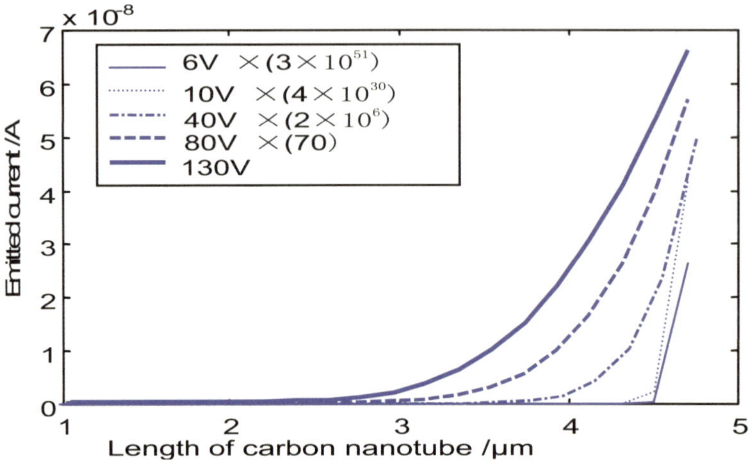


Fig. 4. illustrating the varying contribution of tube length to total current I, for different anode potentials V.

### 2.4 Conclusion

There are lots of field emission experiments about CNTs, but in many cases they just show some field emission characters in specific voltage ranges, thus the result of experiments matching well with F-N curves. Although some experiments get the CNTs field emission character both at low and high electric field intensity, there are no reasonable explanation about them. Based on the F-N tunneling theory, a simple prototype has been established in the paper, which explains well the phenomenon that the slope of F-N curve changes with applied electric field intensity in the process of CNTs field emission. The main reason for increasing of F-N curve slope is considered that the adsorbate plays a big role at the beginning of field emission, and then the field emission of carbon atoms on the surface of CNTs becomes dominated gradually. Meanwhile, as the voltage increasing, the protuberance on the top of CNTs disappears by bombardment of the remained gas particles. In fact, the CNTs field emission is much more complex. First of all, the CNTs itself can be classified into metallic and semiconductor according to the chirality. Secondly, the type of materials adsorbed on the surface and their locations on top of the CNTs are all changed, and the emission area is also a transformable factor. Consequently there will be lots of theoretical and experimental work needed to be carried out if we want to have a more intensive understanding about the process of CNTs field emission.

## 3. The Influence of relative height between cathode and gate on electron transmission efficiency

### 3.1 The key points and technical background

This topic associates with the structure optimization design for a kind field emission display (FED), among the parameters affecting FED performance, the relative height between cathode and gate plays an important role because it dominates FED electron transmission efficiency directly. The electron trajectories as well as their distribution on anode in a large area-full colored FED prototype have been theoretically analyzed by using Monte Carlo and Boundary Element Methods. The result has been used to improve the electron transmission

efficiency via adjusting the vertical distance between cathode and gate, at the same time, the elevation angle of CNT on the influence of transmission efficiency is also discussed.
It is noted that the surface conduction electron emitter displays[1](SCED) proposed by Canon Corporation has simplified production process of FED, and provided a feasible way for FED with large area. According to the technical information, the electron transmission efficiency(the ratio of electron to the anode and electron to the cathode) of this kind is no more than 1%, so we proposed two solutions to try to improve it [9]: First, using CNTs instead of $P_dO$ emitter, thus the electron transmission efficiency can be adjusted by controlling the direction of CNTs , theoretical calculations prove that when the angle between CNTs and gate plane reach 30°, almost all electrons can reach the anode and form the emission current; Meanwhile, giving the symmetrical axis of CNT parallel to the gate plane, we can also improve the electron transmission efficiency according to enlarging the relative height between gate and cathode(as shown in Fig. 5).

### 3.2 The structural and theoretical model of the FED's

Fig.5 is a simplified structural model of one FED pixel, in which the emission cathode consists of CNTs, whose symmetrical axis are parallel to the gate and anode planes, supposing the CNT is 2μm long, and closed by a hemisphere with radius 2.5nm on the top. The distance between anode and gate is 1.5 mm. The side of gate near the CNTs is a quarter arc with radius 0.05μm. By chiral vector, CNTs can be divided into metal and semiconductors which have different band gap. Giving the chiral vectors are uniform distributed, there would be one third of the metal and two thirds of the semiconductor in all type of the carbon tubes [10]. But the experimental results of O.Groning[7] indicate that CNT's field emission characters fit the theoretical model of metal field emission. The experiments of J.M.Bonard[11] demonstrate that, in the case of small current, field emission characters of CNTs fit Fowler-Nordheim formula. So we simplified the CNTs as a metallic cylindrical shell with the end closed by a hemisphere.

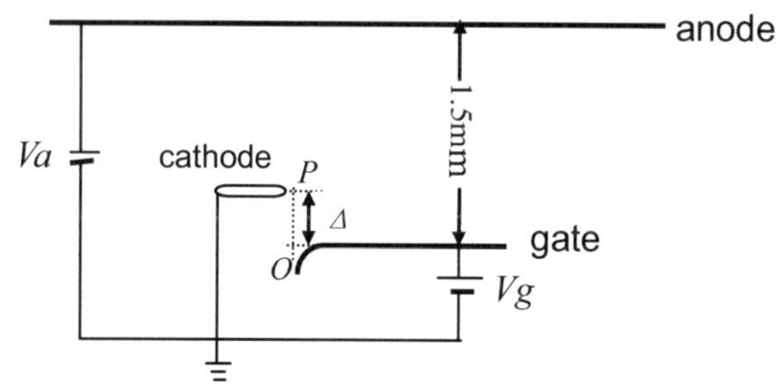


Fig. 5. Model of electric structure

Both the dimension of the electrodes and the distance between them are particularly different (the distance between anode and cathode is 300,000 times than the diameter of CNTs) which brings about the electric field distribution is extremely uneven in the border region, so that the Boundary Element Method (BEM) is an ideal one to calculate the electric distribution on the domain of our interest[12]. 106 units are divided on all the electrodes consisting the boundary of one sub-pixel: cathode( CNT), gate and anode. Considering the quite imbalance of charge density distribution in each electrode, especially the charge density in the cathode is much higher than what in the anode. So that the length of each unit is different in the process of dividing the border, boundary elements tend to put a greater element density where they are needed, such as near the end of electrodes and in regions of high curvature. There are 62 units on the CNTs, 20 units on the gate, and 24 units on the anode. The distribution of initial position, velocity, elevation and intensity of electrons launched by CNT is sampled with Monte Carlo Method.

(1). Localization the initial position of electrons

For an electron launched by emitter, how to decide which discrete unit it is from? For given temperature $T$ and work function $E_{\Phi}$, the higher the field intensity, the more probability of electrons can be emitted. According to Fowler-Nordheim formula, the current density emitted from the cathode can be expressed below [13] [14]:

$$J(T) = \frac{4\pi emkTd}{h^3}\exp(-c)\frac{\pi kT / d}{\sin(\pi kT / d)} \tag{3}$$

In equation (3), $c = \frac{8\pi\sqrt{2m|E_{\varphi}|^3}}{3he\varepsilon}v(y_0)$ $\quad d = \frac{he\varepsilon}{4\pi\sqrt{2m|E_{\varphi}|}t(y_0)}$ $\quad y_0 = \frac{\sqrt{e^3\varepsilon}}{E_{\varphi}}$

Where $e$ is electron charge, $m$ is electron mass, $k$ is Boltzman constant, $h$ is Planck constant, $\varepsilon$ is electric field strength, $E_{\varphi}$ is work function of emitter, $t(y_0)$ is close to 1, $v(y_0)$ is the Nordheim Function, $T$ is the temperature of cathode, When equation (3) is normalized, it can be used as emission probability function for cold field emission, so the normalized emission probability density function $f_1(\varepsilon)$ can be expressed as:

$$f_1(\varepsilon) \approx J(0) / J_M = \frac{1.54\times 10^{-6}\varepsilon^2}{J_M E_{\varphi} t^2(e^3\varepsilon / E_{\varphi})}\exp[-6.83\times 10^7 \frac{E_{\varphi}^{3/2}}{\varepsilon}v(y_0)] \tag{4}$$

In equation (4), $J_M$ is normalized constant. Using expression (4), we can sample the electric density $\varepsilon$ with Rejection Selection Method [15]. In principle, $\varepsilon$ can be any value in [0, ∞], in order to improve sampling efficiency, we just sample in $[1.0, \varepsilon_m]\times 10^7$ V/cm. because for cold cathode, the obvious electron emission occurred only when the electric field intensity of emitter surface is up to 2~3×$10^7$V/cm, while $\varepsilon_m$ is the max electric field intensity on CNT's. After sampling a electric field intensity $\varepsilon$, a comparison of sampling value is made with all the electric field intensities on the nodes, among which the nearest field intensity of the point is, the node is taken as the emitting position, which serve as the initial position of the traced electrons, furthermore, the electric field intensity of these points can be used to calculate the initial acceleration of electrons.

(2) Sampling the initial energies of emitted electrons

The number of electrons with initial energy in $E$~$E+dE$ is:

$$I(E)dE = \frac{4\pi me}{h^3}\cdot d\cdot\exp(-c)\cdot\frac{\exp[(E-E_F)/d]}{\exp[(E-E_F)/kT]+1}dE \tag{5}$$

In equation (5), $E$ is the energy of emitted electron, $E_F$ is the Fermi energy, $I(E)$ is the current density with electron energy of $E$, $I_M$ normalized constant, from equation (5), we can get probability density function $f_2(E)$ of initial energy below:

$$f_2(E) = \frac{I(E)}{I_M} = \frac{4\pi me}{I_M h^3} \cdot d \cdot \exp(-c) \cdot \frac{\exp[(E - E_F)/d]}{\exp[(E - E_F)/kT] + 1} \tag{6}$$

According to expression (6), we can sample initial energy of electrons with Rejection Selection Method too. Because the energy spectrum of field emission electrons is very narrow [16], the electron energy is in the range of 1.5eV (even just 0.5eV [18]) around Fermi level [16] [17], M. J. Fransen, et al[18] have measured the FWHM of CNTs field emission is 0.11~0.70eV, so we will sample of the emission electrons in 4.8~5.3eV for improving the sampling efficiency.

(3) Elevation angle *a* of emitted electrons in each unit obeys Lambert's Law, that is:

$$f_3(\cos\alpha) = \begin{cases} 2\cos\alpha & 0 < \cos\alpha < 1 \\ 0 & else \end{cases} \tag{7}$$

From expression (7), we can sample the elevation angle α of emitted electron.

Assuming T=300K in the cathode emitter, the CNT's work function [11] [15] [16] $E_\varphi$=5.0eV, $t(y_0) = 1$. When we completed the sampling of initial position, initial energy and emission direction according with the above steps, the initial state (position, elevation angle, acceleration and velocity) of electrons were completely determined. Then tracing the trajectories of electrons, among them the electrons reached the gate are called the conduction electrons, and those reached the anode called the emission electrons, which eventually bombard the screen to form image information.

### 3.3 The results and analysis

In the calculation process, let anode voltage Va=4000V, gate voltage Vg=50V. Fig.6 expresses the electric field intensities of every node on the boundary when the distance Δ of CNT's axes above the gate is changed. It is clear that the electric field intensities of CNT's tip can reach strength for field emission; furthermore, the electric field intensity of CNT's tip is three orders of magnitude higher than the electric field intensity on anode. When Δ=0, the minimum distance between CNT's apex and gate is 20.7nm. Fig.7 a, b, c and d show that the distributions of electrons reaching the anode with Δ=300nm, 50nm, 5nm, and 0(assuming in each case, the number of electrons is identical) respectively, while the area under the curve represents the number of electrons reaching the anode. With the increasing of Δ, the electron beam has a tendency of dividing into two beams. But on the whole, focusing of electron beam is rather ideal, the smaller Δ is, the better it focused. When Δ=5nm, the FWHM of electron beam spot is less than 50μm. With Δ changing, the center of the electron beam remains always at about 100μm at the right side of the CNT. This phenomenon must be taken into consideration in the process of the device package.

The relationship of electron transmission efficiency with the vertical distance Δ is shown in table 1, when Δ decreasing from 300nm to 0, the electron transmission efficiency dropped from 50.5% down to 17%. Therefore, the larger *Δ* is, the less constrain the gate would act to the electrons, which means the more electrons can break the restrain of the gate to reach the anode and form the emission current. Fig.8a shows the changes of the actual total number of electrons (logarithm) emitted at the top of CNT with *Δ* changing. It can be seen that , for both *Δ*=300nm and *Δ*=0, the total number of electrons emitted from CNTs is 17 orders of magnitude difference, that is to say, if there are two identical CNTs with *Δ*=300nm and *Δ*=0, the field emission of the CNT which is close to the gate will play a decisive role. Fig.8b shows the change of the maximal strength of electric field on CNT's tip with $\Delta$ changing; No doubt it will directly affect the current density of emission electrons. In fact, in its changing process, the distance between CNT's tip and gate is changing too, so it is difficult to distinguish the main factors whether the vertical distance $\Delta$ or the transactional distance between CNT's tip and gate that affect the electronic transmission efficiency is. So we can let CNT in two specific position for comparison : in Fig.8, if the arc apex $O$ on the left of gates is the origin of coordinates, so at two different emitting points $P_1$(0, 100nm) and $P_2$ (-50nm, 50nm), in both positions, the distances between the emitting points and gate are all 61.8nm., but the electron transmission efficiency are 44.25% and 40.75% respectively, and it proves that one can improve the electronic transmission efficiency by increasing $\Delta$, compared with the experimental results of Canon corporation [15], we can conclude that even if the CNTs and the gate are in the same plane, the electron transmission efficiency can also be increased only by increasing the transversal distance between cathode and gate, but it needs to increase the voltage of the gate at the same time. In order to test the influence of CNT's angle on the electron transmission efficiency, we set the apex of CNT at coordinate (-267.6nm, 1051.25nm) to calculate the transmission efficiencies when the elevation angles equal to 30°and 0° , as a result, the corresponding electron transmission efficiency are 99.7% and 64.7% respectively. It shows that the impact of electron emission direction on electronic transmission efficiency is much greater, because the emission electrons have relative high inertia, so there are more electrons can break the strong confinement of the gate and eventually reach the anode.

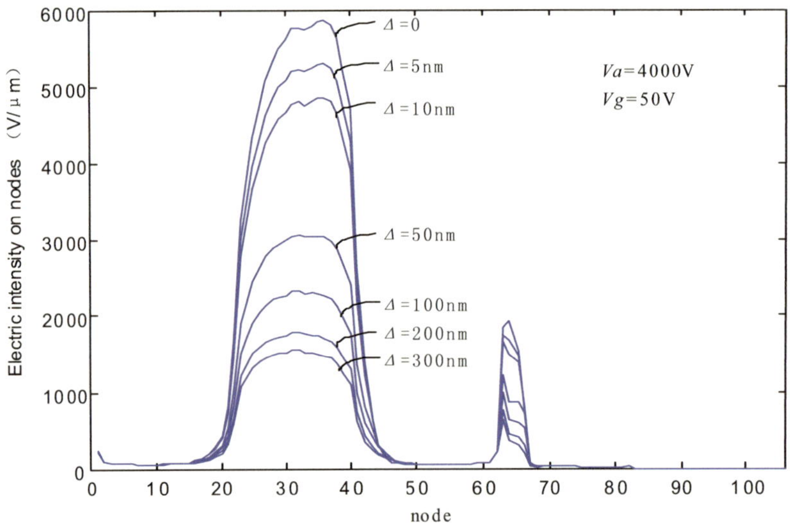


Fig. 6. Relationship of electric intensity on nodes with vertical distance between cathode and gate (the first 62 nodes on tube, the next 20 ones on gate and the last 24 ones on anode)

| $\Delta$(nm) | 300 | 200 | 50 | 5 | 0 |
|---|---|---|---|---|---|
| Efficiency $\eta$ (%) | 50.5 | 45.3 | 44.25 | 30.25 | 17 |

Table 1. Relationship of electron transmission efficiency with $\Delta$

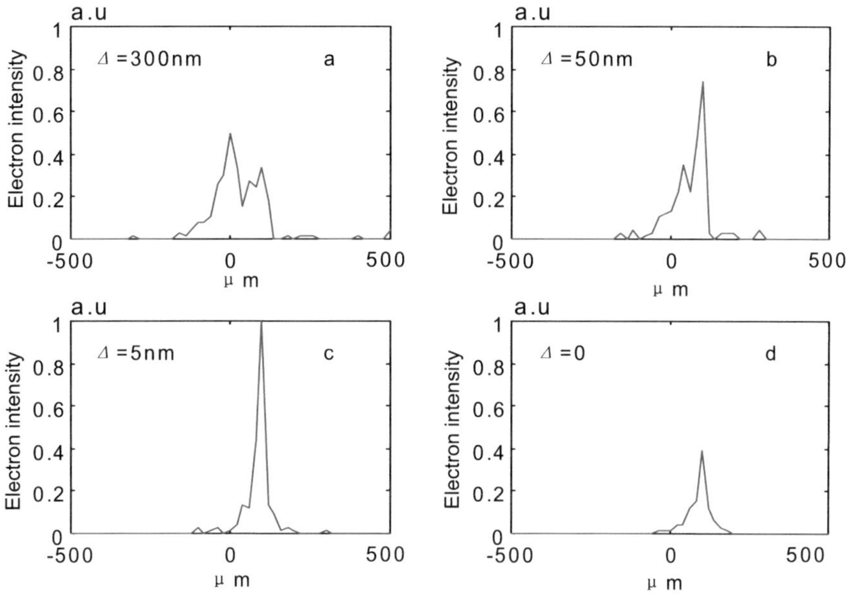


Fig. 7. Distribution of emitted electrons on anode (Va=4000V, Vg=50V, X axis representing the X coordinate of emitted electrons, ranged from -500 to 500μm)

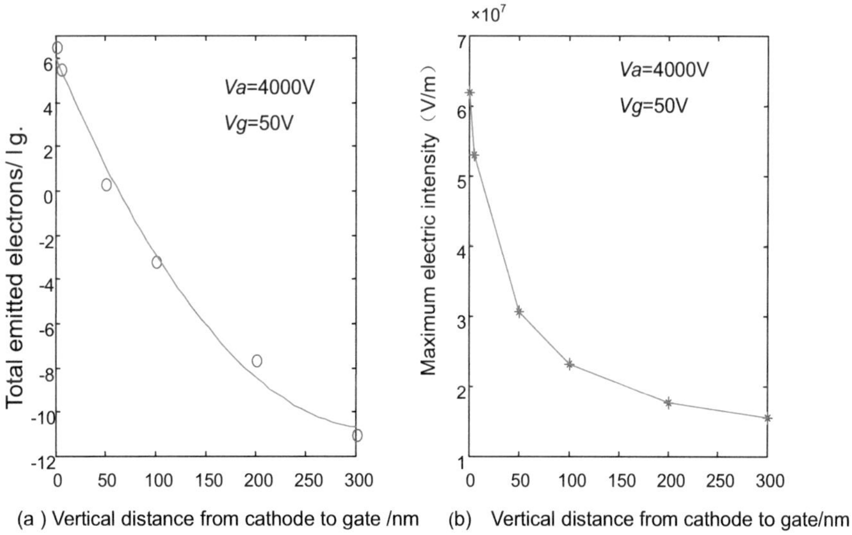


Fig. 8. (a) Relationship of total emitted electrons with the vertical distance Δ (b) Relationship of maximum electric intensity on cathode with Δ

### 3.4 Conclusion

Although the driving voltage of the FED with surface conduction of the Canon corporation is lower (12V), but the distance between cathode and gate is just about 10nm, resulting in the lower electron transmission efficiency, thus limiting the brightness of the display. By increasing the distance between the cathode and gate and the relative height between cathode and gate, especially increasing the elevation angle of CNTs, one can substantially improve the electron transmission efficiency. On the other hand, the focusing performance of electrons reaching anode would be worse, and the gate voltage has to increase, so that we must make an appropriate choice among electron transmission efficiency, driving voltage, resolution and simple arts and crafts.

## 4. Design and experiment on a field emission display prototype based on CNTs

### 4.1 The main points and technical background

Display devices based on carbon materials are considered to be the best choice for field emission large area displays (40″ diagonal or larger). CNTs are capable of emitting high currents (up to 1 A /cm$^2$) at low fields (~5V/μm) [16], and are believed to be ideal candidates for the next generation of field emission flat panel displays and lighting elements. Choi[17] *et al* have demonstrated a fully sealed 4.5 in diode field-emission display using single-wall CNT organic binder. In order to use CNTs as the electron field emitters for large-area displays, it is desirable that an inexpensive substrate such as a lime glass plate can be used for CNTs deposition. A patterned conductive layer needs to be formed on a glass plate before CNT is coated onto the substrate and used as electrode lines. Because of the low melting point of glass and the large mismatch in the coefficient of thermal expansion between metal and glass, carbon deposition cannot be performed at too high a temperature. Previous efforts have been concentrated on controlling the growth process to produce arrays of aligned CNT on patterned substrates and have been successful in some aspects[18~21]. At the same time, various low temperature chemical vapor deposition (CVD) techniques are being studied to achieve low threshold field emission of electrons at a high emission current density [22][23]. Vertically aligned CNTs have been synthesized by plasma- enhanced CVD [1].

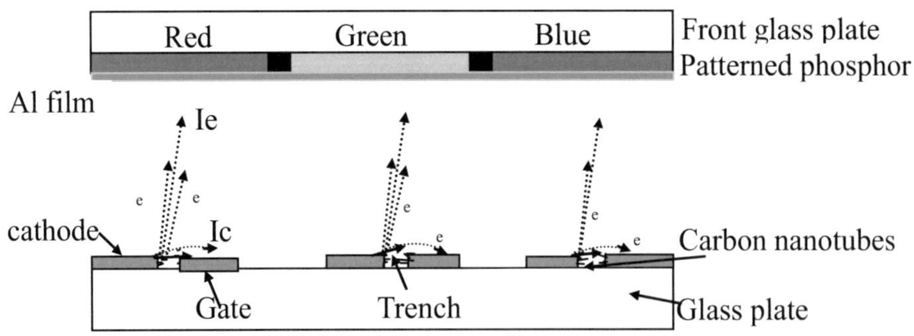


Fig. 9. Schematic structure of a pixel of the plat panel display

Nevertheless, control of a large area synthesis is not easily accessible with such approaches, because this technique requires methods such as e-beam lithography to produce patterned catalyst that is time consuming and expensive for large devices. On the other hand, the complexity of manufacturing gate electrode either on top or beneath of the cathode is unendurable, especially the rigorous requirement of uniform for both the radii of emitter tips and the gate holes, which blocked the development of traditional triode FED. One more efficient way of producing CNT field emission devices is mixing CNT with metal nano-particles in epoxy, results indicate that aligning CNT with the polymer matrix is unnecessary, and the field emission properties comparable to those of aligned CNT can be endurable.

In this part we propose a new type of triode FED prototype that the field emitters are fabricated with CNT organic binders, seeing that the cathode and gate electrode are on the same base plate, this type of FED is easily scalable, and the man stages can be more simplified compared with traditional ones. Besides, it has a high electron transmission ratio of 29.3%, which is well consistent with the results of our theoretical simulation. The proposed triode structure of display is more simple in its fabrication process compared with traditional ones. Our FED prototype possesses a series of good performances such as outstanding brightness, low power consumption et al.

### 4.2 Experiment description

In our experiment, the panel structure of FE-flat panel display is designed as shown in Fig. 9, there are two sets of glass plates: front plate and base plate, CNT binder is between the patterned cathode and gate electrodes on the base plate, and phosphor-coated Al film is on the front glass. The distance between two sets of glass plates is about 1mm, each pixel includes three sub-pixels to realize full color display. The manufacture arts include: firstly, after thoroughly cleaning the substrate with pure water and organic solvent, a pair of device electrodes is formed on the insulating substrate by a proper means , such as vacuum deposition or photolithography; secondly, longitudinal and transverse metal stripes (which are kept insulated with SiO2 at the cross part) are formed to link the device electrodes respectively, thus a series of matrix configuration ( pixel) is formed; thirdly, after the CNT polymer is silk screened between the device electrodes, the polymer was heated at 300°C to 400°C for a little more than 10 minutes, when the organic components are removed, the binder can have a high conductivity to provide electrons to the emitters incessantly, at last, we etch a trench in the middle of the CNT paste with laser beam(YAG). We can see clearly that CNTs are kept on the edge of trench while the nickel nano-particles are melted (see Fig.10b), resulting from their different melting points, and no other than these CNTs consist the virtual emitters. The scanning electron micrograph (SEM) image indicated that not only the slope angles and lengths of CNTs on the edge of trench are not the same, but the CNTs density at different area along the trench is also not uniform. All the factors above can lead the light dots on screen very different, of both its brightness and size. When the density of CNTs reached to certain numbers, the light dots should connect together in a bright line. The trench width is about 0.05mm to 0.37mm depending on the thickness of paste and the diameter and power of laser beam, in our experiment, the trench width is about 200μm, and CNTs diameters are distributed from 20 nm to 40 nm.

The experiments are conducted in a vacuum chamber with the pressure level of $1\times10^{-3}$ Pa. Fig.11 shows the light spots at different anode voltage above the trench, all the light spots are arranged on a line except the top right one, which is the emitting of CNTs on a protuberance formed when pull up the silk screen, only the voltage on anode is high enough to produce electron emitting. in order to confirm this, we reduced the voltage on gate electrode down to zero and found that all the light spots arranged on a line disappeared while the spot on top right existing. In this point, the electro-conductive stripe can also be made with CNT paste as long as the stripe surface is smooth enough, only the voltage on the anode is not high enough to pull out electrons from CNTs on upside of the stripe surface.

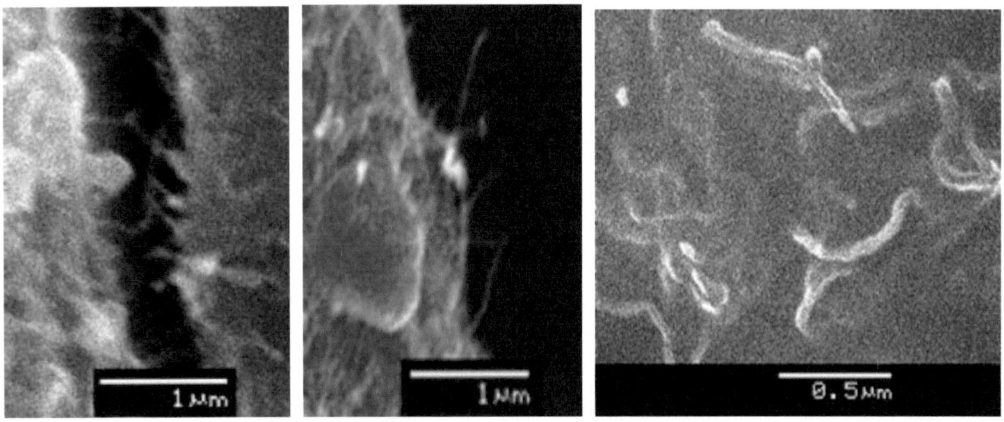


a b c

Fig. 10. Nanotubes on the trench: a). Nanotubes on both edge a trench formed by cool shrinkb). Nanotubes on a trench etched by laser beam c). Nanotubes on the binder surface

Fig. 11 shows the light spots on screen with anode voltage equal to 2.8kV, 2.9kV, 3.0kV and 3.2kV respectively, while the gate voltage keep on 380V. From the emission display we can see that the higher the anode voltage is, the more CNTs take part in emitting electrons and the more electrons reach to anode, therefore, giving higher emission efficiency. In Fig.15c, the total emitting current from CNTs ($I_c$) on the edge of trench is 1.1μA, the emission current arrived at anode ($I_e$) is 0.33μA, the emission current ($I_s$) from CNTs on the protuberance is 0.01μA when the gate voltage is reduced to 0, so the electron transmission ratio is: $\eta = (I_e - I_s)/I_c = 29.3\%$, which is much higher than that of Canon Corporation [24]. In fact, our previous simulation with Monte Carlo method and Boundary Element method has demonstrated that when the slope angle of CNT improves to 30 degree, almost all electrons can reach to anode, which implies a higher electron transmission ratio [25]. Fig. 12 a) shows emission current of CNTs as a function of electric voltage. Fig.12 b) is corresponding to Fowler–Nordheim (F–N) plot of CNTs, indicating a conventional field-emission mechanism. The emission current

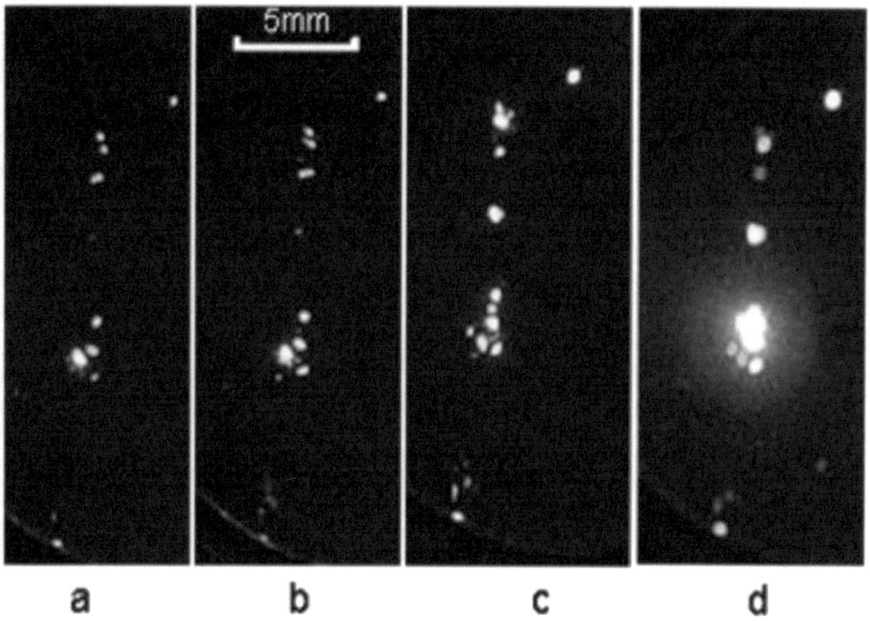


Fig. 11. a, b, c, d are light spots on the screen with different anode voltages 2.8KV, 2.9 KV, 3.0 KV, 3.2 KV respectively(voltage on gate is Vg=380v), The cathode current Ic=1.1μA and the anode current Ie=0.33μA in Fig17c

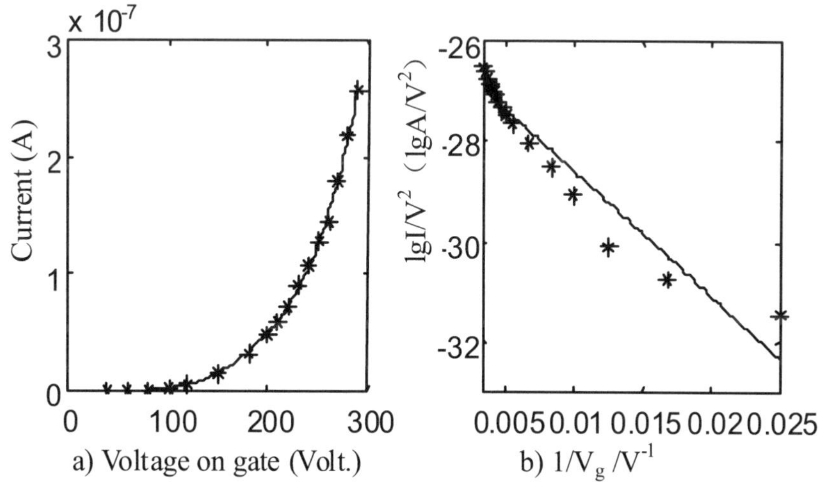


Fig. 12. I-V curve and F-N plot

significantly deviated from F-N behavior in the high-field region, which, we have given a reasonable explanation in other article [26]. We also observed the emission stability of CNTs along trench edge, keeping the voltages on anode and gate fixed, both the total current ($I_c$) and the emitting current ($I_e$) are very stable, and the current fluctuation is not excess to 3% during our one hour's observation.

In summary, a large area, fully colored field emission display panel prototype based upon carbon-CNT emitters was proposed in this research, and there is still a lot of work to do: In our experiment, the trench edge is very irregular, yet the length of CNTs, the wide trench (about 0.2mm) leads to a high gate voltage (380V), the distributing uniformity of CNTs also needs to be improved. We can use a smaller-diameter laser bean to etch CNT paste and before mixing the CNTs by cutting short with chemical or physical methods. To further reduce the gate voltage, single walled CNTs is advisable. To our delight, C.H. Poa[27]et al have mixed high quality single-walled CNT paste suitable for our application. When all these obstacles are removed, we will get higher image quality and lower gate voltages.

## 5. References

[1] Okuda M. Matsutani S., Asai A., et al. Electron trajectory analysis of surface conduction electron emitter displays (SEDS), *SID Intl. Symp. Digest Tech. Papers*, 1998. 523~526

[2] Bonard J. M., Salvetat J. -P., Stöckli T., et al, Field emission from single-wall carbon nanotube films, *Applied physics letters*, 1998, 73(7), 918~920.

[3] Gulyaev Yu. V., Sinitsyn N. I., Torgashov G. V., et al, Work function estimate for electrons emitted from nanotube carbon cluster films, *J. Vac. Sci. Technol. B*, 1997, 15(5), 422~424.

[4] Rinzler A. G., Hafner J. H., Nikolaev P., et al, Unraveling nanotubes: field emission from an atomic wire, *science*, 1995, 269(5230), 1550~1553.

[5] Dean Kenneth A., Allmen Paul Von, Chalamala Babu R., Three behavioral states observed in field emission from single-walled carbon nanotubes, *J. Vac. Sci. Technol. B*, 1999, 17(5), 1959~1969

[6] Maiti A., Brabec C. J., Roland C. M. et al, Growth energetics of carbon nanotubes, *physical review letters*, 1994, 73(18), 2468~2471.

[7] Gröning O., Küttel O. M., Emmenegger Ch., et al. Field emission property of carbon nanotubes, *J. Vac. Sci. Technol. B*, 2001, 18 (2), 665~682

[8] Wadhawan A., Stallcup II R. E., Stephens K. F. et al, Effects of $O_2$, Ar and $H_2$ on the field-emission properties of single-walled and multiwalled carbon nanotubes. *Applied physics letters*, 2001, 79(12), 1867~1869.

[9] Li Ji, Tian Jinshou,Niu Hanben, A full-color field emission display panel based on carbon nanotubes patent No. 02102178.3, 2002.3

[10] Saito R., Fujita M., Dresselhaus G., et al., Electronic structure of chiral graphene tubules. *Appl. Phys. Lett.*, 1992, 60, 2204~2206.

[11] Bonard J.-M., Salvetat Jean-Paul, Stöckli Thomas et al., Why are carbon nanotubes such excellent field emitters? *Ultramicroscopy*, 1998, 73, 7~10.

[12] Li Zhongyuan, Boundary element method applied in electromagnetic field, Beiking Industry College Publish house, 1987. 33~60

[13] Yao Zongxi, Physics-electronics, Xi'an Jiaotong University Publish house, 1991, 145~151.

[14] Xie Zengquan, Wu Quande, Electron emission and electron spectrum, Beiking University Publish house, 1993,75~79.

[15] Choi Won Bong, Lee N.S., Kang J. H., et al., Carbon-nanotubes for full-color field-emission displays, *Jpn. J. Appl. Phys.* 2000,39(5A), 2560~2564.

[16] J. M. Bonard, T. Stoückli, O. Noury, and A. Chaûtelain, Field emission from cylindrical carbon nanotube cathodes : possibilities for luminescent tubes, *Appl. Phys. Lett.*, 2001, 78(18) , 2775~2777.

[17] W. B. Choi, D. S. Chung, J. H. Kang, H. Y. Kim, Y. W. Jin, I. T. Han, Y. H. Lee, J. E. Jung, N. S. Lee, G. S. Park, and J. M. Kim, Carbon nanotubes for fullcolor field emission displays, *Appl. Phys. Lett.* , 1999, 75 (20), 3129~3131

[18] V.I. Merkulov, D.H. Lowndes, Y.Y. Wei, G. Eres, E. Voelkl, Patterned growth of individual and multiple vertically aligned carbon nanofiber, *Appl. Phys. Lett.*, 2000, 76(24), 3555~3557.

[19] M. Chhowalla, K.B.K. Teo, C. Ducati, N.L. Rupesinghe, G.A.J. Amaratunga, A.C. Ferrari, et al., Growth process conditions of vertically aligned carbon nanotubes using plasma enhanced chemical vapor deposition, *J. Appl. Phys.*, 2001,90(10), 5308~5311.

[20] W. C. Hu, D.W Gong, Z. Chen, L. Yuan, K. Saito, C. A. Grimes, P. Kichambare, Growth of well-aligned carbon nanotube arrays on silicon substrates using porous alumina film as a nanotemplate, *Appl. Phys. Lett.*, 2001, 79(19), 3083~3085.

[21] A. Cao, X. Zhang, C.Xu, J. Liang, D. Wu, B.Wei, Thinning and diluting aligned carbon nanotube films for uniform field emission, *Appl. Phys. A*, 2002, 74 (3), 415~418.

[22] B. O. Boskovic, V.B.Govolvko, M. Cantoro, Low temperature synthesis of carbon nanofibres on carbon fiber matrices, *Carbon*, 2005, 43(11), 2643~2648.

[23] S. Hofmann, C. Ducati, J. Robertson, B. Kleinsorge, Low temeprature growth of carbon nanotubes by plasma—enhanced chemical vapor deposition, *Appl. Phys. Lett.*, 2003, 83(1), 135~137.

[24] W. D. Zhang, Y. Wen, W. C. Tjiu, G. Q. Xu, L.M. Gan, Growth of vertically aligned carbon—nanotube array on large area of quartz plate by chemical vapor deposition Growth of vertically aligned carbon nanotube array on large area of quartz plates by chemical vapor deposition, *Appl. Phys. A*, 2002, 74(3), 419~422.

[25] J. S. Tian, J. Li, B. L. Xu, H. B. Niu, Research on the field emission properties of carbon nanotube, *Acta Photonica Sinica*, 2003, 32(12), 1442~1445. ( in Chinese )

[26] J. S. Tian, J. L, B.L. Xu, H.B. Niu, Structural analysis on a field emission display panel based on carbon nanotube, *Acta Photonica Sinica*, 2003, 32(11), 1343~1348. (in Chinese )

[27] C. H. Poa, S. R. P. Silva, P. C. P. Watts, W. K. Hsu, H. W. Kroto, D. R. M. Walton, Field emission from nonaligned carbon nanotubes embedded in a polystyrene matrix, *Appl. Phys. Lett.* 2002, 80(17), 3189~3191.

# Reinforced Thermoplastic Natural Rubber (TPNR) Composites with Different Types of Carbon Nanotubes (MWNTS)

Sahrim Hj. Ahmad[1], Mou'ad.A.Tarawneh[1], S.Y.Yahya[2] and Rozaidi Rasid[1]
[1]*Material Science Programme, School of Applied Physics, Faculty of Science and Technology, Universiti Kebangsaan Malaysia 43600 Bangi, Selangor,*
[2]*Institute sciences, Universiti Teknologi Mara, 40450, Shah Alam, Selangor, Malaysia*

## 1. Introduction

The emergence of thermoplastic elastomers (TPEs) is one of the most important developments in the area of polymer science and technology. TPEs are a new class of material that combine the properties of vulcanized rubbers with the ease of processability of thermoplastics (Abdullah & Dahlan, 1998). Thermoplastic elastomers can be prepared by blending thermoplastic and elastomers at high shear rate. Thermoplastics, for example, polypropylene (PP), polyethylene (PE) and polystyrene (PS), and elastomers, such as ethylene propylene diene monomer (EPDM), natural rubber (NR) and butyl rubber (BR), are among the materials used in thermoplastic elastomer blends.

Blends of natural rubber (NR) and polypropylene (PP) have been reported widely by previous researchers (Abdullah & Dahlan, 1998; Ismail & Suryadiansyah, 2002). According to Abdullah and Dahlan (1998), polypropylene is the best choice for blending with natural rubber due to its high softening temperature (150$^{o}$C) and low glass transition temperature (-60$^{o}$C, is $T_g$ for NR), which makes it versatile in a wide range of temperatures. Even though NR and PP are immiscible, their chemical structure is nearly the same. Thus, stable dispersion of NR and PP is possible. Incompatibility between NR and PP can be overcome by the introduction of a compatibilizer that can induce interactions during blending. Compatibility is important as it may affect the morphology, mechanical and thermal properties of the blends. Among the commonly used compatibilizers are dicumyl peroxide (DCP), m-phenylene bismaleimide (HVA-2) and liquid natural rubber (LNR). Apart from compatibility, mixing torque and curing are interrelated in determining the homogeneity of the TPNR blend (Abdullah & Dahlan, 1998).

A pioneer group of researchers in UKM has studied extensively the utilization of liquid natural rubber (LNR) as a compatibilizer on various natural rubber/polyolefin blends (Ibrahim Abdullah & Sahrim Ahmad, 1999). Liquid natural rubber was produced by photodegradation of natural rubber (NR) in toluene and exposure to ultraviolet for 6 hours (Dahlan, 1998). The LNR has the same microstructure with NR but with a short chain of polyisoprene (different in molecular weight, $M_w$) (Ibrahim, 2002). The $M_w$ for LNR is around

50 000 whereas for NR it is 900 000, new functional new groups such as -OH, C-O and C=O have been formed via oxidation of photochemical sensitization of NR. The LNR with some active terminals like -OH is expected to react with the plastic particles and thereby bond the plastic particles to the NR matrix. There are also successful studies on the NR and various PE with LNR as a compatibilizer (Abdullah et al., 1995; Che Som et al., 1994).
One of the most important aspects to be considered in the thermoplastic elastomer blend is the components ratio. In addition, the processing parameters involving temperature, rotor speed of the mixer and mixing time, play a significant role in producing good blend properties (Abdullah &Dahlan, 1998; Yang et al., 1995). The processing is normally done at a temperature above the melting point of the thermoplastic. The quality of blend reflects its miscibility, where it exhibits better mechanical and thermal properties, as well as homogeneity. The properties obtained can be further enhanced by introducing a variety of additives, such as compatibilizer agents, interfacial agents, cross-linking agentsor fillers to promote miscibility by reducing interfacial tension.
Mechanical blending of PP and NR with the addition of LNR as a compatibilizer was reported to be optimal at a temperature of 175-185oC and a rotor speed of 30-60rpm (Abdullah &Dahlan, 1998). The percentage of LNR used depends on the ratio of NR to PP. For a NR:PP ratio of 30:70 the best physical properties are obtained at 10% LNR (Abdullah & Ahmad, 1992). The compatibilizer helps to induce the interaction between the rubber and plastic interphase and thereby increases the homogeneity of the blend.
Nanometer scale reinforcing particles have attracted considerable attention from polymer scientists. Due to their high aspect ratio (surface/area ratio) and low density, they may be used as substitutes for traditional fibers as fillers in polymer matrices. The excitement that followed the discovery of the possibility of preparing multiwall carbon nanotubes (CNTs) and other carbon nano-structured materials via a catalyst-free process (Ijima S, 1991) inspired scientists regarding a range of potential applications. Depending on the synthesis conditions, nanotubes can be single-walled (a single tubule of 1 nm) or multi-walled (2–50 tubules of 2–50 nm positioned concentrically). The measurements have determined that carbon nanotubes possess excellent mechanical properties (tensile modulus ≈1 TPa (Yu F et al, 2002), tensile strength ≈11–63 MPa (Wagner D et al., 1998)and compressive strength ≈150 GPa (Lourie O et al., 1998)in the direction of the tubule axis. This compressive strength is one order of magnitude higher than that of any other known material. This makes them ideal reinforcing fibers for the manufacture of the next generation of fiber-matrix composite materials (Sinnott B et al., 1998). Conventional microfillers including mica, talc, glass fiber, carbon black, etc. are incorporated in polymer to further improve its mechanical performances. However, additions of a large volume content of microfillers (20–30 wt %) are needed to achieve the desired mechanical properties. The thermal properties of CNT/polymer composites are of particular interest in many applications such as the development of conductive polymer films and electronic components. Because of the high thermal conductivity of CNTs (for SWNTs and MWNTs are 6000 W/m K and 3000 W/m K, respectively) and their extremely high aspect ratio (>1000), the thermal conductivity of CNT-related composites is expected to be very high. This also implies that using only a small amount of CNTs should be able to significantly improve the thermal conductivity properties of polymer-based materials. Compared with SWNTs, MWNTs are more predominantly used as conductive fillers due to their lower cost, better availability and easier dispersability. The addition of a large volume content of microfillers is detrimental to

the processability of the composites. However, only very small amounts of CNTs are required to improve the properties of the polymer. Nanophased matrices based on polymers and carbon nanotubes have attracted great interest because they frequently include superior mechanical, electronic, and flameretardant properties. Different polymer/CNT nanocomposites have been synthesized by incorporating carbon nanotubes (CNTs) into various polymer matrices, such as polyurethane (H.C.Kuan et al., 2005), epoxy (Liao et al., 2004), polypropylene (Li et al., 2004), polyimides (Cai et al., 2004), and polyurethane (Koerner et al., 2005; Kuan et al., 2005).

Currently, the use of composites has primarily been limited to high performance applications like the automotive, aerospace and sporting goods because of low properties. There has been much debate in the literature as to whether this interface region is important in composites and what effect it has on the properties of composites. As far as mechanical and thermal properties are concerned, the main target is to strike a balance of stiffness, strength, toughness, thermal conductivity, specific heat, melting temperature and glass transition temperature. However, with the explosion of the information age and higher demand for materials with superior properties, composites such as TPNR are now making their way into more common applications. Therefore, the most important aspect in the development of reinforced TPNR by carbon nanotubes isto achieve a good combination by incorporating carbon nanotubes into TPNR based composites.Synergistic effects may be achieved in the form of higher mechanical and thermal properties, therefore, making use of this potential will lead to new high performance products and technology that was hitherto not possible.

Despite the considerable number of studies concerning the preparation, characterization and properties of polymer/CNT nanocomposites, no adequate information has been reported on the phenomena behind the property changes resulting from the addition of MWNTsto thermoplastic natural rubber.

In this chapter, thermoplastic natural rubber TPNR reinforced with two types of multi-walled carbon nanotube (MWNTs) nanocomposites were prepared by the melt blending method. Using this method, MWNTs will be dispersed homogeneously in the TPNR matrix in an attempt to increase the mechanical properties of these nanocomposites. The effect of MWNTs on the mechanical properties of TPNR nanocomposites is reported in this chapter.

## 2. Experiment details

Polypropylene, with a density of 0.905 g cm-3, was supplied by Propilinas (M) Sdn. Bhd, natural rubber was supplied by Guthrie (M) Sdn. Bhd, and polypropylene (PP) with a density of 0.905 g/cm3 was supplied by Polipropilinas (M) Sdn. Bhd were used in this research. Maleic anhydride-grafted-polypropylene (MAPP) with a density of 0.95 g/cm3 was supplied from Aldrich Chemical Co., USA. The maleic anhydride content in MAPP was about 0.57%. The chemicals used were sodium hydroxide and sulfuric acid (95-97) % supplied by MERCK, Germany. Liquid natural rubber (LNR) was prepared by the photochemical degradation technique.

In this study two types of MWNTs were used. The first one, MWNTs 1, was supplied by the Chinese academy of science. The second one, MWNTs 2, and was provided by Arkema (GraphistrengthTM C100). Table 1 shows the properties of the two types of multi-walled carbon nanotubes (MWNTs).

Mixing was performed by an internal mixer (Haake Rheomix 600P). The mixing temperature was 180° C, with a rotor speed of 100 rpm and 13 min mixing time. The indirect technique (IDT) was used to prepare nanocomposites, this involved mixing the MWNTs with LNR separately, before it was melt blended with PP and NR in the internal mixer. TPNR nanocomposits were prepared by melt blending of PP, NR and LNR with MWNTs in a ratio of 70 wt% PP, 20 wt% NR and 10wt% LNR as a compatibilizer and 1,3,5 and 7% MWNTs.

| MWNTs | Purity | Length | Diameter | Manufactured |
|---|---|---|---|---|
| MWNTs 1 | >% 95 | 10-30 μm | 8< nm | Catalytic Chemical Vapor Deposition ( CVD) |
| MWNTs2 | >90% | 0.1-10 μm | 10-15 nm. | Catalytic Chemical Vapor Deposition ( CVD) |

MWNTs 1: Fist type of multi-walled carbon nanotubes.

MWNTs 2: Second type of multi-walled carbon nanotubes.

Table. 1 Properties of the two types of multi-walled carbon nanotubes (MWNTs)

The tensile properties were tested using a Testometric universal testing machine model M350-10CT with 5 kN load cell according to ASTM 412 standard procedure using test specimens of 1 mm thickness and a crosshead speed 50 mm $min^{-1}$. At least five samples were tested for each composition, and the average value was reported. The impact test was carried out using a Ray Ran Pendulum Impact System according to ASTM D 256-90b. The velocity and weight of the hammer were 3.5m/s and 0.898kg, respectively. Dynamic mechanical analysis for determining the glass transition temperature, storage and loss modulus was carried out using DMA 8000 (PerkinElmer Instrument), operating in single cantilever mode from -100 to 150 oC at a constant frequency of 1 Hz, with a heating rate of 5 oC/min. The dimensions of the samples were 30 x 12.5 x 3 mm.
The thermal conductivity was measured by a laser flash method. Disk-type samples (12.7 mm in diameter and 1mm in thickness) were set in an electric furnace. Specific heat capacities were measured with a differential scanning calorimeter DSC. Thermal diffusivity ($\lambda$, $Wm_{-1}$ $K_{-1}$) was calculated from thermal diffusivity ($\alpha$, $m^2$ $s_{-1}$), density ($\rho$, g $cm_{-3}$) and specific heat capacity ($C_p$, J $g_{-1}$ $K_{-1}$) at each temperature using the following:

$$\lambda = \alpha.\ \rho.\ C \tag{1}$$

The reference used for the heat capacity calculation was a 12.7mm thick specimen of pyroceram. The reference sample was coated with a thin layer of graphite before the measurement was performed. The thermal conductivity of MWNTs reinforced TPNR matrix composites of all volume fractions was studied from 30°C to 150°C. The morphology of the MWNTs and the composite were examined using a scanning electron microscope (Philips XL 30). The samples were coated with a thin layer of gold to avoid electrostatic charging during examination.

## 3. Results and discussion

### 3.1 Tensile properties

#### 3.1.1 Tensile strength

The tensile strength of TPNR reinforced with two types of MWNTs of different percentages (1%, 3%, 5% and 7%) is shown in Figure 1. Generally, both MWNTs exhibited an increasing trend up to 3wt% content. Further increments in MWNTs content decreased the tensile strength compared to the optimum filler loading.

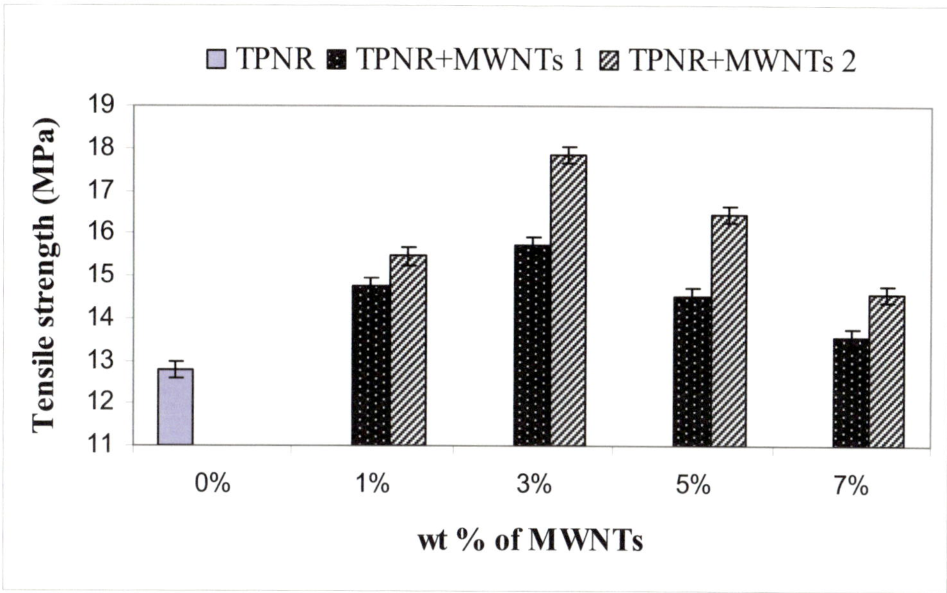


Fig. 1. Tensile strength of TPNR reinforced with two types of MWNTs

From Figure 1,TPNR with MWNTs 1 and MWNTs 2 have optimum results at 3 % wt, which, compared with TPNR, increased by 23% and 39%, respectively. The tensile strength increased radically as the amount of MWNTs concentration increased. The mechanical performance, such as tensile properties, strongly depends on several factors such as the properties of the filler reinforcement and matrix, filler content, filler length, filler orientation, and processing method and condition. The improvement in the tensile strength may be caused by the good dispersion of MWNTs in the TPNR matrix, which leads to a strong interaction between the TPNR matrix and MWNTs. These well-dispersed MWNTs may have the effect of physically crosslinking points, thus, increasing the tensile strength (Bin et al., 2006).

A good interface between the CNTs and the TPNR is very important for a material to stand the stress. Under load, the matrix distributes the force to the CNTs, which carry most of the applied load. When the content of MWNTs is higher, the MWNTs cannot disperse adequately in the TPNR matrix and agglomerate to form a big cluster. This is because of the huge surface energy of MWNTs as well as the weak interfacial interaction between MWNTs and TPNR, which leads to inhomogeneous dispersion in the polymer matrix and negative effects on the properties of the resulting composites that causes a decrease in the tensile strength (Sang et al., 2007).

### 3.1.2 Young's modulus

Figure 5.2 shows the effect of filler content on the tensile modulus of TPNR reinforced by two types of MWNTs. The same trend as for the tensile strength in Figure .1 was observed for the tensile modulus of MWNTs 2, however, for MWNTs 1 it was different. Figure 2 clearly shows that the presence of MWNTs has significantly improved the tensile modulus of the TPNR.

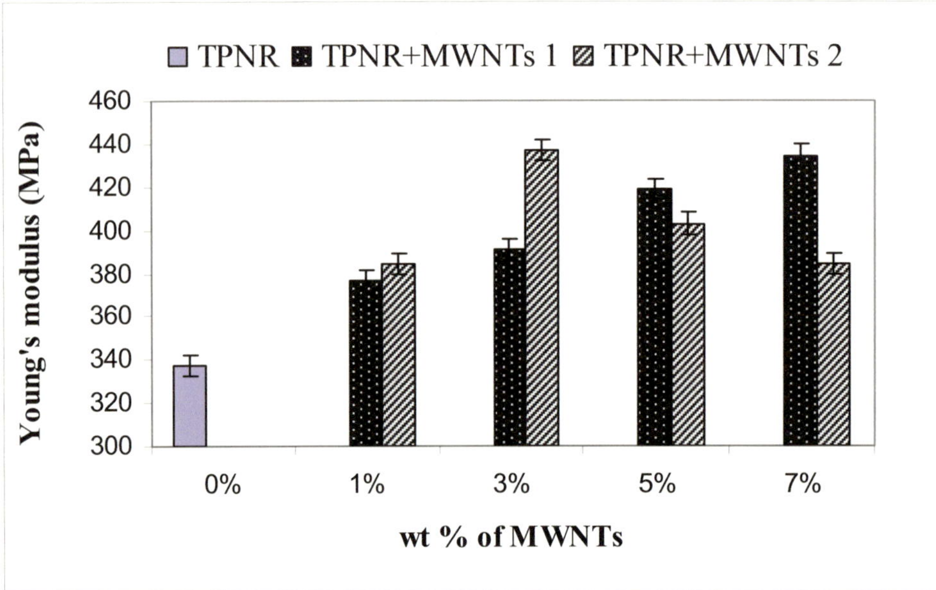


Fig. 2. Young's modulus of TPNR reinforced with two types of MWNTs.

The remarkable increase of Young's modulus with MWNTs 1 content shows a greater improvement than that seen in the tensile strength at high content, which indicates that the Young's modulus increases with an increase in the amount of the MWNTs 1. At 1 wt% of MWNTs the Young's modulus is increased by 11% compared to TPNR. At 3 wt% of MWNTs the increase in the Young's Modulus is about 16%. Further addition of MWNTs from 5 to 7 wt% increased the Young modulus by about 24% and 29%, respectively. The improvement of modulus is due to the high modulus of MWNTs (Treacy et al., 1996).

As depicted in Figure 2 the Young's modulus of MWNTs 2 increased with the increase in the amount of MWNTs. The maximum result was achieved at 3% wt, with an increase of about 30%, which was due to the good dispersion of nanotubes displaying perfect stress transfer (Potschke et al., 2002). As explained before, a reduction in performance occurred at higher filler contents. Initially it increases with filler content and then decreases when exceeding the filler loading limit due to the diminishing interfacial filler-polymer adhesion. It is assumed that aggregates of nanotube ropes effectively reduce the aspect/ratio (length/diameter) of the reinforcement (Lopez Manchado et al., 2005).

### 3.1.3 Elongation at break

The elongation at the break of TPNR with two types of MWNTs is shown in Figure 3. For MWNTs 1, the elongation at break increased with the increase in the amount of MWNTs, at

3% wt the result is the optimum, which increased 29% compared with TPNR. However, the elongation at break of MWNTs 2 decreased with the increase in the amount of MWNTs.
It can be deduced that the reinforcing effect of MWNTs is very marked. As the MWNTs content in the TPNR increases, the stress level gradually increases but at the same time the strain of the nanocomposites decreased. This is because the MWNTs included in the TPNR matrix behave like physical crosslinking points and restrict the movement of polymer chains (Sang et al., 2007). This shows that the inclusion of MWNTs makes the TPNR stronger but more brittle.

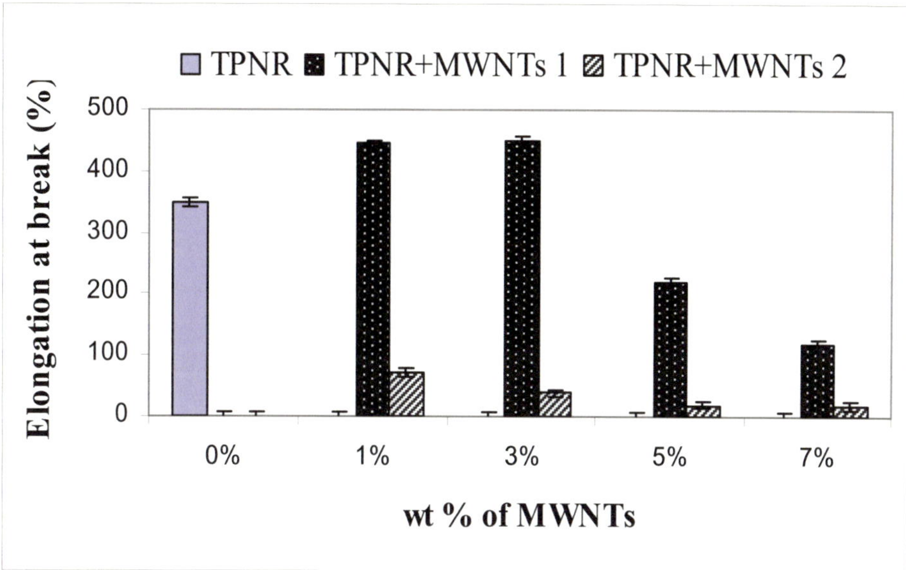


Fig. 3. Elongations at break of TPNR reinforced with two types of MWNTs.

### 3.2 Impact strength

The effect of filler loading on the impact strength of TPNR/MWNTs nanocomposites (MWNTs 1 and 2) is given in Figure 4. It shows that incorporation of MWNTs into TPNR considerably affects the impact strength of TPNR nanocomposites.
The results exhibited better impact strength for MWNT 1 and MWNT 2 at 5 wt% with an increase of almost 70 % and 74%, respectively. This is due to the better dispersion of carbon nanotubes in the matrix, which generated a significant toughening effect on the TPNR/MWNTs nanocomposite. However, when the load is transferred to the physical network between the matrix and the filler, the debonding of the chain segments from the filler surface facilitates the relaxation of the matrix entanglement structure, leading to higher impact toughness.
The low impact energy was attributed to the filler content. This will reduce the ability of reinforced composites to absorb energy during fracture propagation. However, in the case of elastomer-toughened polymer, the presence of the elastomer basically produces stress redistribution in the composite, which causes micro cracking or crazing at many sites, thereby resulting in a more efficient energy dissipation mechanism (Canche-Escamilla et al., 2002).

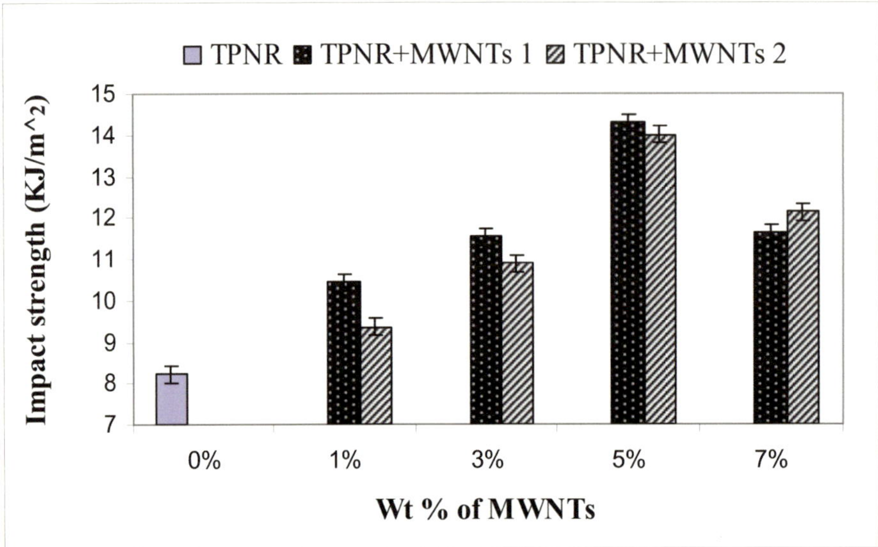


Fig. 4. Impact strength of TPNR reinforced with two types of MWNTs.

Consequently, because of their higher surface energy and large aspect ratio, it will be difficult for the nantubes to disperse in the TPNR when the MWNTs content is higher. This will lead to less energy dissipating in the system due to the poor interfacial bonding and induces micro spaces between the filler and polymer matrix. This causes micro-cracks when impact occurs, which induces easy crack propagation. Therefore, the higher agglomeration of MWNTs can cause the mechanical properties of the composites to deteriorate (Jianfeng Shen et al., 2007).

### 3.3 Dynamic mechanical analysis

#### 3.3.1 Storage modulus

Figure 5 and Figure 6 illustrate the storage modulus of TPNR reinforced with MWNTs 1 and MWNTs 2. As illustrated in these Figures, the storage modulus of the nanocomposites mainly depends on the filler content in the composite. Two consistent variations of the E′ with temperature are observed for all systems, which are a sharp drop around -50°C (±10°C) and a reduction in the E′ drop rate with temperatures above 80°C. The first drop is related to the glass transition region of compatibilised TPNR blends. The reduction in E′ is less severe until the softening point or when the melt region is reached.

The storage modulus of neat TPNR and its nanocomposites decreased with the increasing temperature, as expected. However the most characteristic difference is that the storage modulus E' of TPNR/MWNTs 1 nanocomposites tends to increase with the increase in the amount of MWNTs. This indicates the effectiveness of the MWNTs' reinforcing effect. The maximum values of storage modulus for the first type (MWNTs 1) were obtained at 1wt% and 3wt%, however, for the second type (MWNT 2) it was obtained at 3wt% and 5wt%.

At the same time the storage modulus of two types of MWNTs, below $T_g$, of TPNR/MWNTs is higher compared with composites above $T_g$. The stiffening effect is maximized at temperatures below $T_g$ and becomes greatly reduced at temperatures above $T_g$. Such a phenomenon has also been observed in other well dispersed polymer composite systems

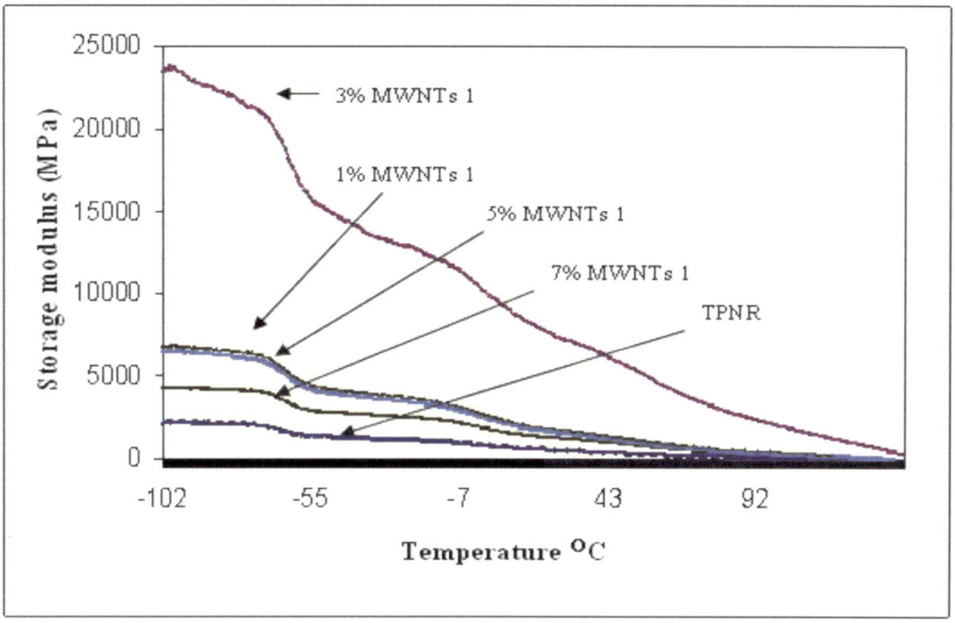


Fig. 5. Storage modulus of TPNR and TPNR/MWNTs 1 nanocomposites.

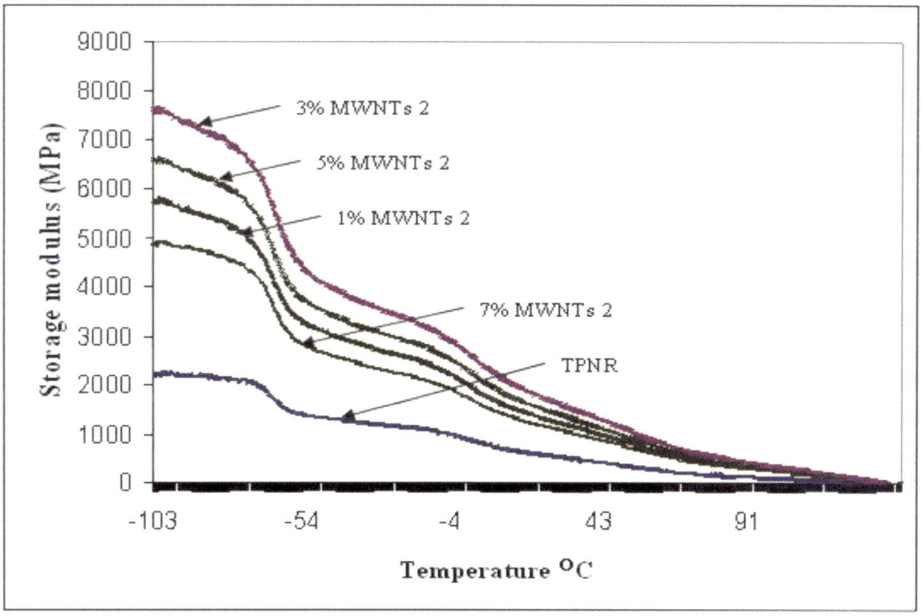


Fig. 6. Storage modulus of TPNR and TPNR/MWNTs 2 nanocomposites.

(Sue et al., 2004). The carbon nanofibers are effective at increasing the stiffness of the prepared nanocomposites, even at relatively small nanofiber loadings. For 1% and 3% of MWNTs this increase should be attributed to the finer dispersion of these nanotubes in the TPNR matrix and the decreased size of the formed aggregates. After 3% of MWNTs 1 and after 5% of MWNTs 2 the storage modulus decreased due to the agglomeration of nanotubes inside the matrix.

### 3.3.2 Loss modulus

In addition, from Figure 7 and Figure 8, the loss modulus E'' of the composites is higher than TPNR at 1% and 3% of MWNTs 1, and 3% and 5% of MWNTs 2, which is because the nanotubes can be more uniformly dispersed in the matrix at this percentage of MWNTs. Therefore, the interfacial area is larger and the interaction between the matrix and nanotubes is stronger.

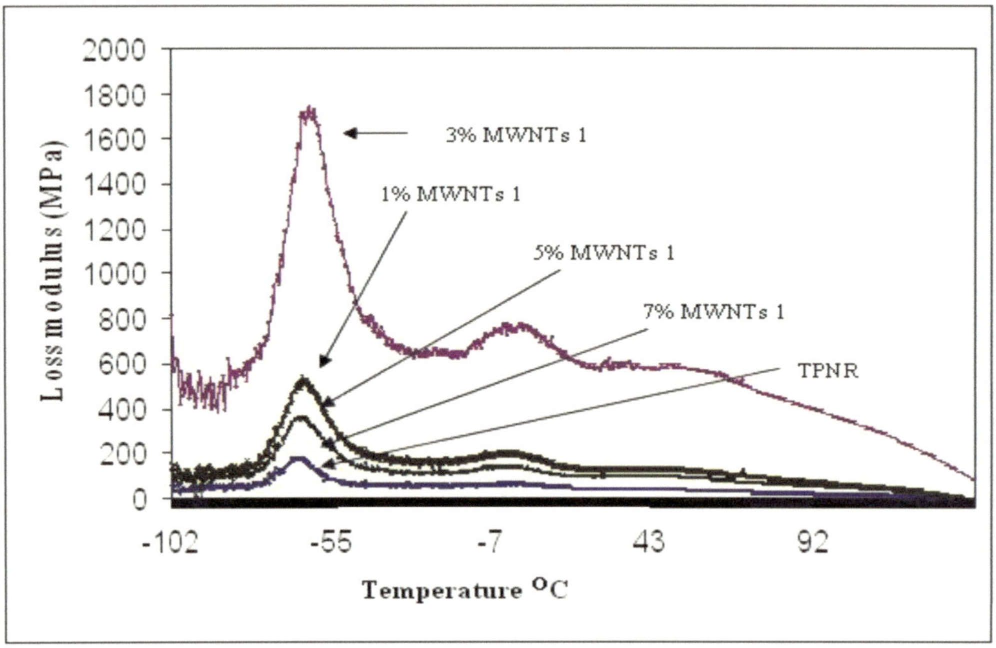


Fig. 7. Loss modulus of TPNR and TPNR/MWNTs 1 nanocomposites.

Since the value of E'' at lower temperatures has a great influence on the toughness of the matrix, it is obvious that MWNTs contribute to the improvement of the toughness of the composites. A mesophase is formed between the matrix and MWNTs. Therefore, when the composites were subjected to external stress, the external energy was dissipated by the friction between MWNTs-MWNTs and MWNTs-TPNR interaction through the mesophase. However, after 3% of MWNTs 1 and 5% of MWNTs 2 the loss modulus E'' had decreased due to agglomeration of MWNTs inside the matrix as mentioned before. Due to the polymer-filler interaction, the adsorption of the polymer chains on the filler's surface reduces the mobility of the macromolecular segments. This transition zone surrounding the nanoparticles exhibits a higher modulus, which gradually reduces with the increasing distance from the filler surface (Wang, 1998).

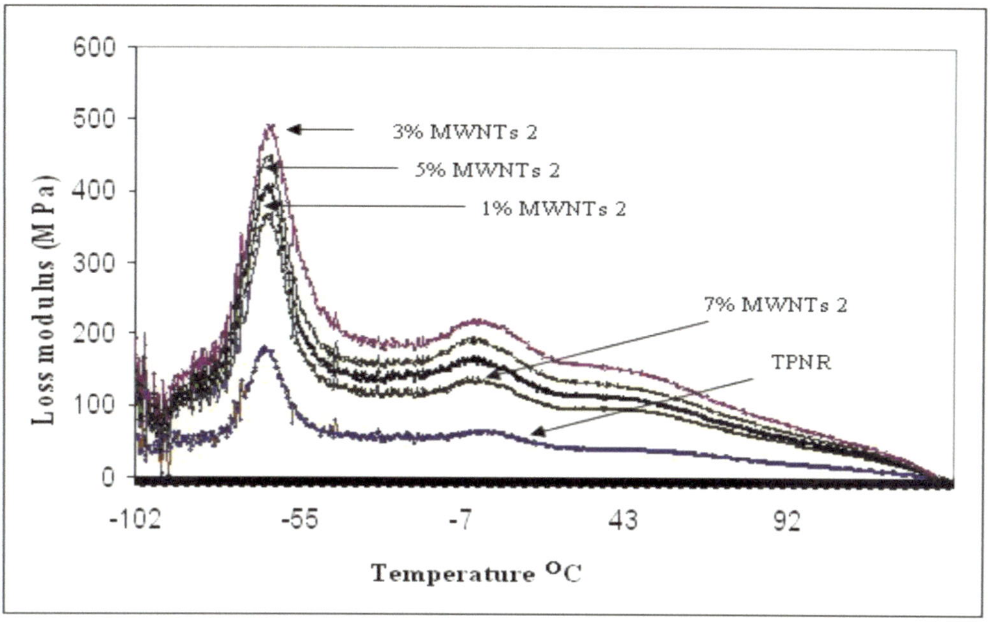


Fig. 8. Loss modulus of TPNR and TPNR/MWNTs 2 nanocomposites.

### 3.3.3 Tan delta

The dynamic mechanical data shows that the glass transition temperature of the TPNR/MWCNTs is affected by the addition of the different amounts of two types of MWNTs, as depicted in Figure 9 and Figure 10.

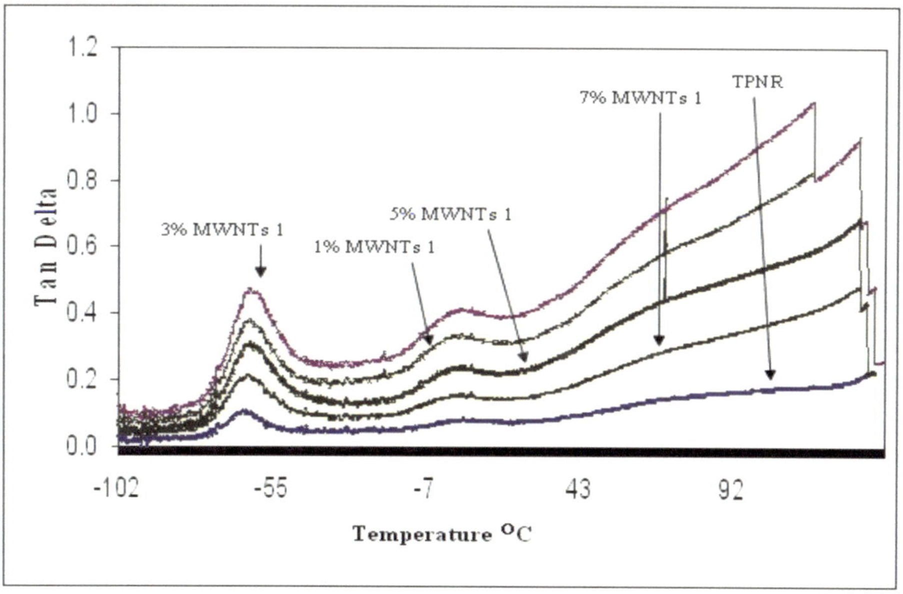


Fig. 9. Tan delta of TPNR and TPNR/MWNTs 1 nanocomposites.

From the figures, the tan δ peak of the nanocomposites is higher than the corresponding temperature of TPNR. As mentioned earlier, usually the $T_g$ of a polymeric matrix tends to increase with the addition of carbon nanotubes. The rise in $T_g$ in any polymeric system is associated with a restriction in molecular motion, reduction in free volume and/or a higher degree of crosslinking due to the interactions between the polymer chains and the nanoparticles, and the reduction of macromolecular chain mobility (Potschke et al., 2002).
With the high amount of MWNTs (after 3% of MWNTs 1 and 5% of MWNTs 2) the $T_g$ drops. This maybe due to the phase separation/agglomeration of MWNTs, which allows the macromolecules to move easily. When the content of MWNTs is higher, the MWNTs congregate, possibly because the intrinsic van der Waals forces occurs, which leads to bubbles and small aggregates. The conglomerations and matrix holes existing in the network of MWNTs may perform as defects, which make the macromolecules move easily, and the $T_g$ of the matrix is decreased.

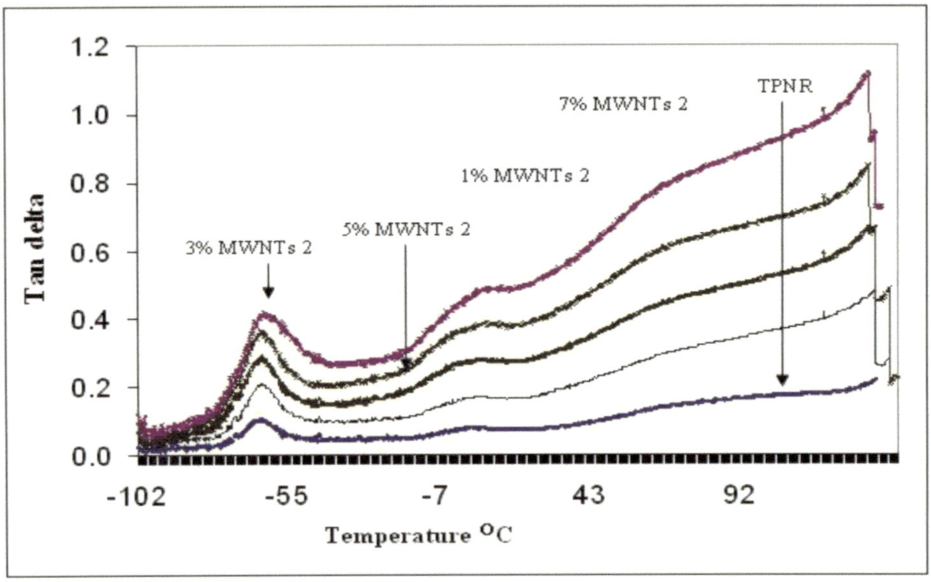


Fig. 10. Tan delta of TPNR and TPNR/MWNTs 2 nanocomposites.

### 3.4 Thermal behavior of TPNR/MWNTS (Two Types OF MWNTS)

#### 3.4.1 Thermal conductivity

To study the effect of MWNTs filler on thermal conductivity, the temperature was varied from (30 - 150) □C. The temperature range was selected based on the melting point of polypropylene (PP), since PP melts above 178 °C (Mustaffa, 1991). The carbon filler loading was from 1%wt to 7%wt for two types of carbon nanotubes. Introducing MWNTs to TPNR can significantly enhance the thermal conductivity of the TPNR matrix, as shown in Figure 11 and Figure 12.
As shown from the figures at 30°C the thermal conductivity of TPNR/MWNTs 1 composites, Thermal conductivity increased 232%, 222%, 17.69% and 18.6% of 1%, 3%, 5% and 7%, respectively, and for TPNR/MWNTs 2, the thermal conductivity increased 80%, 94%, 125% and 108% of 1%, 3%, 5% and 7% respectively, as compared with TPNR at the

same temperature. Thermal transport in the CNT composites includes phonon diffusion in the matrix and ballistic transportation in the filler.

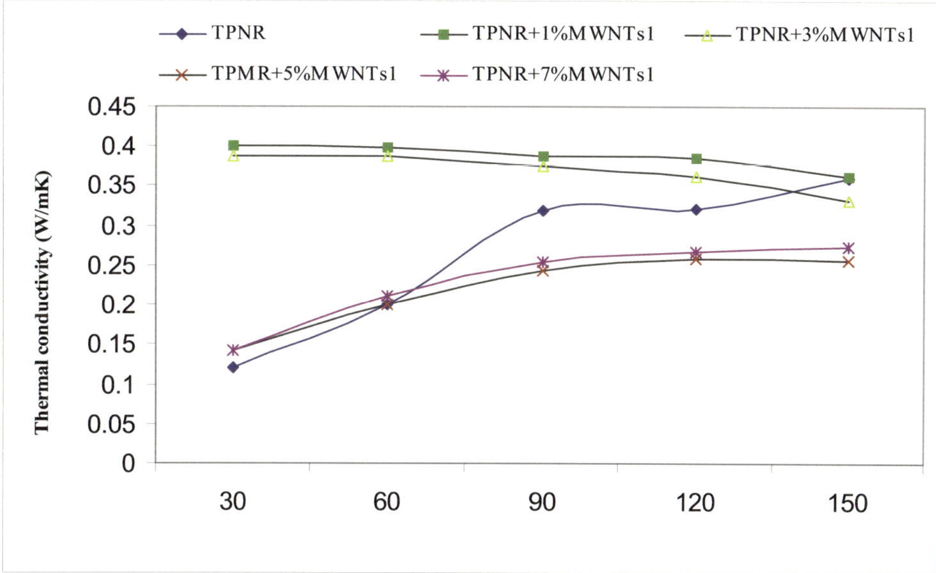


Fig. 11. Thermal conductivity of the composites with different volume fractions of MWNTs 1 (first type) at different temperatures

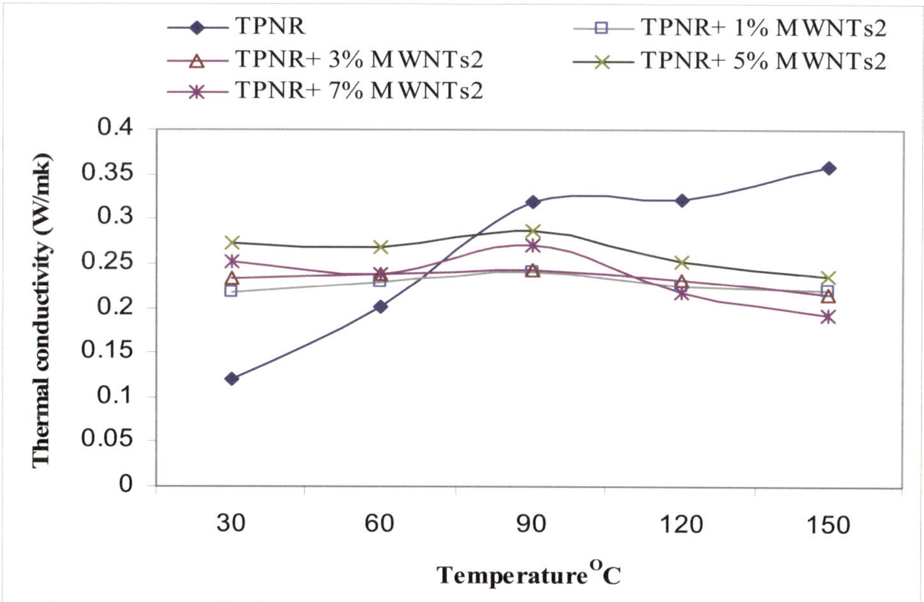


Fig. 12. Thermal conductivity of the composites with different volume fractions of MWNTs 2 (second type) at different temperatures

The improvement in thermal conductivity in MWNTs/TPNR may stem from the improved percolation because of the better dispersion and formation of a network (Kumar et al., 2007). The dispersion of 1wt% and 3wt% MWNTs is better than 5wt% and 7wt% in TPNR for the first type (MWNTs 1), at 5% and 7% the MWNTs 1 agglomerated inside TPNR. Therefore, the large amounts of junctions among the carbon nanotubes form a single conducting path, which is believed to be the reason why the measured thermal conductivity is low. For the second type (MWNT 2) the conductivity at 5wt% and 7wt% is better than 1% and 3%, respectively.

The significant enhancement in the thermal conductivity of CNT nanocomposites is possibly attributed to the kinks or twists of MWNTs. When the phonon travels along the nanotubes, if it meets the kinks or twists it would be blocked at those sites. The existence of such kinks or twists in CNTs would lead to a decrease in the effective aspect ratio of the nanotubes (Nan et al., 2003) when the amount of MWNTs increases, and, thus, the thermal conductivity of MWNTs-TPNR nanocomposites would be reduced. Therefore, the low amount of MWNTs in TPNR could reduce these kinks or twists of MWNTs due to the good dispersion of MWNTs in TPNR, causing the thermal conductivity of the nanocomposites to increase.

As shown in Figure 11 and Figure 12, the thermal conductivity of MWNTs 1 reinforced TPNR matrix composites for all volume fractions studied from 30°C to 150°C is better than MWNTs 2. The effect of temperature on the thermal conductivity is clear from 30°C to 90°C, as shown in the Figures. This is because of the opposing effect of temperature on the specific heat and thermal diffusivity. Eventually, at high temperatures, as the phonon mean free path is lowered, the thermal conductivity of the matrix approaches the lowest limit and the corresponding thermal resistivity approaches the highest limit.

The resistance to phonon movement from one nanotube to another through the junction will hinder phonon movement and, hence, limit the thermal conductivity. The low thermal conductivity could be partly due to the non-uniform diameter and size, the defects in and the nano-scale dimension of MWNTs. However, the numerous junctions between carbon nanotubes involved in forming a conductive path and the exceptionally low thermal conductance at the interface (Yunsheng et al., 2006) are believed to be the main reason for the low thermal conductivity.

The effect of reducing the thermal conductivity is the transfer of phonons from nanotube to nanotube. This transition occurs by direct coupling between CNTs, in the case of the improper impregnated ropes, CNT-junctions and agglomerates, or via the matrix. In all these cases, the transition occurs via an interface and, thus, the coupling losses can be attributed to an intense phonon boundary scattering. At the same time the thermal conductivity decreases with the increase in temperature (if the temperature is near the melting point of the matrix). This indicates that the thermal conductivity of the composites is dominated by the interface thermal transport between the nanotube/matrix or nanotube/nanotube interface. Thus, it is believed that the decreased effective thermal conductivity of the studied composites could be due to the high interface thermal resistance across the nanotube/matrix or nanotube/nanotube interfaces (Ramasamy et al., 2007).

### 3.4.2 Thermal diffusivity

The thermal diffusivity of the TPNR/MWNTs were measured at various temperature points in the range of 30°C-150°C, it decreased with increasing temperature. In addition, the results obtained indicate that the MWNTs content influences the thermal diffusivity values of two types of multi-walled carbon nanotubes, as shown in Figure 13 and Figure 14.

As shown in Figure 13 and Figure 14, as the temperature increased from 30 °C to 150 °C, there is a decrease in the thermal diffusivity. The maximum thermal diffusivity was obtained at 30 °C. However, after 30 °C, there is a sudden drop in diffusivity. The thermal diffusivity of MWNTs 1 dropped as did MWNTs 2 by increasing the test temperature.

The thermal diffusivity of the TPNR/MWNTs decreases with fiber content due to the density of the nanocomposites, which decrease with the increase in the amount of MWNTs (the density of MWNTs less the TPNR density), hence, the density of the samples affects the thermal diffusivity (Kumari et al., 2008). This means that the TPNR containing MWNTs fiber will require a shorter time to be heated or cooled than the TPNR. Above 150 °C, the temperature at which the TPNR starts to melt, a slight variation of the thermal diffusivity with temperature is observed. The thermal diffusivity depends mainly on the mean free path length of the phonons as mentioned before.

As the test temperature goes up, the phonon vibration frequency will be quickened raising the possibility of an increase in collisions. Therefore, the mean free path decreases rapidly, which leads to the rapid decrease of the thermal diffusivity (Ruiying et al., 2004). The decrease in the thermal diffusivity is attributed to the different thermal properties of the individual nanotubes and CNT ropes in the matrix. Therefore, its thermal diffusivity and thermal conductivity are very high under the same temperature.

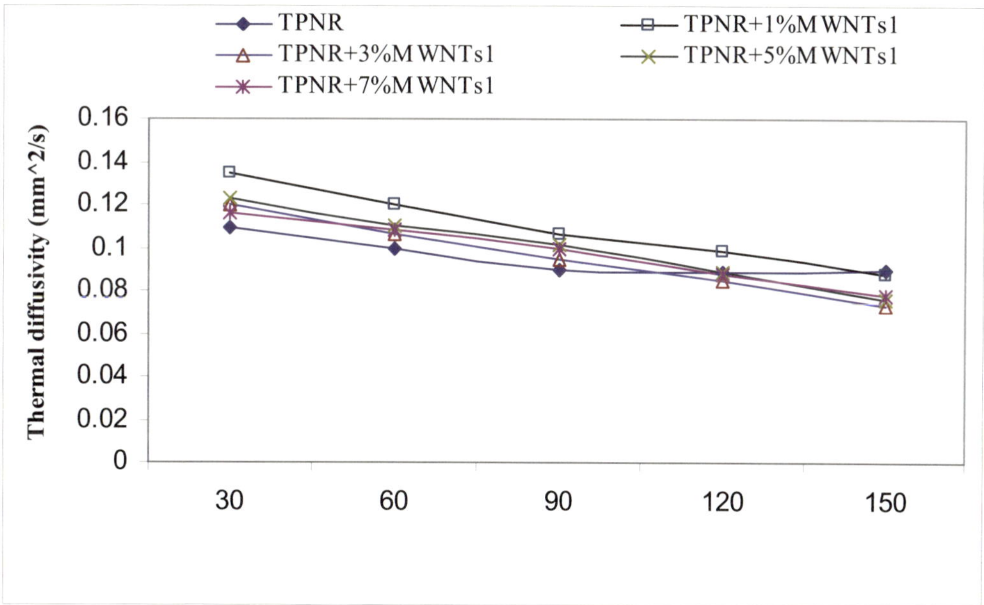


Fig. 13. Thermal diffusivity of the composites with different volume fractions of MWNTs 1 (first type) at different temperatures

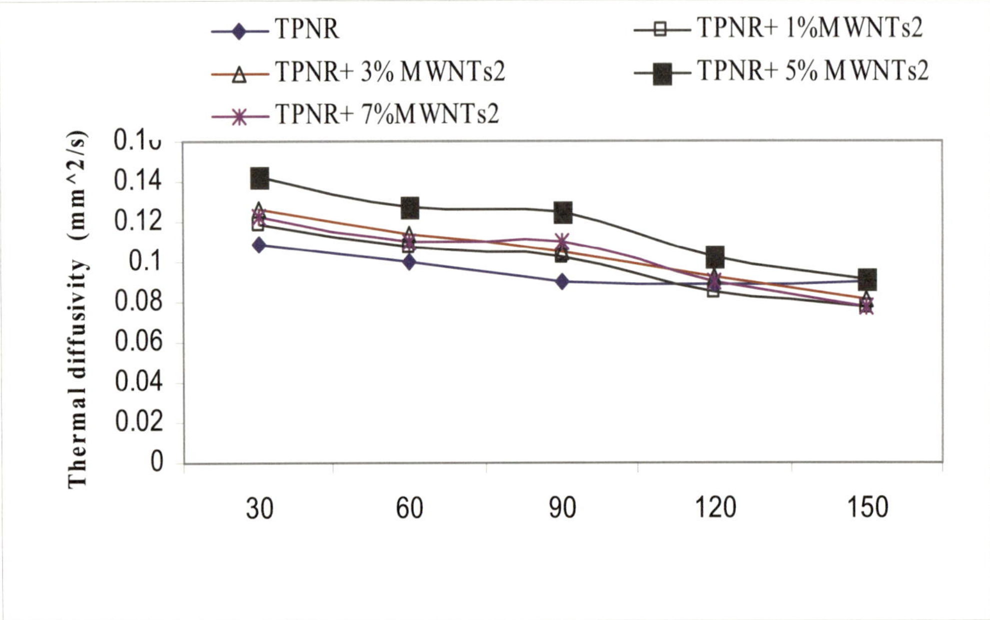


Fig. 14. Thermal diffusivity of the composites with different volume fractions of MWNTs 2 (second type) at different temperature

### 3.4.3 Specific heat

The temperature dependence behavior of the specific heat of the composites is different from that of their thermal diffusivity, which decreased with the temperature. The specific heat of all the measured samples increases linearly with the measured temperature from 30 °C to 150 °C. The addition of MWNTs 1 decreased the specific heat, as expected, from the relatively high specific heat of the matrix, as shown in Figure 15. The specific heat capacity of the composites was influenced by the filler content. It decreased with the increasing filler content, 1% and 3 % have shown high specific heat capacity compared with 5% and 7%. The phonon mean free path is determined by both the phonon-phonon and the phonon-defect interactions. However, its specific heat increases a little with an increase in temperature, while the phonon mean free path decreases, which makes its thermal conductivity first increase and then decrease with temperature.

From Figure 16, the specific heat increased by increasing the MWNTs 2, with a sharp peak at 90°C. This means that the maximum energy is required to change the temperature of material one degree at this temperature. This is clear from the figure of thermal conductivity of MWNTs 2. The most important factor affecting the specific heat is the lattice vibrations or phonons, which means the vibration at this temperature is the maximum, as mentioned before, also the structure of the material. Thus, change in dislocation density, grain size, or vacancies have little effect.

The major mechanism of the specific heat enhancement induced by the CNTs addition is not well understood. It may be related to the multi-walled structure of MWNTs in which the weak interlayer coupling can exhibit anything from 1D to 3D behavior, depending on the detailed value of the radius and number of wall (Benedict et al., 1996). However, the interface between the boundaries of the matrices or CNTs and matrices, or nanotube-nanotube will also affect the heat capacity.

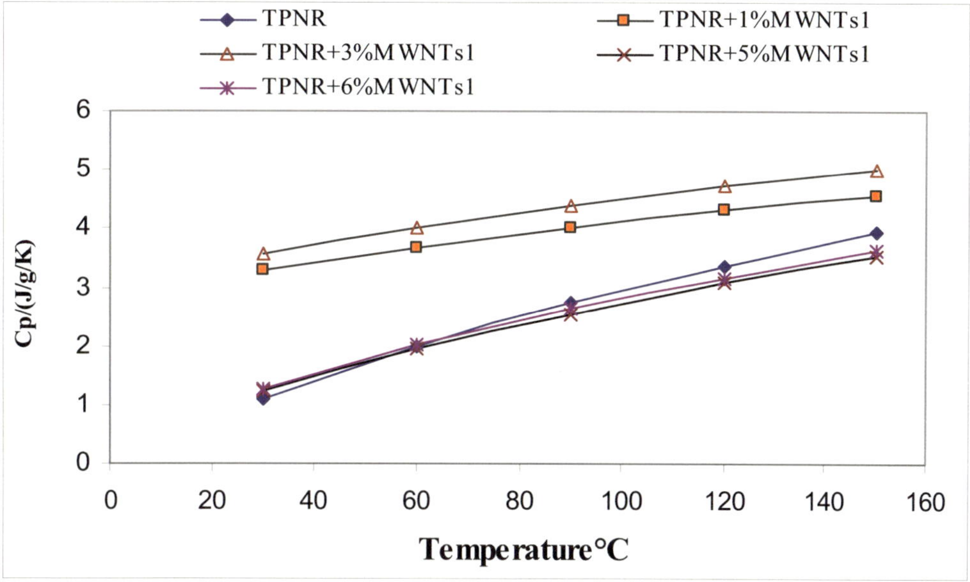


Fig. 15. Specific heat of the composites with different volume fractions of MWNTs 1 (first type) at different temperatures

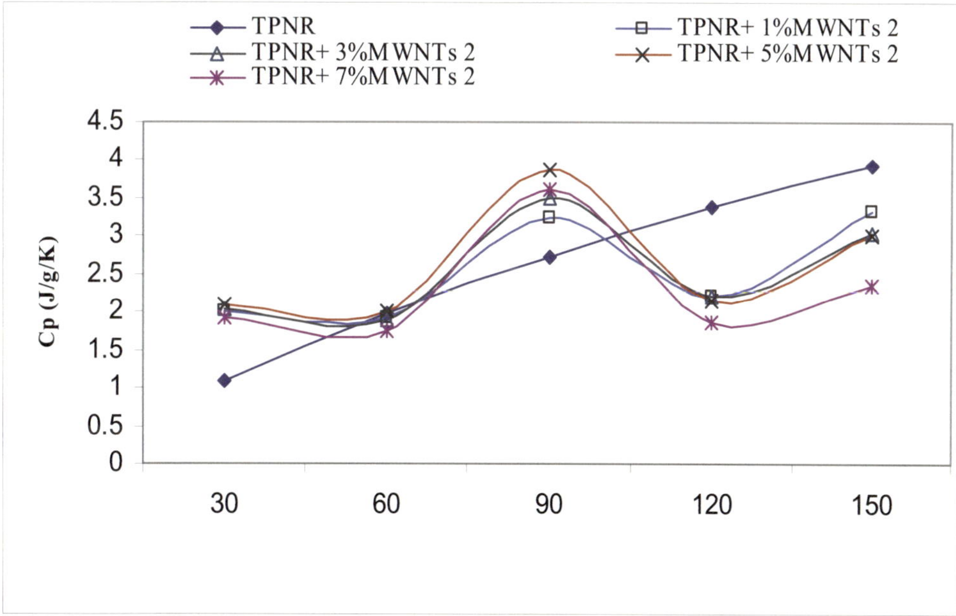


Fig. 16. Specific heat of the composites with different volume fractions of MWNTs 2 (second type) at different temperatures

### 3.4.4 Morphological examination

Figure 17 and Figure 18 show MWNTs (first and second type) as-received, with a large agglomeration of bundles of MWNTs, the carbon nanotubes tend to aggregate to form bundles because of the strong van der Waals forces (many graphene layers wrapped onto themselves). In addition, the Figures show the diameter of the walls of the MWNTs (the diameter of the first type is from 4-7nm, and the diameter of the second type is from 7-15nm). The SEM micrograph of TPNR is shown in Figure 19. This figure shows the TPNR without a filler inside it, and so the surface is smooth.

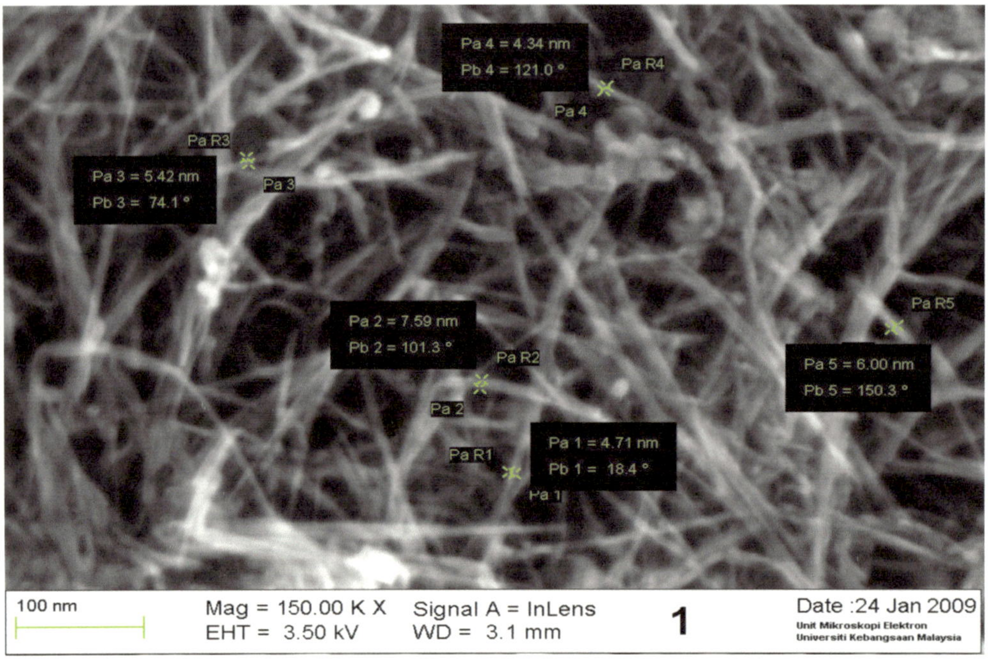


Fig. 17. SEM micrograph of MWNT 1 (first type).

The homogenous dispersion of MWNTs in the composites is confirmed by scanning electron microscopy (SEM). Figure20, Figure 21, Figure 24 and Figure 25, show 1wt% and 3wt% of MWNTs 1 and MWNTs 2, respectively, they are well dispersed as individual tubes in the matrix (the bright dots are the ends of broken MWNTs, indicated by arrows), they also show that the nanotubes that had pulled out from the matrix were still coated with polymer. In addition, the bright spots inside the TPNR, suggestion a strong polymer nanotubes interfacial. The key parameter to improve the composites containing the nanotubes is their uniform dispersion (Sandler et al., 1999; Yoshino et al., 1999). This is often conducted in two steps. One is to reduce the aggregate (the tangled MWNTs) size. The second is to homogenize the individual nanotubes and the aggregates in the matrix. Therefore, strong interfacial adhesion is essential for efficient stress transfer from the matrix to the nanotubes. This supports our observation of the higher efficiency of carbon nanotubes in enhancing the properties of TPNR nanocomposites. Figure 22 and Figure 26 show the SEM image of TPNR with 5wt% of MWNTs 1 and MWNTs 2. They depict an aggregate in TPNR, which observes a large amount of MWNTs that are self-organized in bundles. Low magnification was necessary in Figure 23 (7wt% MWNTs 1) and Figure 27 (7wt% MWNTs 2) to observe the poor dispersion of nanotubes in the TPNR. The small circles in the figures clearly show a large number of unbroken carbon nanotubes, (many zones with very high local MWNTs concentrations), indicating a poor polymer/nanotube adhesion, which contributes to a reduction in the properties of TPNR/MWNTs nanocompsites.

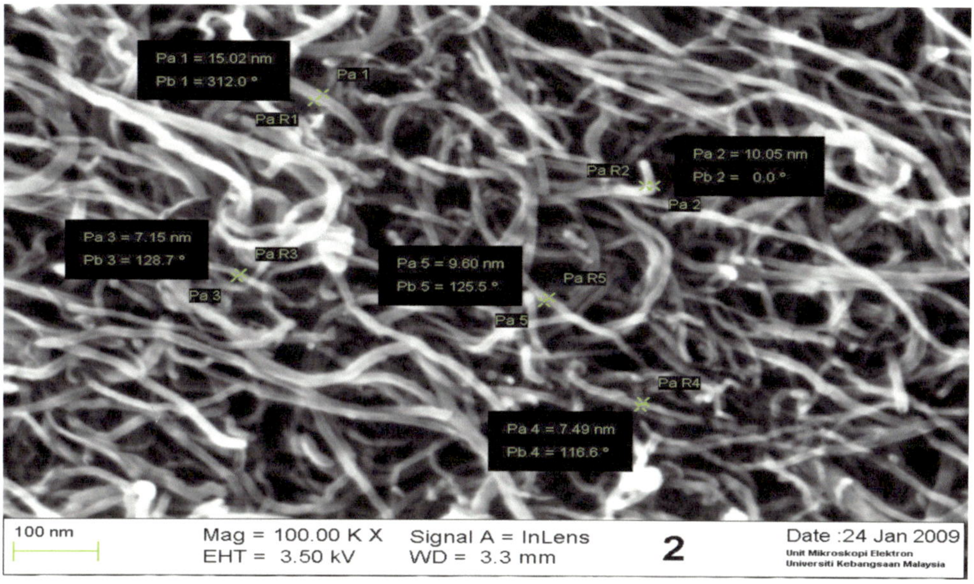


Fig. 18. SEM micrograph of MWNT 2 (second type).

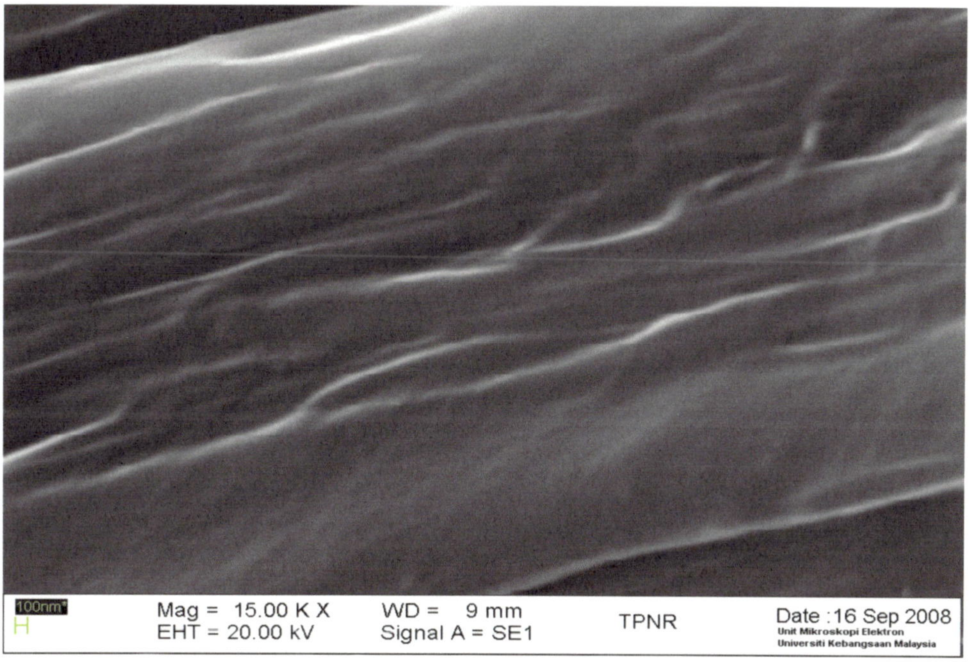


Fig. 19. SEM micrograph of TPNR.

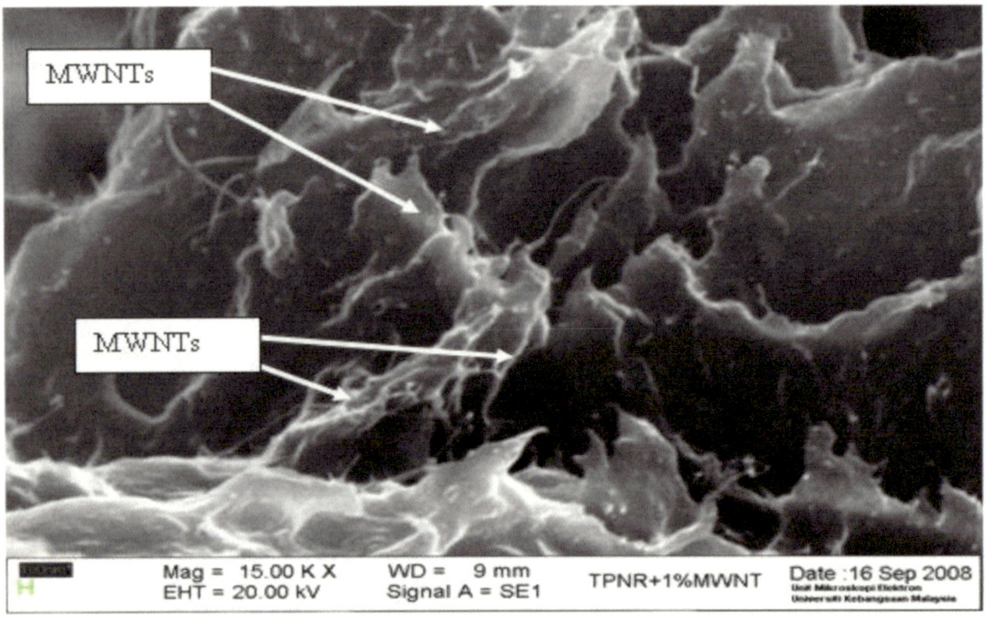


Fig. 20. SEM micrograph of TPNR with 1% MWNTs 1.

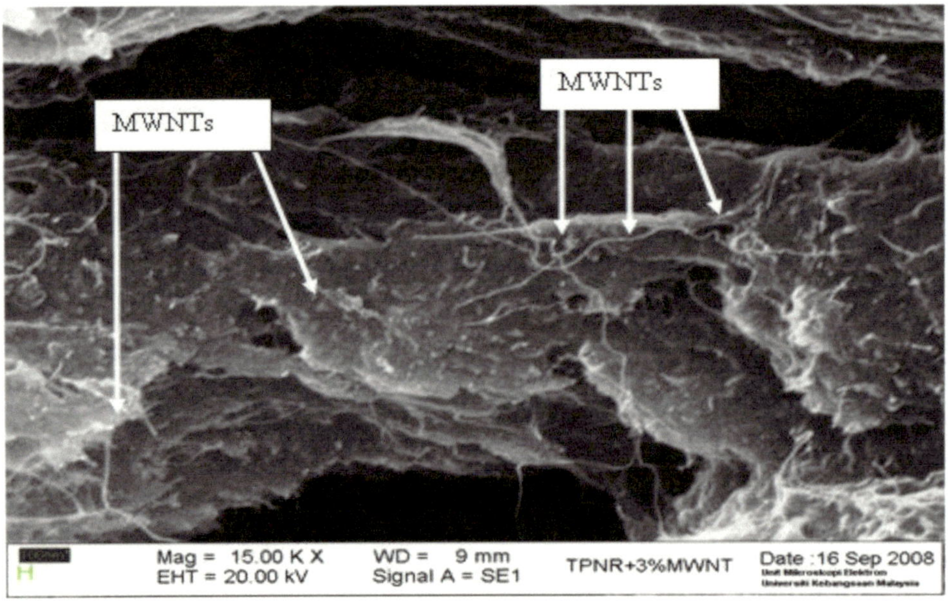


Fig. 21. SEM micrograph of TPNR with 3% MWNTs 1.

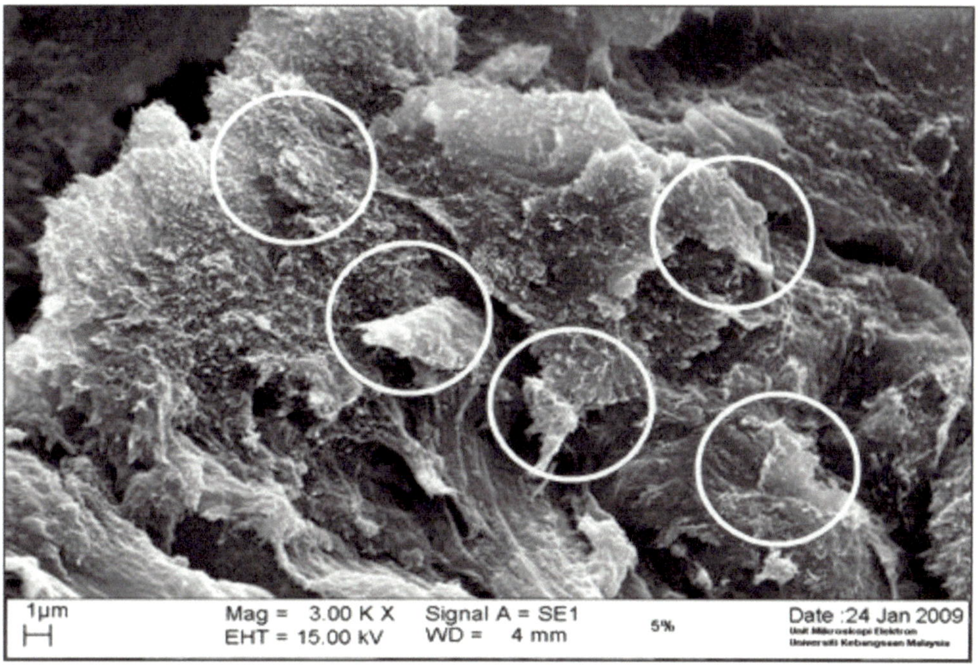


Fig. 22. SEM micrograph of TPNR with 5% MWNTs 1.

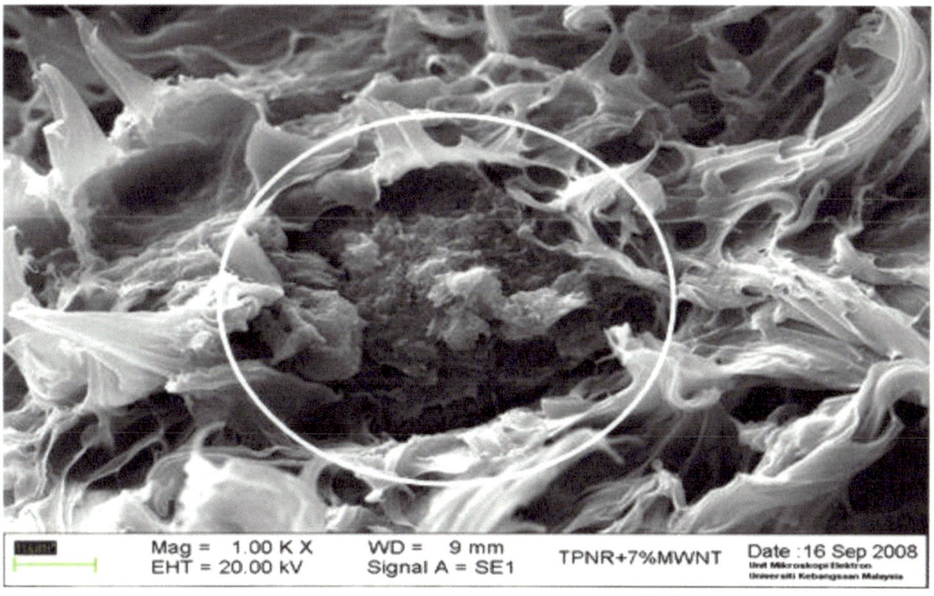


Fig. 23. SEM micrograph of TPNR with 7% MWNTs 1.

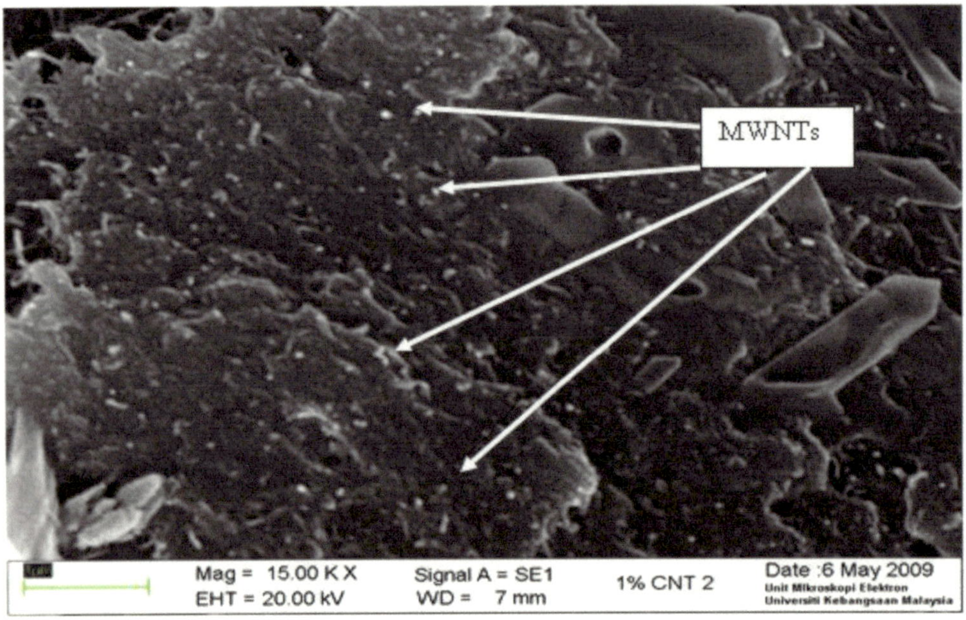


Fig. 24. SEM micrograph of TPNR with 1% MWNTs 2.

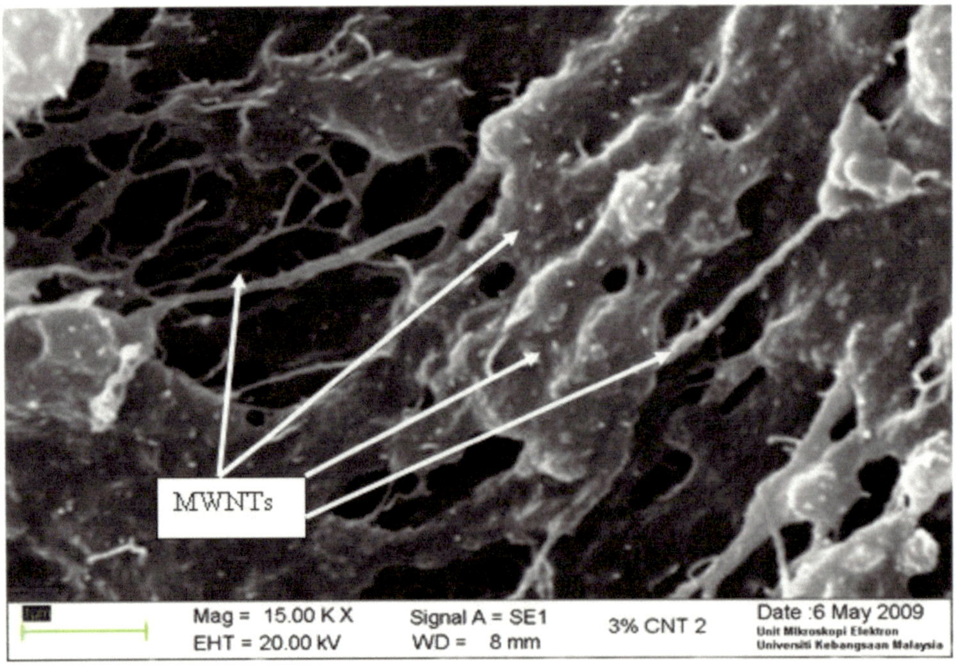


Fig. 25. SEM micrograph of TPNR with 3% MWNTs 2.

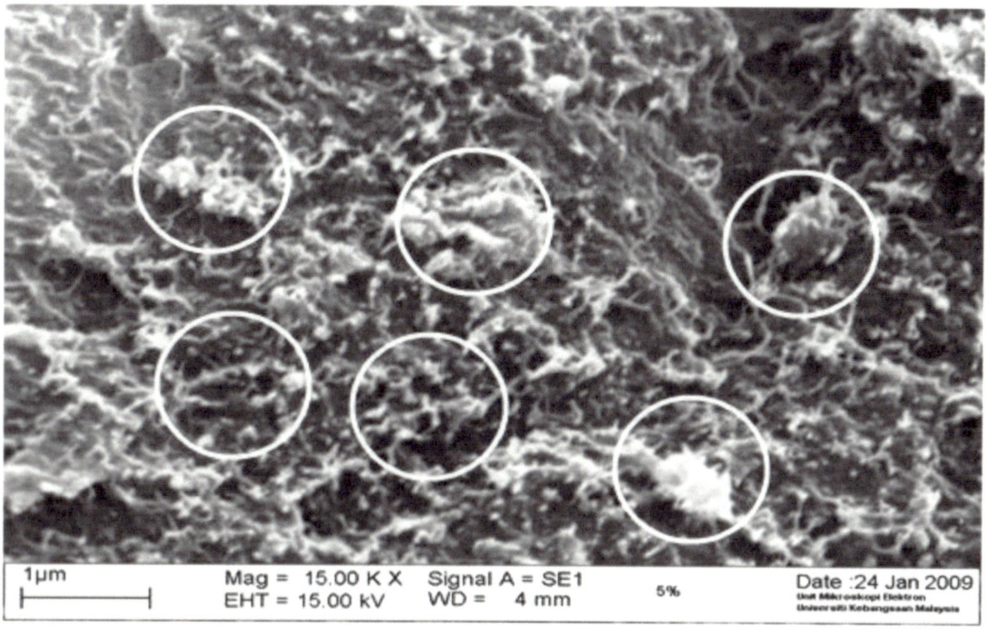


Fig. 26. SEM micrograph of TPNR with 5% MWNTs 2.

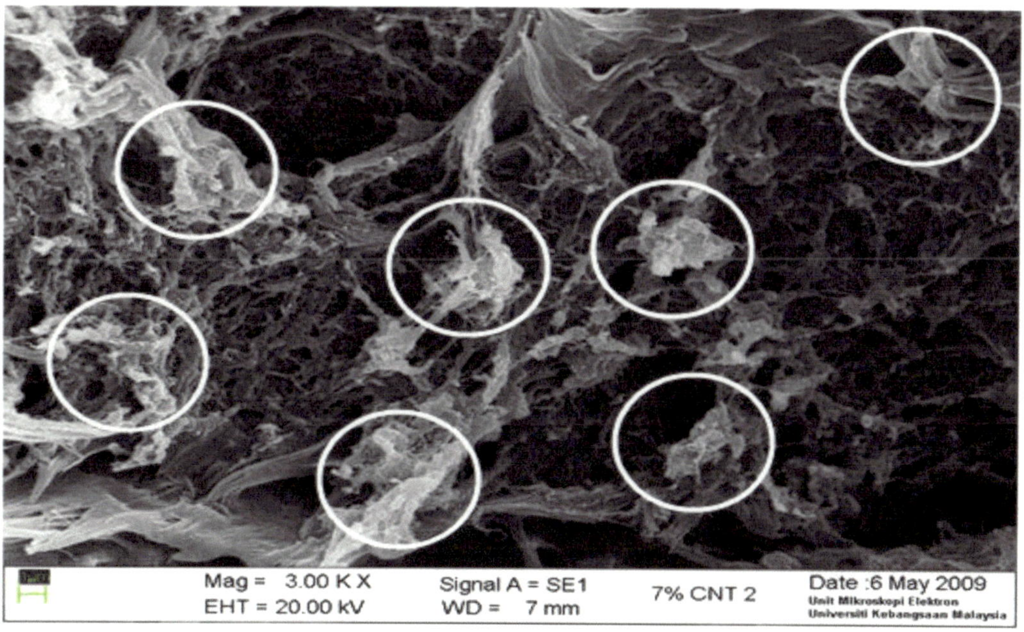


Fig. 27. SEM micrograph of TPNR with 7% MWNTs 2.

## 4. Conclusion

In this work, the MWNTs 1 and 2/TPNR nanocomposites were fabricated and the tensile and properties were measured. The addition of MWNTs in the TPNR matrix improved the mechanical properties. At this percent the tensile strength and elongation at break of MWNTs 1 increased by 23%, and 29%, respectively. The Young's modulus had increased by increasing the content of MWNTs. For MWNTs 2 the optimum result of tensile strength and Young's modulus was recorded at 3% which increased 39%, and 30%, respectively. However, elongation of break decreased by increasing the amount of MWNTs. The results exhibited better impact strength for MWNT 1 and MWNT 2 at 5 wt% with an increase of almost 70 % and 74%, respectively. The reinforcing effect of two types of MWNTs was also confirmed by dynamic mechanical analysis where the addition of nanotubes have increased in the storage modulus E', and the loss modulus E'' also the glass transition temperature (Tg) increased with an increase in the amount of MWNTs. The laser flash technique was used to measure the thermal conductivity, thermal diffusivity and specific heat, from the results obtained. The high thermal conductivity was achieved at 1 wt% and 3 wt% of MWNTs compared with TPNR after 3 wt% it decreased, also the improvement of thermal diffusivity and specific heat was achieved at the same percentage. The homogeneous dispersion of the MWNTs throughout the TPNR matrix and strong interfacial adhesion between MWNTs and matrix as confirmed by the SEM images are considered responsible for the significant mechanical enhancement.

## 5. Acknowledgment

The authors would like to thank Universiti Kebangsaan Malaysia (UKM) for financial support, science fund Grant UKM-OUP-NBT-29-142/2011 and UKM-OUP-FST-2011.

## 6. References

Abdullah, I. & Ahmad, S. (1992). Liquid NR as a compatibilizer in the blending of NR with PP. *Mater. Forum*. Vol.16, No.1, pp.353-357, ISSN 0255-5476

Abdullah, I. & Dahlan, M. (1998). Thermoplastic natural rubber blends. *Prog. Polym. Sci.* Vol.23, No.4 , pp.665-706, ISSN 0079-6700

Abdullah, I., Ahmad, S. & Sulaiman, C.S. (1995). Blending of natural rubber with linear low-density polyethylene. *J. Appl. Polym. Sci.* Vol.58, No.7, pp.1125-1133, ISSN 0021-8995

Benedict, L.X., Louie, S.G. & Cohen, M.L. (1996). Heat capacity of carbon nanotubes. *Solid State Comm*. Vol.100, No.1, pp.177-80. ISSN *0038-1098*

Bin Zhang , Ruowen Fu , Mingqiu Zhang , Xianming Dong , Bin Zhao ,Lichang Wang , Charles U. Pittman Jr. (2006). Studies of the vapor-induced sensitivity of hybrid composites fabricated by filling polystyrene with carbon black and carbon nanofibers. *Composites: Part A*. Vol.37, No. 11, pp.1884–1889. ISSN 1359-835X

Cai, H., Yan, F.Y. & Xue, Q.J. (2004). Investigation of tribological properties of polyimide/carbon nanotube nanocomposites. *Mater. Sci. Eng. A*.Vol.364, No.1-2, pp.94–100. ISSN 0921-5093

Canche-Escamilla, G., Rodriguez-Laviada, J., Cauich-Cupul, J.I., Mendizabal, E., Puig, J.E.&Herrera-Franco, P.J. (2002). Flexural, impact and compressive properties of a rigid-thermoplastic matrix/cellulose fiber reinforced composites. *Compos. Part (A): Appl. Sci. & Manufact.* Vol.33, No.4, pp. 539-549. ISSN 1359-835X

Che Som Sulaiman, Sahrim Hj. Ahmad & Ibrahim Abdullah. (1994). Kajian adunan getah asli/polietilena berketumpatan tinggi dengan getah asli cecair sebagai penserasi. *Sains Malaysiana*. Vol.23, pp. 29-38, ISSN 01266039.

Dahlan Hj. Mohd. (1988). Peranan getah asli cecair dalam penserasi adunan getah asli/polietilena. Tesis Dr. Fal. Universiti Kebangsaan Malaysia, Bangi. Malaysia.

Ibrahim Abdullah & Sahrim Ahmad. (1999). *Process for manufacturing thermoplastic natural rubber (TPNR) using liquid natural rubber (LNR) as a compatibiliser*. Patent no: MY-110733-A.

Ibrahim Abdullah. (2002). *Strategi penyelidikan kimia getah asli*. Bangi: Penerbit Universiti Kebangsaan Malaysia. Malaysia.

Ijima S. Helical microtubules of graphitic carbon. (1991) .*J. Nature*. Vol.354, No. 6348, pp.56-58. ISSN 0028-0836

Ismail, H. & Suryadiansyah. (2002). Thermoplastic elastomers based on polypropylene/natural rubber and polypropylene/recycle rubber blends. *Polym. Test.* Vol.21, No. 4, pp. 389-395. ISSN 0142-9418

Jianfeng Shen, Weishi Huang, Liping Wu, Yizhe Hu, Mingxin Ye. (2007). The reinforcement role of different amino-functionalized multi-walled carbon nanotubes in epoxy nanocomposites, *Composites Science and Technology*. Vol.67, No. 15-16, pp. 3041–3050. ISSN 0266-3538

Koerner, H., Liu, W., Alexander, M., Mirau, P., Dowty, H. & Vaia, R.A. (2005). Deformation-morphology correlations in electrically conductive carbon nanotube- Thermoplastic polyurethane nanocomposites. *Polymer*. Vol.46, No. 42, pp. 4405-4420. ISSN 0032-3861

Kuan, H.C., Ma, C.M., Chang, W.P., Yuen, S.M., Wu, H.H. & Lee, T.M. (2005). Synthesis, thermal, mechanical and rheological properties of multiwall carbon nanotube/waterborne polyurethane nanocomposite. *Compos. Sci. Technol.* Vol.65, No. 11-12, pp.1703–1710. ISSN 0266-3538

Kumar, S., Alam, M.A. & Murthy, J.Y. (2007). Effect of percolation on thermal transport in nanotube composites. *Appl Phys Lett*.Vol. 90, No.10, pp.104105-1-104105-3. ISSN 0003-6951

Kumari, L., T.Zhang, G.D., Li, W.Z., Wang, Q.W., Datye, A. & Wu, K.H.. (2008). Thermal properties of CNT-Alumina nanocompsites, *Composites Science and Technology*. Vol.68, No. 9,pp. 2178-2183. ISSN 0266-3538

Li, C.S., Liang, T.X., Lu, W.Z., Tang, C.H., Hu, X.Q., Cao, M.S. & Liang, J. (2004). Improving the antistatic ability of polypropylene fibers by inner antistatic agent filled with carbon nanotubes. *Compos. Sci. Technol.* Vol.64, No. 13-14, pp.2089–2096. ISSN 0266-3538

Liao, Y.H., Olivier, M.T., Liang, Z.Y., Zhang, C. & Wang, B. (2004). Investigation of the dispersion process of SWNTs/SC-15 epoxy resin nanocomposites. *Mater. Sci. Eng. A*. Vol.385, No. 1-2, pp.175–181. ISSN 0921-5093

Lopez Manchado. M.A, L. Valentine, J. Biagiotti, J.M. Kenny. (2005). Thermal and mechanical properties of single-walled carbon nanotubes-polypropylene composites prepared by melt processing. *Carbon*. Vol.43, No. 7, pp. 1499-1505. ISSN 0008-6223

Lourie O, Cox DM, Wagner HD. (1998). Buckling and Collapse of Embedded Carbon Nanotubes .*J. Phys Rev Lett*. Vol.81, No.8, pp.1638-1641. ISSN 2153-1188

Mustaffa, H.A. (1991). *Sains Bahan Jilid 1*. Kuala Lumpur: Dewan Bahasa dan Pustaka. Malaysia

Nan, C.W., Shi, Z & Lin, Y. (2003). A simple model for thermal conductivity of carbon nanotube-based composites. *Chem Phys Lett*. Vol.375, No.5-6, pp.666-669. ISSN 0009-2614

Potschke P, Fornes TD, Paul DR. (2002). Rheological behavior of multi-walled carbon nanotubes/polycarbonate composites. *Polymer*. Vol.43, No.11, pp. 3247-3255.ISSN 0032-3861

Ramasamy, S., Shuqi, G., Toshiyuki, N. & Yutaka, K. (2007). Thermal conductivity in multi-walled carbon nanotubes/silica-based nanocomposites. *Scripta Materialia*. Vol.56, No.4, pp. 265-268. ISSN 1359-6462

Ruiying, L., Tao, L., Jinsong, L., Hongbo, Z., Zhijun, C. & Guanglai, T. (2004). Thermophysical properties of carbon/carbon composites and physical mechanism of thermal expansion and thermal conductivity. *Carbon*. Vol.42, No. 14, pp. 2887–2895. ISSN 0008-6223

Sandler J, Shaffer MSP, Prasse T, Bauhofer W, Schulte K, Windle AH.(1999). Development of a dispersion process for carbon nanotubes in an epoxy matrix and the resulting electrical properties. *Polymer*. Vol.40, No. 21, pp. 5967-5971. ISSN 0032-3861

Sang Hyun Jin , Young-Bin Park , Kwan Han Yoon, (2007). Rheological and mechanical properties of surface modified multi-walled carbon nanotube-filled PET composite. *Composites Science and Technology*. Vol.67, No. 15-16, pp. 3434–3441. ISSN 0266-3538

Sinnott B, Shenderova OA, White CT, Brenner DW. (1998). Mechanical properties of nanotubule fibers and composites determined from theoretical calculations and simulations .*J. Carbon*. Vol.36, No. 1-2, pp.1-9. ISSN 0008-6223

Sue H-J, Gam KT, Bestaoui N, Spurr N, Clearfield A. (2004). Epoxy nanocomposites based on the synthetic alpha-zirconmium phosphate layer structure. *Chem Mater*. Vol.16, No.2, pp. 242-249. ISSN 0897-4756

Treacy, T.W. Ebbesen, J.M. Gibson. (1996). "Exceptionally high Young's modulus observed for individual carbon nanotubes. *Nature*. Vol.381,No. 6584, pp. 678-680. ISSN 00280836

Wagner D, Lourie O, Feldman Y, Tenne R. (1998). Stress-induced fragmentation of multiwall carbon nanotubes in a polymer matrix .*J. Appl Phys Lett*. Vol.72, No.2, pp.188-190. ISSN 0003-6951

Wang WJ. (1998). Effect of polymer-filler and filler-filler interactions on dynamic properties of filled vulcanizates. *Rubber Chem Technol*. Vol.71, No.3, pp. 520-589. ISSN 0035-9475

Yang, Li-Y, Bigio, D. & Smith, T.G. (1995). Melt blending of LLDPE and PS in a Haake internal mixer. II. Morphology processing relationship. *J. Appl. Polym. Sci.* Vol.58, No.1, pp.129-141, ISSN 0021-8995

Yoshino K, Kajii H, Araki, Sonoda T, Take H, Lee S. (1999). Electrical and optical properties of conducting polymer-carbon nanotube composites. *Fullerene Sci Technol*. Vol.7, No.4, pp. 695-711. ISSN 1064-122X

Yu F, Files BS, Arepalli S, Ruoff RS. (2000) Tensile Loading of Ropes of Single Wall Carbon Nanotubes and their Mechanical Properties. *J. Phys Rev Lett*. Vol.84, No.24, pp.5552-5555. ISSN 2153-1188

Yunsheng, X., Gunawidjaja, R. & Beckry, Abdel-Magid. (2006). Thermal behavior of single-walled carbon nanotube polymer–matrix composites. *Composites A*. Vol.37, No. 1, pp. 114-121. ISSN 1359-835X

# Carbon Nanotubes and Semiconducting Polymer Nanocomposites

Duong Ngoc Huyen
*Hanoi University of Science and Technology, Hanoi*
*Vietnam*

## 1. Introduction

The increasing demand for efficient machines has driven a trend towards the miniaturation of devices and instruments with smaller volume, lesser power consumption but greater performance. The progression relies upon the searching out new desirable materials and the ability of making tiny structures with high precission. However, the development is not so smooth and easy. For instance, current electronic industry based on silicon is very mature and reliable, but it is reaching an unsurmountable barrier of quantum effects as nanoscale approaching. Search for new suitable materials and fabrication methods therefore are being urgent tasks for the near coming development.

Carbon nanotubes (CNTs) and conducting polymers (CPs), newfound materials have shown exceptional characteristics. The coupling of CPs and CNTs furthermore reveals **synergistic effects** which offer an attractive route to create new multifunctional materials with greater potential in application. Envisioned applications from CPs/CNTs systems involve mechanical, thermal, electrical, electrochemical features such as supercapacitors, sensors, organic light emitting diodes (OLEDs), solar cells, electromagnetic absorbers, and, last but not the least, advanced electronic devices.

Conducting polymers commonly are classified as conjugated polymers which consist of alternating single and double bonds along its linear chains (sp2 hybridized structure). The conductivity of the CPs relies on these double bonds which are sensitive to physical or chemical interactions [1-3]. Similarly, CNTs also have sp2 hybridized bonds over the structure. CNTs possess unique structures and exhibit extraordinary electrical, optical, chemical, and mechanical properties, which are somewhat complementary to those of CPs [4-6]. For instance, CNTs have a very long mean free path, ultrahigh carrier mobility, and can be either very good conductors or narrow bandgap semiconductors depending on the chirality and diameter. Mixing up together, both materials show a strong interfacial coupling via dono-acceptor binding and pi-pi interaction [7, 8]. The coupling of CNTs and CPs in a composite has been found to affect their chemical and electronic structure. Beyond a simple combination of their properties, some synergistic effects and new features appear and can develop into applications [9-14]. From a chemical viewpoint of consideration, two possible impacts may take place in a CPs/CNTs system: either the CNTs are functionalized by the CPs or the CPs are modified (doped) with the CNTs. Therefore, either the morphological modifications, electronic interactions, charge transfers or a combination of these effects may occur between the two constituents in the system [15-18]. Due to the nanoscale confinement in the system, the interaction via interfacial bonding is considered to play an essential role in the impacts [19-23]. Morphologically, the interfacial interaction sites on CNTs surface are: (a) defect sites at the tube ends and side walls (b) covalent side wall bindings (c) non-covalent exohedral side wall bindings and (d) endohedral filling (Fig. 1). Three routes have been commonly used for preparation of CPs/CNTs composite: (i) direct mixing (ii) chemical polymerization and (iii) electrochemical synthesis.

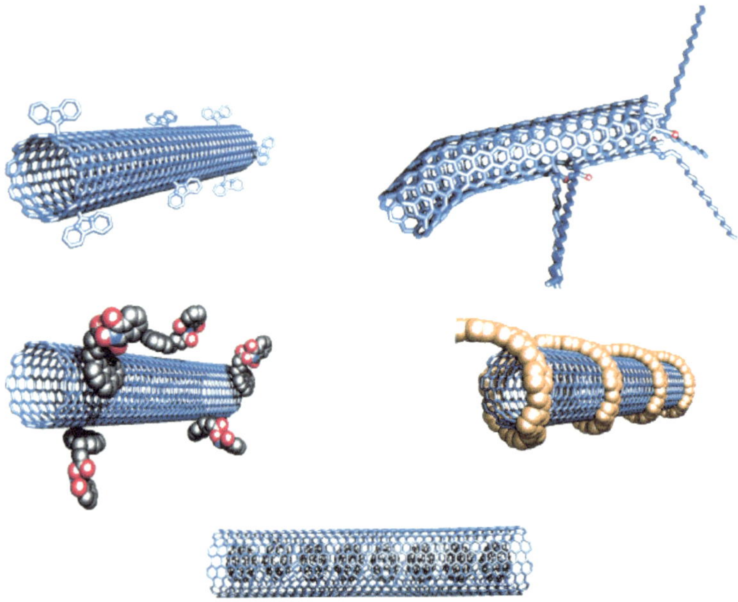

Fig. 1. Possible functional sites CNTs: (a) defect sites at the tube ends and side walls (b) covalent side wall bindings (c) non-covalent exohedral side wall bindings (d), endohedral filling [23].

Back to the development, the study made on the composites of CPs and CNTs relatives started in early 90s of the last century. Since 1992, Heerger *at al.* [24] observed the photovoltaic effects arising from the photoinduced charge transfer at the interface between CPs as donors and a $C_{60}$ film as acceptor although the conversion efficiency was too low (less than 1%). However, the conversion efficiency of CPs based solar cell have been significantly improved (~ 6%) by further use of CNTs derivatives acting as the electron acceptors in the CP/CNTs matrices [25]. The conversion efficiency is still moderate in comparison to that of inorganic systems but the simple in processing and low cost enable CPs/CNTs systems to be promising choice for photovoltaic future. On the other hands, the introduction of CNTs into a polymer matrix improves the electric conductivity, while possibly providing an additional active materials for capacitive energy storage and secondary batteries [26]. Gas sensors fabricated with the PPy/SWNTs nanocomposite has shown sensitivity higher than that of PPy. The improvement is explained as the effect of the increase of specific surface area and anion doping in PPy matrix. The gas sensing improvement in PANi/SWNTs is also shown the same manner [27]. For biosensor applications, it has been demonstrated that the CPs/CNTs nanocomposites are very attractive as tranducers because they provide the best electron transfer and assure a faster ion mass transfer [28]. In some host polymers, the CNTs additive has found to act as a hole-blocking causing a shift of the recombination emission [29, 63]. The interaction between host polymer and CNTs additive is considered to be the main reason accounting for all of these modifications.

Studies on CPs/CNTs systems will further contribute to the fundamental understanding of the nucleating capability of CNTs, epitaxial interaction, and templated crystallization of the polymer at the CNT-polymer interface, and may ultimately lead to more efficient production of bulk nanocomposites. The combination of strong polymer-CNT interaction, nucleation ability of CNTs, CNTs templating of polymer orientation and crystallinity are all features that one can build on to develop high performance composites. With that goal, this chapter tries to provide a look on the development and trend in future research of CPs/CNTs systems.

## 2. Composite preparation

Generally, the composite properties are governed by variety of factors such as the preparation conditions, quantity, contents as well as the nature of its components. With respectto those of CPs/CNTs composite, the **purity** of carbon nanotubes, the **dispersion** and the **interfacial interaction** strength between components are considered to be the essential factors [10, 13]. Carbon nanotubes are created and recombined from carbon sublimation, then naturally accompany with variety of contaminants such as residual catalysts and amorphous carbon phases. In order to obtain purer CNTs, many purification procedures have been proposed to remove these inherent contaminants. Most techniques have been based on chemical and thermal treatment. In the first stage,, strong chemical oxidants, surfactants or burning the unpurified carbon nanotubes were commonly used. However, these techniques have shown some disadvantages such as adding some defects on CNTs surface and poor yield. Later, Davey *et al.* developed a non-destructive and more efficient purification method for that a coiled polymer was used to extract nanotubes from carbonaceous soot [30]. With respect to the mechanical, thermal, electrical, and optical properties, a better dispersion and an increase in interfacial interaction between the CNTs and polymer matrix would lead to even greater improvement. Depending on use purpose, either uniform dispersion or alignment of CNTs in polymer matrix is required. However, due to the Van der Waals attraction, CNTs are held together as bundles and ropes and remain as entangled agglomerates. In order to facilitate a better dispersion of CNTs in polymer matrix, direct mixing, in-situ chemical polymerization or electrochemical polymerizations have been proposed. Direct mixing is actually mechanical/physical means involving solid-state, high shear, solution mixing or melt blending, agitated in a solvent by magnetic stirring, reflux or ultrasonication. Sometimes, additive surfactant or chemical modifications are used to enhance the dispersion of CNTs in the solution. On the other hand, the alignment of nanotube in the polymer matrix is an additive essential factor because it is the only way to optimize the preferred anisotropic behavior of a CNTs composite. The approaches to achieve the alignment of CNTs depend on how nanotubes are introduced into polymer matrices. Carbon arc discharge, composite slicing, film rubbing, chemical vapor deposition, mechanical stretching of CNT-polymer composites and magnetic orientations have been the techniques used for aligning nanotube in composites.

For thermoplastic polymers, melt processing is a common technique. As the dispersion of CNTs in polymer melt is much more difficult than in solution, additive techniques such as mechanical stirring, ultrasonic vibration, melt blending, extrusion and melt spinning are used to enhance the dispersion. For insoluble and thermally unstable polymers, chemical processing is often chosen, which involves *in-situ* polymerization or covalent functionalization of CNTs with polymer molecules. In these methods, CNTs is mixed with monomers and then is chemically polymerized with added oxidant.. Depending on the nature of polymer and the preparation condition, CNTs have been found to play the role of the nucleation sites for polymer growth and crystallized. As a result, the crystallinity and polymerization rate increases when the CNTs are introduced.. The interaction between two entities is a physical factor affecting the polymer orientation and crystallization. This interaction change the morphology and then strongly modify the physical properties of the polymer. The combination of mechanical mixing and *in-situ* polymerization also gives a good dispersion of CNTs in the polymer matrix.
Electrochemical polymerization is another way for the preparation of polymer CNTs composites. A direct electrochemical deposition of monomer on CNTs layers acting as an electrode or an electropolymerization from an aqueous dispersion of monomer and CNTs are common approaches to realize the composites.

## 3. Characterization

Normally, the composites have been characterized by microscopic, spectral, conductivity, mechanical and thermal measurement. Scanning electron microscope (SEM), scanning tunneling microscope (STM), atomic force microscope (AFM), and transmission electron microscope (TEM) are the main techniques used for morphological characterization of the composite. For example, the morphology of the composites could reveal the wrapping of CNTs by conducting polymers in core-shell structure with several modes. UV-Vis, photoluminescence spectra, Raman and FTIR are complementary means used to investigate the chemical structure of the composites. X-ray diffraction, X-ray photoelectron spectral studies made on the composites show the crystallinity, intrinsic oxidation states.

## 4. Application

### 4.1 Electrically conductive composite

The incorporation of CNTs in polymer matrices is made to design electrical properties suitable for variety of applications such as circuit components, electronic products or electrostatic shields. For example, to protect against electrostatic discharge a lower level of conductivity is enough whereas to prevent electromagnetic interference a higher conductivity is needed and for circuit components the level of conductivity must be comparable to those of metals. However, most of composites, although a high weight percentage of CNTs has been used, have only a moderate conductivity below the required level in electronic applications, except for a composite developed by Blanchet-Fincher *et al.* [31]. To prepare a mechanically strong and electrically conductive composite, Niu *et al.* [32] proposed an approach based on solution and melting methods based on polyvinylidene fluoride and CNTs (PVDF/CNTs). The conductivity of this composite has been claimed to be higher than other known polymer composites and can be adjusted by varying the CNTs content as illustrated in Fig. 2. The conductivity achieves a level of CNTs mats with a composite having about 13 wt% of CNTs. Their results shown that the composite prepared by the solution method were better electrical conductors than that by the melting method. Based on Niu route, Charati *et al.* [33] developed a method for manufacturing conductive composites for material-handling, electronic devices (computers, printers etc.) and automotive components that require electrostatic dissipation and electromagnetic shielding. The composite was prepared by blending an organic polymer precursor with SWNTs and polymerizing the polymer precursor to form an organic polymer using an ultrasonicator and shear, elongational forces. They claimed that in this way, at least a portion of the CNTs could be functionalized either at the side wall or hemispherical ends.

In principle, the conductivity of a conducting polymer can be largely tuned by "doping", however highly conductive polymers have been limited in practical applications as they are chemically unstable in use and unsuited for solution or melt processing. Among the conductive polymers, polyaniline (PANi) is known to be chemically stable and readily soluble in environmentally friendly solvents. Protonic acid doped PANi emeradine is the most highly conductive form with conductivity about 5 S/cm, which is well below the $10^2$ S/cm threshold conductivity required for widespread utility in electronics. Blanchet-Fincher *et al.* [31] disclosed a method for the preparation of a composite comprising PANi emeradine salt and CNTs that exhibited electronic conductivities of $10^2$ S/cm while retaining the desirable chemical stability of PANi with only a small wt% of CNTs. They have applied the above composite as a conductive pathway, produced by laser thermal transfer imaging, in electronic circuits. In addition, McElrath *et al.* [34] made a polymer-CNT composite comprising at least one polar polymer with electrical and/or thermal conductivity enhanced six orders higher that of polymer alone. These examples showed the evidences that CNTs acting as effective physical dopants in conducting polymers.

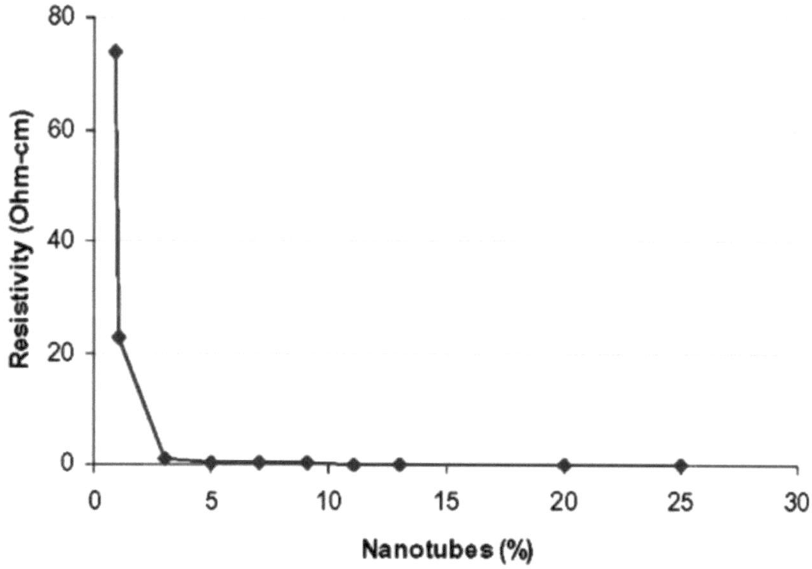


Fig. 2. Resistivity of PVDF/CNT composite as a function of CNT content [32].

### 4.2 Mechanically reinforced composite

With respect to mechanically effective reinforcement, basically required features for composite components are purity, aspect ratio, dispersion, alignment and interfacial stress transfer. Among these, **the interfacial stress transfer**, characterized by the applied shear stress at which the interface fails, is the most important. However, CNTs have atomically smooth non-reactive surfaces, and then lack of interfacial bonding between the CNTs and polymer. In order to increase bonding sites to the polymer chain, chemical modification and functionalization of CNTs such as solution processing, melt processing, and chemical processing were proposed. Quantum mechanics and molecular dynamics calculations supported this approach [35-38]. By chemical functionalizing and integrating CNTs into epoxy polymer, Margrave *et al* [39] found that the composite with 1wt% functionalized CNTs showed an increase of 18% and 24% in tensile strength and modulus respectively over the epoxy composites with unfunctionalized CNTs and a 30% increase in tensile modulus over pure epoxy resin. The enhancement was due to the strong interaction between epoxide rings on the epoxy resin and the functionized groups on CNT walls which were confirmed by infra-red and Raman spectroscopy.

A composite formed by *in-situ* polymerization from the polycondensation of diamine and diacid monomers comprising with dispersed CNTs aligned with rigid-rod polymers [40] showed that the tensile modulus, tensile strength and elongation to break of composite having 10 wt% CNT increased by 20, 60 and 40% respectively. In this composite there was no claim of covalent bond formation at the interfaces of components. The alignment of CNTs could be the reason of the enhancement.

By combining the uniformly dispersed nanotube with polypropylene matrix/solvent mixture and then heating the nanotube/matrix mixture to a temperature above the melting point of the matrix materials, Shambaugh *et al.* [41] recently disclosed a method for producing SWNT-polypropylene composite. The fiber made from composite with 1 wt% of CNTs showed more than 50% increase in tensile strength compared to fiber drawn from matrix materials alone. This substantial increase in strength was believed to be due to a uniform dispersion of CNTs in the matrix materials. In addition, mechanical reinforcements of polymer/CNTs composites have found to be significantly enhanced when high purity nanotubes were exfoliated and well dispersed in the polymer matrix. The above results show that not only the interfacial bonding but also the type of polymer and the conditions used in making the composite are important factors to produce a strong composite.

On the other hand, in some composites, CNTs act as a nucleating agent for polymer crystallization [9, 42-44] and reorientation. For example, SWNTs induced oriented crystallization is observed in PET/SWNTs composites [42]. Two samples, PET and PET/ SWNTs (1wt% loading) are prepared under identical conditions by melt compounding at 270 °C. Both PET and PET/SWNTs samples were made into 0.5 mm thick, 0.4 mm wide strips and subsequently stretched to a draw ratio of four to induce PET and SWNTs orientation. As can be seen from Fig.3, wide-angle X-ray diffraction (WAXD) of both materials shows that PET is oriented in the samples (Fig. 3A and Fig. 3C). However, when the train PET and PET/SWNTs were heated and were recrystallized by cooling down, only PET/SWNTs shows the orientation but not the neat PET (Fig. 3B and Fig. 3D). Individual CNTs also have found to promote polypropylene (PP) crystallization, as a result, thick PP interfacial layer is formed on the nanotube surface [43]. Polypropylene transcrystals were observed on the carbon nanotube surface (Figure 2E), when polymer melt was isothermally crystallized in the temperature range of 118 to 132 °C. In PP/CNTs fiber samples melted and recrystallized spherulite growth is influenced by the presence of CNTs, where the presence of CNTs increase the number of nucleation sites for the polymer and smaller spherulites are observed (Figure 2F to I) even a low CNTs loading (0.1 wt %) [44]. PP/CNTs fibers heated at above polypropylene melting temperature and then cooled down shows that PP/CNTs fibers retain polymer orientation and the average degree of crystallinity is about 80 %. Such a high level of crystallinity is attributed to more complete PP crystallization in dilute solution and PP-CNTs interaction.

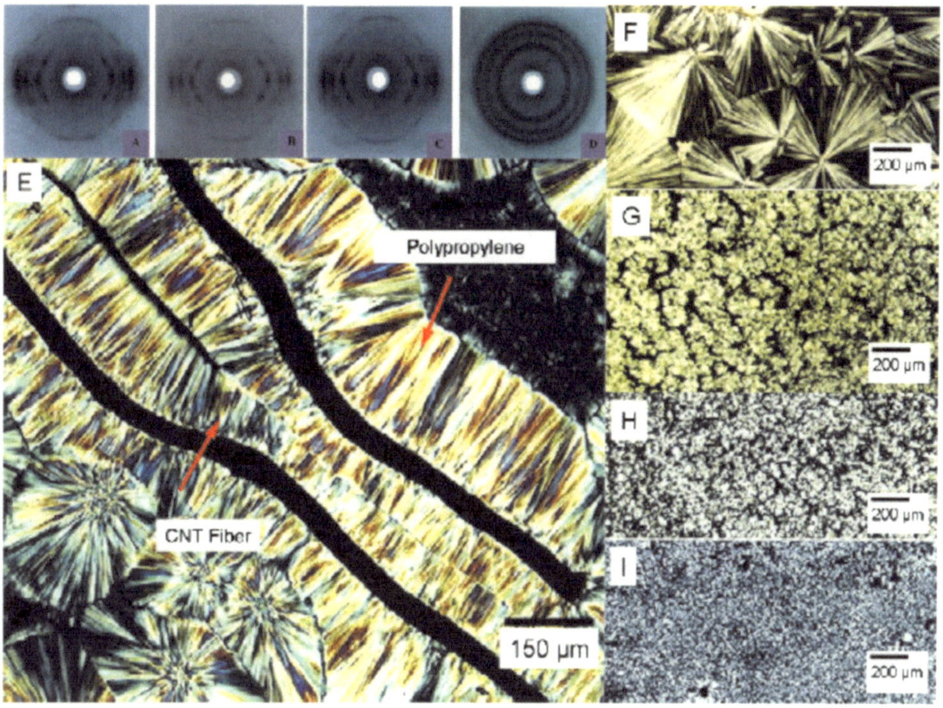


Fig. 3. Wide-angle X-ray diffraction patterns of (A) drawn PET/SWNTs (1.0 wt%) nanocomposite film, (B) PET/SWNTs 1 wt% nanocomposite film after melting to 300 °C and subsequently slowly cooled down to room temperature, (C) drawn neat PET film, (D) drawn neat PET film after melting to 300 °C and subsequently slowly cooled to room temperature [42]. (E) PP transcrystalline interfacial layer on CNTs fibers [43]. The development of transcrystallinity provides the evidence of interaction between PP and CNTs. Optical micrographs with cross-polarizers of spherulite formation in (F) PP, (G) PP/SWNTs (0.1 wt% SWNTs), (H) PP/SWNTs (1 wt% SWNT), and (I) PP/MWNTs (1 wt% MWNTs) bulk samples [44].

In addition, some proteins have been shown to crystallize in an ordered helical fashion on the surfaces of MWNTs. The MWNTs that induce protein crystallization are of a specific size, and protein crystallization occurred consistently throughout the system. The SWNTs have also been shown to induce crystallization and orientation in the sheared polymer melt, and polymer melt containing aligned SWNTs. These observations are important evidences indicating that CNTs can be used as nucleating agents in polymer processing to promote polymer crystallization and orientation.

### 4.3 Electrochemical capacitors

The requirement for energy store and conversion systems has drawn much attention on development of effective energy store devices. Electrochemical capacitor (EC) (or so called supercapacitor) is device which can store the electric field energy in high density by using an electrochemical double-layer with high specific surface-area. With respect to classification, supercapacitors may be distinguished by criteria such as the electrode materials, the electrolyte, or the cell design. Traditionally, the electrode materials are divided into three main groups: carbon based, metal oxides (mainly ruthenium dioxide) and polymeric materials such as PPy, PANi and PTh [45-49]. Polymer-based electrochemical capacitors in particular have attracted much attention due to the polymeric materials possess high capacitive energy density and low cost. However, these kinds of capacitors have shown some disadvantages, namely, lower cycle-life and slow ion transport kinetics because the redox sites in the polymer backbone are not sufficiently stable. As an alternation, the composites of CPs and CNTs (both SWNTs and MWNTs) have been investigated as an electrode materials for supercapacitors [50-52]. The combination of CNTs into a polymer matrix provides an additional active materials for capacitive energy storage. For example, using MWNTs coated with PPy as the active electrode for a supercapacitor assembly, Frackowiak *et al.* report a specific capacitance increase from ca. 50 to 180 F/g demonstrating a synergy between the two components in this composite [53]. The excellent charge storage and transfer capabilities accounted for the improvement are due to the homogeneous coating on the nanotubular materials, high surface area, conductivity, and electrolytic accessibility of the nanoporous structure.

### 4.4 Solar cell

Solarcells are specific devices which transform light into electricity. Depending on the materials and techniques, solarcell may be classified into inorganic or organic (including conducting polymer) based devices. Currently, most solarcells in the market are based on inorganic materials such as Si or CuInGaS(Se) or CdTe. However, due to the production cost and shortage of rare materials organic based solarcell emerges as potential alternative. In principle, photon absorption in the organic-based materials produces primarily bound-state excitons. Some of these charge pairs eventually are dissociated generating free carriers (electrons and holes) which give rise to a current in the materials. The dissociation is faciliated by the potential difference across a polymer-metal junction, provided by agglomeration of excitons near the interface. The dissociation can be further accomplished via electron acceptor impurities [54]. Under illumination, a transfer of electrons to the acceptors will take place and the holes will be preferentially transported through the CPs. This process is known as photo-induced charge transfer. Since the discovery of photo-induced charge transfer, a variety of acceptor materials have been introduced into CPs to produce photovoltaic devices. Amongst of these materials, CNTs have shown to be one of the most effective acceptor materials [24, 25, 55, 56]. Since the efficiency of photo-induced charge generation is dependent on the interface between the two components, the extremely high surface area of CNTs (for purified SWNTs, ~ 1600 $m^2/g$), offers a tremendous opportunity for exciton dissociation. The primary step in these polymer photovoltaic devices is an ultrafast photo-induced electron transfer reaction at the donor-acceptor interface, which results in a metastable charge-separated state. In the case of the oligo(phenylenevinylene)/C60 composite, the quantum efficiency of this step is assumed to be close to one [57, 58]. However, the overall conversion efficiency of these solarcells is limited by the carrier collection efficiency, which is greatly influenced by the morphology of the active film.

SWNT/poly(3-octylthiophene) (P3OT) composites have been used for the fabrication of new photovoltaic devices [25, 26]. P3OT, acting as the photoexcited electron donor, is blended with SWNTs which act as the electron acceptors. In such devices the transferred electrons are transported by percolation paths provided by the addition of SWNTs. It was shown that the internal polymer/nanotube junctions act as dissociation centers, which are able to split up the excitons and also create a continuous pathway for the electrons to be efficiently transported to the negative electrode. This results in an increase of electron mobility, and hence, balances the charge carrier transport to the electrodes. The CP/SWNTs composite represents an alternative class of hybrid organic semiconducting materials that is promising for organic photovoltaic cells with improved performance. Other beneficial properties of SWNTs relevant to polymeric photovoltaic development include composite reinforcement and thermal management. The high Young's modulus and strength/weight ratio of SWNTs could help provide much-needed mechanical stability to large-area thin-film arrays. SWNTs, on the other hand, may provide assistance in thermal management for such arrays, too. Polymer composites doped with as little as 1% wt SWNTs have shown a 70% increase in the thermal conductivity at 40 K [59].

Optical and photovoltaic properties of a composite based on MWNTs and PPV was studied by Curran *et al.* [60]. These authors claim that the helical structure of the polymer chain helps the MWNTs disperse in the polymer solution, and that LEDs made with this composite are five times more stable in air compared with devices without the MWNTs. However, a drastic reduction of the PL efficiency and a change in the vibration structure of the PL spectrum were affected by MWNTs. The reduction of the PL efficiency was understood as an inter-system energy transfer (singlet-triplet), a partial hole transfer from PPV chains to MWNTs and both superposing on an intense scattering and absorption of the exciting light by MWNTs.

### 4.5 Optical limiting devices

Organic light emitting diodes (OLEDs) are one of most interest in recent years for their potential applications in electronic informatic, lighting, display technology. The simplest version of OLED consists of an electroluminescent organic material layer sandwiched between two electrodes. The luminescent emission of OLEDs is due to the radiative recombination of excitons, a process is somewhat opposite to that of solar cell. Fabrication of high efficient OLEDs depends not only on the electronic and the optical properties of the pure organic materials but also on the control of charge transport, holes or electrons through the electroluminescent layers and on the enhancement of charges migration by doping the emissive materials [61, 62]. A proper layer combination in OLEDs can also balance the injected charges in an emissive layer thus increasing the external efficiency. The buffer layer leads to a reduction of the charge injection barrier and an even charge distribution with a large contact area at the electrodes/organics interface. A recent work shows that the dispersion of SWNTs in a host polymer (PmPV, hole conducting) traps the holes in a double emitting organic light emitting diode (DE-OLED) [63]. The device fabricated without SWNTs dispersed in the PmPV has shown a dominant emission near red at 600 nm, which is in the range of the characteristic emission of Nile Red-doped Alq3, while the addition of a small amount of SWNTs enhances a green emission, In addition, the devices fabricated with the polymer/SWNTs composite have shown an increase in the oscillator strength of the green emission with a dominant emission peak near 500 nm, the characteristic emission of PmPV. The shift in the emission indicates that the SWNTs in the PmPV matrix act as a hole-blocking materials that causes a shifting of the recombination region from the Nile Red-doped Alq3 layer to the PmPV composite layer. The addition of CNTs in conducting polymer also has found to modify the electronics properties of polymer composite. For example, OLEDs fabricated with a hole conducting polymers composite dispersed with SWNTs such as PEDOT, poly-carbazole (PVK)),... show a change in EL, PL and I-V data. The modification in electronic structure of the composite originates from the hole trapping nature of SWNTs and SWNT-CP interaction.

### 4.6 Sensors and actuators

Both CPs and CNTs inherit a delocalized $\pi$-electron system over the structure with an "intrinsic" wide band gap. Owing to the interaction with various analytes, the $\pi$-electron system is modulated resulting in physicochemical modifications such as resistance, current, or electrochemical potential/work function. Based on that feature, CPs and CNTs have been the materials widely studied for application in chemical, gas sensors and biosensors, and there are excellent reviews on this topic [2, 3]. With respect to conductivity, the major charge carriers in CPs are polarons and bipolarons arising from variety of chemical and physical doping, charge injection or electronic defects within their pi-orbital backbone. Consequently, through surface interaction the conductivity of a CPs layer is affected by surrounding environment. The feature enables CPs a promising candidate for sensing materials that are superior to the metal oxides counterparts. CNTs also show a similar feature. Kong *et al.* [64] demonstrated chemical sensors based on individual SWNTs. They found that the electrical resistance of a semi-conducting SWNT changed dramatically upon exposure to gas molecules such as $NO_2$ or $NH_3$. From physical view of consideration, the change in conductivity involves the extraction or addition of electrons from valence band of semiconducting CPs and SWNTs (p-type) as a result of gas molecule interaction (physical adsorption). As the $O_2$, $NO_2$ (oxidizing agent) concentration in the environment increases, more electrons are extracted, the conductivity of the materials is increased and vice versa. On the other hand, upon exposure to a reducing agent such as $NH_3$, the conductivity of the materials is decreased. The $NH_3$ gas impact is equivalent to an injection of electron to valence band of *p-type* semiconducting CPs and SWNTs and then reduces the conductivity of the materials. With expectation of achieving synergetic effect, the CP/CNTs composites has been the attractive subject drawn much of attention to develop tiny but ultrasensitive sensor. Some promising. Some works have demonstrated the use of CNT-polymer composite as sensors. Ajayan *et al.* [65] developed a controlled method of producing free-standing nanotube-polymer composite films that can be used to form nanosensor. The nanosensor contains at least one conductive channel comprising an array of substantially aligned carbon nanotubes embedded in a matrix materials (e.g. poly (dimethylsiloxane)). This sensor can be used to determine a real time physical condition of a materials, such as monitoring the physical condition of an airplane wing or chassis while the airplane is in flight. A gas sensor based on poly(*o*-anisidine)–CNTs composite offers a sensitivity of about 28 % compared to a mere 4 % by CNTs alone for the detection of inorganic vapor, HCl [66, 67]. Sensors for the detection of carbon monoxide in the concentration range of 0.01–1000 ppm have been developed [68, 69]. A biosensor based on the composite of pyrene sulfonic acid-functionalized SWNT embedded in PANi has shown two- to sixfold enhancement depending on applied potential, compared to polystyrene sulfonate-doped PANi in the detection of glucose [70]. Another sensor for NADH has been developed with poly(1,2-diaminobenzene)–MWCNT composite in a range from 2 μM to 4 mM [71]. A CNTs composite of poly(anilineboronic acid) functionalized with single-strand DNA (ss-DNA) shows great promise for the detection of dopamine of concentration as low as 40 pM without any interference from ascorbic acid. A biosensor for the detection of chlorine without any interference due to ascorbic acid and uric acid has been developed [72].

Materials such as shape-memory alloys or liquid crystal elastomers exhibit a latent ability to actuate under the right conditions whereas other systems require the blending of two or more materials to impart a new physical response leading to the actuation process. Recently, polymer nanocomposites appeared as the subject of mechanical actuation studies, but most of them concentrated on accentuating the already present features of the host matrix by adding nanotubes. In 2003, Courty *et al.* [73] reported a novel actuator response driven by an electric field due to the presence of MWNTs in nematic elastomer, polysiloxane. They produced a composite materials with embedded and aligned CNTs with an effective dielectric anisotropy, many orders of magnitude higher than in the usual liquid crystals. Koerner *et al.* [74] fabricated MWNT-polydimethyl-siloxane composite that produced a mechanical response to the infrared radiation. They reported that the sample spontaneously contracted and elongated on irradiation. They further showed that the mechanical response is due to photon absorption and not because of the trivial heating of the materials due to irradiation. However, the nature of the actuator mechanism is not known. Ounaies *et al.* [75] developed a technique for making actuating composite materials with polarizable moieties (eg. polyimide) and CNTs. With the aid of an *in-situ* polymerization under sonication and stirring, they achieved an increase of dielectric constant from 4.0 to 31 with a 0.1% volume fraction of SWNTs. The PANi/CNTs composites have received a good deal of attention as smart materials in high-strength actuators which directly convert electrical energy into mechanical energy. Dubbed as artificial muscles, these electrochemomechanical devices can find applications in robotics, optical fiber switches, optical displays, prosthetic devices, microscopic pumps, and antivibration systems. Conducting polymers like PANi and polypyrrole have been extensively investigated for applications in artificial muscles [76-79]. ]. A dual-mode actuation has been reported for a composite containing fibers of chitosan, PANi, and SWCNTs. The activation depends on pH changes and redox polymer reactions. Such dual-mode actuators are expected to have design advantages in the construction of microelectromechanical systems (MEMS) [80]. A hybrid actuator based on cellulose–PANi–CNTs composite has been studied. The effects of the wt % of MWNTs in the composite and sonication time used in its preparation on the actuation behavior have been investigated in detail [81-82].

### 4.7 Thermal conductivity

Carbon based composites with high thermal conductivity have had a number of potential applications, particularly in thermal management such as heat sinking for electronics and motors. In particular, SWNTs are superior to carbon black and carbon fibers because their nano-scale diameter and larger aspect ratio facilitate the formation of extensive network at the same weight loading. Theoretical calculation and measurement show that the thermal conductivity CNTs is much more than that of best thermal conductive metals such as Ag, Cu [83-85]. The thermal conductivity of CP/CNTs nanocomposites although has received lesser attention, initial studies shows that the presence of CNTs in some polymer matrixes has improved the thermal conductivity of the polymer. For example, the thermal conductivity of epoxy/SWNTs composites with only 1 wt% SWNTs enhanced more than 120 % and 70% at room temperature and 40°C, respectively, as compared to epoxy filled with carbon fiber [60, 85]. However, the thermal conductivity of CNTs/epoxy composites seems to be unaffected by increasing CNTs loading in samples [86, 87]. The impact of CNTs on thermal conductivity of polymer/CNTs systems has not been explored as far as the study on the other areas of the composites and needs to be further exploited.

### 4.8 Fuel cell

Fuel cell is an efficient way transforming chemical energy of hydrogen rich compounds to electrical energy. The research in this area has gained momentum since the 80s due to the increased awareness of energy and environmental concerns. Fuel cell are usually characterized by their electrolyte, temperature of operation, transported ion and fuel. The center of the fuel cell is the electrolyte membrane, as it determines the properties needed for the other components. Electrolyte membrance based on conducting polymer has shown some advantages over the other materials due to low operating temperature, high energy density and easy handling of the fuel other than hydrogen. The PANi/CNTs composites can be used as efficient electrocatalytic materials in fuel cell reactions like oxygen reduction and methanol oxidation. The CNTs provides higher surface area and better electronic conductivity while PANi facilitates the electron transfer through the conducting matrix. A PANi-grafted MWNTs composite has shown a 610 mV more positive current onset potential for the two-electron oxygen reduction with a 20-fold enhancement in amperometric current [88]. A poly(*o*- henylenediamine)-MWNTs composite has exhibited a polymer redox-mediated electrocatalytic effect for oxygen reduction with a five-fold enhancement in current and a favorable potential shift of 130 mV compared to the values obtained at a pure poly(*o*-phenylenediamine) electrode [89]. The PANi/CNTs composites can serve as excellent host matrices for metal nanoparticles which can be used as electrode materials in methanol oxidation reaction [90-92]. It has been established that the size of the metal nanoparticles deposited on the composite matrix is smaller than that on PANi. This gives a higher dispersion and better utilization of the metal nanoparticle-impregnated composites, resulting in their high performance and stability. Apart from enhanced electrocatalytic activity, the PANi/CNTs composites consisting of metal nanoparticles have shown a reduced poisoning effect from adsorbed carbon monoxide [91]. A microbial fuel cell with PANi/CNTs composite as anode in 0.1 M phosphate buffer consisting of 5.5 mM of glucose, 2-hydroxy-1,4-naphthoquninone as mediator, and *Escherichia coli* bacteria as the microbial catalyst has been reported to give a cell voltage of 450 mV with a power density of 42 mW $m^{-2}$ [93]. In addition, the polymer nanocomposite membrane resulted in the enhanced thermal, ionic conductivity associated with the lower fuel drag than that of bare polymeric materials. These significant improvements are due to the synergetic combination of component properties. Although still in initial step, the effort. of both industrial and academic activities could be triggered on the development of polymer nanocomposite electrolyte membranes to bring forth the commercialization of fuel cells in near future.

### 4.9 Electromagnetic absorbers

Electromagnetic (EM) shielding by absorption rather than reflection is presently more important for many applications from electronics to military use. Even though metals or metal-coated materials exhibit very high EM shielding efficiency ranging from 40 to 100 dB, they cannot be used as an electromagnetic wave absorbent since their shallow skin depth makes them shield electromagnetic waves mainly through surface reflection. On the other hand, electrically conducting polymers are capable of not only reflecting but also absorbing the electromagnetic waves and therefore exhibit a significant advantage over the metallic materials. Currently, commercial and military applications require high performance absorbing materials with light weight and high strength over a broad frequency band [94]. This could be carried out if one could design and optimize a combination of different CPs components based on their dielectric properties and random scattering effects present due to their respective geometry. For example, EM absorbers with different dielectric properties and thickness were carried out on the base of the polyurethane composite containing carbon nanotubes, carbon fibers and microballoons along with polypyrrole fabric having different surface resistances. It has been shown that both the surface resistance of the PPy fabric and the order in which the composite layers are stacked are critical for the reflection. A PPy fabric composite gave greater than 15 dB reflection loss in the 4–18 GHz frequency range. With proper arrangement, the required bandwidth and better performance can be achieved by using a combination of PPy fabric and composite layer stacks [95, 96].

### 4.10 Other applications

Organic electronics have been a field of most research interests since it exhibits some advantages over inorganic including low-cost and flexible. The development relies on CPs with suitable properties in conductivity, proccecibility, charge mobility, etc. The addition of CNTs in CPs has shown to enhance the conductivity of the nanocomposites and furthermore improve the processing. For example, poly (p-phenlyenevinlene-co-2, 5-dicotoxy-m-phenylenevinylene) (PmPV) with CNTs form a hybrid composite whose conductivity is increased by ten orders of magnitude due to the introduction of CNTs conducting path to the polymer [59, 97]. The addition of SWNTs in PANi doped with dinonylnaphtalene sulfonic acid (DNNSA) creates a highly conducting [31] and the composites show to be a high-resolution printable conductor. Transparent SWNTs film on a polyethylene terephthalate substrate can be used to replace ITO for PANi-based electrochromic devices [98]. Field effect transistors (FETs) based on P3HT using SWNT-contact exhibit three orders of magnitude higher current modulation ($I_{max}$:$I_{min}$) than the metal contacted devices over the same gate voltage $V_{gs}$ equal to −2 to 2 V [99]. Schottky diodes fabricated using composites of high molecular weight PANi and MWNTs produce current levels of significantly higher magnitude than pure PANi devices [100]. However, double linear regions with two different slopes on semi-log I-V curves of these devices were observed. The behavior is explained by the difference in charge transport mechanism: consistent with Ohm's law at lower voltages and with Child's law at higher voltage. Localized defect states arising from PANi and CNTs interactions are considered to be the reason. For optoelectronics, organic materials show a possibility of fast signal processing owing to the greater state change rate and ease of hybridization change. However, the sensitivity to intensity, the long term stability and the thermal electrical conductivity of the materials need to be improved. Composites with CNTs have shown a promising approach for the realization. For instance, an all-optical switch made from polyimide composites consisting about 0.1 wt% SWNTs showed high-sensitivity and an ultra-fast, switching speed exceeding 1 ps for 1.55 μm wavelength light [101]. A organic light emitting diode, fabricated from, PmPV and SWNTs (~0.1wt%), increased oscillator strength in green radiation by about 700% in comparison to that made from PmPV alone [62]. The reason is due to the hole transport blocking by hole traps which were formed in the polymer matrix as SWNTs introduced into the composite. This non-linear optical effect could be employed in developing new light sensitive shielding.

On the other aspect, radiation shielding materials have been developed to protect personnel and equipment from the damaging effects of radiation including galactic cosmic radiation (GCR). Polyethylene (PE) is a CP has been used in space applications for shielding GCR in the low temperature applications. However, transparent composites composed of SWNTs and poly-4-methyl-1-pentene (PMP) exhibit superior strength, optical and thermal properties, and has a melt temperature of 235°C (compared to 136°C for PE) [102]. The PMP is also transparent in the visible region of electromagnetic spectrum and can be modified by doping with an organic dye having phenyl ring. These composites can be employed in thermo-luminescent detection where high energy radiation excite pi-electrons in the phenyl rings and on relaxation to ground state emit photons which can be transported to photodetectors and counted. By this way, the radiation environment of the shielding materials can be continuously monitored.

## 5. Conclusions

This chapter presents a brief summary of the preparative methods, characterization data, and possible applications of conducting polymer/carbon nanotube composites. The electrical, thermal, mechanical and electrochemical properties of the composites in general are intermediate between pure polymer and CNTs but vary depending on the method of preparation, type, purity, content and the dispersion of CNTs in the polymer matrix. In particular, the composite reveals **synergistic effects** and new properties which account for the interaction between CPs and CNTs at nanoscale. The effect offers an attractive route to create new multifunctional materials with great potential inuses involving mechanical, thermal, electrical, electrochemical features. However, the nature of the CP/CNT interaction and its effect on overall properties of the system still are unclear and need to further exploit and develop into practical application.

## 6. Acknowledgement

The work is carried on thanks to the support from Basic Research Project 103 02 103 09 Grant in Aid by National Foundation for Science and Technology Development (Nafosted).

## 7. References

[1] Luis Alcacer, *Conducting Polymers*, D. Reidel Publishing Company, 1987.

[2] Skotheim, T., Elsenbaumer, R., Reynolds, J.,( Eds.); Handbook of Conducting Polymers, 2nd ed.; Marcel Dekker, Inc.: New York, NY, USA, 1998

[3] Skotheim, T., Elsenbaumer, R., Reynolds, J., (Eds.); Handbook of Conducting Polymer 2 Volume Set: Conjugated Polymers: Theory, Synthesis, Properties, and Characterization, Conjugated Polymers: Processing and Applications (2006).

[4] Iijima S., Helical microtubules of graphitic carbon, Nature 354, 56–58 (1991).

[5] Ebbesen, T. W.; Ajayan, P. M., Large-scale synthesis of carbon nanotubes, Nature 358: 220–222 (1992).

[6] Jorio, Ado; Dresselhaus, Gene; Dresselhaus, Mildred S. (Eds.); Carbon Nanotubes: Advanced Topics in the Synthesis, Structure, Properties and Applications, Series: Topics in Applied Physics, Vol. 111, Springer (2008).

[7] H. Zengin, W. Zhou, J. Jin, R. Czerw, J.D.W. Smith, L. Echegoyen, D.L. Carroll, S.H. Foulger, J.H. Ballato, Carbon nanotube doped polyaniline, Adv. Mater., 14, 1480-1483 (2002).

[8] J. E. Huang, X. H. Li, J. C. Xu, H. L Li, Carbon 41, 2731 (2003).

[9] Kwang-Pill Lee, Anantha Iyengar Gopalan and Fernand D.S. Marquis, (Eds.); Functional Composites of Carbon Nanotubes and Applications, Transworld Research Network, Kerala (India) (2009).

[10] Michaela Baibarac and Pedro Gómez-Romero, Nanocomposites Based on Conducting Polymersand Carbon Nanotubes from Fancy Materials to Functional Applications, J. Nanosci. Nanotech., 6 (1), 1–14 (2006).

[11] Breuer, O. and U. Sundararaj, Big returns from small fibers: A review of polymer/carbon nanotube composites. Polymer Composites. 25(6): p. 630-645, (2004).

[12] L. Dai, A. W. H. Mau, Carbon Nanostructures for Adv. Polymeric Composite Materials, Adv. Mater., 13(12) 899-913 (2001).

[13] K. Mylvaganam and L. C. Zhang, Fabrication and Application of Polymer Composites Comprising Carbon Nanotubes, Recent Patents on Nanotechnology, 1, 59-65 (2007).

[14] P. Gajendran and R. Saraswathi, Polyaniline–carbon nanotube composites, Pure Appl. Chem., 80, (11), (2008), 2377–2395.

[15] Y. Min, A.G. MacDiarmid, and A.J. Epstein, The concept of "secondary doping" as applied to polyaniline. Polym Prep 35, 231 (1993).

[16] L. Dai, Polym. Adv. Technol. 10, 357 (1999).

[17] H.Ago, K.Petritch, M.S.P.Shaffer, A.H.Windle, and R.H. Friend, Adv. Mater. 11, 1281 (1999).

[18] G.Z.Chen, M.S.P .Shaffer, D.Coleby, G.Dioxan, W.Zhou, D.J. Fray, and A.H.Windle, Adv. Mater. 12, 522 (2000).

[19] Z.Wei, M.Wan, T.Lin, and L.Dai, Adv. Mater. 15, 136 (2003).

[20] H.Ago, M.S.P .Shaf fer, D.S.Ginger, A.H.W indle, and R.H Friend, Phys. Rev. B 61, 2286 (2000).

[21] W.Feng, X.D.Bai, Y.Q.Lian, J.Liang, X.G.Wang, and K.Yoshino, Carbon 41, 1551 (2003).

[22] M.Baibarac, I.Baltog, C.Godon, S.Lefrant, and O.Chauvet, Carbon 42, 3143 (2004).

[23] A.Hirsch, Angew. Chem. Ind. Ed. 41, 1853 (2002).

[24] N. S. Sariciftci, D. Braun, C. Zhang, V. Srdranov, A. J. Heeger, F. Wudl, Science 258, 1474 (1992).

[25] B. Kraabel, C. H. Lee, D. McBranch, D. Moses, N. S. Sariciftci, A. J. Heeger, Chem. Phys. Lett., 2 13, 389 (1993); N. S. Sariciftci, A. J. Heeger, Int. J. Mod. Phys. B 8, 237 (1994); E.Kymakis and G.A.J.Amaratunga, Appl. Phys. Lett. 80, 112 (2002).

[26] L.S.Schadler, S.C.Giannaris, and P.M.Ajayan, Appl. Phys. Lett. 73, 3842 (1998); H.D.Wagner, O.Lourie, Y.Feldman, and R.Tenne, Appl. Phys.Lett. 72, 188 (1998); D.Qian, E.C.Dick ey, R.Andre ws, and T.Rantell, Appl. Phys.Lett. 76, 2868 (2000).

[27] K.H.An, S.Y .Jeong, H.R.Hwang, and Y.H.Lee, Adv. Mater.16, 1005 (2004); D. N. Huyen, N. D. Chien, J. K. Phys. Soc., 52, 1564 (2008); D. N. Huyen, Proc. Phys., Springer, Berlin 279 (2009).
[28] M.Gao, L.Dai, and G.G.W allace, Electroanal. 15, 1089 (2003). H.Cai, Y.Xu, P.G.He, and Y.Z.Fang, Electroanal. 15, 1864 (2003). S.Carrara, V.Ba vastrello, D.Ricci, E.Sutra, and C.Nicolini, Sensors Actuat. B: Chem. 105, 542 (2005).
[29] D.B.Romero, M.Carrard, W.de Heer, and L.Suppiroli, Adv.Mater. 8, 899 (1996).
[30] Davey, A. Curran, S. Blau, W.: Composition including nanotubes and an organic compound, EP0949199 (1999), US20036576341 (2003).
[31] G.B. Blanchet, C.R. Fincher, and F. Gao, Appl. Phys. Lett. 82, 1290 (2003); Blanchet-Fincher, B.G.: High conductivity polyaniline compositions and uses therefore, US20067033525 (2006).
[32] Niu, C., Ngaw, L., Fischer, A., Hoch, R., Polyvinylidene fluoride composites and methods for preparing same, US20046783702 (2004).
[33] Charati, S.G., Dhara, D., Elkovitch, M., Ghosh, S., Mutha, N., Rajagopalan, S., Shaikh, A. A., Electrically conductive compositions and method of manufacture thereof, S20067026432 (2006).
[34] Roylance ME, McElrath KO, Smith KA, Tiano TM, Composite materials comprising polar polymers and single-wall carbon nanotubes. 2003 (Patent appl. no. 20030216502).
[35] Frankland SJV, Caglar A, Brenner DW, Griebel M, Molecular simulation of the influence of chemical cross-links on the shear strength of carbon nanotube-polymer interfaces, J Phys Chem B, 106, 3046-3048 (2002).
[36] Mylvaganam K, Zhang LC, Nanotube Functionalization and Polymer Grafting: An ab Initio Study, J. Phys Chem B, 108, 15009-15012 (2004).
[37] Mylvaganam K, Zhang LC, Deformation-promoted reactivity of single-walled carbon nanotubes, Nanotechnology, 17, 410-414 (2006).
[38] Mylvaganam K, Zhang LC, Chemical bonding in polyethylene nanotube composites: a quantum mechanics prediction, J. Phys Chem B, 108, 5217-5220 (2004).
[39] Margrave, J. L., Khabashesku, V. N. Zhu, J., Peng, H., Barrera, E.V., Fabrication of carbon nanotube reinforced epoxy polymer composites using functionalized carbon nanotubes, WO05028174A3 (2005).
[40] Kumar, S., Arnold, F. E., Dang, T. D.: US20056900264 (2005).
[41] Shambaugh, R. L, Nanotube/matrix composites and methods of production and use, US 7001556 (2006).
[42] Anoop-Anand, K., K. Agarwal, and R. Joseph, Carbon Nanotubes Induced Crystallization of Poly(ethylene terephthalate). Polymer. 47, 3976-3981 (2006).
[43] Zhang, S., M.L. Minus, L. Zhu, C.P. Wong, and S. Kumar, Polymer Transcrystallinity Induced by Carbon Nanotube,. Polymer 49 (5), 1356-1364 (2008).
[44] Lee, G.W., S. Jagannathan, H.G. Chae, M.L. Minus, and S. Kumar, Carbon Nanotube Dispersion and Exfoliation in Polypropylene and Structure and Properties of the Resulting Composites. Polymer. 49: p. 1831-1840, (2008)
[45] Machida, K., K. Furuuchi, M. Min, and K. Naoi, Mixed proton–electron conducting nanocomposite based on hydrous RuO2 and polyaniline derivatives for supercapacitors. Electrochemistry 72 (6), 402–404 (2004).
[46] Pickup, P.G., C.L. Kean, M.C. Lefebvre, G.C. Li, Z.Q. Qi, and J.N. Shan, Electronically conducting cation-exchange polymer powders: Synthesis, characterization and applications in PEM fuel cells and supercapacitors. J New Mat Elect Syst 3 (1), 21–26 (2000).

[47] Carlberg, J.C. and O. Inganas, Poly(3,4-ethylenedioxythiophene) as electrode material in electrochemical capacitors. J Electrochem Soc 144, L61–L64 (1997).
[48] Arbizzani, C., M. Catellani, M. Mastragostino, and C. Mingazzini, N- and p-doped polydithieno[3,4-B:30,40-D] thiophene: A narrow band gap polymer for redox supercapacitors. Electrochim Acta 40, 1871–1876. (1995).
[49] Fusalba, F., et al., Poly(cyano-substituted diheteroareneethylene) as active electrode material for electrochemcial supercapacitors, Chem Mater 12, 2581 (2000).
[50] Niu, C., et al., High power electrochemical capacitors based on carbon nanotube electrodes. Appl Phys Lett 70, 1480 (1997).
[51] Hughes, M., et al., Electrochemical capacitance of a nanoporous composite of carbon nanotubes and polypyrrole. Chem Mater 14, 1610 (2002).
[52] Jurewicz, K., et al., Supercapacitors from nanotubes-polypyrrole composites. Chem Phys Lett 347, 36 (2001).
[53] Frackowiak, E., et al., Nanotubular materials for supercapacitors, J. Power Sources 97, 822. (2001).
[54] M.Granstrom, K.Petritsch, A.C.Arias, A.Lux, M.R.Anderson, and R.H.Friend, Nature 395, 257 (1998).
[55] N.Camaioni, G.Ridolfi, G.Casalbore-Miceli, G.Possamai, L.Garlaschelli, and M.Maggini, Sol. Energ. Mat. Sol. C 76, 107 (2003).
[56] J.J.M.Halls, K.Pichler , R.H.Friend, S.C.Moratti, and A.B.Holmes, Appl. Phys. Lett. 68, 3120 (1996).
[57] L.Quali, V.V .Krasnik ov, U.Stalmach, and G.Hadziioannou, Adv.Mater. 11, 1515 (1999).
[58] E.Peters, P.A.van Hal, J.Knol, C.J.Brabec, N.S.Saricftci, J.C.Hummelen, and R.A.J.Janssen, J. Phys. Chem. B 104, 10174 (2000).
[59] Biercuk, M.J., M.C. Llaguno, M. Radosavljevic, J.K. Hyun, A.T. Johnson, and J.E. Fischer, Carbon Nanotube Composites for Thermal Management. App Phys Lett., 80(15), 2767-2769, (2002).
[60] S.A.Curran, P.M.Ajayan, W.J.Blau, D.L.Carroll, J.N. Coleman, A.B.Dalton, A.P .Da vey, A.Drury, B.McCarthy, S.Maier, and A.Strevens, Adv. Mater. 10, 1091 (1998).
[61] M.Matsumura, A.Ito, and Y.Miyamae, Appl. Phys. Lett. 75, 1042 (1999).
[62] T.M.Brown, J.S.Kim, R.H.Friend, F.Cacialli, R.Daik, and W.J.Feast, Appl. Phys. Lett. 75, 1679 (1999).
[63] Woo HS, Czerw R, Webster S, Carroll DL, Ballato J, Strevens AE, O'Brien D, Blau WJ., Hole blocking in carbon nanotube–polymer composite organic light-emitting diodes based on poly(m-phenylene vinylene-co-2, 5-dioctoxy-pphenylene vinylene), Appl Phys Lett, 77, 1393-1395 (2000).
[64] Kong J, Franklin NR, Zhou C, Chapline MG, Peng S, Cho K, Dai H, Nanotube Molecular Wires as Chemical Sensors. Science, 287, 622-625 (2000).
[65] Ajayan, P., Lahiff, E., Stryjek, P., Ryu, C. Y., Curran S,: Embedded nanotube array sensor and method of making a nanotube polymer composite, WO04053464A1 (2004).
[66] L. Valentini, V. Bavastrello, E. Stura, I. Armentano, C. Nicolini, J. M. Kenny, Chem. Phys. Lett. 383, 617 (2004).
[67] L. Valentini, J. M. Kenny, Polymer 46, 6715 (2005).
[68] Y. Wanna, N. Srisukhumbowornchai, A. Tuantranont, A. Wisitsoraat, N. Thavarungkul, P. Singjai, J. Nanosci. Nanotechnol. 6, 3893 (2006).
[69] P. Santhosh, K. M. Manesh, A. I. Gopalan, K. P. Lee, Sens. Actuators, B 125, 92 (2007).
[70] E. Granot, B. Basnar, Z. Cheglakov, E. Katz, I. Willner, Electroanalysis 18, 26 (2006).
[71] J. Zeng, X. Gao, W. Wei, X. Zhai, J. Yin, L. Wu, X. Liu, K. Liu, S. Gong, Sens. Actuators, B 120, 595 (2007).

[72] F. Qu, M. Yang, J. Jiang, G. Shen, R. Yu, Anal. Biochem. 344, 108 (2005).
[73] Courty S, Mine J, Tajbakhsh AR, Terentjev EM, Europhys Lett, 64: 654 (2003).
[74] Koerner H, Price G, Pearce NA, Alexander M, Vaia RA, Nat Mater, 3, 115 (2004).
[75] Ounaies, Z., Park, C., Harrison, J. S., Holloway, N. M., Draughon, G. K.: US2006084752 (2006).
[76] K. Kaneto, M. Kaneko, Y. Min. A. G. MacDiarmid, Synth. Met. 71, 2211 (1995).
[77] E. Smela, W. Lu, B. R. Mattes, Synth. Met. 151, 25 (2005).
[78] G. M. Spinks, L. Liu, G. G. Wallace, D. Zhou, Adv. Funct. Mater. 12, 437 (2002).
[79] S. Hara, T. Zama, W. Takashima, K. Kaneto, Polym. J. 36, 151 (2004).
[80] G. M. Spinks, S. R. Shin, G. G. Wallace, P. G. Whitten, I. Y. Kim, S. I. Kim, S. J. Kim. Sens. Actuators, B 21, 616 (2007).
[81] S. Yun, J. Kim, Z. Ounaies, Smart Mater. Struct. 15, N61 (2006).
[82] S. Yun, J. Kim. Synth. Met. 157, 523 (2007).
[83] Berber, S., Y.-K. Kwon, and D. Tomanek, Unusually High Thermal Conductivity of Carbon Nanotubes. Physical Review Letters. 84(20), 4613-4616, (2000).
[84] Kim, P., L. Shi, A. Majumdar, and P.L. McEuen, Thermal Transport Measurements of Individual Multiwalled Nanotubes. Phys. Rev Lett., 87(21), 2155021-2155024, (2001).
[85] Hone, J., M.C. Llaguno, M.J. Biercuk, A.T. Johnson, B. Batlogg, Z. Benes, and J.E. Fischer, Thermal Properties of Carbon Nanotubes and Nonotube-Based Materials.Applied Physics A: Materials Science and Processing. 74, 339-343, (2002).
[86] Gojny, F.H., M.H.G. Wichmann, B. Fiedler, I.A. Kinloch, W. Bauhofer, A.H. Windle, and K. Schulte, Evaluation and identification of electrical and thermal conduction mechanisms in carbon nanotube/epoxy composites. Polymer. 47(6), 2036-2045, (2006).
[87] Moisala, A., Q. Li, I.A. Kinloch, and A.H. Windle, Thermal and electrical conductivity of single- and multi-walled carbon nanotube-epoxy composites, Composites Science and Technology. 66, 1285-1288, (2006).
[88] K. M. Manesh, P. Santhosh, A. I. Gopalan. K. P. Lee. Electroanalysis 18, 1564 (2006).
[89] P. Gajendran, R. Saraswathi, J. Phys. Chem. C 111, 11320 (2007).
[90] Cong H N, Guadarrama V D G, Gautier J L and Chartier P, J., New Mater. Electrochem. Syst. 5, 35 (2002).
[91] P. Santhosh, A. I. Gopalan, K. P. Lee, J. Catal. 238, 177 (2006).
[92] J. Shi, Z. Wang, H. L. Li, J. Mater. Sci. 42, 539 (2007).
[93] Y. Qiao, C. M. Li, S. J. Bao, Q. L. Bao, J, Power Sources 170, 79 (2007).
[94] R.A.Tellakula, V.K.V aradan, T.C.Shami, and G.N.Mathur, Smart. Mater. Struct. 13, 1040 (2004).
[95] Olmedo, L., P. Hourquebie, and F. Jousse, Microwave absorbing materials based on conducting polymers, Adv Mater 5, 373 (1993).
[96] Kim, M.S., et al., PET fabric-polypyrrole composite with high electrical conductivity for EMI shielding. Synth Met 126, 233 (2002).
[97] J.N.Coleman, S.Curran, A.B.Dalton, A.P.Davey, B.McCarthy,W. Blau, and R.C.Barklie, Phys. Rev. B 58, 7492 (1998).
[98] L. Hu, G. Gruner, D. Li, R. B. Kaner, J. Cech, J. Appl. Phys. 101, 1 (2007).
[99] P.Qi, A.Javey, M.Rolandi, Q.Wang, E.Yenilmez, and H.Dai, J. Am. Chem. Soc. 126, 11774 (2004).
[100] P.C.Ramamurthy, W.R.Harrell, R.V.Gregory, B.Sadanadan, and A.A.Rao, Polym. Eng. Sci. 44, 28 (2004).
[101] Zhao, Y., Chen, Y., Zhang, X.-C., Raravikar, N. R., Ajayan, P.M., Lu, T.-M., Wang, G.-C., Schadler, F., Linda, S.: US20046782154 (2004).
[102] Harmon, J. P., Clayton, L. M.: US2005245667 (2005), WO06073454C2 (2006).

23

# Carbon Nanotube-Based Thin Films: Synthesis and Properties

Qiguan Wang[1] and Hiroshi Moriyama[2]
[1]*Research Center for Materials with Integrated Properties. Present Affiliation: School of Materials and Chemical Engineering, Xi'an Technological University*
[2]*Research Center for Materials with Integrated Properties and Department of Chemistry, Toho University,*
[1]*China*
[2]*Japan*

## 1. Introduction

Thin films composed of carbon nanotubes (CNTs) are an emerging class of material with exceptional electrical, mechanical, and optical properties that can be readily integrated into many novel devices.[1–4] These features suggest that CNT films have potential applications as conducting or semiconducting layers in different types of electronic, optoelectronic, and sensor systems. To understand better how to obtain these films and how to fabricate devices using them, film-forming techniques and experimental work that reveals their collective properties are of importance from fundamental and applied viewpoints.

CNTs are a well-known class of material, whose molecular structure can be considered as a series of graphene sheets rolled up in certain directions designated by pairs of integers.[5] In fact, the exceptional electrical, mechanical, optical, chemical, and thermal properties have terms with their unique quasi-one-dimensional structure, atomically monolayered surface, and extended curved $\pi$-bonding configuration.[6–10] For example, with a different chirality and diameter, an individual single-walled nanotube (SWNT) can be either semiconducting, metallic, or semimetallic, and they can be used as active channels in transistor devices because of their high mobilities (up to about 10000 $cm^2V s^{-1}$ at room temperature),[11] or as electrical interconnectors, because of their low resistivities,[12,13] high current-carrying capacities (up to about $10^9$ A $cm^{-2}$),[14] and high thermal conductivities (up to 3500 W $m^{-1}$ $K^{-1}$).[15] With their unique structure, CNTs are stiff and strong, with Young's moduli in the range of 1–2 TPa. Their fracture stresses can be as high as 50 GPa, exhibiting a density-normalized strength 50 times larger than that of steel wires.[16] In addition, the weight-normalized surface area of CNTs can be as high as 1600 $m^2$ $g^{-1}$,[17] thereby rendering them suitable for various sensor applications. CNTs can be used in many areas, ranging from nanoscale circuits,[18,19] to field-emission displays,[20] to hydrogen-storage devices,[21,22] to drug-delivery agents,[23,24] to light-emitting devices,[25,26] thermal heat sinks,[27,28] electrical interconnectors,[29] and chemical/biological sensors.[30]

It should be noted that the electronic features of CNTs are among their most important properties. Because of their high mobilities and ballistic transport characteristics, CNT films have been considered as the best replacement for Si in future devices.[31,32] Although most CNT films show a structure of completely random networks, these films are still attractive in large-area-coverage electronics, such as macroelectronics[33], mechanical flexibility/stretchability, and optical transparency.

## 2. CNT thin-film synthesis

Formation of thin films of CNTs is a necessary step to their fundamental study and use in applications. For the different fabrication techniques, how to control the tube density, the overall spatial layouts, their lengths, and their orientations must be understood, because these parameters significantly influence the collective electrical, optical, and mechanical properties.

### 2.1 Chemical vapor deposition growth

Chemical vapor deposition (CVD) is a direct method to obtain CNT films on solid substrates. Generally, Fe and Co are used as catalysts with CO, ethylene, or ethanol as the feedstock. To prevent the pyrolysis of carbon to form soot,[34] some hydrogen is usually added. Typical processing conditions involve flowing $H_2$ at 400–1000 sccm and CO at 200–1000 sccm with the temperature in the range 600–900 °C under argon.

CNT films formed using the CVD method show high levels of structural perfection, long average tube lengths, high purity, and relative absence of tube bundles. Moreover, the density, morphology, alignment, and position of tubes are also easily controlled in the CVD method. As is well known, the density value (D) is important because of its strong influence on the electrical properties of films. Experimental data demonstrate that the composition and flow rate of the feed gas can be used to control D. Compared with the case of methane, D for films obtained using ethanol as the carbon feedstock significantly increases. This is possibly because of the ability of OH radicals in ethanol to remove amorphous carbon seeds from catalytic sites in the early stages of growth.[35] The nature of the catalyst is also important. The multiple-component catalysts of Fe/Co/Mo [36–38] yield densities higher than those obtained from single Fe nanoparticles, because the former has an increased surface area, pore volume, and catalytic activity. In addition, the concentration of the catalyst, the size,[39–41] composition of the catalyst, growth temperature, pressure, and time can also affect properties such as D, diameter distributions, chiralities, and average tube length.[42]

By using different driving forces from electrical fields,[43,44] laminar flow of feed gas,[45–48] and surface atomic steps,[49,50] as well as anisotropic interactions between CNTs and single-crystalline substrates,[51–53] high alignment can be obtained. For example, electric fields (> 1 V $\mu m^{-1}$) can provide high torques, which are sufficiently large to limit thermal motions of growing CNTs, even with high-temperature growth conditions, thereby yielding field-aligned SWNTs. The degree of alignment is mainly controlled by the surface quality, cleanliness, and the physics of the underlying interactions. With catalysts patterned into small regions on a solid substrate, both perfect levels of alignment and the highest levels of D can be achieved; thus the tubes grow primarily in regions of the substrate with reacted catalyst particles.[54]

Apart from controlling the flow of feed gas, utilization of some templates is another effective method to synthesize aligned CNT films, where the template is usually alumina membranes with regularly distributed pores. In this process, the CVD reactor consists of a quartz tube placed within a tube furnace, in which an alumina template membrane is placed vertically in the CVD reactor, and the reactor temperature is kept at about 670 °C, under argon flow. After flowing ethylene pyrolyzes to yield CNTs on the pore walls as well as thin carbon films on both faces of the membrane, the furnace is turned off and allowed to cool to room temperature. Thus, a parallel array of nanotubes connected together by the carbon surface film can be obtained after dissolution of the alumina template.

Shortly after the discovery of CNTs, several growth methods were developed to synthesize different forms of CNTs in a controlled manner, such as arc discharge,[55] pulsed laser deposition,[56] and catalytic CVD (CCVD).[57] For CCVD, there are several specialized versions, such as hot wire,[58] plasma-enhanced,[59] and template [60] CCVD, which are the most commonly utilized techniques today. Among those methods listed above, CCVD techniques show the great advantage that when applied on prepatterned substrates or catalyst particles, well-aligned CNT films similar to the prepatterned template can be made.[61,62] This feature is essential for applications with special requirements of high thermal conductivity and outstanding mechanical or electrical properties.

Through a later-developed floating catalyst CVD (FCCVD) technique,[63] strong, highly conducting, and large-area transparent SWNT films can be synthesized. In contrast to the typical CVD method, a sublimed mixture of ferrocene/sulfur powder heated to 65–85 °C was used as the catalyst source, and flowed into a reaction zone by a mixture of 1000 sccm argon and 1–8 sccm methane. After 30 min growth, thin films with a thickness of 100 nm formed in the high-temperature zone (over 600 °C) of the quartz tube, which can be easily peeled off. Systematic tests reveal that the electrical conductivity of the CNT films is over 2000 S/cm and the strength can reach 360 MPa, which are both enhanced by more than one order compared with the films made from solution-based processes. It is the long interbundle connections from the firm bondings between CNT bundles that make their conductivity and strength so intriguing.

The next method for obtaining a vertically aligned CNT forest was a plasma-enhanced CVD (PECVD) technique.[64,65] Although a variety of different methods are also currently available, the PECVD process is the only technique that produces perfectly aligned, untangled CNTs. For the PECVD process, there are two main steps. First, the formation of nickel (Ni) catalyst islands on an oxidized (20 nm) silicon substrate through sintering at 650 °C. Second, nanotube growth from these discrete catalyst islands in a DC plasma discharge (bias -600 V) of acetylene and ammonia, at a pressure of 4 Torr. The initial thickness of the Ni catalyst layer controls the nanotube diameter and areal density. The plasma deposition time controls the nanotube height. A typical nanotube forest grown through this process has an areal density of 10 MWNTs per $\mu m^2$, with the vertical MWNTs having a mean diameter of 50 nm and a height of 2 μm.

At present, CVD methods are considered to be well suited for preparation of vertically aligned CNT arrays. The properties of the supporting substrates on which the nanotube films are grown often play a critical role in their applications. Moreover, only a limited variety of substrate materials are suitable for nanotube CVD growth processes, because the typical CVD growth temperature is higher than 600 °C. The interaction between the catalyst and substrate controls growth of nanotubes. Si and quartz wafers are the two substrate materials most commonly used in CVD. By using replication of a growth step and an oxidation step, single-layer or multilayer freestanding nanotube films can be synthesized. In this process, a very low concentration of water vapor can act as a catalyst promoter during nanotube growth, and also as a weak oxidant to etch the nanotube ends after growth. The oxidation step plays the role of an etching process that detaches the nanotube film from the substrate, yielding a freestanding superhydrophobic film. This vertically aligned freestanding CNT film could possibly find practical applications in many devices, such as energy storage, filtration, and the fabrication of superhydrophobic surfaces.

### 2.2 Electrophoretic deposition

The characteristic of inherent insolubility for CNTs has delayed their film formation from a wet method such as drop drying or electrophoretic deposition. After solubility improvement of carbon nanotubes by using two possible approaches of noncovalent [66,67] and covalent functionalization,[68–72] several wet methods have been proposed to prepare thin films of CNTs with controlled morphology and desired function. Compared to the CVD method, the wet methods like drop drying, electrophoretic deposition, Langmuir–Blodgett technique and self-assembling method, which are generally operated at ambient pressure and room temperature, are considered to be simpler and easier, which have attracted more research interest.

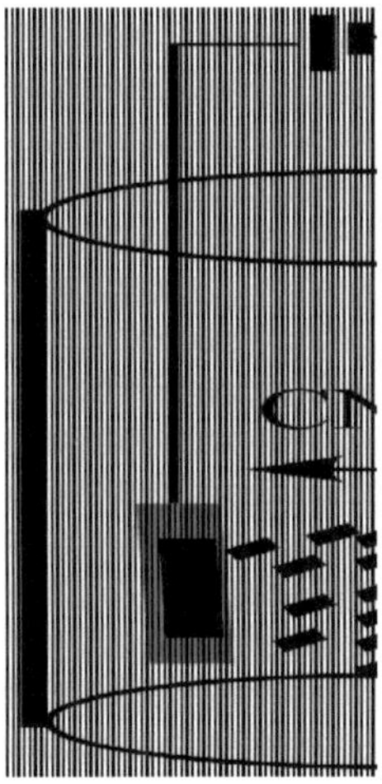

Fig. 1. Schematic diagram of the electrophoresis process for the film deposition of CNTs.

Electrophoretic deposition (EPD) has been shown to be a convenient method of fabricating thin films of CNTs with the desired thickness and excellent macroscopic homogeneity (Figure 1). EPD is fundamentally a combination of two processes, electrophoresis and deposition. In the first step, particles suspended in a liquid are forced to move toward an electrode by applying an electric field. In the second step, the particles collect at the electrode and form a coherent deposited film.[73–75] The applied electric field and deposition time are the crucial parameters to control the CNT deposition yield and thickness. A mathematical model based on Hamaker's law can be used to predict the kinetics of EPD of CNTs. In particular, different local microstructures of CNT deposits lead to variations of Young's modulus and hardness, which are attributed to differences in the packing density of CNTs. The great advantage of the EPD method lies in its simplicity. It is a cost-effective method and offers monolithic or composite coatings with complex shapes and surface patterns. In addition, the deposition rate in EPD is very fast, as much as two orders of magnitude higher than other suspension-based processes, such as slip casting.

By a combination of EPD and fissure formation techniques, a thin film of CNTs was deposited on a Ti substrate from an aqueous mixture of CNT and sodium dodecylsulfate detergent (SDS). This horizontally aligned CNT film can be used as a good field emitter. Assisted by ultrasonic treatment after EPD, oriented SWNT bundles with high density can be formed on a gold electrode by EPD. Applying ultrasonic energy resulted in the deposited SWNT bundles reassembling and orienting normal to the electrode. To get this morphology, experimental parameters, including the gap between the two electrodes, strength of the electric field, and lengths of the CNTs, are important. This combined method may prove useful in the fabrication of CNT-based electron emission devices or fuel-cell electrodes.

### 2.3 Drop drying from solvent

Techniques to form CNT thin films from solution suspensions are attractive because they can be cost-effectively scaled to large areas compatible with various substrates. They generally involve a reliable means of surfactant wrapping, to form stable solutions of CNTs, followed by evaporation of solvent (Figure 2),[76,77] or specific interactions to fabricate a film. However, a major challenge for solution deposition methods is that the low solubility and strong intertube interaction of CNTs make it difficult to obtain robust thin films with uniform moderate-to-high coverage. By means of CNT–substrate chemical interactions, these problems can be reduced to some extent, but they reduce the range of substrates and surfactants that can be used. In addition, these interactions can have adverse effects on CNT properties.

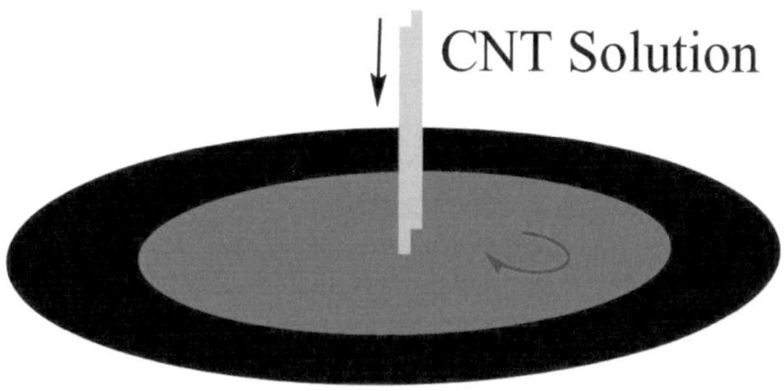


Fig. 2. Schematic diagram of the drop-drying process for the film fabrication of CNTs.

A major advantage of solution methods is that they can yield thin films directly at room temperature using CNTs formed with bulk synthesis procedures, in a manner compatible with patterning techniques such as thermal, piezoelectric, or electrohydrodynamic jet printing.[78,79] A key disadvantage is that the CNTs must first be dispersed in solution suspensions. This step often requires processes including high-power ultrasonication or strong-acid treatments, which degrade the electrical properties and reduce the average lengths of the tubes. In addition, the coated surfactants introduce unwanted organic contaminants for electronic devices.

It is reported that through the addition of liquids that are miscible with the suspending solvent and that also interact with the surfactant, this controlled flocculation (cF) process can actively drive CNTs out of solution in the desired manner. When the fluids are confined close to the surface of a target substrate during mixing, uniform films of CNTs without significant presence of bundles can be produced. Several different methods are expected to help this confinement. In one case, simultaneously introducing methanol and aqueous suspensions of CNTs onto a rapidly spinning substrate can lead to a confined CNT suspension as a thin liquid film close to the substrate surface.[80] The shear flows by spinning help to confine the two liquids vertically and to mix them rapidly, giving uniform coatings of individual or minimally bundled CNTs. Laminar flows in microfluidic channels provide the confinement.[81] The formation of fluids flowing side by side in a microchannel is another effective approach to obtaining thin CNT films. This cF method can form films ranging from monolayer to thick, multilayer coatings by simply increasing the duration of the procedure or the relative amount of CNT suspension.

Processing of CNT-based materials into engineered macroscopic materials is still in its infancy. Evaporation of drops on substrates has been used for solution deposition of CNTs onto nonporous substrates. The moving contact line of a drying drop could be used to form aligned pattern films of CNTs. Evaporation of solvent leads to a local increase in concentration of the suspension, and a very thin gelled crust is formed at the free surface. Crust formation because of solvent mass transfer across an interface qualitatively follows de Gennes' theory. The phenomenon of crusting may be exploited to fabricate thin crusts and coatings of CNTs on substrates.

Solution deposition can also be used to fabricate simple devices such as field-effect transistor (FET) architectures. To form the gate layer, a suspension of CNTs was sprayed onto a supporting substrate to form a dense nanotube network.[82] The suspension consisted of a concentration of around 1 mg/mL of nanotubes in a 1% solution of aqueous SDS. The substrate should be heated after spraying to prevent droplets from forming on the surface and to inhibit flocculation of the nanotubes. Rinsing the substrate by water then removes the SDS. Because the density of the nanotube network in the conducting channel can be tuned by controlling the number of drops of nanotube suspension adsorbed, it is much easier to get reproducible devices than with CVD methods.

If CNTs suspended in a variety of solvents are airbrushed onto a substrate placed on a hot plate at 100–150 °C, nanotube films with a wide range of thicknesses can be obtained by ensuring that all films have a constant, low ohmic resistance.

## 2.4 Langmuir–Blodgett technique

Because chemically solubilized CNTs possess good surface spreading properties at the air/water interface, optically homogeneous thin films have been prepared from the Langmuir–Blodgett (LB) technique (Figure 3). Deposition can be performed in a layer-by-layer (LBL) fashion for more layers either by horizontal lifting or vertical dipping, which allows ready control of the film thickness. This technique exhibits great feasibility. The homogeneous thin films can be synthesized in the presence or absence of assistant polymers employing either horizontal lifting or vertical dipping. By means of the good film-forming properties of poly(*N*-dodecylacrylamide) (PDDA), a stable monolayer of polymer-dispersed chemically solubilized SWNTs can be formed on the water surface, from which thin films of CNTs can be fabricated, by using the LB technique.[83–86] After finely tuning the conditions for the pretreatments and chemical solubilization of SWNTs, fabrication of homogeneous LB films of SWNTs even without using the matrix polymer is possible. More importantly, SWNTs in these films are found to be highly oriented in a specific direction. Polarized absorption spectroscopy and atomic force microscope (AFM) observations demonstrate that the tubes are oriented in the direction of the trough barrier in the case of horizontal lifting or in the dipping direction in the case of vertical dipping. These observations are attributed to compression-induced or flow-induced orientations, respectively, with the latter found to be much stronger than the former. The attainment of homogeneous thin films of SWNTs with a controllable thickness and tube orientation should be an important basis for the future development of their technological applications.

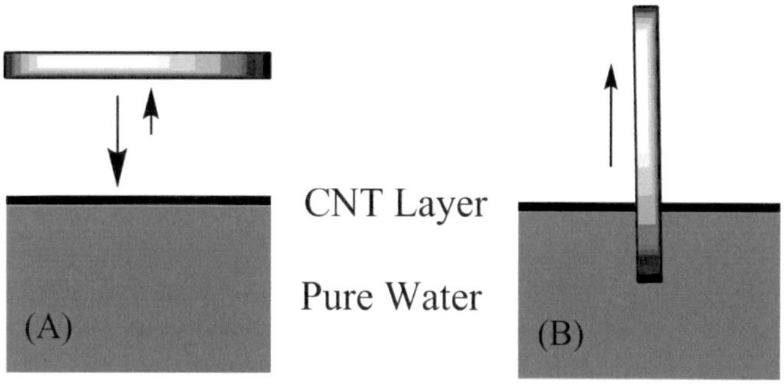


Fig. 3. Scheme for the preparation of Langmuir–Blodgett CNT films by the horizontal-lifting (A) and vertical-dipping (B) methods. Reproduced with permission from Y. Kim, N. Minami, W. Zhu, S. Kazaoui, R. Azumi, M. Matsumoto, *Jpn. J. Appl. Phys.* 2003, 42, 7629. Copyright 2003 The Japan Society of Applied Physics.

### 2.5 Self-assembling method (SAM)

By electrostatic and van der Waals interactions, LBL assembly (Figure 4A) reduces the phase segregation and makes different components highly homogeneous, well dispersed, and interpenetrated.[87] Alternating adsorption of monolayers of components attracted to each other results in a uniform growth of films. Recently, more detailed experiments demonstrated that it can be very successfully applied to the preparation of CNT films. The LBL assembly technique results in a nanotube content in the vicinity of 50%, which is significantly higher than in a typical nanotube composite.

By chemical modification, CNTs can show acid and base groups, which can be treated much as weak polyelectrolytes such as poly(acrylic acid) or poly(allylamine hydrochloride). Based on the LBL assembly method, negatively and positively charged CNTs of CNT–COOH and CNT–$NH_2$ have been alternatively adsorbed onto a substrate to form an all-carbon film, which consists of well-dispersed CNTs.[88] Like other multilayer assemblies, this 100% CNT thin film shows pH-dependent thickness and surface topology, which are characteristics of LBL thin films of weak polyelectrolytes. The surface topology and the inner structure of this thin film are interconnected random network structures with physical entanglements. Sheet resistance and cyclic voltammetry (CV) measurements show that these films are promising electrode materials for high-power and high-energy electrochemical devices.

Self-assembly of true chemical linkages between the molecules and the substrates (Figure 4B) is another low-cost process for formation of functional molecules. Van der Waals interactions between neighboring chemisorbed molecules generally lead to long-distance ordering in the first monolayer. Self-assembly from solution or the gaseous state can result in very good coverage. By using this SAM method, an ordered CNT film with a perpendicular orientation can be prepared on gold surfaces. The as-grown nanotubes were first chemically functionalized by thiol groups. The ordered assembly of CNTs was accomplished by their spontaneous chemical adsorption on gold via Au–S bonds. The adsorption kinetics of the nanotubes was very slow in comparison with conventional alkane thiols. The adsorption rate varied inversely with tube length. The nanotubes tend to form bundles as the adsorption propagates, following a "nucleation adsorption mechanism." Functionalized CNTs perpendicular to the surface can be assembled on various substrates via a predesigned bonding nature. For example, carboxylic acid-terminated CNTs assemble on an amino-terminated silicon and silver surface via electrostatic interaction or chemical bonding such as the surface condensation reaction.[89]

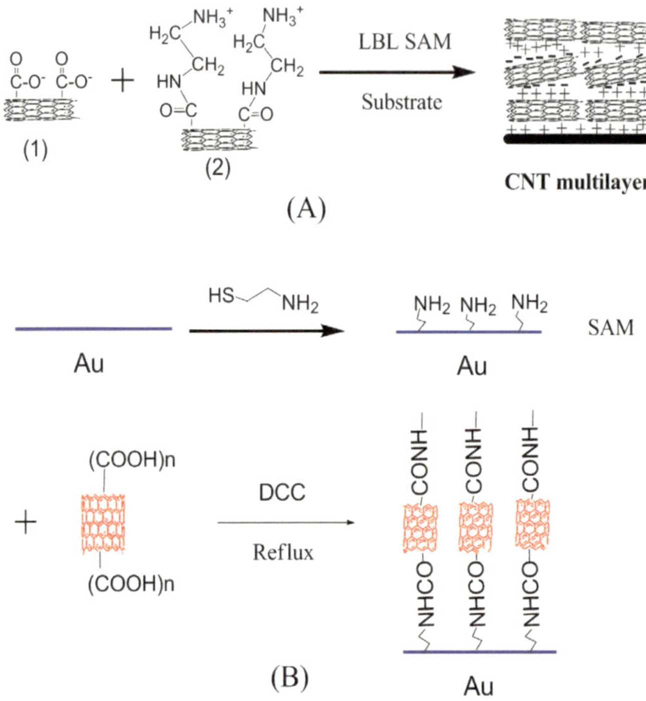


Fig. 4. Scheme for the preparation of SAM CNT films by the electrostatic-interaction (A) and chemical-bonding (B) methods.

With the assistance of an electric field, highly aligned CNT thin films can be fabricated by the chemical assembly approach.[90] With increase in the electric field, the assembling kinetics of CNTs is remarkably speeded up, and the packing density can even exceed the saturated density of the conventional assembly method by a factor of four. The molecular dynamics simulation results illustrated the alignment of CNTs with their long axes along the electric flux in solution, leading to the increase in packing density and efficiency. Under a d.c. electric field, the CNTs are aligned with their long axes along the electric flux and drift toward the anode substrate with higher velocity, leading to the increase in packing density by overcoming the steric hindrance of the "giant" CNTs and the effective decrease in assembling time.

### 2.6 Electropolymerization

Electropolymerization of polymerizable monomers is an effective method for assembling polymer films on electrode surfaces, and the electropolymerization of monomer-functionalized giant metal nanoparticles, for example, pyrrole-capped Au nanoparticles,[91] was reported to yield a two-dimensional (2-D) nanoparticle array. Recently, the preparation of *p*-mercaptoaniline-capped CdS nanoparticles and their electropolymerization on a Au electrode in a monolayer assembly has been successfully achieved.

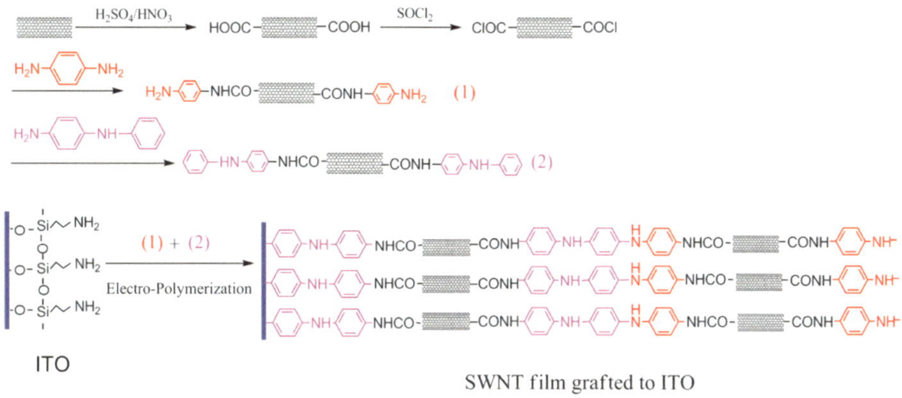


Fig. 5. Scheme for the preparation of CNT films by using the electropolymerization method.

Similarly, a CNT film can be generated by using the electropolymerization method, if CNTs were correctly modified by some polymerizable groups such as phenylamine. The oxidized multiwalled nanotube was functionalized with *p*-phenylenediamine, which gave functional groups on the surface. In our group, an SWNT film on indium–tin oxide (ITO) was prepared by electropolymerizing the *N*-phenyl-1,4-phenylenediamine-modified SWNT in aqueous solution (Figure 5). The CNT film showed a homogeneous structure (Figure 6), where the SWNTs were interconnected to form a dense film, which provided another path for ease of preparing CNT film with good mechanical properties.

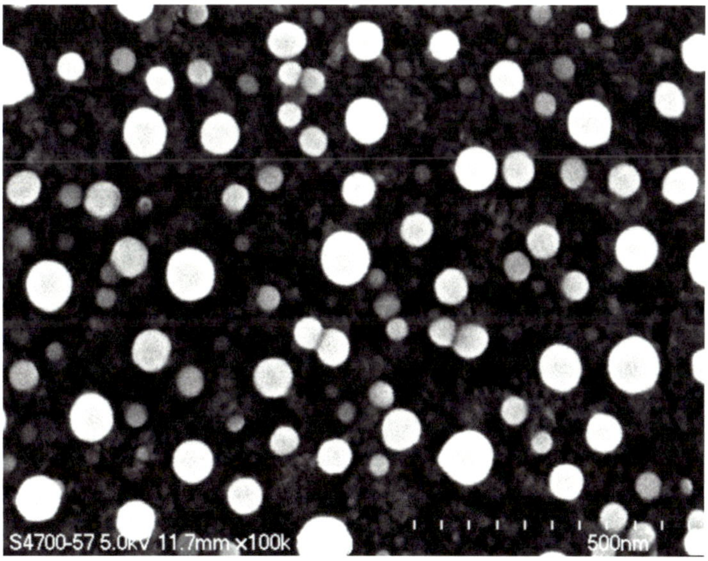


Fig. 6. SEM images of CNT films prepared by using the electropolymerization method.

### 2.7 Vacuum-filtering method

Probably, compared with the other methods, the vacuum-filtering method (Figure 7) is regarded as the simplest process for the fabrication of ultrathin, transparent, optically homogeneous, electrically conducting films composed of pure CNTs.[92] This process is quite simple, containing three steps: vacuum filtering a dilute, surfactant-based suspension of purified nanotubes onto a filtration membrane to form a homogeneous film on the membrane, then washing away the surfactant with purified water to allow film formation of pure CNT, followed by dissolving the filtration membrane in solvent. From the above, this filtration method has several advantages: (i) as the nanotubes accumulate, the generated filter cake acts to impede the permeation rate, which can tune the local permeation rate and associated deposition rate automatically. Therefore, homogeneity of the films is guaranteed. (ii) Under vacuum, the nanotubes tend to lie straight, gaining maximum overlap and interpenetration within the film as they accumulate. This yields maximum electrical conductivity and mechanical integrity throughout the films. (iii) The film thickness is controlled, with nanoscale precision, by the nanotube concentration and volume of the suspension filtered.

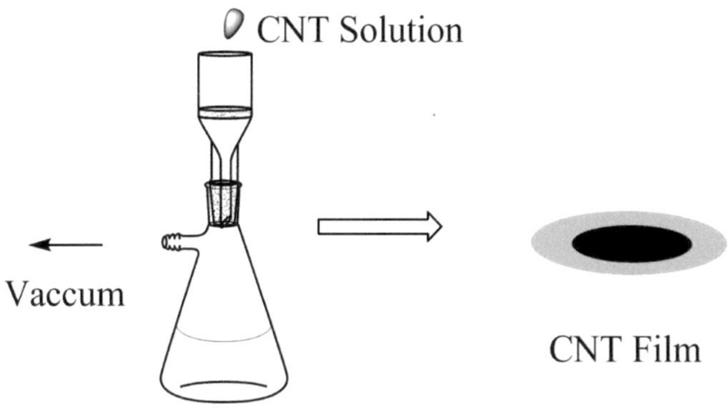


Fig. 7. Scheme for the preparation of CNT films by using the vacuum-filtering method.

In addition, through a simple postdeposition method, the conductivity of the obtained CNT film can be improved via exposure to nitric acid and thionyl chloride. Sheet resistance of CNT thin films is decreased by a factor of five via exposure to thionyl chloride. This enhancement in transport properties upon $SOCl_2$ treatment is related to the formation of acyl chloride functionalities. The obtained CNT films are more flexible than ITO, and can be a replacement for those expensive semiconductors.

## 3. CNT-based thin films

### 3.1 Nonconjugated molecule–CNT thin films

Nonconjugated molecule–CNT thin films are generally prepared by two approaches: solution casting and LBL assembly. The driving forces include electrostatic interactions,[93–95] hydrogen bonding,[96,97] charge-transfer interactions,[98] and coordination bonding.[99]

For construction of CNT composites, LBL assembly allows excellent control of thickness and composition and diminished phase segregation compared with other methods.[100] So far, all LBL composites containing CNTs have been constructed via electrostatic interactions[101] or hydrogen bonding.[102] However, covalent cross-linking between SWNT and polymer is needed for strengthened composite films. By using some special reactions such as covalent linkage under UV irradiation, the electrostatic LBL film of SWNT–poly(sodium 4-styrenesulfonate) (PSS) and a diphenylamine-4-diazoresin can be converted to a cross-linked film. The SWNT–PSS was 55 wt% SWNT.[103] Apart from the increase in mechanical strength, the resistance of the film toward etching by polar solvents increased significantly after irradiation.

By using spin coating with a mixture that consists of a solvent with low volatility, transparent electrically conductive films of CNTs and thermoplastic polymer poly(methyl methacrylate) (PMMA) can be obtained, which may replace ITO.[104]

For the LBL process, poly(diallyldimethylammonium chloride) (PDDA) can be used as a model for preparation of polymer/CNT films.[105] A clean hydroxy-bearing silicon wafer is first dipped into a 1 wt% aqueous solution of PDDA for some time, such as 10 min, and the wafer rinsed with deionized water, then dried with nitrogen. Then, the PDDA-treated wafer is placed horizontally, face down, into a dispersion of purified CNTs in dimethylformamide (DMF) for 100 min, removed, rinsed with DMF, and dried with nitrogen. The CNT-terminated film is then dipped into a 1 wt% aqueous solution of PDDA in 1.0 M NaCl for 10 min, followed by rinsing with deionized water and drying with nitrogen. The addition of 1.0 M NaCl to the PDDA was required for uniform film growth as attempts to form films with only 1 wt% PDDA resulted in little sequential adsorption. Studies on polyelectrolyte multilayer films have shown that the addition of salt causes a dramatic increase in the amount of polyelectrolyte deposited. Atomic force and scanning electron microscopies indicated that the adsorbed CNTs were mostly in the form of 5-10 nm bundles and that uniform substrate coverage occurred. Absorbance spectrophotometry confirmed that the adsorption technique resulted in uniform film growth.

In most recent reports on CNT/ isotactic polypropylene (iPP) nanocomposites, the melt blending technique has been employed,[106] which provides a very simple preparation method. However, some of the drawbacks associated with melt-compounding methods include high energy cost, risk of filler deterioration during processing, and a generally poor dispersion quality. Solution mixing provides an alternative preparation method; however, it requires the use of organic solvents and is limited to relatively small quantities. To overcome the above defects, a novel latex-based method was developed, by which CNT/polypropylene films were prepared through the incorporation of CNTs into a polypropylene matrix. In addition to being versatile and environmentally friendly, latex technology allows for the achievement of high dispersion qualities. Moreover, it can be easily extended to any matrix polymer with a latex form. It allows the preparation of high-performance lightweight CNT/iPP films, while overcoming the drawbacks of conventional processing methods.

By solution casting from dilute solutions, interpenetrating networks of entangled CNTs and polystyrene (PS) chains were prepared in thin films.[107] The CNTs were first surface grafted with PS chains to provide good compatibility and steric hindrance against reaggregation of the CNTs in the solution phase. The CNTs dispersed quite well in PS–toluene solutions. The dispersion of the nanotubes was uniform, extending globally to form a percolated network, capable of withstanding deformation of more than 25% without fracture. Experimental data show that micronecking of the fracture precursor of crazing was strongly suppressed, which leads to the enhancement of mechanical properties.

Conjugated macromolecules such as poly(*p*-phenyleneethynylene)s (PPEs) can be used to noncovalently functionalize and solubilize CNTs. Using PPE, the resulting SWNT solubilized in chloroform can produce a homogeneous SWNT–polycarbonate (PC) composite solution by mixing with a PC solution.[108] After the solution is cast on a glass dish and dried very slowly, a free-standing film can be peeled from the substrate. The infrared photoresponse in the electrical conductivity of SWNTs is dramatically enhanced by embedding SWNTs in the electrically and thermally insulating polymer matrix.

An insulating polymer surface can be used as a guide for the deposition of two-dimensional networks of CNTs. For example, the CNT solution was cast and dried on the surface of electrospun polyamide 11 (PA11) nanofiber films, which can manufacture transparent and electrically conductive thin films. Multiple deposition cycles lead to increased coverage and conductivity.[109]

Also, by a facile method of spray coating, CNT/silane compound hybrid films at a silane sol concentration of 70 wt% were achieved. In addition, the wettability of the transparent, conductive films can be varied from superhydrophobicity to superhydrophilicity by varying the chemical functionality of the silane sol. The stable CNT/silane sol solution was prepared based on the intermolecular interactions between the hydroxyl groups of the CNTs and the silanol groups of the silane sol.[110] This CNT-based film may provide a wide range of applications in the development of self-cleaning coatings for optoelectronics, transparent film heating, electrostatic discharging, and electromagnetic interference shielding.

CNTs/PMMA composite films showing anisotropic electrical transport properties can be fabricated using the electric-field-assisted thermal annealing method.[111] Because of the alignment of the SWNT along the electric field direction, the electric-field-assisted thermal annealing of octadecylamine-functionalized SWNT/PMMA films induces an increase in the composite transverse conductivity by several orders of magnitude and a decrease in the lateral conductivity.

### 3.2 Conjugated polymer–CNT thin films

Because of the strong π–π interactions, CNTs are easily dispersed into conjugated polymer solutions. So, for the synthesis of conjugated polymer–CNT thin films, the solution-casting method is very applicable. For example, CNT/polythiophene (P3HT) films can be fabricated using a very simple spin-casting technique. The resulting film is regarded as a high-performance chemical sensor.[112]

Among the conjugated polymers, P3HT has attracted much research interest because the high-molecular-weight P3HT forms very stable dispersions. Based on solution casting, a free-standing, light-pink-colored SWCNT/P3HT film is readily released from the glass slide substrate as soon as it is dipped into deionized water. This free-standing film exhibits good electrical properties, comparable with commercial ITO and PEDOT/PSS systems.

The highly aromatic pyrenyl group is known to interact strongly with the basal plane of graphite via π-stacking, and also strongly interacts with the sidewalls of SWNTs in a similar manner.[113] Based on this interaction, a CNT film was prepared containing a dye, *N*-(1-pyrenyl)maleimide (PM), and a functionalized SWNT-conjugated polymer, poly(3-octylthiophene) (P3OT), from drop or spin casting. The photoresponse was improved by functionalizing the SWNT with dye molecules. The short-circuit current was found to increase by more than an order of magnitude compared with the SWNT–polymer diode without dye. The increase in short-circuit current is probably because of efficient transfer of holes by dye molecules to P3OT at the dye/polymer interface and the rapid transfer of the generated electrons to the SWNTs at the dye/nanotube interface.

Studies have shown that addition of small amounts of conjugated polymer to nanotube dispersions enables straightforward fabrication of uniform network films by spin coating. After treatment with thionyl chloride, electrodes have significantly decreased sheet resistances. For example, adding a minimal quantity of P3AT or poly[2-methoxy-5-(2-ethylhexyloxy)-1,4-phenylene vinylene] (MEH–PPV) to CNT dispersions is sufficient to disperse the nanotubes for spin coating onto glass or PET substrates, to fabricate a transparent conducting film with a uniform CNT network. The technique provides an easy, reliable, scalable, plastics-compatible method for fabricating flexible transparent electrodes directly from solution onto the substrate of interest.[114]

Polybenzimidazole (PBI) has been shown individually to dissolve/disperse SWNTs in *N,N*-dimethylacetamide (DMAc).[115] By casting these dispersions, SWNTs/PBI composite films were successfully fabricated on substrates without macroscopic aggregation. The addition of SWNTs to PBI does not reduce the thermal stability of the matrix film, and the mechanical properties of the PBI film were reinforced by ca. 50% with only 0.06 wt% addition of the SWNTs because of the $\pi$–$\pi$ interaction between the PBI and the sidewalls of the SWNTs.

In the case of CNTs, the hydrophobic part of the poly(4-vinylpyridine) (PVP) chain can be bound to the CNTs' surface via hydrophobic and other intermolecular interactions (e.g., $\pi$-stacking interactions) to form a stable CNT/PVP composite. By using this feature, a CNT/PVP/PB composite film was synthesized by casting CNTs wrapped with PVP on gold electrodes followed by electrochemical deposition of PB, which was shown to act as an amperometric biosensor, because of the remarkable synergistic effect of the CNTs and PB.[116]

Electrochemical codeposition is another concise chemical method to prepare conjugated polymer–CNT thin films based on their respective electrochemical properties. By this method, homogeneous nanocomposites of CNT–polyaniline (PANI) resulted.[117] For this, the CNTs should be functionalized in advance via polymerizable groups. This also helped to disperse the nanotubes in aniline. The combination of PANI with CNTs would offer an attractive composite support material for an electrocatalyst to enhance its activity and stability based on morphological modification or electronic interaction between two components.

### 3.3 Pure carbon thin films from fullerene and CNT

$C_{60}$ and CNTs, as novel all-carbon $\pi$-electron systems, have increasingly invited exploration for preparing their composite films from both fundamental and practical points of view. It is known that clusters of $C_{60}$ [118–120] and carbon nanostructures [121] can be deposited electrophoretically onto electrodes to form a film. In this manner, the clusters of $C_{60}$ and SWNT were attached to electrodes such as FTO/$SnO_2$ (FTO represents F-doped tin oxide) to form a film of $(C_{60} + \text{SWNT})_m$. Under application of a high d.c. electric field (200 V for 120 s), the clusters of $C_{60}$ and functionalized SWNT move toward the positively charged electrode. With increasing time of deposition, the FTO/$SnO_2$ electrode turns brown for $(C_{60})_m$ and $(C_{60} + \text{SWNT})_m$, or black for $(\text{SWNT})_m$. The time to reach a maximum absorbance increases in the order $(\text{f-SWNT})_m < (C_{60} + \text{f-SWNT})_m < (C_{60})_m$, because of the faster deposition of $(\text{f-SWNT})_m$ than of $(C_{60})_m$. The difference in the mobilities of the clusters leads to inhomogeneous structures in the deposited composite film of $C_{60}$ and f-SWNT. The composite film exhibited an incident photon-to-photocurrent efficiency as high as 18% at 1400 nm under an applied potential of 0.05 V vs. SCE. The photocurrent generation efficiency is the highest value among CNT-based photoelectrochemical devices in which CNTs are deposited onto electrodes electrophoretically, electrostatically, or covalently.[122]

For the preparation of $C_{60}$ and CNTs, one of the key challenges is to overcome the high aggregation tendency of these nanoscale carbon spheres and fibers. A $C_{60}$–CNT composite film was created by CV.[123] Briefly, purified MWNTs and $C_{60}$ (MWNTs/$C_{60}$ = 2:1) with a total amount of about 1 mg were dispersed in 10 mL toluene in an ultrasound bath for 30 min to give a 0.1 mg $mL^{-1}$ suspension. A volume of 15 mL of the suspension was cast directly onto a glassy carbon (GC) electrode surface and the solvent was allowed to evaporate at room temperature. This $C_{60}$–MWNT film electrode was subjected to potential scanning in acetonitrile solution containing 0.1 M tetrabutylammonium hexafluorophosphate ($TBAPF_6$) between 0.0 and -2.0 V (vs. Ag/AgCl). The resultant $C_{60}$–MWNT film electrode was then washed with acetonitrile several times to remove the electrolytes, and then dried at room temperature. The uniform composite films show reversible redox behavior, which is similar to that of $C_{60}$ dissolved in organic solution but is very different from those of either $C_{60}$ films, CNT films, or peapod films. It is presumed that these novel properties come from the covalent anchorage of $C_{60}$ to the CNTs in a uniform fashion.

Using the above method, hemoglobin (Hb) was embedded into $C_{60}$–CNT film.[124] Experimental results demonstrated that $C_{60}$–CNT films can facilitate the direct electron transfer of Hb much more effectively than bare CNT films. This is attributed to the faster electron-transfer kinetics on the $C_{60}$–CNT film from the roles of electron mediator and protein docking site played by $C_{60}$, which is finely dispersed on the MWNT surfaces. In this way, Hb can transfer electrons to and from the electrode more easily through $C_{60}$ in the $C_{60}$–CNT nanocomposite film. The obtained Hb/$C_{60}$–MWNT film was shown to act as a new biochemical sensor for the reduction of $O_2$.

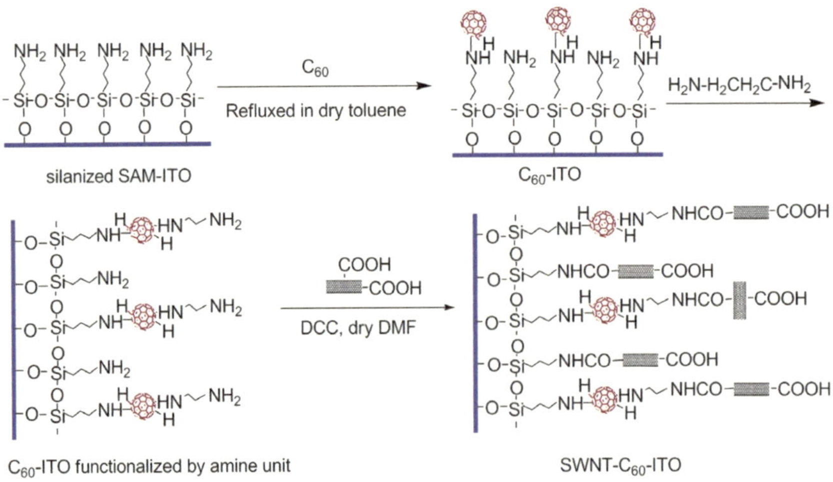


Scheme 1. Schematic illustration of the preparation route to $C_{60}$–SWNT ultrathin film grafted onto ITO step-by-step. Reproduced with permission from Ref. [125]. Copyright 2009 American Chemical Society.

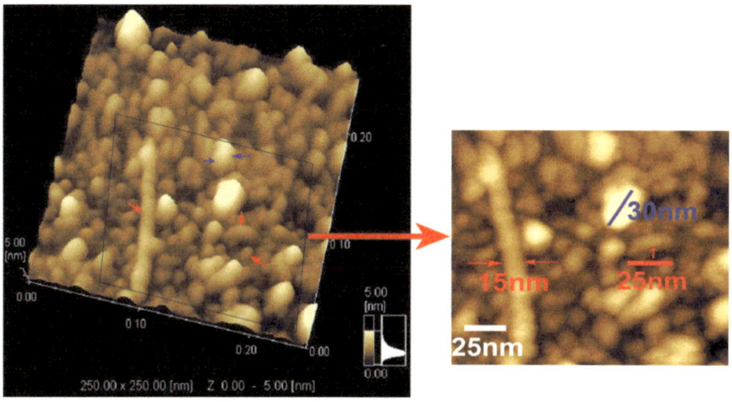


Fig. 8. Typical AFM images of SWNT–$C_{60}$–ITO. Reproduced with permission from Ref. [125]. Copyright 2009 American Chemical Society.

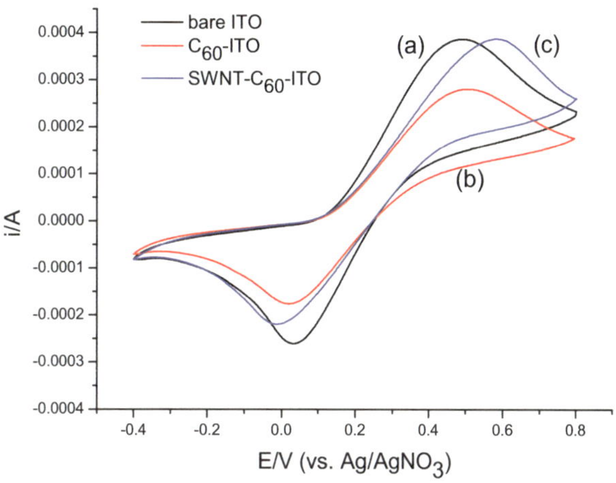


Fig. 9. Cyclic voltammograms of bare ITO (a), $C_{60}$–ITO (b), and SWNT–$C_{60}$–ITO (c) in acetonitrile with 3 mM ferrocene as an internal probe. Reproduced with permission from Ref. [125]. Copyright 2009 American Chemical Society.

By using a step-by-step method, we prepared homogeneous ultrathin films composed of [60]-fullerene ($C_{60}$) and SWNTs, grafted to the functional surface of an alkylsilane SAM on an ITO substrate with an ITO–$C_{60}$–SWNT sequence using amine addition across a double bond in $C_{60}$ followed by amidation coupling with acid-functionalized SWNTs (Scheme 1).[125] AFM images of the resulting composite film showed two-component ball-tube microstructures with high-density coverage, where $C_{60}$ was homogeneously distributed in the SWNT forest (Figure 8). The attachment of SWNTs to the residual amine units in the SAM on the ITO substrate (SAM–ITO) as well as on the $C_{60}$ sphere results in the $C_{60}$ molecules in the aggregated clusters being more separately dispersed, which forms a densely packed composite film as a result of the π–π interaction between the $C_{60}$ buckyballs and the SWNT walls. It was found using ferrocene as an internal redox probe that the oxidative and reductive processes at the film–solution surface were effectively retarded because of obstruction from the densely packed film and the electronic effect of SWNT and

$C_{60}$ (Figure 9). In addition, the electrochemical properties of $C_{60}$ on SAM–ITO plates observed by CV were significantly modified by chemical anchorage using SWNTs (Figure 10). X-ray photoelectron spectroscopy (XPS) analysis also indicated the successful grafting of $C_{60}$ and SWNT. The XPS chemical shift of the binding energy showed the presence of electronic interactions between $C_{60}$, SWNT, and ITO components. Such a uniformly distributed $C_{60}$–SWNT film may be useful for future research in electrochemical and photoactive nanodevices.

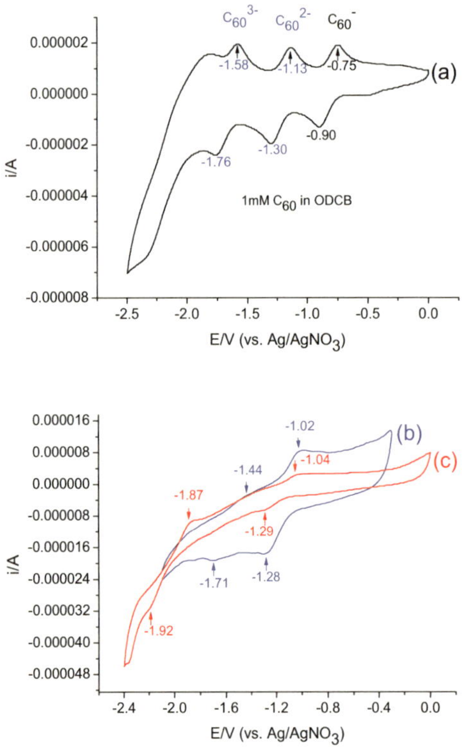


Fig. 10. Cyclic voltammograms of (a) 1 mM $C_{60}$/ODCB solution, (b) $C_{60}$–ITO, and (c) SWNT–$C_{60}$–ITO in $CH_3CN$. Reproduced with permission from Ref. [125]. Copyright 2009 American Chemical Society.

## 4. Properties

### 4.1 Mechanical strength

Similar to other engineering materials, the strength of macroscale SWNTs in film is dominated by the stress-transfer mechanism rather than the strength of individual CNTs. A 200-nm-thick film exhibits high tensile strength and good toughness. The tensile strength is 360 MPa, which is 30 and 10 times higher than typical bulky paper and sheets from oleum,[126] respectively; the density-normalized stress is 280 MPa/(g/cm$^3$). The Young's modulus is about 5 GPa. Compared with the theoretical strength of individual SWNTs (37 GPa), the film strength is two orders lower. As is known, despite the high stiffness and strength of individual SWNTs, slippage between nanotube surfaces reduces the prospect of using SWNT bundles as reinforcing material in composites.[127,128] To resolve the "slipping problem", several routes have been proposed, such as reducing the bundles' diameters, bridging adjacent tubes by electron-beam irradiation,[129] or prolonging the contact length between tubes;[130] however, none of them have proved feasible at the macroscale.

By using a homemade microextensiometer set inside the SEM chamber, the in situ morphologies of CNT can be observed when the films are extended. The changes in morphology show that when the strain is far below the strain-to-failure, the "meshes" in the networks extend continuously and homogeneously. With increasing strain, stress concentrations occur at weak points and become more and more severe. Close to the strain-to-failure, extension mainly occurs at the breaking point, where meshes are destroyed and the remaining tubes completely align. Once the breaking point develops, the concentrated stress will split the films rapidly. The maximum extension ratio of the basic unit is 33%, far higher than the typical strain-to-failure of films (10%). From the above, the mechanical property of the macroscopic films is dominated by the basic units, meshes, rather than straight bundles, and the load is homogeneously transferred to the whole film through shared bundles of adjacent meshes.

Mechanical characterization of the CNT films was provided by nanoindentation tests. Similar load–depth data were obtained for loads of 10 mN and 1 mN. The values of Young's modulus vary significantly, from 7.7 to 77.7 GPa for the 1 mN load, and from 70.0 to 157.8 GPa for the 10 mN load. Hardness varies from 0.15 to 1.19 GPa and from 0.5 to 2.12 GPa, respectively. Probably, the broad ranges come from a network of rods that are very rigid in tension but flexible in bending,[131] and are probed at the same length scale as the network features.

It is also worth noting that the mechanical properties of individual nanoscale objects are difficult to measure directly; indeed, nanotubes are particularly heterogeneous, both in dimensions and internal perfection, giving rise to significant variation from one nanotube to another. In fact, the response is controlled by a small number of nanotubes, and is susceptible to local variations in microstructure.

### 4.2 Thermal response

As a transparent conducting coating, thin films of CNTs have outstanding performance as a thermal interface layer for heat dissipation in high-density electronic packaging.[132] Because CNT film is composed of a network of individual CNTs and CNT bundles, the thermal and electrical resistances are dominated by the intertube junctions,[133–136] which depend strongly on chemical modification of the SWNTs and the film-preparation technology. In general, the relative contribution of the electron and phonon components of the thermal conductivity can be evaluated on the basis of the Lorenz number, $L = \kappa / \sigma T$. The Lorenz number for the purified SWNT film is close to $7 \times 10^{-6}$ W $\Omega/K^2$ at temperatures between 50 and 300 K, which corresponds to a ratio of the electron-to-phonon contribution to the thermal conductivity of 1 to 100, which is further decreased to 1 to 10 000 in the case of the as prepared SWNT network.

The electrical resistance at the junction of two metallic SWNTs was found to be 200 kΩ, contact of two semiconducting SWNTs showed a junction resistance of 500 kΩ, while contact of metallic and semiconducting SWNTs provided the most resistive junction (> 10

MΩ) because of the Schottky barrier.[36] The heat conductance at the intertube junctions, $G_J$, was evaluated theoretically not to exceed $10^{-9}$ W/K.[137,138] The average junction electrical resistance can be evaluated at about $R_J = 10^6\ \Omega$. Experimental data show a Lorenz number for individual cross-junctions at 300 K of $L_J = 3 \times 10^{-6}\ W\ \Omega/K^2$, close to the value obtained for purified SWNT film and two orders of magnitude higher than the pure electronic value, $L_e = 2.4453 \times 10^{-8}\ W\ \Omega/K^2$. That is to say, heat transport across the intertube junction is dominated by the phonon component. Both electrical and thermal transport in SWNT networks are dominated by intertube junctions. It should be noted that, in the case of an SWNT network embedded in a polymer matrix, stronger suppression of both electrical conductivity and larger Lorenz numbers ($10^{-2}\ W\ \Omega/K^2$) were observed.

### 4.3 Electrical conductivity

Although the axial conductivity of an SWNT rope can reach 10000–30000 S/cm, conductivity in films or networks is usually one or two orders lower. For CNT films, sheet resistance is the result of three distinct contributions. The first is from the CNTs themselves. Many inherent factors have an effect on the electronic properties of nanotubes, including diameter, chirality, defect, curvature, and local environment.[139] As a result, their inhomogeneous distribution complicates the conductivity of the films. The second component is the existence of some barriers at intertube junctions.[140] Electron transport via the hopping mechanism through the intertube junctions is predominant in the conductivity of CNT films. Finally, the additional resistances introduced during the fabrication process of CNT films also contribute to sheet resistance, such as residual surfactant.

For transparent conductive thin films fabricated through a procedure based on the filtration method, the sheet resistance has varying degrees of improvement after the multistep purification process. After removing the mixed cellulose ester (MCE) filtration membrane, L-SWNT films ("laser" nanotubes) present the lowest sheet resistances, while those of H-SWNT ("HiPCO" nanotubes) films show the highest. The sheet resistance of A-SWNT ("arc-discharge" nanotubes) films is close to that of L-SWNT films, because of the same range of diameters and lengths. The high resistance of H-SWNT films arises from their much smaller diameter and length compared with those of L-SWNTs or A-SWNTs.[141]

The conductivity of the SWNT films features a sharp jump of several orders of magnitude, attributed to a typical electrical phenomenon dealing with the formation of a network of conductive particles in terms of percolation theory.[142] Percolation is a statistical geometric theory that has established the universality of the exponents in the power law dependence of geometrical parameters. In plain terms, for SWNT films just above the percolation threshold, sheet resistance reduces dramatically with the increase in film thickness, while in the region far from the threshold, sheet resistance decreases inversely with film thickness, as expected for constant conductivity.

After washing off the surfactants, the electrical conductivity of the SWNT film coatings was improved further by treatment with various acids. Upon treatment with acids, Geng et al.[143] observed a fivefold increase in the electrical conductivity of SWNT thin films that had been made using a surfactant-based dispersion and had been washed to remove residual surfactant. They proposed that the acid removed residual surfactant molecules adsorbed on the surface of the nanotubes, leading to better contact between the nanotubes, densification of the films, and improvement in overall electrical conduction properties.

### 4.4 Electrochemical properties

CNTs have a high electrochemically accessible area of porous tubes, as well as good electronic conductance and good electrocatalytic activity, which give CNTs enormous potential as components of nanoscale electronic devices and biosensors, particularly for the CNT films fabricated on electrodes.

A potential application of the electrochemical active CNT film is the electrocatalytic activity toward $O_2$ reduction in alkaline media.[144] These properties essentially suggest that the CNTs are a potential candidate for development of effective, low-cost, and environmentally benign nonplatinum alkaline air electrodes for energy conversions. For example, the CNT multilayer films on GC electrodes, developed by the LBL method, based on the electrostatic interaction between positively charged poly(diallyldimethylammonium chloride) (PDDA) and negatively charged and shortened MWNTs, show remarkable electrocatalytic activity for $O_2$ reduction in alkaline media.

Because the diameters and carrier densities of SWNTs are comparable to the sizes and surface-charge densities of biomacromolecules, SWNTs can serve as ultrasensitive transducers in biosensors based on chem-resistor or transistor structures.[145,146] A more generalized and reliable approach to achieve specific detection involves direct chemical functionalization of the SWNTs. Noncovalent approaches are generally preferred as they do not degrade the intrinsic electrical properties of the SWNTs.[147]

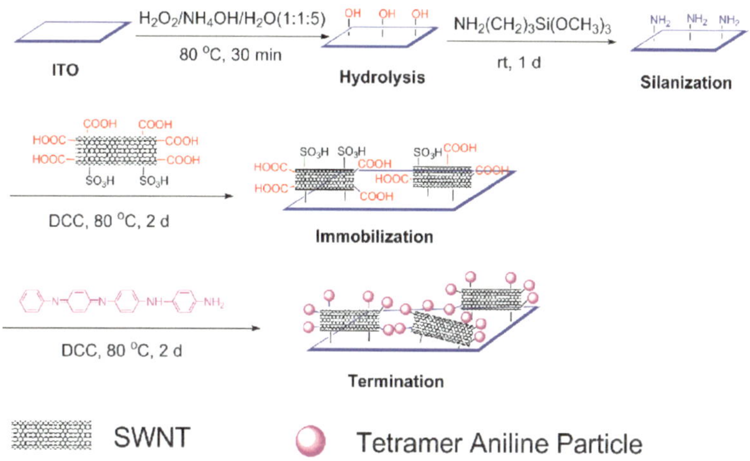


Scheme 2. Schematic illustration of the preparation route for attaching a functionalized SWNT layer onto an ITO endcapped by a tetramer aniline group. Reproduced with permission from Ref. [149]. Copyright 2009 Chemical Society of Japan.

SWNTs chemically assembled on functional monolayer-coated Au substrates show quasireversible CV features, indicating that, although directly linked to the insulating monolayer, the assembled SWNTs allow electron exchange between the gold electrode and the redox couple in solution. Electron tunneling between assembled SWNTs and the underlying gold substrate is involved in the charge-transfer process. The insulating monolayer between the gold substrate and the SWNTs acts as an electron-tunneling barrier. The high electron-transfer efficiency for the electrodes was ascribed to the large $\pi$-conjugated system within SWNTs, which enables SWNTs to accept or donate electrons, and to the efficient through-bond tunneling between the gold electrode and SWNTs, which can be described by the apparent tunneling resistance.[148]

In our group, SAMs of SWNTs covalently attached to a (3-aminopropyl)trimethoxysilane-modified ITO surface (SAM–ITO) were prepared from a soluble SWNT, which was safely obtained via a two-step process assisted by microwave irradiation (Scheme 2).[149]

It has been reported that SWNTs can be quickly functionalized under the assistance of UV or microwave irradiation,[150,151] plasma or ozone treatment.[152,153] A two-step method was developed to prepare soluble functionalized SWNTs assisted by a microwave oven in our group. Compared with the preparation under higher pressure,[151] the two-step approach allowed for a safer and easier operation. The FT–IR data of the soluble functionalized SWNTs showed a strong stretching mode of the –COOH groups from the SWNT backbone, and a weaker peak attributed to the asymmetric $SO_2$ stretching mode of the acid sulfonate (–$SO_2OH$) group, which implied that most of the functionalized carbon atoms on the SWNT backbone were carboxylated, with the remainder being sulfonated.

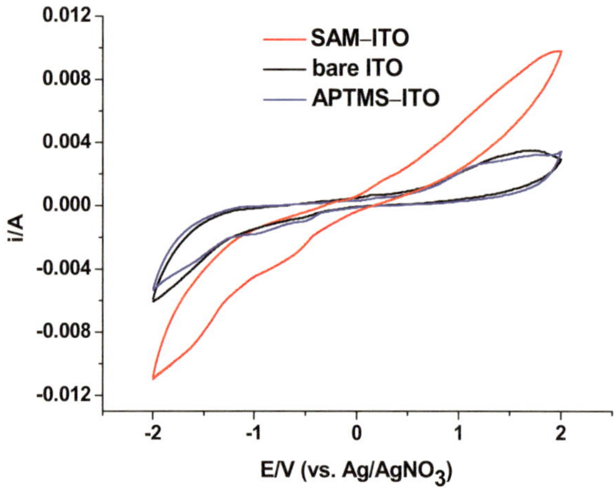


Fig. 11. CV traces of bare ITO, (3-aminopropyl)trimethoxysilane-modified ITO (APTMS–ITO) and SWNT-functionalized ITO (SAM–ITO) in $CH_3CN$ with 0.1 M tetrabutylammonium perchlorate (TBAP) as the supporting electrolyte. Scan rate = 0.05 V/s. Reproduced with permission from Ref. [149]. Copyright 2009 Chemical Society of Japan.

Cyclic voltammograms of CNT thin films self-assembled on ITO-coated glass by the coupling reaction of the amine groups with the carboxyl groups from the soluble SWNTs showed a higher capacitor charging current than in the bare ITO plates, as shown by the curves in Figure 11, which was attributed to the presence of SWNTs, resulting in an increase in the active electrochemical components.

To evaluate the stability in water of the water-soluble SWNTs layer, the SAM–ITO electrode was successively scanned for five cycles from –0.3 to 0.9 V at a rate of 0.05 V/s in a 1.0 M $H_2SO_4$ aqueous solution. Surprisingly, the CV data of the SAM–ITO electrodes (Figure 12a) demonstrated that oxidation occurred at 0.42 and 0.56 V and reduction occurred at 0.24 V, which showed a lower stability than in an organic TBAP /acetonitrile solution. Similar electrochemical reactions in aqueous solution, associated with surface oxygen complexes and increasing defect densities of the carbon nanotubes, have been reported previously, and the redox peaks have been assigned recently.[154,155] This is reasonable, considering that these peaks were also assigned to redox reactions involving defects and sidewalls of soluble functionalized SWNTs.

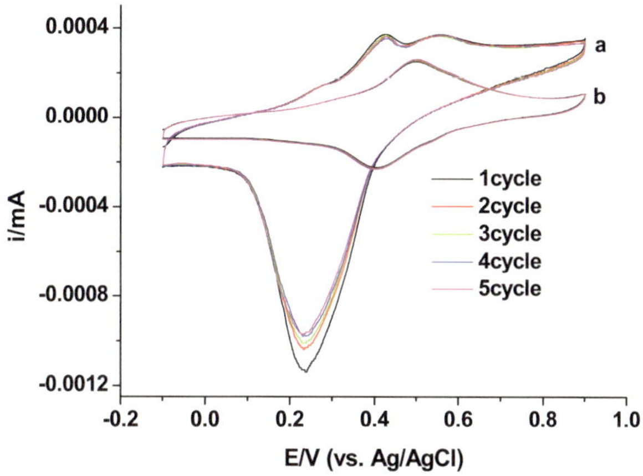


Fig. 12. CV traces of (a) SWNT SAM–ITO and (b) SWNT SAM–ITO endcapped by tetramer aniline groups in an aqueous 1.0 M $H_2SO_4$ solution. Scan rate = 0.05 V/s. Reproduced with permission from Ref. [149]. Copyright 2009 Chemical Society of Japan.

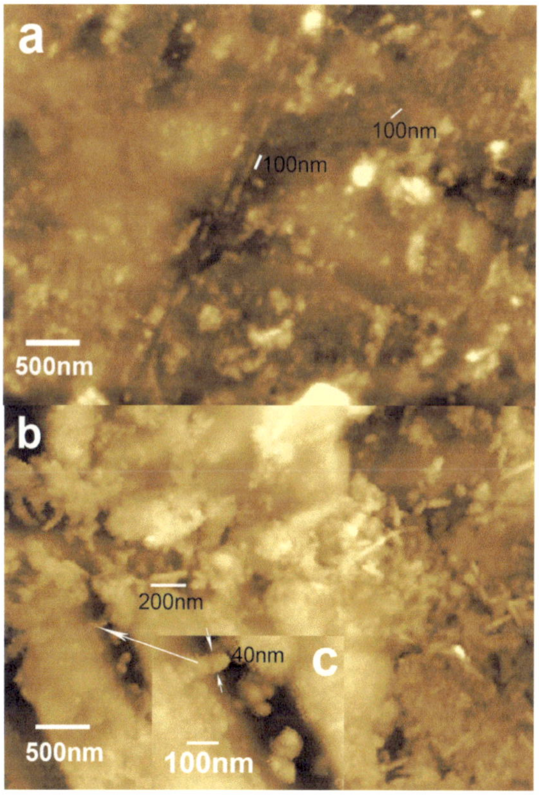


Fig. 13. Tapping mode AFM images of (a) SAM–ITO and (b) tetramer aniline-terminated ITO surface. Insert (c) shows the high-resolution image of an SWNT surface after being endcapped by tetramer aniline. Reproduced with permission from Ref. [149]. Copyright 2009 Chemical Society of Japan.

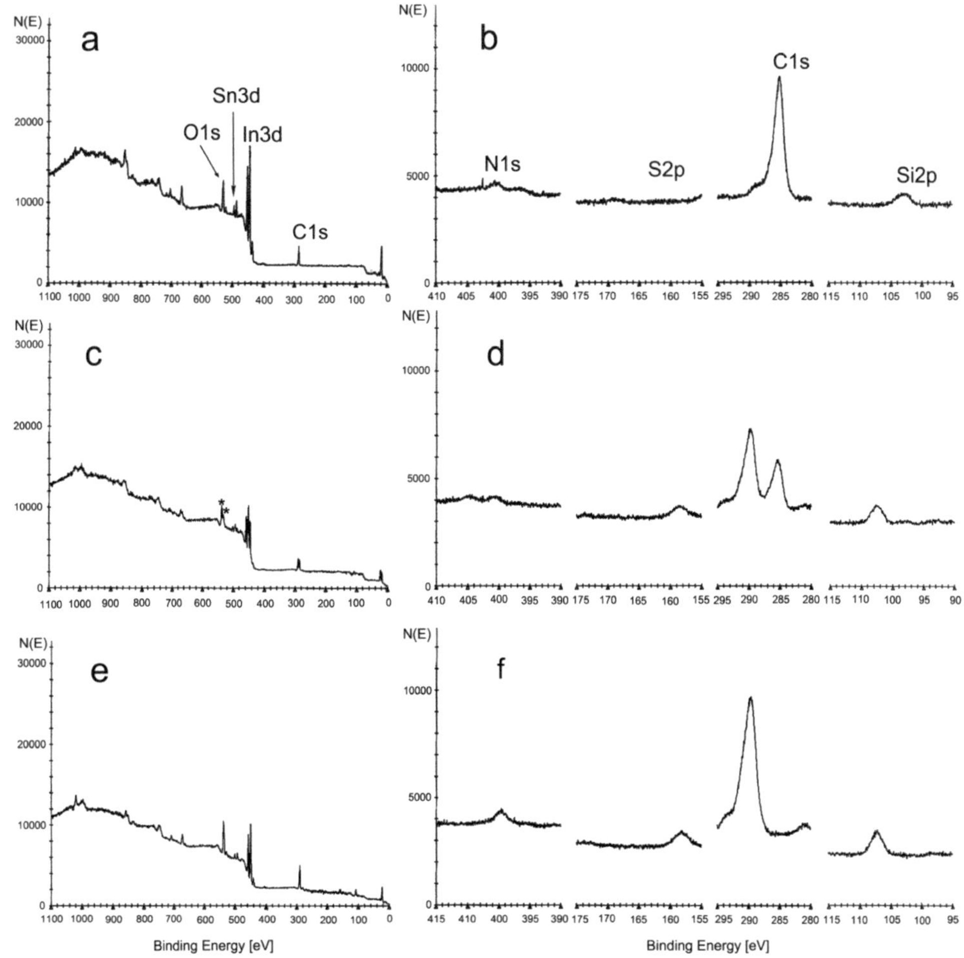


Fig. 14. The XPS wide-scan spectra of: (a) pristine ITO, (c) SAM–ITO, (e) tetramer aniline-terminated ITO surface, and (b), (d), and (f) their corresponding high resolution N 1s, S 2p, C 1s, and Si 2p spectra, respectively. Reproduced with permission from Ref. [149]. Copyright 2009 Chemical Society of Japan.

However, the CV data recorded for tetramer aniline end-capped SAM–ITO (Figure 12b) showed two reversible redox couples occurring at 0.50 and 0.40 V under the same conditions, which were assigned to the oxidation and reduction of the tetramer aniline between the leucoemeraldine and emeraldine oxidation states. Moreover, this was very reproducible, indicating that the stability of the SAM–ITO was improved after being end-capped by the tetramer aniline molecules.

From the AFM images in Figure 13, the SAM–ITO plate showed a monolayer with tube-like particle topography, implying the successful incorporation of functionalized SWNTs onto the ITO surface. The average length of the functionalized SWNTs is around 100 nm, with average diameter of around 20 nm. The morphology of the obtained SWNT layer was significantly changed after being end-capped using tetramer aniline groups (Figure 13b), which showed a rougher SWNT surface compared with that of SWNT–ITO (see arrow label in Figure 13c). From the AFM image shown in Figure 13b, the average length of the tetramer aniline terminated SWNTs is around 200 nm with an average diameter of around 40 nm, which showed a significant increase compared with that of the pure SWNT layer, because of the presence of a tetramer aniline particle layer linked with functionalized SWNTs by chemical bonds.

The XPS analysis showed the oxygen-containing defects on functionalized SWNT walls were covered and protected by tetramer aniline groups after end-capping, as shown in Figure 14. In addition to the intense and sharp features of the In 3d, Sn 3d, and O 1s peaks, the spectrum of the bare ITO plate exhibited weak C 1s and Si 2p peaks, which correspond to minor surface carbon contamination and the silicon from the glass, respectively (Figures 14a and 14b).[156] After attachment of the soluble functionalized SWNT, the peak intensity of the In 3d, Sn 3d, and O 1s peaks decreased, and the Si 2p peaks increased, because of the linked APTMS agent between the ITO layer and the functionalized SWNT (Figures 14c and 14d). In addition, the observed S 2p feature showed the presence of acid sulfonate ($-SO_2OH$) groups on the functionalized SWNTs. Aside from the C–C/ C–H peak at 285.3 eV, an additional photoemission representing the higher binding energy bands indicated the presence of carbon atoms bonded to other functional groups. The binding energy peak occurring at 290 eV was attributed to C=O and O–C=O groups from the functionalized SWNTs. The two-peak feature of O 1s peaks (denoted by asterisks in Figure 14c) shown in the wide-scan spectra compared to that of the bare ITO plate, also offers evidence for different chemical bonded states that correspond to COOH groups on the functionalized SWNT surface. However, after end-capping the residual carboxyl acid groups by the tetramer aniline terminated by amine groups, the intensity of the C 1s and O 1s features in the higher binding energy bands was significantly decreased to form a shoulder peak, as shown in Figures 14e and 14f. In addition, from the corresponding peak areas, the atomic content of the C 1s and N 1s peaks was enhanced compared to that of the In 3d or Sn 3d peaks. This is reasonable, considering that the enriched particle assembled by end-terminated tetramer aniline molecules tightly covered the sidewalls of the soluble functionalized SWNT via $\pi$–$\pi$ interactions, where only the photoemission of the C 1s and N 1s peaks assigned to the tetramer aniline backbone was detected. This results in an increase of the N 1s and C 1s peaks in the lower binding energy bands, and a decrease in the C 1s and O 1s peaks in the higher binding energy bands.

As discussed above, the conjugate layer, assembled by the end-terminated tetramer aniline molecules, tightly covered the sidewalls of the soluble functionalized SWNTs via covalent bonding and $\pi$–$\pi$ interactions, together with the doping effect between tetramer aniline and the sulfonic groups on the SWNT surface, which made the SWNT layer more stable in the CV scans in acidic aqueous solutions. It is expected that this stable electroactive SWNT layer will find a wealth of applications in nanocomposite architectures.

## 5. References

[1] R. H. Baughman, A. A. Zakhidov, W. A. de Heer, Science 2002, 297, 787.
[2] P. M. Ajayan, O. Stephan, C. Colliex, D. Trauth, Science 1994, 265, 1212.
[3] P. Calvert, Nature 1999, 399, 210.
[4] T. X. Liu, I. Y. Phang, L. Shen, S. Y. Chow, W. D. Zhang, Macromolecules 2004, 37, 7214.
[5] P. M. Ajayan, Chem. Rev. 1999, 99, 1787.

[6] P. Avouris, Acc. Chem. Res. 2002, 35, 1026.
[7] M. Ouyang, J. L. Huang, C. M. Lieber, Acc. Chem. Res. 2002, 35, 1018.
[8] V. N. Popov, Mater. Sci. Eng. R 2004, 43, 61.
[9] Q. Cao, J. A. Rogers, Adv. Mater. 2009, 21, 29
[10] H. J. Dai, Acc. Chem. Res. 2002, 35, 1035.
[11] X. J. Zhou, J. Y. Park, S. M. Huang, J. Liu, P. L. McEuen, Phys. Rev. Lett. 2005, 95, 146805.
[12] C. T. White, T. N. Todorov, Nature 1998, 393, 240.
[13] B. M. Quinn, S. G. Lemay, Adv. Mater. 2006, 18, 855.
[14] Z. Yao, C. L. Kane, C. Dekker, Phys. Rev. Lett. 2000, 84, 2941.
[15] E. Pop, D. Mann, Q. Wang, K. Goodson, H. J. Dai, Nano Lett. 2006, 6, 96.
[16] E. W. Wong, P. E. Sheehan, C. M. Lieber, Science 1997, 277, 1971.
[17] M. Cinke, J. Li, B. Chen, A. Cassell, L. Delzeit, J. Han, M. Meyyappan, Chem. Phys. Lett. 2002, 365, 69.
[18] S. E. Thompson, S. Parthasarathy, Mater. Today 2006, 9, 20.
[19] P. Avouris, Z. H. Chen, V. Perebeinos, Nat. Nanotechnol. 2007, 2, 605.
[20] W. B. Choi, D. S. Chung, J. H. Kang, H. Y. Kim, Y. W. Jin, I. T. Han, Y. H. Lee, J. E. Jung, N. S. Lee, G. S. Park, J. M. Kim, Appl. Phys. Lett. 1999, 75, 3129.
[21] L. Schlapbach, A. Zuttel, Nature 2001, 414, 353.
[22] A. S. Arico, P. Bruce, B. Scrosati, J. M. Tarascon, W. van Schalkwijk, Nat. Mater. 2005, 4, 366.
[23] Z. Liu, M. Winters, M. Holodniy, H. J. Dai, Angew. Chem., Int. Ed. 2007, 46, 2023.
[24] M. Prato, K. Kostarelos, A. Bianco, Acc. Chem. Res. 2008, 41, 60.
[25] M. Freitag, J. C. Tsang, J. Kirtley, A. Carlsen, J. Chen, A. Troeman, H. Hilgenkamp, P. Avouris, Nano Lett. 2006, 6, 1425.
[26] J. Chen, V. Perebeinos, M. Freitag, J. Tsang, Q. Fu, J. Liu, P. Avouris, Science 2005, 310, 1171.
[27] K. Kordas, G. Toth, P. Moilanen, M. Kumpumaki, J. Vahakangas, A. Uusimaki, R. Vajtai, P. M. Ajayan, Appl. Phys. Lett. 2007, 90, 123105.
[28] T. Iwai, Y. Awano, Fujitsu Sci. Tech. J. 2007, 43, 508.
[29] G. F. Close, S. Yasuda, B. Paul, S. Fujita, H. S. P. Wong, Nano Lett. 2008, 8, 706.
[30] S. N. Kim, J. F. Rusling, F. Papadimitrakopoulos, Adv. Mater. 2007, 19, 3214.
[31] P. Avouris, J. Chen, Mater. Today 2006, 9, 46.
[32] J. Appenzeller, Proc. IEEE 2008, 96, 201.
[33] R. H. Reuss, B. R. Chalamala, A. Moussessian, M. G. Kane, A. Kumar, D. C. Zhang, J. A. Rogers, M. Hatalis, D. Temple, G. Moddel, B. J. Eliasson, M. J. Estes, J. Kunze, E. S. Handy, E. S. Harmon, D. B. Salzman, J. M. Woodall, M. A. Alam, J. Y. Murthy, S. C. Jacobsen, M. Olivier, D. Markus, P. M. Campbell, E. Snow, Proc. IEEE 2005, 93, 1239.
[34] J. P. Edgeworth, N. R. Wilson, J. V. Macpherson, Small 2007, 3, 860.
[35] S. Maruyama, R. Kojima, Y. Miyauchi, S. Chiashi, M. Kohno, Chem. Phys. Lett. 2002, 360, 229.
[36] G. Y. Zhang, D. Mann, L. Zhang, A. Javey, Y. M. Li, E. Yenilmez, Q. Wang, J. P. McVittie, Y. Nishi, J. Gibbons, H. J. Dai, Proc. Natl. Acad. Sci. U.S.A. 2005, 102, 16141.

[37] Y. Murakami, S. Chiashi, Y. Miyauchi, M. H. Hu, M. Ogura, T. Okubo, S. Maruyama, Chem. Phys. Lett. 2004, 385, 298.
[38] Q. Cao, S.-H. Hur, Z.-T. Zhu, Y. Sun, C. Wang, M. A. Meitl, M. Shim, J. A. Rogers, Adv. Mater. 2006, 18, 304.
[39] G. S. Duesberg, A. P. Graham, M. Liebau, R. Seidel, E. Unger, F. Kreupl, W. Hoenlein, Nano Lett. 2003, 3, 257.
[40] Y. M. Li, W. Kim, Y. G. Zhang, M. Rolandi, D. W. Wang, H. J. Dai, J. Phys. Chem. B 2001, 105, 11424.
[41] K. M. Ryu, A. Badmaev, L. Gomez, F. Ishikawa, B. Lei, C. W. Zhou, J. Am. Chem. Soc. 2007, 129, 10104.
[42] B. B. Wang, S. Lee, X. Z. Xu, S. H. Choi, H. Yan, B. Zhang, W. Hao, Appl. Surf. Sci. 2004, 236, 6.
[43] E. Joselevich, C. M. Lieber, Nano Lett. 2002, 2, 1137.
[44] Y. G. Zhang, A. L. Chang, J. Cao, Q. Wang, W. Kim, Y. M. Li, N. Morris, E. Yenilmez, J. Kong, H. J. Dai, Appl. Phys. Lett. 2001, 79, 3155.
[45] S. M. Huang, M. Woodson, R. Smalley, J. Liu, Nano Lett. 2004, 4, 1025.
[46] S. M. Huang, X. Y. Cai, J. Liu, J. Am. Chem. Soc. 2003, 125, 5636.
[47] Z. Jin, H. B. Chu, J. Y. Wang, J. X. Hong, W. C. Tan, Y. Li, Nano Lett. 2007, 7, 2073.
[48] S. M. Huang, B. Maynor, X. Y. Cai, J. Liu, Adv. Mater. 2003, 15, 1651.
[49] A. Ismach, L. Segev, E. Wachtel, E. Joselevich, Angew. Chem. Int. Ed. 2004, 43, 6140.
[50] A. Ismach, D. Kantorovich, E. Joselevich, J. Am. Chem. Soc. 2005, 127, 11554.
[51] C. Kocabas, S. H. Hur, A. Gaur, M. A. Meitl, M. Shim, J. A. Rogers, Small 2005, 1, 1110.
[52] S. Han, X. L. Liu, C. W. Zhou, J. Am. Chem. Soc. 2005, 127, 5294.
[53] L. Ding, D. N. Yuan, J. Liu, J. Am. Chem. Soc. 2008, 130, 5428.
[54] C. Kocabas, M. Shim, J. A. Rogers, J. Am. Chem. Soc. 2006, 128, 4540.
[55] T. W. Ebbesen, P. M. Ajayan, Nature 1992, 358, 220.
[56] S. Bandow, S. Asaka, Y. Saito, A. M. Rao, L. Grigorian, E. Richter, P. C. Eklund, Phys. Rev. Lett. 1998, 80, 3779.
[57] K. Mukhopadhyay, A. Koshio, T. Sugai, N. Tanaka, H. Shinohara, Z. Konya, J. B. Nagy, Chem. Phys. Lett. 1999, 303, 117.
[58] H. Yokomichi, F. Sakai, M. Ichihara, N. Kishimoto, Thin Solid Films 2001, 395, 253.
[59] L. C. Qin, D. Zhou, A. R. Krauss, D. M. Gruen, Appl. Phys. Lett. 1998, 72, 3437.
[60] G. W. Meng, Y. J. Jung, A. Cao, R. Vajtai, P. M. Ajayan, Proc. Natl. Acad. Sci. U.S.A. 2005, 102, 7074.
[61] B. Q. Wei, R. Vajtai, Y. Jung, J. Ward, R. Zhang, G. Ramanath, P. M. Ajayan, Nature 2002, 416, 495.
[62] Y. J. Jung, B. Q. Wei, R. Vajtai, P. M. Ajayan, Nano Lett. 2003, 3, 561.
[63] B. D. Yao, N. Wang, J. Phys. Chem. B 2001, 105, 11395.
[64] Z. F. Ren, Z. P. Huang, J. W. Xu, J. H. Wang, P. Bush, M. P. Siegel, P. N. Provencio, Science 1998, 282, 1105.
[65] K. B. K. Teo, M. Chhowalla, G. A. J. Amaratunga, W. I. Milne, D. G. Hasko, G. Pirio, P. Legagneux, F. Wyczisk, D. Pribat, Appl. Phys. Lett. 2001, 79, 1534.
[66] J. Chen, M. A. Hamon, H. Hu, Y. Chen, A. M. Rao, P. C. Eklund, R. C. Haddon, Science 1998, 282, 95.

[67] J. Chen, A. M. Rao, S. Lyuksyutov, M. E. Itkis, M. A. Hamon, H. Hu, R. W. Cohn, P. C. Eklund, D. T. Dolbert, R. E. Smalley, R. C. Haddon, J. Phys. Chem. B 2001, 105, 2525.
[68] S. Banerjee, T. Hemraji-Benny, S. S. Wong, Adv. Mater. 2005, 17, 17.
[69] P. J. Boul, J. Liu, E. T. Mickelson, L. M. Ericson, I. W. Chiang, K. A. Smith, D. T. Colbert, R. H. Hauge, J. L. Margrave, R. E. Smalley, Chem. Phys. Lett. 1999, 310, 367.
[70] V. Georgakilas, K. Kordatos, M. Prato, D. M. Guldi, M. Holzinger, A. Hirsch, J. Am. Chem. Soc. 2002, 124, 760.
[71] J. L. Delgado, P. de la Cruz, F. Langa, A. Urbina, J. Casado, N. J. T. Lopez, Chem. Commun. 2004, 1734.
[72] D. Tasis, N. Tagmatarchis, A. Bianco, M. Prato, Chem. Rev. 2006, 106, 1105.
[73] A. R. Boccaccini, J. Cho, J. A. Roether, B. J. C. Thomas, J. E. Minay, M. S. P. Shaffer, Carbon 2006, 44, 3149.
[74] O. O. Van der Biest, L. J. Vandeperre, Annu. Rev. Mater. Sci. 1999, 29, 327.
[75] P. Sarkar, P. S. Nicholson, J. Am. Ceram. Soc. 1996, 79, 1987.
[76] T. V. Sreekumar, T. Liu, S. Kumar, L. M. Ericson, R. H. Hauge, R. E. Smalley, Chem. Mater. 2003, 15, 175.
[77] L. Hu, D. S. Hecht, G. Grüner, Nano Lett. 2004, 4, 2513.
[78] J. U. Park, M. Hardy, S. J. Kang, K. Barton, K. Adair, D. K. Mukhopadhyay, C. Y. Lee, M. S. Strano, A. G. Alleyne, J. G. Georgiadis, P. M. Ferreira, J. A. Rogers, Nat. Mater. 2007, 6, 782.
[79] K. Kordas, T. Mustonen, G. Toth, H. Jantunen, M. Lajunen, C. Soldano, S. Talapatra, S. Kar, R. Vajtai, P. M. Ajayan, Small 2006, 2, 1021.
[80] M. A. Meitl, Y. X. Zhou, A. Gaur, S. Jeon, M. L. Usrey, M. S. Strano, J. A. Rogers, Nano Lett. 2004, 4, 1643.
[81] J. U. Park, M. A. Meitl, S. H. Hur, M. L. Usrey, M. S. Strano, P. J. A. Kenis, J. A. Rogers, Angew. Chem. Int. Ed. 2006, 45, 581.
[82] E. Artukovic, M. Kaempgen, D. S. Hecht, S. Roth, G. Grüner, Nano Lett. 2005, 5, 757.
[83] E. Pop, D. Mann, Q. Wang, K. Goodson, H. J. Dai, Nano Lett. 2006, 6, 96.
[84] M. F. Yu, O. Lourie, M. J. Dyer, K. Moloni, T. F. Kelly, R. S. Ruoff, Science 2000, 287, 637.
[85] Z. W. Pan, S. S. Xie, L. Lu, B. H. Chang, L. F. Sun, W. Y. Zhou, G. Wang, D. L. Zhang, Appl. Phys. Lett. 1999, 74, 3152.
[86] P. Poncharal, Z. L. Wang, D. Ugarte, W. A. de Heer, Science 1999, 283, 1513.
[87] G. Decher, Science 1997, 277, 1232.
[88] S. W. Lee, B. S. Kim, S. Chen, Y. Shao-Horn, P. T. Hammond, J. Am. Chem. Soc. 2009, 131, 671.
[89] Z. Liu, Z. Shen, T. Zhu, S. Hou, L. Ying, Z. Shi, Z. Gu, Langmuir 2000, 16, 3569.
[90] Z. Chen, Y. Yang, Z. Wu, G. Luo, L. Xie, Z. Liu, S. Ma, W. Guo, J. Phys. Chem. B 2005, 109, 5473.
[91] H. B. Yildiz, R. Tel-Vered, I. Willner, Adv. Funct. Mater. 2008, 18, 3497.
[92] Z. C. Wu, Z. H. Chen, X. Du, J. M. Logan, J. Sippel, M. Nikolou, K. Kamaras, J. R. Reynolds, D. B. Tanner, A. F. Hebard, A. G. Rinzler, Science 2004, 305, 1273.
[93] N. A. Kotov, MRS Bull. 2001, 26, 992.
[94] G. Decher, J. D. Hong, Makromol. Chem. Macromol. Symp. 1991, 46, 321.
[95] G. Decher, J. D. Hong, J. Schmitt, Thin Solid Films 1992, 210, 831.

[96] W. B. Stockton, M. F. Rubner, Macromolecules 1997, 30, 2717.
[97] L. Y. Wang, Z. Q. Wang, X. Zhang, J. C. Shen, L. F. Chi, H. Fuchs, Macromol. Rapid Commun. 1997, 18, 509.
[98] Y. Shimazaki, M. Mitsuishi, S. Ito, M. Yamamoto, Langmuir 1997, 13, 1385.
[99] J. K. Mwaura, D. L. Thomsen, T. Phely-Bobin, M. Taher, S. Theodoropulos, F. Papadimitrakopoulos, J. Am. Chem. Soc. 2000, 122, 2647.
[100] A. A. Mamedov, N. A. Kotov, M. Prato, D. M. Guldi, J. P. Wicksted, A. Hirsch, Nature Mater. 2002, 1, 190.
[101] J. H. Rouse, P. T. Lillehei, Nano Lett. 2003, 3, 59.
[102] S. H. Qin, D. Q. Qin, W. T. Ford, J. E. Herrera, D. E. Resasco, Macromolecules 2004, 37, 9963.
[103] S. H. Qin, D. Q. Qin, W. T. Ford, J. E. Herrera, D. E. Resasco, S. M. Bachilo, R. B. Weisman, Macromolecules 2004, 37, 3965.
[104] L. M. Clayton, A. K. Sikder, A. Kumar, M. Cinke, M. Meyyappan, T. G. Gerasimov, J. P. Harmon, Adv. Funct. Mater. 2005, 15, 101.
[105] J. H. Rouse, P. T. Lillehei, Nano Lett. 2003, 3, 59.
[106] H. E. Miltner, N. Grossiord, K. Lu, J. Loos, C. E. Koning, B. Van Mele, Macromolecules 2008, 41, 5753.
[107] N. Grossiord, J. Loos, L. van Laake, M. Maugey, C. Zakri, C. E. Koning, J. Hart, Adv. Funct. Mater. 2008, 18, 3226.
[108] B. Pradhan, K. Setyowati, H. Liu, D. H. Waldeck, J. Chen, Nano Lett. 2008, 8, 1142.
[109] M. Havel, K. Behler, G. Korneva, Y. Gogotsi, Adv. Funct. Mater. 2008, 18, 2322.
[110] J. T. Han, S. Y. Kim, J. S. Woo, G. Lee, Adv. Mater. 2008, 20, 3724.
[111] L. Valentini, S. B. Bon, J. M. Kenny, Macromol. Mater. Eng. 2008, 293, 867.
[112] H. Gu, T. M. Swager, Adv. Mater. 2008, 20, 4433.
[113] A. Ikeda, K. Nobusawa, T. Hamano, J. Kikuchi, Org. Lett. 2006, 8, 5489.
[114] J. Geng, B. S. Kong, S. B. Yang, S. C. Youn, S. Park, T. Joo, H. T. Jung, Adv. Funct. Mater. 2008, 18, 2659.
[115] M. Okamoto, T. Fujigaya, N. Nakashima, Adv. Funct. Mater. 2008, 18, 1776.
[116] J. Li, J. Qiu, J. Xu, H. Chen, X. Xia, Adv. Funct. Mater. 2007, 17, 1574.
[117] Z. Zhu, Z. Wang, H. Li, Appl. Surf. Sci. 2008, 254, 2934.
[118] Y. P. Sun, Nature 1993, 365, 398.
[119] H. Imahori, J. Mater. Chem. 2007, 17, 31.
[120] T. Hasobe, Y. Kashiwagi, M. A. Absalom, J. Sly, K. Hosomizu, M. J. Crossley, H. Imahori, P. V. Kamat, S. Fukuzumi, Adv. Mater. 2004, 16, 975.
[121] T. Hasobe, S. Fukuzumi, P. V. Kamat, Angew. Chem. Int. Ed. 2006, 45, 755.
[122] T. Umeyama, N. Tezuka, M. Fujita, S. Hayashi, N. Kadota, Y. Matano, H. Imahori, Chem.-Eur. J. 2008, 14, 4875.
[123] H. Zhang, L. Fan, Y. Fang, S. Yang, Chem. Phys. Lett. 2005, 413, 346.
[124] H. Zhang, L. Fan, S. Yang, Chem.-Eur. J. 2006, 12, 7161.
[125] Q. Wang, H. Moriyama, Langmuir 2009, 25, 10834.
[126] X. F. Zhang, T. V. Sreekumar, T. Liu, S. Kumar, J. Phys. Chem. B 2004, 108, 16435.
[127] P. M. Ajayan, L. S. Schadler, C. Giannaris, A. Rubio, Adv. Mater. 2000, 12, 750.
[128] M. F. Yu, B. S. Files, S. Arepalli, R. S. Ruoff, Phys. Rev. Lett. 2000, 84, 5552.
[129] A. Kis, G. Csanyi, J. P. Salvetat, T. N. Lee, E. Couteau, A. J. Kulik, W. Benoit, J. Brugger, L. Forro, Nature Mater. 2004, 3, 153.

[130] D. Qian, W. K. Liu, R. S. Ruoff, Compos. Sci. Technol. 2003, 63, 1561.
[131] M. Olek, K. Kempa, S. Jurga, M. Giersig, Langmuir 2005, 21, 3146.
[132] A. Yu, M. E. Itkis, E. Bekyarova, R. C. Haddon, Appl. Phys. Lett. 2006, 89, 133102.
[133] G. Grüner, J. Mater. Chem. 2006, 16, 3533.
[134] L. Hu, D. S. Hecht, G. Grüner, Nano Lett. 2004, 4, 2513.
[135] E. Bekyarova, M. E. Itkis, N. Cabrera, B. Zhao, A. Yu, J. Gao, R. C. Haddon, J. Am. Chem. Soc. 2005, 127, 5990.
[136] J. Hone, M. C. Llaguno, N. M. Nemes, A. T. Johnson, J. E. Fischer, D. A. Walters, M. J. Casavant, J. Schmidt, R. E. Smalley, Appl. Phys. Lett. 2000, 77, 666.
[137] S. Shenogin, L. P. Xue, R. Ozisik, P. Keblinski, D. G. Cahill, J. Appl. Phys. 2004, 95, 8136.
[138] N. Shenogina, S. Shenogin, L. Xue, P. Keblinski, Appl. Phys. Lett. 2005, 87, 133106.
[139] M. Ouyang, J. L. Huang, C. M. Lieber, Acc. Chem. Res. 2002, 35, 1018.
[140] M. S. Fuhrer, J. Nygard, L. Shih, M. Forero, Y. G. Yoon, M. S. C. Mazzoni, H. J. Choi, J. Ihm, S. G. Louie, A. Zettl, P. L. McEuen, Science 2000, 288, 494.
[141] E. J. Garboczi, K. A. Snyder, J. F. Douglas, M. F. Thorpe, Phys. Rev. E 1995, 52, 819.
[142] Y. Wang, C. Di, Y. Liu, H. Kajiura, S. Ye, L. Cao, D. Wei, H. Zhang, Y. Li, K. Noda, Adv. Mater. 2008, 20, 4442.
[143] H. Z. Geng, K. K. Kim, K. P. So, Y. S. Lee, Y. Chang, Y. H. Lee, J. Am. Chem. Soc. 2007, 129, 7758.
[144] M. Zhang, Y. Yan, K. Gong, L. Mao, Z. Guo, Y. Chen, Langmuir, 2004, 20, 8781.
[145] B. L. Allen, P. D. Kichambare, A. Star, Adv. Mater. 2007, 19, 1439.
[146] G. Grüner, Anal. Bioanal. Chem. 2006, 384, 322.
[147] R. J. Chen, Y. Zhang, D. Wang, H. Dai, J. Am. Chem. Soc. 2001, 123, 3838.
[148] P. Diao, Z. Liu, J. Phys. Chem. B 2005, 109, 20906.
[149] Q. Wang, H. Moriyama, Bull. Chem. Soc. Jpn. 2009, 82, 743.
[150] M. L. Sham, J. K. Kim, Carbon 2006, 44, 768.
[151] Y. B. Wang, Z. Iqbal, S. Mitra, J. Am. Chem. Soc. 2006, 128, 95.
[152] N. P. Zschoerper, V. Katzemaier, Y. Vohrer, M. Haupt, C. Oehr, T. Hirth, Carbon 2009, 47, 2174.
[153] K. Peng, L. Q. Liu, H. Li, H. Meyer, Z. Zhang, Carbon 2011, 49, 70.
[154] M. J. Bleda-Martínez, D. Lozano-Castelló, E. Morallón, D. Cazorla-Amorós, A. Linares-Solano, Carbon 2006, 44, 2642.
[155] J. G. Zhou, C. Booker, R. Y. Li, X. T. Zhou, T.-K. Sham, X. L. Sun, Z. F. Ding, J. Am. Chem. Soc. 2007, 129, 744.
[156] K. R. Kissell, K. B. Hartman, P. A. W. Van der Heide, L. J. Wilson, J. Phys. Chem. B 2006, 110, 17425..